9th Symposium on Frequency Standards and Metrology (9FSM 2023)

Journal of Physics: Conference Series Volume 2889

Kingscliff, Australia
16 - 20 October 2023

Editor:

Michael E. Tobar

ISBN: 979-8-3313-1025-7
ISSN: 1742-6588

Printed from e-media with permission by:

Curran Associates, Inc.
57 Morehouse Lane
Red Hook, NY 12571

Some format issues inherent in the e-media version may also appear in this print version.

Institute of Physics
Dirac House, Temple Back
Bristol BS1 6BE UK

Phone: 44 1 17 929 7481
Fax: 44 1 17 920 0979

techtracking@iop.org

Additional copies of this publication are available from:

Curran Associates, Inc.
57 Morehouse Lane
Red Hook, NY 12571 USA
Phone: 845-758-0400
Fax: 845-758-2633
Email: curran@proceedings.com
Web: www.proceedings.com

TABLE OF CONTENTS

Journal of Physics: Conference Series **2889** (2024) 011001 doi:10.1088/1742-6596/2889/1/011001

The Symposia on Frequency Standards and Metrology

9th Symposium on Frequency Standards and Metrology 2023

Journal of Physics: Conference Series
jpcs.iop.org
Volume XXX, 2024

9th Symposium on Frequency Standards and Metrology 2023

16–20 October 2023, Kingscliff, Australia

Editor

Michael E Tobar
The University of Western Australia

Journal of Physics: Conference Series **2889** (2024) 011001 doi:10.1088/1742-6596/2889/1/011001

Welcome to Country

We respectfully acknowledged the Bundjalung people, the Traditional Custodians of the land on which this event took place. We recognized their enduring connection to land, waters, and culture, and we paid our respects to their Elders past, present, and emerging. We extended that respect to all Aboriginal and Torres Strait Islander peoples who were present.

Journal of Physics: Conference Series **2889** (2024) 011001 doi:10.1088/1742-6596/2889/1/011001

Preface

The 9th Symposium on Frequency Standards and Metrology [1] was held from the 16th to 20th of October at the Mantra on Salt Beach Resort in Kingscliff, New South Wales, Australia. The Symposium included a keynote lecture plus another 47 invited oral presentations and 105 invited poster contributions continually displayed over the week with two dedicated poster sessions. The Symposium is the most recent one of a series [2] that was initially organized by Jacques Vanier in 1971 in Forêt Montmorency in Quebec, Canada. The next ones took place in 1976 in Copper Mountain, USA, organized by Helmut Hellwig, 1981 in Aussois, France (Claude Audoin) [3], 1988 in Ancona, Italy (Andrea De Marchi) [4], 1995 in Woods Hole, USA (James Bergquist) [5], 2001 in St Andrews, UK (Patrick Gill) [6], 2008 in Pacific Grove, USA (Lute Maleki) [7], and in 2015 Potsdam, Germany (Fritz Riehle) [8]. The symposium series serves as an international discussion forum on precision frequency standards throughout the electromagnetic spectrum and associated metrology. It focuses on the fundamental scientific aspects of the latest ideas, results and applications in relation to these frequency standards. During the seven years after the last symposium very significant progress has occurred in various associated fields.

The 9th Symposium, focusing on frequency standards and related metrology, was held at an historically significant time, as metrology recently underwent a radical change in the definitions of the SI base units. In 2019, these units were redefined in terms of physical constants, with time playing a central role, as the definition of the second, based on the ground state hyperfine frequency of the caesium atom, now influences the definition of all base units (except the mole). The symposium series remains a crucial meeting, with many laboratories conducting studies on optical frequency standards, a promising area with the potential to replace microwave frequency standards for a possible redefinition of the second. This could significantly impact the SI unit system. The realisation of the second based on caesium standards appears to have reached its accuracy limit at the level of 10^{-16}. In contrast, the use of optical radiations allows for the realisation of frequency standards with accuracy at the level of 10^{-18} and beyond. Related to this, in 2016, the Consultative Committee for Time and Frequency (CCTF) published the first roadmap towards redefining the second. By 2020, the CCTF established a task force comprising about 40 experts from around the world to update this roadmap in preparation for a potential redefinition of the second in the coming years. The fundamentals of this process, along with the criteria that must be met and the challenges involved, are detailed in an open-source paper that fully supports the presentation by Tavella given at the 9th Symposium on Frequency Standards and Metrology [9].

The Symposium was a truly international event with 197 attendees from UK, Singapore, USA, France, Germany, China, Italy, Poland, Korea, Japan, Canada and Australia. It included invited talks from world experts, and more than one hundred invited poster presentations. Topics included: Precision Tests on Fundamental Physics, Precision and Low Noise Signal Generation and Techniques, Molecular, Atomic, Ion, and Nuclear Clocks, Time and Frequency Transfer, Precision and Quantum Metrology with Atoms, Photons and Phonons, Miniature, Portable, and Space Systems, SI definition, Clocks, and Time Scales. The keynote presentation was presented by the winner of the 2023 Shaw Prize in Astronomy, Professor Matthew Bailes from Swinburne University Melbourne on "Millisecond Radio Pulsars:

Journal of Physics: Conference Series **2889** (2024) 011001 doi:10.1088/1742-6596/2889/1/011001

Nature's clocks in the sky". This was followed by the founder of the symposium series, Jacques Vanier, discussing the influence of frequency standards on the SI unit system.

The Symposium excursion was to Tropical Fruit World, which saw the attendees experience a wide range of native fruits and fauna as well as different fruits from all around the world. At the Symposium dinner the joint best poster prize, was presented to Josipa Madunic from FEMTO-ST, Besancon France, entitled 'Microfabricated surface-electrode ion trap for frequency metrology', and Hiutat Antony Yang from the University of Tokyo, Japan, entitled "Development of Transportable Optical Lattice Clock". There was also an award for the most interesting paper award as voted by attendees, which was presented to Emma Paterson from the University of Western Australia, entitled "The Twisted Anyon Cavity Resonator as a Potential Dark Matter Detector and Sensing Device".

The week before the Symposium we held the associated Summer School on Frequency Standards, Precision and Quantum Measurement, held at the Gold Coast in Queensland on the 9th to 13th of October 2023. The summer school was targeted for students, early career researchers, and those professionals who want to learn more on the subject. Twelve experts from all over the world came to deliver tutorials on topics, which included, optical lattice clocks, frequency combs, optical and microwave metrology, atomic precision and quantum metrology, vapor cells, quantum sensing, time keeping, nuclear clocks, highly charged ion clocks, mathematical and statistical methods for time and frequency standards, UTC and the role of primary frequency standards, quantum metrology with photons, low noise and frequency stable oscillators, time transfer methods and molecular clocks. Approximately 50 participants came from many parts of the world including France, USA, UK, Germany, Singapore, China and Australia. Attendees were invited to present a poster on the topic of their research on the afternoon of Wednesday the 11th of October, with awards for the best student poster and the best early career researcher poster. The winners of the poster session were Martina Matusko from FEMTO-ST, Besancon France, for best student poster, entitled "Superradiance on ytterbium clock transition for frequency metrology", and Dr Elizaveta Klantsataya from Adelaide for best poster for a postdoc, entitled "A Laser-Cooled Optical Beam Clock". The excursion on Thursday 12th of October was at the nearby Currumbin Wildlife Sanctuary, where everyone was introduced to a range of Australian animals. Special thanks to the major sponsors, the IEEE and EQUS for supporting this event.

Finally, I would like to express my cordial thanks to the sponsors, the Governing Board, the International Steering Committee, the Local Organizing Committee, the session chairs and the reviewers who helped with the realization of the Symposium and the preparation of these proceedings.

The Editor

Michael E Tobar

Journal of Physics: Conference Series **2889** (2024) 011001 doi:10.1088/1742-6596/2889/1/011001

References

[1] See https://www.qdmlab.com/9fsm2023

[2] Le Journal de Physique Colloques - Third Symposium on Frequency Standards and Metrology, J.Phys. Colloques 42 No. C8 (Decembre 1981) http://jphyscol.journaldephysique.org/en/articles/jphyscol/abs/1981/08/contents/contents.html

[3] Proc. 4th Symposium on Frequency Standards and Metrology, Ancona, 1988 ed A De Marchi (Springer Verlag Berlin ISBN 3-540-50818-X)

[4] Proc. 5th Symposium on Frequency Standards and Metrology, Woods Hole, 1995 ed J C Bergquist (World Scientific Singapore ISBN 981-02-2527-X)

[5] Proc. 6th Symposium on Frequency Standards and Metrology, St Andrews, 2001 ed P Gill (World Scientific New Jersey ISBN 981-02-4911-X)

[6] Proc. 7th Symposium on Frequency Standards and Metrology, Pacific Grove, 2008 ed L Maleki (World Scientific New Jersey ISBN – 10 981-283-821-X)

[7] Proc. 8th Symposium on Frequency Standards and Metrology, ed Fritz Riehle, Journal of Physics: Conference Series, 723 (2016)

[8] D Wineland, 2016 The evolution of the Frequency Standards and Metrology symposium and its physics J. Phys. Conf. Ser. 723

[9] N. Dimarcq, et al, “Roadmap towards the redefinition of the second” Metrologia 61, 012001, DOI 10.1088/1681-7575/ad17d2 (2024).

Committees

Governing Board

- Jacques Vanier (Canada)
- Andrea de Marchi (Italy)
- James C. Bergquist (USA)
- Patrick Gill (Great Britain)
- Lute Maleki (USA)
- Fritz Riehle (Germany)

Symposium Chair

- Michael Tobar (Australia)

Symposium Secretariat

- Angela Bird (Australia)

International Steering Committee

- Michael Biercuk (Australia)
- Sébastien Bize (France)
- Davide Calonico (Italy)
- Scott Diddams (USA)
- Liz Donley (USA)
- Pierre Dubé (Canada)
- Victor Flambaum (Australia)
- Kurt Gibble (USA)
- Patrick Gill (Great Britain)
- Tetsuya Ido (Japan)
- Hidetoshi Katori (Japan)
- John Kitching (USA)
- Jeroen Koelemeij (Netherlands)
- Nikolay Kolachevsky (Russia)
- Long-Shen Ma (China)
- Helen Margolis (Great Britain)
- Ekkehard Peik (Germany)
- Fritz Riehle (Germany)
- Patrizia Tavella (France)
- Thomas Udem (Germany)
- Peter Wolf (France)
- Michael Wouters (Australia)
- Jun Ye (USA)
- Nan Yu (USA)

Local Organising Committee

- Linda Barbour (Australia)
- Jeremy Bourhill (Australia)
- Warwick Bowen (Australia)
- Scott Foster (Australia)
- Maxim Goryachev (Australia)
- Tingrei Tan (Australia)
- Virginia Tobar (Australia)
- Katrina Tune (Australia)
- Robert Williams (Australia)
- Magdalena Zych (Australia)

Journal of Physics: Conference Series **2889** (2024) 011001 doi:10.1088/1742-6596/2889/1/011001

Symposium Photographs

Beautiful Salt Beach, Kingscliff photo courtesy Tara Fortier

After a Welcome to Country, participants took part in an Indigenous Smoking Ceremony for goodwill, cleansing and to ward off bad spirits.

Journal of Physics: Conference Series **2889** (2024) 011001 doi:10.1088/1742-6596/2889/1/011001

Monday 16 October, 2023

The Mayor of Tweed, Chris Cherry, welcomed the 9FSM Symposium to the town of Kingscliff.

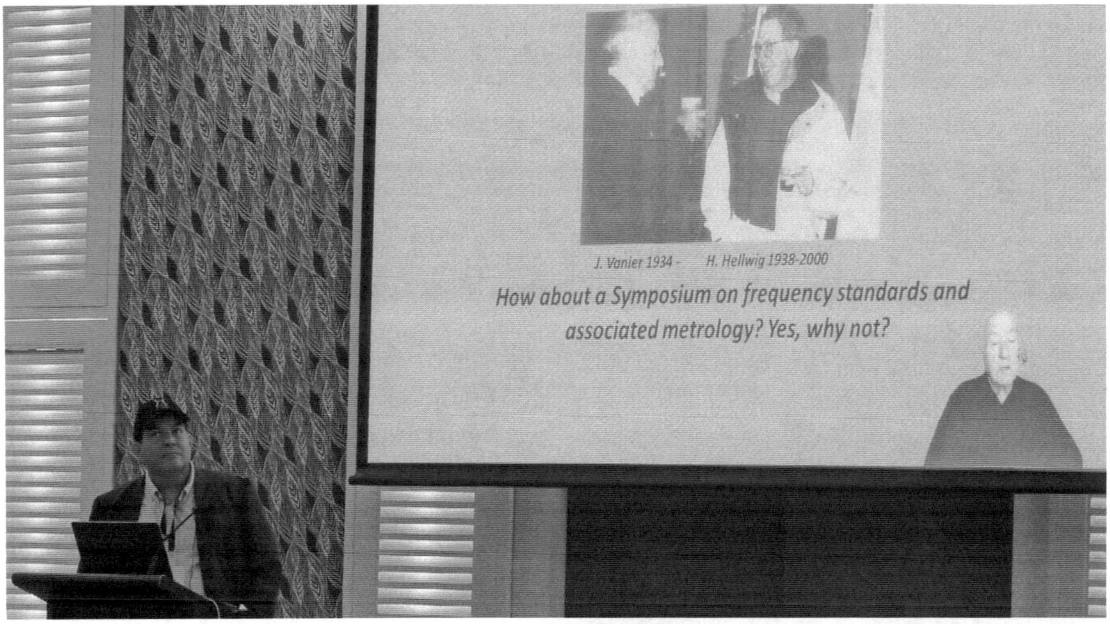

The 2023 Symposium Chair, Mike Tobar welcomed delegates and introduced Jacques Vanier, organiser of the inaugural FSM Symposium in 1971. Jacque's video presentation 'Atomic frequency standards, physical constants and metrology' described the ideology behind establishing the symposia on frequency standards and the ambitious foresight of including 'metrology' in the symposia title.

Keynote speaker Matthew Bailes of Swinburne University of Technology. His address was entitled 'Millisecond Radio Pulsars: Nature's clocks in the sky'.

Journal of Physics: Conference Series **2889** (2024) 011001 doi:10.1088/1742-6596/2889/1/011001

Journal of Physics: Conference Series **2889** (2024) 011001 doi:10.1088/1742-6596/2889/1/011001

Tuesday 17 October, 2023

Journal of Physics: Conference Series **2889** (2024) 011001 doi:10.1088/1742-6596/2889/1/011001

Wednesday 18 October, 2023

Tropical Fruit World

Journal of Physics: Conference Series **2889** (2024) 011001 doi:10.1088/1742-6596/2889/1/011001

Banquet dinner and poster awards

Pre-dinner drinks and group photo

Banquet icebreaker games

Sponsors and Governing Board

 IOP Publishing
Journal of Physics: Conference Series **2889** (2024) 011001 doi:10.1088/1742-6596/2889/1/011001

Joint poster prize: Josipa Madunic

Joint poster prize: Hiutat Antony Yang

Most interesting paper: Emma Paterson

Thursday 19 October, 2023

doi:10.1088/1742-6596/2889/1/011001

Friday 20 October, 2023

Journal of Physics: Conference Series **2889** (2024) 011001 doi:10.1088/1742-6596/2889/1/011001

Sponsors

Australian Research Council Centre of Excellence for Engineered Quantum Systems (EQUS), Australia

---

National Institute of Standards and Technology
US Department of Commerce (NIST), USA

National Physical Laboratory, United Kingdom

Commonwealth Scientific and Industrial Research Organisation (CSIRO), Australia

---

Australian Government Defence

IEEE Ultrasonics, Ferroelectrics, and Frequency Control Society, USA

Helmholtz Fonds e.V., Germany

The University of Western Australia

Quantum Technologies and Dark Matter Research Lab, Australia

ARC Centre of Excellence for Dark Matter Particle Physics, Australia

---

NSW Government
Office of the Chief Scientist & Engineer, Australia

Australian Government
Department of Industry, Science and Resources
National Measurement Institute

Journal of Physics: Conference Series **2889** (2024) 011001 doi:10.1088/1742-6596/2889/1/011001

Participants of the 9FSM, 2024

Michel Abgrall	France
Ben Allen	United Kingdom
Anne Amy-Klein	France
Kyle Arnold	Singapore
Matthew Bailes	Australia
Mark Baker	Australia
Linda Barbour	Australia
Murray Barrett	Singapore
Scott Beattie	Canada
Julian Berengut	Australia
Christopher Billington	Australia
Angela Bird	Australia
Sébastien Bize	France
Klaus Blaum	Germany
Andy Boes	Australia
Nicolas Bourbeau Hebert	Australia
Jeremy Bourhill	Australia
Dmitry Budker	Germany
Eric Burt	United States
Luigi Cacciapuoti	Netherlands
Ashlee Caddell	Australia
Davide Calonico	Italy
Claudio Eligio Calosso	Italy
James Camparo	United States
William Campbell	Australia
Christian Chardonnet	France
Jingbiao Chen	China
Ziyang Chen	China
Chin-wen Chou	United States
Taylor Christie	Australia
Roman Ciurylo	Poland
Robert Crew	Australia
Anne Curtis	United Kingdom
Benjamin Dix-Matthews	Australia
Elizabeth Donley	United States
Carter Fairhall	Australia
Sichen Fan	China
Fang Fang	China
Yanying Feng	China
Graeme Flower	Australia
Tara Fortier	United States
Alex Frost	Australia
Hao Gao	China
Guillaume Gauthier	Australia
Stephen Gensemer	Australia
Moji Ghadimi	Australia
Kurt Gibble	United States
Jonathan Gillot	France
Jacinda Ginges	Australia
Michele Giunta	Germany
Maxim Goryachev	Australia
Hua Guan	China
Xiaolei Guan	China
Lichung Ha	United States
Hidekazu Hachisu	Japan
James Hanssen	United States
Yara Hariri	France
Elrina Hartman	Australia
Michael Hatzon	Australia
Bryan Hemingway	United States
Ian Hill	United Kingdom
Ashby Hilton	Australia
Hyun-Gue Hong	Republic of Korea
Yao Huang	China
Matthew Hummon	United States
Nils Huntemann	Germany
Tetsuya Ido	Japan
Eugene Ivanov	Australia
Andrey Jarmola	United States
Nakarin Jayjong	Singapore
Julian Jee	Australia
Qianqian Ji	China
Weichen Jia	China
Yanyi Jiang	China
Haifeng Jiang	China
Douglas Jones	United Kingdom
Seji Kang	Republic of Korea
Mark Kasevich	United States
Hidetoshi Katori	Japan
Mona Kempkes	Germany
Attila Kinali	Germany
John Kitching	United States
Joshua Klose	Germany
Michael Kriele	Australia
Alexander Kuhl	Germany
Sang-Bum Lee	Republic of Korea
Won-Kyu Lee	Republic of Korea
Michael Lee	Singapore
Yuanhao Li	China
Yige Lin	China

Journal of Physics: Conference Series **2889** (2024) 011001 doi:10.1088/1742-6596/2889/1/011001

Jérôme Lodewyck	France	Sarah Scholten	Australia
Tim Lücke	Germany	Adam Sharp	Australia
Andrew Ludlow	United States	Tiantian Shi	China
Andre Luiten	Australia	Yeshpal Singh	United Kingdom
Longsheng Ma	China	Sabrina Slimani	Australia
Josipa Madunic	France	Ben Sparkes	Australia
Lute Maleki	United States	Robert Spero	United States
Claude Marceau	Canada	Uwe Sterr	Germany
Helen Margolis	United Kingdom	Erik Streed	Australia
Andrey Matsko	United States	Yabei Su	China
Martina Matusko	France	Krzysztof Szymaniec	United Kingdom
William McGehee	United States	Ting Rei Tan	Australia
Shawn McSorley	Australia	Patrizia Tavella	France
Tanja Mehlstäubler	Germany	Michael Tobar	Australia
Gerard Milburn	Australia	Virginia Tobar	Australia
Maverick Millican	United States	Thomas Udem	Germany
Arnan Mitchell	Australia	Eduardo Villasenor	Australia
Atsuo Morinaga	Japan	Andrew Wade	Australia
Piotr Morzyński	Poland	Zhiyang Wang	China
Mehran Mossammaparast	United States	Yanhui Wang	China
Takashi Muramatsu	Japan	Ying-Ju Wang	United States
Tyler Neely	Australia	Shengzhe Wang	China
Montana Nelligan	Australia	Shiguang Wang	China
Nils Nemitz	Japan	Xiumei Wang	China
Rachel Offer	Australia	Sarah Watzdorf	Australia
Scott Papp	United States	Jayden Webster	Australia
Sonali Parashar	Australia	Timm Wegehaupt	Germany
Sang Eon Park	Republic of Korea	Andrew White	Australia
Young-Ho Park	Republic of Korea	Sam White	Australia
Emma Paterson	Australia	Robert Williams	Australia
Ekkehard Peik	Germany	Michael Wouters	Australia
Steven Peil	United States	Atsushi Yamaguchi	Japan
Joseph Pham	Australia	Peiqiang Yan	China
Paul-Eric Pottie	France	Hiutat Antony Yang	Japan
Qichen Qin	Singapore	Yuan Yao	China
Franklyn Quinlan	United States	Jun Ye	United States
Aaron Quiskamp	Australia	Nan Yu	United States
Fritz Riehle	Germany	Dai-Hyuk Yu	Republic of Korea
Erling Riis	United Kingdom	Tanya Zelevinsky	United States
Benjamin Roberts	Australia	Ya Zhang	Australia
Jun Ruan	China	Zhao Zhang	Singapore
Enrico Rubiola	France	Jia Zhang	China
Steven Samuels	Australia	TongBao Zhang	China
Jordan Scarabel	Australia	Jianwei Zhang	China
Sascha Schediwy	Australia	Qi Zhao	Singapore
Piet Schmidt	Germany	Cindy Zhao	Australia
Robert Scholten	Australia		

Journal of Physics: Conference Series **2889** (2024) 011001 doi:10.1088/1742-6596/2889/1/011001

Peer review statement

All papers published in this volume of Journal of Physics: Conference Series have been peer reviewed through processes administered by the proceedings Editor. Reviews were conducted by expert referees to the professional and scientific standards expected of a proceedings journal published by IOP Publishing.

Journal of Physics: Conference Series **2889** (2024) 011002 doi:10.1088/1742-6596/2889/1/011002

Peer Review Statement

All papers published in this volume have been reviewed through processes administered by the Editors. Reviews were conducted by expert referees to the professional and scientific standards expected of a proceedings journal published by IOP Publishing.

1. **Type of peer review:** Single Anonymous
2. **Conference submission management system:** Morressier
3. **Number of submissions received:** 70
4. **Number of submissions sent for review:** 69
5. **Number of submissions accepted:** 62
6. **Acceptance Rate (Submissions Accepted / Submissions Received × 100):** 88.6 %
7. **Average number of reviews per paper:** 1.04
8. **Total number of reviewers involved:** 15
9. **Contact person for queries:**
   Name: Michael Tobar
   Email: michael.tobar@uwa.edu.au
   Affiliation: The University of Western Australia - The University of Western Australia Department of Physics

Pulsars as Clocks

Matthew Bailes

Swinburne University of Technology, PO Box 218 Hawthorn, Vic, 3122, Australia

E-mail: mbailes@swin.edu.au

Abstract. Radio pulsars are naturally occurring flywheels in space that come equipped with a radio transmitter emitting a beam at a fixed rotation phase of the neutron star. The pulsars slow down due to the loss of energy associated with a magnetic braking torque, and can be thought of as an accurate clock with a predictable, but steady, slow-down rate. By recording when the radio pulsar's pulse strikes the Earth, it is possible to both use the pulsar as a clock, and as an experimental device that enables them to be used to explore a large range of physical and astronomical phenomena. In this paper I will explain how radio pulsars can be used to tell the time, what they have revealed about the Universe and nature's physical laws, and their limits, one of which is the gravitational background (GWB) caused by supermassive black hole binaries in the Universe. This paper is aimed at non-astronomers and closely follows the invited plenary talk delivered at the 9th Symposium on Frequency Standards and Metrology.

1. Introduction

For millennia the rotation of the Earth with respect to the Sun provided the most accurate time-keeping available to humans on long timescales. To the layperson, it might seem remarkable that the number of days in a year was known to reasonably high accuracy even in ancient Roman times, but it arguably involved one of the first applications of astronomy, numeracy, some patience, and a similar principle to that of pulsar timing, where many rotation cycles are counted over many years to deduce an average. Farmers needed to know the timing of the seasons to judge when to plant their crops. This was ascertained by learned scholars counting the nights between the repeated appearance of the same constellations. Julius Caesar decreed that a year would be 365.25 days, and the Julian calendar was adopted during his reign. Although ancient astronomers couldn't necessarily determine stellar positions very accurately in just one year, even if they got it wrong by a day or few, it didn't matter too much, as over decades the error in the number of the days in a year declined linearly and led to reasonably accurate estimates for the length of a year.

The Earth has a large moment of inertia and spins in the vacuum of space where the braking torque on it is quite small. Although its rotation speed is influenced by many subtle factors, such as the tides raised by the Moon and the Sun, its precession and even atmospheric effects, the rotation period of the Earth is remarkably stable. It possesses variations in the length of a day of only milliseconds per century, or a period derivative of $O(7\times10^{-13})$ s/s, or about 7×10^{-8} rotations per rotation.

Maritime navigation provided the impetus for improved terrestrial time-keeping, as when time on an accurate chronometer could be compared with the apparent locations of the stars, the longitude of an observer could be determined. The engineer Harrison's clocks were one of the main reasons Cook's voyages were so successful in the late 1700s, as Harrison had reduced the friction in the mechanics of the clock, and how to compensate for changes in temperature[1], leading to unambiguous determination of Cook's longitude. Much of government early investment in astronomy was linked to its practical applications such as navigation.

2. Discovery of Radio Pulsars

Following the Second World War radio astronomy flourished, in part driven by the rapid evolution of radar technology during the conflict. At first the radio sky was poorly understood and this was not aided by the limited resolution of radio telescopes (about a degree) and their analogue data recording systems (literally pen and paper chart recorders). Karl Janksy[2] had discovered that our own Milky Way galaxy emitted radio waves along its plane in the 1930s, and motivated by this discovery the remarkable amateur astronomer Grote Reber undertook the first (privately funded) mapping of radio emission across the sky[3] by building his own radio telescopes. This revealed a concentration of radio emission along the galactic plane with a peak near the galactic centre. Radio astronomy had its first golden era in the 1960s, when large facilities were constructed in both hemispheres enabling access to the entire radio sky.

The first evidence for super-massive black holes was provided by the discovery of quasars[4], incredibly luminous beacons with very small angular scales, in 1963. Quasars are now accepted to be fueled by supermassive black holes that are powered by the in-fall of ionised gas generating powerful radio jets.

In an attempt to study quasar dimensions, Hewish and his PhD student Jocelyn Bell built a new instrument to study the rapid quasar "twinkling" or scintillation in the interplanetary medium, and built one of the first radio telescopes with a high time resolution instrument attached. Their efforts were rewarded when Bell serendipitously discovered the first radio pulsar because of its regular radio pulses leaving behind a "bit of scruff" on the chart records, a sign of regular impulsive radiation[5]. Radio pulsars are neutron stars, the collapsed-cores of once-massive stars. Like the Earth they exhibit regular rotation, but unlike the Earth it is extremely rapid. The first pulsar pulsed just once every 1.33 s but now the most rapidly-spinning pulsar known rotates once just every 1.39595 milliseconds[6]. With masses some 500,000 times that of the Earth ($\sim 1.4\ M_\odot$), but radii of only ~ 10 km pulsars are remarkable flywheels, and only weakly braked by their rotating magnetic fields. Their estimated moments of inertia are $I = 10^{38}$ kg m^2 and their inferred magnetic field strengths range from about 10^8-10^{14} G. The discovery of pulsars was recognised by the awarding of the Nobel prize in Physics in 1974.

3. Pulsar Properties

In the magnetospheres of radio pulsars, charged particles are subjected to enormous accelerations above their polar caps which accelerate them to near the speed of light. Powerful magnetic fields collimate the flow of the charged particles that emit synchrotron emission that is visible in the radio spectrum. The radio pulsar beams are anchored by the pulsar's rotating magnetic field, causing a lighthouse-like appearance as they spin.

Pulsars are characterised by their spin periods P and period derivatives $\dot{P}$ and often plotted in the P-$\dot{P}$ diagram (see Fig 1). Two distinct types of pulsars arise. The "normal" pulsars are thought to be the results of supernovae, and although they are born spinning quickly (~ 50 Hz), their $\sim 10^{12}$ G fields cause them to slow down to periods closer to a second within a few hundred thousand years. By the time they have slowed to a few seconds most cease radio emission and disappear, usually forever. A second group are descendants of the normal pulsars, but have benefitted by the presence of a stellar companion. When the companion swells up at the end of its life the pulsar accretes mass and angular momentum. The mass accretion onto the neutron star weakens the magnetic field strength, and also spins up the pulsar to sometimes millisecond spin periods. These are usually known as the "millisecond pulsars" or MSPs. They are consistent

Journal of Physics: Conference Series **2889** (2024) 012001 doi:10.1088/1742-6596/2889/1/012001

rotators and have exceptionally weak braking torques, hence they find applications as precise clocks. For a magnetic dipole of field strength B spinning with a rotation period P, the period

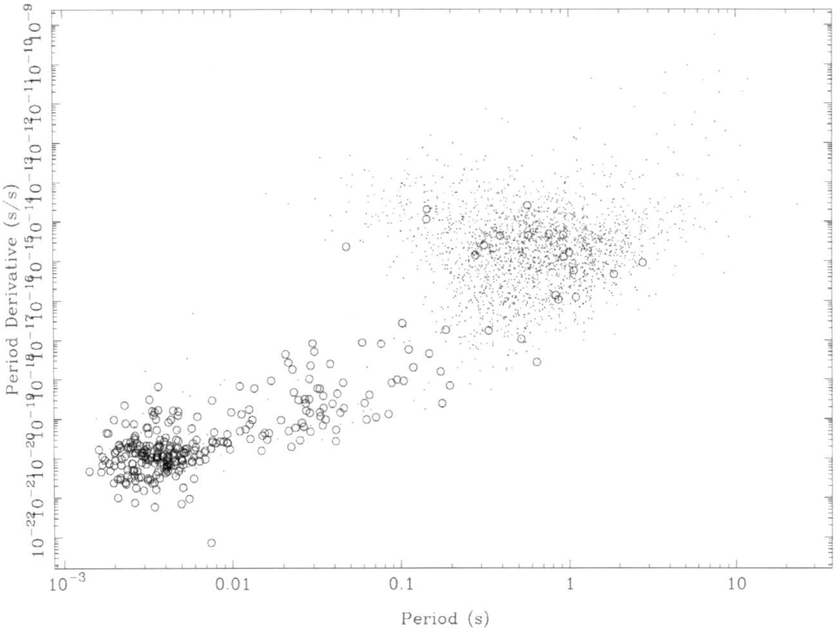

Figure 1. The pulsar period-period derivative or "$P - \dot{P}$" diagram. Circles depict binary pulsars. Young pulsars are born in the top-left hand side of the diagram and lose their kinetic energy over millions of years. Pulsars can be recycled by the accretion of matter from a companion and the high binary fraction amongst the most rapidly spinning pulsars is consistent with the recycling hypothesis.

derivative $\dot{P}$ will be

$$\dot{P} = \frac{B^2}{P} \frac{8\pi^2 R^6 \sin^2 \alpha}{3c^3 I} \tag{1}$$

where R is the radius of the neutron star, α the offset angle between the spin and magnetic field axis, c is the speed of light and I is the moment of inertia of a neutron star. In the case of a 10^8G neutron star with an orthogonally-aligned magnetic field a 5 ms pulsar would possess a period derivative of just $2{\times}10^{-21}$ s/s. In a billion years its period would only increase by about 60μs making it almost immortal (the Universe is only about 13 billion years old). Pulsars' extreme densities (a billion tons/cc cf Earth 6g/cc) and weak braking torques make them far superior to the Earth as a rotational frequency standard with a fractional increase in period of only 4×10^{-19} turns/turn.

Although a pulsar's mean pulse profile is very stable, its individual pulses are not. This gives rise to pulse "jitter", meaning that a pulsar is only a useful clock if many thousands or even hundreds of thousands of pulses are recorded in an observation. This can be seen in Figure 3, which displays 173 consecutive pulses from the bright millisecond pulsar PSR J0437–4715 and the pulsar's mean profile. The pulsar with the lowest known jitter is PSR J2241–5236. In a one hour integration the mean profile only suffers from about 4 ns of phase jitter[7] although its timing precision is often signal-to-noise limited. Many MSPs are an order of magnitude worse, although this doesn't preclude them being exceptional time standards that can find use in a variety of experiments.

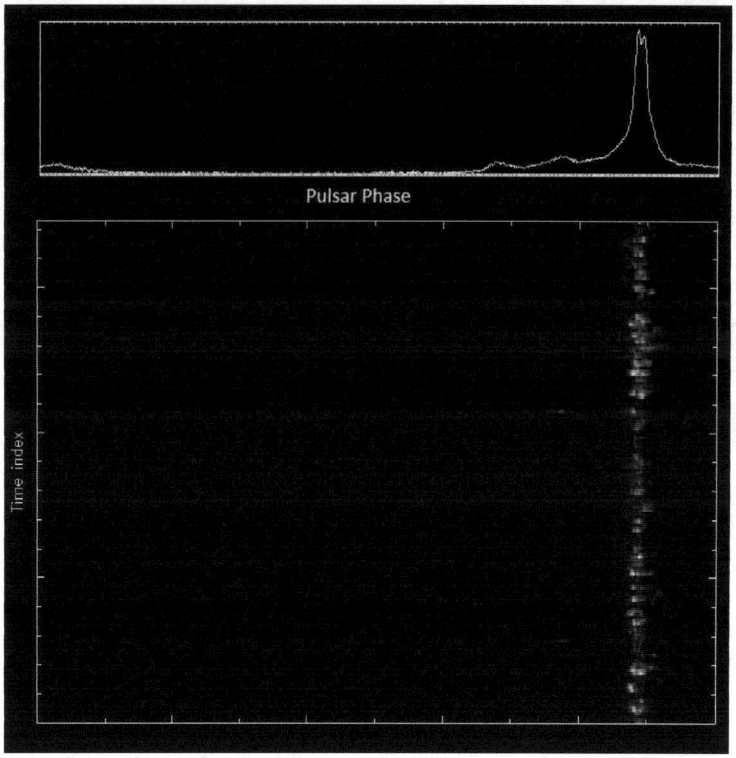

Figure 2. The mean pulse profile of the 5.75 ms PSR J0437–4715 after 1 second (upper panel) and a greyscale depicting the energy as a function of pulse phase for 173 consecutive pulses (lower panel).

Unlike an atomic clock, pulsars have no spin period that can be deduced a priori, nor is their spin period derivative possible to assert without measurement. Although the fractional error in a pulsar's spin can be remarkable, its phase is affected by the pulsar's imperfectly known proper motion, parallax and changing accelerations of the Sun and pulsar in the Milky Way. Some MSPs also contain easily visible red noise that makes their spin phase wander arbitrarily. These tend to have higher magnetic fields. Occasionally MSPs can experience sudden changes to their braking torques, resulting in poor extrapolation of their timing solutions[8].

In Figure 3 the pulsar timing residuals for the millisecond pulsar J1744–1134 are shown over a four year interval observed with the MeerKAT Pulsar Timing Array [9]. PSR J1744–1134 was discovered at the Parkes radio telescope and first published in 1997[10]. On Modified Julian Day (MJD) 55636 it had a spin period of 0.004074545942316071(4) seconds [11]. In Figure 3 the exact number of pulsar rotations between the first and last observation is 32,526,518,537. With a 100 ns typical error per epoch the fractional error in the period is deduced to about 1 part in 10^{15}, hence the remarkable precision in the published values.

The International Pulsar Timing Array is attempting to share data and methodologies to detect the stochastic background of gravitational waves being generated by the coalescence of supermassive black hole binaries [12]. In a recent study[13], developed the timescale TT(IPTA16), which although still inferior to atomic time standards, provides a useful check on them. Sadly our imperfect knowledge of the position of the Earth with respect to the Solar System Barycentre (SSB) requires constant attention to ensure the pulsars can be used as clocks on decadal timescales. Regardless, there appears to be a "common red noise" present in the millisecond pulsars, which may either represent an inherent limit to their stability, or the presence of a gravitational wave background (GWB)[14].

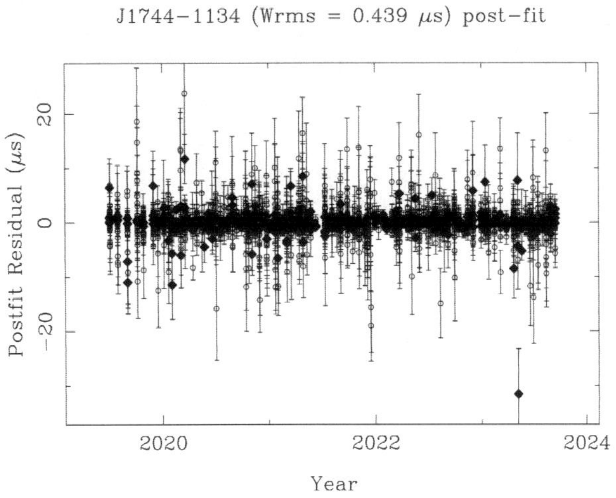

Figure 3. Pulsar timing residuals for the solitary millisecond pulsar PSR J1744–1134 after subtracting the motion of the Earth around the Sun and a model of the pulsar's dispersion measure, rotation, period derivative, position, parallax and proper motion with the MeerKAT pulsar timing array. The weighted rms timing residual is just 436 nanoseconds, and there are 32 arrival times at each epoch. A weighted average leads to a timing precision of better than 100 ns per epoch.

4. Pulsar Timing Applications

4.1. Dispersion

The speed of the pulsar's radio waves is affected by free electrons in the Galaxy and the observing frequency. Changes in the integrated column density cause changing delays that must be corrected by observing over large fractional bandwidths. With a model of the free electron content of the Galaxy we can ascribe a distance to each pulsar. The integrated column density of free electrons along the line of sight to the pulsar is referred to as the dispersion measure or DM. Changes to the DM present a barrier to using pulsars as clocks unless it is corrected for. With a model of the free electron density of the galaxy pulsar positions can be ascribed. These are shown in Figure 4.

All of the known pulsars are either in our galaxy or the Magellanic Clouds. In 2007 a new source of impulsive radio emission was discovered in the form of Fast Radio Bursts (FRBs) that enabled the first estimates of the extragalactic DM [15]. FRBs exhibit similar dispersion properties to the radio pulsars but are at cosmological distances and more isotropic in their distribution. When traced to host galaxies, they enable the determination of the baryonic content of the Universe. For a recent review see reference [16].

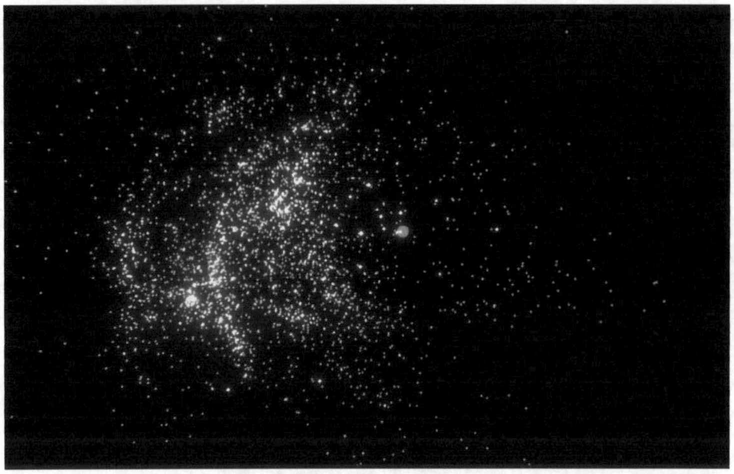

Figure 4. Pulsar positions in our Galaxy as ascribed by their dispersion measures and a model of free electrons in the galaxy. In this figure the centre of our Galaxy is depicted by the blue sphere. The Sun is near the clump of pulsars about 25% of the way from the left of the Figure and about 33% of the way up from the bottom. The high pulsar density near the Sun is due to a selection effect. The colour of the pulsars depicts their dispersion measure, with red pulsars possessing low dispersion and the blue high dispersion. Pulsars are "missing" from the far side due to a combination of factors including the inverse square law, radio pulse scattering and high sky background temperatures.

4.2. Relativistic Gravity

Some rare pulsars are in tight orbits where their velocities can exceed 0.1% of the speed of light, and their trajectories are affected by the curvature of space time as described by Einstein's General theory of Relativity (GR). Although the Earth takes a year to orbit the sun, some pulsars complete almost ten orbits of their companion neutron stars in a single day. This enables us to witness the shrinking of the orbit due to the emission of gravitational waves. The first astronomers to discover a binary neutron star[17] won the 1993 Nobel Prize in physics for the eventual proof that the orbit shrinks by 3 mm per orbit as predicted by GR[18]. One of the most relativistic systems is the celebrated double pulsar J0737-3039A/B with an orbital period just over 2.4 h. It exhibits many relativistic effects, including time dilation and gravitational redshift, relativistic advance of periastron, orbital decay and Shapiro delay. In a recent paper [19] the mass of the pulsar was determined to be 1.33819(1) $M_\odot$. The orbital period derivative is $-1.247929(8) \times 10^{-12}$ ss^{-1}, mainly due to the emission of gravitational waves[20]. Pulsars like this provided the impetus for the construction of the Advanced LIGO interferometer that discovered gravitational waves. In 2017 LIGO saw two neutron stars merge in the event known as GW170817[21]. At the time of the merger the orbital period of the two neutron stars was just 1 millisecond! Less than two seconds later a burst of gamma-rays struck the Earth. Within 24 hours a new star appeared in the sky at the location of the merger in what is described as a kilonova. This was arguably the most observed source in modern astrophysics[22]. A large fraction of the r-process elements such as gold may be produced in such events.

General relativity predicts that light passing a massive body will experience an extra time delay due to the compression of space time. In Figure 5 we show the effect of this on the time delays seen from the binary pulsar J1909–3744 as it orbits its white dwarf companion. The white dwarf has a mass of just 0.2 $M_\odot$ but the high orbital inclination gives rise to a clearly visible Shapiro delay that can be seen once the highly circular orbit is subtracted from the timing residuals. The orbit of this pulsar is one of the most perfectly measured circles in the Universe. Its orbital eccentricity is only $e = 7\times10^{-8}$ in an orbit with a projected semi-major axis of $a = 1.89799$ light seconds. The difference between the semi-major and semi-minor axes is only $0.5ae^2 = 1.4$ microns in an orbit over a million km in diameter!

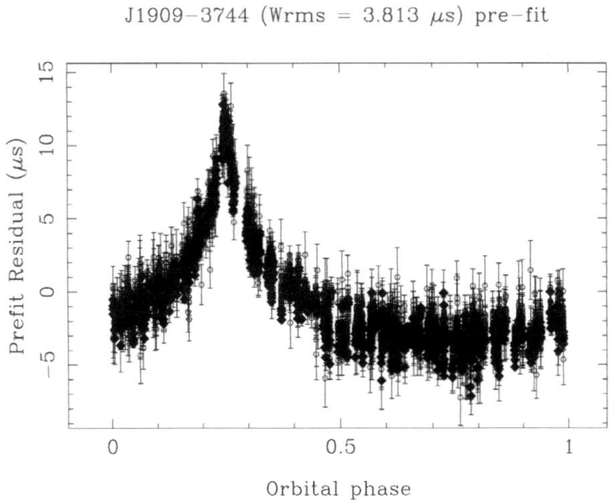

Figure 5. The effects of Shapiro delay on the timing residuals of the binary pulsar PSR J1909–3744 as seen with the MeerKAT radio telescope. At an orbital phase of 0.25 the pulsar passes behind the 0.2 $M_\odot$ white dwarf companion at an inclination angle of about 87°.

4.3. Stellar Evolution and Neutron Star Masses

Counting and timing radio pulses in a pulsar binary offers an insight into their orbital geometry. Systems in which the last star to evolve exploded remain highly eccentric, whilst those that accrete matter from a dying companion are often extremely circular due to tidal forces. Many binary pulsars have orbital eccentricities approaching unity, meaning they were almost unbound by the explosions that produced them. This places constraints on the mass ejected and kicks imparted to neutron stars at birth which appear to often be in excess of several 100 km s^{-1}. On the other hand, the binary pulsar PSR J1909–3744's orbit was circularised by the mass transfer, and the difference between the semi-major and semi-minor axes is negligible.

Systems favourably inclined permit the derivation of their orbital orientations and pulsar masses via the Shapiro delay. The most massive neutron stars appear to be nearly twice the mass of the Sun [23]. This has implications for the equation of state of nuclear matter.

4.4. The Search for a Gravitational Wave Background

Many collaborations are now searching for a gravitational wave background (GWB) generated by coalescing supermassive black hole binaries in the Universe by looking at the arrival times of millisecond pulsars.

Gravitational waves stretch and squeeze space time, and induced correlated arrival times for closely-separated pulsars on the sky, and a well-defined mean correlation coefficient as a function of angular separation - the so-called Hellings and Downs curve. The mean correlation coefficient is only 0.5 for closely-spaced pulsars, as although the Earth is common to the pulsar-earth pairs, each was experiencing their own GWB displacement when the pulse left the pulsars.

The GWB places a fundamental limit on how well even a perfect pulsar could be timed. In 2023 the four major pulsar timing array collaborations all reported 2-4 sigma "detections" of the GWB by exploring the correlation of their arrival times between many pairs of pulsars and it suggested a GWB amplitude of order a few $\times 10^{-15}$ [14, 24, 25, 26]. This limits the fractional timing stability they can achieve as does our knowledge of the solar system barycentre. Inverting the problem, MSPs can be used to infer the masses of the major planets [27].

5. Conclusions

Although the timing of pulsars is limited to a few-ten nanoseconds at any epoch, over long timescales their stability is impressive for clocks that are provided freely by the Universe. Modern pulsar astronomy demands reference to time standards accurate to less than 5-10 nanoseconds so that timing precision is not compromised. The most precisely timed pulsars are sensitive to the effects of a gravitational wave background and a handy check of the validity of modern time standards.

5.1. Acknowledgments

I am grateful to Dr Chris Flynn for assistance with the figures and to the MeerTime collaboration for data used in the plots.

6. References

[1] R. H. Brown, *Man and the stars* (Oxford University Press, 1978).
[2] K. G. Jansky, *Nature* **132**, 66 (1933).
[3] G. Reber, *Astrophysical Journal* **100**, 279 (1944).
[4] M. Schmidt, *Nature* **197**, 1040 (1963).
[5] A. Hewish, S. J. Bell, J. D. H. Pilkington, P. F. Scott, R. A. Collins, *Nature* **217**, 709 (1968).
[6] J. W. T. Hessels, *et al.*, *Science* **311**, 1901 (2006).
[7] A. Parthasarathy, *et al.*, *MNRAS* **502**, 407 (2021).
[8] J. W. McKee, *et al.*, *MNRAS* **461**, 2809 (2016).
[9] M. T. Miles, *et al.*, *MNRAS* **519**, 3976 (2023).
[10] M. Bailes, *et al.*, *Astrophys. J* **481**, 386 (1997).
[11] D. J. Reardon et al., *MNRAS* **507**, 2137 (2021).
[12] G. Hobbs, *Classical and Quantum Gravity* **27**, 084013 (2010).
[13] G. Hobbs, *et al.*, *MNRAS* **491**, 5951 (2020).
[14] D. J. Reardon, *et al.*, *Astrophys. J. Lett.* **951**, L6 (2023).
[15] D. R. Lorimer, M. Bailes, M. A. McLaughlin, D. J. Narkevic, F. Crawford, *Science* **318**, 777 (2007).
[16] M. Bailes, *Science* **378**, abj3043 (2022).
[17] R. A. Hulse, J. H. Taylor, *Astrophys. J. Lett.* **195**, L51 (1975).
[18] J. H. Taylor, J. M. Weisberg, *Astrophys. J.* **345**, 434 (1989).
[19] H. Hu, *et al.*, *A&A* **667**, A149 (2022).
[20] M. Kramer, *et al.*, *Physical Review X* **11**, 041050 (2021).
[21] B. P. Abbott, *et al.*, *Physical Review Lett.* **119**, 161101 (2017).
[22] B. P. Abbott, *et al.*, *Astrophys. J. Lett.* **848**, L12 (2017).
[23] H. T. Cromartie, *et al.*, *Nature Astronomy* **4**, 72 (2020).
[24] G. Agazie, *et al.*, *Astrophys. J. Lett.* **951**, L8 (2023).
[25] EPTA Collaboration, *et al.*, *A&A* **678**, A50 (2023).
[26] H. Xu, *et al.*, *Research in Astronomy and Astrophysics* **23**, 075024 (2023).
[27] R. N. Caballero, *et al.*, *MNRAS* **481**, 5501 (2018).

Journal of Physics: Conference Series **2889** (2024) 012002 doi:10.1088/1742-6596/2889/1/012002

Atomic frequency standards, physical constants, and metrology

Jacques Vanier*
Université de Montréal, Montréal, Canada,
*Retired
E-mail: vanier_jacques@videotron.ca

Abstract. The paper describes the physics behind the change that has taken place recently of the nature of the International System of Units, commonly called SI. This change took place in 2019 in the definition of the units forming the basis of the System: those base units, some of them defined originally by means of laws and artefacts, are now all defined in terms of physical constants. An important aspect of this fundamental change is the role played by the base unit, the second, associated with frequency and time measurements. Its definition in terms of the ground state hyperfine frequency of the ^{133}Cs atom plays now a central role entering into the definition of all base units, except the mol. The paper shows how this is done and consequently shows the importance of the developments of atomic frequency standards for obtaining the greatest precision and accuracy in the field of metrology.

1. Introduction

In 2019 the International System of Units, commonly known as the SI, has undergone a radical transformation that began several years ago. At the 26th General Conference on Weights and Measures (CGPM) held in November 2018, it was decided that from May 20, 2019, the SI base units would all be defined by means of physical constants[1].

The SI is essentially based on the introduction of seven fundamental quantities by means of which any experimental physical measurement made on entities within the universe can be expressed in terms of seven base units representing those quantities. Thus, seven measurement base units appear to be sufficient for that purpose. In fact, the universe is not fundamentally characterized by base units. Physics describes the universe by means of laws expressed in the language of mathematics that the human being has invented. However, those laws need experimental verifications, and this is done by means of measurements, expressed in numbers. Those numbers are related to each other within the physical laws, and those laws are correlated and must be coherent. This is done by means of units. Those units are the kilogram (kg), the metre (m), the second (s), the ampere (A), the kelvin (K), the candela (cd) and the mole (mol), representing respectively the quantities: mass, length, time, electric current, temperature, light intensity, and quantity of matter. Other units, qualified as derived, have been introduced through physical rules, connecting those fundamental units to each other, the connexion factor being unity. For example, the unit of force called newton (N) is expressed by the combination $kg \cdot m/s^2$; the unit of energy, called joule (J), is expressed by the combination $N \cdot m$ or even by the combination $kg \cdot m^2/s^2$. The definition of the base units, before May 20, 2019, are given in Appendix 1. The relationships between certain selected derived units and the base units are given in Appendix 2 by means of a diagram connecting those units.

It appears that a system of units resulted from identified needs, among other activities, in the exchange of material goods, local and international trade, industry, machining, and, in particular, in the field of science. Its application in these various sectors, in order to standardize relations and the exchange of goods and data, requires the construction and maintenance of basic standards representing those units in accordance with their definition. In addition, those standards must be available and easy to use. For example, in the past, the metre, defined as 1 / 10,000,000 of a quarter of the Earth's meridian, was not easily achievable and to say the least, applicable in daily measurements. In order to have a standard of length easily accessible and capable of being replicated, a single physical standard was manufactured, namely a metal bar (platinum-iridium) representing, as best as possible, a practical realization of the definition of the metre. Similarly, the kilogram, by its definition, could be achieved using a determined volume of water, i.e., a litre, a standard then based at the same time on the unit of length. However, it could vary with the temperature and the purity of the water used. As in the case of the metre, a physical standard representing as best as possible that definition of the kg was manufactured, that is, a small metallic cylinder also made of platinum and iridium. Copies of these two primary standards, kept with great care at the International Bureau of Weights and Measures in Sèvres, near Paris, France, were produced and distributed to various countries greatly facilitating the unification of measurements across the planet. This kind of activity, extended to all possible types of physical measurements, led to the introduction of the metric system of measurements, later called the International System of Units, the SI. However, it is evident that standards based on artifacts are subject to variation over time. For example, comparisons between copies of the international standard kilogram (called International Prototype Kilogram, IPK) have shown that, due to environmental causes which are difficult to establish, their relative mass varied over the years, the IPK itself not being immune from those fluctuations even if it is kept isolated with great care. Those fluctuations can be of the order of 25 micrograms. Likewise, the standard metre and its copies were subject to fluctuations.

For its part, the time unit, the second, is subject to a particular type of problem. We are familiar with the concept that the length of a day on earth, the interval between two passages of the sun at the zenith, is 86,400 seconds. We could easily use this concept to define the second and establish its unit. However, astronomers have found that the Earth's rotation period shows significant fluctuations, due for example to tides, motion of ice packs at the pole, and even changes in the direction of the polar axis. It was generally accepted that the period of translation of the earth around the sun should be very stable. Therefore, a more accurate definition was possible. Indeed, until 1967, the second was defined as: 1/31 556 925.975 of the tropical year 1900. However, it is obvious that this definition introduced a major problem of measurement and its practical realization in the laboratory.

As has been suggested by many scientists in the past, a solution to these problems, which seems most appropriate, exists in the use of standards provided directly by nature through, for example, universal laws or immutable intrinsic properties of atoms. Those laws introduce physical constants, which in our mathematical language appear to be characteristics of the universe. Those constants, as demonstrated experimentally with great precision, do not seem to vary with space and time. If we could define the units in terms of physical constants, or internal properties of the atomic world, we could establish criteria for the construction of primary standards based on those laws. There would no longer be any need to have physical artifacts to represent the SI units. The units could then be realized independently in any laboratory on the planet. It is this new philosophy that is now being advocated.

2. The base units

Two quantities have appeared relatively early in the development of standards as being connected to frequency standards. Time is obvious, being a quantity that affects human life on a continuous basis: a pendulum in oscillation is a frequency standard or, by definition of frequency, a time standard, and has been used to construct clocks in various configurations for centuries. Other mechanical classic types of various natures such as those using springs or solid sate devices using quartz crystals have been developed. However, those devices could not be used as standards to define the unit of time. Their period of oscillation is arbitrary, and their frequency is not stable. They need to be calibrated in terms of the base unit definition such as the earth rotation or translation. However, as said above, that approach leads to difficulties. Fortunately, in the 1950's the atomic clock was invented and an electronic transition within an atom was used as a pendulum. Similarly, the use of a spectral line in krypton, with a fixed frequency or wavelength, was used to implement a length standard. In both cases it was thus essentially the use of an oscillation phenomenon within an atom, a basic atomic property. In the case of length, it assumed that the speed of light had very permanent properties, and later, it was the speed of light itself that was used to define the unit of length. Let us look at those two units as they were triggers in the introduction of the new SI.

The time unit: the second [s]

The first practical frequency standard, which appeared to fulfill the requirement, that is using an atomic property for the definition of base units, and having promising characteristics as a standard of time, was implemented in 1950's. It was an instrument, now well known, using the ground state hyperfine frequency of a ^{133}Cs atoms as pendulum. The most interesting property of that instrument is its reproducibility. Indeed, it was quickly shown that the operating frequency of independently constructed devices in various laboratories was the same with an accuracy of the order of 10^{-13}. In addition, this frequency had long-term stability of the same order of magnitude. No laboratory device had such properties. Precise measurements were made to determine the exact value of this frequency in relation to the definition of the astronomical second described above. Given the accuracy of the device in question and the limited accuracy of the astronomical measurement, it was quickly concluded that the caesium atomic clock, easily accessible and relatively easy to build by any serious laboratory, could serve as a basic standard for the definition of the second. It was readily done in 1967.The frequency adopted for the hyperfine transition of the ground state of Cs^{133} which seemed in best agreement with astronomical measurements of the second is:

$$\Delta\nu_{Cs} = \nu_{hf}(Cs) = 9\ 192\ 631\ 770\ \text{Hz} \tag{1}$$

This frequency has been adopted as an exact value. We can qualify Eq. 1 as a definition. We thus have as an identity:

$$1\text{Hz} = \frac{\Delta\nu_{Cs}}{9\ 192\ 631\ 770} \tag{2}$$

Since Hz is the inverse of the second, we have:

$$1\ \text{s} = \frac{9\ 192\ 631\ 770}{\Delta\nu_{Cs}} \tag{3}$$

This expression says that if $\Delta\nu_{Cs}$ is accepted as the value said above, the equation becomes an identity, and 1 s becomes the duration of 9,192,631,770 periods of this radiation. (CGPM 1967). This is now the definition of the base unit second. The primary standard of time is thus the Cs atomic clock using the physical constants that determine the hyperfine structure of the ^{133}Cs ground state. The Cs hyperfine frequency is not really a physical constant as in the case of the speed of light or the electric charge of an electron but is assumed to be the result of interactions totally controlled by physical laws themselves determined by physical constants, assumed immutable. Many other frequency standards using different atoms with a particular frequency have been developed. Their frequency has been determined exactly relative to the frequency of Cs. They can be used as very precise secondary standards.

Journal of Physics: Conference Series **2889** (2024) 012002 doi:10.1088/1742-6596/2889/1/012002

The length unit: the metre [m]

Another striking example is the metre, which was defined for a long time as the length of a metal bar artefact as mentioned above. It was later replaced by a length equivalent to 1,650,763.73 wavelengths of the light emitted by a Krypton-86 atom in a specific electronic transition (11th CGPM). However, during the 1960's, following the invention of the laser, numerous experiments were carried out which allowed the use of certain lasers in the field of distance measurement with increased precision. The wave emitted by a laser is coherent, which makes measurements easier. That work also made it possible to stabilize the frequency of these lasers and thus determine this frequency as a function of the time standard based on Cs. Indeed, we have the fundamental relationship:

$$\lambda \nu = c \tag{4}$$

where c is the physical constant characterizing the speed of light in space, independent of the speed of the reference frame in which the wave propagates. It is therefore easy to realize that, during a measurement, the frequency being determined exactly, the wavelength can also be determined exactly. It is also possible to directly link the wavelength to the standard metre artefact by means of interference phenomenon. Indeed, if we know ν, such a measurement, using the standard metre as a basis, leads to the determination of the speed of light. The accuracy of a measurement of this nature, based on the second defined using an atomic clock, is much greater than that obtained with measurements of c by other methods. We conclude that the speed of light is an appropriate physical constant to define the metre. However, the determination of c using the standard metre is limited in accuracy by the nature of the standard artefact itself. In practice, therefore, its value is fixed with an accuracy which reflects the precision reached in the measurement. The following value is used as a definition:

$$c = 299{,}792{,}458 \text{ m / s} \tag{5}$$

From such a definition, we can write:

$$1\text{m} = \frac{c}{299\,792\,458}\text{s} \tag{6}$$

and from the definition of the second we have:

$$1\text{ m} = \frac{9\,192\,631\,770}{299\,792\,458}\frac{c}{\Delta\nu_{Cs}} = 30.663\,318\ \frac{c}{\Delta\nu_{Cs}} \tag{7}$$

where c and $\Delta\nu_{Cs}$ are both exact quantities. In principle, we can therefore use this relation to define the metre, as being the distance traveled by light in 3.3356411 nanoseconds. This is what was decided in 1983 by the General Conference of Weights and Measures (CGPM). This makes the metre a unit that is function of the time unit definition by means of an atomic clock based on the Cs ground state hyperfine frequency.

Further steps: Josephson effect and Quantum Hall effect

Following those two definitions we can examine the state of the global SI, now depending on two physical constant parameters, the hyperfine frequency of Cs and the speed of light. We have abandoned the metre standard artefact as well as the astronomical reference for time. This is in part unfortunate since time was based on astronomical phenomena. However, it remains a very special entity of the universe, which appears to have been created at its birth and to which we attach a lot of importance even though we do not understand exactly its nature.

Are there other physics constant that we can attach to the other base units of the SI? It is not obvious at all. What could be done for units of mass and electricity? It does not seem that there is any connection between mass and electric charge. Let us look at that. Two phenomena discovered in solid state physics come to the rescue. The first is a phenomenon that occurs when two superconductive materials are separated by a thin layer of insulation: a structure like a diode is created. However, when the device is exposed to microwave radiation, that structure has a discontinuous voltage-current (*I-V*) characteristic. Quantum jumps are observed in the behavior of the voltage-current function as shown in Figure 1.

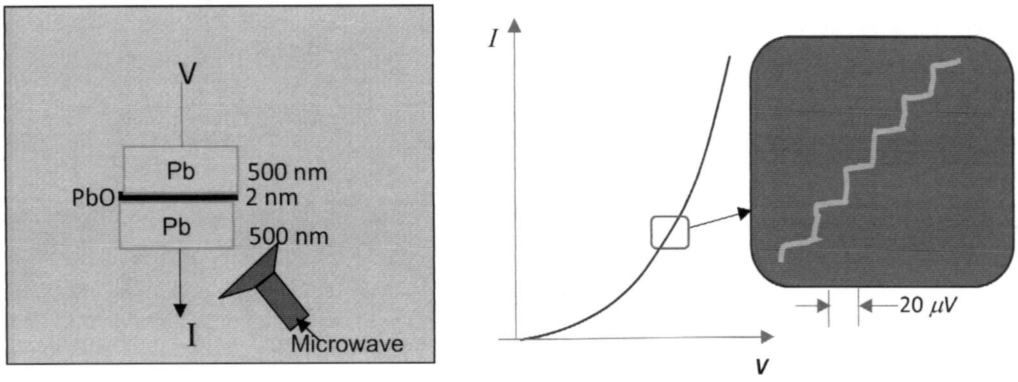

Figure 1. Josephson effect: schematic representation of superconducting junction and approximate *I-V* response.

As determined theoretically, the *I-V* curve has fixed voltage plateaus at values:

$$V = \nu \frac{h}{2e} n_J = K_J \nu n_J \tag{8}$$

where ν is the frequency of the applied radiation, n_J identifies the number of the plateau observed, h is Planck's constant, and e is the charge of the electron. The $h/2e$ ratio is known as the Josephson constant, K_J, from the name of the scientist who discovered this effect. With $\nu = 10$ GHz, the plateau $n_J = 100$ appears at V = ~2 mV. One can build units in which a large number of such junctions are placed in series. A plateau of the order of one volt can then be realized. Therefore, it is possible to measure the value of an unknown voltage by comparison with the voltage developed by a system using this effect. We can therefore determine experimentally an electrical tension in terms of physical constants h and e by comparison with a Josephson junction. It is readily observed that the voltage given by Eq. 8 is a function of the microwave frequency applied and thus is related also to the time base unit defined earlier.

The second phenomenon is another quantum effect, called quantum Hall effect, which has been studied by von Klitzing in semiconductor materials at very low temperature. In this case, it has been shown that the so-called Hall resistance of a material varies by plateaus as a function of the intensity of the applied magnetic field. In fact, it is shown with great precision that the resistance associated with Hall effect in the material in question is given by the expression:

$$R = \frac{h}{e^2} n_K = K_K n_K \tag{9}$$

where h is Planck's constant, e is the charge of the electron and n_K identifies the plateau, The ratio h/e^2 is called the von Klitzing constant and has a value of 24.5…kilo-ohms. It is therefore possible, by means of measurements using this phenomenon, to express an unknown electrical resistance in terms of fundamental constants. It is thus possible to enter the SI by means of derived units such as resistance and electromotive force and in practice it is possible to use these effects to determine a resistance and an electromotive force. This is shown in Figure 2.

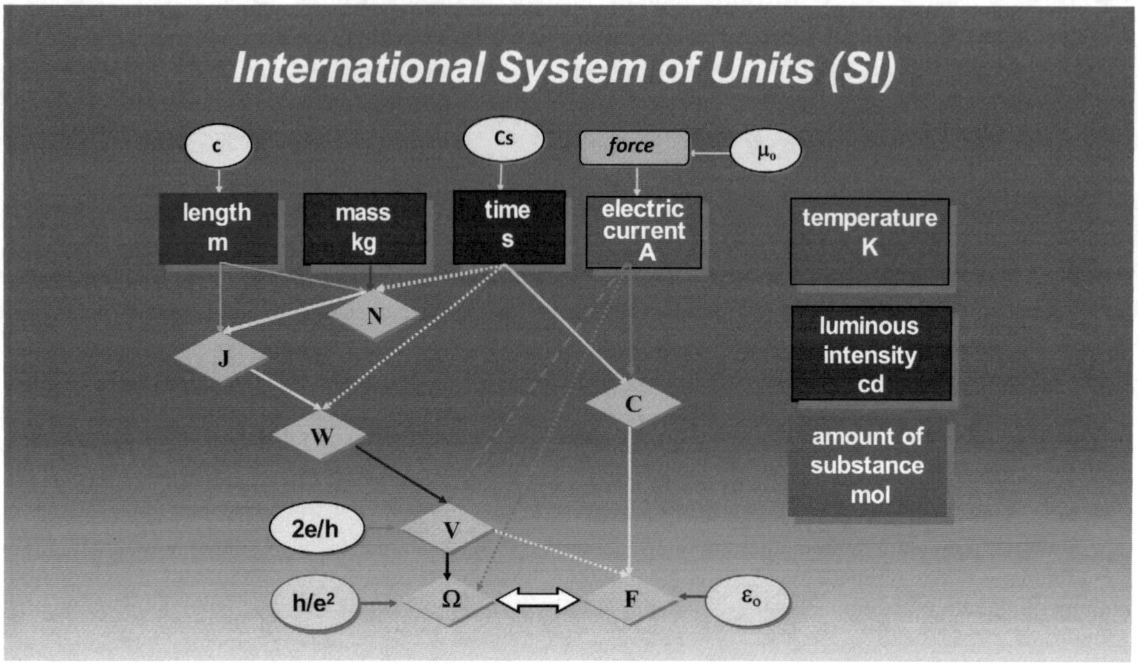

Figure 2. International system of units with length and time base units defined from physical constants. Only a few derived units important for the present discussion are shown (See ref. 2).

It should be mentioned that here is also a possibility to define the impedance associated with a capacitance C by means of the electric constant ε_o. This is done through the so-called calculable capacitance made of 4 symmetrical rods of 1 m length[2] . In the case of perfect symmetry, the capacitance of each pair of rods is given exactly by:

$$C_o = \varepsilon_o(ln2/\pi) \qquad [F/m] \tag{10}$$

It effectively provides a method of evaluating the electric constant ε_o through the path using e^2/h offered in the figure. It is clear, however, that coherence must be maintained throughout since various paths are possible.

As can be observed, the electric current base unit, the ampere A, could in principle be defined from the derived units, volt, and ohm, by means of Ohm's law. However, the ampere is already defined as the current causing a force of 2 x 10^{-7} Newton per metre created by the magnetic field generated between two infinite wires 1 metre apart. The system must stay coherent at that level too, that is coherent with previous definitions. In such a case, the classical method used to implement experimentally the base unit in the laboratory is necessarily to construct a system that can be used to determine such a force.

Time as a central unit

At this stage, as can be seen readily, time acts as a central unit in the SI. It enters into the Josephson effect; it plays an important role in the definition of the metre. Consequently, we can draw a diagram making explicit its interest as a central quantity, and creating some unification, a basic goal searched for in physics[3]. This is shown in Figure 3.

This is accomplished by realizing that the metre and the volt both are functions incorporating time in their definitions. Using quantum Hall effect, it is then possible to define the ampere by means of ohm's law and finally obtain a definition of the kilogram by means of a device measuring a force, in accordance with the original definition of the ampere. The whole system, electrical and mechanical, is then totally determined by physical constants. We will see below that the approach has some hidden difficulties. However, these difficulties can be resolved, and the SI can be integrated totally in terms of those physical constants.

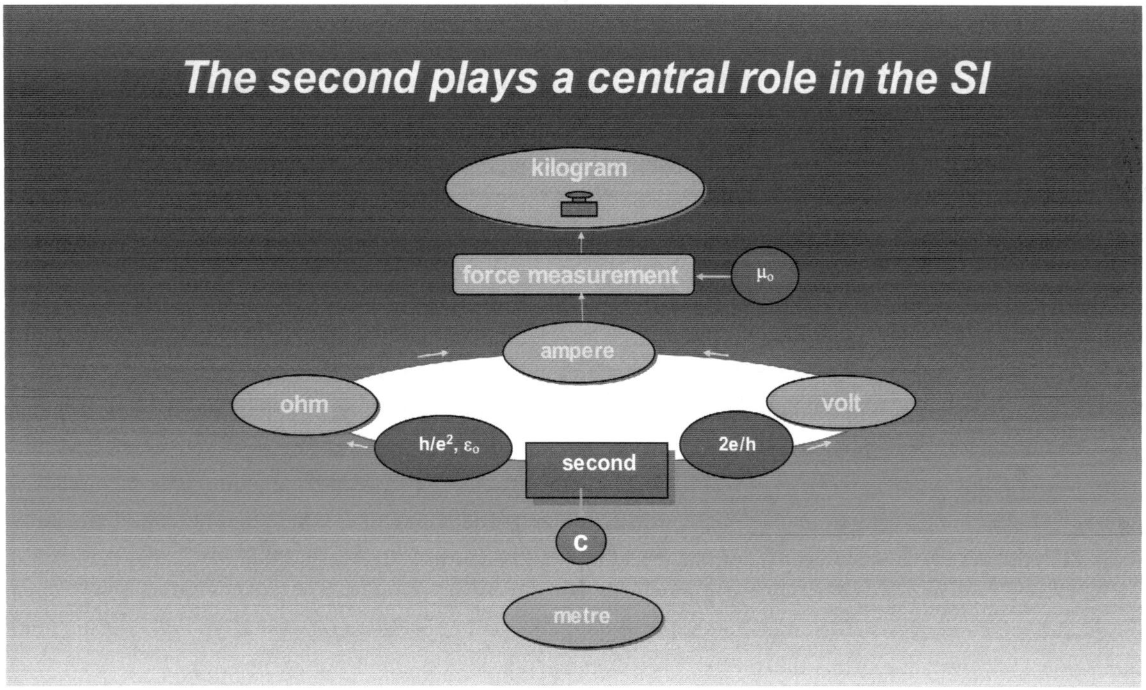

Figure 3. Time plays a central role in the SI, coupling electric and mechanical units together (See Ref. 3).

The mass unit, the kilogram (kg), and the unit of electric current, the ampere [A]

As emphasized above, however, the unit of electric current is effectively isolated through its definition in terms of a force. We need a method to measure that force. The best one known is a balance working in the earth's gravitational field, which is called appropriately an ampere balance: it measures that force against the force exerted by gravity on a kg primary standard. This can in principle be done with coils for the purpose of increasing the size of the magnetic force involved, which effectively is very small. However, the technique is rather inaccurate because it relies on the exact determination of the whole geometry of the structure used. Fortunately, it is possible to avoid that problem by means of a special balance developed by B. Kibble at NPL, UK[4]. It is most interesting as the technique developed solves at the same time the question of the use of an artefact as the primary standard of mass. Let us look at that approach, the final result allowing the use of physical constants for defining the SI base units and the Cs hyperfine frequency as a common basic entity for six of those units.

The force is obviously produced by the magnetic field created by the current. But how do you measure that force? As just said, the simplest idea is to build an apparatus which could compare the force developed by a magnetic field of induction ***B*** on a coil carrying a current ***I***, with the force produced on a mass by the gravitational terrestrial field, this force being equal to mg where g is the acceleration of gravity and m is the mass (not to be confused with the previous unit metre). We immediately see that this method couples the two quantities: electric current and mass. This causes a problem of consistency and priority. If the standard kilogram is used as a reference, the ampere becomes essentially a unit derived from the kilogram. The technique, however, leads to a problem. Part of that problem originates from the complexity of the geometric calculations involved in determining the magnetic field and its interaction with the coil representing the wire.

However, we can clearly see an even more serious problem in this approach. The Newton, defined in terms of the kilogram, makes the ampere, measured in this way, a function of an artefact which can very well vary over time. So, in principle, this definition causes a fundamental problem. However, we could define the ampere in a different way. In fact, the ampere can be defined as the flow of a charge of 1 coulomb per second, therefore from another derived unit. In the past, the coulomb was defined from the ampere, as shown in Fig. 2. But we know that the electric current is created by a flow of electrons. If we know well the electric charge of the electron, can we not simply define the ampere from this concept as a certain flow of electrons per second? It is possible, but first we must assess the charge of the electron independently and find a method to count electrons.

On the other hand, we could also look for a relationship between the fundamental constants which would allow the definition of the kilogram as a function of those constants. However, in any practical realization of a standard based on this new definition, it is necessary that this standard be compatible with the definition used until then and represent its value as accurately as possible. We saw previously that this was possible for the second and the metre. The new definitions of these units remained compatible with their original definition by measurements on the original standards to the limit of possible accuracy at the time of measurement. A new definition of the kilogram should allow the realization of a practical standard that can be evaluated according to the original artefact.

The solution to the dilemma relating to the definition of the ampere according to a force appeared in the last decades thanks to the realization of a modified ampere balance as well as to discoveries made in the field of solid-state physics outlined earlier, the Josephson and Quantum Hall effects. The new balance is the Kibble balance, mentioned earlier[4]. It was designed and constructed so as to avoid the obligation of geometric measurements and calculations when evaluating the force generated by the magnetic field on a coil carrying a current I. Rather than doing those complex calculations, the balance allows, as illustrated in Figure 4, the evaluation of the magnetic field by means of the voltage induced in the coil when it is set in motion in the same field at speed $\mathbf{v}$. Two measurements are made and the system is fully characterized, based on fundamental constants, which are then used to assess either the kilogram or the ampere. The first measurement, illustrated in Figure 4a, consists of the evaluation of the force exerted by an applied radial magnetic induction B_r on the coil made of a wire of length L. This force is given by the expression:

$$F = IB_r L \tag{11}$$

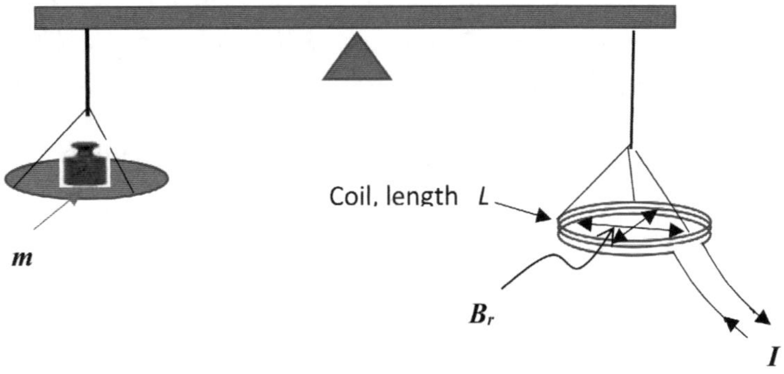

Figure 4a. Kibble balance. First measurement: comparison of the two forces, magnetic and gravitational.

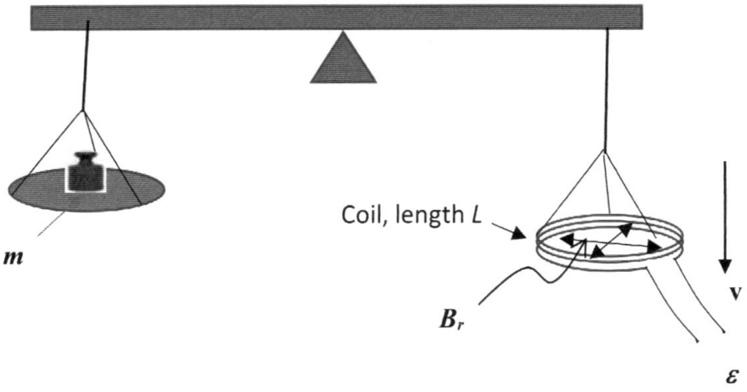

Figure 4b. Second measurement: evaluation of the B_r field by means of the induced voltage ε when the coil made of a wire of length L is set in motion at speed v.

The current I in the coil is adjusted so that an equilibrium of the balance is obtained. The force developed is then equal to mg where g is the acceleration of gravity at the place where the measurement is made. Consequently:

$$mg = IB_rL \tag{12}$$

In the second measurement, as illustrated in Figure 4b, the coil is set in motion at speed v and the emf ε developed at the end of the wire forming this coil is given by:

$$\varepsilon = \mathrm{v}B_rL \tag{13}$$

The speed v, with unit m/s, can be measured and expressed as a function of the meter and the second defined above. The value of the product B_rL is therefore determined by the induced voltage ε and combining Eqs. 12 and 13 we obtain a unique relationship between mass m and current I:

$$m = \frac{\varepsilon}{\mathrm{v}g} I \tag{14}$$

This equation can be written $m\mathrm{v}g = \varepsilon I$, having watt as unit. For this reason, the device is often called a watt balance. We can take m as the mass of the standard IPK, and we have an expression which determines I as a function of the standard kilogram, the relation being based only on parameters which we can, in principle, easily measure, voltage ε, speed v and acceleration of gravity g. But, as mentioned above, this standard is an artifact. In addition, the voltage ε is expressed as a function of a derived unit, the volt which contains the current I in its definition. There is therefore a problem of consistency between units which depend on each other. Fortunately, the two phenomena of a quantum origin mentioned earlier allow us to solve this problem.

We can first determine the electromotive force ε in terms of fundamental constants h and e by comparison with a Josephson junction. We can then use quantum Hall effect to determine the resistance of the wire of length L, thus determining the current through Ohm's Law. We can therefore use these two properties to determine the values of the emf ε and current I in Eq. 14. We obtain, after some algebra:

$$m = \frac{n_{J1} n_{J2} \nu_{J1} \nu_{J2}}{n_K} \frac{1}{4g\mathrm{v}} h \tag{15}$$

where subscripts 1 and 2 identify the two independent measurements, I and ε. This equation being established, there is thus a relationship between h and m using a somewhat sophisticated laboratory instrument. This instrument allows either a determination of h or the measurement of an unknown mass. If the mass m is the artefact, prototype of the standard kilogram, we then have a determination of the value of Planck's constant, all the terms of Eq. 15 being determined by precise measurements. Note that this is made possible through two quantum effects using the constants e and h. The experimental value of h obtained is [5]:

$$h = 6.626\ 069\ 57 \times 10^{-34}\ \text{J s} \tag{16}$$

We note that the units of h are those of *action* or angular momentum used in mechanics, that is J s. The joule J is a unit of energy which can be expressed in $N\,m$, the Newton itself being expressed in kg m s^{-2} (see Appendix 2). Combining these various units, we obtain J s = kg m^2 s^{-1} as the unit of h.

The kilogram [kg]

We can therefore write:

$$h = 6.626\ 0\ 69\ 57 \times 10^{-34}\ \text{kg m}^2\ \text{s}^{-1} \tag{17}$$

Inverting this relationship, it follows that:

$$1\ \text{kg} = \frac{h}{6.626\ 069\ 57\ x\ 10^{-34}}\ \text{m}^{-2}\ \text{s} \tag{18}$$

Using the definitions for the metre and the second introduced previously, we thus obtain the unit of the mass quantity, the kilogram, as:

$$1\text{kg} = 1.475\ 5214 \times 10^{40} \frac{\Delta\nu_{Cs}}{c^2}\ h \tag{19}$$

An important advantage of this new definition is that the primary standard artefact of the kilogram is no longer necessary. It could disappear and be reconstituted again to its original value, on the basis of h and other physical constants in Eq.19. Putting this definition into practice can be done using the Kibble balance, which is of course an uneasy task, but provides a means of realizing the unit without artefact. However, given the accuracy of the copies of the primary standard available in many laboratories (~ 25 micrograms), it is obvious that these standards can still be used for measurements which do not require high accuracy. They can always be recalibrated. We note that the Cs ground state hyperfine frequency enters in the definition of the base unit, the kilogram.

The ampere [A]

In the previous paragraphs, we showed that with a somewhat sophisticated ampere balance we could very well connect the ampere and the kilogram by means of measurements of speed, tension, and resistance, the two latter determined by quantum effects, the Josephson effect, and the Quantum Hall effect.

In a coherent system, basic and derived units are linked together by well-known rigid rules (see Appendix 2). For example, the coulomb electrical charge unit, C, is equal to the ampere current unit, A, multiplied by the unit second s. We can thus write:

$$1\ \mathrm{C} = 1\ \mathrm{A\ s} \tag{20}$$

This equation can also be written, multiplying both sides by a well chosen number as in the following equation:

$$1.602\ 176\ 565\ 10^{-19}\ \mathrm{C} = 1.602\ 176\ 565\ 10^{-19}\ \mathrm{A\ s} \tag{21}$$

We have chosen, as multiplier, the well-known electronic charge equal to:

$$e = 1.602\ 176\ 565\ 10^{-19}\ \mathrm{C} \tag{22}$$

Inverting Eq. 21, we thus have:

$$1\ \mathrm{A} = \frac{e}{1.602\ 176\ 565 \times 10^{-19}}\mathrm{s}^{-1} \tag{23}$$

Using the definition of the second we have:

$$1\ \mathrm{A} = 6.789\ 687\ \times 10^{8}\ \Delta\nu_{Cs}\ e \tag{24}$$

Consequently, we have a definition of the ampere as a function of the electronic charge, *e*, another physical constant, as we wanted earlier. By this definition, the Ampere is the electric current which corresponds to the flow of 1 / 1.602 176 565 x 10^{-19} elementary charges per second. However, the simplest way to put this definition into practice is to use ohm's law, $V = RI$, where R and V are determined using Josephson and Quantum Hall effects. We note again that the cesium ground state hyperfine frequency enters again in the definition of the base unit, the ampere.

The temperature unit: kelvin (K)

In the *SI* used earlier, the unit of temperature was called kelvin, *K*, defined as the fraction 1 / 273.16 of the thermodynamic temperature of the triple point of water. In the new system, the kelvin is defined as being related to Boltzmann's constant, *k*. Indeed, in thermodynamics this constant is given by the equation:

$$k = 1.380\ 648\ 8 \times 10^{-23}\ \mathrm{J\ K}^{-1}. \tag{25}$$

We use the coherent connection between the energy unit, the joule *J*, and the basic units as we did before. We can then write the product J K^{-1} as being kg m^2 s^{-2} K^{-1}. We replace J K^{-1} in Eq. 25 by this equivalent combination of units, reverses this equation and replaces the various units by their definitions as we did previously. We obtain:

$$1\,\mathrm{K} = \frac{1.380\,6489 \times 10^{-23}}{(6.626\,070\,15 \times 10^{-34}) \times (9\,192\,631\,770)} \frac{\Delta\nu_{Cs} h}{k} \tag{26}$$

or:

$$1\,\mathrm{K} = 2.266\,665\,3 \frac{\Delta\nu_{Cs}\, h}{k} \tag{27}$$

The effect of this new definition is to match the Kelvin to a change in thermal energy kT equal to 1.380 648 9 x 10^{-23} *J.* We see that in the earlier definition, the triple point of water was established as exact. Now k is an exact value. Therefore, this new definition implies that the triple point of water must be determined experimentally. We note again that the cesium ground state hyperfine frequency enters in the new definition of the base unit kelvin.

Light intensity: the candela

In the original |SI system, the candela, cd, is defined as being the light intensity in a given direction from a source which emits monochromatic radiation of frequency 540 x10^{12} Hz and whose energy intensity in the direction of emission is 1/683 watt by steradian (sr). This concept is related to the human eye's response to light. The new definition of the candela is based on these concepts and the relative luminous efficacy of the human eye. The luminous efficacy K_{cd} is defined as being 683 lumen per watt, where the lumen, lm, is the unit derived from the candela expressed in the form (see Appendix 2):

$$1\ \mathrm{lm} = 1\ \mathrm{cd\ sr} \tag{28}$$

Thus, we have:

$$K_{cd} = 683\ \mathrm{lm\ W}^{-1} = 683\ \mathrm{cd\ sr\ W}^{-1} \tag{29}$$

We reverse this definition, and we get:

$$1\mathrm{cd} = \frac{K_{cd}}{683}\ \mathrm{W\ sr}^{-1} \tag{30}$$

The watt is expressed in the form (see Appendix 2):

$$\mathrm{W} = \mathrm{kg\ m^2\ s^{-3}} \tag{31}$$

Using the new definitions of the various units introduced above and the fact that the steradian has no unit, we finally obtain:

$$\mathrm{cd} = \frac{1}{683} \frac{1}{6.626\,069\,57 \times 10^{-34}} \frac{1}{(9\,192\,631\,770)^2} K_{cd}\, h\,(\Delta\nu_{Cs})^2 \tag{32}$$

or:

$$1\ \mathrm{cd} = 2.614\,830\ \times 10^{10} K_{cd}\, h\,(\Delta\nu_{Cs})^2 \tag{33}$$

This new definition of the candela is none other than the light intensity in a given direction from a source which emits at a frequency of 540 x 10^{12} Hz and which has a radiation intensity of (1/683) watt per steradian. We note again that the definition includes the ground state hyperfine frequency of the cesium atom.

The amount of matter: the mole [mol]

In the original definition of the mole (see Ap. 1), the molar mass of carbon 12 was set at 0.012 kg / mol, and the mole contained 6.022 141 29 x 10^{23} atoms. The new definition fixes this number as an exact value called N_A, the Avogadro constant. We thus have :

$$N_A = 6.022\ 141\ 29 \times 10^{23}\ \text{mol}^{-1} \tag{34}$$

We can reverse this expression and we get:

$$1\ \text{mol} = \frac{6.062\ 141\ 29 \times 10^{-23}}{N_A} \tag{35}$$

The consequence of this definition is that the mole is now defined as the quantity of matter in a system which contains 6.022 141 29 x 10^{23} elementary entities. The molar mass of carbon 12 is no longer exact and must be determined experimentally.

The constants

The number of physical constants used to define the base units is therefore 7. They are given in Table 1 with the constants defined as:

$\Delta\nu_{Cs}$: ground state hyperfine frequency of resonance of the of the cesium atom

c: speed of light h: Planck constant e: elementary charge

k: Boltzmann constant N_A: Avogadro constant K_{cd}: luminous efficacy constant

In the new SI, the values of these constants are considered to be exact.

Quantity	Base unit and symbol	Constants, exact defined value
Time t	second: s	$\Delta\nu_{Cs}$= 9 192 631 770 s^{-1}
Length l	metre: m	c = 299 792 458 m s^{-1}
Mass m	kilogram: kg	h = 6.626 069 57x 10^{-34} J s
Electric Current I or i	ampere: A	e =1.602 176 565 x 10^{-19} C
Temperature T	kelvin: K	k = 1.380 648 8 x 10^{-23} J K^{-1}
Quantity de matter n	mole: mol	N_A=6.022 141 29 x 10^{23} mol^{-1}
Luminous intensity I_v	candela: cd	K_{cd}= 683 lmW^{-1}

Table 1. Physical constants used in the definition of the basic units of the SI[5].

Summary of new definitions of base units

In the new **SI**, base units' definitions are given in the form of equations:

Second

$$1\,s = \frac{9\ 192\ 631\ 770}{\Delta\nu_{Cs}} \tag{36}$$

Metre

$$1\,m = 30.663\ 318\ \frac{c}{\Delta\nu_{Cs}} \tag{37}$$

Kilogram

$$1kg\ = 1.475\ 5214 \times 10^{40}\,\frac{\Delta\nu_{Cs}}{c^2}\ \ h \tag{38}$$

Ampere $$1\,A = 6.789\,687 \times 10^{8}\,\Delta\nu_{Cs}\, e \tag{39}$$

Kelvin $$1\,K = 2.266\,665\,3\,\frac{\Delta\nu_{Cs}\,h}{k} \tag{40}$$

Candela $$1\,cd = 2.614\,830 \times 10^{10}\,K_{cd}\,h\,(\Delta\nu_{Cs})^2 \tag{41}$$

Mole $$1\,mol = \frac{6.062\,141\,29 \times 10^{-23}}{N_A} \tag{42}$$

3. Reflections on the new SI

The new SI system is therefore entirely defined according to seven physical constants whose value are declared to be exact. This obviously has important consequences for our interpretation of the relationships between the units of measurement and those physical constants. We can represent these relations by Figure 5.

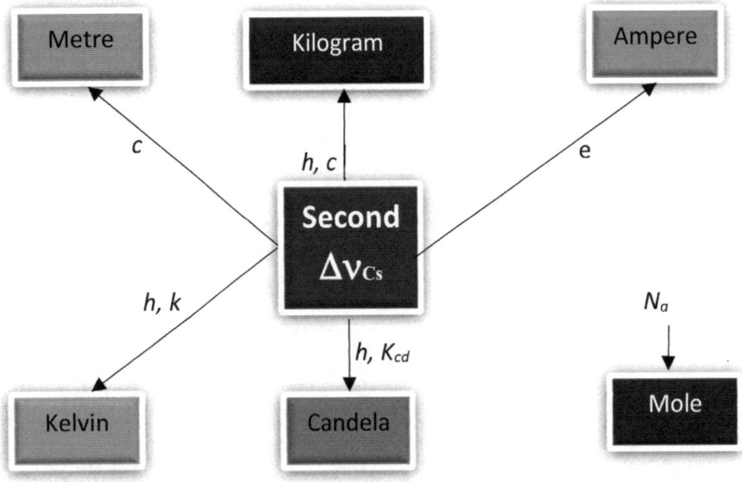

Figure 5. Diagram making explicit the important role played by the unit second and the Cs atomic clock in the definition of base units in the new (2019) International System of Units.

A major point that emerges is that the definitions are not independent of each other. The second plays a central role in the system. It appears in the definition of all the base units, except the mole. We live in a dynamic world and time, although we cannot define it exactly, plays a crucial role. The second is the unit of measurement for which we can make a most accurate standard: the Cs atomic clock. We also note that time being involved in the base units, all derived units are also function of the time unit. While the universe appears to be quantized, time remains a continuous quantity whose origin seems to be the big bang.

We must recognize that relativity introduces time as a relative quantity that varies with the velocity of the frame of reference when measured from a rest frame of reference. In a rest frame, we refer to time as *proper* time. It is in that context that the second is defined and it is in that frame that new definitions apply. When motion is involved, time is function of velocity[6], but the speed of light remains constant as determined in Michelson-Morley experiments. On the other hand, time is also a function of the gravitational potential in which the clock is situated. This was measured very accurately with H masers[7]. In a singularity like in a black hole, it does not appear that time exists reinforcing the concept that time has its origin at the big bang singularity.

On the other hand, the physical constants that are used have relationships with each other and are not independent of each other. In electromagnetism, two fundamental constants are introduced in particular, the magnetic constant μ_o and the electric constant ε_o. In electromagnetism, the magnetic constant was introduced having the exact value $4\pi \times 10^{-7}$ henry per metre. The speed of light c obtained in the same theory is:

$$c = \frac{1}{\sqrt{\mu_o \epsilon_o}} \tag{43}$$

In the definition of the metre, we have given c an exact value and, in this context, the constant ε_o should have therefore an exact value since μ_o has an exact value. In the new system of units, the ampere is defined using the constants, e and $\Delta\nu_{Cs}$, and the magnetic force, in the Kibble balance, is now determined by the constants e and h. The constant μ_o is not used. Therefore, ε_o is no longer defined as exact. It is the elementary charge that is now defined. It follows that, in the new system, ε_o and μ_o must be determined experimentally.

We also note that the fine structure constant is established as:

$$\alpha = \frac{e^2}{2\pi\varepsilon_0 hc} \qquad \text{(no units)} \tag{44}$$

and has a value known very precisely as:

$$\alpha = 7.297\,352\,566\,4\,(17) \times 10^{-3} \tag{45}$$

This constant therefore fixes a relation between the constants ε_o, e, h, c. It is important to mention that there exist questions on the constancy of fundamental constants, particularly α. However, that constant seems to be invariable over time to a level of the order of 10^{-15}, which makes the physical constants used invariable to that level (see Ref. 8 for details).

It should also be recalled, as we mentioned it earlier, that ε_o can also be linked to the meter by means of Lampard's theorem which applies to the calculable capacity made of 4 symmetrical metal rods, which is characterized by a capacity per metre given by Eq. 10: the resulting capacitance can serve as a standard impedance and can be linked to the ohm, itself linked to the volt and the ampere as well as to e and h through the quantum Hall effect and the Josephson effect.

It follows from this discussion that the constants introduced to define the base units, in addition to first having a value that keeps the system consistent with the previous definitions, must be consistent with each other when they are determined through various physical laws. The new definitions of the kilogram and the ampere illustrate this requirement very well. On the one hand, the Kibble balance allows us to measure h well as a function of the original kilogram artefact. Thereafter h is used in determining the kilogram through the units that characterizes h. We defined the ampere through the charge of the electron and the definition of the second, even if at the beginning the ampere was used in the expression of the force in the Kibble balance.

Finally, we must remember that the derived units are linked to the basic units by relationships that do not include a numerical parameter: the coulomb is 1 ampere-second; the volt is 1 watt per ampere, the joule is 1 newton-metre; etc. Consequently, the derived units are also described by equations including physical constants like the base units. We are thus in the presence of an ensemble in which all units can be considered on an equal footing. In the realization of standards representing the base units, it is therefore possible to imagine much simpler systems which can be applied to particular circumstances and fully comply with the definitions of base units. In fact, the established definitions will always be valid, and the achievement of standards may depend only on the imagination of the scientist involved. Their value will not vary since the fundamental constants used are defined as exact.

Appendices

Appendix 1. Definition of the basic units of the SI system before May 2019

second [s]

The second is the duration 9,192,631,770 periods of radiation corresponding to the transition between the two hyperfine levels of the ground state of the cesium 133 atom.

metre [m]

The metre is the length of the path traveled in a vacuum by light in a period of 1/299 792 458 seconds.

kilogram [kg]

The kilogram is the unit of mass; it is equal to the mass of the international prototype of the kilogram.

ampere [A]

The ampere is the intensity of a constant current which maintained in two parallel conductors of infinite length of negligible circular section placed at a distance of 1 metre from each other in a vacuum, would produce between the conductors a force of 2×10^{-7} newton per metre of length.

kelvin [K]

The kelvin, unit of temperature is the fraction 1 / 273.16 of the thermodynamic temperature of the triple point of water.

mole [mol]

The mole is the amount of matter in a system that contains as many elementary units as there are atoms in 0.012 kilogram of carbon 12.

candela [cd]

The candela is the luminous intensity in a given direction, of a source which emits a monochromatic radiation of frequency 540×10^{12} hertz and whose energy intensity in this direction is 1/683 watt per steradian.

Supplementary units

radian (rad): Measurement of the plane angle whose apex is the center of a circle and subtended by an arc whose length is equal to that of the radius.

steradian (sr): unit of measurement of a solid angle whose apex is at the center of a sphere, and which delimits a spherical surface area equal to that of a square whose sides are equal to the radius.

There are other related derivative units accepted for the protection of human health. They are outside the context of this analysis.

Appendix 2. Relation between selected derived units and base units.

There are many ways in which connections between base and derived units can be represented graphically. The literature contains several of those representations introduced particularly by institutes involved in the field of metrology. The following Figure A2 represents a combination of such representations and provides the relation between based units and selected derived units in a self-evident way. It should be noted that all units are now based on fundamental constants through mathematical relations.

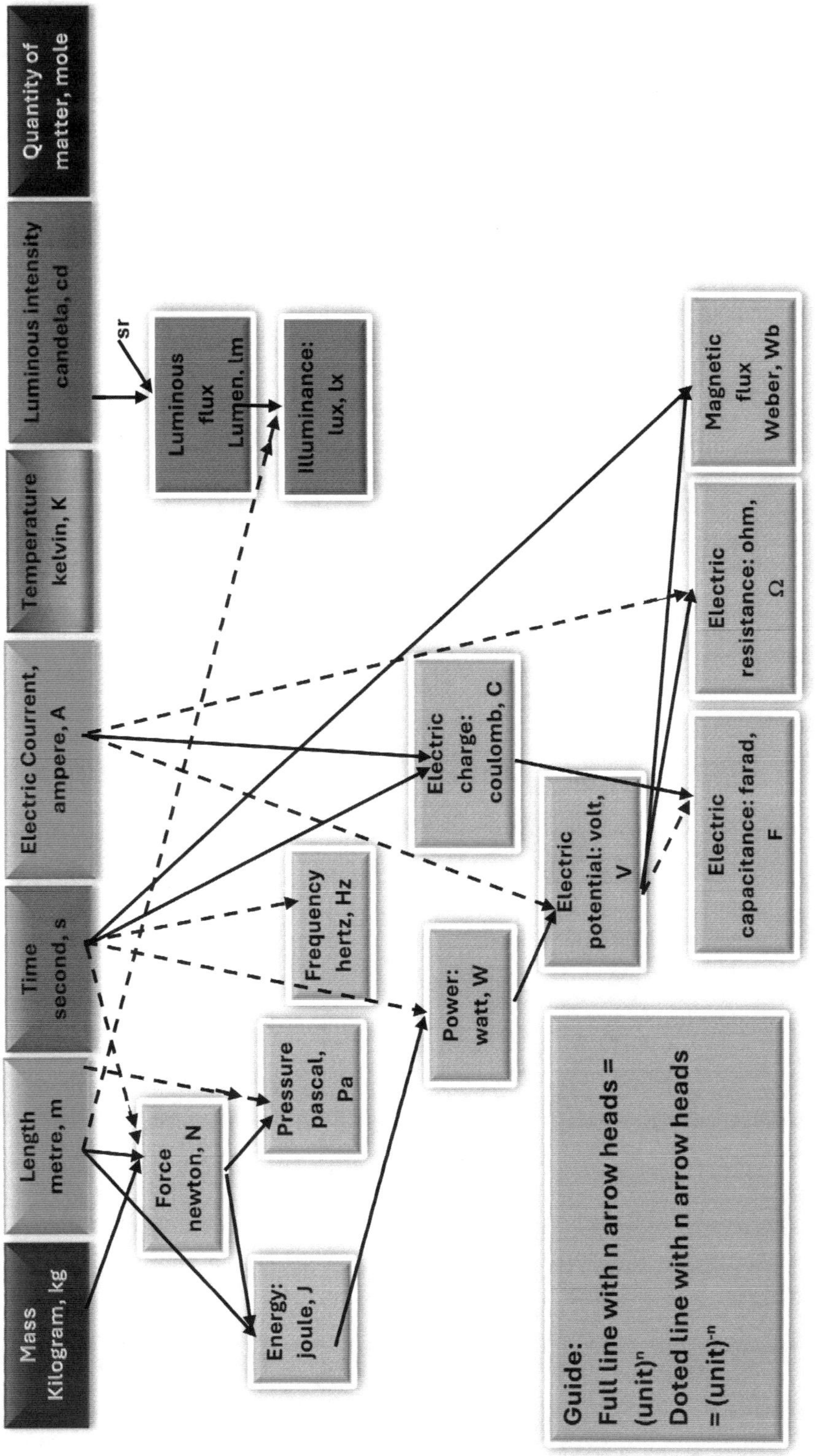

Figure A2. Relation between selected derived and base units used in the international system of units (SI) and defined in the main text.

Acknowledgement

I wish to thank Cipriana Tomescu for her encouragement during the writing of this text, and Thomas Udem for instructive discussions.

Notes and references

1) Several articles and reports, in particular the Brochure International System of Units, SI, BIPM , 9th edition 2019, have been used as the basis for writing this text.
2) Thompson A.M. and Lampard D.G. *A New Theorem in Electrostatics and its Application to Calculable Standards of Capacitance.* Nature 177, 888 (1956).
3) Vanier J. *Atomic frequency standards: Basic physics and impact on metrology*, Proc. Int. School of Physics "Enrico Fermi" Course CX LVI, Quinn T.J. S. Leschiutta, P. Tavella, IOS Press Amsterdam, 2001. See also Vanier J. and Tomescu C. *The Quantum Physics of Atomic Frequency Standards, recent developments*, 2016, Taylor and Francis, CRC Press.
4) Kibble B. P. Robinson I. A. Belliss J. H. *A realization of the SI Watt by the NPL Moving-coil Balance,* Metrologia 27 (4) pp 173-192, 1990.
5) 9th SI BIPM Brochure 2018. A very precise determination of the value of *h* was recently carried out using this method by B.M. Wood et al. Metrologia, 54, 3, 2017. The result is: $h = 6.626\,070\,133\,(60) \times 10^{-34}\,J\,s$, close to the recommended value.
6) Hafele J.C. and Keating, *Around the World Atomic Clocks: Predicted Relativistic Time Gains; Observed Relativistic Time Gains*, Science **177**, 166 1972.
7) Vessot RFC et al. *Test of relativistic gravitation with a space -borne hydrogen maser*, Phys. Rev. Lett. 45, 2081, 1980
8) Marion H. Pereira Dos Santos F. Abgral M. Zhang S. Sortais Y. Bize S. Maksimovic, I. Calonico D. Grunert J. Mandache C. Lemonde, P Santarelli G. Laurent Ph. Clairon A. and Salomon C. *A Search for Variation of Fundamental Constant using Atomic Fountain Clocks,* Phys. Rev. Lett. 90,15 150801(2003)

Multi-messenger astronomy in the new physics modality with GPS constellation

Arko P. Sen[1]

Kalia Pfeffer[1]

Paul Ries[2]

Geoff Blewitt[1]

Andrei Derevianko[1]

[1]University of Nevada, Reno, NV 89511, USA

[2]Jet Propulsion Laboratory, California Institute of Technology, 4800 Oak Grove Dr., Pasadena, CA

E-mail: `andrei@unr.edu`

Abstract.

We explore a novel, exotic physics, modality in multi-messenger astronomy. We are interested in exotic fields emitted by the mergers and their *direct* detection with a network of atomic clocks. We specifically focus on the rubidium clocks onboard satellites of the Global Positioning System. Bursts of exotic fields may be produced during the coalescence of black hole singularities, releasing quantum gravity messengers. To be detectable such fields must be ultralight and ultra-relativistic and we refer to them as exotic low-mass fields (ELFs). Since such fields possess non-zero mass, the ELF bursts lag behind the gravitational waves emitted by the very same merger. Then the gravitational wave observatories provide a detection trigger for the atomic clock networks searching for the feeble ELF signals. ELFs would imprint an anti-chirp transient across the sensor network. ELFs can be detectable by atomic clocks if they cause variations in fundamental constants. We report our progress in the development of techniques to search for ELF bursts with clocks onboard GPS satellites. We focus on the binary neutron star merger GW170817 of August 17, 2017. We find an intriguing excess in the clock noise post LIGO gravitational wave trigger. Potentially the excess noise could be explained away by the increased solar electron flux post LIGO trigger.

1. Introduction

Since the initial discovery of gravitational waves (GW) by LIGO in 2015 [1], there have been multiple detection of GWs. Most of these GWs result from mergers of black hole binaries. However, on August 17, 2017, a new class of GW events was discovered: the binary neutron star merger GW170817 [2]. That was the first astrophysical source detected in both the GW and multiwavelength electromagnetic radiation modalities and such is referred to as the multi-messenger astronomy [3]. The messengers came from a host galaxy 40 megaparsecs (~100 million light-years) away. Ref. [4] extended the gravitational and electromagnetic modalities of multi-messenger astronomy to exotic, beyond the Standard Model of elementary particles, fields. We start by summarizing salient points of the concept paper [4] and then report on our progress in the search for such exotica in Global Positioning System (GPS) atomic clock datastreams.

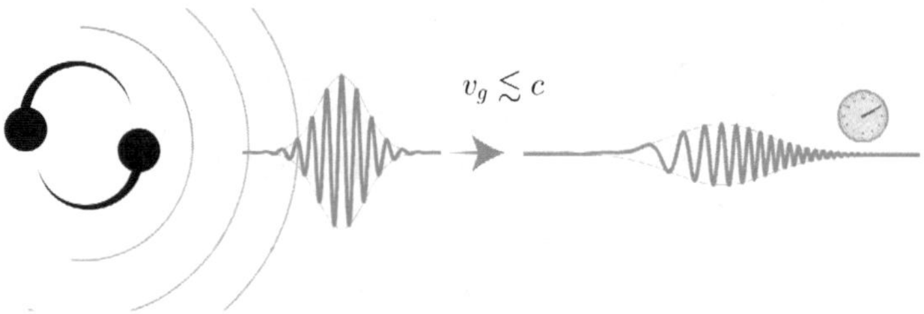

Figure 1. An emitted ELF burst propagates with the group velocity $v_g \lesssim c$ to the detector, lagging behind the GW burst. Since the more energetic ELF components propagate faster, the arriving ELF wavepacket exhibits a characteristic frequency anti-chirp. From Ref. [4].

Fig. 1 summarizes the idea of multi-messenger astronomy with the addition of exotica modality. A merger emits both the GW and an exotic low-mass fields (ELFs) bursts. Since such fields possess non-zero mass, the burst propagates at a group velocity v_g smaller than the speed of light. Thereby, the ELF pulse lags behind the GW burst. Because of the dispersion, the ELF burst tends to spread out as it travels. More energetic components of the ELF wave-packet travel faster, imprinting a universal anti-chirp signature in the atomic clock data. We refer the reader to Ref. [4] for technical details and to Ref. [5] for a more qualitative and informal discussion.

As to the exotic physics modality, we focus on ultralight (yet non-zero mass) bosonic fields as the messengers. We assume that bursts of such ELFs could be generated by cataclysmic astrophysical events such as black hole or neutron star mergers [6, 7]. Considering that much of the underlying physics remains to be discovered, a pragmatic approach [4] is based on energy arguments: based on the uncertainties for the LIGO events, $\Delta E \sim M_\odot c^2$ of energy can be released in the form of ELFs from a black hole binary and $\Delta E \sim 0.1\, M_\odot c^2$ in a neutron star merger. These are extreme (compared to terrestrial experiments) amounts of energy potentially released in the ELF modality. Atomic clocks are sensitive to gentle perturbations of energy levels by coherent, classical waves. Due to copious numbers of emitted ELFs this condition is indeed satisfied all the way to the sensors. Notice that the atomic clocks sensitivities are such that we do not need require $\Delta E \sim M_\odot c^2$ for the ELFs to be detectable. Much smaller fractions of the solar mass would suffice for LIGO detectable events.

There is a wide variety of speculative scenarios [4] for coherent ELF production in the mergers. One particularly intriguing scenario is due to quantum gravity. Much of the underlying physics of coalescing singularities in black hole mergers remains unexplored as it requires understanding of the as yet unknown theory of quantum gravity [8]. Then the ELF burst would emerge from the merger as a quantum gravity messenger.

Relativistic massive fields satisfy the energy-momentum dispersion relation

$$\omega(k) = \sqrt{(ck)^2 + \Omega_c^2}, \tag{1}$$

where the Compton frequency $\Omega_c = mc^2/\hbar$ depends on the ELF mass m. The energy and momentum of ELF quanta are given by $\varepsilon = \hbar\omega$ and $p = \hbar k$, where k is the wave number. Since we require a close time correlation (say delay less than a week) between the arriving GW and ELF bursts and the vast burst travel distances, the ELFs must be necessarily relativistic, i.e. their group velocity $v_g \approx c$.

A propagation of the initially emitted Gaussian ELF pulse (of central frequency ω_0 and of pulse duration τ_0) leads to the solution [4] for a field at the clock a distance R away from the progenitor

$$\phi(R,t) \approx \frac{1}{R}\left(\frac{c\Delta E}{2\pi^{3/2}\omega_0^2\tau}\right)^{1/2} \exp\left(-\frac{(t-t_s)^2}{2\tau^2}\right) \times \cos\left(\omega_0(t-t_s) - \frac{\omega_0}{4\delta t}(t-t_s)^2\right), \quad (2)$$

where $t_s = t_{GW} + \delta t$ is the time of arrival of the center of the pulse at the sensor. It is offset by $t_{GW} = R/c$, the time of arrival of the GW messenger. ΔE is the total energy released in the ELF channel. The signal duration τ and GW-ELF time delay δt are related as

$$\tau \approx 2\delta t/(\omega_0\tau_0).$$

Note that the instantaneous ELF frequency is time-dependent, $\omega(t) = (1-(t-t_s)/(2\delta t))\omega_0$, exhibiting a frequency "anti-chirp" at the sensor. The slope of the anti-chirp is given by

$$d\omega/dt = -1/(\tau\tau_0) = -\omega_0/(2\delta t).$$

This reflects a qualitative fact that the more energetic (or higher frequency) components have a higher phase velocity ω/k. The waveform (2) is shown in Fig. 1.

A GPS clock sampling rate in our dataset is 1 Hz, see Sec. 4. This limits ω_0 from above. These frequencies fix energies $\varepsilon_0 = \hbar\omega_0$ of detectable ELFs to below 10^{-14} eV. Practically, we are limited to a duration of the pulse at the sensor to about a month; this limits ω_0 from below. In fact, our GPS data set, see Sec. 5.2, is effectively limited to $\sim$ a day, requiring that $\omega_0 \gtrsim 2\pi/\text{day}$, i.e. above $\sim 10^{-19}$ eV. Moreover, it is desirable to observe at least several oscillations, so the anti-chirp can be observed. To summarize,

$$10^{-19}\,\text{eV} \lesssim \varepsilon_0 \lesssim 10^{-14}\,\text{eV}. \quad (3)$$

Now we know how an ELF pulse looks like at a clock. How can we detect them? The most obvious way how the clock frequencies can be affected is through the variation of fundamental constants, discussed in the following section.

2. Transient variation in fundamental constants

ELFs can couple to atomic clocks through interactions that lead to the variation of fundamental constants. We are interested in the phenomenological Lagrangian for the SM-ELFs interactions

$$\mathcal{L} = \hbar c\left(-\sum_f \Gamma_f m_f c^2 \bar{\psi}_f\psi_f + \frac{\Gamma_\alpha}{4}F_{\mu\nu}F^{\mu\nu}\right)\phi^2. \quad (4)$$

Here ψ_f are the SM fermion fields (of masses m_f) and $F_{\mu\nu}$ is the electromagnetic Faraday tensor. Γ_X are coupling constants (with $X = \alpha, m_f$) quantifying the strength of the quadratic interaction between the scalar ϕ and SM fields. We focus on the interactions that are quadratic in the ELF field ϕ for a number of reasons, like the U(1) invariance for complex-valued fields and the fact that the current limits on the couplings being much weaker for the quadratic couplings than for the linear ones.

The interaction Lagrangian (4) leads to the effective redefinition of fundamental constants (FCs): fermion masses and the fine structure constant,

$$m_f^{\text{eff}}(\mathbf{r},t) = m_f\left[1 + \Gamma_f \hbar c\phi^2(\mathbf{r},t)\right], \quad (5)$$

$$\alpha^{\text{eff}}(\mathbf{r},t) = \alpha\left[1 + \Gamma_\alpha \hbar c\phi^2(\mathbf{r},t)\right]. \quad (6)$$

The variations in FCs modify clock frequencies and, thereby, the apparent time as measured by the clock. In general, the sensitivity of an oscillator resonance frequency ν to the variation in FC X can be quantified by the coefficient

$$\kappa_X = \frac{\partial \log \nu}{\partial \log X}. \quad (7)$$

In atomic clocks, the clock frequency is determined by frequencies of both the atomic transition and the local oscillator (LO), see Sec. 3. The LOs in GPS clocks are quartz oscillators with their fundamental mode frequencies depending on FCs. We can introduce the LO sensitivity coefficient κ_X^{LO}, as in Eq. (7). Moreover, in GPS we compare the satellite and terrestrial clocks, so, in general, we have to operate in terms of differential sensitivity coefficients, K_X, that take sensitivities of individual clocks into account.

As the ELFs are ultra-relativistic, their travel time from satellites to terrestrial atomic clock is about 0.1 s, an order of magnitude smaller than our sampling time interval, $\Delta_t = 1$ s. Here we used the fact that the GPS constellation diameter $D_{\mathrm{GPS}} \approx 5 \times 10^4$ km. Moreover, all the clocks would experience approximately the same value of the field if the ELF wavelength $\lambda \approx 2\pi\hbar c/\varepsilon_0 \gg D_{\mathrm{GPS}}$. For the range (3) of detectable ELF energies, the corresponding wavelengths map into the range $10^5 \lesssim \lambda \lesssim 10^{10}$ km. This means that for $\varepsilon_0 \lesssim 10^{-13}$ eV, $\lambda \gg 10^4$ km and the entire GPS constellation measures nearly the same ELF phase and can be considered as a composite yet point-like quantum sensor. In the opposite limit of $\lambda \ll 10^5$ km, field values at different satellite clocks differ and if we were to take an average over all the clocks at a given time, the ELF signal would appear as an excess noise.

The fractional frequency excursion due to quadratic interaction (4) reads

$$s(t) = \frac{\delta\nu(t)}{\nu_{\mathrm{clock}}} = \Gamma_{\mathrm{eff}}\hbar c \phi(t)^2 , \tag{8}$$

where ν_{clock} is the unperturbed or the nominal clock frequency, $\phi(t)$ is the ELF field at the sensor (2), and $\Gamma_{\mathrm{eff}} \equiv \sum_X K_X \Gamma_X$ is the effective coupling constant expressed in terms of the differential sensitivity coefficients, K_X.

The sensitivity coefficients K_X depend on the internal composition and design of the clock and data taking rate. Our GPS clock phase (or bias) data, see Sec. 5.2, is sampled every second. These clock biases are the differences between the satellite and the fixed reference terrestrial atomic clock times as measured by the clocks. We will focus on ^{87}Rb microwave GPS clocks and fix an active hydrogen maser located in Kourou, French Guiana (station ID: KOUR) as a reference clock. In the following section, we discuss the operation of such atomic clocks.

If the characteristic time scale of the transient variation of FCs is longer than the servo loop time constant, $K_X = \kappa_X^{\mathrm{atom}}$, otherwise $K_X = \kappa_X^{\mathrm{LO}}$. From Ref. [9, 10, 11], we have $\kappa_\alpha^{\mathrm{Rb}} \approx 4.34$, $\kappa_{m_e}^{\mathrm{Rb}} = 2$, $\kappa_{m_q}^{\mathrm{Rb}} \approx -0.069$, $\kappa_\alpha^{\mathrm{H}} \approx 4$, $\kappa_{m_e}^{\mathrm{H}} = 2$, $\kappa_{m_q}^{\mathrm{H}} \approx -0.150$, $\kappa_\alpha^{\mathrm{LO}} = 2$ and $\kappa_{m_e}^{\mathrm{LO}} = 3/2$, where m_e and m_q are the electron and average up and down quark masses respectively. As discussed in Sec. 4, our data is acquired in the regime when $K_X = \kappa_X^{\mathrm{atom}}$. Thereby, the effective coupling constant for the Rb-H pairwise comparison becomes

$$\Gamma_{\mathrm{eff}} = 0.34\,\Gamma_\alpha + 0.081\,\Gamma_{m_q} , \tag{9}$$

where the dependence on Γ_{m_e} cancels out. Notice that the ELF signal would disappear in the data if we were to compare clocks of the same kind such as Rb-Rb, since $\Gamma_{\mathrm{eff}} \equiv 0$; this is a powerful test of systematic effects.

3. Atomic clocks of GPS network

GPS satellites host either ^{87}Rb or ^{133}Cs clocks. We focus on the Rb clocks because of their abundance (24 out of total 26 clocks are Rb in our data set for the GW170817 event) and because their frequency noise is primarily white [12]. The ^{87}Rb $5S_{1/2}$ electronic ground state splits into a manifold of two, $F = 1$ and $F = 2$, hyperfine levels. The clock transition is between two $M_F = 0$ B-field insensitive Zeeman sub-levels of the hyperfine manifold with the nominal frequency $\nu_{\mathrm{Rb}} \approx 6.835$ GHz. The frequency of a quartz oscillator is locked to ν_{Rb} through a servo-loop operation, driving a frequency synthesizer to generate a 10.23 MHz digital clock. This then drives the rest of the electronics generating the GPS microwave signals which are transmitted on carrier waves precisely at 154×10.23 MHz = 1.57542 GHz (L1 band) and 120×10.23 MHz = 1.22760 GHz (L2 band) [13]. (Actually, the frequency of the digital clock is set slightly lower to allow for the mean combined relativistic effects of time dilation and gravitational red shift, so that the clock frequency is precisely 10.23 MHz according to an observer at the gravitational potential at mean sea level on Earth.) The satellites broadcast microwave signals with a carrier phase that is commensurate with the on-board atomic clock system, through a phase locked loop frequency synthesizer. The phases of the microwave signals are further measured by the terrestrial receivers and through the data processing are interpreted as a measured GPS clock time, see Sec. 4. To estimate the effect of the FC variations, we need to gain a deeper understanding of how the GPS signal is generated, and how it propagates to the terrestrial receivers. We start with describing the operation of on-board Rb clocks in Sec. 3.1 and then, in Sec. 3.2, the terrestrial H-masers.

3.1. Rb atomic clocks

The GPS Rb atomic clocks are vapour-cell type [14, 15]. Such clocks include several packages which function together to ensure the frequency locking process. The Rb atoms are a part of the physics package (PP) which is primarily comprised of a ^{87}Rb discharge lamp, a filter cell containing ^{85}Rb isotope, a resonance cell containing ^{87}Rb, and a photo-detector [14]. The resonance vapor cell is placed in a microwave cavity and a solenoid generating quantizing magnetic field [15]. The PP is integrated with supporting electronic circuits which control temperature of the discharge lamp, filter cell and resonance cell and also control the solenoid coil's current supply [14].

Notice the use of two isotopes of Rb: ^{87}Rb and ^{85}Rb with nuclear spins $I = 3/2$ and $I = 5/2$ respectively. The ^{87}Rb lamp emits on the transitions from the $5\,^2P_{1/2}$, $5\,^2P_{3/2}$ levels to the ground state manifold (D1 and D2 lines). The emitted light passes through a filter cell containing the ^{85}Rb isotope. The fortuitous coincidence in the spectral line frequency associated with the $F = 2$ hyperfine level of ^{87}Rb with those associated with $F = 3$ hyperfine level of ^{85}Rb leads to the absorption of the $F = 2$ spectral lines of D1 and D2 emissions of ^{87}Rb from the discharge lamp by ^{85}Rb in the filter cell [16].

The filtered light is used for quantum state preparation via optical pumping to the $F = 2$ clock state. When this $F = 2$ state is subjected to a microwave radiation, the microwaves drive the population between the $F = 2$ and $F = 1$ clock levels. As a result the intensity of the light coming through the filter cell depends on the detuning of the microwave frequency from $\nu_{\rm Rb}$ [15, 16]. The difference in light intensity is measured by the photodetector. This is used by a servo-loop circuit to generate an error signal which further controls the frequency of the LO that determines the microwave frequency. The magnitude of the error signal is the difference between the microwave frequency and $\nu_{\rm Rb}$.

At the LO heart in GPS clocks are quartz crystals. Quartz is a piezo-electric material and when an electrical field is applied across the crystal, it undergoes mechanical deformations and vibrates at a fundamental frequency of the crystal that remains stable over relatively short timescales. The vibration modes depend on the geometry of the crystal. The relevant sensitivity coefficients to the variation of FCs was given in Sec. 2.

The servo-loop time constant $\tau_{\rm servo}$ determines the duration of the fastest perturbation that can be recorded by the clock. Military operators manually adjust $\tau_{\rm servo}$ and its precise value is not available to us at a given epoch, however it is known to be within 0.01 s to 0.1 s [17, 18]. This is sufficient to suppress the effect of occasional frequency jumps in the quartz oscillator down to the level of $< 10^{-11}$ s in accumulated clock phase [14]. Our dataset, described in Sec. 5, is sampled every second, giving enough time for LO frequency to be steered to the Rb atom frequency. This explains our choice of sensitivity constants in Eq. (9). It is worth pointing out that the electronics itself can be affected by the varying FCs; we presume that all the derived frequencies follow the Rb atom frequency at a sufficiently fast time scale, so that the sampled phases of the L1 and L2 GPS signals by the terrestrial GPS receivers are a faithful representation of the behavior of the Rb atomic frequency.

3.2. H-maser atomic clock

The terrestrial reference clock used for satellite clocks bias data (see Sec. 5.2), is an active H-maser T4Science atomic clock housed in the station named KOUR, located in Kourou, French

Guiana [19, 20]. Below we focus on the general principles of H-maser operations [21]. In these clocks, the quartz oscillator is locked to the transition frequency between hyperfine levels attached to the ground state of the hydrogen atom, $\nu_{\rm H} \approx 1.420$ GHz.

Such a clock uses a beam of atomic hydrogen, obtained by dissociating of molecular hydrogen H_2 in a discharge chamber. The atomic beam passes through an inhomogeneous state-selecting magnetic field, which focuses hydrogens in the $F = 1$ states into a Teflon-coated quartz storage bulb. This "magic" coating preserves atomic coherence for $\sim 10^6$ consecutive collisions with the bulb walls, critically enabling long interrogation times.

The bulb is placed within a radiofrequency (RF) cavity tuned to the transition frequency, $\nu_{\rm H}$ between $F = 1, M_F = 0$ and $F = 0, M_F = 0$ states of the H-atom [22, 23]. As the H-atoms undergo the transition from the $F = 1$ state down to the $F = 0$ state, the emitted microwave photons stimulate transitions in other atoms [24], hence the maser qualifier in these clocks. Cascade of stimulated transitions produce a highly coherent oscillation in the RF cavity. This results in a microwave signal to which the quartz oscillator is phase-locked.

The RF cavity is enveloped by magnetic shields to attenuate the ambient fields. An inner solenoid coil creates a uniform magnetic field to separate the different magnetic sublevels. There are two types of interactions that occur in the RF cavity: the atomic relaxation process that affects the population in $F = 1$ state and the process that leads to the decay of atomic phase coherence that affect the frequency stability [23]. The major purpose of the bulb is to spatially confine the atoms, enabling interaction with the in-phase components of the RF magnetic field in the cavity [23].

For longer sustenance of the oscillation either a sufficiently high H-atom density is to be maintained or the RF cavity is to be interrogated with external RF radiation at $\nu_{\rm H}$. Atomic clocks based on the former process is termed as active H-maser clocks, whereas clocks based on the latter are passive H-maser clocks. In active H-masers the cavity undergoes self-sustained oscillation. Both types of H-masers eventually drive a quartz oscillator which outputs a reference signal with a high stability and a low drift rate by control voltage adjustment [25]. The principal circuit of H-maser atomic clock has a crystal oscillator, an isolation amplifier circuit, an up and down-conversion circuit, a digital servo, and a frequency synthesizer circuit. There is an auxiliary circuit system that includes a constant temperature and a high-voltage source circuit, and a constant current source [25]. In passive H-maser clocks the interrogation microwave field is produced by an auxiliary frequency generator. H-masers have a good short-term stability, $\sigma_y(\tau = 1\text{ s}) < 10^{-12}$, reaching a noise floor of $\approx 10^{-15}$ after about an hour of averaging [24].

The cavity frequency is stabilized by a servo loop. The value of the time constant of the servo-loop response is in the order of 1 s [26]. As with the Rb clocks, the response to a sudden change in FCs is that of a quartz for time-scales much shorter than the servo-loop time constant. For time scales $\gg$ than the servo-loop time constant, it is the response of the hydrogen hyperfine splitting that matters. Once again, we ignore the complicated response of electronic circuits to the FC variations, as we believe, such a response is only important for very short time scales. These considerations govern our choice of sensitivity constants to FC variations in Sec. 2.

4. GPS data processing

As discussed in Sec. 3, GPS satellites broadcast microwave signals with a carrier phase that is commensurate with the on-board atomic clock system, through a phase locked loop frequency synthesizer. The carrier phase is sampled at regular intervals by a global terrestrial network of GPS receiver stations. Each GPS receiver measures the beat phase between the incoming carrier phase signal mixed with an internal replica signal, which is locked in phase to the receiver's internal quartz oscillator. Thus the beat phase measures the difference between time of transmission according to each GPS satellite clock, and the time of reception according to each receiver clock.

The Jet Propulsion Laboratory (JPL) routinely analyzes carrier phase data every 30 seconds from the global GPS network [27]. For the special purpose of this ELF search, JPL processed data at 1Hz across multiple days from an identical global 30-station network. Data processing includes modeling the time of flight of the carrier signal traveling at the phase velocity of light, accounting for relativistic Shapiro delay, rotational wind-up of circularly polarized phase, atmospheric refraction, and deterministic station motion caused by tidal and surface loading deformation. For each station, a linear combination of carrier phases at two different carrier frequencies is computed such that it eliminates the effect of ionospheric refraction n_{ion} assuming that $(n_{\text{ion}} - 1) \propto 1/\omega^2$ [28]. Note that this elimination applies even if fundamental constants are perturbed. The linear combination conserves the satellite clock variations, which for GPS is designed to be identical on carrier signals of both frequencies. Automatic data editing algorithms allow for arbitrary variation of clock phase, and so unusual clock behavior is not a determining factor in data outlier deletion [29].

Parameter biases in the model are simultaneously estimated using a square-root estimation filter. Those parameters include all the satellite and receiver clock biases every second, with the exception of one high quality receiver clock which defines the reference coordinate time. The receiver reference clock is selected such that it is synchronized to a stable H-maser frequency standard. Other biases that are important to estimate but are not explicitly used in our subsequent analysis include station coordinates, satellite orbit parameters, atmospheric delays and their variation, and carrier phase biases associated with an ambiguity in the integer number of cycles, which are subsequently determined precisely and fixed [28, 30].

For the neutral atmosphere, the signal delay is non-dispersive at microwave frequencies, thus it is independent of frequency variations in the carrier signal that may be caused by FC variations. However, in principle there could be a contribution of FC variations to delay arising from changes in refractive index. Variations in refractive index due to weather are accounted for using the GPT2 model [31], which was based on fits to numerical weather models of the European Center for Medium-Range Weather Forecasts.

Any residual variations in refractive index, such those caused by lack of spatio-temporal resolution in the GPT2 model, are estimated as part of the multi-parameter bias estimation by the square root information filter. Specifically, the variations are estimated as a random walk in the zenith delay and in two gradient parameters for each station. These parameters are separable from satellite clock variations owing to the so-called "mapping function," that is the modeled variation of atmospheric slant thickness with elevation angle of the satellite [31]. Such estimation should also absorb variation in refractive index caused by the FC variations we are seeking, leaving the imprint of clock frequency variations intact. The assumption is that any refractive index variations caused by ELFs do not greatly exceed the level of variations expected from weather.

The resulting data set used in our analysis is a 1-second time series of atomic clock biases from each of the GPS satellites that have Rb clocks. All of these biases are relative to the clock of the selected reference station. We also have access to receiver clock biases, however most GPS stations only use quartz oscillators without an external atomic frequency standard, and so are of limited use. Other useful metadata include satellite positions and velocities.

One aspect of the GPS system that we aim to exploit is that, in the co-rotating Earth-fixed reference frame of the global GPS station network, the satellite positions are designed to repeat approximately every sidereal day [32, 33]. The precise repeat time must account for the precession of each satellite's orbit, caused by Earth's oblateness, and the orbit's eccentricity, inclination, and semi-major axis. This precession is slightly different for each satellite, resulting in a repeat period of that varies in the range 1 day minus 245 ± 5s [33]. This repeat of the orbit geometry causes systematic errors such as ground multipath reflections to also repeat, hence errors in the carrier phase measurement tend also to repeat. Therefore, there is the potential to enhance signal to noise by comparing measured carrier phase time series from one repeat cycle to the next, as in the common mode error subtraction techniques. Such methods have proven to greatly reduce scatter in geodetic GPS positioning every second, with accuracy at the few mm level [33], therefore signal to noise ratio in clock bias data should similarly be improved.

5. Search for ELF signal and preliminary results

5.1. Effect of ELFs on the clock output

The Rb-locked LO frequency is measured by a narrow band circuit whose signal can be related to a sinusoidal wave [34]. The output voltage of such a circuit can be expressed as

$$V(t) = [V_0 + \varepsilon(t)] \sin [2\pi \nu_{\text{clock}} t + \varphi(t)] \,, \tag{10}$$

where V_0 and ν_{clock} are nominal peak voltage amplitude and the clock frequency respectively and $\varepsilon(t)$ and $\varphi(t)$ are the deviation from the nominal peak voltage and nominal clock phase [34, 35]. Usually the oscillator frequency remains constant for a short time interval but it can drift over a relatively larger period of time. The instantaneous frequency is conventionally defined as the first derivative of the phase, which is

$$\nu(t) = \nu_{\text{clock}} + \frac{1}{2\pi}\frac{d\varphi(t)}{dt}. \tag{11}$$

The extra phase term can be due to either technical noise or due to the ELF signal. Given a sampling time interval Δ_t, the off-set phase is explicitly

$$\begin{aligned} \varphi(t_j) &= 2\pi \left[\int_{t_{j-1}}^{t_j} \nu(t')dt' - \nu_{\text{clock}} \int_{t_{j-1}}^{t_j} dt' \right] \\ &\approx 2\pi[\nu(t_j) - \nu_{\text{clock}}]\Delta_t \,, \end{aligned} \tag{12}$$

where $t_j = j\Delta_t$; $j = \overline{1, N_t}$ and N_t being to total number of data points.

Note that the deviation from the nominal clock frequency is $\delta\nu(t_j) = \nu(t_j) - \nu_{\text{clock}}$. Then the fractional clock frequency excursions are

$$\frac{\delta\nu(t_j)}{\nu_{\text{clock}}} \approx \frac{1}{2\pi\Delta_t}\frac{\varphi(t_j)}{\nu_{\text{clock}}}, \tag{13}$$

to be matched to the sought ELF waveform (8).

5.2. Data analysis and results

In our GPS.ELF group, we pursue the analysis of data from the GPS atomic clocks. One can think of the GPS constellation as the largest human-built $\sim 50,000$ km-aperture sensor array. Combined with other satellite positioning constellations and terrestrial clocks, the total number of sensors $N_c \sim 100$. Employing networks is crucial for distinguishing ELF signals from technical noise. Furthermore, by having baselines with the diameter of the Earth or larger, one can resolve the sky position of the ELF source. This is a critical feature for multi-messenger astronomy that enables correlation with other observations of the progenitor. However, an ELF feature (say leading edge) propagation time across the GPS constellation is 0.2 s. The publicly available GPS data is sampled every $\Delta_t = 30$ s. Our GPS.ELF group has developed techniques to generate more frequently sampled, $\Delta_t = 1$ s, GPS datastreams. Still the limited time resolution makes tracking the leading edge of the ELF pulse across the GPS network challenging. Nevertheless, clock networks can still act collectively, gaining $\sqrt{N_c}$ in sensitivity and vetoing signals that do not affect all the sensors in the network.

We work with the GPS clock bias data for the day of GW170817 event detection and the day before. The satellite clock biases are relative to a fixed reference terrestrial clock. This is a non-stationary time series, dominated by random walk noise. This discrete clock bias data stream $d^{(0)}_{a,j}$, sampled every $\Delta_t = 1$ s, can be expressed, for the ath satellite clock, in units of time as

$$d^{(0)}_{a,j} = \frac{\varphi_a(t_j)}{\omega_a} - \frac{\varphi_r(t_j)}{\omega_{r0}}, \tag{14}$$

where the index $j = \overline{1, N_t}$, enumerates the epochs. N_t is the number of points sampled during the orbital period, as discussed below. The frequencies ω_a and ω_{r_0} are the nominal clock angular frequency of a satellite a and the reference terrestrial clocks. The time dependent discrete phases of the satellite and reference clocks are denoted by $\varphi_a(t_j)$ and $\varphi_r(t_j)$ respectively. The reference terrestrial clock for this work is a hydrogen maser at the KOUR station located in Kourou, French Guiana, see Sec. 3.2.

Although the proposal paper [4] has used the formalism of excess noise statistic and time-frequency decomposition, here we carry out a simplified analysis of the excess noise to also include a possibility of high-frequency/short wavelength ELFs, see Sec. 1.

As discussed in Sec. 4, the orbital geometry of the satellites repeats relative to observer on Earth, leading to repeating multipath errors. To account for the multipath errors, the GPS satellite orbital period based on the inertial Earth-fixed reference frame is set approximately to 24 hours minus 247 s = 86,153 s. The sampling interval of 1 s thus leads to the total number of points per day $N_t = 86,153$. To whiten the data, we employ the first order differencing of the discrete clock bias data, the so-called discretized pseudo-frequency [36],

$$d^{(1)}_{a,j} = d^{(0)}_{a,j+1} - d^{(0)}_{a,j}\,, \tag{15}$$

where a indexes individual datastreams and j spans over sampling grid. The differencing reduces the total number of discrete pseudo-frequency points to $N \equiv N_t - 1$. We remove the frequency drift and the offset for each clock a

$$u^{(1)}_{a,j} = d^{(1)}_{a,j} - f_{a,j}\,, \tag{16}$$

by fitting the $d^{(1)}_{a,j}$ data to the straight line of the form, $f_{a,j} = m_a t_j + b_a$, with m_a and b_a being fitting coefficients. The weighted average of $u^{(1)}_{a,j}$ over the network of N_c satellite clocks is

$$\bar{d}^{(1)}_j = \frac{\sum_a^{N_c} u^{(1)}_{a,j}/\sigma_a^2}{\sum_a^{N_c} 1/\sigma_a^2}\,, \tag{17}$$

where

$$\sigma_a^2 = \frac{1}{N-1}\sum_{j=1}^{N}\left(u^{(1)}_{a,j} - \left\langle u^{(1)}_a\right\rangle\right)^2, \tag{18}$$

$$\left\langle u^{(1)}_a\right\rangle = \frac{1}{N}\sum_{j=1}^{N} u^{(1)}_{a,j}\,. \tag{19}$$

Here σ_a^2 and $\left\langle u^{(1)}_a\right\rangle$ are the variance and mean of $u^{(1)}_{a,j}$ respectively for $N_c = 24$ Rb atomic clocks.

Further, the network weighted average pseudo-frequency as calculated using Eq. (17), is partitioned into M windows with each window consisting of $N_w = N/M$ points. The standard deviation in each window is calculated as

$$\sigma_w = \sqrt{\frac{1}{N_w - 1}\sum_{k=1}^{N_w}\left(\bar{d}^{(1)}_{w,k} - \left\langle \bar{d}^{(1)}_w\right\rangle\right)^2}\,, \tag{20}$$

with

$$\left\langle \bar{d}^{(1)}_w\right\rangle = \frac{1}{N_w}\sum_{k=1}^{N_w} \bar{d}^{(1)}_{w,k}\,, \tag{21}$$

where w enumerates the windows and k spans points in a given window.

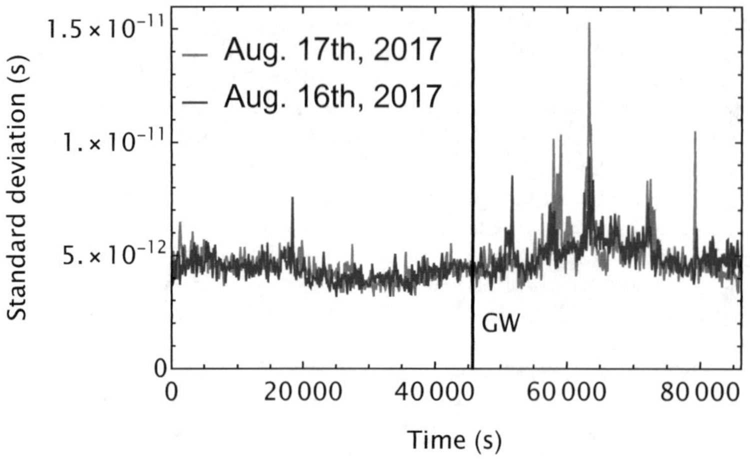

Figure 2. Standard deviations of pseudo-frequency for the GPS network average of Rb clocks for August 16 and 17, 2017. GW trigger detected at 45,682 s from the start of the day of August 17, 2017 (GPST), is shown with the black line. In computing standard deviation, we partitioned each day dataset into $M = 712$ windows, each containing $N_w = 121$ data points.

In Fig. 2, the standard deviation of $\bar{d}_j^{(1)}$ is graphed with respect to time in seconds from the start of a day, based on GPS Time (GPST) scale. Note that the difference between GPST and Coordinated Universal Time (UTC) time scale is 18 s for the year 2017. There are two traces: for the day before the GW170817 LIGO trigger (August 16, 2017) and for the day of the event (August 17, 2017). Comparing the two days, we see that the clock network exhibits excess noise level post the GW trigger shown by a black vertical line. In particular, we observe several spikes in the standard deviation of $\bar{d}_j^{(1)}$ occuring after the GW LIGO trigger (on August 17, 2017). The spike with the highest magnitude is delayed by 17,600 s (≈ 4.9 hours), from the GW LIGO trigger. Although we see the spike with a high magnitude in both the days' data, the spike post GW trigger is larger by a factor of about 2.3. This factor was computed by taking the average standard deviation for both days (at $\approx 4.7 \times 10^{-12}$ s) as the floor. Generically, one would expect both spikes of approximately the same magnitude due to repeating multipath errors in geodetic receivers. It is plausible that this extra noise is due to the putative ELF signal. To quantify this statement additional statistical analysis is required, in particular by finding the optimal size of the window N_w and characterizing the clock statistics. Notice, however, that we found that the solar electron flux was elevated during this period, see Fig. 3, perhaps negating the ELF hypothesis. This is discussed in the following section.

6. Systematic effects due to solar activity

The expulsion of solar energetic particles into the near-Earth space environment during periods of enhanced solar activity can be responsible for compromising GPS radio communication and positioning accuracy. Enhanced solar activity is defined as periods of time in which the solar corona exhibits increased magnetic field deformation. The acceleration of energetic particles from the solar corona, with the total energy release of about 10^{24} to 10^{32} ergs, typically characterizes a solar flare. Coronal mass ejections (CME) often accompany solar flares, where charged plasma particles with an average velocity ranging from 500 to 800 km/s are ejected into the environment [37].

Energetic particles emitted during solar activity events are introduced to Earth's magnetosphere and can precipitate to the ionosphere through acceleration along magnetic field lines. The energetic electrons and protons can also be trapped within the Van Allen radiation belts, accessible to GPS satellites. Exposure to high energetic particle fluxes within

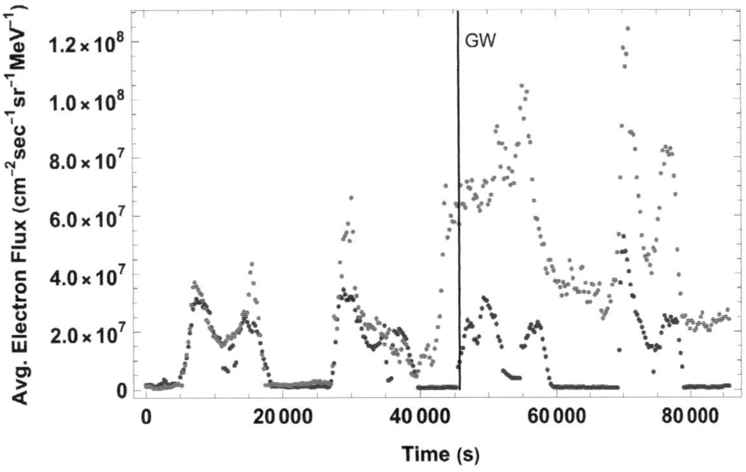

Figure 3. The average electron flux depicts a period of elevated electron flux that occurs after the GW event, GW170817. The GW LIGO event occurred at 45,682 s on August 17, 2017 (GPST) and is marked with a black vertical line. The elevated electron flux continues onto the following two days, maintaining the same order of magnitude.

the radiation belts can induce malfunctions to the satellites' electric power system [38]. The precipitation of energetic particles to the ionosphere can introduce ionization density variations, resulting in fluctuations to the amplitude and phase of transmitted signals. This phenomenon is known as ionospheric scintillation. GPS transmitted signals are vulnerable to strong ionospheric scintillation associated with enhanced electron content [39].

The Los Alamos National Laboratory has released energetic particle data recorded by the Combined X-Ray Dosimeter (CXD) aboard 21 operational GPS satellites. The CXD is equipped with high-energy and low-energy particle sensors that can record electron fluxes with an energy of 1 MeV to above 5 MeV and proton fluxes with an energy of 16 MeV to above 60 MeV [40]. An increased sunspot number associated with enhanced solar activity can correspond to elevated energetic particles introduced to the GPS orbit environment. Increased energetic particle exposure can be represented by spikes in electron and proton flux measurements. The daily sunspot number for August of 2017 indicates elevated values occurring at the end of the month.

We examined the network average proton and electron fluxes from the CXD aboard 17 GPS satellites for 16 August and 17 August, 2017, c.f. Fig. 2. The plot of the average electron flux is shown in Fig. 3. We observe a period of elevated electron flux that occurs after the GW event, GW170817. The proton flux exhibited no obvious spike or irregularity before this GW event. The spike in the network weighted average of the pseudo-frequency depicted in Fig. 2 could be related to elevated energetic particle exposure. Notice that the solar electron flux leads to a charge build-up on a satellite that could randomly discharge after reaching a certain level [41], which could potentially cause frequency jumps in the quartz oscillator [14, 18], modifying GPS microwave signals. This elevated electron flux starts around the time of the LIGO trigger. One of the intriguing explanations might be that the GWs perturb the Sun causing the solar activity.

7. Conclusions

The ELF proposal [4] has introduced a novel, exotic physics, modality in multi-messenger astronomy. In this report we outlined some practical considerations for the ELF search with the network of atomic clocks on board GPS satellites. We have generated 1 Hz clock bias data specifically for this analysis. Our search method focused on the excess noise in the clock network weighted average pseudo-frequency. There is an intriguing excess in the clock noise post

LIGO gravitational wave trigger for GW170817 binary neutron star merger of August 17, 2017. Potentially it could be explained away by the increased solar electron flux during the observed spike in the clock noise. It is also intriguing that solar activity is greatly enhanced around the time and after the LIGO trigger.

Acknowledgments
This work was supported in part by the U.S. National Science Foundation grant PHY-2207546, by the Heising-Simons Foundation, and by the NASA Grant 80NSSC23M0104.

References

[1] B Abbott *et al* (2016) *Physical Review Letters* **116** 061102 ISSN 0031-9007 URL https://link.aps.org/doi/10.1103/PhysRevLett.116.061102

[2] B Abbott *et al* (2017) *Phys. Rev. Lett.* **119** 161101 ISSN 0031-9007 (*Preprint* 1710.05836) URL https://link.aps.org/doi/10.1103/PhysRevLett.119.161101

[3] B Abbott *et al* (2017) *The Astrophysical Journal Letters* **848** L13 URL https://dx.doi.org/10.3847/2041-8213/aa920c

[4] C Dailey *et al* (2021) *Nature Astronomy* **5** 150–158 ISSN 2397-3366 (*Preprint* 2002.04352) URL http://www.nature.com/articles/s41550-020-01242-7

[5] Derevianko A (2023) (*Preprint* 2305.17138) URL http://arxiv.org/abs/2305.17138

[6] Bini D, Geralico A and Ortolan A (2017) *Phys. Rev. D* **95**(10) 104044 URL https://link.aps.org/doi/10.1103/PhysRevD.95.104044

[7] Baumann D, Chia H and Porto R (2019) *Phys. Rev. D* **99** 044001 ISSN 2470-0010 URL https://doi.org/10.1103/PhysRevD.99.044001 https://link.aps.org/doi/10.1103/PhysRevD.99.044001

[8] Loeb A (2018) (*Preprint* 1805.05865) URL https://arxiv.org/abs/1805.05865

[9] W Campbell *et al* 2021 *Phys. Rev. Lett.* **126**(7) 071301 URL https://link.aps.org/doi/10.1103/PhysRevLett.126.071301

[10] V Flambaum *et al* 2004 *Phys. Rev. D* **69**(11) 115006 URL https://link.aps.org/doi/10.1103/PhysRevD.69.115006

[11] B Roberts *et al* 2018 *Phys. Rev. D* **97** 083009 ISSN 2470-0010 URL http://arxiv.org/abs/1803.10264 https://link.aps.org/doi/10.1103/PhysRevD.97.083009

[12] Danzy F and Riley W 1989 Stability test results for gps rubidium clocks *Proceedings of the Annual Precise Time and Time Interval (PTTI) applications and Planning Meeting (21st), Held in Redondo Beach, California on November 28-30, 1989* URL https://apps.dtic.mil/sti/pdfs/ADA224769.pdf

[13] Blewitt G (2015) GPS and space-based geodetic methods *Treatise on Geophysics* (Oxford: Elsevier) pp 307–338 URL http://linkinghub.elsevier.com/retrieve/pii/B9780444538024000609

[14] Riley W (1981) A rubidium clock for GPS *Proceedings of the 13th Annual Precise Time and Time Interval Systems and Applications Meeting* URL https://apps.dtic.mil/sti/pdfs/ADA494259.pdf

[15] Camparo J 2007 *Physics Today* **60** 33–39 ISSN 0031-9228 (*Preprint* https://pubs.aip.org/physicstoday/article-pdf/60/11/33/16696572/33_1_online.pdf) URL https://doi.org/10.1063/1.2812121

[16] Y Guo *et al* (2022) *AIP Advances* **12** 045216 ISSN 2158-3226 URL https://doi.org/10.1063/5.0086523

[17] Griggs E, Kursinski E R and Akos D 2015 *Radio Science* **50** 813–826 (*Preprint* https://agupubs.onlinelibrary.wiley.com/doi/pdf/10.1002/2015RS005667) URL https://agupubs.onlinelibrary.wiley.com/doi/abs/10.1002/2015RS005667

[18] Dupuis et al R 2008 *Frequency Control Symposium, 2008 IEEE International* 655–660

[19] International GNSS Service (IGS) File Access URL https://files.igs.org/pub/station/log/kour_20230405.log

[20] International Terrestrial Reference Frame, (ITRF) URL https://itrf.ign.fr/en/station/KOUR-97301M210

[21] Major F G 2007 *The Quantum Beat:Principles and Applications of Atomic Clocks* 2nd ed (Springer-Verlag)

[22] Kleppner D, Goldenberg H M and Ramsey N F 1962 *Phys. Rev.* **126**(2) 603–615 URL https://link.aps.org/doi/10.1103/PhysRev.126.603

[23] Vessot R F C 2005 *Metrologia* **42** S80 URL https://dx.doi.org/10.1088/0026-1394/42/3/S09

[24] Lombardi M 2002 *Time and Frequency* pp 783–801 URL https://www.researchgate.net/publication/269420574_Time_and_Frequency

[25] Hu W, Shuai T, Xie Y, Chen P, Pei Y, Zhao Y and Wang R 2023 *Sensors* **23** ISSN 1424-8220 URL https://www.mdpi.com/1424-8220/23/22/9202

[26] Audoin, C 1981 *Rev. Phys. Appl. (Paris)* **16** 125–130 URL https://doi.org/10.1051/rphysap:01981001603012500, https://hal.science/jpa-00244902/document

[27] W Bertiger *et al* 2020 *Advances in Space Research* **66** 469–489 ISSN 0273-1177 URL https://www.sciencedirect.com/science/article/pii/S0273117720302532

[28] Blewitt G 1989 *J. Geophys. Res.* **94** 10187 ISSN 0148-0227 URL http://dx.doi.org/10.1029/JB094iB08p10187 http://doi.wiley.com/10.1029/JB094iB08p10187

[29] Blewitt G 1990 *Geophys. Res. Lett.* **17** 199–202 ISSN 00948276 URL http://dx.doi.org/10.1029/GL017i003p00199 http://doi.wiley.com/10.1029/GL017i003p00199

[30] Bertiger et al W 2010 *Journal of Geodesy* **84** 327–337 URL https://doi.org/10.1007/s00190-010-0371-9

[31] J Böhm *et al* 2015 *GPS Solutions* **19** 433–441 URL https://doi.org/10.1007/s10291-014-0403-7

[32] D Agnew *et al* 2007 *GPS Solutions* **11** 71–76 URL https://doi.org/10.1007/s10291-006-0038-4

[33] K Larson *et al* 2007 *Journal of Geophysical Research: Solid Earth* **112** (*Preprint* https://agupubs.onlinelibrary.wiley.com/doi/pdf/10.1029/2006JB004367) URL https://agupubs.onlinelibrary.wiley.com/doi/abs/10.1029/2006JB004367

[34] D Howe *et al* 1981 Properties of Signal Sources and measurement methods *Proceedings of the 35th Annual Symposium on Frequency Control* URL https://tf.nist.gov/general/pdf/554.pdf

[35] Riley W and Howe D 2008 Handbook of frequency stability analysis URL https://tsapps.nist.gov/publication/get_pdf.cfm?pub_id=50505

[36] B Roberts *et al* (2017) *Nature Comm.* **8** 1195 ISSN 2041-1723 URL http://www.nature.com/articles/s41467-017-01440-4

[37] Duffin R T, White S M, Ray P S and Kaiser M L 2015 *Journal of Physics: Conference Series* **642** 012006 URL https://dx.doi.org/10.1088/1742-6596/642/1/012006

[38] Chen Y, Carver M R, Morley S K and Hoover A S 2021 Determining ionizing doses in medium earth orbits using long-term gps particle measurements *2021 IEEE Aerospace Conference (50100)* pp 1–21 URL https://ieeexplore.ieee.org/document/9438516

[39] Kintner P M, Ledvina B M and de Paula E R 2007 *Space Weather* **5** (*Preprint* https://agupubs.onlinelibrary.wiley.com/doi/pdf/10.1029/2006SW000260) URL https://agupubs.onlinelibrary.wiley.com/doi/abs/10.1029/2006SW000260

Journal of Physics: Conference Series **2889** (2024) 012003 doi:10.1088/1742-6596/2889/1/012003

[40] J Distel *et al* 1999 The combined X-ray dosimeter (CXD) on GPS Block IIR satellites Tech. rep. Tech. Rep. LA-UR-99 URL https://permalink.lanl.gov/object/tr?what=info:lanl-repo/lareport/LA-UR-99-2280

[41] Frederickson A, Mullen E, Brautigam D, Kerns K, Robinson P and Holman E 1991 *IEEE Transactions on Nuclear Science* **38** 1614–1621 URL https://ieeexplore.ieee.org/document/124153

Journal of Physics: Conference Series **2889** (2024) 012004 doi:10.1088/1742-6596/2889/1/012004

Development of a molecular Hg_2 clock to investigate fundamental physics

M. Witkowski, A. Linek, M. Zawada, R. Ciuryło

Institute of Physics, Faculty of Physics, Astronomy and Informatics, Nicolaus Copernicus University in Toruń, ul. Grudziądzka 5, 87-100 Toruń, Poland.

E-mail: rciurylo@umk.pl

Abstract. We describe the development of an experimental system based on ultra-cold Hg atoms. We will use it to explore search possibilities for new interactions beyond the Standard Model. For this proposal, we will use two-color photoassociation and molecular clock transitions, and we will test the linearity of King's plot with clock transitions in ultra-cold mercury.

1. Introduction

We introduce a novel molecular sensor to study fundamental interactions, focusing on clock transitions within a Hg-Hg system. Our project implements optical Feshbach resonances in systems involving Hg_2 or Hg-alkali systems [1, 2], with the ultimate goal of constructing a Hg_2 optical molecular clock [3, 4]. This tool has the potential to push limits for fundamental research by achieving unprecedented advancements in terms of precision and accuracy.

We will employ Hg_2, one of the heaviest two-atom molecules to experimentally investigate new hadron-hadron interactions. The interatomic potential of Hg-Hg is relatively well-characterized [5], especially compared to other heavy molecules like Yb_2. Nonetheless, to mitigate the impact of theoretical imperfections, we will leverage quantum defect theory [6] and measure rovibronic bound states near the dissociation threshold within the ground electronic configuration [7]. This approach has recently emerged as a promising method for probing new interactions [8].

The spectroscopy of Hg_2 will be conducted in gas samples cooled to microkelvin temperatures within a dipole trap. Experimental results from one- and two-color photoassociations (PA) involving various isotopologues of Hg_2 will provide essential insights into quantum defects that describe near-threshold bound states, with and without electron excitation. Mapping these bound states within the Hg_2 system will allow us to select the appropriate isotopologue to realize the optical molecular clock [3]. To achieve the electronic ground state of Hg_2, a two-stage experiment will be employed. Firstly, a weakly bound state of Hg_2^* (near the molecule dissociation limit) will be created through photoassociation. Subsequently, the electronic ground-state molecule will be formed via stimulated Raman adiabatic passage (STIRAP). The realization of an optical molecular clock will be based on the scenarios proposed in Ref. [3]. The photoassociation spectra will be referenced to optical atomic frequency standards, for sub-kHz level accuracy. The formation of ultra-cold Hg_2 molecules through photoassociation will significantly advance our understanding of fundamental physics.

2. Search for new interactions

Non-Newtonian correction to the gravity can be parametrized by the Yukawa-type potential [9, 10]

$$V_{NNG}(r) = -\alpha G m_1 m_2 \frac{e^{-r/\lambda}}{r}, \tag{1}$$

where λ is the interaction range, and α is the interaction coupling constant, m_1 and m_2 are masses of interacting bodies separated by distance r, G is the gravitational constant. The additional interaction can also be seen as an appearance of the so-called fifth force due to the exchange of light bosons that couple nucleons [9, 10, 11]. The corresponding fifth force potential has a similar form

$$V_{5F}(r) = -N_1 N_2 \frac{g^2 \hbar c}{4\pi} \frac{e^{-r/\lambda}}{r}, \tag{2}$$

where N_1 and N_2 are atomic mass numbers, $\lambda = \hbar/(Mc)$ is determined by the mass M of an exchange particle, g^2 is a dimensionless parameter describing the coupling strength between nucleons and the new particle. Conversion between α and g^2 is: $\alpha \approx g^2 \hbar c/(4\pi G m_P^2) \approx 1.347 \times 10^{37} g^2$, where c is speed of light, $\hbar$ is reduced Planck constant, m_P is proton mass.

In our approach, the new Yukawa-type interaction is to be searched by comparing the measured energies of the near-threshold bound states in the ground electronic configuration of Hg_2 with the theoretical prediction. The theory involved relies on *ab initio* calculations of the energies of the molecular bound states near the dissociation threshold [7] with or without the new Yukawa-type interaction with properly adjusted quantum defect techniques [7, 12]. This approach has been recently proposed as a promising tool for searching for new interactions [8].

The experimental realization requires the Hg_2 to be in a weakly bound state of its electronic ground state. The required state will be created by a two-color photoassociation followed by a stimulated Raman adiabatic passage [13]. It allows the next implementation of the molecular clock scheme [3] to improve measurement accuracy further. Constraints on g^2 coupling strength between nucleons and new possible particle field and on α describing the magnitude of non-Newtonian gravity correction as a function of Yukawa range parameter λ, were determined from Yb_2 photoassociation [8]. Even better predictions are expected with Hg_2 [5], especially because van der Walls interaction in this system is significantly weaker than in other molecules composed of popular clock atoms such as Sr or Yb.

The other scenario is to look into electron-quarks interaction induced by Higgs boson coupling [14]. For this purpose, we plan to explore King plot [15] for transitions in Hg atom. Linearity of Kings plot is based on the assumption that the isotopic shift for i-th transition in isotopes A' and A is given by the following equation:

$$IS_i^{A'A} = \frac{A' - A}{A'A} MS_i + E_i \delta < r^2 >^{A'A}, \tag{3}$$

where MS_i is mass shift term, $\delta < r^2 >^{A'A}$ is the difference between the nuclear charge mean-square radii of isotopes, E_i is the electronic field shift term:

$$E_i = \frac{Ze^2}{6\epsilon_0} \Delta\rho_{\rm el}(0), \tag{4}$$

Z is the atomic number, e is the charge of the electron, ϵ_0 is the vacuum permittivity, and $\rho_{\rm el}(0)$ is a difference of electron density at nuclei in the upper and lower states of i-th transition. The normalized isotopic shifts:

$$\zeta_i^{A'A} = \frac{A'A}{A' - A} IS_i^{A'A}, \tag{5}$$

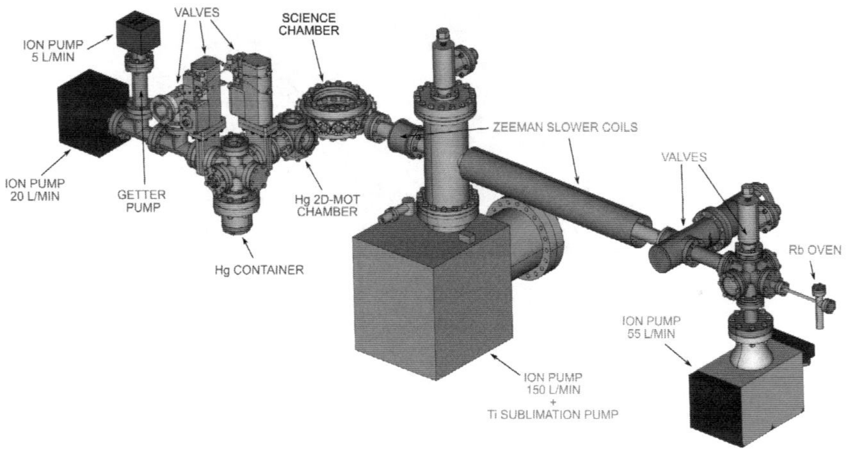

Figure 1. Schematic of the vacuum assembly for cooling and trapping Rb and Hg atoms with labelled key elements. The scientific chamber will be replaced with a new one, better suited for researching on ultracold Hg_2 molecules, as shown in Fig. 2.

for two transitions i and j in the same atom fulfill the linear relation [15]:

$$\zeta_i^{A'A} = \zeta_j^{A'A}\frac{E_i}{E_j} + MS_i - MS_j\frac{E_i}{E_j}. \tag{6}$$

Up to date, we tested linearity of King's plot in Hg for two transitions: $6s6p\ ^3\mathrm{P}_2 - 6s7s\ ^3\mathrm{S}_1$ [16] and $6s^2\ ^1\mathrm{S}_0 — 6s6p\ ^3\mathrm{P}_1$ [17, 18]. However, exploration of clock transitions present in Hg will allow us to significantly improve accuracy of such tests [19, 20] and look for new physics beyond Standard Model [14, 21, 22].

3. Development of experimental setup

The above scientific goals require adaptation of our experimental setup depicted in Fig. 1 and described in [1]. The setup will be enhanced by adding an ion detector, which will provide an alternative mechanism for molecule detection besides the commonly used method of atomic loss detection. To facilitate precise measurements of Hg clock transitions $^1\mathrm{S}_0$–$^3\mathrm{P}_0$ and $^1\mathrm{S}_0$–$^3\mathrm{P}_2$, the system will be equipped with relevant clock laser beams at 227 nm and 266 nm and the magic-wavelength optical lattice at 363 nm. Piezoelectric elements will control the in-vacuum lattice mirrors. To ensure proper beams' polarization, the lattice and clock beams will be introduced vertically through vacuum windows mounted at Brewster's angle. The spectral narrowing of both clock lasers will be achieved with dual-wavelength high-finesse cavity [23]. Additionally, repumping laser beams from Hg clock states $^3\mathrm{P}_0$ (405 nm) and $^3\mathrm{P}_2$ (546 nm) will be introduced into the system. To improve vacuum conditions, an ion-getter pump will be added to the main chamber. Furthermore, the inner wall of the chamber will be coated with a layer of gold, providing an additional mechanism for mercury atom pumping.

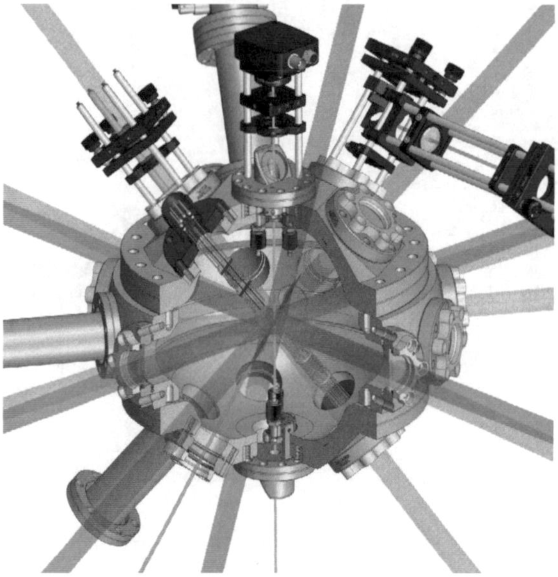

Figure 2. Vacuum chamber for cooling and trapping Hg atoms. The vertical beam constitutes an optical lattice operating at the magic wavelength at 363 nm. Cooling ($^1S_0$–$^3P_1$ at 254 nm) and repumping laser beams from Hg clock states $^3P_0$ (405 nm) and $^3P_2$ (546 nm) are also introduced into the system. To ensure proper polarization of the clock beams and the optical lattice, the entrance windows of the vertical beams are oriented at the Brewster angle. In the horizontal plane, optical dipole trap and STIRAP beams will be inserted into the chamber. A layer of gold provides an additional mechanism for pumping mercury atoms.

Acknowledgments
This project 20FUN01 TSCAC has received funding from the EMPIR programme co-financed by the Participating States and from the European Union's Horizon 2020 research and innovation programme. This work was supported by the National Science Centre, Poland, Project no. 2021/42/E/ST2/00046 and 2021/41/B/ST2/00681.The measurements were performed as a part of research done at the National Laboratory FAMO (KL FAMO) in Toruń, Poland. KL FAMO in Toruń is supported by the subsidy of the Ministry of Education and Science.

References
[1] Witkowski M, Nagórny B, Munoz-Rodriguez R, Ciuryło R, Żuchowski P S, Bilicki S, Piotrowski M, Morzyński P and Zawada M 2017 *Opt. Express* **25** 3165–3179 URL `https://opg.optica.org/oe/abstract.cfm?URI=oe-25-4-3165`
[2] Borkowski M, Muñoz Rodriguez R, Kosicki M B, Ciuryło R and Żuchowski P S 2017 *Phys. Rev. A* **96**(6) 063411 URL `https://link.aps.org/doi/10.1103/PhysRevA.96.063411`
[3] Borkowski M 2018 *Phys. Rev. Lett.* **120**(8) 083202 URL `https://link.aps.org/doi/10.1103/PhysRevLett.120.083202`
[4] Kondov S, Lee C H, Leung K, Liedl C, Majewska I, Moszynski R and Zelevinsky T 2019 *Nature Physics* **15** 1118–1122 URL `https://doi.org/10.1038/s41567-019-0632-3`
[5] Krośnicki M, Strojecki M, Urbańczyk T, Pashov A and Koperski J 2015 *Physics Reports* **591** 1–31 ISSN 0370-1573 URL `https://www.sciencedirect.com/science/article/pii/S0370157315003348`
[6] Mies F H 1984 *The Journal of Chemical Physics* **80** 2514–2525 ISSN 0021-9606 URL `https://doi.org/10.1063/1.447000`
[7] Kitagawa M, Enomoto K, Kasa K, Takahashi Y, Ciuryło R, Naidon P and Julienne P S 2008 *Phys. Rev. A* **77**(1) 012719 URL `https://link.aps.org/doi/10.1103/PhysRevA.77.012719`
[8] Borkowski M, Buchachenko A A, Ciuryło R, Julienne P S, Yamada H, Kikuchi Y, Takasu Y and Takahashi Y 2019 *Scientific Reports* **9** 14807 URL `https://doi.org/10.1038/s41598-019-51346-y`

[9] Fayet P 1996 *Classical and Quantum Gravity* **13** A19–A31 URL https://dx.doi.org/10.1088/0264-9381/13/11A/004

[10] Adelberger E G, Heckel B R and Nelson A E 2003 *Annual Review of Nuclear and Particle Science* **53** 77–121 URL https://doi.org/10.1146/annurev.nucl.53.041002.110503

[11] Knapen S, Lin T and Zurek K M 2017 *Phys. Rev. D* **96**(11) 115021 URL https://link.aps.org/doi/10.1103/PhysRevD.96.115021

[12] Borkowski M, Buchachenko A A, Ciuryło R, Julienne P S, Yamada H, Kikuchi Y, Takahashi K, Takasu Y and Takahashi Y 2017 *Phys. Rev. A* **96**(6) 063405 URL https://link.aps.org/doi/10.1103/PhysRevA.96.063405

[13] Gaubatz U, Rudecki P, Schiemann S and Bergmann K 1990 *The Journal of Chemical Physics* **92** 5363–5376 ISSN 0021-9606 URL https://doi.org/10.1063/1.458514

[14] Delaunay C, Ozeri R, Perez G and Soreq Y 2017 *Phys. Rev. D* **96**(9) 093001 URL https://link.aps.org/doi/10.1103/PhysRevD.96.093001

[15] King W H 1963 *J. Opt. Soc. Am.* **53** 638–639 URL https://opg.optica.org/abstract.cfm?URI=josa-53-5-638

[16] Rayman M D, Aminoff C G and Hall J L 1989 *J. Opt. Soc. Am. B* **6** 539–549 URL https://opg.optica.org/josab/abstract.cfm?URI=josab-6-4-539

[17] Witkowski M, Kowzan G, Munoz-Rodriguez R, Ciuryło R, Żuchowski P S, Masłowski P and Zawada M 2019 *Opt. Express* **27** 11069–11083 URL https://opg.optica.org/oe/abstract.cfm?URI=oe-27-8-11069

[18] Linek A, Morzyński P and Witkowski M 2022 *Opt. Express* **30** 44103–44117 URL https://opg.optica.org/oe/abstract.cfm?URI=oe-30-24-44103

[19] Miyake H, Pisenti N C, Elgee P K, Sitaram A and Campbell G K 2019 *Phys. Rev. Res.* **1**(3) 033113 URL https://link.aps.org/doi/10.1103/PhysRevResearch.1.033113

[20] Counts I, Hur J, Aude Craik D P L, Jeon H, Leung C, Berengut J C, Geddes A, Kawasaki A, Jhe W and Vuletić V 2020 *Phys. Rev. Lett.* **125**(12) 123002 URL https://link.aps.org/doi/10.1103/PhysRevLett.125.123002

[21] Flambaum V V, Geddes A J and Viatkina A V 2018 *Phys. Rev. A* **97**(3) 032510 URL https://link.aps.org/doi/10.1103/PhysRevA.97.032510

[22] Berengut J C, Budker D, Delaunay C, Flambaum V V, Frugiuele C, Fuchs E, Grojean C, Harnik R, Ozeri R, Perez G and Soreq Y 2018 *Phys. Rev. Lett.* **120**(9) 091801 URL https://link.aps.org/doi/10.1103/PhysRevLett.120.091801

[23] Linek A, Muñoz-Rodriguez R, Zawada M and Witkowski M 2023 *2023 Joint Conference of the European Frequency and Time Forum and IEEE International Frequency Control Symposium (EFTF/IFCS)* pp 1–2 URL https://ieeexplore.ieee.org/document/10272089

Journal of Physics: Conference Series **2889** (2024) 012005 doi:10.1088/1742-6596/2889/1/012005

Atomic Clock Ensemble in Space

L. Cacciapuoti[1], A. Busso[1], R. Jansen[1], S. Pataraia[1], T. Peignier[1], S. Weinberg[1], P. Crescence[2], A. Helm[2], J. Kehrer[2], S. Koller[2], R. Lachaud[2], T. Niedermaier[2], F.-X. Esnault[3], D. Massonnet[3], D. Goujon[4], J. Pittet[4], A. Perri[4], Q. Wang[4], S. Liu[5], W. Schaefer[5], T. Schwall[5], I. Prochazka[6], A. Schlicht[7], U. Schreiber[7], P. Laurent[8], M. Lilley[8], P. Wolf[8], C. Salomon[9]

[1] European Space Agency, ESTEC, Noordwijk, The Netherlands
[2] Airbus Defence and Space, Friedrichshafen, Germany
[3] CNES, Toulouse, France
[4] Safran Timing Technologies SA, Neuchâtel, Switzerland
[5] Timetech, Stuttgart, Germany
[6] Czech Technical University in Prague, Prague, Czech Republic
[7] Technical University of Munich, Munich, Germany
[8] SYRTE, Observatoire de Paris-PSL, CNRS, Sorbonne Université, LNE, Paris, France
[9] Laboratoire Kastler Brossel, ENS-PSL, Sorbonne Université, Collège de France, CNRS, Paris France

E-mail: `Luigi.Cacciapuoti@esa.int`

Abstract. The Atomic Clock Ensemble in Space (ACES) mission is developing high performance clocks and links for space to test Einstein's theory of general relativity. From the International Space Station, the ACES payload will distribute a clock signal with fractional frequency stability and accuracy of 1×10^{-16} establishing a worldwide network to compare clocks in space and on the ground. ACES will provide an absolute measurement of Einstein's gravitational redshift, it will search for time variations of fundamental constants, contribute to test topological dark matter models, and perform Standard Model Extension tests. Moreover, the ground clocks connected to the ACES network will be compared over different continents and used to measure geopotential differences at the clock locations.

After solving some technical problems, the ACES flight model is now approaching its completion. System tests involving the laser-cooled Cs clock PHARAO, the active H-maser SHM and the on-board frequency comparator FCDP have measured the performance of the clock signal delivered by ACES. The ACES microwave link MWL is currently under test. The single-photon avalanche detector of the optical link ELT has been tested and will now be integrated in the ACES payload.

The ACES mission concept, its scientific objectives, and the recent test results are discussed here together with the major milestones that will lead us to the ACES launch.

1. The ACES mission

Atomic Clock Ensemble in Space [1, 2, 3] is a mission using high performance clocks in space and on the ground to test the foundations of general relativity, contribute to the generation and distribution of global time scales, and perform chronometric geodesy experiment.

The ACES payload (see Fig. 1) will be installed onboard the International Space Station (ISS), at the external payload facility of the Columbus module. It includes two high stability and

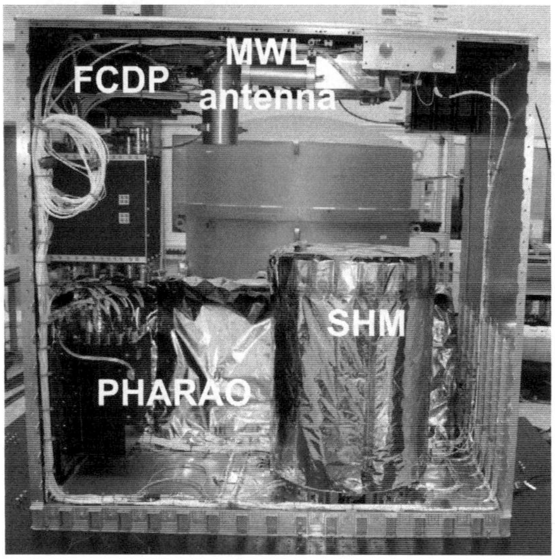

Figure 1. The ACES payload during integration. The vacuum chamber that will be used to test the payload is visible in the background.

accuracy atomic clocks: PHARAO, acronym for Projet d'Horloge Atomique par Refroidissement d'Atomes en Orbite, is a primary frequency standard using laser-cooled Cs atoms; SHM, the ACES flywheel oscillator, is an active Space Hydrogen Maser.

PHARAO [4] is specified to a fractional frequency stability of $1\times10^{-13}/\sqrt{\tau}$ for $1\ \text{s} < \tau < 10^6$ s, where τ is the integration time expressed in seconds, and an accuracy of $1-2\times10^{-16}$ (see Fig. 2). The 100 MHz signals generated by PHARAO and SHM enter the Frequency Comparison and Distribution Package (FCDP), which compares the phase of the two clocks and distributes the ACES clock signal to the MicroWave Link (MWL) electronics. The phase comparison measurement between PHARAO and SHM is also the error signal of a phase locked loop (PLL) used to steer the local oscillator of PHARAO on the SHM clock signal on a time scale of 1-2 s (short-term servo-loop). A second loop, a frequency locked loop operating on a time scale of 100 s or longer (long-term servo-loop), will then use the frequency measurements performed in the PHARAO clock with respect to the Cs reference to correct SHM output against temperature variations and long-term drifts. The two loops lock the onboard clocks together thus defining the ACES frequency reference, which will have the short-term stability of SHM and the long-term stability and accuracy of the Cs clock PHARAO. At this point, the clock signal enters the MWL electronics where the ACES atomic time scale is built and used for the operation of the two onboard links. MWL is a 2-way, 3-frequency link [5, 6, 7], with an up-link in the Ku-band and two down-links in the Ku and S-band. Here, the 100 MHz clock signal is upconverted, modulated by a PN-code carrying the time information, and sent towards ground users via the two antennas placed on the top panel of the ACES payload (see Fig. 1) and facing the Earth. A distributed network of MWL ground terminals located at prominent metrology institutes worldwide (SYRTE, PTB, Wettzell Observatory, and NPL in Europe; JPL and NIST in the United States; NICT in Japan) will be the physical interface to connect ground clocks to ACES. Space-to-ground and ground-to-ground comparisons of clocks will be enabled by ACES. Moreover, a transportable terminal will be available to calibrate fixed MWL ground stations for time transfer experiments and to perform measurement campaigns on request at other metrology institutes. MWL is also responsible for time tagging the time-of-arrival of the laser pulses

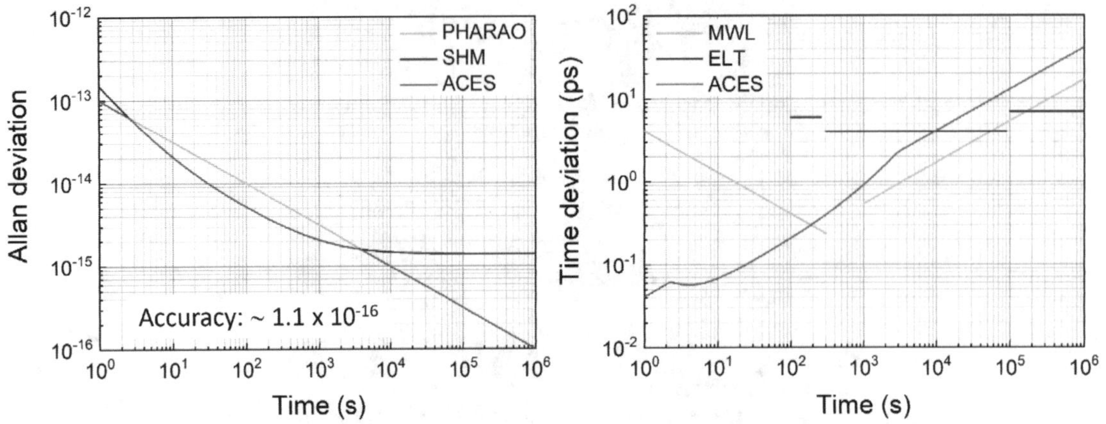

Figure 2. (Left) Stability requirement of the ACES clocks expressed in Allan deviation. (Right) Time deviation specifying the stability of the ACES links for space-to-ground clock comparisons.

received by the ELT (European Laser Timing) optical link. A network of ground-based Satellite Laser Ranging (SLR) stations connected to atomic clocks is completing the ground segment needed for the ELT experiment. SLR stations will be coordinated by the Technical University of Munich within the International Laser Ranging Service (ILRS), to which ACES has already been proposed as an official ranging target. Wettzell SLR is the ACES primary station, which, together with Graz, Herstmonceux, Potsdam and Zimmerwald, has already been calibrated for time transfer experiments with ELT. Other stations can join, provided that they comply with the ISS safety standards. Finally, a dedicated GNSS receiver will provide precise orbit data of the ACES payload, important to operate the ACES time & frequency links and precisely locate the PHARAO clock to perform redshift measurements.

The space-to-ground clock de-synchronization measurements delivered by MWL and ELT are the primary data product of the ACES mission. They are used to compare clocks both space-to-ground and ground-to-ground to test Einstein's theory of general relativity, and develop applications in time & frequency metrology, and geodesy.

In the fundamental physics domain, ACES will perform a gravitational redshift test on the Earth-to-ISS baseline. Differently from other redshift tests, the modulation of the redshift effect between perigee and apogee is quite small due to the reduced eccentricity of the ISS orbit and therefore marginally sufficient for a competitive measurement. Therefore, ACES will rely on the accuracy of the PHARAO clock and of Cs fountain clocks on the ground to perform an absolute measurement of the frequency difference between the space and the ground clock. The ACES performance for a redshift test along the ISS orbit has been studied in [8]. With PHARAO uncertainty at $1-2\times10^{-16}$ and a network of up to 8 ground stations, an uncertainty of 2-3 ppm is expected to be reached after 20 days of measurement, a factor 70 improvement on the historical GPA experiment [9, 10], and a factor 10 better than the more recent redshift test based on the GALILEO 5 and 6 satellites [11, 12]. The characterization of the PHARAO accuracy budget to $1-2\times10^{-16}$ will require the complete mission duration to minimize the statistical uncertainty on the evaluation of the frequency shifts due to cold collisions and distributed cavity phase.

One of the most important aspects of ACES is its network potential. The stability specified for the ACES links for space-to-ground clock comparisons is reported in Fig. 2. However, ACES can also be seen as a relay satellite connecting clocks in space and on the ground in a worldwide network. Ground clocks can be compared when in common view with the ACES

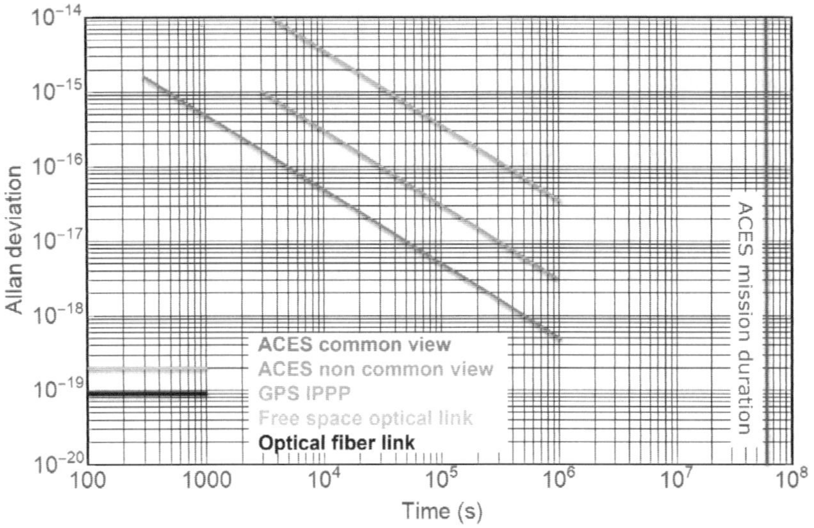

Figure 3. ACES MWL performance for the comparison of two ground clocks in common view (red) and non-common view (green) [3] compared to the GPS IPPP performance (blue) [13]. The typical stability of optical links through a fiber (black) or in free space (yellow) are also reported [14].

payload, or across continents in the non-common view configuration. In the first case, the noise of the onboard clock is in common-mode and cancels when taking the difference of the two simultaneous space-to-ground comparisons. As shown in Sec. 3, both MWL and ELT long-term stability remains below 1 ps. Therefore, the performance of the common view comparison, now depending on the link noise only, allows reaching a fractional frequency uncertainty of 1×10^{-17} after a few days of integration time. In a non-common view comparison, the SHM clock will act as a flywheel oscillator keeping the time across continents. The uncertainty in the comparison of the two ground clocks will now depend on both the link noise and the time error accumulated by SHM during the deadtime. Due to the extra noise introduced by the clock error, non-common view comparisons will require about 5 days to reach the 1×10^{-17} uncertainty level. Figure 3 shows the ACES performance for common view (red) and non-common view (green) comparisons of two ground clocks [3], together with the GPS IPPP performance (blue) [13] and the typical stability of optical links though a fiber (black) and in free space (yellow) [14].

Optical atomic clocks have experienced a tremendous progress in the last decades [15, 16, 17, 18, 19] and ACES will be able to exploit their potential connecting them in a worldwide network. Cross comparisons of atomic clocks based on different transitions will allow testing the time stability of the fine structure constant, the quark mass, and the electron mass [20, 21]. In a similar way, the ACES network of atomic clocks will also be used to search for dark matter and place boundaries on topological dark matter models [22, 23, 24]. Topological dark matter can be represented by a scalar field that couples to the three fundamental constants thus inducing fluctuations of atomic transition frequencies. The comparison of clocks based on different atomic transitions will allow to place limits on the three energy scales, which are defining the coupling of the dark matter field to the three fundamental interactions. The simultaneous observation with several clocks compared along different baselines will provide ways to confirm any observation above the sensitivity threshold and control systematic errors. Moreover, the screening effect of the dark matter field due to the Earth mass will be reduced to about 0.06 on the space clock

Journal of Physics: Conference Series **2889** (2024) 012005 doi:10.1088/1742-6596/2889/1/012005

PHARAO, several order of magnitude less than the 10^{-7} shielding factor experienced by ground clocks [25]. Therefore, the PHARAO clock, which is sensitive to all three fundamental constants, will assume a key role in the hunt for dark matter.

The ACES network of ground clocks will also be used for chronometric geodesy experiments [19, 26, 27, 28] on a worldwide scale. The measurement performed at the Tokyo Skytree Tower has demonstrated the possibility of resolving the geopotential difference between two transportable optical lattice clocks at the level of 5×10^{-18} in relative frequency [19]. ACES intercontinental comparisons of atomic clocks will extend this method to a global scale by providing a resolution on the geoid height better than 10 cm. The coverage offered by ACES will thus complement the results of the CHAMP, GRACE, and GOCE missions providing fixes with a very high spatial resolution to constrain existing Earth gravity models.

In the time & frequency metrology domain, the ACES microwave and optical links will also be used to synchronize clocks on the ground. While MWL will provide 100 ps accuracy in ground-to-ground time transfer experiments, the optical link ELT is expected to reach even better performance, down to the 50 ps level. Finally, by connecting optical clocks over four continents to a fractional frequency uncertainty of 1×10^{-17} or better, the ACES links will help the process towards the realization of a new definition of the SI second based on an optical frequency standard. Optical clocks and space links are clearly the next generation technology for the distribution of time and frequency on a worldwide scale.

The ACES mission will be commanded and controlled from the ACES Users Support and Operations (USOC) center, which will be located in CADMOS, at the CNES campus in Toulouse (FR). From there, we will be able to send commands to the ACES payload and to the MWL ground terminals, and to receive the science and housekeeping telemetry. CADMOS is also the central hub where all the ACES data will be stored and analysed to generate quick-look products for the monitoring and control of the flying instruments. From there, the data will be accessed by the data analysis centers and further downstream by the scientific community. SYRTE is hosting the MWL data analysis center and is responsible for providing full performance space-to-ground clock de-synchronization measurements and higher level data products. The Technical University of Munich will be the only interface between the ACES USOC and SLR stations contributing to the ELT experiment, also responsible for the ELT data analysis. Finally, orbitography products will be generated by the German Space Operations Center (GSOC), which is located in Oberpfaffenhofen (DE).

ACES will be launched to the ISS on a Space X Dragon vehicle carried by a Falcon 9 rocket. After docking to the ISS, the robotic arm will extract the ACES payload from the Dragon trunk and install it at the lower balcony of the Columbus module, with the MWL antennas looking towards the Earth. The first 6 months on orbit will be dedicated to commissioning activities, which include switch-on, temperature control optimization, instruments tuning, and calibrations. During this phase, the performance of the ACES clock signal and of the time & frequency links will be optimized in preparation for the routine science phase. ACES will then be operated in nominal mode for the next 2 years, during which the PHARAO systematic frequency shifts will be measured with increasing precision and the links continuously used to compare the ACES clock signal to ground clocks.

2. The ACES clock signal

The ACES clock signal combines the medium-term frequency stability of SHM with the long-term stability and accuracy of the PHARAO clock.

PHARAO is a primary frequency standard designed to operate under weightlessness conditions [4]. Cs atoms are laser cooled in an optical molasses to a temperature of $1-2$ μK. The cooling process produces a cloud of 15-20 mm diameter containing 5×10^8 atoms after a loading time of about 1 s. Using the moving molasses technique, the atomic cloud is launched along the

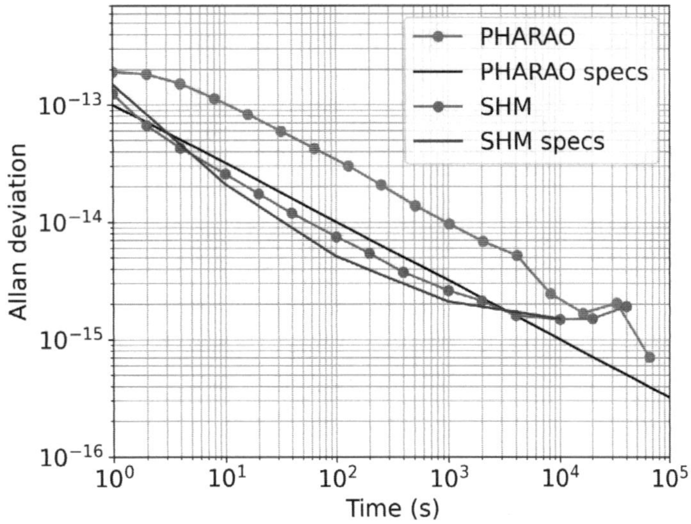

Figure 4. Fractional frequency stability of PHARAO in autonomous mode and of SHM in stand-alone configuration during ground tests. In microgravity, the 5.6 Hz linewidth of the PHARAO resonance will be reduced by one order of magnitude, enabling the specified stability indicated by the black line [29]. The plots include the noise contribution from the reference H-maser. After removing it, SHM stability becomes compatible with the specified performance.

PHARAO tube where it crosses two microwave cavities in resonance with the clock transition at 9.192631770 GHz (definition of SI-second). In the first microwave cavity, the atoms are prepared in the initial clock state; in the second, they are interrogated on the clock transition in a Ramsey sequence. At the end of the tube, the atomic population in the two hyperfine levels of the Cs ground state is selectively measured by fluorescence detection. The cooling and loading process, the launch, the preparation, the interrogation, and detection phases have been optimized. Figure 4 shows the fractional frequency stability of the PHARAO clock when operated in autonomous mode. In this configuration, the frequency of the 100 MHz oscillator of PHARAO is continuously measured against the Cs frequency reference and used in a feedback loop to stabilize the 100 MHz clock output. On the ground, Cs atoms are launched against gravity with a velocity of 3.54 m/s. The limited transit time between the two interrogation zones of the PHARAO Ramsey cavity defines the width of the central fringe of the Ramsey resonance, which can be 5.6 Hz, limiting the fractional frequency stability of the clock to $3.3 \times 10^{-13}/\sqrt{\tau}$, where τ is the integration time expressed in seconds. In space, under free fall conditions, Cs atoms will be launched very slowly and, thanks to a narrower resonance of 0.2-0.5 Hz, a frequency stability of $1.1 \times 10^{-13}/\sqrt{\tau}$ will be reached. The frequency accuracy of the PHARAO clock has been evaluated several times in the past years [4, 29]. The PHARAO accuracy budget, measured on the ground to $1 - 2 \times 10^{-15}$, has also been checked by performing an absolute frequency measurement between the PHARAO clock at Airbus premises in Friedrichshafen (DE) and the Cs fountain clocks operated by SYRTE in Paris (FR) through a GPS link. The frequency measurement was found compatible with zero within 1.5×10^{-15}. The major contributions to the PHARAO accuracy on the ground arise from the collisions between cold atoms and the distributed phase of the microwave interrogation field. The distributed cavity phase shift is strongly increased by the power imbalance between the two pulses of the Ramsey cavity due to the gravity deceleration. In space, with a narrower linewidth, a much lower atomic density, and a well balanced Ramsey interaction, it will be possible to characterize PHARAO accuracy to $1 - 2 \times 10^{-16}$. PHARAO has completed its acceptance test and it is integrated in the ACES payload.

SHM is an active H-maser designed for space. Despite the significantly reduced mass, volume, and power consumption (46 kg, 70 l, 93 W), SHM frequency stability is very close to the one of active H-masers on the ground. SHM has a crucial role on the ACES payload. It is the stable frequency reference used in space to characterize the PHARAO systematic frequency shifts and determine its accuracy budget. Moreover, it is the ACES flywheel oscillator that, distributed by FCDP, drives the MWL electronics, where the ACES atomic time scale is built to be sent to ground users. After the failure of the ion pump, SHM went through a major refurbishment activity to replace the getter material and activate it. SHM has now completed its acceptance tests and it has recently been integrated in the ACES payload. Its frequency stability has been characterized against a reference H-maser in terms of Allan deviation (see Fig. 4). After removing the contribution of the reference H-maser the clock stability becomes 7.9×10^{-14} at 1 s of integration time, 2.0×10^{-14} at 10 s, 6.2×10^{-15} at 100 s, 2.2×10^{-15} at 1000 s, reaching the 1×10^{-15} level at 10^4 s, compatible with the Allan deviation specified for SHM.

The ACES clock signal is generated from PHARAO and SHM after closing the ACES servo-loops. In this configuration, the 100 MHz clock signal from SHM has the short-term to medium-term frequency stability of the active H-maser and the long-term stability and accuracy of the Cs clock PHARAO. The Allan deviation of SHM and PHARAO when the ACES servo-loops are closed is shown in Fig. 5. The two curves overlap to a large extent after 40 s of integration time. The short-term servo loop forces PHARAO to follow the SHM clock. The discrepancy that can be observed between PHARAO and SHM below 40 s is due to the noise of the FCDP phase comparator, which is transferred to the PHARAO 100 MHz clock signal by the short-term servo-loop. This noise averages down as $1/\tau$, where τ is the integration time, and it becomes negligible already after 40 s. For integration times longer than 100 s, the long-term servo loop forces the SHM clock signal to follow the PHARAO stability curve (see also Fig. 4).

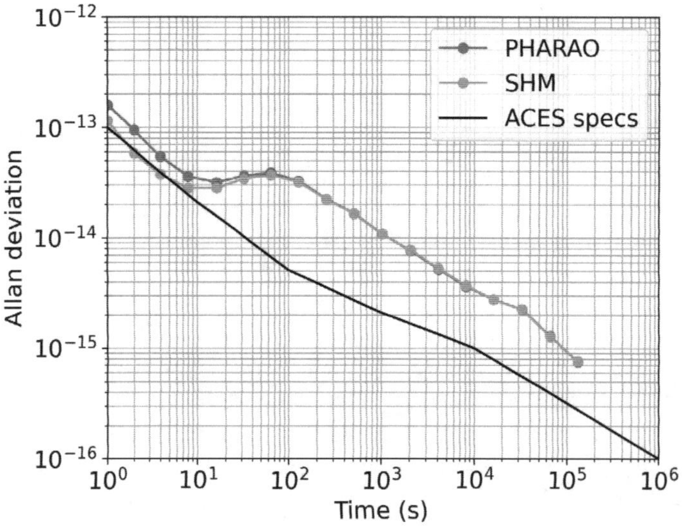

Figure 5. Fractional frequency stability of PHARAO and SHM when the short-term and long-term servo loops are closed. After 100 s, the long-term servo loop forces the SHM clock signal to follow PHARAO stability curve (see also Fig. 4).

The ACES servo-loops are further optimized during the final system tests, which involve the complete ACES payload. During this campaign, another mode of operation, called the PHARAO evaluation mode, will also be tested. In the evaluation mode, the long-term servo-loop is kept open with the PHARAO clock continuously measuring the frequency offset between SHM and the Cs resonance. This configuration avoids any frequency offsets or instabilities that might be introduced by the long-term servo-loop on the PHARAO clock and allows to characterize any perturbations injected on the PHARAO local oscillator through the short-term servo-loop (e.g. AM/PM conversion effects at the ACES phase comparator FCDP). With PHARAO in evaluation mode, the SHM clock signal distributed by ACES to ground users is freely running and the long-term stability and accuracy of the PHARAO clock are not available in real-time. However, they can be retrieved in post-processing from the measurements of the frequency difference between PHARAO and SHM provided in the telemetry of the PHARAO clock.

3. The ACES time & frequency transfer links

MWL is a 2-way 3-frequency link in the microwave domain [5, 6, 7]. The 2-way link in the Ku-band is used to efficiently remove the tropospheric time delay and the first-order Doppler effect, and finally to measure the space-to-ground clock de-synchronization. In addition, the symmetry between space and ground transceiver will allow reducing phase perturbations introduced by the signal dynamics (e.g. tracking loop errors or AM to PM conversion effects). The second downlink in the S-band is used with the Ku-band downlink to determine the Total Electron Content (TEC) of the ionosphere and correct the ionospheric time delay. The phase of the Ku-band carrier is modulated (BPSK) by a Pseudo-Random Noise (PN) code at 100 MChip/s. This modulation, coherently generated from the ACES clock, superimposes unambiguous time tags on the carrier signal, thus defining the ACES time scale. The high chip rate ensures high resolution code phase measurements, which are important to identify the cycle of the carrier. In addition it helps reducing multipath effects. 4 receiver channels are able to simultaneously compare the ACES clock signal with 4 atomic clocks on the ground. In each receiver channel, a Delay Locked Loop (DLL) acquires the PN code and tracks it. The carrier of the incoming signal is extracted after the correlator and tracked in a PLL with respect to the local replica. Every 80 ms, the space and ground transceivers provide a code and carrier phase measurement, which determines the time difference between a given transition of the PN code and zero crossing of the carrier in the received signal and the same code transition and carrier zero crossing in the signal locally generated by the transceiver, coherently with the local clock. From the combination of the phase measurements performed by the MWL transceivers in space and on the ground, we extract the MWL scientific products, which include the space-to-ground clock de-synchronization, the Total Electron Content (TEC) of the ionosphere, and the combination of the round trip pseudorange and tropospheric delay.

To test the stability of code and carrier phase measurements, we connect the MWL flight transceiver to the ground terminal one using cables. Both transceiver are connected to the same clock signal (common-clock test) and are operated under static conditions with a transmit signal of -108 dBm, corresponding to the power level expected at reception when the ISS reaches its maximum elevation above a ground station in Europe. Figure 6 shows the stability of code and carrier phase measurement expressed in time deviation. Code phase measurements reach a sub-ps noise floor after about 1000 s of integration time. With such performance, the cycle of the carrier can be resolved in less than 10 s of integration time. However, the ultimate stability is obtained by using carrier phase measurements, which reach a time deviation of 60 fs after 10 s and remain well below 1 ps on the long term.

MWL tests in the presence of signal dynamics are currently ongoing. In this case, the MWL flight segment transceiver is connected to the ground terminal through an RF system that generates the Ku-band uplink, the Ku-band downlink, and the S-band downlink signals with

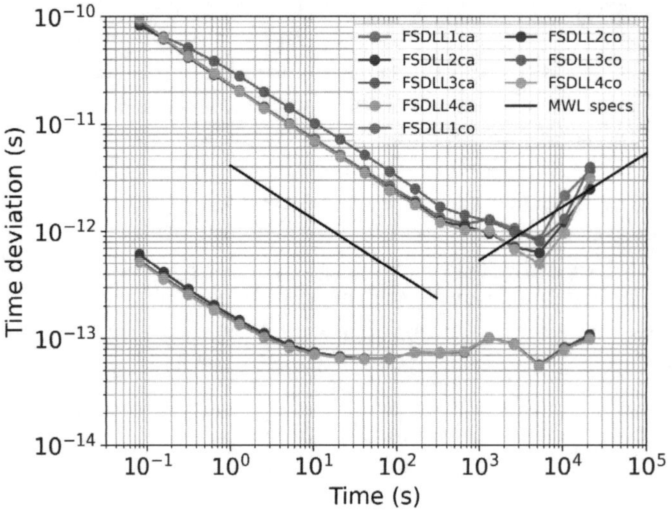

Figure 6. Time stability of code (co) and carrier (ca) phase measurements for the 4 DLL receivers of the MWL flight segment electronics.

the dynamical behavior (in Doppler frequency and amplitude) expected along the ISS orbit. Such tests are important to determine the link sensitivity to AM/PM conversion effects, group delays, and tracking loop error. To evaluate the flight transceiver performance in the presence of signal dynamics, we repeatedly execute three passes with low (16°), medium (36°) and high elevation (84°) in parallel for all 4 receiving DLLs. As the pass starts, MWL DLLs enter the lock acquisition mode. In this phase, the frequency of the code and carrier oscillators is steered in open loop to follow the incoming signal. The signal is correlated in the DLL against the local replica of the PN-code while the range is scanned, searching for the correlation peak. At this point the delay locked loop is closed and the code oscillator tracks the frequency variations of the incoming signal. After acquisition of the 1PPS (Pulse Per Second) marker, also the PLL on the carrier is closed. At this point, both code and carrier phase measurements are correctly tracking the incoming signal. The 3-pass sequence was repeated several times for a total of 200 passes. All 4 receiver channels acquired lock correctly, albeit with different performance. DLL 1 and 3 entered the full tracking mode in less than 50 s, corresponding to an elevation of 10°. Closing the loop on the incoming signal with DLL 2 and 4 takes longer, requiring further optimization of the loop parameters. Figure 7 (left) shows the typical profile of code and carrier phase measurements during a pass. The code phase is non-ambiguous and results from the sum of the integrated Doppler frequency of the incoming signal and the initial de-synchronization between the MWL flight electronics and the RF system generating the dynamic signals. On the contrary, carrier phase measurements can only be resolved up to a constant offset after determining the integer number of carrier cycles with respect to code phase measurements. The large phase variations due to the integrated Doppler frequency can be cancelled by calculating the phase difference between two DLLs. Figure 7 (right) shows the stability of code and carrier phase differences between the 4 DLLs of the MWL flight transceiver. After removal of the phase variations due to the Doppler effect, the time stability reaches levels compatible with the flicker floor measured during static tests (see Fig. 6). This analysis is useful to infer the performance of the end-to-end link in the presence of signal dynamics, however it does not reflect the ultimate performance of MWL. Indeed, on the one hand, the phase difference between two

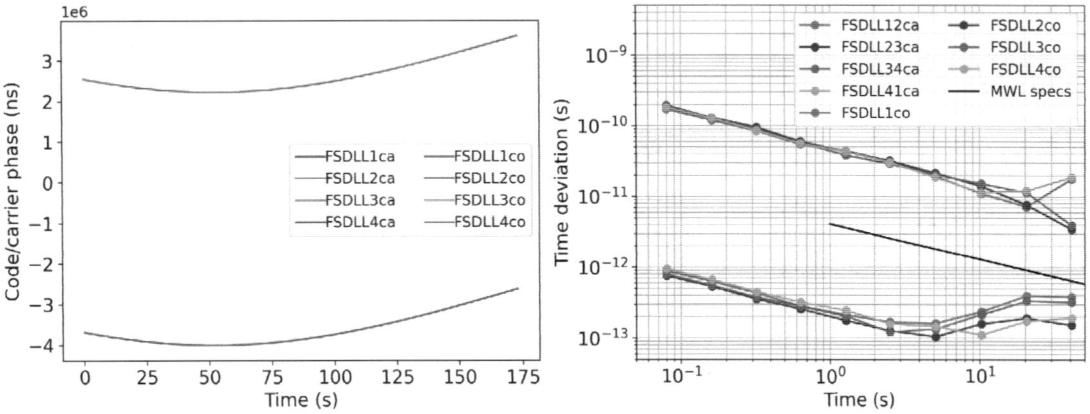

Figure 7. (Left) Evolution of code and carrier phase measurement during the simulated ISS pass. (Right) Time stability of the difference of code (co) and carrier (ca) phase measurements for the 4 DLLs of MWL flight segment electronics.

DLLs removes the correlated instrumental noise of the receiving channels that would naturally enter the comparison between the space and the ground clock; on the other hand, signal dynamics effects like AM/PM conversion, tracking look errors, and group delay dependence on the Doppler frequency, which are DLL specific and need to be calibrated, are removed only partially.

End-to-end tests in the presence of signal dynamics are continuing, together with the calibration of signal dynamics effects. In addition, the temperature sensitivity of code and carrier phase measurements, the sensitivity to supply voltage, and clock input power will be determined. Soon, all the elements to correctly combine the phase measurements from the flight and ground transceivers in the 2-way formula will be at hand and we will be able to assess the ultimate performance of MWL on ISS-representative passes.

ELT is an optical link that uses short laser pulses exchanged between a satellite laser ranging (SLR) station on the ground and ACES to determine the de-synchronization between ACES and the SLR station clock. The SLR station ranges the ISS by firing laser pulses with a duration of about 40-60 ps. The laser pulse emission event is time stamped on the ground in the local clock time scale (t_{start}). Once it reaches the ISS, the pulse is detected by the ELT Single-Photon Avalanche Detector (SPAD) and time stamped in the ACES time scale, which is generated in MWL (t_{space}). At the same time, the onboard corner cube reflector redirects the echo return to the SLR station where its time of arrival is recorded (t_{stop}). From the three time tags, it is possible to calculate the space-to-ground clock de-synchronization as

$$\tau_g(t) - \tau_s(t) = \frac{t_{start} + t_{stop}}{2} - t_{space} + \tau_{rel} + \tau_{atm} + +\tau_{geom} \quad (1)$$

where $\tau_{g,s}(t)$ is the proper time on the ground and in space at coordinate time t, and τ_{rel}, τ_{atm}, and τ_{geom} are the delays due to relativistic effects, propagation in the atmosphere, and geometrical ties.

The ELT detection chain comprises the SPAD unit and the electronics for time tagging the detection events, which is integrated in MWL [30]. The temperature stability of the detector package was tested between -55°C to +60°C, showing a temperature sensitivity of 1 ps/K over the entire temperature range. The optical-to-electrical detection delay was also characterized with an uncertainty of 20 ps [31]. To test the stability of the ELT flight hardware, we connect

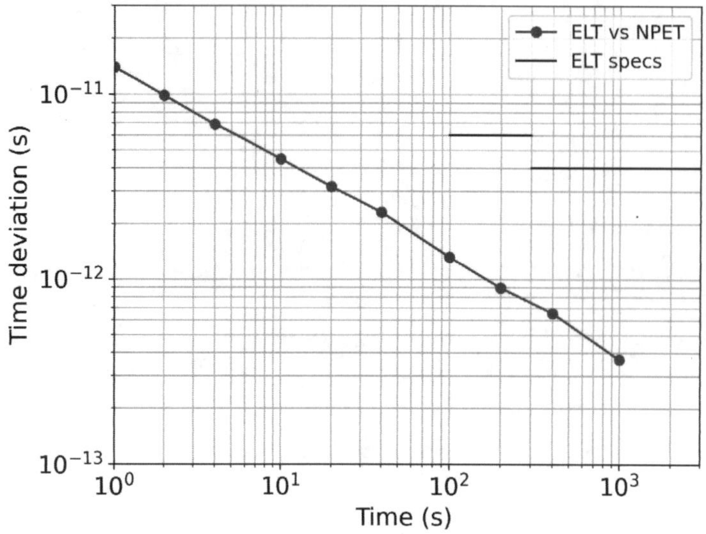

Figure 8. Time stability of the ELT detection chain. The SPAD is operated in dark count mode and the detection events are measured in parallel by the MWL time tagging board and an external event timer, both referenced to a common clock. In this experiment, the SPAD unit of ELT is gated with 100 Hz repetition rate.

the SPAD flight unit in parallel to MWL and to an external time tagging unit [31]. The time scales of both MWL and the external event timer are referenced to the same clock signal. With the SPAD aperture closed by a light-blocking cap and the detector operated in dark count mode, we measure the difference between the detection events recorded in MWL and the external event timer. Figure 8 shows the time deviation of the time differences obtained by gating the SPAD unit with 100 Hz repetition rate. The measurements show sub-ps stability already after 300 s of integration time, well below the performance specified for the ELT optical link.

ELT operation relies on existing SLR stations with a stable timing infrastructure connected to atomic clocks. To reach the specified link stability and accuracy for time transfer experiments, the optical ground stations need to be characterized and their delays measured. A first calibration campaign of the European SLR stations (Wettzell, Graz, Herstmonceux, Potsdam, and Zimmerwald) has already been completed in 2016 to demonstrate the feasibility of the ELT experiment [32]. SLR station delays were monitored up to 1 year, showing peak-to-peak variations at the 20 ps level. A new calibration campaign will be performed before the ACES mission launch.

There is no major limitation to the number of SLR station that can participate to the ELT experiment, provided that they comply with the ISS safety standards for laser emission.

4. Conclusions

The ACES payload is currently completing the qualification tests. The PHARAO and SHM clocks, the frequency comparator FCDP, the onboard GNSS receiver, the power distribution unit, and the onboard computer are all integrated in the ACES payload, and currently tested at system level to measure the performance of the servo-loops and of the ACES clock signal. The MWL flight electronics is continuing its qualification tests before final integration. MWL calibrations will be soon established together with the characterization of the link delays for time transfer experiments. After completion of the MWL calibration campaign, the flight hardware will enter the final performance tests where, in end-to-end configuration with the MWL ground terminal electronics, the link performance will be established under a signal dynamics representative of the ISS orbit. After the integration of MWL, the ACES payload will be in its final configuration. The clocks and the links will then be commanded and controlled in a flight representative operational environment for the final acceptance tests before flight. ACES will be ready for flight in the last quarter of 2024.

Acknowledgments

The authors wish to acknowledge the contribution of the ACES science team members, the ACES project team at ESA, the ACES teams at Airbus Defence and Space, Timetech, Safran, and the PHARAO team at SYRTE and CNES.

References

[1] Cacciapuoti L, *et al.* 2020 *Eur. Phys. J.* D **74** 164
[2] Laurent P, Massonnet D, Cacciapuoti L, and Salomon C 2015 *C. R. Phys.* B **4731** 459
[3] Cacciapuoti L and Salomon C 2009 *Eur. Phys.J. Spec. Top.* B **172** 57
[4] Laurent P, *et al.* 2006 *Appl. Phys.* B **84** 683
[5] Meynadier F, Delva P, Le Poncin-Lafitte C, Guerlin C, and Wolf P 2018, *Classical Quantum Gravity* **35** 35
[6] Hess MP, Kehrer J, Kufner M, Durand S, Hejc G, Frühauf H, Cacciapuoti L, Much R, and Nasca R *2011 Joint Conference of the IEEE International Frequency Control and the European Frequency and Time Forum Proceedings* (San Francisco, CA, 2011) p. 26
[7] Duchayne L, Mercier F, and Wolf P 2009 *Astron. Astrophys.* **504** 653
[8] Savalle E, Guerlin C, Delva P, Meynadier F, Le Poncin-Lafitte C, and Wolf P 2019 *Classical Quantum Gravity* **36** 245004
[9] Vessot RFC, *et al.* 1980 *Phys. Rev. Lett.* **45** 2081
[10] Vessot RFC and Levine MW 1979 *Gen. Relativ. Gravit.* **10** 181
[11] Delva P, *et al.* 2015 *Phys. Rev. Lett.* **121** 231101
[12] Herrmann S, *et al.* 2015 *Phys. Rev. Lett.* **121** 231102
[13] Petit G, Loyer S, and Perosanz F *2017 European Frequency and Time Forum* (IEEE, 2017) p. 784
[14] Dix-Matthews BP, *et al.* 2021 *Nature Communications* **12** 515
[15] Nicholson T, *et al.* 2015 *Nat. Commun.* **6** 6896
[16] McGrew WF, *et al.* 2018 *Nature* **564** 87
[17] Brewer SM, Chen JS, Hankin AM, Clements ER, Chou CW, Wineland DJ, Hume DB, and Leibrandt DR 2019 *Phys. Rev. Lett.* **123** 33201
[18] Bothwell, Kedar TD, OelkerE, Robinson JM, Bromley SL, Tew WL, Ye J, and Kennedy CJ 2019 *Metrologia* **56** 065004
[19] Takamoto M, Ushijima I, Ohmae N, Yahagi T, Kokado K, Shinkai H, and Katori H 2020 *Nature Photonics* **14** 411
[20] Dzuba VA, Flambaum VV, and Webb JK 1999 *Phys. Rev. Lett.* **82** 888
[21] Dzuba VA, Flambaum VV, and Webb JK 1999 *Phys. Rev. A* **59** 230
[22] Stadnik YV and Flambaum VV 2015 *Phys. Rev. Lett.* **114** 161301
[23] Stadnik YV and Flambaum VV 2015 *Phys. Rev. Lett.* **115** 201301
[24] Derevianko A and Pospelov M 2014 *Nature Phys* **10** 933
[25] Schkolnik V, *et al.* 2023 *Quantum Sci. Technol.* **8** 014003
[26] Takano T, Takamoto M, Ushijima I, Ohmae N, Akatsuka T, Yamaguchi A, Kuroishi Y, Munekane H, Miyahara B, and Katori H 2016 *Nat. Photonics* **10** 662
[27] Grotti J, *et al.* 2018 *Nat. Phys.* **14** 437
[28] Lion G, Panet I, Wolf P, Guerlin C, Bize B, and Delva P 2017 *J. Geod.* **91** 597
[29] Laurent P, *et al.* 2020 *Metrologia* **57** 055005
[30] Schreiber U, Prochazka I, Lauber P, Hugentobler U, Schafer W, Cacciapuoti L, and Nasca R 2010 *IEEE Trans. Ultrason. Ferroelectr. Freq. Control* **57** 728
[31] Prochazka I, Blazej J, and Kodet J 2011 *Metrologia* **48** L13
[32] Prochazka I, Kodet J, Blazej J, Kirchner G, Koidl F, and Wang P 2017 *Adv. Space Res.* **59** 2466
[33] Prochazka I, Blazej J, and Kodet J 2018 *Rev. Sci. Instrum.* **89** 056106

Journal of Physics: Conference Series **2889** (2024) 012006 doi:10.1088/1742-6596/2889/1/012006

Laser-cooling Cadmium Bosons and Fermions with Near Ultraviolet Triplet Excitations

Kurt Gibble

Department of Physics, The Pennsylvania State University, University Park, PA, 16802 USA

E-mail: kgibble@psu.edu

Abstract. Cadmium is laser-cooled and trapped with excitations to triplet states with UVA light, first using only the 67 kHz wide 326 nm intercombination line and subsequently, for large loading rates, the 25 MHz wide 361 nm $^3P_2 \rightarrow {}^3D_3$ transition. Eschewing the hard UV 229 nm $^1S_0 \rightarrow {}^1P_1$ transition, only small magnetic fields gradients, less than 6 G cm^{-1}, are required enabling a 100% transfer of atoms from the 361 nm trap to the 326 nm narrow-line trap. All 8 stable cadmium isotopes are straightforwardly trapped, including two nuclear-spin-$\frac{1}{2}$ fermions that require no additional repumping. We observe evidence of $^3P_2$ collisions limiting the number of trapped metastable atoms, report isotope shifts for ^{111}Cd and ^{113}Cd of the 326 nm $^1S_0 \rightarrow {}^3P_1$, 480 nm $^3P_1 \rightarrow {}^3S_1$, and 361 nm $^3P_2 \rightarrow {}^3D_3$ transitions, and measure the ^{114}Cd 5s5p $^3P_2 \rightarrow$ 5s5d $^3D_3$ transition frequency to be 830 096 573(15) MHz.

1. Introduction

Strontium, ytterbium and other atoms with two valence electrons are attractive for state-of-the-art quantum sensors such as optical-lattice atomic clocks and atom interferometers [1–8], for quantum gas experiments [9–14], and for quantum computation, information and simulation [15–20]. Yb, Ca, Cd and Hg have more than three stable spin 0 isotopes, enabling tests of fundamental physics through precise measurements of isotope shifts (ISs) [21–27]. The broad applicability of alkaline-earth-like atoms to high-coherence measurements is facilitated by their nearly forbidden clock and other intercombination transitions, from the singlet $^1S_0$ ground state to triplet $^3P_{J=0,1,2}$ states, as well as fast laser cooling, on the broad electric dipole allowed $^1S_0 \rightarrow {}^1P_1$ transition for most atoms (Ca, Sr, Yb, Mg, Cd and Zn). These atoms have clock transitions with mHz natural linewidths, their $^1S_0 \rightarrow {}^3P_1$ intercombination transitions have sub-MHz linewidths, which allow laser cooling to μK temperatures and lower, and their paired electron spins provide a natural insensitivity to perturbations from external magnetic fields. For the currently most accurate optical lattice clocks, based on Sr and Yb, one of the leading systematic errors is the frequency shift due to blackbody radiation (BBR) [1,2,4]. An attractive feature of Cd and Hg optical lattice clocks is their order of magnitude smaller sensitivities of their clock transitions to BBR than the Sr and Yb clock transitions [3,4,28,29].

Previous laser cooling of Cd and Hg has used hard UV laser light, exciting the broad 91 MHz wide 229 nm $^1S_0 \rightarrow {}^1P_1$ Cd transition [28,30,31] and the 1.3 MHz wide 254 nm $^1S_0 \rightarrow {}^3P_1$ Hg intercombination transition [3,4,32]. These UVC sources are subject to degradation of non-linear crystals and optical coatings, making long-term operation, especially autonomous operation, challenging [3,32]. Here, we demonstrate large loading rates and efficient laser-cooling of Cd

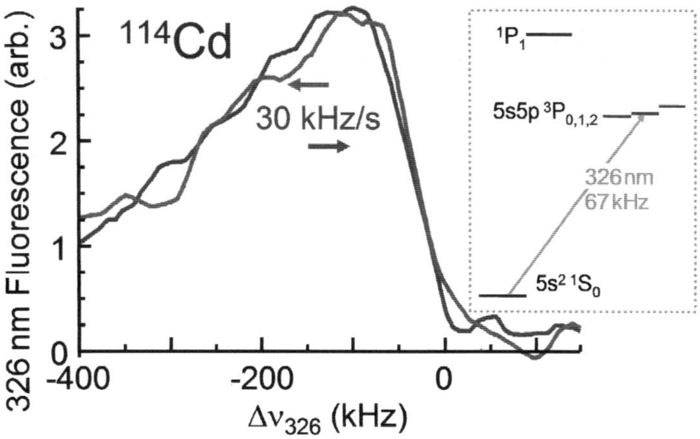

Figure 1. Fluorescence of a Cd MOT that uses only the 67 kHz wide ${}^1S_0 \rightarrow {}^3P_1$ 326 nm intercombination transition (see inset) to capture and trap atoms from an effusive source at 133 C. The blue (red) curve is the average of four frequency scans at $+(-)30\,\mathrm{kHz\,s^{-1}}$ and shows a sharp blue edge near the ${}^1S_0 \rightarrow {}^3P_1$ resonance.

using only near UV excitations to triplet states, with 326 nm and 361 nm light, in addition to two low-power blue optical pumping lasers, importantly, without having used 229 nm ${}^1S_0 \rightarrow {}^1P_1$ UVC light [30,31,33,34]. The multilevel lower state structure of the 361 nm broad-line magneto-optic trap (MOT) does not require the usual large magnetic field gradients of ${}^1S_0 \rightarrow {}^1P_1$ MOTs and provides polarization gradient cooling, facilitating a 100% transfer of atoms to the 326 nm narrow-line MOT.

We first demonstrate laser-cooling and trapping of approximately 10^4 Cd atoms from an effusive source using only the 67 kHz wide 326 nm ${}^1S_0 \rightarrow {}^3P_1$ intercombination transition [35,36] (see Fig. 1). While the narrow transition linewidth enables low Doppler cooling temperatures [28], in the absence of stimulated processes [37], the slow ${}^3P_1$ spontaneous emission limits the rate at which 326 nm photons can slow Cd atoms as they cross the laser beams. This is the narrowest transition used to capture atoms from a room-temperature (or hotter) source into an atom trap [38–40].

Next, we dramatically increase the laser-cooling capture rate using the dipole-allowed ${}^3P_2 \rightarrow {}^3D_3$ transition. For Cd, this transition is in the near UV, 361 nm in Fig. 2a), and has a reasonably broad 25 MHz linewidth. With other alkaline-earth atoms, this transition has been used in conjunction with the singlet transition to cool atoms to increase MOT and lattice loading fractions [41–43] and for high signal-to-noise detection [42]. Excitations from ${}^3P_J$ to other states, such as ${}^3P_0 \rightarrow {}^3D_1$, have been used to cool to sub-recoil temperatures [44,45]; and ${}^3P_{0,1} \rightarrow {}^3S_1$, to increase the number of Sr atoms loaded into a magnetic trap, and subsequently an intercombination MOT [46]; and ${}^3P_1 \rightarrow {}^1D_2$ and ${}^3P_1 \rightarrow {}^1S_0$, to increase the scattering rate for a Ca intercombination MOT [47,48].

Exciting the 361 nm UVA transition, we efficiently capture room temperature atoms and load a metastable ${}^3P_2 \rightarrow {}^3D_3$ MOT with all 6 of the stable nuclear-spin-zero Cd bosonic isotopes, ^{106}Cd to ^{116}Cd. We further show that no additional lasers are needed for hyperfine pumping to efficiently capture both nuclear-spin $I = \frac{1}{2}$ fermionic isotopes, ^{111}Cd and ^{113}Cd, in the metastable MOT. We observe evidence of ${}^3P_2$ collisional loss, report the ISs of three transitions, the 326 nm ${}^1S_0 \rightarrow {}^3P_1$, the 480 nm ${}^3P_1 \rightarrow {}^3S_1$ and the 361 nm ${}^3P_2 \rightarrow {}^3D_3$ transition, and measure the absolute frequency of the 361 nm ${}^3P_2 \rightarrow {}^3D_3$ transition, improving its previous uncertainty

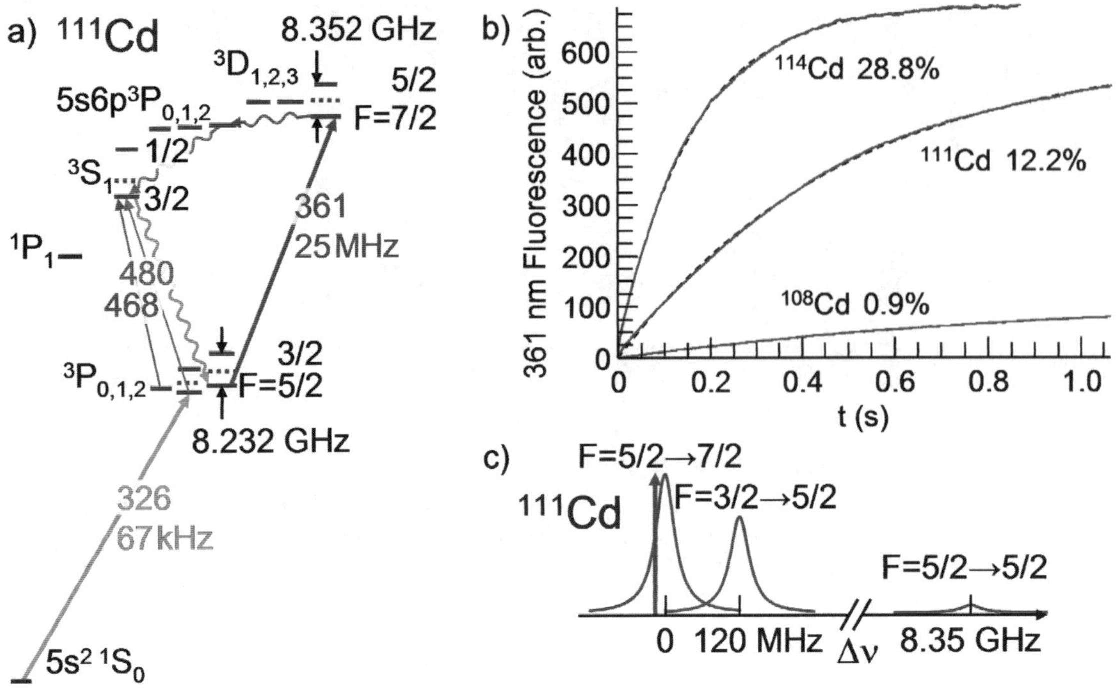

Figure 2. a) Energy levels of Cd fermions (solid) and bosons (dotted), where the annotated transition wavelengths are in nanometers. Atoms are excited to $^3P_1$ and optically pumped to $^3P_2$ via $^3S_1$ with 480 nm and 468 nm light. The 361 nm $^3P_2 \rightarrow {}^3D_3$ MOT transition is nearly closed, with a slow 11.4 μm and 1.4 μm cascade to 5s6p $^3P_2$ and $^3S_1$. The fermion hyperfine splittings from Table 1 are not depicted to scale and those of the $^1P_1$, $^3D_{1,2}$ and 5s6p $^3P_{1,2}$ states are omitted. b) Exciting the 25 MHz wide 361 nm $^3P_2 \rightarrow {}^3D_3$ transition efficiently captures the 8 Cd bosonic and fermionic isotopes. The faster saturation of the ^{114}Cd and ^{111}Cd loading than that for the 0.9% abundant ^{108}Cd suggests that $^3P_2$ collisional loss limits the number of atoms trapped. c) To trap ^{111}Cd and ^{113}Cd, the 361 nm laser is red-detuned from the $^3P_2$ $F = \frac{5}{2} \rightarrow {}^3D_3\, F = \frac{7}{2}$ MOT transition. The inverted ^{111}Cd and ^{113}Cd fermion hyperfine structure and the larger hyperfine splitting of $^3D_3$ than that of $^3P_2$ in Table 1 enable convenient repumping (and cooling) of $^3P_2\, F = \frac{3}{2}$ atoms via the $^3P_2\, F = \frac{3}{2} \rightarrow {}^3D_3\, F = \frac{5}{2}$ transition, from which the 361 nm laser is further red-detuned by 120 MHz for ^{111}Cd, and 125 MHz for ^{113}Cd.

by a factor of 20. This $^3P_2 \rightarrow {}^3D_3$ trapping may be applied to other alkaline-earth-like atoms, for example, Hg, where it can be challenging to maintain sufficient 254 nm power to capture many atoms in a $^1S_0 \rightarrow {}^3P_1$ MOT [3, 32].

Table 1. ^{111}Cd, and ^{113}Cd, hyperfine splittings of Fig. 2a) in MHz [35, 49–52].

A	$^3S_1$	$^3P_1$	$^3P_2$	$^3D_3$
111	−11626.2(13)	−6185.72(2)	−8232.341(2)	−8352.04(9)
113	−12162.9(14)	−6470.79(2)	−8611.586(4)	−8736.88(9)

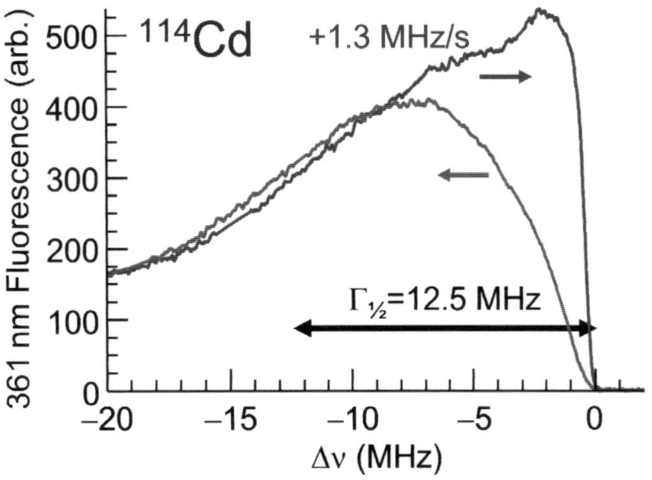

Figure 3. 361 nm metastable MOT fluorescence as the laser frequency is tuned across the $^3P_2 \rightarrow {}^3D_3$ resonance.

Table 2. Isotope shifts of Cd I fermions relative to ^{114}Cd, in MHz. We measure the 326 nm $^1S_0 \rightarrow {}^3P_1\,F = \frac{3}{2}$, the 361 nm $^3P_2\,F = \frac{5}{2} \rightarrow {}^3D_3\,F = \frac{7}{2}$ and the 480 nm $^3P_1\,F = \frac{3}{2} \rightarrow {}^3S_1\,F = \frac{3}{2}$ hyperfine transitions and calculate the transition centers-of-mass using the $^3P_1$, $^3P_2$, $^3S_1$ and $^3D_3$ hyperfine splittings in Table 1.

A	326 nm	361 nm	480 nm
111	861.6(10)	−227.4(10)	−334.2(11)
113	370.8(10)	−75.3(10)	−135.0(11)

2. Cadmium Intercombination and Metastable MOT's

2.1. Narrow-line 326 nm intercombination MOT

Trapping a detectable number of Cd atoms using only the 67 kHz wide 326 nm intercombination line in Fig. 1 is an important first step to bootstrap to the large loading rates of the 361 nm $^3P_2 \rightarrow {}^3D_3$ MOT. Our effusive source of Cd atoms with natural isotopic abundances has a 1.2 cm diameter, is 2.2 cm from the center of the MOT, and is heated to 120 C to 142 C. We begin with 50 to 150 mW of 326 nm light that is split into three retro-reflected beams with e^{-2} diameters of 8.5 mm to form a MOT with a quadrupole field gradient of 0.5 G cm^{-1} [53]. To increase the velocity capture range, an acousto-optic modulator (AOM) frequency modulates (FM) the 326 nm light at 50 kHz with a peak-to peak amplitude of 8.6 MHz [28]. We trap the six stable bosonic and fermionic isotopes, ^{110}Cd to ^{116}Cd, which have abundances greater than 7.5%. The number of trapped ^{114}Cd atoms versus frequency in Fig. 1 shows a sharp blue edge [54], which we use to measure isotope shifts [27]. The ^{111}Cd and ^{113}Cd fermions are trapped using their $^1S_0\,F = \frac{1}{2} \rightarrow {}^3P_1\,F = \frac{3}{2}$ hyperfine components and their ISs, given in Table 2, are consistent and more precise, as are the bosonic ISs [27], with other recent measurements [34]. To trap ^{106}Cd and ^{108}Cd, which have 1.25% and 0.89% abundances, we use the 361 nm $^3P_2 \rightarrow {}^3D_3$ MOT, described next, to increase the loading rate.

2.2. 361 *nm Metastable MOT*

With cold atoms trapped in the 326 nm intercombination MOT, it is straightforward to successively add and frequency tune the lasers for the 480 nm $^3P_1 \rightarrow {}^3S_1$ optical pumping, the 361 nm $^3P_2 \rightarrow {}^3D_3$ metastable MOT, and the 468 nm $^3P_0 \rightarrow {}^3S_1$ optical pumping [see Fig. 2a)]. The optical pumping to $^3P_2$ is well saturated with 100 μW of power in 12.6 mm diameter 468 nm and 8.7 mm diameter 480 nm laser beams. The 480 nm $^3P_1 \rightarrow {}^3S_1$ resonance can be easily found as this pumping extinguishes the 326 nm MOT fluorescence. We use 50 to 200 mW of 361 nm light that is intensity controlled and frequency shifted by an AOM before splitting into three retro-reflected MOT beams with e^{-2} diameters of 10.4 mm. The number of atoms depends weakly on the optimal MOT quadrupole gradient of 5.7 G cm^{-1}, notably smaller than the of order 200 G cm^{-1} gradients of $^1S_0 \rightarrow {}^1P_1$ singlet MOT's [28, 30, 31, 38]. Here, the 326 nm FM depth is larger, 16.3 MHz, consistent with the larger capture velocity of the 25 MHz wide 361 nm $^3P_2 \rightarrow {}^3D_3$ MOT [55]. From the saturation of the 361 nm fluorescence versus a pulsed intensity of the MOT beams, the number of trapped ^{114}Cd atoms is greater than 6×10^6. We trap all 6 spin 0 bosonic isotopes, which have no hyperfine structure [dashed levels in Fig. 2a)], including ^{106}Cd and ^{108}Cd with 1% abundances [Fig. 2b)]. As discussed below, the loading behavior in Fig. 3 for increasing frequency sweeps (blue curve), as well as the different loading time constants in Fig. 2b), suggests that $^3P_2$ collisional loss limits the number of trapped metastable atoms.

In addition to the 6 Cd bosonic isotopes, the 361 nm $^3P_2 \rightarrow {}^3D_3$ metastable MOT efficiently captures ^{111}Cd [Fig. 2b)] and ^{113}Cd nuclear-spin-$\frac{1}{2}$ fermions. Whereas additional repumping lasers are normally required for $J \rightarrow J+1$ MOT transitions with $J, I \neq 0$, especially for large nuclear spins [43, 46], no additional laser frequencies are required here for Cd. This is enabled by the negative nuclear magnetic moments of ^{111}Cd [Fig. 2a)] and ^{113}Cd and their $I = \frac{1}{2}$ nuclear spins. Each $J \neq 0$ state has only two hyperfine components, $F = J \pm I$, from the nuclear magnetic dipole interaction, with no electric quadrupole shifts for $I \leq \frac{1}{2}$. Table 1 gives the Cd hyperfine splittings and we highlight that the $^3D_3$ hyperfine splitting for ^{111}Cd(^{113}Cd) is 120(125) MHz larger than the $^3P_2$ hyperfine splitting [51, 52]. Combined with a negative nuclear magnetic moment, which yields the "inverted" ($^3S_1$, $^3P_1$,) $^3P_2$ and $^3D_3$ hyperfine structure, the $^3P_2\, F = \frac{3}{2} \rightarrow {}^3D_3\, F = \frac{5}{2}$ transition in Fig. 2c) is therefore 120 MHz blue detuned from the $^3P_2\, F = \frac{5}{2} \rightarrow {}^3D_3\, F = \frac{7}{2}$ metastable MOT cycling transition. As a result, atoms lost from the cycling transition to $^3P_2\, F = \frac{3}{2}$ are naturally repumped, cooled and trapped via the $^3P_2\, F = \frac{3}{2}, \rightarrow {}^3D_3\, F = \frac{5}{2}$ transition by the red detuned 361 nm MOT beams. The nuclear magnetic moment and resulting hyperfine splittings of ^{113}Cd in Table 1 are the same, within 5%, as for ^{111}Cd, allowing similarly large loading rates of ^{113}Cd as in Fig. 2b).

The low temperatures and small quadrupole magnetic field of the 361 nm metastable MOT facilitates an essentially complete transfer of captured atoms to the 326 nm intercombination MOT in Fig. 4. This compares to 20% to 30% direct transfer efficiencies from singlet to narrow-line Ca and Sr MOTs [47, 56]. Here, at $t = 0$, the 480 nm $^3P_1 \rightarrow {}^3S_1$ optical pumping light is extinguished. The 361 nm fluorescence decays in 0.7 ms as atoms no longer populate the $^3P_2$ state, e.g., through cascaded spontaneous emission from $^3D_3 \rightarrow 5s6p\, {}^3P_2 \rightarrow {}^3S_1 \rightarrow {}^3P_1 \rightarrow {}^1S_0$, and may then be captured by the FM-broadened 326 nm intercombination MOT. Resuming the 480 nm $^3P_1 \rightarrow {}^3S_1 \rightarrow {}^3P_2$ optical pumping with variable delays shows that 99.6(4)% of the atoms trapped in the metastable MOT are transferred to the intercombination MOT in Fig. 4b) and that it has a 1.2 s lifetime. Notably, this is significantly longer than the ≈ 0.5 s 361 nm MOT loading time in Fig. 4a) - if background gas collisions limited the lifetimes of both MOTs, the smaller trapping forces of the 326 nm MOT would normally yield a shorter lifetime.

Here, fine structure changing $^3P_2$ collisions, energy-pooling collisions or light-assisted collisions may shorten the loading time in Fig. 4a) and the number of atoms in Fig. 2b) in the 361 nm metastable MOT [57–59]. In Fig. 2b), the initial 361 nm MOT loading rates for ^{114}Cd, ^{111}Cd and ^{108}Cd are approximately proportional to their isotopic abundances while the

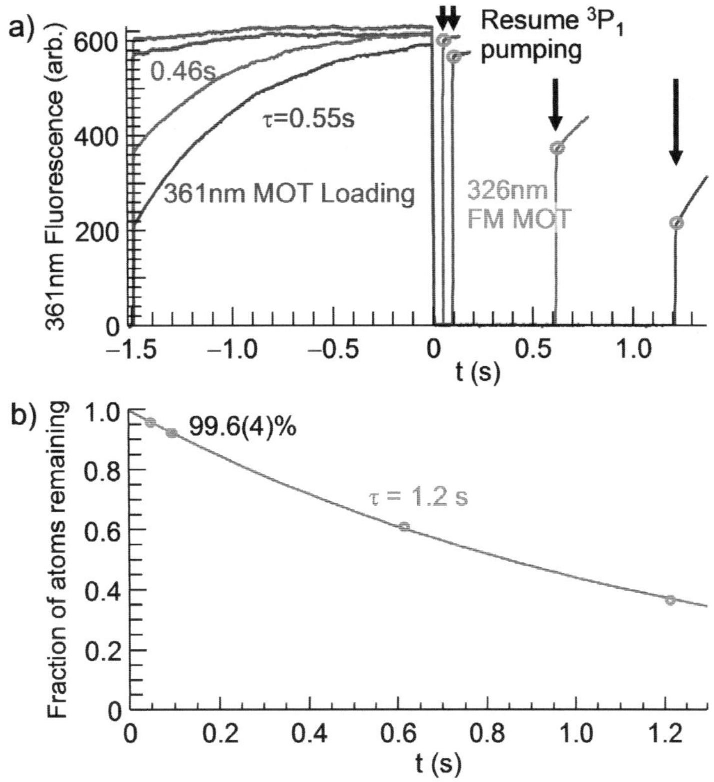

Figure 4. a) Inhibiting the 480 nm $^3P_1 \rightarrow {}^3S_1 \rightarrow {}^3P_2$ optical pumping in Fig. 2a) for a variable time transfers atoms from the metastable MOT to the the narrow-line MOT and back to the metastable MOT. b) A fit of the fraction of atoms remaining in the narrow-line MOT versus time shows 99.6(4)% of the atoms are captured and the lifetime is 1.2 s, well longer than the ≈ 0.5 s loading time of the metastable MOT. During the narrow-line MOT capture, the magnetic field gradient decreases from 5.7 G cm^{-1} to 0.6 G cm^{-1}.

asymptotic numbers of trapped atoms is limited more quickly for the abundant isotopes, in spite of a higher background Cd pressure for the ^{108}Cd data. Further supporting that $^3P_2$ collisions may limit the number of atoms trapped, the 361 nm fluorescence in Fig. 3 (blue curve) abruptly flattens as the metastable MOT light frequency increases beyond -7 MHz of resonance, and then the fluorescence suddenly increases at -3 MHz before falling rapidly to 0 at resonance. From images of the trap, the MOT becomes unstable with large atom numbers at these small detunings, as expected from repulsive radiation pressure forces [60]. Interestingly, here, this MOT instability appears to allow a larger number of trapped atoms with a lower density.

Figure 5 shows the 361 nm metastable MOT fluorescence when the 468 nm $^3P_1 \rightarrow {}^3S_1$ optical pumping light is extinguished for $\tau_e = 4$ ms to 14 ms. This transfers cold $^3P_2$ atoms to $^3P_0$, e.g., via decay from $^3D_3$ to 5s6p $^3P_2$ in Fig. 2a). These $^3P_0$ atoms are not trapped and begin to slowly leave the detection region, yielding less fluorescence just after they are repumped to $^3P_2$ at $t = 0$. Because the atoms are recaptured in much less than the MOT loading time, they do not exit the MOT laser beams volume during this time. After the 468 nm light is extinguished, the 326 nm excitation to $^3P_1$ and the 480 nm pumping from $^3P_1$ to $^3S_1$ to $^3P_2$, continues to load the 361 nm metastable MOT, albeit at a lower rate because some $^3P_1$ atoms excited to $^3S_1$ decay to $^3P_0$ instead of $^3P_2$. Nonetheless, for $\tau_e = 4$ ms, the 361 nm MOT fluorescence at $t = 4$ ms is

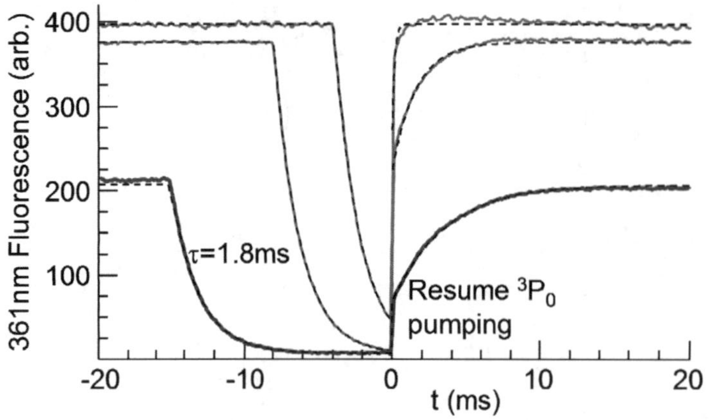

Figure 5. Atoms accumulate in the $^3P_0$ excited clock state with a time constant of 1.8 ms after the 468 nm $^3P_0 \rightarrow {}^3S_1$ optical pumping in Fig. 2a) is inhibited. The black dashed curves are exponential fits to the red, green and blue curves. For the red curve, after $^3P_0$ atoms are repumped to $^3P_2$, more atoms are detected at $t = 4$ ms than at $t = -4$ ms, when the optical pumping is turned off. This suggests that $^3P_2$ collisional loss limits the number of atoms in the metastable MOT. Here, the cycle repeats every 68 ms, which traps fewer atoms for the 14 ms $^3P_0$ holding time (blue curve).

greater than at $t = -4$ ms, consistent with a lower loss rate for atoms in $^3P_0$ than in $^3P_2$, and the dominant $^3P_2$ loss being fine structure changing collisions. For $\tau_e = 14$ ms in Fig. 5 (blue curve), the smaller number of trapped atoms is due to the short cycle time of 68 ms.

3. Laser System

To generate the four laser sources in Fig. 2a) [27, 61, 62], we begin with a 1083 nm extended-cavity diode laser (ECDL) that seeds a fiber amplifier, which is frequency doubled to 542 nm with LBO in a resonant cavity. We generate 361 nm UVA light for the metastable MOT with doubly resonant sum-frequency generation (SFG) of the 542 nm and 1083 nm light. The 326 nm UVA light for the intercombination MOT [27, 28, 61–63] is made in a second doubly resonant BBO cavity for SFG of the 542 nm light and 820 nm light from an ECDL and tapered amplifier. Finally, the low power blue 468 nm and 480 nm optical pumping light is generated with single-pass SFG of 1083 nm with 823 nm and 862 nm light in periodically-poled lithium niobate (PPLN) waveguides. The 1083 nm ECDL is frequency stabilised with a tunable offset to a ring reference-cavity, which is also used to monitor the 820 nm, 823 nm and 862 nm ECDL frequencies [27]. Saturated absorption of molecular I_2 provides an absolute frequency reference at 542 nm.

4. Isotope Shift and Absolute Frequency Measurements

We use the sharp blue-edge of the MOT fluorescence in Fig. 3 to measure the isotope shifts [27] and the absolute frequency of the $^3P_2 \rightarrow {}^3D_3$ transition. Relative to the the R(83)28-0 I_2 a_7 hyperfine component, the ^{114}Cd $^3P_2 \rightarrow {}^3D_3$ transition frequency is $\nu_{114} = \frac{3}{2}\nu_{R(83)28-0\,a_7} + 261(3)$ MHz, where a model of a number of I_2 absolute frequencies gives $\nu_{R(83)28-0\,a_7} = 553\ 397\ 541(10)$ MHz [64, 65]. This improves the previous uncertainty [66] by a factor of 20, yielding a 5s5d $^3D_3$ energy of 59 515.990 cm^{-1} for natural Cd, the value of [66], after a likely misprint is corrected [67]. The previous 300 MHz uncertainty contributed to some difficultly initially finding this broad transition - changing the frequency of our 361 nm light by more than the 10 MHz tuning range of the AOM requires changing the 820 nm laser frequency and the dispersion of the 326 nm doubly-resonant SFG cavity to preserve the narrow-line MOT [61]. We use the blue MOT edge to measure the isotope shifts for the Cd bosons [27] and fermions, which are given in Table 2. We similarly report the 480 nm $^3P_1 \rightarrow {}^3S_1$ ISs from the optical pumping rate of the metastable MOT fluorescence, as in [27], which reported these ISs for Cd bosons.

5. Conclusions

Cadmium has a number of attractive properties for atomic clocks and interferometers and quantum gas experiments. These include narrow linewidth clock and other intercombination transitions, a small sensitivity of its clock transition to BBR, small sensitivities to external magnetic fields, a level structure that requires minimal repumping for laser cooling, and six bosonic isotopes and two nuclear-spin-$\frac{1}{2}$ fermionic isotopes. We demonstrate that all 8 Cd isotopes can be efficiently laser-cooled without exciting the hard UVC 229 nm $^1S_0 \rightarrow {}^1P_1$ transition. We begin by loading a MOT using only the Cd 67 kHz wide 326 nm $^1S_0 \rightarrow {}^3P_1$ UVA intercombination transition, the narrowest transition used to capture room temperature atoms. We dramatically increase the laser-cooling capture rate using the dipole-allowed 361 nm $^3P_2 \rightarrow {}^3D_3$ UVA transition to load a metastable MOT, with no additional repumping required for the Cd fermions.The $J = 2$ or $F = \frac{5}{2}$ substate structure of $^3P_2$ lower level of the metastable Cd MOT, along with the favorable $^3P_2 \rightarrow {}^3D_3$ hyperfine structure, yields the usual MOT behaviors of alkali atoms, in contrast to those of alkaline-earth singlet and intercombination MOTs. The $^3P_2 \rightarrow {}^3D_3$ metastable Cd MOT avoids cumbersomely large magnetic field gradients, provides low temperatures, and enables a 100% transfer of atoms to the narrow-line 326 nm MOT for subsequent cooling to microkelvin temperatures [28]. We observe evidence of collisional loss in the metastable MOT, report an absolute frequency of the $^3P_2 \rightarrow {}^3D_3$ transition, reducing its uncertainty by a factor of 20, as well as its isotope shifts and those of the 326 nm and 480 nm transitions [27] for the ^{111}Cd and ^{113}Cd fermions.

Capturing atoms with $^3P_2 \rightarrow {}^3D_3$ metastable MOTs may also efficiently collect atoms from an atomic beam that is optically pumped to $^3P_2$ and slowed with the $^3P_2 \rightarrow {}^3D_3$ transition, or a beam slowed by low-power $^1S_0 \rightarrow {}^1P_1$ light, avoiding the $^1S_0 \rightarrow {}^1P_1$ MOT and its required large magnetic field gradients [38–40]. Beyond Cd, these techniques may be particularly useful for Hg [3, 32], requiring only low power UVC light, as well as for Sr [43] and Yb, and possibly Zn [68], Ca and Mg [42]. Avoiding UVC light, or reducing its required power, is advantageous for long-term operation of future Zn, Cd and Hg clocks and other quantum sensors, particularly in remote environments.

Acknowledgments

We gratefully acknowledge stimulating and helpful conversations with U. Sterr, S. Gupta, A. Yamaguchi, H. Katori, A. Kramida, and E. Tiemmann and contributions of D. Schussheim [61, 62].This material is based upon work supported by the U.S. National Science Foundation under award No. 2012117.

References

[1] Ludlow A D, Boyd M M, Ye J, Peik E and Schmidt P O 2015 *Rev. Mod. Phys.* **87**(2) 637–701 https://link.aps.org/doi/10.1103/RevModPhys.87.637

[2] Beloy K *et al* 2021 *Nature* **591** 564–569 ISSN 1476-4687 https://doi.org/10.1038/s41586-021-03253-4

[3] McFerran J J, Yi L, Mejri S, Di Manno S, Zhang W, Guéna J, Le Coq Y and Bize S 2012 *Phys. Rev. Lett.* **108**(18) 183004 https://link.aps.org/doi/10.1103/PhysRevLett.108.183004

[4] Ohmae N, Bregolin F, Nemitz N and Katori H 2020 *Opt. Express* **28** 15112–15121 http://www.opticsexpress.org/abstract.cfm?URI=oe-28-10-15112

[5] Tarallo M G, Mazzoni T, Poli N, Sutyrin D V, Zhang X and Tino G M 2014 *Phys. Rev. Lett.* **113**(2) 023005 https://link.aps.org/doi/10.1103/PhysRevLett.113.023005

[6] Abe M *et al* 2021 *Quantum Science and Technology* **6** 044003 https://dx.doi.org/10.1088/2058-9565/abf719
[7] Tino G M 2021 *Quantum Science and Technology* **6** 024014 https://dx.doi.org/10.1088/2058-9565/abd83e
[8] Bandarupally S, Tinsley J N, Chiarotti M and Poli N 2023 *Journal of Physics B: Atomic, Molecular and Optical Physics* **56** 185301 https://dx.doi.org/10.1088/1361-6455/acf3bf
[9] Takasu Y, Maki K, Komori K, Takano T, Honda K, Kumakura M, Yabuzaki T and Takahashi Y 2003 *Phys. Rev. Lett.* **91**(4) 040404 https://link.aps.org/doi/10.1103/PhysRevLett.91.040404
[10] Kraft S, Vogt F, Appel O, Riehle F and Sterr U 2009 *Phys. Rev. Lett.* **103**(13) 130401 https://link.aps.org/doi/10.1103/PhysRevLett.103.130401
[11] Stellmer S, Tey M K, Huang B, Grimm R and Schreck F 2009 *Phys. Rev. Lett.* **103**(20) 200401 https://link.aps.org/doi/10.1103/PhysRevLett.103.200401
[12] de Escobar Y N M, Mickelson P G, Yan M, DeSalvo B J, Nagel S B and Killian T C 2009 *Phys. Rev. Lett.* **103**(20) 200402 https://link.aps.org/doi/10.1103/PhysRevLett.103.200402
[13] Cazalilla M A and Rey A M 2014 *Reports on Progress in Physics* **77** 124401 https://dx.doi.org/10.1088/0034-4885/77/12/124401
[14] Chen C C, González Escudero R, Minář J, Pasquiou B, Bennetts S and Schreck F 2022 *Nature* **606** 683–687 ISSN 1476-4687 https://www.nature.com/articles/s41586-022-04731-z
[15] Gorshkov A V, Hermele M, Gurarie V, Xu C, Julienne P S, Ye J, Zoller P, Demler E, Lukin M D and Rey A M 2010 *Nature Physics* **6** 289–295 ISSN 1745-2481 https://doi.org/10.1038/nphys1535
[16] Daley A J 2011 *Quantum Information Processing* **10** 865 ISSN 1573-1332 https://doi.org/10.1007/s11128-011-0293-3
[17] Pagano G, Scazza F and Foss-Feig M 2019 *Advanced Quantum Technologies* **2** 1800067 https://onlinelibrary.wiley.com/doi/abs/10.1002/qute.201800067
[18] Madjarov I S, Covey J P, Shaw A L, Choi J, Kale A, Cooper A, Pichler H, Schkolnik V, Williams J R and Endres M 2020 *Nature Physics* **16** 857–861 ISSN 1745-2481 https://doi.org/10.1038/s41567-020-0903-z
[19] Bloch I, Dalibard J and Nascimbène S 2012 *Nat. Phys.* **8** 267–276 ISSN 1745-2481 https://doi.org/10.1038/nphys2259
[20] Schäfer F, Fukuhara T, Sugawa S, Takasu Y and Takahashi Y 2020 *Nature Reviews Physics* **2** 411–425 ISSN 2522-5820 https://doi.org/10.1038/s42254-020-0195-3
[21] Counts I, Hur J, Aude Craik D P L, Jeon H, Leung C, Berengut J C, Geddes A, Kawasaki A, Jhe W and Vuletić V 2020 *Phys. Rev. Lett.* **125**(12) 123002 https://link.aps.org/doi/10.1103/PhysRevLett.125.123002
[22] Ono K, Saito Y, Ishiyama T, Higomoto T, Takano T, Takasu Y, Yamamoto Y, Tanaka M and Takahashi Y 2022 *Phys. Rev. X* **12**(2) 021033 https://link.aps.org/doi/10.1103/PhysRevX.12.021033
[23] Figueroa N L, Berengut J C, Dzuba V A, Flambaum V V, Budker D and Antypas D 2022 *Phys. Rev. Lett.* **128**(7) 073001 https://link.aps.org/doi/10.1103/PhysRevLett.128.073001
[24] Hur J *et al* 2022 *Phys. Rev. Lett.* **128**(16) 163201 https://link.aps.org/doi/10.1103/PhysRevLett.128.163201
[25] Solaro C, Meyer S, Fisher K, Berengut J C, Fuchs E and Drewsen M 2020 *Phys. Rev. Lett.* **125**(12) 123003 https://link.aps.org/doi/10.1103/PhysRevLett.125.123003
[26] Zheng X, Dolde J, Lochab V, Merriman B N, Li H and Kolkowitz S 2022 *Nature* **602** 425–430 https://doi.org/10.1038/s41467-023-40629-8
[27] Ohayon B, Hofsäss S, Padilla-Castillo J E, Wright S C, Meijer G, Truppe S, Gibble K and Sahoo B K 2022 *New Journal of Physics* **24** 123040 https://dx.doi.org/10.1088/1367-2630/acacbb
[28] Yamaguchi A, Safronova M S, Gibble K and Katori H 2019 *Phys. Rev. Lett.* **123**(11) 113201 https://link.aps.org/doi/10.1103/PhysRevLett.123.113201
[29] Dzuba V A and Derevianko A 2019 *Journal of Physics B: Atomic, Molecular and Optical Physics* **52** 215005 https://dx.doi.org/10.1088/1361-6455/ab4434
[30] Brickman K A, Chang M S, Acton M, Chew A, Matsukevich D, Haljan P C, Bagnato V S and Monroe C 2007 *Phys. Rev. A* **76**(4) 043411 https://link.aps.org/doi/10.1103/PhysRevA.76.043411
[31] Kaneda Y, Yarborough J M, Merzlyak Y, Yamaguchi A, Hayashida K, Ohmae N and Katori H 2016 *Opt. Lett.* **41** 705–708 https://opg.optica.org/ol/abstract.cfm?URI=ol-41-4-705
[32] Lavigne Q, Groh T and Stellmer S 2022 *Phys. Rev. A* **105**(3) 033106 https://link.aps.org/doi/10.1103/PhysRevA.105.033106
[33] Tinsley J N, Bandarupally S, Penttinen J P, Manzoor S, Ranta S, Salvi L, Guina M and Poli N 2021 *Opt. Express* **29** 25462–25476 https://opg.optica.org/oe/abstract.cfm?URI=oe-29-16-25462
[34] Hofsäss S, Padilla-Castillo J E, Wright S C, Kray S, Thomas R, Sartakov B G, Ohayon B, Meijer G and Truppe S 2023 *Phys. Rev. Res.* **5**(1) 013043 https://link.aps.org/doi/10.1103/PhysRevResearch.5.013043
[35] Lurio A and Novick R 1964 *Phys. Rev.* **134**(3A) A608–A614 https://link.aps.org/doi/10.1103/PhysRev.134.A608
[36] Czajkowski M, Bobkowski R and Krause L 1989 *Phys. Rev. A* **40**(8) 4338–4343 , this more recent measurement suggests a 3.0(1) μs lifetime and a 53(2) kHz linewidth https://link.aps.org/doi/10.1103/PhysRevA.40.4338

[37] Norcia M A, Cline J R K, Bartolotta J P, Holland M J and Thompson J K 2018 *New Journal of Physics* **20** 023021 https://dx.doi.org/10.1088/1367-2630/aaa950

[38] Kawasaki A, Braverman B, Yu Q and Vuletic V 2015 *Journal of Physics B: Atomic, Molecular and Optical Physics* **48** 155302 , report capturing Yb using only its 184 kHz 556 nm intercombination transition. https://dx.doi.org/10.1088/0953-4075/48/15/155302

[39] Kuwamoto T, Honda K, Takahashi Y and Yabuzaki T 1999 *Phys. Rev. A* **60**(2) R745–R748 https://link.aps.org/doi/10.1103/PhysRevA.60.R745

[40] Lee J, Lee J H, Noh J and Mun J 2015 *Phys. Rev. A* **91**(5) 053405 https://link.aps.org/doi/10.1103/PhysRevA.91.053405

[41] Grünert J and Hemmerich A 2002 *Phys. Rev. A* **65**(4) 041401 https://link.aps.org/doi/10.1103/PhysRevA.65.041401

[42] Friebe J *et al* 2011 *New Journal of Physics* **13** 125010 https://dx.doi.org/10.1088/1367-2630/13/12/125010

[43] Akatsuka T, Hashiguchi K, Takahashi T, Ohmae N, Takamoto M and Katori H 2021 *Phys. Rev. A* **103**(2) 023331 https://link.aps.org/doi/10.1103/PhysRevA.103.023331

[44] Zhang X, Beloy K, Hassan Y S, McGrew W F, Chen C C, Siegel J L, Grogan T and Ludlow A D 2022 *Phys. Rev. Lett.* **129**(11) 113202 https://link.aps.org/doi/10.1103/PhysRevLett.129.113202

[45] Chen C C, Siegel J L, Hunt B D, Grogan T, Hassan Y S, Beloy K, Gibble K, Brown R C and Ludlow A D 2024 *Phys. Rev. Lett.* **133**(5) 053401 https://link.aps.org/doi/10.1103/PhysRevLett.133.053401

[46] Barker D S, Reschovsky B J, Pisenti N C and Campbell G K 2015 *Phys. Rev. A* **92**(4) 043418 https://link.aps.org/doi/10.1103/PhysRevA.92.043418

[47] Curtis E A, Oates C W and Hollberg L 2001 *Phys. Rev. A* **64**(3) 031403 https://link.aps.org/doi/10.1103/PhysRevA.64.031403

[48] Sterr U, Degenhardt C, Stoehr H, Lisdat C, Schnatz H, Helmcke J, Riehle F, Wilpers G, Oates C and Hollberg L 2004 *Comptes Rendus Physique* **5** 845–855 ISSN 1631-0705 fundamental metrology https://www.sciencedirect.com/science/article/pii/S1631070504001525

[49] Frömmgen N *et al* 2015 *The European Physical Journal D* **69** 164 ISSN 1434-6079 https://doi.org/10.1140/epjd/e2015-60219-0

[50] Brimicombe M S W M, Stacey D N, Stacey V, Hühnermann H, Menzel N and Bleaney B 1976 *Proceedings of the Royal Society of London. A. Mathematical and Physical Sciences* **352** 141–152 https://royalsocietypublishing.org/doi/abs/10.1098/rspa.1976.0168

[51] Faust W, McDermott M and Lichten W 1960 *Phys. Rev.* **120**(2) 469–469 https://link.aps.org/doi/10.1103/PhysRev.120.469

[52] Chantepie, M and Lecluse, Y 1971 *J. Phys. France* **32** 415–419 https://doi.org/10.1051/jphys:01971003205-6041500

[53] The MOT is loaded for 400 ms and the atom fluorescence is integrated for 16.67 ms after the laser's FM is inhibited and its intensity and detuning are ramped [62]. A subsequent background, also integrated for 16.67 ms after the atoms are cleared with a blue laser detuning, is subtracted. The direction of the MOT magnetic field gradient is then reversed and its background, from two more integrations, is additionally subtracted, giving an updated number of trapped atoms every 907 ms.

[54] Walhout M, Megens H J L, Witte A and Rolston S L 1993 *Phys. Rev. A* **48**(2) R879–R882 https://link.aps.org/doi/10.1103/PhysRevA.48.R879

[55] Gibble K E, Kasapi S and Chu S 1992 *Opt. Lett.* **17** 526–528 https://opg.optica.org/ol/abstract.cfm?URI=ol-17-7-526

[56] Loftus T H, Ido T, Ludlow A D, Boyd M M and Ye J 2004 *Phys. Rev. Lett.* **93**(7) 073003 https://link.aps.org/doi/10.1103/PhysRevLett.93.073003

[57] Hansen D and Hemmerich A 2006 *Phys. Rev. Lett.* **96**(7) 073003 https://link.aps.org/doi/10.1103/PhysRevLett.96.073003

[58] Traverso A, Chakraborty R, Martinez de Escobar Y N, Mickelson P G, Nagel S B, Yan M and Killian T C 2009 *Phys. Rev. A* **79**(6) 060702 https://link.aps.org/doi/10.1103/PhysRevA.79.060702

[59] Kelly J F, Harris M and Gallagher A 1988 *Phys. Rev. A* **38**(3) 1225–1229 https://link.aps.org/doi/10.1103/PhysRevA.38.1225

[60] Walker T, Sesko D and Wieman C 1990 *Phys. Rev. Lett.* **64**(4) 408–411 https://link.aps.org/doi/10.1103/PhysRevLett.64.408

[61] Schussheim D and Gibble K 2018 Laser system to laser-cool and trap cadmium: towards a cadmium optical lattice clock *Frontiers in Optics / Laser Science* (Optical Society of America) p LTh1F.2 http://www.osapublishing.org/abstract.cfm?URI=LS-2018-LTh1F.2

[62] Schussheim D T and Gibble K 2023 *Review of Scientific Instruments* **94** 085101 ISSN 0034-6748 https://doi.org/10.1063/5.0157330

[63] Manzoor S, Tinsley J N, Bandarupally S, Chiarotti M and Poli N 2022 *Opt. Lett.* **47** 2582–2585 https:

//opg.optica.org/ol/abstract.cfm?URI=ol-47-10-2582
[64] Tiemann E 2022 Private Communication
[65] Bodermann B, Knöckel H and Tiemann E 2002 *The European Physical Journal D* **19** 31–44 https://doi.org/10.1140/epjd/e20020052
[66] Burns K and Adams K B 1956 *J. Opt. Soc. Am.* **46** 94–99. From their measured wavelength, it is likely that their calculated 5s5d $^3D_3$ energy was 59 515.990 cm^{-1} [67]. https://opg.optica.org/abstract.cfm?URI=josa-46-2-94
[67] Kramida A 2022 Private Communication
[68] Büki M, Röser D and Stellmer S 2021 *Appl. Opt.* **60** 9915–9918 https://opg.optica.org/ao/abstract.cfm?URI=ao-60-31-9915

Ultralow phase noise optoelectronic oscillator

Dongliang Yan[1], Jun Ge[1] and Dan Liu[1,2]

[1] Beijing Institute of Radio Metrology and Measurement, Beijing, China

[2] E-mail: liuan1014@163.com

Abstract. The optoelectronic oscillator(OEO)'s main technological advantage over the traditional microwave oscillator is its ultra-low phase noise property. In this article, we have been using a high-power low-RIN laser, a long fiber loop, and an ultra-low noise optical connection to generate an OEO with a phase noise of -162.5dBc/Hz@10kHz at a frequency of 10 GHz to investigate the optoelectronic oscillator's single-loop structure. More significant factors on OEO have been explored in this work, including residual phase noise of the optoelectronic link, linear operating state of the optical link, optical power input to the photodetector, and photodetector bias voltage.

1. Introduction

Low phase noise microwave oscillators are widely used in radar, navigation, communications, metrology, deep space exploration, etc[1]-[4]. Its phase noise directly affects the detection accuracy of radar, the positioning accuracy of navigation systems, the bit error rate of high-speed communications, the transmission accuracy of measurement standards, and the limited distance of deep space exploration[5]. However, as the frequency of the traditional microwave oscillator increases, the phase noise worsens by 20log$N$, where N is the multiplication factor, which is due to the generation process of the traditional microwave oscillator.

In the field of low phase noise frequency generation technology, researchers have proposed a variety of various technological solutions, and the superior technical specifications reached in these solutions are as follows. One is the low-temperature sapphire microwave frequency generation system[6]-[8], the advantage of this system is that the short-term frequency stability is quite excellent, but the phase noise is not as good as in other approaches., and because this frequency generation system is exceedingly sophisticated, huge, and requires a strict environment, it is hard to receive advances in technological applications. The second is the use of ultra-stable laser sources combined with low phase noise optical frequency comb technology program[9], the advantages of this technologist are not only its superior short-term frequency stability, but phase noise performances are also very good, which is the technological solution that produces the best phase noise indicator, but the disadvantages of this program are also obvious, including the large size, expensive, so the same does not have engineering advantage. Another is based on the photoelectric oscillator structure to generate low phase noise frequency source[10]-[15], this scheme produces a frequency source of phase noise are superior at offset frequencies from 100Hz to 1MHz, and the expense is lower, smaller in size, very appropriate for large-scale engineering applications. The paper released by OEWAVES, USA, offers the best technical specification of the worldwide published photoelectric oscillator, with a phase noise of -163dBc/Hz@6kHz at a frequency of 10GHz[11]-[12]. The photoelectric oscillator produced in this work has a phase noise of -162.5dBc/Hz@10kHz at a frequency of 10GHz.

Journal of Physics: Conference Series **2889** (2024) 012007 doi:10.1088/1742-6596/2889/1/012007

2. Technology principle

The current photoelectric oscillator structure principle covers single-loop, double-loop, and multi-loop structures, with the single-loop structure still being the best loop structure in terms of phase noise performance. The block diagram below depicts the single-loop structure employed in this paper. The laser's wavelength is 1550nm, the output power is 70mW, the electro-optical modulator's insertion loss is 3dB, the frequency range is 12GHz, the fiber ring is 10km single-mode fiber, the photodetector's saturation output power is 10mW, the gain of the low-phase noise microwave amplifier is 33dB, the insertion loss of the directional coupler is 1dB, and the center frequency of the bandpass cavity filter is 10GHz.

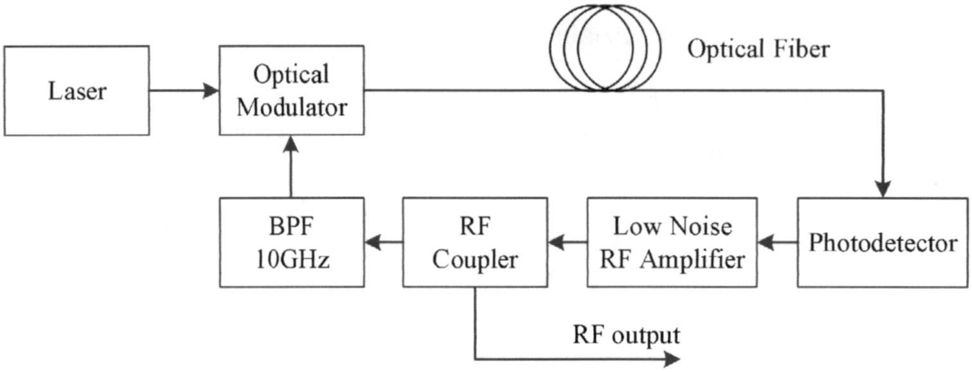

Figure 1. Block diagram of the single-loop OEO.

This is a classic single-loop structure OEO, three components make up the majority of the phase noise of its RF signal output, one is the residual phase noise of the photoelectric link, which consists primarily of the photoelectric detector and low phase noise microwave amplifier, here, this paper does not measure the noise of each component individually, because it eventually has to work as a whole, so the noise of the overall photoelectric link is measured. The other is the impact of the laser RIN, theoretically, the laser RIN after the photodetector PD converted to phase noise will be suppressed by 30dB~50dB. The third is the dispersion noise introduced by the long-distance fiber and the non-linear effect caused by the high-power input laser, the wavelength used in this paper is 1550 nm, and the dispersion noise of a single-mode fiber at this wavelength is much larger than that of a fiber at 1310 nm. This wavelength has the advantage of not requiring a 1310nm laser, electro-optical modulator, or photodetector; in addition, 1550nm optical equipment is more adaptive and less expensive.

3. The main factors affecting OEO phase noise

3.1. Additional phase noise in optical links

Because the residual phase noise of the optoelectronic link is a significant source of noise, it must be evaluated first when exploring ultra-low phase noise optoelectronic oscillators. The measurement block diagram of the residual phase noise of the optoelectronic link is displayed in Figure 2, and the phase noise is presented in Figure 3. It shows that the phase noise at the carrier frequency of 10GHz is -162dBc/Hz@10kHz, which may be decreased by a further 3dB due to comparison measurement.

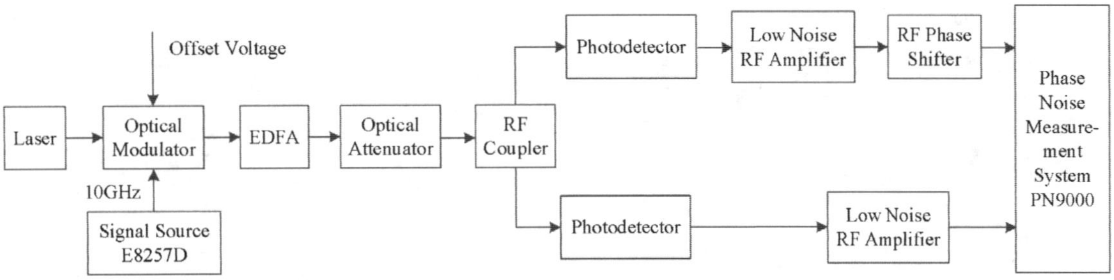

Figure 2. Block diagram for optical link residual phase noise measurement.

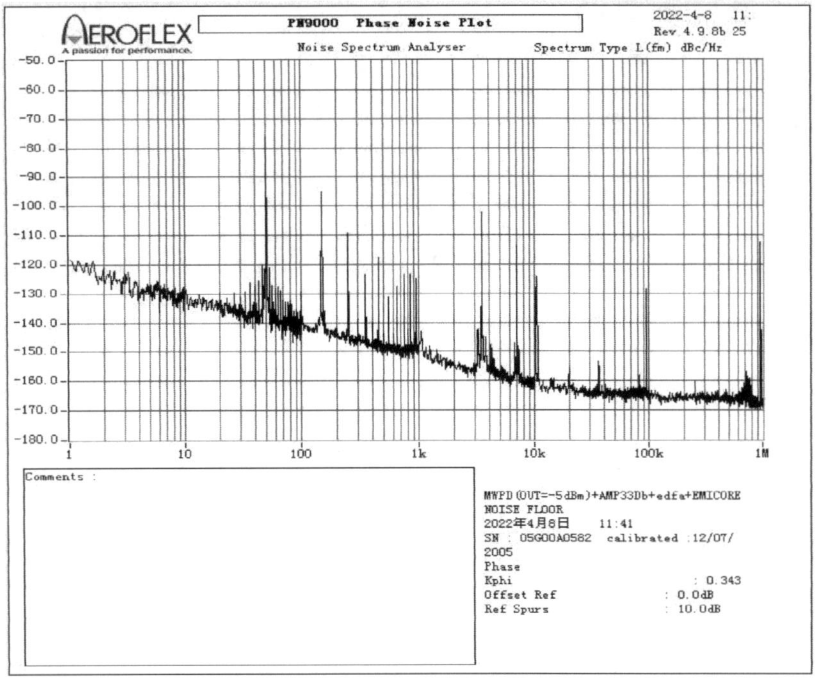

Figure 3. Optical link noise floor.

3.2. Linear operating state of the optical link

The functioning state of the optical link is another essential aspect influencing the phase noise of the OEO under the assumption of assuring superior additional phase noise of the optical link. The optical power input to the photodetector and the photodetector bias voltage are the two most important parameters influencing the optical link in the linear working range.

Because the OEO phase noise's turning point offset frequency is 10 kHz, the emphasis is on phase noise performance at that frequency. A set of data conditions is the maximum optical power input to the photodetector is 12.3dBm, the photodetector bias voltage is 8.22V, 8.72V, and 9V, phase noise curve as shown in Figure 4~6, another set of data conditions is the maximum optical power input to the photodetector is 12.6dBm, the photodetector bias voltage is 8.72V and 9V, respectively, and the phase noise curves are shown in Figure 7~8.

A comparison of the graphs above shows that under the condition that the photodetector receives the same optical power, adjusting the bias voltage of the photodetector can effectively improve the phase noise, and the photodetector has the best working bias point, the photodetector used in this paper is in the range of maximum input optical power 12.3dBm~12.6dBm, and the best bias working point is 8.72V, compared to other voltage bias points, the phase nosie is (1~2)dB superior.

Journal of Physics: Conference Series **2889** (2024) 012007 doi:10.1088/1742-6596/2889/1/012007

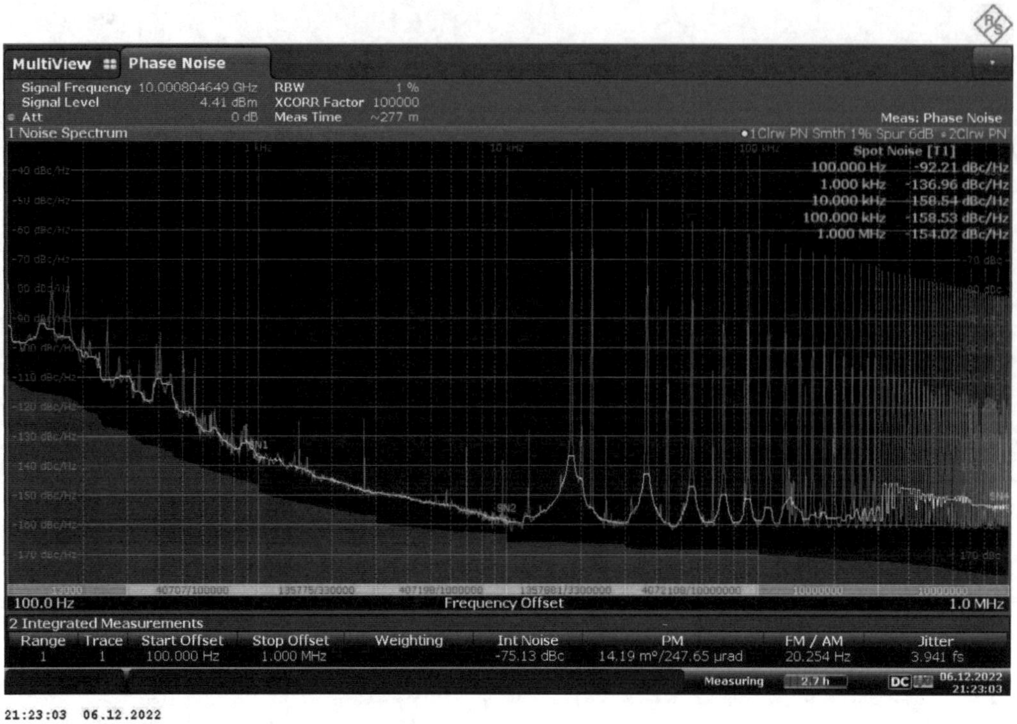

Figure 4. The PD's maximum input power is 12.3dBm, and its bias voltage is 8.22V.

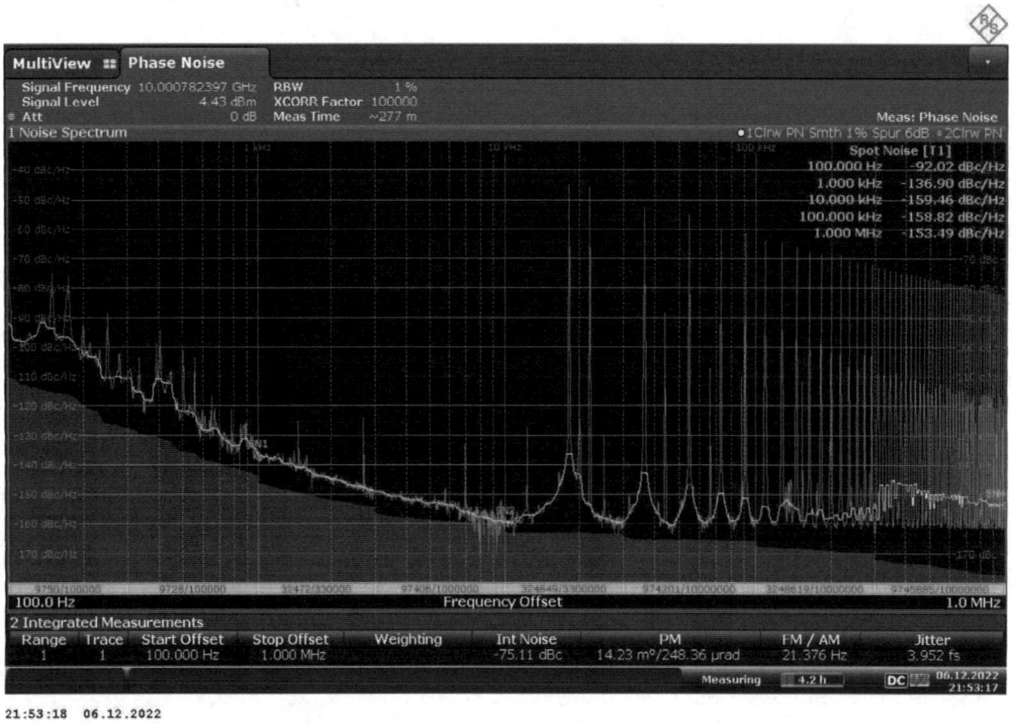

Figure 5. The PD's maximum input power is 12.3dBm, and its bias voltage is 8.72V.

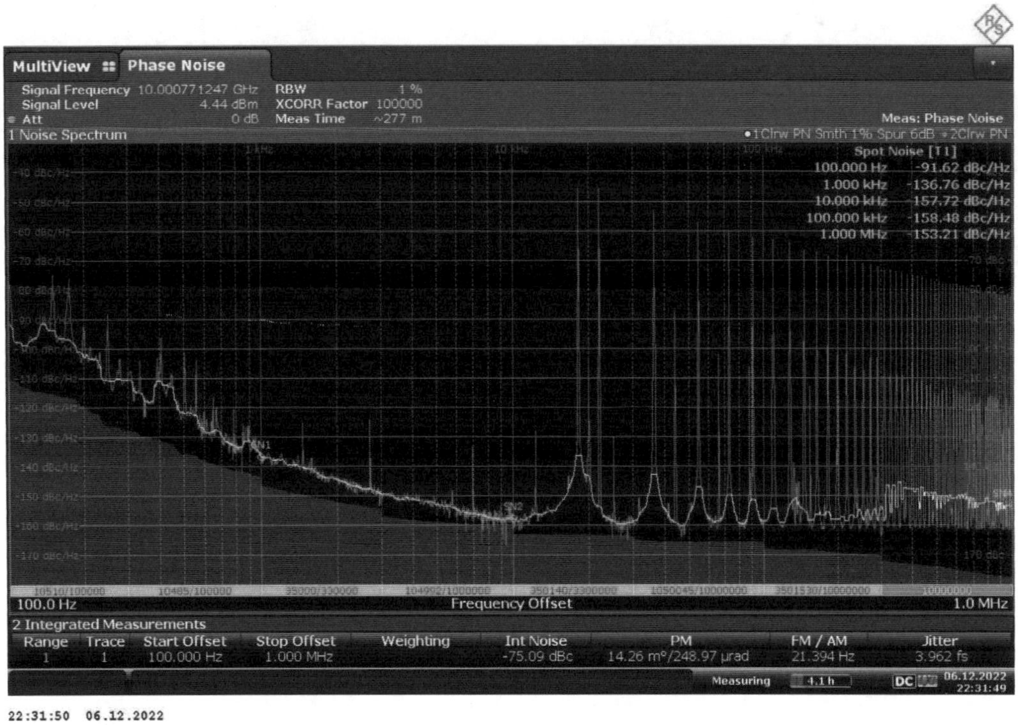

Figure 6. The PD's maximum input power is 12.3dBm, and its bias voltage is 9V.

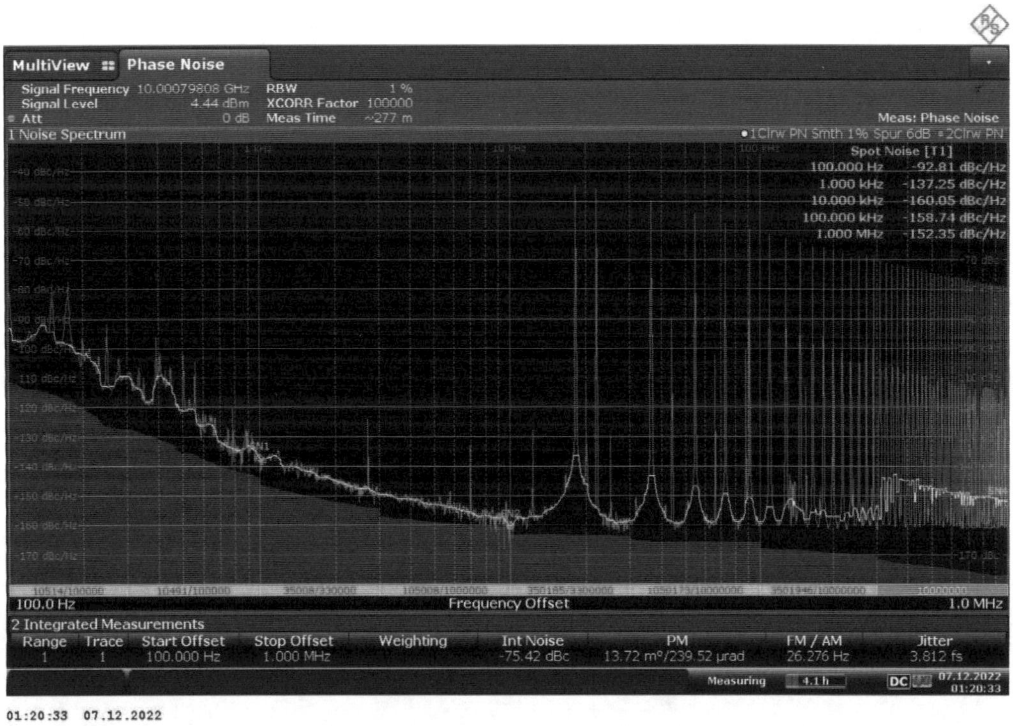

Figure 7. The PD's maximum input power is 12.6dBm, and its bias voltage is 8.72V.

Journal of Physics: Conference Series **2889** (2024) 012007 doi:10.1088/1742-6596/2889/1/012007

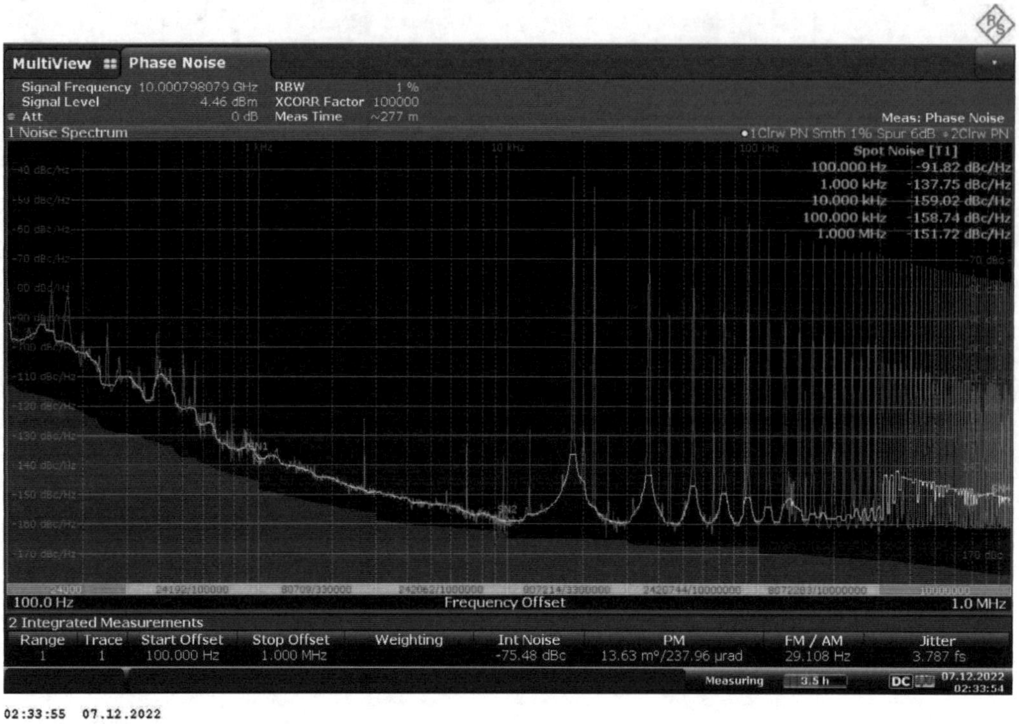

Figure 8. The PD's maximum input power is 12.6dBm, and its bias voltage is 9V.

In the experiment, it was found that the operating current of the photodetector is an important factor affecting the phase noise performance of the OEO. If the photodetector's initial current is too low, the output signal's signal-to-noise ratio will be inadequate, the phase noise will be undesirable, and if the photodetector's initial current is too great, the noise introduced by the dark current will be considerable. Therefore, the current of the photodetector has an ideal working point as well. The following takes a set of data for analysis and explanation, the conditions of the data are the maximum optical power input to the photodetector is 12.8dBm, the bias voltage of the photodetector is 9V, the working current is 6mA, 7mA, and 9mA, the phase noise curve as shown in Figure 9~11. The comparison of the figures shows that the best working current under the given circumstances is 7mA.

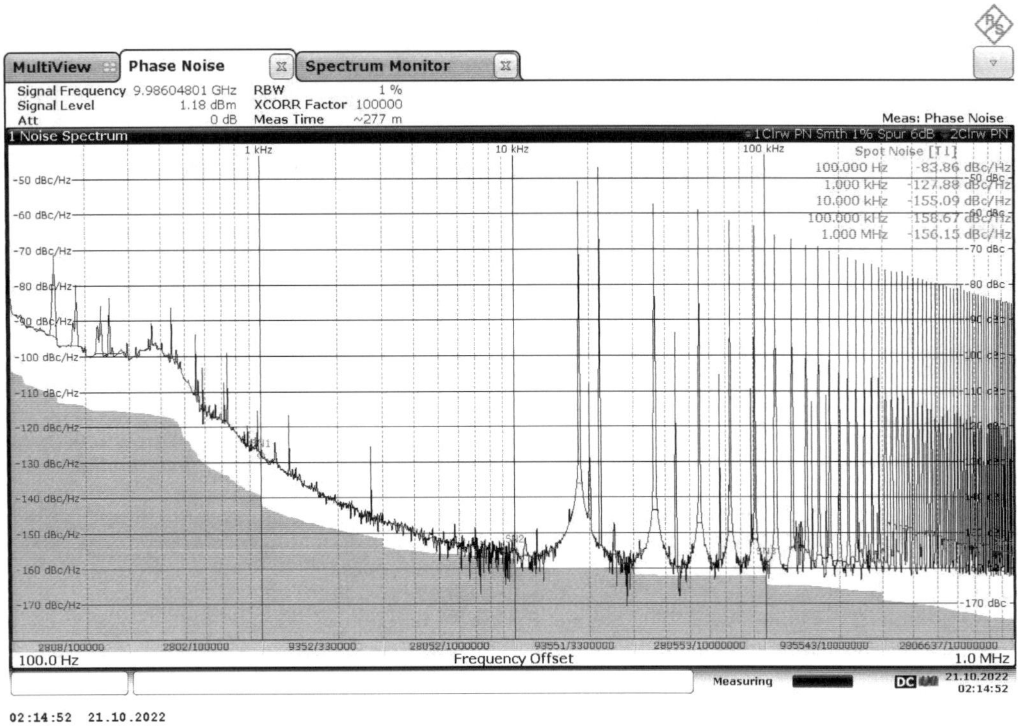

Figure 9. PD's operating voltage is 9V and its operating current is 6mA.

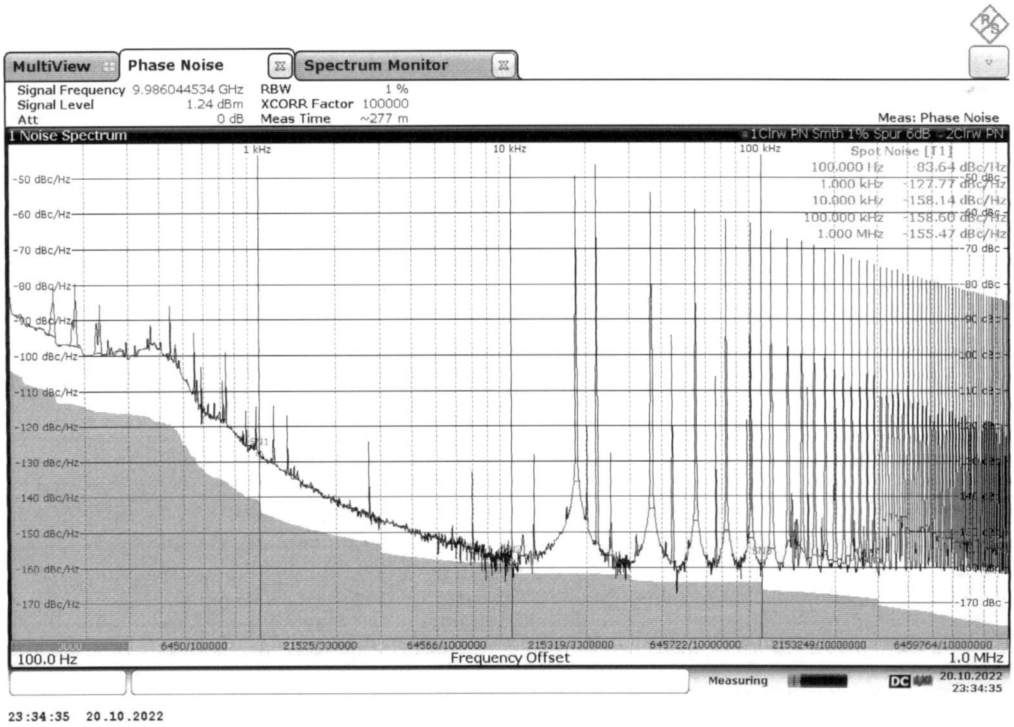

Figure 10. The PD's operating voltage is 9V and its operating current is 7mA.

Journal of Physics: Conference Series **2889** (2024) 012007 doi:10.1088/1742-6596/2889/1/012007

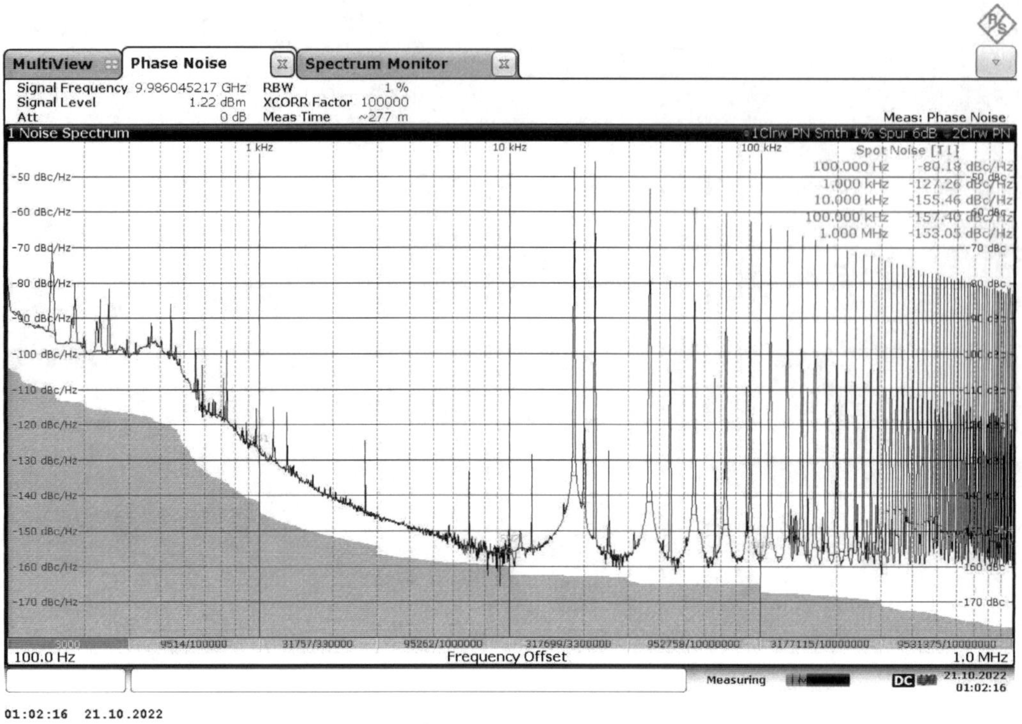

Figure 11. PD's operating voltage is 9V and its operating current is 9mA.

4. Photoelectric oscillator performance

Based on the single-loop design of Figure 1, the researched ultra-low phase noise optoelectronic oscillator has a phase noise of -162.5 dBc/Hz@10 kHz at a frequency of 10 GHz, as shown in Figure 12.

Journal of Physics: Conference Series **2889** (2024) 012007 doi:10.1088/1742-6596/2889/1/012007

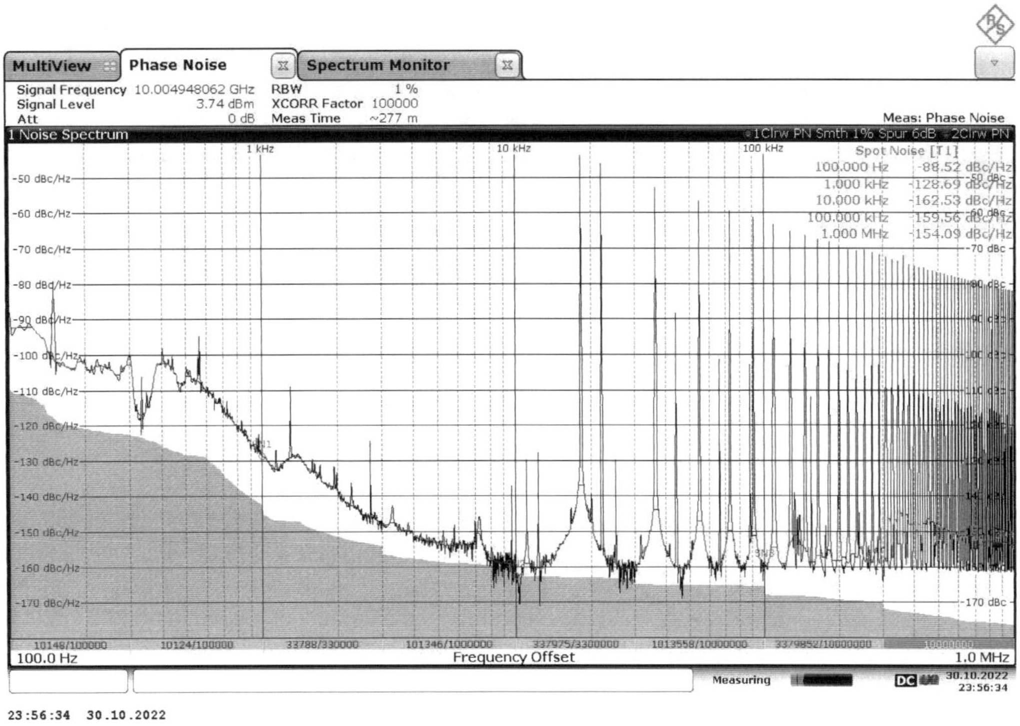

Figure 12. Ultra-low phase noise opto-oscillator phase noise.

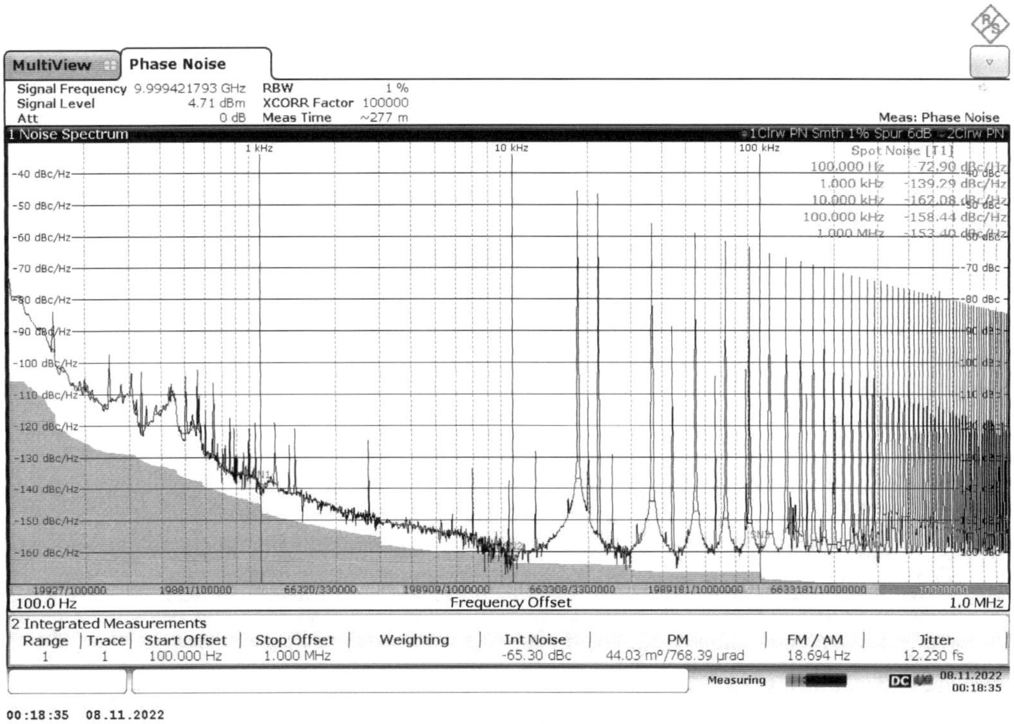

Figure 13. Improved phase noise.

In our experiments, we discovered a strong tendency of phase noise deterioration induced by external factors such as temperature, airflow from air conditioning, and vibration, particularly in the near-carrier frequency less than 10 kHz offset frequency area. So the phase noise is significantly improved when the optoelectronic oscillator is positioned in a chassis, compared to Figure 12, the phase noise is optimized by about 10 dB at 1 kHz frequency offset, as illustrated in Figure 13.

5. Conclusion

This paper demonstrated that the ophotoelectric oscillator's single-loop structure can generate microwave signals with ultra-low phase noise, which is much better than that of traditional microwave oscillators, and that its simple structure and low cost make it suitable for large-scale engineering applications.

References

[1] Padgett, J. E., Gunther, C. G., Hattori, T.. *Overview of wireless personal communications*. IEEE Communications Magazine, vol 33, no. 1, pp. 28-41, 2002.

[2] Hall, J. R., Hastrup, R. C., Bell, D. J.. *Microwave systems applications in deep space telecommunications and navigation: space exploration initiative architectures*. IEEE Transactions on Microwave Theory & Techniques,vol. 40, no. 6, pp. 1171-1178, 1992.

[3] Gubler, H., Hiller, M.. *The use of microwave fmcw radar in snow and avalanche research*. Cold Regions Science & Technology, vol 9, no 2, pp. 109-119, 1984.

[4] Peng, Z., Li, C.. *Portable microwave radar systems for short-range localization and life tracking*: A review, 2019.

[5] Gliserin, A. , Apolonski, A. , Krausz, F., Baum, P.. *Compression of single-electron pulses with a microwave cavity*. New Journal of Physics, vol. 14, no. 7, pp. 2565-2573, 2012.

[6] Mann, A. G., Sheng, C., Luiten, A. N.. *Cryogenic sapphire oscillator with exceptionally high frequency stability*. IEEE Transactions on Instrumentation and Measurement, vol. 50, no. 2, pp. 519-521, 2001.

[7] Hartnett, J. G. , Nand, N. R. , & Lu, C. *Ultra-low-phase-noise cryocooled microwave dielectric-sapphire-resonator oscillators with 1 x 10^-16 frequency instability*. Applied Physics Letters, vol. 100,no. 18, pp. 1, 2012.

[8] Calosso, C. E., Vernotte, F., Giordano, V., Fluhr, C., Dubois, B., Rubiola, E.. *Frequency stability measurement of cryogenic sapphire oscillators with a multichannel tracking dds and the two-sample covariance*. IEEE Transactions on Ultrasonics, Ferroelectrics, and Frequency Control, vol. 66, no. 3, pp. 616-623, 2019.

[9] Millo J, Abgrall M, Lours M, et al. *Ultra-low noise microwave generation with fiber-based optical frequency comb and application to atomic fountain clock*. Applied Physics Letters, vol. 94:141105, 2009.

[10] X.S. Yao and Lute Maleki. *Optoelectronic Microwave Oscillator*. J. Opt. Soc. Am. B., vol. 13, no. 8, pp. 1725-1735, Aug. 1996.

[11] Eliyahu D, Seidel D, Maleki L. *Phase noise of a high performance OEO and an ultra low noise floor cross-correlation microwave photonic homodyne system*. IEEE International Frequency Control Symposium, 2008.

[12] Eliyahu D, Seidel D, Maleki L. *RF Amplitude and Phase-Noise Reduction of an Optical Link and an Opto-Electronic Oscillator*. IEEE Transactions on Microwave Theory & Techniques, vol. 56, no. 2, pp. 449-456, 2008.

[13] X. S. Yao and Lute Maleki. *Dual Microwave and Optical Oscillator*. Opt. Letters, vol. 22, no. 24, pp. 1867-1869, Dec. 1997.

[14] X.S. Yao and Lute Maleki. *Multi-loop Optoelectronic Oscillator*. IEEE, J. of Quant. Electron., vol 36, pp. 79-84. Jan. 2000.

[15] W. Zhou and G. Blasche. *Injection-locked dual opto-electronic oscillator with ultra-low phase noise and ultra-low spurious level*. IEEE Transactions on Microwave Theory and Techniques, vol. 53, pp. 929-933, 2005.

Acknowledgments
The authors would like to express their gratitude to Beijing MWPHOTON Technology Co., Ltd. for supplying the excellent photodetector used in this experiment.

Journal of Physics: Conference Series **2889** (2024) 012008 doi:10.1088/1742-6596/2889/1/012008

Frequency comb for $^{88}Sr^{+}$ clock frequency comparison with Cs fountain clock at the NRC

Claude Marceau, Scott Beattie, Wojciech Pakulski, and Marina Gertsvolf

1. National Research Council Canada, 1200 Montreal road, Ottawa, Ontario, K1A 0R6 Canada

E-mail: claude.marceau@nrc-cnrc.gc.ca

Abstract. We describe a polarization maintaining erbium fibre optical frequency comb that was used to link NRC-FCs2 to an ultrastable laser, generating ultrastable microwaves (MW) that are shown to significantly shorten the fountain clock averaging time. A detailed experimental setup for our next measurement linking NRC $^{88}Sr^{+}$ clock to the NRC-FCs2 Cs fountain clock is also presented.

1. Introduction

The redefinition of the SI second in terms of optical atomic transitions is expected to happen within the next few years. The International Committee for Weights and Measures has issued requirements for candidate clocks to be considered for the redefinition [1, 2, 3]. Among these requirements is continuity: the candidate clock frequencies need to be measured against current primary frequency standards and the measurements need to be mainly limited by the systematic uncertainty of the Cs clocks. NRC has recently commissioned its Cs fountain clock NRC-FCs2 [4]. NRC also has a candidate clock for the redefinition, the $^{88}Sr^{+}$ single ion clock, which was used recently to make an absolute frequency measurement with an uncertainty of 4.3×10^{-16} [5].

In this work, we present the portable saturable absorber-based fibre combs made for this clock comparison [6, 7]. We also present the setup that we intend to use for the $^{88}Sr^{+}$ single ion clock measurement campaign against NRC-FCs2. The ultrastable probe laser [8] of the $^{88}Sr^{+}$ clock at 674 nm will serve as the local oscillator of both clocks. This scheme will also be used to facilitate the re-evaluation of the systematic biases of NRC-FCs2.

2. Transportable erbium fibre comb

An earlier generation of this comb was described in [6]. This transportable polarization-maintaining erbium fibre optical frequency comb was developed in a collaboration between NIST and NRC [7]. The oscillator mode-locks passively with the use of a semiconductor saturable absorber mirror (SESAM). Modulation of the 1480 nm oscillator pump power is used to stabilize the oscillator carrier-envelope offset frequency (f_0). Both a slow and a fast piezoelectric transducer (PZT) glued to the oscillator fibre are used to phase-lock one tooth of the comb to an optical frequency reference (f_B). Temperature control of the aluminum housing using thermoelectric cooling is also implemented to keep the slow PZT within range.

Journal of Physics: Conference Series **2889** (2024) 012008 doi:10.1088/1742-6596/2889/1/012008

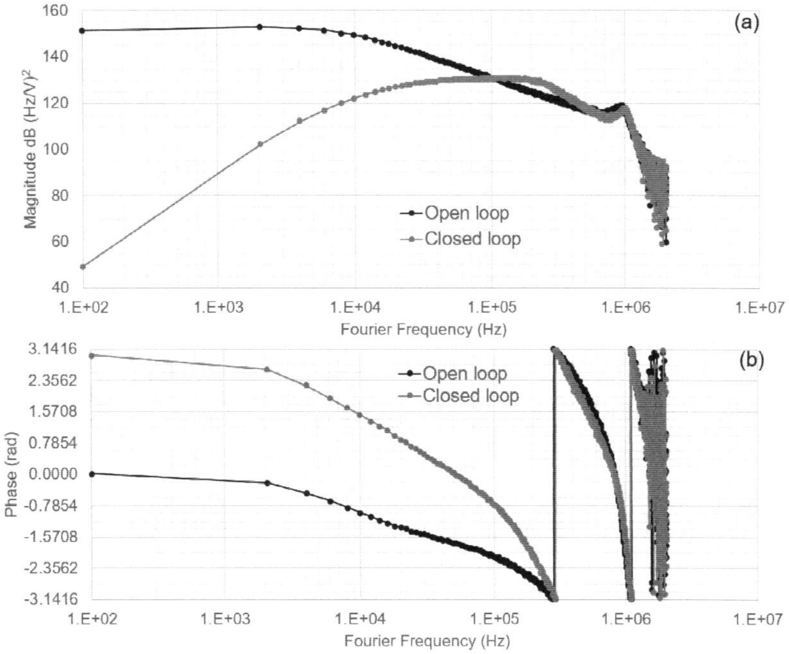

Figure 1. f_0 transfer function, measured from the fibre comb controller voltage modulation input to f_0. (a) Magnitude. (b) Phase. The phase reaches $-\pi$ around 300 kHz. The unity gain bandwidth is around 100 kHz. The oscillator pump diode current is controlled to lock f_0.

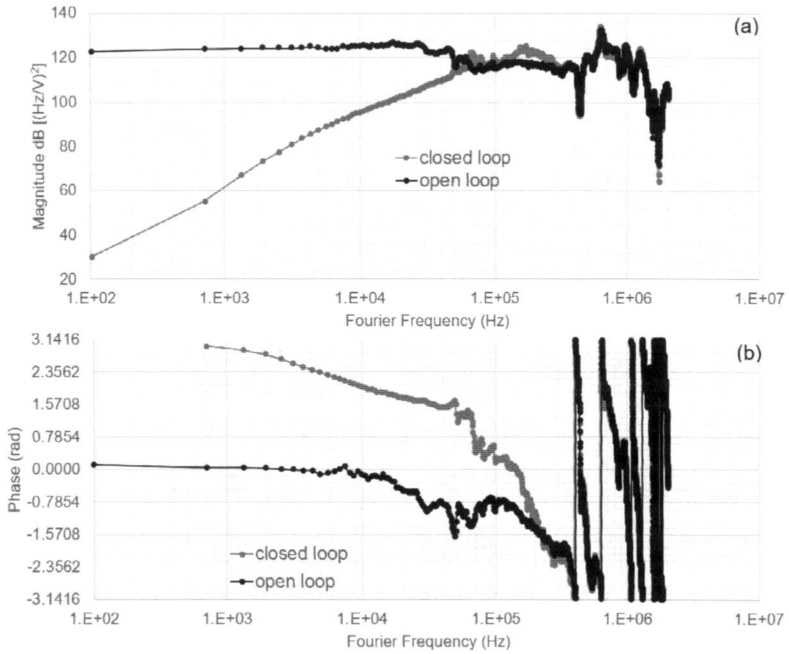

Figure 2. f_B transfer function, measured from the fibre comb controller voltage modulation input to f_B at 193 THz. (a) Magnitude. (b) Phase. The phase reaches $-\pi$ around 400 kHz. The unity gain bandwidth is around 60 kHz. The fast PZT response is dominant above 1 kHz.

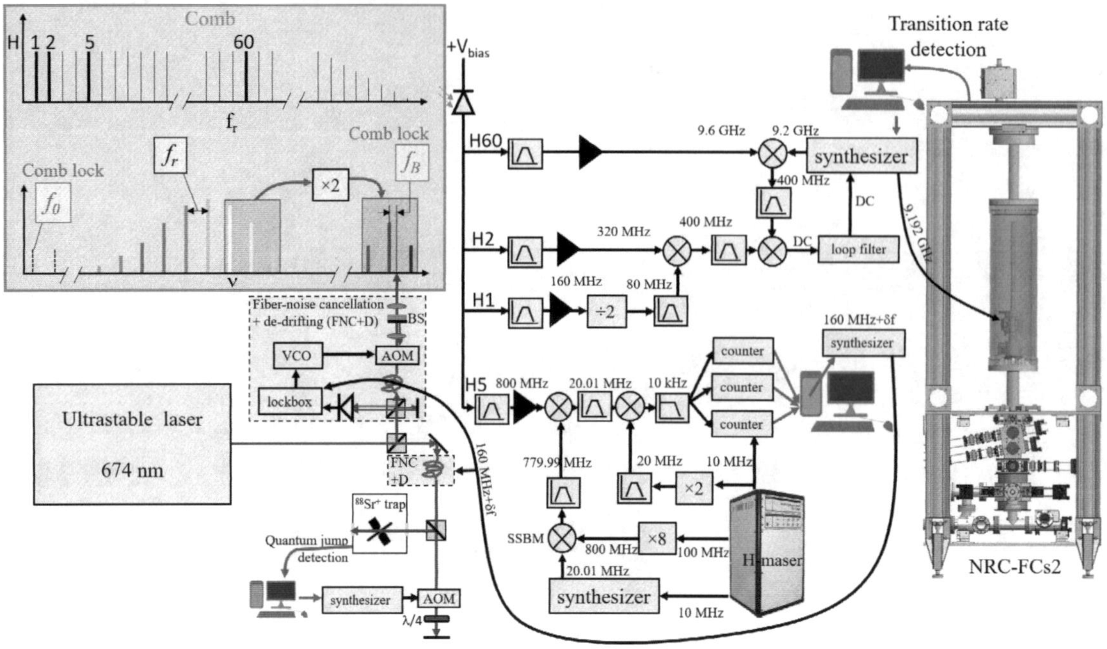

Figure 3. $^{88}Sr^+$ clock measurement scheme against NRC-FCs2 Cs fountain clock. Legend: AOM: acousto-optic modulator; SSBM: Single sideband mixer; FNC+D: Fibre-noise cancellation + de-drifting.

An amplifier is spliced to the oscillator and is pumped by two 980 nm diode lasers. The output of the amplifier is sent to a highly-nonlinear fibre, which broadens the spectrum beyond one octave. A frequency-doubling waveguide (1960 nm → 980 nm) is employed to produce an f_0 signal at 980 nm.

The f_0 transfer function (comb controller input voltage → frequency) is shown in Fig. 1. The optical beat f_B transfer function (comb controller input voltage → frequency) is shown in Fig. 2. The digital phase-locked loop (PLL) bandwidths are approximately 100 kHz for f_0, 60 kHz for f_B at 1550 nm (193 THz) and 50 kHz for f_B at 674 nm (445 THz). The Allan deviations ($\sigma_y(\tau = 1s)$) of the optical locks are 1.2×10^{-18} and 6.4×10^{-18} at 193 THz nm and 445 THz, respectively.

3. $^{88}Sr^+$ single ion clock measurement

The setup to measure the $^{88}Sr^+$ clock frequency against NRC-FCs2 where ultrastable MW from the locked comb is used as the local oscillator of the MW synthesizer is shown in Fig. 3. We use the fibre-noise cancellation modules to de-drift the probe laser over long averaging times (τ >100 s). The error signal to control the probe laser de-drifting can come from the $^{88}Sr^+$ clock itself or from a comparison with a hydrogen maser. This scheme takes advantage of the ultrastable RF/MW that are generated when one tooth of the fibre comb is locked to an ultrastable laser. The 60th harmonic of the comb at 9.6 GHz was chosen for this phase locking scheme because of its reduced photodetection noise compared to the fundamental (160 MHz). The fountain synthesizer (model CS-1, SpectraDynamics) has an output available at 9.2 GHz, which we mix with the 60th harmonic. The resulting beat at 400 MHz is then mixed with the sum of the second harmonic of the repetition rate (320 MHz) and half the repetition rate (80 MHz) to yield a DC error signal to close the main loop. The de-drifted ultrastable laser at 674 nm acts as the

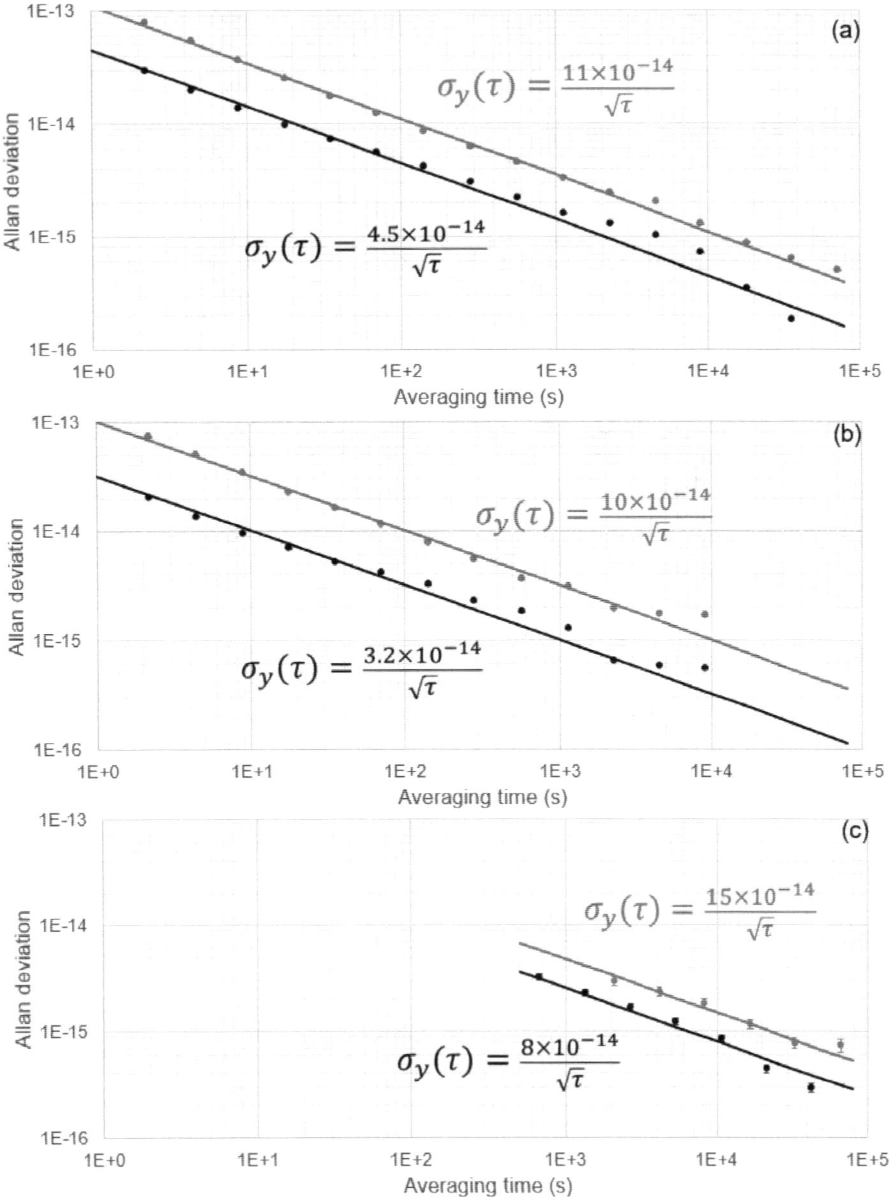

Figure 4. Stability of NRC-FCs2 Cs fountain clock using two different local oscillators. (a) Low atomic density mode. (b) High density mode. (c) Clock mode: an extrapolation to zero density is used to correct the bias introduced by cold collisions. The local oscillator is (red circles) NRC's best active hydrogen maser VM1; (black circles) the ultrastable laser with a slow drift correction calculated from the maser.

local oscillator of both the $^{88}Sr^+$ clock and the Caesium fountain clock NRC-FCs2. It allows a direct measurement of the $^{88}Sr^+$ clock frequency and faster averaging than the usual local oscillator of NRC-FCs2, a hydrogen maser.

4. Ultrastable MW generation

The Allan deviation $\sigma_y(\tau)$ of the NRC-FCs2 fountain clock against the VM1 (Vremya VCH-1003M) hydrogen maser and against the ultrastable 445 THz probe laser via the comb are shown in Fig. 4. The stability of the clock against its local oscillator is compared using the setup of Fig. 3, but without operating the $^{88}Sr^+$ clock. We observe in Fig. 4 (b) a short term stability improvement of 3.1 with the ultrastable laser in high atomic density mode, which should result in a reduction of averaging time by a factor of 10 for a given level of accuracy. This was used to speed up the systematic uncertainty re-evaluation of NRC-FCs2.

References

[1] Riehle F 2015 *Comptes Rendus Physique* **16** 506–515 ISSN 1631-0705 the measurement of time / La mesure du temps URL `https://www.sciencedirect.com/science/article/pii/S1631070515000638`

[2] Riehle F, Gill P, Arias F and Robertsson L 2018 *Metrologia* **55** 188

[3] URL `https://www.bipm.org/en/cgpm-2022/resolution-5`

[4] Beattie S, Jian B, Alcock J, Gertsvolf M, Hendricks R, Szymaniec K and Gibble K 2020 *Metrologia* **57** 035010 URL `https://dx.doi.org/10.1088/1681-7575/ab7c54`

[5] Jian B, Bernard J, Gertsvolf M and Dubé P 2022 *Metrologia* **60** 015007 URL `https://dx.doi.org/10.1088/1681-7575/aca615`

[6] Sinclair L C, Deschênes J D, Sonderhouse L, Swann W C, Khader I H, Baumann E, Newbury N R and Coddington I 2015 *Review of Scientific Instruments* **86** 081301 ISSN 0034-6748 (*Preprint* `https://pubs.aip.org/aip/rsi/article-pdf/doi/10.1063/1.4928163/14051684/081301_1_online.pdf`) URL `https://doi.org/10.1063/1.4928163`

[7] Marceau C, Beattie S and Gertsvolf M 2022 *2022 Joint Conference of the European Frequency and Time Forum and IEEE International Frequency Control Symposium (EFTF/IFCS)* pp 1–2

[8] Dubé P, Madej A A, Bernard J E, Marmet L and Shiner A D 2009 *Appl. Phys. B* **95** 43–54 URL `https://doi.org/10.1007/s00340-009-3390-6`

Low-noise optical frequency divider for precision measurement

Yanyi Jiang, Haosen Shi, Yuan Yao, Hongfu Yu, and Longsheng Ma

State Key Laboratory of Precision Spectroscopy, East China Normal University, Shanghai, China

E-mail: yyjiang@phy.ecnu.edu.cn

Abstract. We describe the development of an accurate optical frequency divider based on a Ti:Sapphire optical frequency comb. The division instability and uncertainty of the optical frequency divider are demonstrated to be 10^{-18} at 1 s averaging time and 3×10^{-22}, respectively. The ability of coherence transfer is also demonstrated by resolving a hertz-level-linewidth spectroscopy.

1. Introduction

Since the 8th symposium, significant progress has been made in optical frequency standards, low noise optical oscillators, and optical frequency combs. Both the long-term frequency instability and the frequency uncertainty of optical atomic clocks achieve ~1×10^{-18} [1,2]. The short-term frequency instability of optical oscillators frequency-stabilized to cryogenic optical cavities approach 4×10^{-17} [3]. The uncertainties induced by optical frequency combs in optical frequency ratio measurement approach 1.4×10^{-21} [4]. Moreover, optical atomic clocks based on different species can be compared with an uncertainty at the 10^{-18} level [5].

These impressive advances in optical frequency metrology have drawn wide attention since many precision measurements succeed in gaining higher sensitivity or accuracy by transforming other quantities into a frequency. For example, the constancy of the fine structure constant has been tested at a new limit of $1.0(1.1) \times 10^{-18}$/yr by measuring the frequency ratio between the $^2S_{1/2}(F=0)\rightarrow{}^2D_{3/2}(F=2)$ electric quadrupole (E2) and the $^2S_{1/2}(F=0)\rightarrow{}^2F_{7/2}(F=3)$ electric octupole (E3) transition of $^{171}Yb^+$ over time [6]. Moreover, the gravitational redshift has been tested with an uncertainty of $(1.4 \pm 9.1) \times 10^{-5}$ using two ground-based optical clocks [7], comparable to that measured in space. Although the measurement sensitivities or uncertainties have been pushed to unprecedented levels, there are growing interests to push these limits to even lower levels. Search for dark matter, tests of relativity, and detection of gravitational wave anticipate even more precise optical frequency ratio measurement beyond 10^{-18} $(\tau/s)^{-1/2}$, where τ is the averaging time [8,9].

In pursuit of this goal, we need to keep reducing the frequency instabilities and uncertainties of optical atomic clocks and optical frequency combs as well. In several talks of the 9th symposium, researchers have set their goals to achieve optical atomic clocks with uncertainties toward 10^{-21}. In this paper, we will introduce our recent effort on reducing the uncertainty in optical frequency ratio measurement down to 10^{-22} level, by using an optical frequency divider based on a Ti:Sapphire optical frequency comb. Meanwhile, we will also demonstrate other features of the optical frequency divider, such as the ability to transport between labs, multi-channel optical frequency division and long-term operation.

2. Optical frequency divider based on a Ti:Sapphire comb

When using optical frequency combs to link the frequency gap between two optical atomic clocks or c.w. lasers in optical frequency ratio measurement or coherence transfer, the frequency noise of combs will be added onto the result. In order to achieve a low frequency noise and uncertainty, we choose a comb based on a Ti:Sapphire mode-locked laser. Since the gain medium of a Ti:Sapphire mode-locked laser is a Ti:Sapphire crystal with a length at the millimeter level, it suffers less frequency noise compared to fiber mode-locked lasers. Meanwhile, its spectra after broadening are usually in the range of 0.5 μm to 1.1 μm, where most optical atomic clocks or their fundamental light lies in, as shown in Fig. 1(a). Besides, Ti:Sapphire combs have a larger repetition rate (f_r) in the range of 0.5–10 GHz compared to fiber combs, and thus it has more power in each comb line to obtain beating signals against optical atomic clocks with a good signal to noise ratio. Figure 1(b) shows the beat notes of our Ti:Sapphire comb against optical atomic clocks and c.w. lasers.

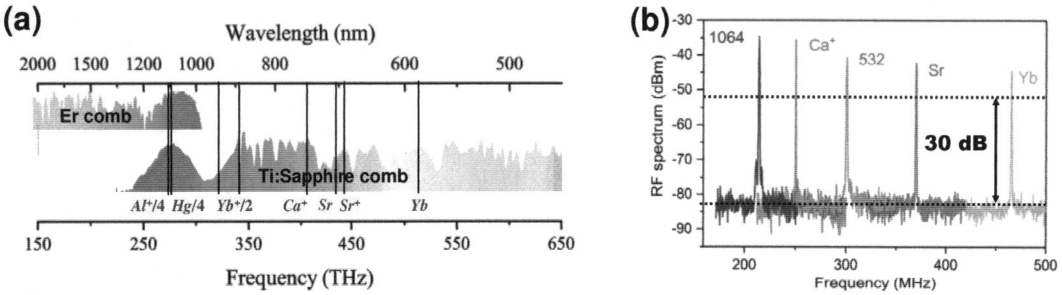

Figure 1: (Colour online) (a) Typical spectrum of a Ti:Sapphire comb after broadened in a nonlinear fiber. (b) Beat note of optical atomic clocks or c.w. lasers against our Ti:Sapphire comb in a resolution bandwidth (RBW) of 300 kHz.

However, easy operation and long continuous operation are the weakness of Ti:Sapphire combs. We cooperated with Albrecht Bartels from Novanta Europe GmbH to customize a commercial Ti:Sapphire mode-locked laser, which can be continuously mode-locked for over a month, limited by our observation time [10].

Although the frequency noise of combs can be largely eliminated by phase locking to ultra-stable lasers, it requires high servo bandwidth of combs. However, combs lose phase lock easily in the present of external perturbance. Therefore, we stabilize the Ti:Sapphire comb to a RF clock such as a hydrogen maser or a rubidium clock. We also customize the commercial Ti:Sapphire mode-locked laser by installing both a fast piezo and a slow one inside the laser cavity, which allows us to adjust the cavity length (or f_r) with moderate bandwidth and tuning range. By combining a fast analog feedback to pump current and a slow digital feedback to both an intracavity wedge and the pump power of the Ti:Sapphire mode-locked laser as well, the carrier envelope offset frequency (f_0) of the comb can be stabilized for days. With all these improvements, we extend the continuous operation time of the frequency-stabilized Ti:Sapphire comb to five days [10].

However, when stabilizing the comb to a RF clock, the 1 s-frequency instability of comb lines is no better than 10^{-13}. To eliminate the comb frequency noise, transfer oscillator scheme is adopted [11]. As shown in Fig. 2, first the beat notes between two lasers with frequency of ν_1 and ν_2 against their nearby comb lines are detected

$$f_{b1} = \nu_1 - \nu_N = \nu_1 - (Nf_r \pm f_0), \quad (1)$$

$$f_{b1}^* = \nu_{N+1} - \nu_1 = (N+1)f_r \pm f_0 - \nu_1, \quad (2)$$

$$f_{b2} = \nu_2 - \nu_M = \nu_2 - (Mf_r \pm f_0), \quad (3)$$

where N and M are integers, and ν_N (ν_M) is the frequency of the N-th (M-th) comb line. In order to understand easily, we use the plus sign before f_0 in the following text.

To remove f_0 from the above beating signals, we mix them with the signal of f_0, which is detected using 1f-2f self-referencing interferometer. Then the outputs of the mixers are divided in direct digital synthesizers (DDSs), whose divisors (m and n) are set to meet $M/m=N/n$. Then the outputs of the DDSs are compared on a third mixer. Thus we have

$$f_b = \frac{f_{b2}+f_0}{m} - \frac{f_{b1}+f_0}{n} = \frac{\nu_2 - Mf_r}{m} - \frac{\nu_1 - Nf_r}{n} = \frac{\nu_2}{m} - \frac{\nu_1}{n}. \quad (4)$$

We can phase lock f_b to a RF signal. Thereby we can fix the frequency relation between ν_1 and ν_2. In order to make the frequency ratio between ν_1 and ν_2 independent of any external RF frequency standard, we use the transfer oscillator scheme again to obtain an optically-referenced RF time base at about 10 MHz [4], which is

$$f_{\text{time}} = \left[\frac{f_{b1}+f_0}{k} + \frac{f_{b1}^*-f_0}{l}\right]\frac{1}{k_1} = \left[\frac{\nu_1 - Nf_r}{k} + \frac{(N+1)f_r - \nu_1}{l}\right]\frac{1}{k_1} = \frac{\nu_1}{k_1}\left(\frac{1}{k} - \frac{1}{l}\right), \quad (5)$$

where k and l are the divisors of another two DDSs, which satisfy $N/k = (N+1)/l$, and k_1 is a divisor of a DDS to make f_{time} close to 10 MHz. Then, we use f_{time} to generate a RF signal of $f_{\text{tune}} = f_{\text{time}}/k_2 = \nu_1/K$, where $1/K = (1/k - 1/l)/(k_1k_2)$.

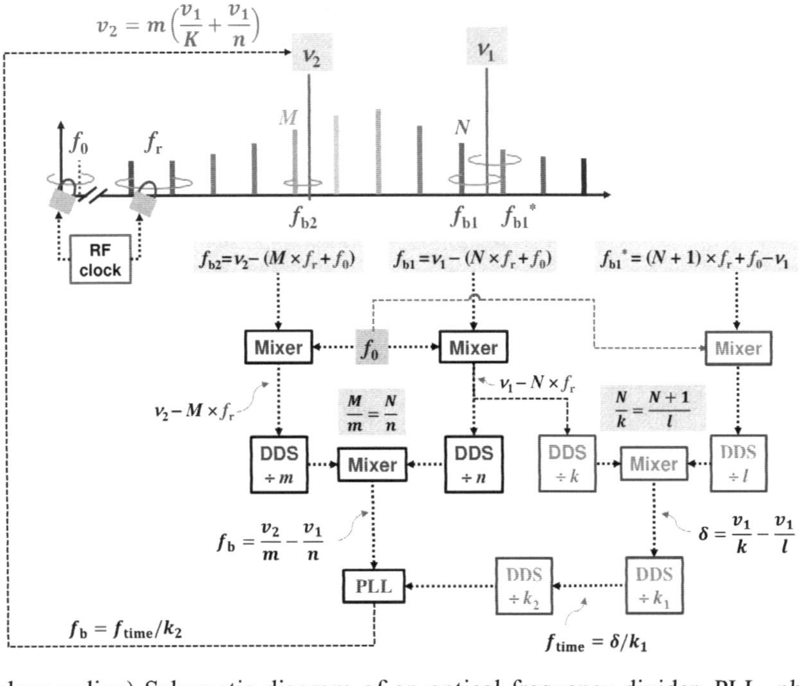

Figure 2: (Colour online) Schematic diagram of an optical frequency divider. PLL, phase lock loop. DDS, direct digital synthesizer.

As long as f_b is phase-locked to f_{tune}, then

$$\nu_2 = \frac{m}{n}\nu_1 + \frac{m}{K}\nu_1 = \frac{\nu_1}{R}, \quad (6)$$

where the divisor R is only determined by the divisors of the DDSs. When we phase lock f_b to f_{tune} by adjusting ν_2, ν_2 is phase-locked to ν_1 with a pre-set ratio of R, similar to a frequency divider. If we count the frequency of f_b on a frequency counter with the time base of f_{time}, then we can obtain the frequency ratio between ν_1 and ν_2.

The frequency ratio between ν_1 and ν_2 is tested by comparing an independent frequency divider realized by second harmonic generation. As shown in Fig. 3(a), we divide the frequency of a cavity-stabilized laser at 1064 nm ($\nu_{c\text{-}1064}$) [12] to a 532 nm laser ($\nu_{t\text{-}532}$), $\nu_{t\text{-}532} = \nu_{c\text{-}1064} \div R_{set}$ with R_{set} = 0.499 999 219 876 815 223 597 90 or 0.500 000 106 518 137 671 678 87 on different days. The 532 nm laser is second harmonic generated from a second 1064 nm laser ($\nu_{t\text{-}1064}$), that is $\nu_{t\text{-}532} = \nu_{t\text{-}1064} \div R_{SHG}$ with $R_{SHG} = 0.5$. By measuring the beating frequency between $\nu_{c\text{-}1064}$ and $\nu_{t\text{-}1064}$, we can check the accuracy and noise of R_{set} or R_{SHG}.

Figure 3(b) shows the measured instability of R_{set} induced in optical frequency division when the comb is frequency stabilized to a Rb clock (blue squares), a hydrogen maser (red dots) or an ultra-stable laser (purple solid line). The 1 s-instability of R_{set} is below 10^{-17}, and the lowest 1 s-instability of 6×10^{-19} is achieved when the 1 s-frequency instability of comb lines is $\sim 1 \times 10^{-15}$. As we can see in the figure, the frequency noise of the comb slightly affects the division noise. The long-term instability of R averages down to 10^{-22} when the comb is frequency-stabilized to a hydrogen maser, mainly due to the long-term operation of the Ti:Sapphire comb as well as an improved air sealed enclosure and robust mechanics. Moreover, we evaluate the agreement between the set value of R_{set} and a measured value from a known R_{SHG}, which is $(-1.9 \pm 3.0) \times 10^{-22}$. Therefore, we can conclude that the optical frequency divider can divide the frequency of the state-of-the-art optical atomic clocks without degrading their performance.

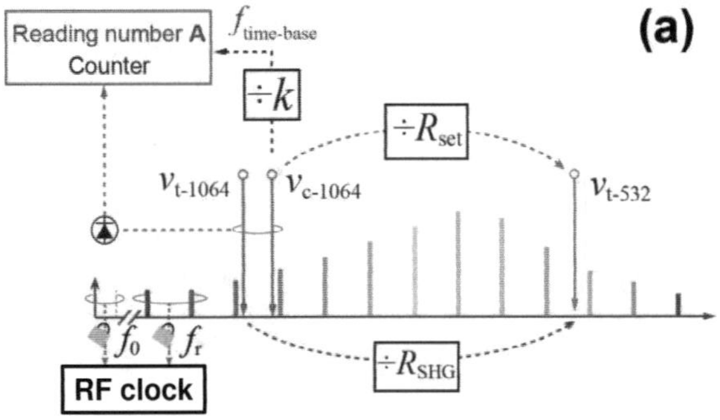

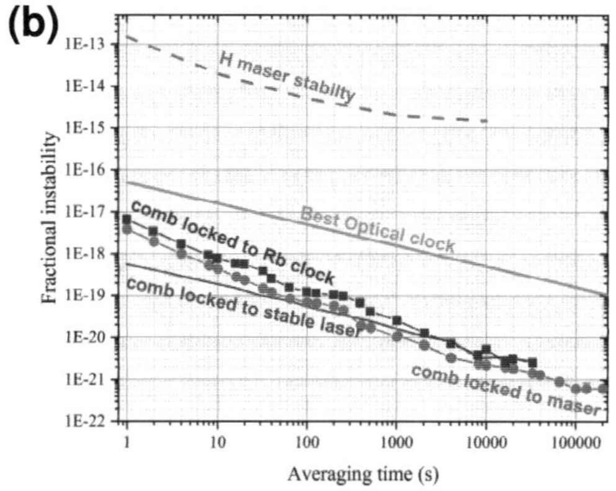

Figure 3: (Colour online) Division noise measurement. (a) Schematic diagram. (b) Fractional instability of R when a comb is frequency-stabilized to a Rb clock (blue squares), a hydrogen maser (red dots) or an ultra-stable laser (purple solid line). The frequency instabilities of typical optical atomic clocks and hydrogen maser are shown with a red solid line and a grey dashed line, respectively.

We also tested the ability of coherence transfer using the optical frequency divider. The experimental setup is shown in Fig. 4. A 1064 nm laser is frequency-stabilized to the resonance of a 10 cm-long cubic cavity, which is tested to have a 1 s-frequency instability of 6×10^{-16} and a laser linewidth better than 0.4 Hz [12]. Using the optical frequency divider when the Ti:Sapphire mode-locked laser is stabilized to a rubidium clock with a 1 s-frequency instability of 10^{-11}, the frequency of this cavity-stabilized laser is converted to a laser at 578 nm. The error signal is generated as shown in Fig. 2, and it is feed back to the current of the 578 nm laser to close the loop. Therefore, the frequency of the 578 nm laser is $\nu_{578} = \nu_{ref} / R_x$, with R_x a preset and adjustable division ratio. Although the frequency jitter of the comb lines is measured to be on the MHz level, the 578 nm laser is expected to inherit the coherence of the 1064 nm reference laser.

The remaining part of the phase-locked 578 nm laser light is delivered to Yb atoms for spectroscopy via a piece of polarization maintaining optical fiber with fiber noise cancelled [13]. Laser cooling and trapping of Yb atoms is described in detail in Ref. [14]. Atoms of ^{171}Yb isotope are slowed and trapped into a two-stage magneto-optical trap, and then they are loaded into an optical lattice with a trap depth of $210E_r$ (E_r is recoil energy) and a cross-section diameter of 110 μm. During state preparation, laser light at 556 nm optically pumps the atoms to either one of the two $^1S_0$ Zeeman sub-states. After probing the clock transition of Yb atoms, the populations in the $^1S_0$ and the $^3P_0$ states are measured in sequence using shelving detection. We obtain the Rabi spectrum of Yb atoms by stepping R_x. With a 400-ms atomic interrogation time, 27 spectra of the $^1S_0$ ($m_F = +1/2$) – $^3P_0$ ($m_F = +1/2$) transition with an average spectral linewidth of 2 Hz are obtained. In spite of the large frequency jitter of the comb, the 578 nm laser inherits the coherence from the cavity-stabilized 1064 nm laser, demonstrating the ability of optical frequency synthesis for high precision spectroscopy. It shows that the immunity to the comb frequency noise can be as large as 6 orders of magnitude.

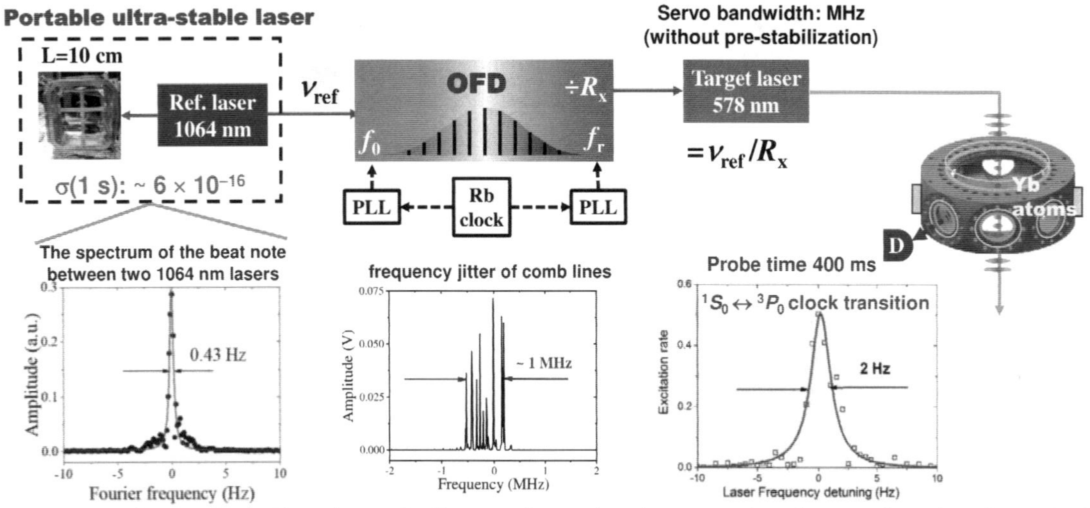

Figure 4: (Colour online) The schematic diagram for testing the spectral purity transfer using the optical frequency divider. Although the frequency jitter of the comb lines is on the MHz level, the coherence of the cavity-stabilized laser at 1064 nm is faithfully transferred to 578 nm, which is then used to resolve a spectroscopy with a linewidth of 2 Hz.

3. Multi-channel optical frequency division and ability of transport

In the above optical frequency divider, we use f_0 and the beating signals of reference and target lasers against their nearby comb lines to generate an error signal for locking the target laser frequency with precise frequency ratio. The error signal is independent on the frequency stability of the optical frequency comb. Therefore, we can realize multi-channel optical frequency division from a same optical reference or independent optical references. The outputs of optical frequency division can be in the optical region or even in the microwave region.

Multi-channel optical frequency division has advantages in multi-task applications. Moreover, it enables frequency comparison among ≥3 optical atomic clocks in a closed loop. When using a single frequency reference by multi-channel optical frequency division, the frequency noise of local oscillators can be common and be rejected in a synchronous clock comparison, largely reducing the statistical noise in optical frequency ratio measurement [15]. This operation mode is commonly useful for frequency comparison among optical lattice clocks. Moreover, when dividing the frequency of an optical atomic

Journal of Physics: Conference Series **2889** (2024) 012009 doi:10.1088/1742-6596/2889/1/012009

clock to a local oscillator of a second optical clock, differential spectroscopy can be performed to reduce the statistical noise in optical frequency ratio measurement, especially for single-ion optical clocks [16].

For daily operation, the optics of the optical frequency divider is sealed in an aluminum box. This sealed box together with the Ti:Sapphire mode-locked laser and a 1064 nm c.w. laser are put on a platform with a size of 600 mm × 900 mm. All the electronics for comb frequency stabilization, optical frequency division for four channels are put in a 19-inch rack. Figure 5 shows the photo of the whole experimental setup. To test the ability to transport the system between laboratories, we packed the optics platform in a wood equipment box, and we took the electrics out from the rack and packed them in delivery boxes. They were transported in a van from Shanghai to NIM (National Institute of Metrology) in Beijing (1300 km away). After we unpacked the equipment, reconnected all the cables and turned on the lasers, broadened rainbow spectrum was observed. The optical frequency divider started to work after optimizing the SNRs of f_0 and the beat signals by slightly tuning the steering mirror in front of the coupling lens to correct minor deviation of the output beam pointing of the Ti:Sapphire mode-locked laser. The performance of the optical frequency divider is almost the same before and after the transportation [17].

Figure 5: (Colour online) Photos of the optical frequency divider and the transportation from Shanghai to Beijing.

4. Conclusion and outlook

We introduce an optical frequency divider based on a Ti:Sapphire optical frequency comb, which is able to measure optical frequency ratios with an uncertainty at the 10^{-22} level. The instability of the divisor R at 10^{-18} or less can be obtained in optical frequency ratio measurement or frequency conversion, which supports optical frequency division of modern optical clocks without degrading their performance. The optical frequency divider can support multi-channel optical frequency division and meanwhile it can be transported from one place to another.

Low noise and accurate optical frequency dividers will play an important role in frequency comparison among optical atomic clocks, which is the prerequisite for the redefinition of the SI second. Meanwhile, they will also be the key to the realization of optical frequency synthesizers, which can generate optical waves or microwaves at any desired frequency with the same excellent frequency stability and accuracy of optical clocks for their exciting applications.

Acknowledgments

This work was supported by the Natural Science Foundation of China (11927810, 91636214) and the National Key R&D Program of China (2022YFB3904001).

References

[1] McGrew W F, Zhang X, Fasano R J, Schäffer S A, Beloy K, Nicolodi D, Brown R C, Hinkley N, Milani G, Schioppo M, Yoon T H and Ludlow A D 2018 *Nature* **564** 87
[2] Oelker E, Hutson R B, Kennedy C J, Sonderhouse L, Bothwell T, Goban A, Kedar D, Sanner C, Robinson J M, Marti G E, Matei D G, Legero T, Giunta M, Holzwarth R, Riehle F, Sterr U and Ye J 2019 *Nat. Photonics* **13** 714
[3] Matei D G, Legero T, Häfner S, Grebing C, Weyrich R, Zhang W, Sonderhouse L, Robinson J M, Ye J, Riehle F and Sterr U 2017 *Phys. Rev. Lett.* **118** 263202
[4] Yao Y, Jiang Y, Yu H, Bi Z and Ma L 2016 *Natl. Sci. Rev.* **3** 463
[5] Beloy K *et al.* 2021 *Nature* **591** 564
[6] Lange R, Huntemann N, Rahm J M, Sanner C, Shao H, Lipphardt B, Tamm Chr, Weyers S and Peik E 2021 *Phys. Rev. Lett.* **126** 011102
[7] Takamoto M, Ushijima I, Ohmae N, Yahagi T, Kokado K, Shinkai H and Katori H 2020 *Nat. Photoics.* **14** 411
[8] Arvanitaki A, Huang J and Van Tilburg K 2015 *Phys. Rev. D* **91** 015015
[9] Kolkowitz S, Pikovski I, Langellier N, Lukin M D, Walsworth R L and Ye J 2017 *Phys. Rev. D* **94** 124043
[10] Yang G, Shi H, Yao Y, Yu H, Jiang Y, Bartels A and Ma L 2021 *Chin. Opt. Lett.* **19** 121405
[11] Telle H R, Lipphardt B and Stenger J 2002 *Appl. Phys. B* **74** 1
[12] Chen X, Jiang Y, Li B, Yu H, Jiang H, Wang T, Yao Y and Ma L 2020 *Chin. Opt. Lett.* **18** 030201
[13] L. S. Ma, P. Jungner, J. Ye, and J. L. Hall 1994 *Opt. Lett.* **19** 1777
[14] Yao Y, Li B, Yang G, Chen X, Hao Y, Yu H, Jiang Y and Ma L 2021 *Photon. Res.* **9** 98
[15] Nemitz N, Ohkubo T, Takamoto M, Ushijima I, Das M, Ohmae N and Katori H 2016 *Nat. Photonics* **10** 258
[16] Kim M E, McGrew W F, Nardelli N V, Clements E R, Hassan Y S, Zhang X, Valencia J L, Leopardi H, Hume D B, Fortier T M, Ludlow A D and Leibrandt D R 2023 *Nat. Phys.* **19** 25
[17] Shi H, Yao Y, Li B, Wang C, Yu H and Ma L 2023 *APL Photonics* **8** 100802

Journal of Physics: Conference Series **2889** (2024) 012010 doi:10.1088/1742-6596/2889/1/012010

Techniques on Crystal Oscillator Vibration Compensation

M Mossammaparast, P Mullin, L Ronchetti, R Koehler and N Tiches

Quantic Wenzel, Austin, TX 78758, USA

mehran.moss@quanticwenzel.com

Abstract. Oven Controlled Crystal Oscillators (OCXOs) have been employed for precision timing and frequency distribution and synchronization applications, such as telecommunication, instrumentation, and test equipment in today's Commercial, Military, Scientific, Space and Low Earth Orbit markets. Since 1978, Wenzel Laboratory has been researching and developing outstanding performance oscillators to provide the lowest phase noise, as low as -190dBc/Hz, and highest short-term stability for these markets. In this paper, various compensation techniques developed at Quantic Wenzel are discussed and compared. Quantic Wenzel employs such techniques to greatly improve the stability of its OCXOs in various mild to harsh vibration environments. Consequently, this lends itself to provide quieter performance in multiple applications, whether OCXOs are exposed to micro-vibration environments caused by disturbance sources onboard spacecraft or noisy environments such as a jet aircraft landing or taking off on a carrier. An example of a compact microcontroller-based vibration compensated OCXO developed at Quantic Wenzel will be presented. Such digitally controlled OCXOs can also be employed to correct for thermal drift or errors in GPS location when GPS signals become unavailable in telecom and navigation systems.

1. Introduction

An OCXO is a precision timing circuit based on a high-Q quartz resonator enclosed in a temperature-controlled oven designed to minimize the effects of external temperature fluctuations and maintain crystal and oscillator circuits at a stable temperature. OCXOs can provide good short-term stability and clean frequency references for Atomic Frequency Standards (AFS), whether Rb [1], Cs or Hydrogen Maser atomic clocks [2]. By locking the OCXO reference to atomic resonance, short-term stability of the reference is combined with the long-term stability of an atomic resonance to enable novel tests of fundamental laws of physics which require extreme levels of precision and accuracy.

OCXO temperature stabilization and vibration isolation are two critical factors for femtosecond level stability. The OCXO phase noise performance is one of the key parameters considered when used as a reference in highly stable application requirements [3]. Specifically, dynamic phase noise performance is critical where minute vibrations could affect short-term stability such as from a ground-based rotating platform, an aircraft cruising altitude, or a satellite in orbit.

The precise time/frequency distribution is of significant concern in the field of radio astronomy array applications. In the Square Kilometer Array (SKA) project the frequency stability and phase noise performance of a H-maser clock needs to be distributed by fiber and then recovered at each dish site. This synchronization frequency signal needs to be distributed to hundreds of receivers in a star-shaped topology. The coherence requirements to support astronomical observations requires short term phase stability of better than $1\mathrm{x}10^{-12}$ at $\tau = 1$sec interval [4]. High stability phase locked OCXOs are required between H-masers at transmit and receiver sites. In this case, the short-term stability and high frequency offset phase noise performance of the OCXO is used when phase locked to the H-maser [4].

Similarly in autonomous driving systems, using doppler radars, frequency accuracy and stability of the reference clock play essential roles in determining accuracy and probability of moving target detection, such as objects or pedestrian on the road [5]. Unless vibration induced noise, such as road conditions, vehicle acceleration or braking situation, is accounted for, it can detrimentally affect the OCXO frequency shifts in phase and degrade the performance of the radar receiver's frequency measurement system.

As such, Quantic Wenzel employs two 5MHz Blue Tops Ultra Low Noise (BTULN) reference oscillators for measuring Allan Deviation (ADEV) of its high-performance products, using Microchip 53100A phase noise analyzers. The references typically have a stability of better than $3x10^{-13}$ at $\tau = 1$ sec and static phase noise under -125dBc/Hz at 1Hz frequency offset. The single sideband (SSB) phase noise plot of a BTULN oscillator is shown below.

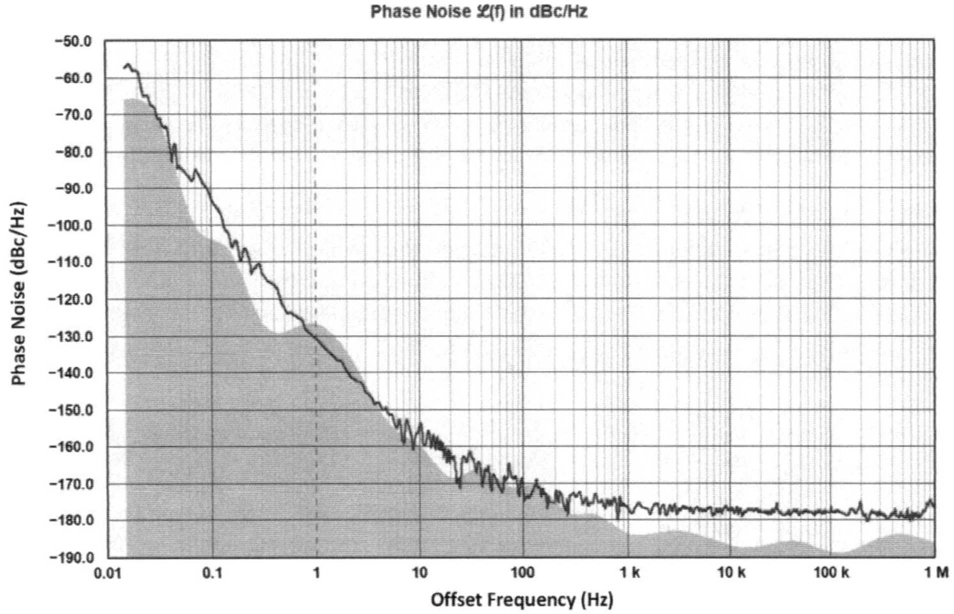

Trace	DUT Freq	Input Amplitude	dBc/Hz at 1 Hz
5 MHz ULN 02864	5.0 MHz	14 dBm	−130.8

Figure 1. SSB Phase Noise of a Quantic Wenzel 5MHz BTULN, -130dBc/Hz at 1Hz offset frequency.

2. Dynamic Phase Noise

It is known [6] that when an oscillator of frequency, f_0, is subjected to random vibration, its resonant frequency shift, Δf, is proportional to the magnitude and dependent on the direction of vibration and can be written as, $\Delta f = f(\vec{a}) - f_0 = (\vec{\Gamma}.\vec{a})f_0$, where $\vec{\Gamma}$ and $\vec{a}$ are the vibration sensitivity and the peak applied vibration vectors respectively. As a result, the associated dynamic phase noise, $\mathcal{L}(f_v)$, expressed in dB, can be determined as, $\mathcal{L}(f_v) = 20\log\left(\frac{\vec{\Gamma}\cdot\vec{a}f_0}{2f_v}\right)$ dBc/Hz.

For a sinusoidal vibration, $\vec{a}$, is the peak applied voltage level in units of g and $\mathcal{L}(f_v)$, is expressed in units of dBc. Under random vibration, the power is randomly distributed over a range of frequencies and as a result, the vibration is expressed in terms of Power Spectral Density (PSD). For a random vibration, $\vec{a} = \sqrt{2PSD}$ is expressed in units of g^2/Hz.

As an example, if a 100MHz (F1), $1x10^{-10}$/g OCXO is subjected to a random vibration level of 0.01 g^2/Hz in x-direction, its dynamic phase noise, $\mathcal{L}(f_v)$, is shown in Table 1.

Table 1. SSB Phase Noise of a 100MHz OCXO vs 10GHz multiplied OCXO at offset frequency of 10Hz

	f_0 (MHz)	$\vec{\Gamma}_x$ (1/g)	$\vec{a}$ (g^2/Hz)	f_v (Hz)	$\mathcal{L}(f_v)$ (dBc/Hz)
F1	100	$1x10^{-10}$	0.01	10	-83
F2	10,000	$1x10^{-10}$	0.01	10	-43

It is noteworthy that for the same vibration sensitivity, $\vec{\Gamma}$, when an OCXO is multiplied to X-band range, say 10GHz (F2) in an EW radar for example, phase noise levels (or sidebands in case of sinusoidal vibration) increase by 20log(N) or in this case, 40dB (See Table 1). This can be a major limiting factor in X-Band radar systems when an OCXO signal source is multiplied to X-band frequencies. Due to phase noise level rising, weak targets can be buried under the radar noise signature and remain undetected.

3. Vibration Compensation Techniques

Reducing OCXO sensitivity to vibration begins with crystal selection; low phase noise and low-g sensitive crystals (as low as 0.1ppb/g) are essential for success. In addition, a further reduction in OCXO g-sensitivity is possible by sensing the vibration affecting the crystal and compensating for the effect.

3.1. Passive Vibration Isolation

Passive vibration isolation systems can be implemented to limit the system vibration sensitivity at vibration frequencies above the system resonant frequency (See Figure 2). This type of vibration isolation is a simple but effective method of vibration mitigation and can provide a high level of roll-off above the system resonant frequency [7].

Figure 2. Passive Isolation Oscillators.

Figure 3 below shows the dynamic phase noise of an OCXO utilizing a passive isolation system to reduce the induced random vibration above 35Hz. However, vibration isolation may not be a viable solution for minimizing vibrations beyond the natural resonant frequency (typically below 100Hz) of the isolated system. The design challenge, in this case, is to attain a low system resonant frequency that is outside of the required phase noise frequency offsets.

In addition, for wide environmental temperature ranges, the vibration isolation/damping effect required for the design, may be affected at low or high temperatures. Passive vibration isolation may also cause excessive thermal isolation if the isolated OCXO is not effectively thermally connected to the attached chassis. When used properly, passive isolators can provide significant attenuation of high frequency vibration and prevent shock-induced damage to the isolation system. In a passively compensated Citrine OCXO, isolator resonance can be optimized in the range of 30Hz to 70Hz [8].

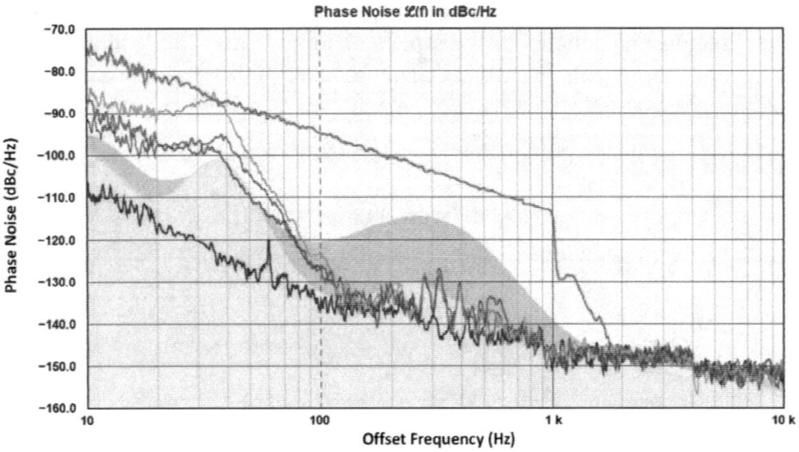

Trace	DUT Freq	Input Amplitude	dBc/Hz at 100 Hz
VXO Static PN	100.0 MHz	14 dBm	-135.2
VXO Uncompensated	100.0 MHz	14 dBm	-94.9
VXO Passive Isolation X-Axis	100.0 MHz	13 dBm	-127.3
VXO Passive Isolation Y-Axis	100.0 MHz	13 dBm	-126.4
VXO Passive Isolation Z-Axis	100.0 MHz	13 dBm	-123.1

Figure 3. An example of a Passive Vibration Isolation depicting Static and Dynamic Phase Noise of a 100MHz OCXO.

3.2. Active Vibration Compensation

Quantic Wenzel has developed a compact real-time active vibration compensation (RTC) which is a sensor-based measurement system designed to compensate for different types of vibration induced in an OCXO, whether sinusoidal, random, or both [8]. This type of compensation is microcontroller based (referred to as μOCXO) and could be broadband by design, restricted only by the bandwidth and type of sensors employed. Figure 4 indicates the effect of active vibration compensation when a hard-mounted OCXO was subjected to a random vibration profile of $0.002g^2/Hz$. In this case, it can be said that the g-sensitivity of the crystal has been reduced by an order of magnitude, having an "effective" g-sensitivity of <0.01ppb/g.

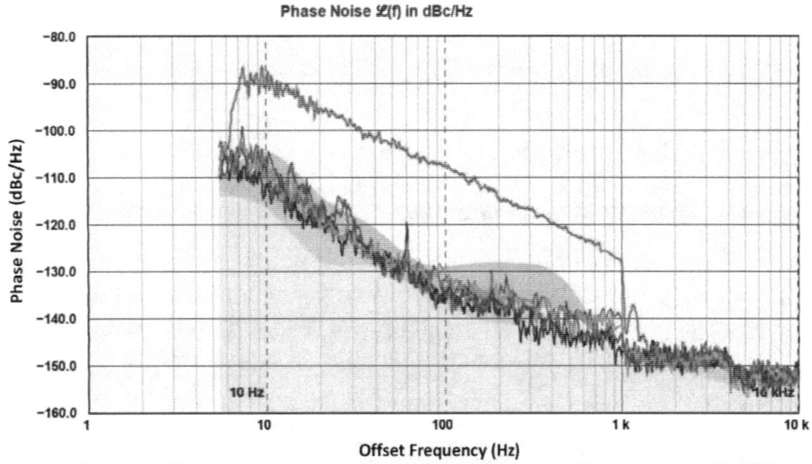

Notes	DUT Freq	Input Amplitude	dBc/Hz at 100 Hz	RMS Jitter
VXOM Static PN	80.0 MHz	9 dBm	-135.2	2.1E-14 s
VXOM DPN Uncompensated	80.0 MHz	9 dBm	-107.4	3.1E-13 s
VXOM X-Axis DPN	80.0 MHz	9 dBm	-131.7	3.1E-14 s
VXOM Y-Axis DPN	80.0 MHz	9 dBm	-135.0	2.8E-14 s
VXOM Z-Axis DPN	80.0 MHz	9 dBm	-130.9	2.6E-14 s

Figure 4. An example of an Active Vibration Compensation depicting Static and Dynamic Phase Noise of an μOCXO when subjected to $0.002g^2/Hz$ random vibration.

Journal of Physics: Conference Series **2889** (2024) 012010 doi:10.1088/1742-6596/2889/1/012010

According to a NASA survey, spacecraft micro-vibrations are small-amplitude mechanical vibrations due to dynamic interactions, in the ranges of micro-g to milli-g, typically occurring at low frequencies up to a few hundred Hz [9]. The RTC technique, in this case, is very effective in reducing these micro vibrations between 0.01sec to 1sec. A plot of active compensation is shown in Figure 5 below:

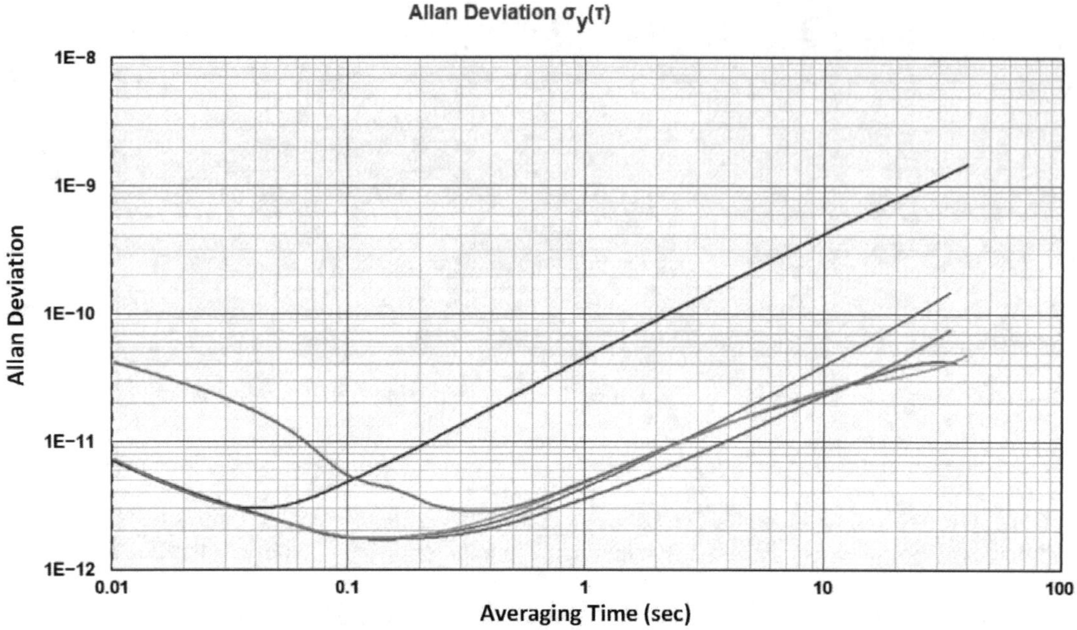

Notes	DUT Freq	Input Amplitude	ADEV at 0.01s
VXOM Static PN	80.0 MHz	9 dBm	7.14E−12
VXOM DPN Uncompensated	80.0 MHz	9 dBm	4.23E−11
VXOM X-Axis DPN	80.0 MHz	9 dBm	7.32E−12
VXOM Y-Axis DPN	80.0 MHz	9 dBm	7.30E−12
VXOM Z-Axis DPN	80.0 MHz	9 dBm	7.44E−12

Figure 5. ADEV of an 80MHz μOCXO in static and dynamic environment (profile 0.002g^2/Hz), Pink Trace: Vibe compensation disabled, Green/Red/Light Blue Traces: Vibe compensation enabled.

To integrate and implement a digital RTC technique in an extremely low phase noise OCXO requires careful design consideration. Sensor inherent delay and time variant processes present technical challenges that must be overcome to integrate an effective compensation system.

An advantage of implementing a digital compensation system is that it allows 3-axis calibration to be conducted automatically, especially useful in a production environment. In such a system, the micro controller can be leveraged to integrate a digital phase locked loop to lock the OCXO to an external reference as well as conducting automatic frequency aging correction. In addition, a wide temperature range of -40°C to +85°C can be supported.

3.3. Active & Passive Vibration Compensation

A hybrid active and passive vibration compensation system can provide higher degrees of vibration isolation and compensation (See Figure 6). The active compensation can significantly attenuate the resonant peak caused by the passive vibration isolation system. The combination of active and passive vibration guarantees ultra-low phase noise performance and a stable frequency output even when μOCXO is exposed to severe vibrations.

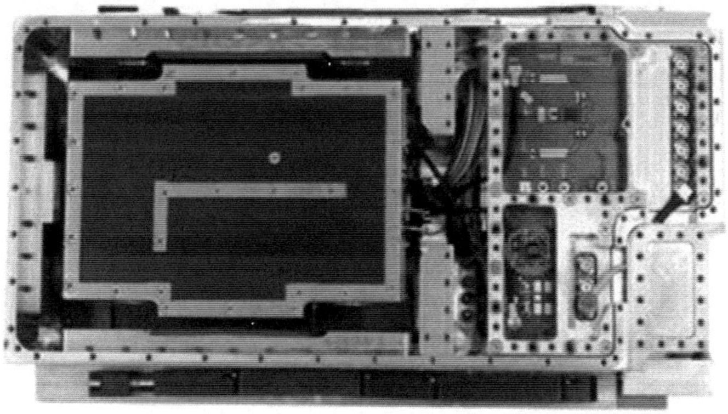

Figure 6. 3U VPX Hybrid Vibration Compensation Oscillator at 100MHz and 3GHz.

Figure 7 indicates the improvement achieved using a hybrid vibration compensation, technique when the μOCXO module was subjected to a random vibration profile of $0.02g^2/Hz$.

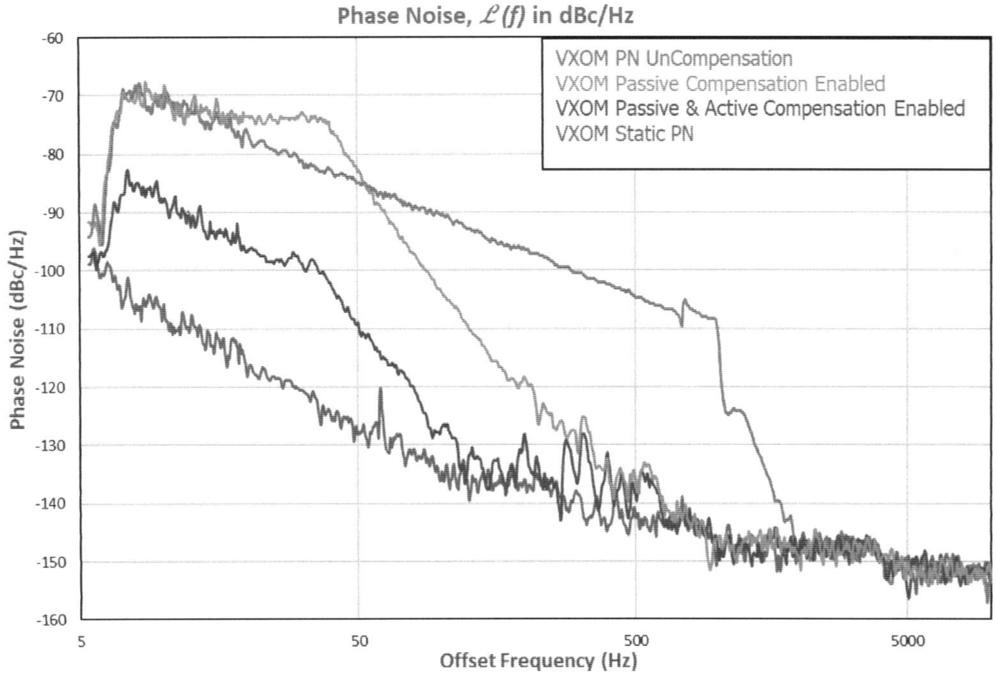

Figure 7. The effect of vibration compensation on the μOCXO phase noise performance. Blue Trace: OCXO static phase noise plot, Red Trace: OCXO phase noise plot under random vibration - Uncompensated, Green Trace: OCXO phase noise plot with passive vibration isolation, Violet Trace: OCXO phase noise plot with hybrid vibration compensation enabled.

This microcontroller based OCXO can also be readily used to digitally phase lock to a Rb based oscillator at < 0.1Hz bandwidth such as the Safran ruggedized mRO-50 Rubidium atomic clock. A typical block diagram for such configuration is shown in Figure 8.

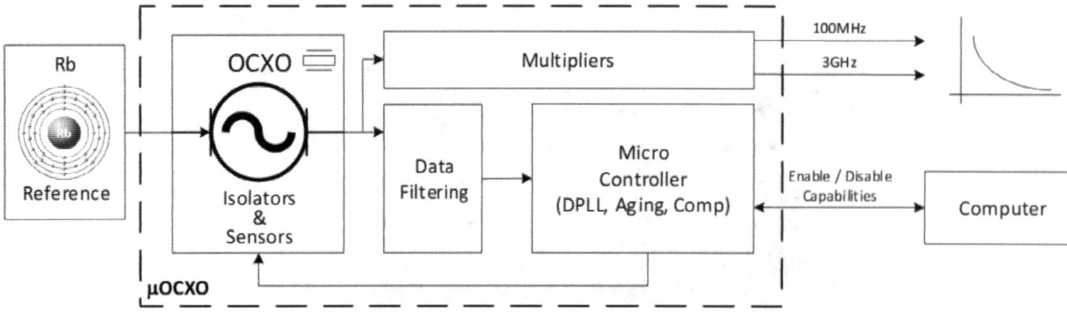

Figure 8. Multiplied µOCXO Locked to Rubidium.

4. Conclusion

The active compensation technique is an effective scheme in reducing the vibration induced phase noise of quartz based OCXOs. Combined with passive vibration isolation it can achieve unprecedented levels of phase noise in mild to harsh vibration and noisy environments.

Acknowledgments

The authors would like to thank Charles Wenzel, Mike Sawicki, and William Walsh for their valuable suggestions, support, and assistance in editing the manuscript.

References

[1] Chen, Z. et al. 2012 'A low phase noise microwave source for atomic spin squeezing experiments in 87rb', *Review of Scientific Instruments*, 83(4). doi:10.1063/1.3700247.

[2] Mushtaq Ahmed et al. 2008 'The Brazilian time and frequency atomic standards program', *Annals of the Brazilian Academy of Sciences*, 217-252, doi:10.1590/S0001-37652008000200002.

[3] Francois, B. et al. 2015 'Simple-design ultra-low phase noise microwave frequency synthesizers for high-performing Cs and Rb vapor-cell atomic clocks', *Rev. Sci. Instrum.* **86**, doi:10.1063/1.4929384.

[4] Wang, B. et al. 2015 'Square Kilometre Array Telescope—Precision Reference Frequency Synchronization via 1f-2f Dissemination', *Sci. Rep.* **5**, doi: 10.1038/srep13851.

[5] F. Hau, F. Baumgärtner and M. Vossiek, 2017 'Influence of vibrations on the signals of automotive integrated radar sensors', *IEEE MTT-S International Conference on Microwaves for Intelligent Mobility (ICMIM)* , 159-162, doi: 10.1109/ICMIM.2017.7918881.

[6] Hati, A., Nelson, C. and Howe, D. 2009 'Vibration-induced PM noise in oscillators and its suppression', *Aerial Vehicles* [Preprint]. doi:10.5772/6476.

[7] Wenzel Associates, A. 2012 '1 to 650 Mhz ultra low noise ocxos', *Microwave Journal*. Available at: https://www.microwavejournal.com/articles/17872-1-to-650-mhz-ultra-low-noise-ocxos (Accessed: 28 January 2024).

[8] Narayan Propato, Mehran Mossammaparast and Patrick Mullin, 2022 'Improving Oscillator Dynamic Phase Noise with Passive Vibration Isolation and Accelerometer-Based Vibration Compensation', *Microwave Journal*, Available: https://www.microwavejournal.com/articles/37979-improving-oscillator-dynamic-phase-noise-with-passive-vibration-isolation-and-accelerometer-based-vibration-compensation (Accessed: 28 January 2024).

[9] C. Dennehy, 2018 'Spacecraft Micro-Vibration: A survey of problems, experiences, potential solutions, and some lessons learned', *NASA Technical Reports Server* (NTRS), Available: https://ntrs.nasa.gov/citations/20180006315.

[10] Wenzel Associates, Inc. 2023 'Golden VHF Bootstrap Oscillator' *Quantic Wenzel*, Available: https://www.quanticwenzel.com/model/golden-vhf-bootstrap/ (Accessed: 28 January 2024).

Journal of Physics: Conference Series **2889** (2024) 012011 doi:10.1088/1742-6596/2889/1/012011

High-Temperature Chip-Scale Atomic Clock

Li-Chung Ha*, Jay Noble, Bruce Newell, Robert Lutwak
Microchip Technology, Frequency and Time Systems, 34 Tozer Rd., Beverly, MA 01915

* lichung.ha@microchip.com

Abstract

We present the development of alkali-alloy-based vapor cells for operating a cesium-based Chip-Scale Atomic Clock (CSAC) at environmental temperatures up to 105 °C. Leveraging Raoult's Law for a mixture of potassium and cesium metal, we have demonstrated vapor pressure suppression equivalent to 25 °C, allowing a CSAC to operate at higher temperatures with minimal degradation of performance or power consumption.

Introduction

Chip-Scale Atomic Clocks (CSAC) enable atomic-quality timing in portable battery-powered applications. The Microchip's (formerly Symmetricom) SA.45s CSAC was developed, with DARPA funding, in 2000-2010, and was described at FSM 7, in 2008[1]. Over 100,000 model SA.45s CSACs have been manufactured since the product release in 2011. The improved timing accuracy that CSAC provides to battery-powered devices has enabled new applications and network architectures for navigation, communications, and remote sensing.

The operating temperature range of the SA.45s is limited by the operating temperature of the vapor cell, typically 90-95 °C, which sets the upper operating temperature of the SA.45s to 85 °C. While this is suitable for most applications, several key applications, notably in the automotive and industrial sectors, remain unserved by CSAC, due to its inability to operate in a 105 °C environment.

The limitation to the upper operating temperature of the vapor cell is fundamental. At cell temperatures above 95 °C, the cesium vapor density is so high that the width of the atomic resonance line is significantly broadened, due to collisional decoherence, such that the CSAC Allan deviation (ADEV) frequency stability is degraded. At temperatures exceeding 110 °C, such as is necessary for operation in a 105 °C ambient, the cesium density is so high that the vapor cell becomes opaque to light transmission and clock operation is impossible.

This paper focuses on our efforts to achieve CSAC operating temperatures above 105 °C for automotive and industrial applications. We review the basic architecture of the CSAC and the approach of using alkali metal mixtures to permit operation of the vapor cell at elevated temperatures. We present experimental data that demonstrates clock performance, including frequency stability, temperature sensitivity and aging rate.

Journal of Physics: Conference Series **2889** (2024) 012011 doi:10.1088/1742-6596/2889/1/012011

Microchip CSAC

The Microchip SA.45s CSAC provides Allan deviation short-term frequency stability of $\sigma_y(\tau) < 3x10^{-10}\ \tau^{-1/2}$ over an environmental temperature range of -40 °C to 80 °C. As shown in Fig.1, the core of the physics package is an integrated stack of the laser, vapor cell and photodiode. The core is supported by a pair of polyimide membranes, which provide both thermal insulation and mechanical stability. The vapor cell is a 2 mm x 2 mm x 2 mm cube, fabricated by anodically-bonding 250 μm-thick Pyrex onto each side of a 1.5 mm thick silicon cell body. The cell bore is a 1.5 mm diameter x 1.5 mm long cylinder, containing cesium and an Ar/N_2 temperature-compensated buffer gas mixture[2,3].

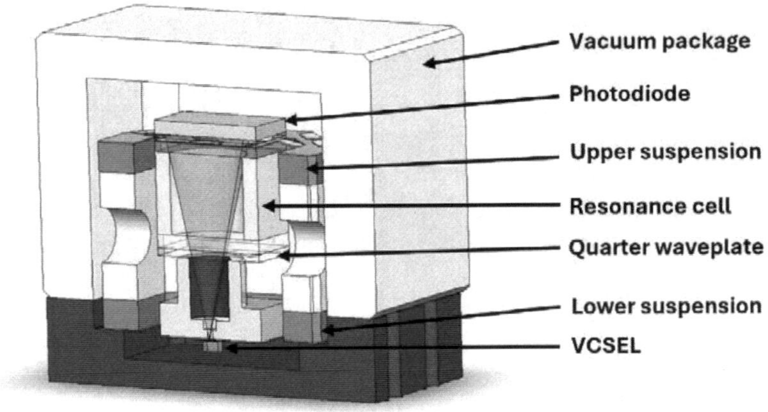

Fig.1 SA.45s CSAC physics package. The thermally stabilized core contains VCSEL laser, vapor cell, and photodiode. The polyimide suspensions provide thermal insulation of the core assembly, and electrical traces on the membranes transmit signals between the core and the outer structure with minimum thermal conduction. The physics package is sealed under vacuum for additional thermal isolation from the environment.

Within the vapor cell, the buffer gas minimizes collisional broadening due to cesium-cesium and cesium-wall collisions[4], at the expense of cesium-buffer gas collisions. The clock linewidth can be estimated by:

$$\gamma_2^{clock} = \gamma_2^{wall} + \gamma_2^{buffer\ gas} + \gamma_2^{Cs\text{-}Cs} \qquad [Eq.1]$$

where γ_2^{wall} is the line broadening due to cesium collisions with the walls, $\gamma_2^{buffer\ gas}$ is the broadening due to cesium-buffer gas collisions, and $\gamma_2^{Cs\text{-}Cs}$ is the broadening due to cesium-cesium collisions.

In order to maintain stable thermal control, the physics core assembly temperature (T_{CORE}) must be stabilized 5-10 °C higher than the highest expected environmental temperature (T_{ENV}), e.g. T_{CORE} = 90 °C for operation up to T_{ENV} = 80 °C. Emerging automotive and industrial applications require CSAC operation in T_{ENV} up to 105 °C, which requires $T_{CORE} \cong 115$ °C. The consequence of operating the vapor cell at an elevated temperature is the higher alkali vapor density which results in a higher collisional broadening as shown in Fig.2(a). With the Cs-Cs collisions being the dominating mechanism for line broadening, 25 °C increase of T_{CORE} results in 5X degradation of the clock linewidth and, thereby, the short-term stability of the CSAC.

A second consequence of increased cesium vapor density is higher optical density, which results in a lower optical transmission through the gas cell and thereby lower signal-to-noise ratio of the atomic resonance signal. Together, these factors result in significant (>10X) degradation of CSAC short-term stability. Empirically, we have also observed that the clock resonance line shape becomes asymmetric at elevated cell temperatures due to inhomogeneous AC Stark shift across the cell from the optically thick cesium vapor.

First proposed by the French chemist François-Marie Raoult in 1887, the partial pressure of a vapor in equilibrium above the surface of an ideal liquid mixture is suppressed in proportion to its molar fraction in the solution[5]. Raoult's Law has been foundational to understanding strengths of molecular bonds in chemical mixtures which causes the deviations from ideality in mixtures. We leverage Raoult's Law, in a near-ideal mixture of highly similar alkali metals, to reduce the cesium partial pressure at elevated T_{CORE}. Rather than pure cesium metal, the cell is filled with a cesium-potassium alloy[6]. The presence of potassium reduces the cesium partial pressure in proportion to the molar fraction of cesium in the alloy according to

$$p = p'_{Cs}\, x_{Cs} + p'_{K}\, x_{K} \qquad \text{[Eq.2]}$$

where p is the total vapor pressure, p'_{Cs} and p'_{K} are the saturated vapor pressures of cesium and potassium respectively, and x_{Cs} and x_K are the molar ratios of cesium and potassium in the alloy.

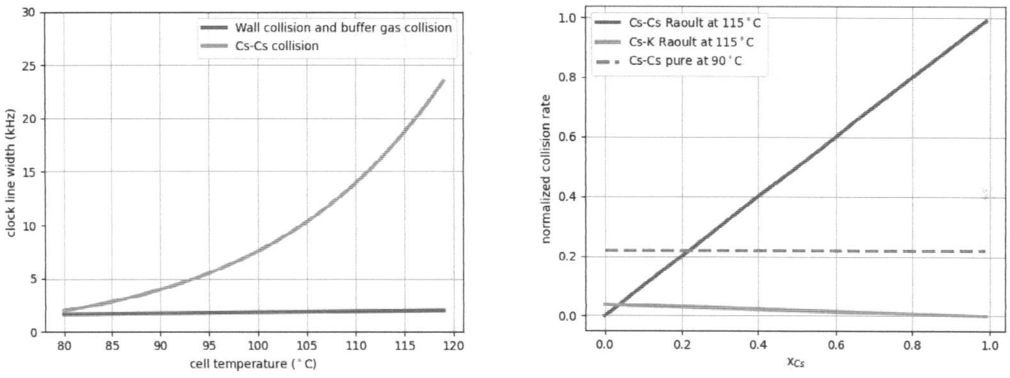

(a) (b)

Fig.2 (a) Linewidth contributions in the SA.45s vapor cell at different temperatures. (b) Normalized collisional broadening contribution from Cs-Cs collisions and Cs-K collisions at different mixing ratios. To achieve ΔT = 25 °C at 115 °C, the target x_{Cs} is around 0.2. The broadening due to Cs-K collisions is an order of magnitude smaller than that of Cs-Cs collisions.

Introducing potassium into the cell adds an additional linewidth-broadening mechanism due to collisions between cesium and potassium. However, due to potassium's higher melting point, the potassium vapor pressure is much lower than the cesium vapor pressure and, as shown in Fig.2 (b), the Cs-K collisional rate[7] is negligible compared to the Cs-Cs collisions. At a molar fraction of x_{Cs} = 0.2, the linewidth contributed by Cs-Cs collisional broadening for the Raoult cell at 115 °C is comparable to the linewidth of a pure Cs cell at 90 °C. Therefore, the introduction of potassium in the cell mitigates the collisional broadening of the clock line when operating at a higher temperature.

Frequency Stability

We built 10 SA.45s CSACs with Cs-K Raoult cells at $x_{Cs} \approx 0.2$, which reduces the Cs vapor pressure by the equivalent of 25 °C. Fig.3 shows the Allan deviation of seven of these devices, operating with $T_{CORE} > 110$ °C and exhibiting short-term stability of $\sigma_y(\tau = 1 \text{ sec}) = 6\text{-}12 \times 10^{-11}$. The stability performance is superior to that of the standard SA.45s, which typically exhibits $\sigma_y(\tau = 1 \text{ sec}) = 1\text{-}2 \times 10^{-10}$ at $T_{CORE} = 85\text{-}95$ °C.

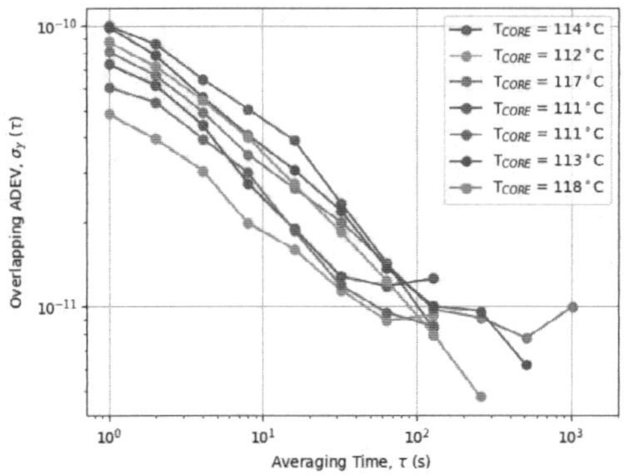

Fig.3 ADEV measurements of seven high-temperature CSACs with $x_{Cs} \approx 0.2$ and $T_{CORE} > 110$ °C.

Temperature Sensitivity

For a typical SA.45s CSAC, the normalized frequency varies by about 4×10^{-9} peak-to-peak over an environmental temperature range of -10 °C to +70 °C, as shown in blue in Fig.4(a). The temperature sensitivity is primarily caused by the variation of temperature gradient across the core, however, this behavior is sufficiently repeatable that digital compensation reliably reduces the total peak-to-peak sensitivity to $< 1 \times 10^{-9}$, as shown in orange. The test results of one prototype high-temperature CSAC are shown in Fig.4(b), with x_{Cs} equivalent to $\Delta T = 25$ °C and $T_{CORE} = 119$°C. The high-temperature unit shows peak-to-peak sensitivity of 1.2×10^{-9} over the T_{ENV} range of -40 °C to 105 °C (blue), which is reduced to $< 2 \times 10^{-10}$ with digital compensation (orange). Across all of the hot-CSAC prototypes, the peak-to-peak temperature sensitivity over $T_{ENV} = -40$ °C to 105 °C was superior to that of a standard CSAC operating over the narrower range of -10 °C to +70 °C. These results confirm both temperature sensitivity and repeatability of the high-temperature prototypes.

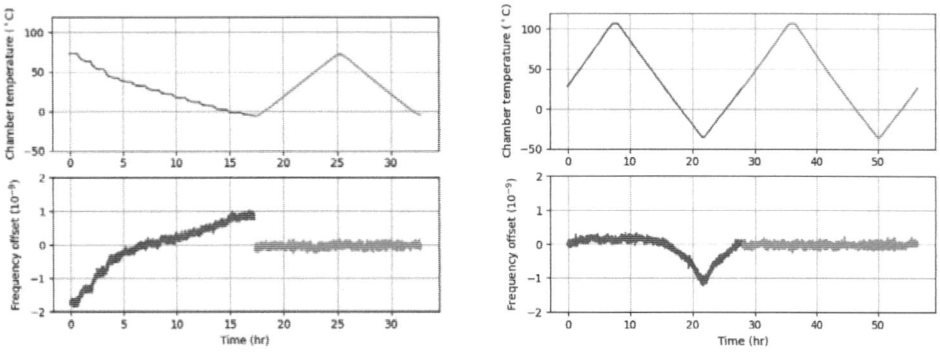

(a) (b)

Fig.4 Total peak-to-peak frequency sensitivity over temperature with and without digital compensation in blue and orange, respectively. (a) SA.45s CSAC with T_{CORE} = 100°C. (b) High-temperature CSAC with T_{CORE} = 119°C.

Clock Aging

Long-term frequency aging of the SA.45s is dominated by migration of the excess cesium in the vapor cell, which may obfuscate the cell windows, resulting in long-term variation of the intensity of the interrogation light and, thereby, drift of the clock frequency due to the AC Stark effect ("light shift"). The early frequency aging of a standard SA.45s CSAC and a high-temperature CSAC prototype is shown in Fig.5. The standard SA.45s CSAC shows an initial aging rate of $9x10^{-9}$ per month, which settles to $9x10^{-10}$ per month after 100 days of operation. In contrast, the hot-CSAC prototype shows roughly 10X superior aging rate on all time scales since the migration rate is inhibited by the presence of potassium.

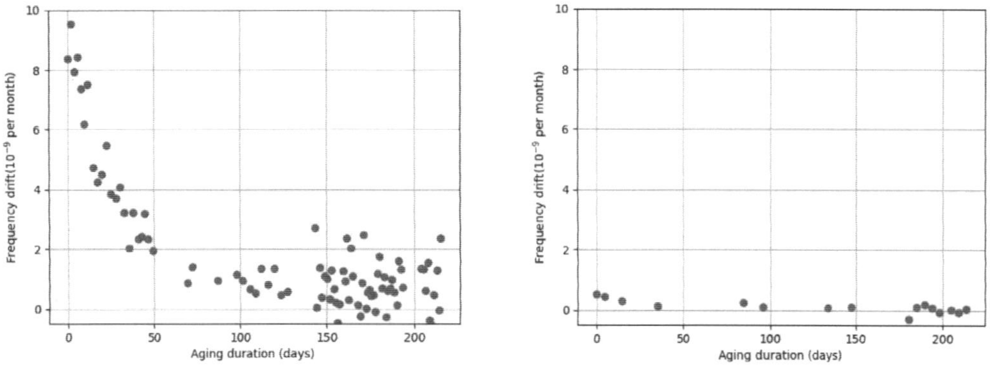

(a) (b)

Fig.5 Early aging behavior of (a) SA.45s CSAC and (b) High-temperature CSAC. We observe a lower aging rate for the CSAC constructed with Raoult cell. To resolve the lower aging rate with a higher precision, more raw data points are used for the calculation.

Vapor Cell Lifetime

Our target Raoult ratio for operation at T_{CORE} = 115 °C is x_{Cs} = 0.2, which reduces the cesium vapor pressure by 5X, thereby similar to the vapor pressure of pure cesium at T_{CORE} = 90 °C. We confirmed that the suppression ratio remains constant over the temperature range from 60 °C to 150 °C.

To investigate time-evolution of the molar ratio, we performed an accelerated lifetime test of a Cs-K resonance cell optimized for ΔT = 25 °C. The cell was stored at 250 °C to accelerate the depletion process. The cell was periodically removed from the oven and its optical absorption was measured at a cell temperature of 90 °C, as shown in Fig.6. Compared with the cell filled with pure cesium, with a typical optical density of 1.2, we observe an absorption signal suppressed to 0.25. The optical absorption depth remains constant as the alkali alloy is depleted, indicating that the mixing ratio remained constant for the lifetime of the resonance cell.

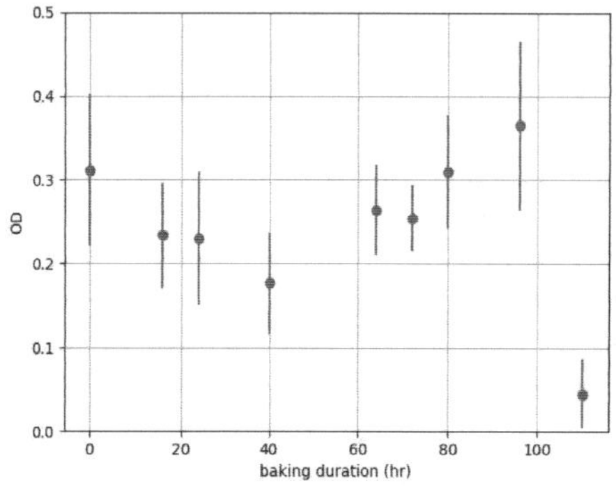

Fig.6 Accelerated lifetime test of a Cs-K cell. The cell was stored in a 250 °C furnace for 120 hours and taken out periodically to check the cesium optical absorption at 90 °C. The data represent the extracted optical densities from those measurements. The optical density remained constant until shortly before the alkali was depleted, indicating that the suppression ratio stayed constant for the lifetime of the vapor cell.

Conclusion

We have developed vapor cells, based on Raoult mixtures of alkali metals, which support operation of atomic clocks in environmental temperatures above 105 °C. With a Cs/K molar ratio of x_{Cs} = 0.2, the cesium vapor pressure is reduced by an equivalent ΔT = 25 °C. We have evaluated short-term stability, temperature sensitivity, and frequency aging and found them to be comparable or superior to SA.45s CSAC performance at lower temperatures. The results show promise for deployment of CSACs in higher-temperature automotive and industrial environments.

Journal of Physics: Conference Series **2889** (2024) 012011 doi:10.1088/1742-6596/2889/1/012011

References

[1] R. Lutwak, A. Rashed, M. Varghese, G. Tepolt, J. LeBlanc, M. Mescher, D. K. Serkland, K. M. Geib, G. M. Peake, "CSAC - The Chip-Scale Atomic Clock," Proceedings of the 7th Symposium Frequency Standards and Metrology, 453-462 (2008).

[2] O. Kozlova, S. Guérandel, and E. de Clercq, "Temperature and pressure shift of the Cs clock transition in the presence of buffer gases: Ne, N_2, Ar," Phys. Rev. A **83**, 062714 (2011).

[3] L.-C. Ha, X. Zhang, N. Dao and R. Overstreet, "D1 line broadening and hyperfine frequency shift coefficients for ^{87}Rb and ^{133}Cs in Ne, Ar and N_2," Phys. Rev. A **103**, 022826 (2021).

[4] N. Cyr, M. Tetu, M. Breton, "All-Optical Microwave Frequency Standard: A Proposal," IEEE Trans. Instrum. Meas. **42**, 640-649 (1993).

[5] F.-M. Raoult, "General law of vapor pressures of solvents," Proceedings of the Academy of Sciences, **104**, 1430 (1886).

[6] E. E. Shpil'rain, A. M. Belova, V. I. Shkermontov, A. G. Mozgovoi, S. N. Skovorod'ko, "Experimental investigation of the pressure of saturated vapors of liquid alloys of alkali metals at high temperatures: K-Cs system," High Temperature, **39**, 670 (2001).

[7] A. Dalgarno and M. R. H. Rudge, "Spin-change cross-sections for collisions between alkali atoms," Proceedings of the Royal Society of London. Series A, Mathematical and Physical Sciences **286**, 519-524 (1965).

Journal of Physics: Conference Series **2889** (2024) 012012 doi:10.1088/1742-6596/2889/1/012012

Towards Compact, Robust and Highly Stable Optical Frequency References for Space Applications

T Wegehaupt[1,2], M Gohlke[3], F Kuschewski[1], M Oswald[1,2], K Abich[1], T Alam[1], T Blomberg[1], J Bischof[1], A Bußmeier[1], N Röder[1], J Wüst[1,4], J Grosse[2], J Sanjuan[1,5], T Schuldt[1] and C Braxmaier[1,4]

[1]German Aerospace Center (DLR), Institute of Quantum Technologies, Ulm, Germany
[2]University of Bremen, Center of Applied Space Technology and Microgravity (ZARM), Bremen, Germany
[3]German Aerospace Center (DLR), Institute of Space Systems, Bremen, Germany
[4]University of Ulm, Institute of Microelectronics, Ulm, Germany

E-mail: timm.wegehaupt@dlr.de

Abstract. Future space missions from different fields of applications, like global navigation satellite systems (GNSS), next generations of gravity missions or tests of fundamental physics will rely on optical frequency references. While laboratory setups of various optical frequency reference technologies already demonstrate frequency stabilities orders of magnitude better than current microwave references for space applications, another step must be accomplished for high performance optical references in space. One promising technology for this is Doppler-free spectroscopy of molecular iodine, for which we have already developed several iterations of laboratory setups with frequency instabilities down to the 10^{-15} level for integration times $> 100\,\mathrm{s}$. Based on the experience gained from this laboratory setups, we are now working on an iodine reference that will become the first optical clock in space as part of the COMPASSO mission. It will be delivered to the international space station in 2026 for a mission time of 2 years. In addition to the iodine development, we are currently working on a next iteration of a high-finesse optical cavity designed with emphasis on space compatibility to meet the requirements of the next generation of gravity missions. Both technologies together, the iodine spectroscopy unit and the high-finesse optical cavity, should form a hybrid setup suitable for new GNSS architectures.

[5]Present address: A&M University, Aerospace Engineering & Physics, College Station, TX 77843

1 Introduction

State-of-the-art optical frequency references such as ion clocks have demonstrated frequency stabilities at the 10^{-18} level and below [1]. With these new technologies, many applications will benefit from either time or distance measurements. In terrestrial applications, optical references are already replacing microwave references, but for space applications, optical references are not yet as widely used because reliability, size, weight, and power budgets are key considerations in addition to the performance. One technology already being used for space missions is optical cavities. Optical cavities can offer high short-term stability, small volumes, low power consumption and comparable low complexity. This makes them suitable for communication applications or space based laser-interferometry applications such as the future gravitational wave detector LISA (Laser Interferometer Space Antenna) or the LRI (Laser Ranging Interferometer) on present and future gravity field missions [2–4]. With current frequency stabilities at the 10^{-17} level on short timescales for terrestrial applications and 10^{-15} for space applications, there is still a need for development. In addition, cavities have the disadvantage of being relative references, which limits their suitability to applications where long-term stability or absolute frequency knowledge is required only to a certain level. Here, absolute frequency references are the technology of choice, based on the principle that a laser is stabilized to an atomic or molecular transition. This results in high long-term stability and absolute frequency knowledge, but at the cost of reduced short-term stability and more complex setups. At the DLR Institute of Quantum Technologies, in cooperation with the Universities of Bremen and Ulm, frequency references covering both, short- and long-term stability are currently under development with a focus on space compatibility. The demonstrated frequency stabilities of two laboratory setups are depicted and compared to operational space clocks in Fig. 1.

2 Iodine references

One well known - and compared to ion clocks less complex technology of an absolute reference - is Doppler-free spectroscopy of molecular iodine. This technology has demonstrated frequency stabilities at the 10^{-15} level for integration times $> 100\,\mathrm{s}$ in laboratory experiments, been flown on a sounding rocket and is currently being further developed to be launched within the COMPASSO project to the ISS Bartolomeo platform in 2026, becoming the first optical clock in space [5, 7–9].

Over the last decades, several iterations were carried out to reach this goal, with a selection of them shown in Fig. 2. Molecular iodine has a rovibronic transition at around 532 nm with hyperfine components spanning over several hundred MHz and a natural line-width of each transition of about 300 kHz. These transitions are well studied and its absolute frequency was determined with around 1 kHz accuracy, which is at the same range as the reproducibility of the locking scheme.[10–12]

The locking principle of choice for high performance setups is known as Modulation Transfer Spectroscopy (MTS). Two counter propagating laser beams are sent on the same optical path but in opposite directions through a gas cell containing molecular iodine. A strong pump beam burns a so called 'Bennet-hole' into the velocity distribution of the gas. The counter propagating probe beam then addresses only molecules moving perpendicular to the beam, hence eliminating the Doppler shift. The MTS signal (comparable with an error signal in the widely used Pound-Drever-Hall (PDH) technique for cavity locking) is then generated by phase modulating the pump beam and demodulation of the absorption signal with the same frequency.

2.1 EBB-Design

After several modular breadboard experiments a new setup on Elegant-Breadboard (EBB) level was realized using a Nd:YAG 1064 nm Laser with an internally doubled second output at 532 nm. The 532 nm light is then used for generation of probe and pump beams whereby the required modulation is applied accordingly with an acusto-optic modulator (AOM). The spectroscopy unit of the EBB setup utilizes an OHARA Clearceram-Z HS board with a size of 550 mm × 250 mm × 50 mm as basis to achieve a minimum level of mechanical deformation. All optical components are made out of fused silica and bonded by a space-qualified two-component epoxy, thus resulting in a semi-monolithic optical assembly. As reference cell, a triple pass 300 mm cell is used. The MTS signal is detected by a balanced noise cancelling (NC) detector and used for a feedback loop for laser frequency stabilization. In addition, a correction for Residual Amplitude Modulation (RAM) as well as intensity stabilization for probe and pump beam are integrated.

This setup demonstrated a frequency stability in Allan deviation of below 1×10^{-14} at integration times of 1 s and below 4×10^{-15} for integration times between 10 s and 1000 s.

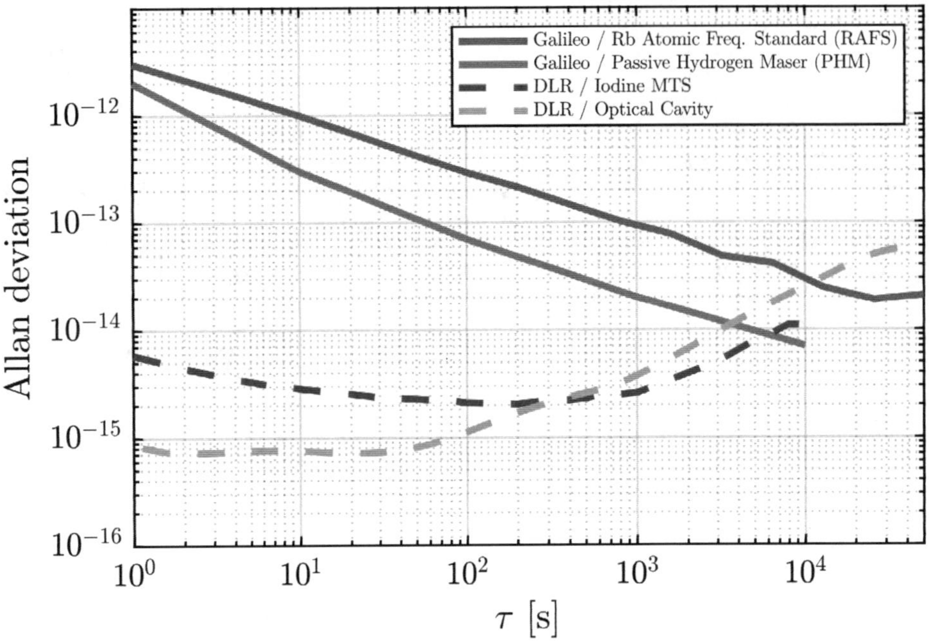

Figure 1: Relative frequency stability of two laboratory setups that are being further developed towards space suitability and are, therefore, compared here with two current space clocks. The Iodine spectroscopy reference is using Modulation Transfer Spectrscopy (MTS) and its frequency stability is determined over a record time of 76 000 s [5]. The optical cavity setup was characterized against a duplicated system over a record time of 170 000 s and subtracted by a linear detrend with a slope of about 100 mHz/s [6].

2.2 EM-Design

Upon a further development, a setup on Engineering Model (EM) level was realized and thus further optimized with respect to compactness, mechanical-, and thermal-stability. The core of the optimization was a special nine-pass gas cell with dimensions of 100 mm × 100 mm × 30 mm mounted on a fused silica breadboard with dimensions of 380 mm × 180 mm × 40 mm only, which corresponds to a reduction in size of about 50 %. Fused silica is utilized in this new iteration after considering that such high mechanical stability for the breadboard is not necessary. In addition, the gas cell and optics are all made out of fused silica and thus a material match between the optical components and the baseplate is given. All optical components are integrated using adhesive bonding technology. The spectroscopy was subjected to thermal cycling (−20°C to 60°C) and vibrational testing with test levels of up to 30 g for sine vibration and 25.1 g_{rms} for random vibration. A frequency stability of 5×10^{-15} for integration times $\tau > 1000$ s was achieved before and after the tests.

2.3 ADVANTAGE

In order to take the next step into space applications, a new iteration based on the knowledge of the EBB and EM developments was designed within the ADVANTAGE project (Advanced Technologies for Navigation and Geodesy). Here, one main objective was further development towards a fully integrated physics box combining the spectroscopy unit with the laser system in a compact and robust package. At the same time, the temperature requirements of this package were set to 15 °C ± 5 °C. Within this range, an active control system guarantees the operation of the spectroscopy board within 22 °C ± 0.1 °C. As laser source a Nd:YAG laser provides 1064 nm light, which is then coupled into a

Journal of Physics: Conference Series **2889** (2024) 012012 doi:10.1088/1742-6596/2889/1/012012

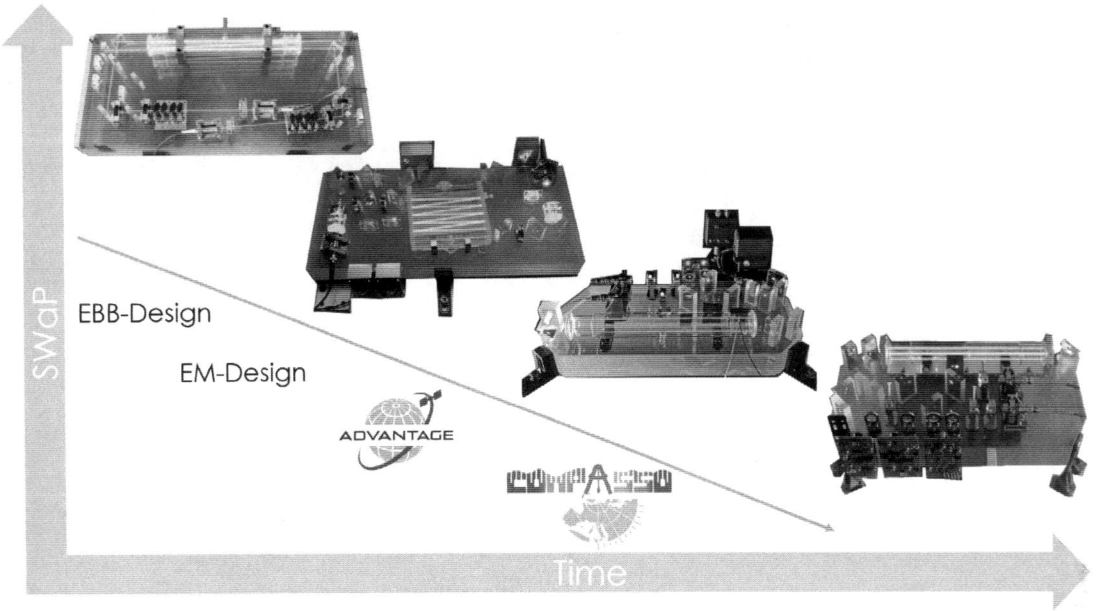

Figure 2: Qualitative illustration of the further development in Size-, Weight and Power-Budget of the iodine spectroscopy iterations.

fiber and splitted into probe and pump beam. The pump beam is guided to an AOM for RAM correction and intensity-stabilization and then phasemodulated by an electro-optic phase modulator (EOM) for modulation transfer spectroscopy (MTS). Also, the AOM creates a frequency offset between probe and pump beam. The intensity of the probe beam is controlled by an electro-optic amplitude modulator (EOAM). Both modulated beams are then guided to second harmonic generators (SHG) to create 532 nm light. All modulators are fiber-coupled. This principle was intended as a baseline for future designs using an External Cavity Diode Laser (ECDL).

2.4 COMPASSO

With the upcoming COMPASSO mission, an iodine clock will become the first optical clock in space. COMPASSO is a mission of the German Aerospace Center with the aim to demonstrate high-performance quantum optical technologies in space. The mission includes two laser-based absolute iodine references, a frequency comb, and a laser communication and ranging terminal (LCRT). The mission will be delivered to the International Space Station (ISS) in 2026 for a mission time of 2 years. Primary goals include the technology demonstration and performance evaluation of the two iodine references and a frequency comb referenced to a GNSS disciplined oven-controlled crystal oscillator (OCXO), while secondary goals also include the transfer of time and frequency via the LRCT. [9]
The iodine references are built on heritage of the previously introduced iterations and a CAD model of the COMPASSO-EM spectroscopy unit is shown in Fig. 3 with it's zerodur breadboard illustrated in yellow and one part of the thermal shield in red. The COMPASSO breadboard dimensions are 300 mm × 150 mm × 35 mm with a honeycomb structure milled into its body. The spectroscopy uses the same 4-pass layout as in ADVANTAGE but with a shorter 200 mm gas cell. All components are mounted by adhesive bonding to achieve the rigidity for spaceflight and are arranged to achieve a crossed-polarized beam distribution scheme. This, on one hand, adds the complexity of using polarizing optics in space, but on the other hand, allows for an efficient distribution and dumping of the beams. The board is mounted with adhesive bonded titanium holders to a base-plate and encapsulated by an actively controlled thermal shield. The spectroscopy and laser system are then integrated together into one exterior housing, forming the Iodine Spectroscopy Unit (ISU). A CAD model of the overall iodine reference consisting of ISU and Iodine control electronics (ICE) is shown in Fig. 4.

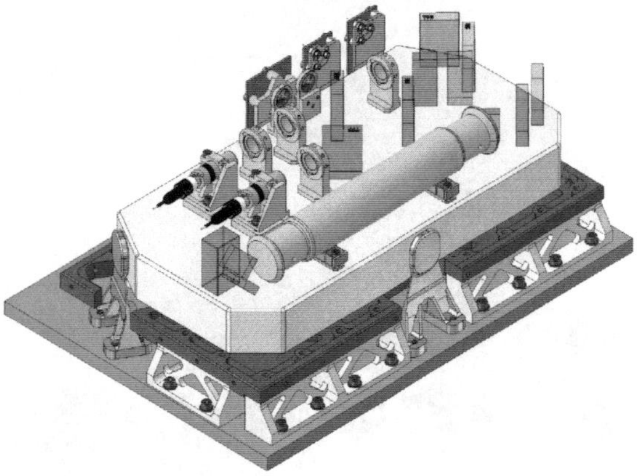

Figure 3: CAD of the COMPASSO-EM optical bench.

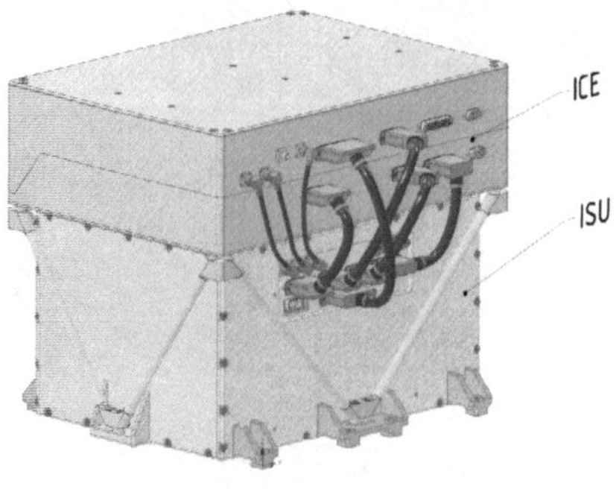

Figure 4: CAD of the COMPASSO-EM iodine reference including the Iodine-Control-Electronics (ICE) and the Iodine-Spectroscopy-Unit (ISU).

3 Optical cavities

For applications where no absolute frequency references are required, optical cavities are well suited due to less complexity, small dimensions and low power consumption. Here, the further development of high finesse optical cavities towards space compatibility is the research field of the DLR Quantum Metrology group. One goal is the development of an optical cavity with respect to the requirements of future gravity missions, global navigation satellite systems, and special relativity tests in space. The baseline for all cavity setups is the cube design of the National Physical Laboratory of Great Britain (NPL), that is shown in Fig. 5. [13]

The first setup on EBB-level was designed as a technology demonstrator of a space based Kennedy-Thorndike experiment within the BOOST project [14]. The requirements for this setup were quite challenging for a cavity, as the setup should provide a frequency stability of 1×10^{-14} on 90 min integration time. This is realized by covering an 8.7 cm cavity in a five folded thermal shield, where the outer shield is actively stabilized by peltier-elements to the pre-characterized cavity's zero crossing point. A theoretical model based on [15] served as the basis for the thermal design. The system is placed in a vacuum chamber and optical breadboards are rigidly mounted to it, containing in-coupling optics on one side and components for active intensity stabilization on transmission side. This setup demonstrated a frequency stability of 7.5×10^{-16} for $\tau < 90$ s and fulfilled the required 1×10^{-14} on 90 min. Furthermore, with this setup an extensive noise study and characterization was performed to allow for improvements and simplifications for subsequent experiments. [6, 16]

With this knowledge and investigations, a new compact setup is under development focusing on a compact and robust setup on EM-level that should fulfill the requirements for the next generations of gravity missions and, as a hybrid system that is further described in section 4, the requirements of a proposed architecture of future global navigation satellite systems.[17]

Therefore, a 5 cm cube cavity made out of ultra low expansion (ULE) glass with fused silica mirrors and crystalline coatings is covered by two thermal shields, while a customized 6-axis collimator mount is directly contacted to the outer shield to improve the in-coupling stability in rough environments. The 6-axis collimator mount is made out of Ti6Al4V and gets aligned by additional fine adjustment screws. The six sides of the outer shield are actively temperature stabilized by using kapton heaters and NTC sensors. Decoupled from the cavity shield, an optical breadboard surrounds the translation stage and carries all required optical and electro-optical components beside the laser. This includes components for PDH laser stabilization, intensity stabilization and RAM correction. It should be noted here that the reduction to a minimum of free beam components is made possible by the use of an optical circulator. All components are fiber coupled by polarization maintaining fibers and all connections are spliced. Together with the cavity itself, the breadboard is enclosed in a vacuum chamber, creating a fiber-coupled robust assembly with a minimum amount of free-beam optics. First performance validations are ongoing and a picture of the experimental setup without the vacuum chamber is shown in Fig. 6.

Figure 5: Picture of a 5 cm cube cavity with crystalline mirror coating and compensation rings mounted in a tetrahedral configuration.

Figure 6: Compact cavity setup based on a cubic cavity enclosed by two actively controlled thermal shields. All electro-optical components required for this setup beside the laser are fiber coupled and placed on a breadboard next to the cavity.

4 Hybrid lock

With a combination of both previously introduced technologies, work on a so called 'hybrid lock' is ongoing and was first demonstrated in [18]. The hybrid lock offers the advantages of both technologies and thus resulting in an optical frequency reference offering the long-term stability, reproducibility and absolute accuracy of an iodine reference, with the short term stability of an optical cavity. The principle of the first step is analogous to the iodine lock described in section 2, while a portion of the laser light is then locked to the resonance of an optical cavity by offset sideband locking. The control parameters of both, the iodine lock and cavity lock, define the cutoff frequency until the laser follows the iodine transition and then the cavity takes over, as shown in Fig. 7. For the next step, a more compact setup is currently under development where both parts, iodine spectroscopy unit and cavity setup are especially designed for being part of a hybrid lock. In addition, theoretical work on the scalability and thus possible simplifications of the principle is ongoing. Depending on actual mission requirements, the principle can become an alternative for future absolute frequency determination of optical cavities, where the full capabilities of the hybrid lock do not need to be exploited, and thus size, weight, and power can be further reduced.

5 Conclusion

We presented our current progress in the further development of optical frequency references for space applications. Multiple iterations of laboratory iodine references were evolved providing frequency stabilities up to 4×10^{-15} for long integration times. Now, following on from this development, iodine spectroscopy units are currently being built that will become the first optical clocks in space as part of the COMPASSO project and will be sent to the ISS in 2026. In addition, we presented our work on a compact and robust high finesse optical cavity setup, which is also being developed with an emphasis on space compatibility. Based on the knowledge of an extensively characterized laboratory setup as well as theoretical models, a setup with a minimum of free beam optics has been designed. After the ongoing performance validation, the next step will be to integrate this setup into a hybrid lock consisting of a compact iodine reference and a compact cavity setup.

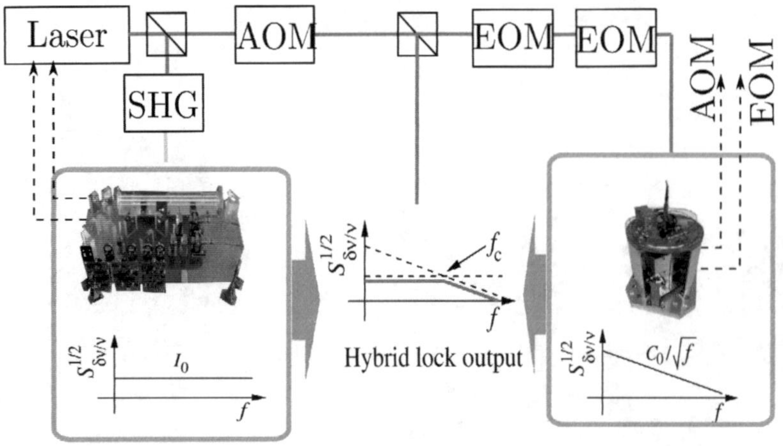

Figure 7: Schematic of the hybrid lock concept showing the laser lock to an iodine spectroscopy unit and the subsequent offset sideband lock to an optical cavity. The first EOM is bridging the frequency gap between the iodine transition and the cavity resonance by creating sidebands for a so-called offset sideband lock. The second EOM is then responsible for the phase modulation for the PDH technique. An additional AOM, even before the EOMs, acts as fast frequency actuator to generate the short- and long term stable laser which simultaneously provides absolute frequency knowledge.

References

[1] Brewer S M, Chen J S, Hankin A M, Clements E R, Chou C W, Wineland D J, Hume D B and Leibrandt D R 2019 *Phys. Rev. Lett.* **123**(3) 033201

[2] Sheard B, Heinzel G, Danzmann K, Shaddock D, Klipstein W and Folkner W 2012 *Journal of Geodesy* **86** 1083–1095

[3] Dionisio S, Anselmi A, Bonino L, Cesare S, Massotti L and Silvestrin P 2018 The "next generation gravity mission": challenges and consolidation of the system concepts and technological innovations *2018 SpaceOps Conference* p 2495

[4] Amaro-Seoane P, Audley H, Babak S, Baker J, Barausse E, Bender P, Berti E, Binetruy P, Born M, Bortoluzzi D *et al.* 2017 *arXiv preprint arXiv:1702.00786*

[5] Schuldt T, Döringshoff K, Kovalchuk E, Keetman A, Pahl J, Peters A and Braxmaier C 2017 *Applied Optics* **56**

[6] Sanjuan J, Abich K, Gohlke M, Resch A, Schuldt T, Wegehaupt T, Barwood G P, Gill P and Braxmaier C 2019 *Opt. Express* **27** 36206–36220

[7] Schuldt T, Gohlke M, Oswald M, Wüst J, Blomberg T, Peters A, Wicht A, Lezius M, Voss K, Krutzik M, Herrmann S and Braxmaier C 2021 *GPS Solutions* **25**

[8] Döringshoff K, Schuldt T, Kovalchuk E, Stühler J, Braxmaier C and Peters A 2017 *Applied Physics B: Lasers and Optics* **123**

[9] Kuschewski F, Wüst J, Oswald M, Blomberg T, Gohlke M, Bischof J, Boac A, Alam T, Bußmeier A, Abich K *et al.* 2024 *GPS Solutions* **28** 10

[10] Arie A and Byer R L 1993 *JOSA B* **10** 1990–1997

[11] Nevsky A Y, Holzwarth R, Reichert J, Udem T, Hänsch T, Von Zanthier J, Walther H, Schnatz H, Riehle F, Pokasov P *et al.* 2001 *The Hydrogen Atom: Precision Physics of Simple Atomic Systems* 576–588

[12] Hong F L, Ishikawa J, Zhang Y, Guo R, Onae A and Matsumoto H 2004 *Optics communications* **235** 377–385

[13] Webster S and Gill P 2011 *Optics letters* **36** 3572–3574

[14] Gürlebeck N, Wörner L, Schuldt T, Döringshoff K, Gaul K, Gerardi D, Grenzebach A, Jha N, Kovalchuk E, Resch A *et al.* 2018 *Physical Review D* **97** 124051

[15] Sanjuan J, Gürlebeck N and Braxmaier C 2015 *Optics Express* **23** 17892–17908

[16] Wegehaupt T, Sachit S, Müller V, Heinzel G, Braxmaier C and Grosse J 2023 Mode number determination of optical cavities for next generations of gravity missions *2023 Joint Conference of the European Frequency and Time Forum and IEEE International Frequency Control Symposium (EFTF/IFCS)* (IEEE) pp 1–4

[17] Wegehaupt T, Sanjuan J, Gohlke M, Schuldt T and Braxmaier C 2022 Optical cavity setup for future hybrid lock concept *2022 Joint Conference of the European Frequency and Time Forum and IEEE International Frequency Control Symposium (EFTF/IFCS)* (IEEE) pp 1–4

[18] Sanjuan J, Abich K, Blümel L, Gohlke M, Gualani V, Oswald M, Wegehaupt T, Schuldt T and Braxmaier C 2021 *Optics Letters* **46** 360–363

Journal of Physics: Conference Series **2889** (2024) 012013 doi:10.1088/1742-6596/2889/1/012013

Preliminary characterization of a surface electrode Paul trap for frequency metrology

Josipa Madunic[a], Lucas Groult[1], Bachir Achi[2], Thomas Lauprêtre[3], Alan Boudrias[a], Pierre Roset[a], Valérie Soumann[a], Yann Kersalé[b], Moustafa Abdel Hafiz[b] and Clément Lacroûte[a]

[a]Université de Franche-Comté, CNRS, institut FEMTO-ST, 26 rue de l'Epitaphe, 25000 Besançon, France

[b]SUPMICROTECH, CNRS, institut FEMTO-ST, 26 rue de l'Epitaphe, 25000 Besançon, France

E-mail: `clement.lacroute@femto-st.fr`

Abstract.
We are developing a single-ion optical clock based on a surface-electrode (SE) trap that we will operate with $^{171}Yb^{+}$ ions on the electric quadrupole transition at 435.5 nm. We present heating rate measurements performed with a prototype SE trap. We also introduce a new, micro-fabricated SE trapping chip using silicon on insulator technology. Electric tests were performed under ultra-high vacuum using a testing chip, including breakdown voltages measurements and flashover detection. We present suitable trapping parameters for this chip, as well as a road-map for improving its design.

1. Introduction

Atomic clocks are ubiquitous in modern society, and impact various domains such as telecommunications, navigation, energy distribution, etc. Optical atomic clocks are pushing atomic clocks to never-explored territories, reaching systematic fractional uncertainties of 10^{-18} and below [1, 2, 3, 4]. Such fractional frequency resolution enables a range of new applications, within time and frequency (TF) metrology, fundamental physics [5, 6], geodesy [7, 8, 9]. Several demonstrations of on-field measurements were performed in the last few years [10, 11], and portable optical clocks are being developed [12, 13, 14, 15, 16, 17].

In parallel to the development of optical clocks, trapped-ions based quantum information processing has also made tremendous progress in the past few years, leading to a number of research and industrial achievements [18, 19, 20]. Surface-electrodes (SE) ion traps have been proposed as the base of future quantum computers, with capabilities to control, route and coherently couple hundreds of qubits [21, 22]. Technological developments have recently lead to SE ion traps that embedded photonic integrated chips. Such trap use gratings to outcouple light from underneath the trapping electrodes and have been demonstrated with several ionic

[1] now at Heriot Watt univerity, Edinburgh, Scotland.
[2] now at Thales AVS/MIS, Vélizy-Villacoublay, France.
[3] now at Laboratoire de Physique des Lasers, Université Sorbonne Paris Nord-Institut Galilée, Villetaneuse, France

Journal of Physics: Conference Series **2889** (2024) 012013 doi:10.1088/1742-6596/2889/1/012013

species [23, 24, 25]. Such gratings can also be used to collect single-ions fluorescence and are compatible with non-destructive measurements on trapped ions [26].

We are investigating the use of a SE ion trap in the context of optical frequency metrology [27]. We have developed a prototype SE trap, which we characterized experimentally [28]. In this article, we summarize the main results of the characterization of this prototype trap and we present an upgraded trapping chip designed and fabricated for optical frequency metrology. We then present initial electric characterization of the chip, including radio frequency (RF) electrodes capacitance and breakdown voltages measurements.

2. Prototype surface-electrodes trap

This section summarizes results obtained with a prototype trap that are presented in more details in [28]. A simple, 5-wire geometry SE trap was designed and produced on a printed-circuit board (PCB). It was fabricated externally with commonly used materials to serve as a preliminary test-bed for a future SE trap-based ion clock.

Fig. 1 shows the prototype trap mounted on its ceramic support, along with its filtering capacitors and SD-format connections. RF voltages with amplitude V_{RF} and frequency Ω_{RF} are applied to the central, rectangular asymmetric electrodes that define the longitudinal axis of the trap. The RF electrodes widths of 600 and 1200 µm induce a so-called pseudo-potential with a trapping distance of about 500 µm. DC voltages are applied to the remaining electrodes in order to close the trap and to cancel DC electric fields at the trap minimum location. The resulting total potential is the combination of the RF pseudo-potential and the DC electrodes potentials, and is close to a harmonic trap with a low longitudinal secular motion frequency and higher transverse secular motion frequencies (see for instance [21, 29] for details).

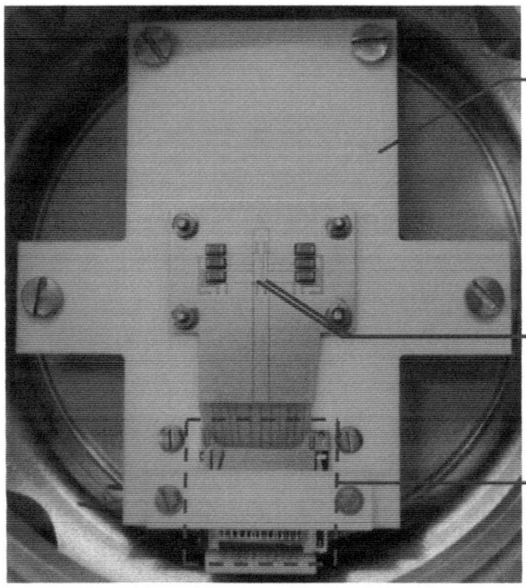

Figure 1. Photograph of the prototype SE trap used in [28]. The trap is held on a MACOR support and electrically connected through a standard SD-connector. It is housed in a vacuum chamber and faces a CF100 window for fluorescence detection.

Ions are loaded in the trap from a neutral atom beam generated with a commercial dispenser, using two-photon ionization at 399 nm and 370 nm [30]. We obtained an average trap lifetime of 1200 s with cooling lasers on, with trapping frequencies of 85 kHz, 320 kHz and 350 kHz along the longitudinal and transverse directions. The trap heating rate was estimated using a combination of the Doppler-recooling techniques and ion fluorescence imaging. The results obtained by fitting the width of the ion fluorescence spot are shown Fig. 2. We observed a heating rate of 8000 phonons/s in all directions, which we think could be related to technical noise on the experiment as well as a non-negligible background pressure at the trap location. Combining the heating rate value with preliminary excess micro-motion measurements, we estimated that a fractional frequency stability of $5\times10^{-14}\tau^{-1/2}$ and trap-related frequency shifts of order 10^{-16} could be obtained with such a trap [28].

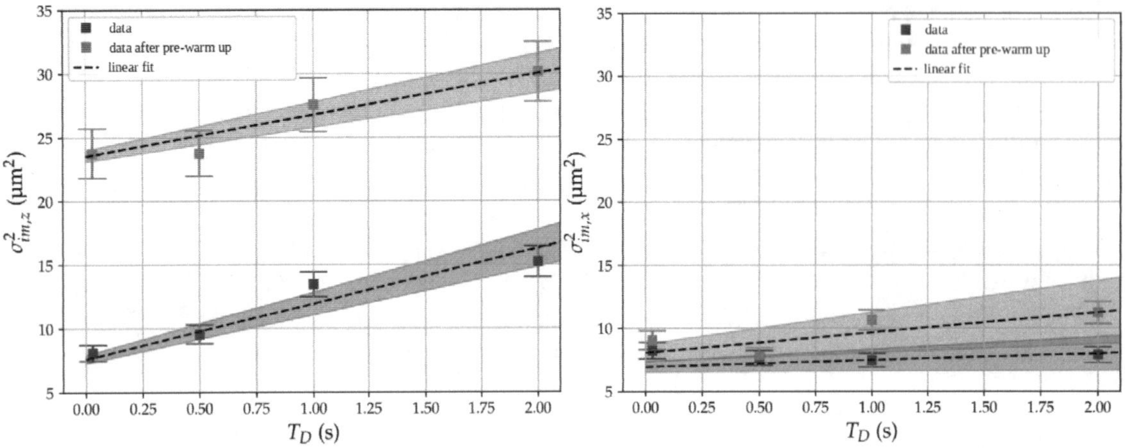

Figure 2. Heating rate measurements performed using the prototype chip [28]. The ion fluorescence image square width σ^2_{im} is plotted versus the dark time T_D along the longitudinal (x) and transverse (z) directions. Left: axial heating rate measured with (red) and without (blue) an initial pre-warm up time. Right: transverse heating rate measured with (red) and without (blue) an initial pre-warm up time.

This prototype and the related results were the basis of the design of a new trapping chip, with the main objectives being to enhance both the trapping time and heating rate, and to reduce the related fractional-frequency shifts.

3. New chip design

The prototype trap used in [28] was affected by a number of technical limitations, including:

- large inter-electrode spacing (120 µm) that results in exposed dielectrics. These areas can host free charges that distort the trapping potential;
- non-ultra-high-vacuum (UHV) dedicated materials that cause a high local pressure in the trap vicinity, reducing the single-ions lifetime via high-energy collisions;
- in-plane conductors routes to the connector, which can also distort the trapping potential.

We have therefore designed and fabricated a new trapping chip. Silicon is rarely used as a substrate for SE ion traps, because of its high dieletric RF losses. However, doped silicon can be used as an electrode material [31]. Silicon offers many advantages as a material for microfabrication:

- it is a streamlined material in any MEMS or CMOS microfabrication facility;
- it can easily be etched into high aspect ratios geometries, enabling low inter-electrode spacing and large electrodes height;
- it can be used on a number of low-RF losses substrates, including SiO_2;
- it is inexpensive thanks to its very large use.

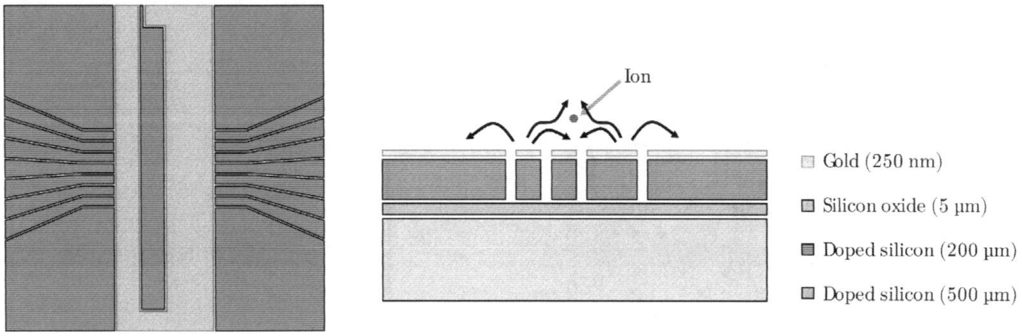

Figure 3. Silicon on insulator SE trap design. Left: geometrical scheme of the trapping electrodes. The blue, asymmetric central electrode is the RF electrode. All remaining gray electrodes are DC. Right: transverse cut of the trapping chip. Yellow: gold-coating of the electrodes. Dark gray: doped silicon electrodes. Blue: SiO_2 insulation layer. Light gray: doped silicon ground plane. Arrows: electric field lines. An additional 500 µm doped Si layer is not shown, see text for details.

3.1. Microfabricated chip geometry and fabrication process

We have designed a trapping geometry based on a 5-wire geometry similar to the prototype trap, shown Fig. 3. With 600 µm and 1200 µm wide RF electrodes, the trapping distance of 500 µm is unchanged. We added control electrodes, which enable finer displacement of the trap location in the longitudinal direction, as well as the generation of multiple trap locations. This opens the possibility of multi-ions experiments.

The chip has been fabricated using deeply reactive ion etching (DRIE) on a silicon on insulator (SOI) wafer. The initial wafer consists of a 200 µm-thick doped-silicon device layer, a 5 µm-thick buried oxyde (box) SiO_2 layer and a 500 µm-thick handle layer (see Fig. 3). The handle is doped in order to serve as an electrical ground plane.

In order to increase the chip overall height so that its surface rises above the chip carrier, the wafer handle side is bonded to a 500 µm thick doped silicon wafer. This results in a total 1.2 mm chip height, which raises the electrodes surfaces about 100 µm above the chip carrier surface (commercial ceramic pin grid array).

The electrodes are gold coated, and etched through the device layer using DRIE, resulting in 200 µm high electrodes with 20 µm inter-electrodes distance. The resulting 1/10 aspect ratio ensures that the ion is shielded from the box by the electrodes.

Fig. 4 shows a picture of the finalized microfabricated chip. It is glued inside a ceramic pin grid array carrier (CPGA) with conductive epoxy. Electrodes are wire-bonded to the CPGA pads. 220 nF capacitors are connecting all DC electrodes to ground in order to prevent RF leakage from the RF to the DC electrodes. The chip is a 8-mm long square, while the CPGA length is one inch. The overall size of the setup will be determined by the vacuum chamber volume as well as the optical setup and control electronics. Using traditional components, an overall size of a few hundred liters can be achieved, see for instance [17]. Using dedicated photonic integrated chips, digital electronics [32] and a compact vacuum cell [33], this could be reduced to the liter scale.

3.2. Trapping parameters

Trapping parameters were derived to operate the clock in a Doppler-free regime. This can be ensured by operating the clock laser and ion trap in the resolved-sideband, recoil-free regime

Figure 4. 5-wire surface-electrode trap microfabricated at FEMTO-ST. The trap is embedded in a commercial, UHV compatible ceramic chip carrier. Filter capacitors prevent RF leakage to the DC electrodes.

[34]. The resolved-sideband condition is $\Gamma \ll \omega$, where Γ is the clock transition width and ω is the secular motional frequency along the laser direction. Recoil-free clock laser interrogation is ensured when $E_{\text{recoil}} \ll \hbar\omega$ where $E_{\text{recoil}} = \hbar\omega_{\text{recoil}}$ is the recoil energy associated with the clock laser photons. For the quadrupole transition of Yb^+, $\Gamma \approx 2\pi\times3$ Hz and $\omega_{\text{recoil}} \approx 2\pi\times6$ kHz; for typical radial trapping frequencies $\omega \gtrsim 2\pi\times100$ kHz, both the resolved-sideband regime and recoil-free interrogation are ensured.

We devised two sets of trapping parameters, to be able to reach either $\omega > 2\pi\times100$ kHz or $\omega \geq 2\pi\times500$ kHz. These parameters are given in table 1. Both would enable working in the resolved-sideband regime. However, higher secular frequencies are preferred in order to reduce the decoherence caused by heating during interrogation [28]. Set (1) would therefore be more favorable in that regard, but as shown in the next section, the required RF voltage might be hard to reach. Set (2) represents a compromise to reach the Doppler-free regime with more realistic voltages.

Table 1. Proposed trapping parameters. Ω_{RF} is the trap RF drive frequency, V_{RF} is the RF voltage amplitude, $V_{\text{DC}}^{\text{max}}$ max indicates the maximum required DC voltage, q is the transverse RF stability parameter.

Set	Ω_{RF}	V_{RF}	$V_{\text{DC}}^{\text{max}}$	ω	Trap depth	q	η
1	4.9 MHz	300 V	7 V	500 kHz	325 meV	0.28	0.11
2	3 MHz	100 V	7 V	310 kHz	95 meV	0.26	0.14

4. New chip preliminary tests results

In order to trap single ions, we apply high radiofrequency voltages to the RF electrode using a simple circuit. This circuit forms an RLC resonator with the chip, which is decoupled from the source by a transformer. The trap voltage is read on a capacitor bridge with a division ratio of 100. The loaded quality factor is measured as the ratio $f_0/\Delta f$, where f_0 is the resonant frequency and Δf is the frequency interval where the voltage transmission is greater than the maximum over $\sqrt{2}$. Values in the range of 60-70 are usually measured measured.

4.1. Electrical characterization

Determining the capacitive and resistive loads formed by the trapping electrodes within an ion trap system, represented as an RC circuit, is crucial for constructing a highly accurate experiment. Determining the capacitance of the chip enables us to set the value of the resonator inductance for the desired RF frequency Ω_{RF}. It also gives information on the amount of power dissipated by the resonator and the heat generated in the process. The value of resistance sets the quality factor and RF voltage gain across the resonator, and also contributes to the overall power dissipation.

The presented trapping chip is modeled as a capacitor and resistor in series, with typical measured values of 80 to 100 pF for capacitance and around 3 Ω for resistance. These values were determined through a series of measurements using a vector network analyzer (VNA) on 3 different chips. A calibration of the VNA is carried out for each measurement. In this setup, the VNA measures the S_{11} parameters. These parameters were converted to express the capacitance and resistance of the chip between the RF electrode and ground plane. The dissipated power in the resonator is given by: $P = (C \cdot \Omega_{\mathrm{RF}} \cdot V_{RMS})^2 \cdot R$ [35]. Assuming a total resistance of 3 Ω, a trap capacitance of 100 pF, an RF trap frequency of 4 MHz and a voltage of 200 V amplitude, the dissipated RF power is around 0.4 W, which is significant as compared to other SE trap-based resonators in the literature and could lead to heating of the chip. Indeed, when the trap is operated with a voltage amplitude of 200 V, an increase in pressure at the level of 10^{-12} mbar is observed.
Table 2 summarizes the contributions to the overall capacitance and resistance of various components, including the chip, wire bonding and inductor in the setup, for one of the tested devices, with resonant frequency of 3.9 MHz. All values were measured using the VNA except for the total quality factor which was directly measured as $\frac{f_0}{\Delta f}$. The quality factor of the chip and inductor have been calculated from the measured values of resistance, capacitance and inductance.

Table 2. Overview of different contributions to overall capacitance and resistance. Capacitances, resistances and quality factor of the resonance are given in the table. The capacitance of the wires and inductors are considered to be negligibly small compared to the capacitance of the chip.

	Chip	Wire bonding and carrier	Inductor	RF source	Total
Capacitance (pF)	107	-	-	-	107
Resistance (Ω)	3	0.6	0.5	2.2	6.3
Q	127	na	312	na	60

The directly measured Q value almost perfectly matches the value calculated from the losses measured by the VNA $Q = 1/(2\pi f_0 C R)$. The main contribution to the losses comes from the trapping chip itself. This is mostly explained by the rather low conductivity of doped silicon compared to metals. We are currently investigating ways to enhance this design and reduce both ohmic and capacitive losses, which can be performed with two main changes:

- increasing the insulator layer thickness would decrease the chip capacitance, which would result in a reduction in power dissipation and an increase in RF voltage breakdown value.
- reducing the total resistance of the doped silicon elements, using wafers with the highest possible conductivity as well as by optimizing the electrodes geometry. This will necessitate finite-element modeling of the chip in order to account for RF ohmic losses in the trap.

4.2. Breakdown voltage tests
SE traps require high RF voltages, mainly dictated by the mass of the ion, electrodes configuration and the desired trapping frequencies [36, 37]. However excessive voltages applied to the

electrodes can lead to RF breakdown [38, 39]. The impact of the breakdown damage to the surface of the electrode depends on the location, scale, and physical structure of the damage. Concerns arise in the design of micrometer-scale ion traps due to the relatively lower breakdown threshold at RF frequency [40]. This issue is particularly significant for surface-electrode ion traps, which exhibit a shallower trap depth compared to conventional linear Paul traps. The need for a deep well depth and high trapping efficiency is critical. Hence, the design of microtrap structures that allow for a high applied voltages is a favorable condition. This requires a high inter-electrode separation.
Breakdown can also occur between the RF electrodes and the ground plane. In surface ion traps, where high RF voltages between the electrodes and the ground plane are desirable, the dielectric layer separating the two metal layers should be thick to maximize the breakdown voltage. In our case, the thickness of the dielectric layer (SiO_2) is 5 µm, limited by the commercially available SOI wafer, which adds constraints to the voltages that can be applied to the electrodes. This limitation makes achieving ion trapping with the desired stability parameters challenging.

In order to measure the typical voltage the RF electrodes can handle without breakdown, several flashover tests were conducted in a test vacuum chamber with pressures lower than 6×10^{-11} mbar. The experimental setup is shown in Fig. 5. A 1:1 imaging system, with a CCD camera, was used to observe the surface of the chips. The camera, cooled to -80°C, captures images with a 3-second acquisition time every 4.07 seconds.

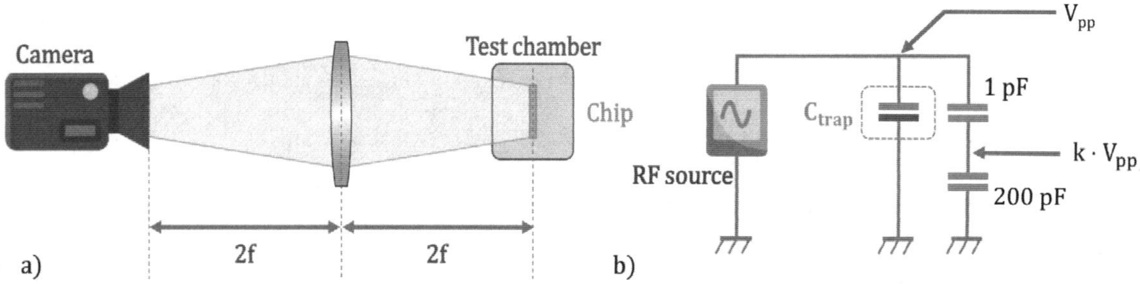

Figure 5. a) The chip is placed in the test chamber at a pressure of 6×10^{-11} mbar. Throughout the test, a voltage ramp is applied, and data acquisition is carried out using a CCD camera. b) Experimental setup schematic for supplying and reading the trap voltage with the chip in a vacuum chamber. The voltage is supplied using an RLC circuit resonator and monitored through a capacitor bridge with a ratio of k=1/200 with respect to the trap voltage V_{pp}.

The measurement was conducted for a duration of two hours. 10 minutes voltage ramps followed by a 20-minute wait at the resonant frequency and a quality factor measurement are applied to the RF electrode. From 150 V peak-to-peak on the RF electrode, reflections of flashes occurring at the edge of the chip started to be observed (see Fig. 6, left). However, voltages of several hundred volts peak-to-peak were reached without clear breakdown observed on the voltage monitor, Fig. 6, right. Flashing events observed on the camera signal were accompanied by a pressure increase of around 10^{-11} mbar, as shown in Fig. 6, right. After this voltage was reached, flashes are visible in nearly all acquired images, and their intensity increases with higher voltage. They stopped spontaneously before the end of the experiment, at 200 V peak-to-peak (Fig. 6, right).

Following the conclusion of these experiments, the trapping parameters specified in Table 1

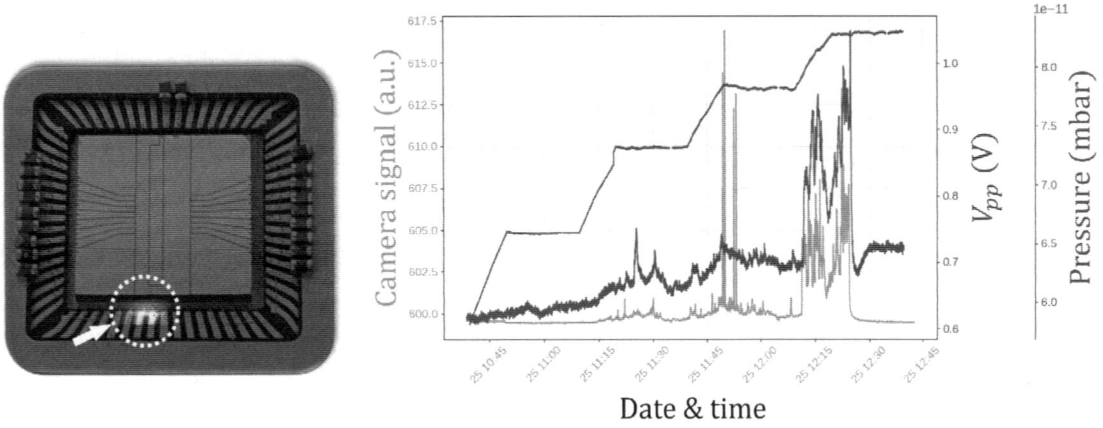

Figure 6. Left: Depiction of the chip surface with the reflection from the flashover highlighted on the chip carrier pads within a white circle. Right: Corresponding flashover data. The camera signal integrated over a region of interest centered on the flashing region is represented in green, pressure data in blue, and the voltage read on the monitor in pink.

were formulated. The first configuration of parameters is not achievable with the existing design; however, the second configuration might be possible.
After the measurements, the chip was inspected using the SEM microscope, and no clearly visible damage was observed on the electrodes at the location of the flashing events. The results of these tests have shown that flashover is most likely to occur at voltages of 150 V peak-to-peak. The obtained results will guide the design of an upgraded version of the chip, with a thicker layer of insulator between the electrodes and the ground plane.

5. Conclusion

We are investigating the use of silicon-based SE traps for optical frequency metrology. Using a SE trap combines the advantages of offering a very reduced footprint and the possibility to integrate optical as well as electrical functions. Silicon is the material of choice for micro-fabrication, and the realization of silicon-based, high-performance SE traps for optical atomic clocks would be a key technological advance towards industrial, transportable optical atomic clocks.

In this article, we have summarized experiments performed with a prototype SE trap, and presented a new micro-fabricated trapping chip that underwent initial electric characterization. Further improvements in the design are already identified and planned for reaching higher values of V_{RF} and ω. This new chip is compatible with UHV pressures and trapping configurations in the resolved-sidebands regime. Future experiments will include heating rate measurements as well as coherence time measurements on the optical clock transition at 435.5 nm, for integration in a transportable ion clock.

Acknowledgments

The authors would like to thank Philippe Abbé and Yannick Gruson from the TF department at FEMTO-ST for their technical help with chip connections and support and RF management. We would also like to thank Nicolas Passilly from the MN2S department and William Daniau from the TF department at FEMTO-ST for their advice on the chip microfabrication process.

This work has been supported by the EIPHI Graduate School (contract "ANR-17-EURE-0002"), by the ANR-10-LABX-48-01 First-TF and ANR-11-EQPX-0033 Oscillator-IMP, by the

Région Bourgogne Franche-Comté and by the French RENATECH network and its FEMTO-ST technological facility.

References

[1] Huntemann N, Sanner C, Lipphardt B, Tamm Chr and Peik E 2016 *Physical Review Letters* **116** 063001
[2] Brewer S M, Chen J S, Hankin A M, Clements E R, Chou C W, Wineland D J, Hume D B and Leibrandt D R 2019 *Physical Review Letters* **123** 033201
[3] Kim K, Aeppli A, Bothwell T and Ye J 2023 *Physical Review Letters* **130** 113203
[4] Srinivas R, Löschnauer C M, Malinowski M, Hughes A C, Nourshargh R, Negnevitsky V, Allcock D T C, King S A, Matthiesen C, Harty T P and Ballance C J 2023 *Physical Review Letters* **131** 020601
[5] Derevianko A 2016 *Journal of Physics: Conference Series* **723** 012043
[6] Safronova M S, Budker D, DeMille D, Kimball D F J, Derevianko A and Clark C W 2018 *Reviews of Modern Physics* **90** 025008
[7] Flury J 2016 *Journal of Physics: Conference Series* **723** 012051
[8] Lion G, Panet I, Wolf P, Guerlin C, Bize S and Delva P 2017 *Journal of Geodesy* **91** 597–611
[9] McGrew W F, Zhang X, Fasano R J, Schäffer S A, Beloy K, Nicolodi D, Brown R C, Hinkley N, Milani G, Schioppo M, Yoon T H and Ludlow A D 2018 *Nature* **564** 87
[10] Grotti J, Koller S, Vogt S, Häfner S, Sterr U, Lisdat C, Denker H, Voigt C, Timmen L, Rolland A, Baynes F N, Margolis H S, Zampaolo M, Thoumany P, Pizzocaro M, Rauf B, Bregolin F, Tampellini A, Barbieri P, Zucco M, Costanzo G A, Clivati C, Levi F and Calonico D 2018 *Nature Physics* **14** 437–441
[11] Takamoto M, Ushijima I, Ohmae N, Yahagi T, Kokado K, Shinkai H and Katori H 2020 *Nature Photonics* **14** 411–415
[12] Poli N, Schioppo M, Vogt S, Falke S, Sterr U, Lisdat C and Tino G M 2014 *Applied Physics B* **117** 1107–1116
[13] Cao J, Zhang P, Shang J, Cui K, Yuan J, Chao S, Wang S, Shu H and Huang X 2017 *Applied Physics B* **123** 112
[14] Koller S B, Grotti J, Vogt St, Al-Masoudi A, Dörscher S, Häfner S, Sterr U and Lisdat Ch 2017 *Physical Review Letters* **118** 073601
[15] Hannig S, Pelzer L, Scharnhorst N, Kramer J, Stepanova M, Xu Z T, Spethmann N, Leroux I D, Mehlstäubler T E and Schmidt P O 2019 *Review of Scientific Instruments* **90** 053204
[16] Kong D H, Wang Z H, Guo F, Zhang Q, Lu X T, Wang Y B and Chang H 2020 *Chinese Physics B* **29** 070602
[17] Stuhler J, Abdel Hafiz M, Arar B, Bawamia A, Bergner K, Biethahn M, Brakhane S, Didier A, Fortágh J, Halder M, Holzwarth R, Huntemann N, Johanning M, Jördens R, Kaenders W, Karlewski F, Kienle F, Krutzik M, Lessing M, Mehlstäubler T E, Meschede D, Peik E, Peters A, Schmidt P O, Siebeneich H, Tamm C, Vogt E, Wicht A, Wunderlich C and Yu J 2021 *Measurement: Sensors* **18** 100264
[18] Pino J M, Dreiling J M, Figgatt C, Gaebler J P, Moses S A, Allman M S, Baldwin C H, Foss-Feig M, Hayes D, Mayer K, Ryan-Anderson C and Neyenhuis B 2021 *Nature* **592** 209–213
[19] Auchter S, Axline C, Decaroli C, Valentini M, Purwin L, Oswald R, Matt R, Aschauer E, Colombe Y, Holz P, Monz T, Blatt R, Schindler P, Rössler C and Home J 2022 *Quantum Science and Technology* **7** 035015
[20] Zhu D, Kahanamoku-Meyer G D, Lewis L, Noel C, Katz O, Harraz B, Wang Q, Risinger A, Feng L, Biswas D, Egan L, Gheorghiu A, Nam Y, Vidick T, Vazirani U, Yao N Y, Cetina M and Monroe C 2023 *Nature Physics* **19** 1725–1731
[21] Chiaverini J, Blakestad R B, Britton J, Jost J D, Langer C, Leibfried D, Ozeri R and Wineland D J 2005 *Quantum Information and Computation* **5** 419–439
[22] Seidelin S, Chiaverini J, Reichle R, Bollinger J J, Leibfried D, Britton J, Wesenberg J H, Blakestad R B, Epstein R J, Hume D B, Itano W M, Jost J D, Langer C, Ozeri R, Shiga N and Wineland D J 2006 *Physical Review Letters* **96** 253003
[23] Mehta K K, Bruzewicz C D, McConnell R, Ram R J, Sage J M and Chiaverini J 2016 *Nature Nanotechnology* **11** 1066–1070
[24] Mehta K K, Zhang C, Malinowski M, Nguyen T L, Stadler M and Home J P 2020 *Nature* **586** 533–537
[25] Niffenegger R J, Stuart J, Sorace-Agaskar C, Kharas D, Bramhavar S, Bruzewicz C D, Loh W, Maxson R T, McConnell R, Reens D, West G N, Sage J M and Chiaverini J 2020 *Nature* **586** 538–542
[26] Saha U, Siverns J D, Hannegan J, Prabhu M, Quraishi Q, Englund D and Waks E 2023 *Physical Review Applied* **19** 034001
[27] Lacroûte C, Souidi M, Bourgeois P Y, Millo J, Saleh K, Emmanuel Bigler, Boudot R, Giordano V and Kersalé Y 2016 *Journal of Physics: Conference Series* **723** 012025
[28] Lauprêtre T, Achi B, Groult L, Carry É, Kersalé Y, Delehaye M, Hafiz M A and Lacroûte C 2023 *Applied Physics B* **129** 37 ISSN 1432-0649

[29] House M G 2008 *Physical Review A* **78** 033402
[30] Lauprêtre T, Groult L, Achi B, Petersen M, Kersalé Y, Delehaye M and Lacroûte C 2020 *OSA Continuum* **3** 50–57
[31] Britton J, Leibfried D, Beall J A, Blakestad R B, Wesenberg J H and Wineland D J 2009 *Applied Physics Letters* **95** 173102–173102–3
[32] Pruttivarasin T and Katori H 2015 *Review of Scientific Instruments* **86** 115106
[33] Wilpers G, See P, Gill P and Sinclair A G 2013 *Applied Physics B* **111** 21–28
[34] Ludlow, A D and Boyd, M M and Ye, J and Peik, E and Schmidt, P O 2015 *Reviews of Modern Physics* **87** 637–701
[35] Marcus D Hughes Bjoern Lekitsch J A B and Hensinger W K 2011 *Contemporary Physics* **52** 505–529
[36] McLoughlin J J, Nizamani A H, Siverns J D, Sterling R C, Hughes M D, Lekitsch B, Stein B, Weidt S and Hensinger W K 2011 *Physical Review A* **83** 013406
[37] Hong S, Lee M, Cheon H, Kim T and Cho D i D 2016 *Sensors* **16**
[38] Bautista-Salvador A, Zarantonello G, Hahn H, Preciado-Grijalva A, Morgner J, Wahnschaffe M and Ospelkaus C 2019 *New J. Phys.* **21** 043011
[39] Sterling R C, Hughes M D, Mellor C J and Hensinger W K 2013 *Applied Physics Letters* **103** 143504
[40] Stick D, Hensinger W K, Olmschenk S, Madsen M J, Schwab K and Monroe C 2006 *Nature Phys* **2** 36–39

Journal of Physics: Conference Series **2889** (2024) 012014 doi:10.1088/1742-6596/2889/1/012014

DSAC and next generation high stability, long life trapped mercury ion frequency standards

E.A. Burt, T.A. Ely, and R.L. Tjoelker

Jet Propulsion Laboratory, California Institute of Technology, Pasadena, CA*

Corresponding author's e-mail address: eric.a.burt@jpl.nasa.gov

Abstract. The methods of trapping and cooling of atoms and ions have been transformative for atomic clocks due to the reduction, and in some cases elimination, of major systematic frequency shifts. Continuously operating atomic clocks based on trapped mercury ions have existed for decades but until recently have been restricted to terrestrial applications. The recently completed Deep Space Atomic Clock (DSAC) mission demonstrated the first trapped ion clock operation in space. Here we review DSAC as well as follow-on improvements towards the realization of high stability, long life Hg ion atomic clocks for foreseen ground and flight application.

1. Introduction

Ion trapping has had a major impact on achieved atomic clock stability and accuracy because of its elimination of atom interactions with the surrounding vacuum chamber walls [1-5]. Laser-cooling of atoms and ions to milli-Kelvin temperatures and below, have further reduced other systematics but does introduce the complexity of laser systems [6]. At JPL we have had a long-standing research effort to develop trapped ion clocks that operate near room temperature with a noble buffer gas and no lasers. Clock stability achievements include short-term stabilities as low as $2x10^{-14}/ t^{1/2}$ [4] and long-term drift of < $3x10^{-17}$/day – sufficient to provide a single-clock representation of UTC [7]. In addition to high stability performance, the 40.5 GHz clock transition has low sensitivity to the environment and the technology lends itself to field applications requiring lower SWaP [8, 9]. In 2011, NASA initiated the Deep Space Atomic Clock (DSAC) Technology Demonstration Mission (TDM) with the goal of demonstrating ion clock operation and performance in space. The DSAC demonstration mission was successfully executed between 2019 and 2021.

In section 2, we briefly review the DSAC demonstration mission [10, 12]. In Section 3 we present some more recent DSAC instrument life analysis. In Section 4 we describe some upcoming ground and spaceflight applications and summarize several frequency standard technologies with potential to meet needed stability, reliability, and volume requirements. In Section 5 we describe recent work to develop a DSAC follow-on ion clock that is smaller and more manufacturable. This new version will also further increase instrument life. Core clock components are being developed for dual use to address both future

* The research was carried out at the Jet Propulsion Laboratory, California Institute of Technology, under a contract with the National Aeronautics and Space Administration (80NM0018D0004). © 2024, California Institute of Technology. Government sponsorship acknowledged. The technology demonstration mission was supported in part by the NASA Space Technology Mission Directorate (STMD) and in part by the NASA Space Communications and Navigation Directorate (SCAN). The authors are with the Jet Propulsion Laboratory, California Institute of Technology, Pasadena CA 91109 USA (e-mail: eric.a.burt@jpl.nasa.gov).

ground (e.g. as an alternative to hydrogen masers) and spaceflight needs (e.g. GNSS or Lunar/Mars Precision Navigation and Timing).

2. DSAC Mission

The DSAC mission was a culmination of many decades of research at JPL on trapped ion atomic clocks. The DSAC demonstration clock design did not include cryogenics, a microwave cavity, or lasers. Instead, it operated at near room temperature, used simple travelling wave microwave components, and used a plasma discharge deep UV light source. The maturity and robust operability of each of these enabled launch into and operation demonstration in space.

On the ground, the DSAC instrument demonstrated a short-term fractional frequency stability of $1.5 \times 10^{-13}/\sqrt{\tau}$ [11]. The DSAC instrument was launched into low Earth orbit in 2019 for a planned 2-year demonstration [17]. Mission and space operation results are well documented [10, 12]. After launch and several months of host spacecraft and instrument subsystem functional validation, the ion clock stability was characterized over several long uninterrupted runs of approximately 60 days in duration. In each case, DSAC performed without incident and the run was only terminated by space craft power availability. To determine DSAC performance, a GPS receiver was included as part of the mission payload and enabled GPS frequency comparisons to UTC. In addition to clock noise, measurements included orbit determination noise and GPS noise so these measurements should be considered an upper bound on actual clock performance.

In summary, the DSAC demonstration clock operated for 2 years in space. It achieved a fractional frequency stability in space of 1.5×10^{-13} at one second, a long-term stability of 3×10^{-15} for averaging times greater than a day, a time deviation of only 4 ns at 23 days (no drift removal), and an estimated drift of $3.0(0.7) \times 10^{-16}$ per day. Among the most stable space clocks currently in use, each of these characteristics established a new space clock performance standard by at least an order of magnitude [13, 14, 15]. The DSAC clock performed well in the space environment, due to low sensitivities of the 40.5 GHz clock transition to variations in radiation, temperature, and magnetic fields. For a detailed discussion of these DSAC environmental sensitivity, see [10, 11].

In addition to demonstrating the first operation of a trapped-ion atomic clock in space, the project also demonstrated the clock's utility for applications requiring autonomy such as one-way navigation, whereby signal delay times are measured in-situ making autonomous deep-space probe radio navigation possible [16].

3. DSAC Clock Life Estimate

The DSAC demonstration clock operated in low Earth orbit between June 2019 and November 2021. As built and flown, the demonstration clock life was estimated to be greater than 5 years [10]. The three leading life limiting mechanisms were mercury depletion, neon buffer gas depletion, and plasma discharge light source life. Since the trap load rate is proportional to the background neutral mercury vapor pressure and electron emission current, with the emission current fixed mercury depletion was monitored by periodically measuring the load rate. These measurements show how the load rate is evolving. The lifetime of the instrument against this mechanism is determined by the requirement that the load rate remain much faster than the trap lifetime. In DSAC, the trap lifetime was measured to be greater than 60 days due to the extremely good vacuum quality. Thus, even very slow load times are viable. Fig. 1a shows two ion loading measurements separated by 3 years giving exponential 1/e times of 10 and 22 minutes respectively. The slowly increasing load time can be extrapolated to a useful instrument life due to mercury depletion of 7 years.

The DSAC instrument includes both a quadrupole and a multipole trap [11]. Neon buffer gas depletion is determined by monitoring the trap shuttling efficiency. Shuttling is the process of moving ions back and forth between the two traps. Due to space-charge repulsion and the different confinement characteristics of each trap, moving ions between traps can lead to ion heating and is heavily dependent on the buffer gas to prevent ions from becoming too energetic to be trapped (ion loss). Thus, shuttling efficiency (the fraction of ions remaining trapped after one round-trip shuttle) taken together with pressure calibrations made on the ground, can be a good measure of neon buffer gas pressure. Fig. 1b shows ion number decay rates as a function of the number of shuttles measured at three different times separated by several years. The three measurements together with ground calibrations indicate that the neon pressure, which started at about 5×10^{-5} torr, will not reach a lower threshold for quadrupole trap operation of 1×10^{-7} torr for 8-9 years, thereby setting the instrument lifetime due to this mechanism.

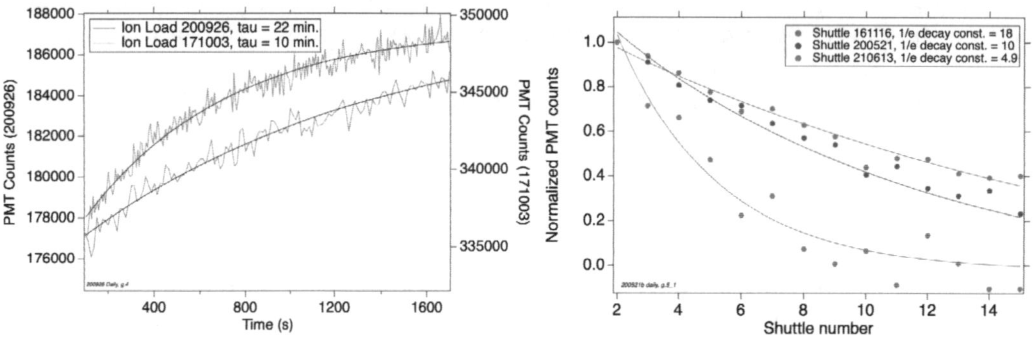

Figure 1. Lifetime measurements: a) on the left is shown DSAC ion trap loading in 2017 (green trace) and approximately 3 years later in 2020 (red trace). b) on the right is shown DSAC ion number decay rate as a function of number of shuttles as measured in 2016 (red), 2020 (blue) and 2021 (green).

The third life-limiting mechanism is degradation in the optics system, which is dominated by the lifetime of the plasma discharge light source ("lamp"). This is determined by loss of metallic mercury inside the lamp to the lamp envelope or recombination to mercury oxide. This loss rate is determined by monitoring the background light levels as a function of time at constant lamp temperature (lamp output is dependent on temperature so only light levels taken at the same lamp temperature are compared). Fig. 2 shows background light levels as a function of lamp temperature at two different times separated by about one year. The difference in light level of about 15% can be extrapolated forward to a sufficient light level for 7-years, however it is possible for mercury to go below the level required to sustain a plasma before the lower light level is reached. Given that the lamp had already operated for a total of approximately 4 years (including ground operation), an estimated total life of 5 years is conservative.

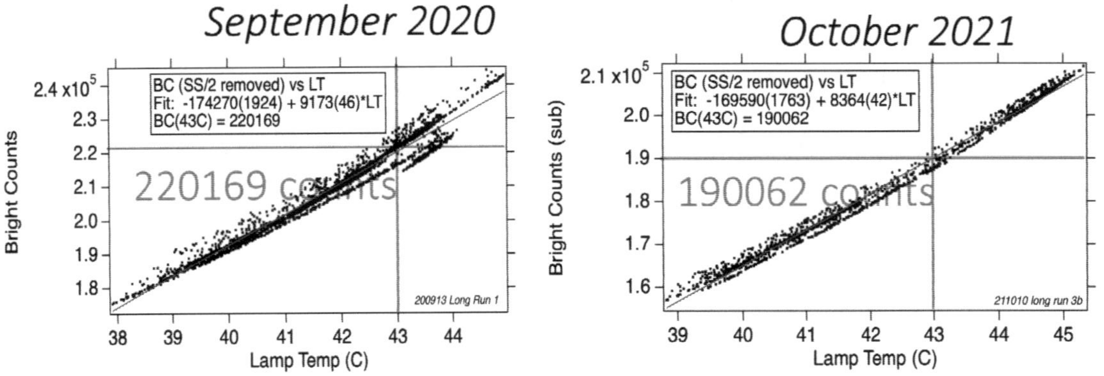

Figure 2. DSAC background light levels as a function of lamp temperature measured at two times (September 2020 and October 2021) separated by about 1 year. The levels are compared at the same lamp temperature of 43 ℃. The change from approximately 220,000 counts to 190,000 counts shown gives a decrease of about 15% over that time period.

4. Next generation high performance trapped ion clocks for ground and space

An essential application of atomic clocks is timekeeping and navigation. There are two ways that clocks can be used for navigation in space. In two-way navigation, a signal is sent from the Earth to a satellite, which acts as a transponder reflecting the signal back to Earth. The round-trip travel time as determined by a precise clock on the ground gives the radial distance to the space craft. This combined with local gravitational models and VLBI measurements gives both the radial and angular position [17]. The second method, referred to as one-way navigation, places the precise clock on-board the satellite enabling navigational calculations to be made in-situ.

In addition to deep space navigation, navigation on Earth using various GNSS systems requires high-performance atomic clocks in each of the GNSS satellites. For GNSS applications it has been shown that a clock performance of 1×10^{-13} at one second and 1×10^{-15} at one day is sufficient and that better stability currently offers little advantage [18]. In addition to the performance for averaging times of 1 day or less, lower long-term drift, for example below 1×10^{-15} per day, would provide a higher level of autonomy (current GNSS clocks with drifts $>1\times10^{-14}$ per day require solutions to be updated several times a day to avoid undesirable positioning errors).

Traditionally, hydrogen masers have been used as a ground reference clock in two-way navigation. Hydrogen masers have proved challenging to put in space with the same level of performance achieved on the ground [19]. Trapped ion clocks with maser-like stability offer a possible lower-SWaP alternative on the ground and have been demonstrated in space as described in Section 2. Responding to the needs of both potential space and ground high-performance clock applications, we are targeting a performance of 1×10^{-13} at one second and 1×10^{-15} at one day with drift below 1×10^{-15} per day in the next generation trapped ion clocks.

However, stability performance is only one metric in a larger trade space that also requires low SWaP and high reliability/instrument lifetime. To capture this three-dimensional space, Fig. 3 shows achieved stability at 1000 seconds (an averaging time of particular interest to navigation) vs. the average time between interactions required to keep the clock running. In the Figure, marker size is also proportional to current clock volumes. Fig. 3 includes representative stability, reliability, and volume properties of many existing and developing frequency standard technologies. See Appendix 1 for a description of each clock included in this figure as well as references.

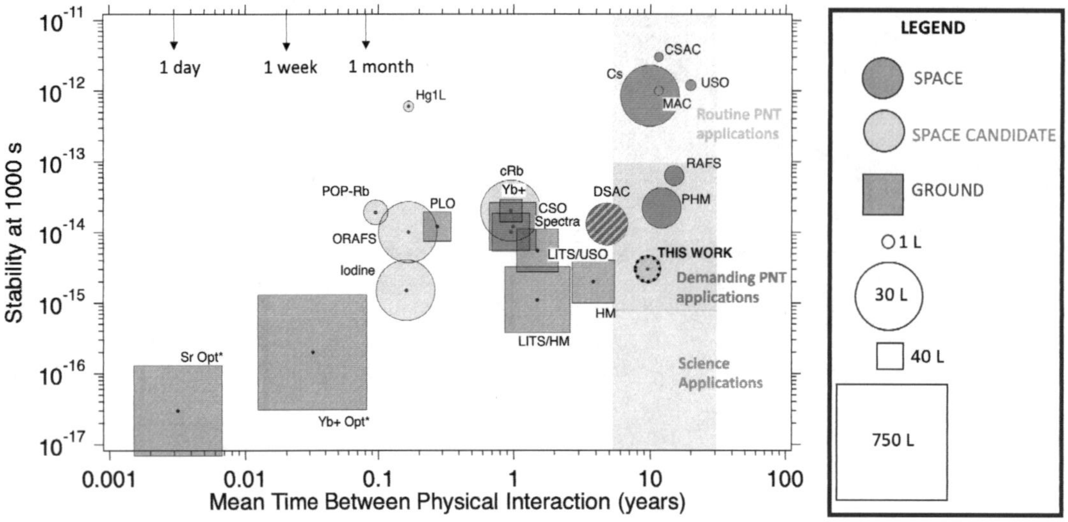

Figure 3. A selection of currently operating ground and space clocks showing their stability at 1000 s averaging time vs. the mean time between interaction required to keep the clock operating (in some cases this is the instrument lifetime). The size of each marker (the diameter for the circles or the length of the side of the squares) is proportional to the volume of the clock with the circles representing clocks currently operating in space (dark blue), clocks that are either space qualified or have direct space heritage, but are not yet operating in space (light blue) and squares representing current ground clocks (these are on a different size scale as shown in the legend). The cross hatching for DSAC indicates that it has operated in space, but is not currently doing so. The marker, "This Work" refers to the follow-on work described here. See appendix 1 for more details on this figure.

As is evident in Fig. 3, only a few clock technologies are close to meeting all three metrics required for demanding Precision Navigation and Timing (PNT). This graph is only a snapshot in time and several of the newer instruments represented are likely to move down and to the right over time. The DSAC symbol is cross hatched indicating that it has flown in space, but that it is not currently doing so. The dashed circle represents the current target of our next-generation trapped mercury ion clock, which has not yet flown but has DSAC heritage. A recent effort to demonstrate a small, 1-liter mercury ion clock is also included in the Figure [20]. An important message to take away from this graphic is that there is no one clock that meets all requirements for all applications. Some routine PNT applications may be able to tolerate lower performance and benefit by the extremely low SWaP of a CSAC or a USO. On the other end of the spectrum, some scientific applications requiring the best possible stability may be able to tolerate a lesser degree of autonomy.

5. Future clock performance and SWaP goals

At the conclusion of the DSAC mission, NASA initiated a Technology Maturation Task to address better performance, lower SWaP, and improved manufacturability of the JPL trapped mercury ion technology. As previously mentioned, for the next generation trapped ion clock, we have chosen a stability target of 1×10^{-13} at one second and 1×10^{-15} at 1 day. In addition, we have set a SWaP goal of 10 L, 10 kg and 34 W, which will fit well within the current GPS clock footprint. We are also targeting a lifetime goal of > 10 years. These properties are improvements from DSAC, which was 19 L, 19 kg, 54 W, and > 5 years life. Fig. 4 shows a relative size comparison for DSAC (the quartz Local Oscillator is not shown) and the next generation (with integrated quartz Local Oscillator). The next generation's lower SWaP, higher stability, and life performance will be useful for GNSS, but is also designed to fill various ground applications. For example, the GPS footprint chosen can easily be converted to a 3U rack mount configuration. With performance similar to that of an active hydrogen maser [21] and in a smaller package the next version trapped ion clock may also be useful as an alternative to maser technology with the lower SWaP enabling simpler configuration logistics.

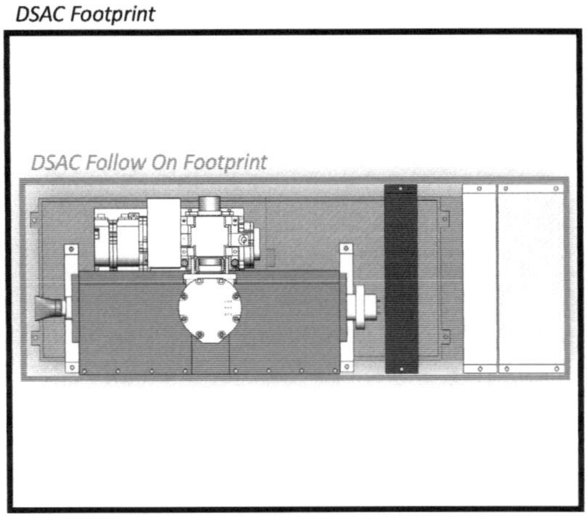

Figure 4. A comparison between the DSAC instrument footprint as viewed from above with 19 L and 19 kg (black outline) to the next generation trapped ion clock as viewed from above with 10 L and 10 kg (green outline).

While the DSAC follow-on clock will be implemented with an internal USO as the LO, future clock variants could easily be adapted to an advanced LO. Some scientific applications such as space VLBI [22] can benefit from significantly better short-term performance. While optical clocks may be the ultimate solution for these types of measurements, they are not yet mature enough to be flown. The trapped ion technology has always had the ability to take advantage of different LO architectures: from small lower-performing crystal oscillators all the way to LO's based on optical to microwave generation [35, 41]. Our next generation instrument will have an external LO input allowing flexibility in this choice. With an optimized signal-to-noise (SNR), a trapped ion clock coupled with a space-hardened optical-to-microwave LO (OLO) could achieve $3x10^{-15}$ at one second and $2x10^{-14}/\tau^{1/2}$ in the long term (see Fig. 5). Both the comb and ultra-stable laser components of an OLO have already flown in space, making the ensemble space viable. Fig. 5 shows the estimated performance of a practical, room temperature bulk resonator cavity OLO compared to a conventional USO. An OLO coupled with either the $1x10^{-13}/\tau^{1/2}$ stability described here (red trace) or a trapped ion standard with a fully optimized SNR enabling $2x10^{-14}/\tau^{1/2}$ (blue trace) would result in a clock system with exceptional short- and long-term stability.

Journal of Physics: Conference Series **2889** (2024) 012014 doi:10.1088/1742-6596/2889/1/012014

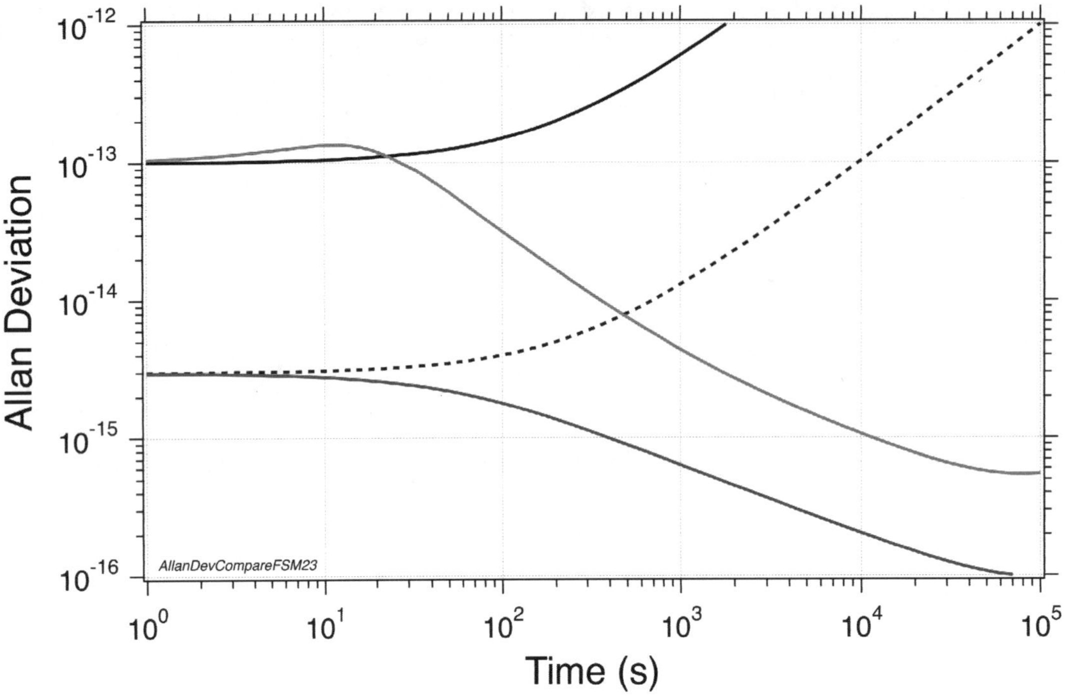

Figure 5. The red trace shows the target Allan deviation of next generation high performance trapped ion clocks of $1x10^{-13}/t^{1/2}$ coupled with a USO. An estimate of the Allan deviation possible using the best short-term stability of $2x10^{-14}/t^{1/2}$ achieved with this technology coupled to an OLO is shown in the blue trace. For comparison, the estimated Allan deviation for a practical Optical Local Oscillator or OLO (dashed black) and a typical USO are also shown (solid black).

6. Conclusions

Trapped ion frequency standards are poised to have impact in upcoming ground and space applications that require high stability, reliable autonomous operation, low environmental sensitivity, and lower SWaP. We have summarized demonstration of a trapped ion atomic clock in space, which operated for 2 years from 2019 through 2021. The clock's 1-second stability determined by the USO was $1.5x10^{-13}$ and long-term drift of $< 3x10^{-16}$/day was measured (with no drift-removal performed on the data). This level of stability corresponds to a time-deviation of less than 4 ns at 23 days, thereby demonstrating the technology's importance to applications requiring autonomy. We have also estimated the as-flown instrument life by extrapolating measurements taken on the three leading life-limiting mechanisms: mercury depletion, neon buffer gas depletion, and plasma discharge light source aging. Each of these was individually found to be in excess of 5 years. Finally we discussed future developments towards a higher-performance, lower-SWaP version of this technology and described how these fit into the broader atomic clock technology and PNT application context. This next generation of mercury ion clock is expected to have broader ground and flight clock impact.

Appendix 1

The clocks shown in Fig. 3 belong to one of three categories: 1) space clocks: clocks currently operating in space, 2) space candidate clocks: clocks that have been space qualified, or have significant space heritage, but are not currently operating in space, and 3) ground clocks.

Space clocks include, the Chip Scale Atomic Clock (CSAC) [23], a space microwave cesium beam tube atomic clock (Cs) [24], the GPS rubidium microwave atomic frequency standard (RAFS) [25], the Galileo passive hydrogen maser space microwave clock (PHM) [26], an Ultra-Stable quartz Oscillator (USO) [27], the trapped mercury ion microwave Deep Space Atomic Clock (DSAC) [10] (operated in space from 2019 through 2021), and the Miniature Atomic Clock (MAC) [28], and a laser-cooled rubidium space clock (cRb) [29] (operated in space from 2016 through 2018).

Candidate space clocks include, a Pulsed Optically Pumped rubidium microwave clock (POP-Rb) [30], a laser-cooled cesium microwave clock (Spectra) [31], a DSAC follow-on clock with lower SWaP and improved performance [reported here and indicated as a dashed line in the figure], an ytterbium ion microwave clock (Yb+) [32], an iodine optical clock (Iodine) [33], a 1-liter trapped mercury ion microwave clock (Hg1L) [34], a space-based Optical-to-microwave Local Oscillator (OLO-s) [35], and a space active hydrogen maser [36].

Ground clocks include, a trapped mercury ion microwave clock operating with a USO LO (LITS/USO) [37], an optical rubidium atomic frequency standard (ORAFS) [38], a trapped mercury ion microwave clock operating with a maser LO (LITS/HM) [39], a hydrogen maser (HM) [21], a cryogenic sapphire oscillator (CSO) [40], an Optical to microwave Local Oscillator operating on the ground (OLO) [41], a strontium optical clock (Sr Opt) [42], and an Ytterbium ion optical clock (Yb+ Opt) [43],

References

[1] H. Dehmelt, Bull. Am. Phys. Soc. **18**, 1521 (1973)

[2] Cutler, L. S., Giffard, R. P. & McGuire, M. D. "A trapped mercury 199 ion frequency standard". Proc. 13th Annual Precise Time and Time Interval (PTTI) Applications and Planning Meeting, 563–577 (Institute of Navigation, 1981).

[3] Prestage, J.D., Dick, G.J., and Maleki, L. "New ion trap for frequency standard applications" J. Appl. Phys. **66** (3), 1 August 1989.

[4] R.L. Tjoelker, J.D. Prestage, L. Maleki, "Record atomic frequency standard stability with mercury in a linear ion trap", TDA Progress Report 42-126, August 15, 1996

[5] Tjoelker, R.L., et al. "A mercury ion frequency standard engineering prototype for the NASA deep space network." *Proc. Of the 50th annual IEEE International Frequency Control Symposium*, pp. 1073-1081 (1996)

[6] D.J. Wineland, "Trapped Ions, Laser Cooling, and Better Clocks," Science **226**, 395 (1984)

[7] E.A. Burt, W. A. Diener, and R. L. Tjoelker, "A compensated multi-pole linear ion trap mercury frequency standard for ultra-stable timekeeping," IEEE Trans. Ultrason., Ferroelectr., Freq. Control, vol. 55, no. 12, pp. 2586–2595, Dec. 2008.

[8] J. D. Prestage, et al., "Compact microwave mercury ion clock for deep-space applications," in Proc. Joint IEEE Freq. Control Symp. Eur. Freq. Time Forum, Geneva, Switzerland, May 2007, pp. 1113–1115.

[9] R. L. Tjoelker et al., "Mercury atomic frequency standard development for space-based navigation and timekeeping," in Proc. 43rd Precise Time and Time Interval (PTTI), Long Beach, CA, USA, Nov. 2011, pp. 293–304.

[10] E.A. Burt, J.D. Prestage, R.L. Tjoelker, et al., "Demonstration of a trapped-ion atomic clock in space," Nature **595**, 43 (2021)

[11] Tjoelker, R.L., et al., "Deep Space Atomic Clock (DSAC) for a NASA Technology Demonstration Mission," *IEEE Trans. On Ultrasonics, Ferroelectrics, and Frequency Control* **63**, 1034 (2016)

[12] T. Ely et al, "Deep Space Atomic Clock Technology Demonstration Mission Results"; 2022 IEEE Aerospace Conference (AERO), March 2022.

[13] Lutwak, R., Emmons, D., Garvey, R.M., and Vlitas, P. "Optically pumped cesium-beam frequency standard for GPS-III," *Proc. 33rd Ann. Precise Time and Time Interval (PTTI)*, pp. 19-30 (2001)

[14] Riley, W.J. "Rubidium atomic frequency standards for GPS block IIR," *Proc. 22nd Ann. Precise Time and Time Interval (PTTI)*, pp. 221-230 (1990)

[15] Droz, F, *et al*. "Space Passive Hydrogen Maser - Performances and lifetime data," *Proc. 2009 IEEE International Frequency Control Symposium Joint with the 22nd European Frequency and Time forum,* pp. 393-398 (2009)

[16] Seubert, J., Ely, T., and Stuart, J. "Results of the deep space atomic clock deep space navigation analog experiment," *Proc. AAS/AIAA Astrodynamics Specialist Conference*, in press, page numbers awaiting publication (2020)

[17] Ely, Todd. "Advancing Navigation, Timing, and Science with the Deep Space Atomic Clock." In *SpaceOps 2014 Conference*. Pasadena, CA: American Institute of Aeronautics and Astronautics, 2014. https://doi.org/10.2514/6.2014-1856.

[18] Jaduszliwer, Bernardo, and James Camparo. "Past, Present and Future of Atomic Clocks for GNSS." *GPS Solutions* 25, no. 1 (January 3, 2021): 27. https://doi.org/10.1007/s10291-020-01059-x.

[19] [Goujon, D., P. Rochat, P. Mosset, D. Boving, A. Perri, J. Rochat, N. Ramanan, et al. "Development of the Space Active Hydrogen Maser for the ACES Mission." In *EFTF-2010 24th European Frequency and Time Forum*, 1–6, 2010. https://doi.org/10.1109/EFTF.2010.6533730.].

[20] T. M. Hoang, et al. "Micro Mercury Trapped Ion Clock Prototypes with 10^{-14} Frequency Stability in 1-Liter Packages." Scientific Reports 13, no. 1 (June 30, 2023).

[21]"Ground-based active hydrogen maser frequency standard" Microchip MHM2020 data sheet https://www.microchip.com/en-us/products/clock-and-timing/components/atomic-clocks/atomic-system-clocks/mhm-2020-hydrogen-masers

[22] Gurvits, Leonid I. "Space VLBI: From First Ideas to Operational Missions." *Advances in Space Research*, High-resolution space-borne radio astronomy, 65, no. 2 (January 15, 2020): 868–76. https://doi.org/10.1016/j.asr.2019.05.042.

[23] "Chip Scale Atomic Clock" Microchip CSAC-SA65 data sheet, https://www.microchip.com/en-us/product/csac-sa65

[24] "Space qualified cesium beam atomic clock" Cash, P., P. Vlitas, D. Emmons, R. Lutwak, and E. Watts. "Space-Qualified Microprocessor-Controlled Cesium Atomic Clock," 820–25, 2010. http://www.ion.org/publications/abstract.cfm?jp=p&articleID=9198.

[25] "GPS rubidium atomic frequency standard" Excelitas RAFS data sheet, https://www.excelitas.com/product/space-qualified-rubidium-atomic-frequency-standard-clocks

[26] "Galileo passive hydrogen maser" Leonardo PHM data sheet, https://electronics.leonardo.com/en/products/phm-1

[27] "Quartz Ultra-Stable Oscillator" Accubeat USO data sheet, https://www.accubeat.com/uso

[28] "Miniature Atomic Clock", Microchip MAC-SA5X data sheet, https://www.microsemi.com/product-directory/embedded-clocks-frequency-references/5570-miniature-atomic-clock-mac-sa5x

[29] "Chinese cold rubidium atomic frequency standard", Liu, Liang, De-Sheng Lü, Wei-Biao Chen, Tang Li, Qiu-Zhi Qu, Bin Wang, Lin Li, et al. "In-Orbit Operation of an Atomic Clock Based on Laser-Cooled 87Rb Atoms." *Nature Communications* 9, no. 1 (July 24, 2018): 2760. https://doi.org/10.1038/s41467-018-05219-z.

[30] "Pulsed optically pumped rubidium atomic frequency standard" Micalizio, S., F. Levi, C. E. Calosso, M. Gozzelino, and A. Godone. "A Pulsed-Laser Rb Atomic Frequency Standard for GNSS Applications." *GPS Solutions* 25, no. 3 (April 23, 2021): 94. https://doi.org/10.1007/s10291-021-01136-9.

[31] "SpectraDynamics laser-cooled rubidium atomic frequency standard", SpectraDynamics cRb-CLOCK data sheet, https://spectradynamics.com/products/crb-clock/

[32] "Trapped Yb+ ion microwave atomic frequency standard", Zhang, Xianli, Jay Noble, Jonathan Tallant, Hyunwook Park, Michelle Nguyen, Jackie Ellett, Dan Boschen, et al. "Progress on a Tactical Ytterbium Microwave Ion Clock," 1–9, 2023. https://doi.org/10.33012/2023.18685.

[33] "Iodine vapor cell optical atomic frequency standard", Roslund, Jonathan D., Arman Cingöz, William D. Lunden, Guthrie B. Partridge, Abijith S. Kowligy, Frank Roller, Daniel B. Sheredy, et al. "Optical Clocks at Sea." arXiv, August 23, 2023. https://doi.org/10.48550/arXiv.2308.12457.

[34] "1-liter trapped mercury ion atomic frequency standard", Hoang, Thai M., Sang K. Chung, Thanh Le, Sehyun Park, Sung-Jin Park, J. Gary Eden, Christopher Holland, et al. "Micro Mercury Trapped Ion Clock Prototypes with 10^{-14} Frequency Stability in 1-Liter Packages." *Scientific Reports* 13, no. 1 (June 30, 2023): 10629. https://doi.org/10.1038/s41598-023-36411-x.

[35] "Space-based photonic local oscillator – optical to microwave generation", A. Matsko, W. Zhang, private communication.

[36] "Space-based active hydrogen maser frequency standard" Goujon, D., P. Rochat, P. Mosset, D. Boving, A. Perri, J. Rochat, N. Ramanan, et al. "Development of the Space Active Hydrogen Maser for the ACES Mission." In *EFTF-2010 24th European Frequency and Time Forum*, 1–6, 2010. https://doi.org/10.1109/EFTF.2010.6533730.

[37] "Linear Ion Trap atomic frequency Standard with USO LO" Tjoelker, R.L., C. Bricker, W. Diener, R.L. Hamell, A. Kirk, P. Kuhnle, L. Maleki, et al. "A Mercury Ion Frequency Standard Engineering Prototype for the NASA Deep Space Network." In *Proceedings of 1996 IEEE International Frequency Control Symposium*, 1073–81, 1996. https://doi.org/10.1109/FREQ.1996.560296.

[38] "Optical Rubidium Atomic Frequency Standard", Lemke, Nathan D., Kyle W. Martin, River Beard, Benjamin K. Stuhl, Andrew J. Metcalf, and John D. Elgin. "Measurement of Optical Rubidium Clock Frequency Spanning 65 Days." *Sensors* 22, no. 5 (January 2022): 1982. https://doi.org/10.3390/s22051982.

[39] "A Compensated Multi-Pole Linear Ion Trap Mercury Frequency Standard for Ultra-Stable Timekeeping." Burt, Eric A., William A. Diener, and Robert L. Tjoelker. *IEEE Transactions on Ultrasonics, Ferroelectrics, and Frequency Control* 55, no. 12 (December 2008): 2586–95. https://doi.org/10.1109/TUFFC.2008.975.

[40] "Cryogenic Sapphire Oscillator", Dick, G.J., R.T. Wang, and R.L. Tjoelker. "Cryo-Cooled Sapphire Oscillator with Ultra-High Stability." In *Proceedings of the 1998 IEEE International Frequency Control Symposium (Cat. No.98CH36165)*, 528–33, 1998. https://doi.org/10.1109/FREQ.1998.717949.

[41] "Ground-based photonic local oscillator – optical to microwave generation", Fortier, T. M., M. S. Kirchner, F. Quinlan, J. Taylor, J. C. Bergquist, T. Rosenband, N. Lemke, et al. "Generation of Ultrastable Microwaves via Optical Frequency Division." *Nature Photonics* 5, no. 7 (July 1, 2011): 425–29. https://doi.org/10.1038/nphoton.2011.121.

[42] "Strontium Optical Lattice Optical Clock", Ohmae, Noriaki, Masao Takamoto, Yosuke Takahashi, Motohide Kokubun, Kuniya Araki, Andrew Hinton, Ichiro Ushijima, et al. "Transportable Strontium Optical Lattice Clocks Operated Outside Laboratory at the Level of 10−18 Uncertainty." *Advanced Quantum Technologies* 4, no. 8 (2021): 2100015. https://doi.org/10.1002/qute.202100015.
[43] Stuhler, J., M. Abdel Hafiz, B. Arar, A. Bawamia, K. Bergner, M. Biethahn, S. Brakhane, et al. "Opticlock: Transportable and Easy-to-Operate Optical Single-Ion Clock." Measurement: Sensors 18 (December 1, 2021): 100264. https://doi.org/10.1016/j.measen.2021.100264

Journal of Physics: Conference Series **2889** (2024) 012015 doi:10.1088/1742-6596/2889/1/012015

Next-Generation Chip Scale Atomic Clocks

John Kitching, Matt Hummon, William McGehee, Ying-Ju Wang, and Susan Schima

Time and Frequency Division, National Institute of Standards and Technology, MS 847, 325 S. Broadway, Boulder CO 80305

john.kitching@nist.gov

Abstract. We describe three independent research directions focussed on the development of next-generation chip-scale atomic clocks, which combine small size, low power consumption and manufacturability with high frequency stability. The use of optical transitions in microfabricated vapor cells improves both short- and long-term frequency stability to near 10^{-13} at the cost of the added complexity of a chip-scale optical frequency comb. Chip-scale atomic beam microwave clocks have been successfully demonstrated and offer the potential for microsecond-level timing accuracy over one week. And finally, Sr vapor cells have been developed with lifetimes exceeding 250 hours and residual gas pressures below 1 Torr.

1. Introduction

Chip scale atomic clocks [1-3], developed throughout the 2000s and released as a product in 2011 [7], have surpassed 95,000 units in sales [9] and have established themselves as a significant component of the commercial atomic timing market. With a power consumption of under 200 mW and a timing stability of 1 μs over 1 day under controlled environmental conditions [7], this unique combination of properties enables their use in application spaces where GPS is unavailable and battery operation is needed. Chip-scale atomic clocks are finding widespread use in underwater timing for oil and gas exploration [10,11], have been flown in space [12,13] and can improve the performance of satellite-based navigation systems [14,15]. Efforts are underway to reduce production cost, potentially enabling use in an even broader set of applications.

The key technical elements of these clocks are the use of silicon micromachining to fabricate the vapor cell [16], a low-power diode laser as the light source [17] and a thermal suspension system to isolate the physics package from the environment [18]. Together, these elements allow clock operation with significantly lower power consumption [19] than is possible in conventional designs [20].

The commercial success of this technology, as well as new applications requiring a higher level of timing precision, has motivated the search for improvements across a range of parameters, but most importantly in the long-term stability. Chip-scale atomic clocks exhibit a fractional frequency drift of up to 3×10^{-10}/mo [7], limiting their use in applications where accurate time is required over extended periods. In addition, vapor cell clocks possess no intrinsic frequency accuracy due to the large shift of the clock frequency created by the presence of the buffer gas. Periodic recalibrations are therefore required to maintain timing accuracy.

In this Proceedings, we review efforts in our group over the last few years to develop next-generation chip-scale atomic clocks that retain compact form factors, low power consumption and a high degree of manufacturability while simultaneously achieving improved frequency stability, especially over long integration times. Compact optical clocks [5,8,21] have been demonstrated based on the two-photon optical transition in Rb at 778 nm, which incorporate microresonator frequency combs to divide the optical frequency down to the microwave domain. A microfabricated microwave atomic beam clock was also developed that shows promise for microsecond-level timing stability over one week. Finally, we have demonstrated Sr vapor cells, with lifetimes exceeding several hundred hours at elevated temperature.

2. Compact optical clocks

Following the development of the optical frequency comb in the late 1990s and early 2000s [22], tremendous progress has been made in the field of optical clocks, which now achieve fractional frequency uncertainties near 10^{-18}, a factor of 100 better than the best microwave clocks [23]. This improvement is largely driven by the ~ $10^5 \times$ larger carrier frequency, which leads to correspondingly higher resonance quality factors. Optical frequency combs have been implemented using Kerr nonlinearities in on-chip whispering-gallery mode optical resonators [24,25], which has opened the door to the possibility of chip-scale optical clocks.

The two-photon (2-hv) transition in Rb at 778 nm has been investigated for many decades [26-29] as a high-performance optical frequency reference and is identified by the BIPM as a secondary representation of the SI meter [30]. This transition has the advantages of a reasonably narrow linewidth (300 kHz), a small light shift, and accessibility in a simple vapor cell configuration.

2.1. A compact 2-hv optical clock

A compact Rb 2-hv optical clock using a pair of toroidal microresonator frequency combs is shown in Fig. 1(a) [5]. The Rb 2-hv transition is excited using a distributed Bragg reflector laser at 778 nm. The resonance is detected by monitoring the 420 nm fluorescence generated by the atoms decaying to their ground state via a cascade transition through the $6P_{3/2}$ level (see Fig. 1(b)). Locking the DBR laser to the 2-hv transition generated a short-term frequency stability on the optical reference of $4.4 \times 10^{-12}/\sqrt{\tau}$, with a flicker floor near 10^{-13}, as shown in Figure 1(c).

The frequency division is accomplished using two microresonator combs: a small Si_3N_4 resonator driven at 1540 nm produces an octave-spanning comb with a spacing of 1 THz that is self-referenced to control the carrier offset frequency; and a larger SiO_2 resonator driven at 1556 nm produces a comb with a mode spacing of 22 GHz, which spans the frequency gap between two adjacent comb teeth in the 1 THz comb. Phase locks between the comb teeth stabilize the carrier envelope offset frequency for the comb pair. A phase lock of one comb tooth to the Rb DBR laser allows coherent division of the atomically stabilized light at 778 nm down to 22 GHz without adding any additional instability as shown by the overlap of the microwave and optical reference data points in Fig. 1(c).

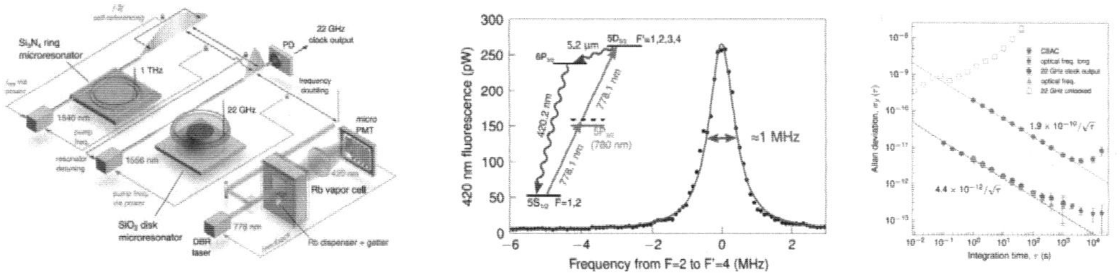

(a) (b) (c)

Fig. 1 (a) Chip-scale optical clock incorporating a Rb 2-hv optical reference and a pair of microresonator optical frequency combs to perform the frequency division. (b) Fluorescence signal from Rb 2-hv optical reference. (c) Allan deviation of the chip-scale optical clock compared to a conventional (microwave) chip-scale atomic clock. (From Ref. [5])

Full integration and manufacturability of such optical clocks is a major remaining challenge. The clock optics require multiple materials platforms, Si_3N_4, SiO_2, and InGaAsP if the lasers are to be included on-chip, as well as frequency doubling capability. Substantial optical power from the lasers may also be required (100 mW at 1540 nm and 160 mW at 1556 nm) due to losses associated with coupling light on and off the chips. In addition, a more highly integrated Rb 2-hv reference would need to be developed, perhaps using photonic integrated circuits [31].

2.2. *High-performance 2-hv optical reference*

The short-term stability of the optical clock in the previous section was limited by intermodulation effects, in which FM noise on the DBR laser at twice the FM modulation frequency used to lock the laser is coupled down to DC. By replacing the DBR laser with an extended-cavity laser with a narrower linewidth, this source of instability could be suppressed as shown in Fig. 2 [8]. In this case, the short-term instability could be reduced to $1.8 \times 10^{-13}/\sqrt{\tau}$, with about half of this instability coming from photon shot noise and the other half due to the intermodulation effect. A flicker floor below 2×10^{-14} was observed out to an integration time of one hour.

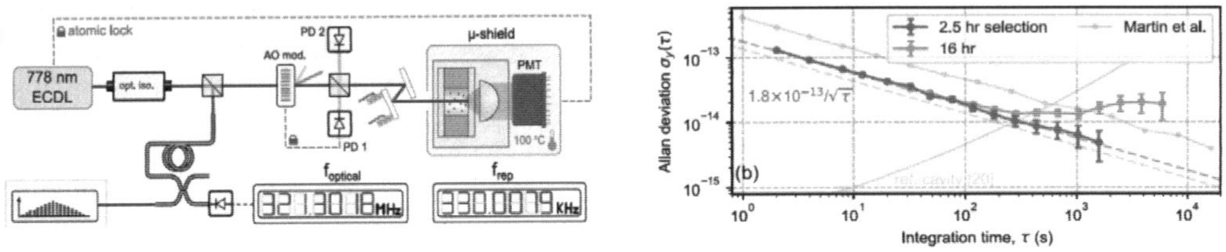

(a) (b)

Fig. 2 (a) Rb 2-hv optical clock operated with an external cavity laser. (b) improved short-term frequency stability is obtained compared to operation with a DFB laser due to the reduction of intermodulation instability caused by laser FM noise. (From Ref. [8])

3. Chip-scale atomic beam clocks

Atomic beam clocks have been a workhorse in precision timing since the 1950s [32,33] and commercial versions of these with beam tube lengths on the order of 15 cm achieve accuracies below 10^{-12} and long-term stabilities below 10^{-13}. Such a clock, if implemented at the chip scale with low power consumption, could therefore be an ideal complement to chip-scale vapor cell clocks even with the degraded stability and accuracy implied by the shorter tube length.

3.1. *Clock design*

A chip-scale atomic beam clock physics package is shown in Fig 3 [4]. The device is 25 mm in length, 19 mm wide and 5 mm thick, and contains a source region containing a thermal vapor of Rb, channels etched in silicon to create a collimated beam [34], and drift region in which the atoms are interrogated. Rb is created in the source region from a Rb_2MoO_4/AlZr pill activated to produce elemental Rb by a high-power laser after the chamber is sealed [35]. Non-evaporable getters pump most gases in the drift region and graphite is used to reduce the background Rb gas pressure in the drift region. The beam flux at a temperature of 90 °C is approximately 5×10^{11} at/s at the output of the collimator.

Journal of Physics: Conference Series **2889** (2024) 012015 doi:10.1088/1742-6596/2889/1/012015

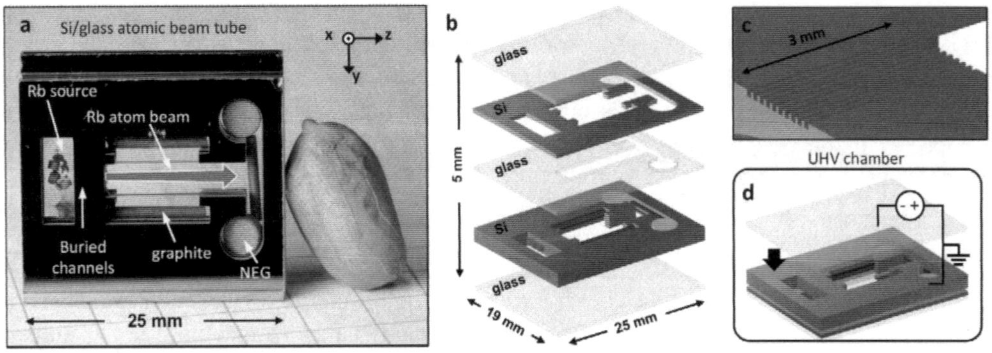

Fig. 3 (a) Chip-scale atomic beam clock photograph showing the Rb source region and the atomic beam propagation direction. (b) Device structure. (c) Channels etched in Si that collimate the atomic beam. (d) Fabrication process in which glass is anodically bonded to the top silicon surface to form an evacuated package. (From Ref. [4])

3.2. Frequency stability

Atoms in the atomic beam are excited using Ramsey coherent population trapping [36,37]. The atomic florescence detected in the second Ramsey zone shows typical thermal beam Ramsey fringes with a width of 12 kHz, as shown in Fig. 4(a), determined by the transit time of the atoms between the zones. Stabilization of a microwave oscillator using this signal results in a sort-term instability of $1.2 \times 10^{-9}/\sqrt{\tau}$ out to 100 s, as shown in Fig. 4(b). While this stability is quite poor (worse even than the stability of a commercial chip-scale atomic clock), significant improvements are anticipated with future work, for example, through improving the collection efficiency of the florescence light. It is expected that the long-term stability at integration times beyond one day, will be considerably better than a vapor cell CSAC if systematic effects such as the AC Stark shift and Rb-Rb collision shift can be adequately controlled.

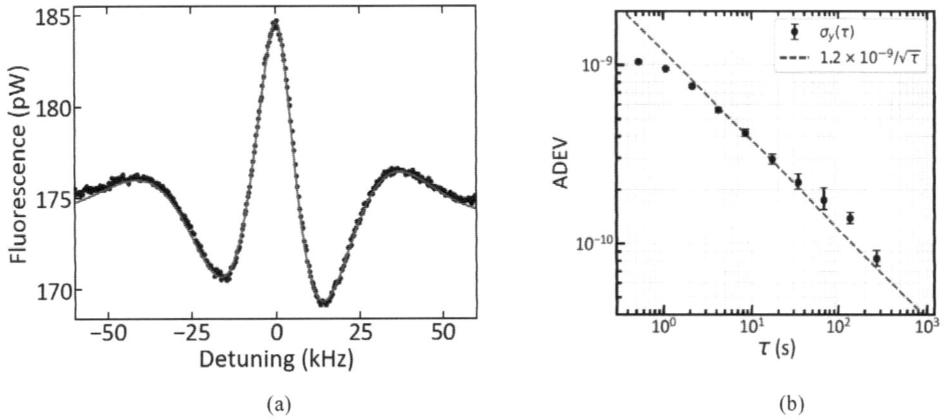

Fig. 4 (a) Ramsey fringes generated by coherent population trapping in a microfabricated atomic beam clock with a Ramsey zone of 1 cm. (b) Clock stability. (From Ref. [4])

4. Strontium vapor cells

Alkaline earth atoms are of high interest for optical frequency references due to the long coherence times of some transitions in these species. These atoms are the basis for some of the most stable and accurate optical lattice clocks currently under development [23]. Alkaline earth atoms have transitions that span a broad range of linewidths, from several tens of megahertz to the millihertz range. This makes them interesting for references based on vapor cell or atomic beams [38], where transit time broadening in centimetre-scale geometries results in minimum linewidths in the range of 10 kHz - 100 kHz. The low vapor pressure of alkaline earth atoms means that temperatures near 300 °C are required for optical thicknesses approaching unity. At this elevated temperature alkaline earth atoms are known to damage most glasses typically used as windows. As a result, fairly complex experimental systems such as hollow cathode lamps or heat pipes are traditionally used.

4.1. Cell fabrication

Sr vapor cells, shown in Fig. 5(a) were fabricated using anodic bonding processes similar to those used to make alkali vapor cells. The interior windows of the cell were coated with Al_2O_3 [39] to prevent interaction of the Sr with the glass as shown in Fig. 5(b). Sr metal was inserted into the cell cavity in a dry nitrogen environment and the filled cavity was quickly transferred into an anodic bonding system, which was subsequently evacuated for sealing under vacuum.

4.2. Absorption spectroscopy

After fabrication, absorption spectroscopy was carried out at temperatures above 300 °C to confirm the presence of Sr vapor in the cell. A typical absorption line in a counterpropagating geometry is shown in Fig. 5(c). A key question is the extent to which the Al_2O_3 coating protects the cell windows over extended periods. This was assessed by measuring the transmission through the vapor cell (at 461 nm) over many days; when alkaline earth atoms react with the glass walls, the glass becomes opaque, and the optical transmission is reduced. No measurable change in the transmission was observed over 250 hours at a temperature above 300 °C.

The width of the narrow Doppler-free feature near the line centre was used to assess the pressure of residual gases in the cell, since such gases broaden optical transitions in alkaline earth atoms. By measuring the linewidth of the transition at 461 nm as a function of optical power, and extrapolating to zero power, an upper limit on the residual gas pressure was established of ~ 50 Pa (0.4 Torr), consistent with zero. Further measurements on narrower transitions (for example of the 7.5 kHz-wide transition at 689 nm) are expected to provide more stringent limits on the presence of any residual gas.

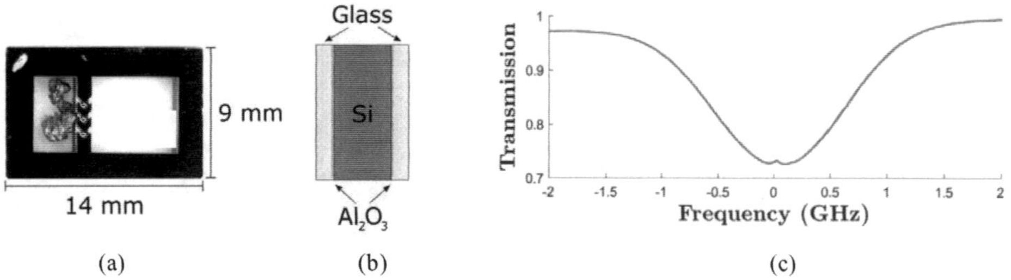

Fig. 5 (a) Photograph of a microfabricated Sr vapor cell. (b) Cell design: Al_2O_3 coatings on the interior surfaces of the vapor cell protect the cell windows from the corrosive Sr vapor. (c) Absorption spectrum at 461 nm at a temperature near 320 °C with counterpropagating optical fields. A Doppler-free feature is observed at the line centre. (From Ref. [6])

5. Conclusion

Next-generation chip-scale atomic clocks based on optical transitions in atoms or atomic beams promise improved frequency stability at both short and long integration times while retaining the low power consumption and manufacturability of the original technology. The proof of principle demonstrations of such clocks described here provide a foundation for future work to enhance the engineering and manufacturing maturity. While considerable work is still needed to achieve product prototypes,

Journal of Physics: Conference Series **2889** (2024) 012015 doi:10.1088/1742-6596/2889/1/012015

especially with regard to thermal engineering for low power consumption, such clocks would likely make considerable impact on real-world applications if successfully commercialized.

Distribution Statement A – Approved for Public Release, Distribution Unlimited. This research was developed with funding from the Defense Advanced Research Projects Agency (DARPA).

References

[1] Knappe S, Shah V, Schwindt P D D, Hollberg L, Kitching J, Liew L A, and Moreland J 2004 *Appl. Phys. Lett.* **85**, 1460
[2] Lutwak R et al. 2007 *Proc. Precise Time and Time Interval (PTTI) Meeting*, Long Beach, CA, 269
[3] Kitching J 2018 *Applied Physics Reviews* **5**, 031302
[4] Martinez G D, Li C, Staron A, Kitching J, Raman C, and McGehee W R 2023 *Nat. Commun.* **14**, 3501
[5] Newman Z L et al. 2019 *Optica* **6**, 680
[6] Pate J M, Kitching J, and Hummon M T 2023 *Opt. Lett.* **48**, 383
[7] Lutwak R 2011 *Proc. Precise Time and Time Interval (PTTI) Meeting*, Long Beach, CA, 207
[8] Newman Z L, Maurice V, Fredrick C, Fortier T, Leopardi H, Hollberg L, Diddams S A, Kitching J, and Hummon M T 2021 *Opt. Lett.* **46**, 4702
[9] Cash P, Krzewick W, Machado P, Overstreet K R, Silveira M, Stanczyk M, Taylor D, and Zhang X 2018 *Proc. 2018 European Frequency and Time Forum (EFTF)*, 65
[10] Gardner A T and Collins J A 2012 *Proc. MTS/IEEE Oceans Conference*, Hampton Roads, VA, 1
[11] Gardner A T and Collins J A 2016 *Proc. MTS/IEEE Oceans Conference*, Monterey, CA, 1
[12] Lezius M et al. 2016 *Optica* **3**, 1381
[13] CAPSTONE Mission, 2023 https://advancedspace.com/capstone-30aug23-update/
[14] Krawinkel T and Schön S 2015 *Proc. ION GNSS*, 2867
[15] Krawinkel T and Schon S 2016 *GPS Solutions* **20**, 687
[16] Liew L A, Knappe S, Moreland J, Robinson H, Hollberg L, and Kitching J 2004 *Appl. Phys. Lett.* **84**, 2694
[17] Chantry P J, Liberman I, Verbanets W R, Petronio C F, Cather R L, and Partlow W D 1996 *Proc. IEEE Int. Freq. Cont. Symp.*, 1002
[18] Mescher M J, Lutwak R, and Varghese M 2005 *Proc. Solid-State Sensor, Actuator and Microsystems Workshop*, Seoul, Korea, 311
[19] Lutwak R et al. 2004 *Proc. Precise Time and Time Interval (PTTI) Meeting*, Washington, DC, 339
[20] Vanier J and Audoin C 1992 *The Quantum Physics of Atomic Frequency Standards* (Bristol: Adam Hilger)
[21] Maurice V et al. 2020 *Opt. Exp.* **28**, 24720
[22] Diddams S A et al. 2000 *Phys. Rev. Lett.* **84**, 5102
[23] Ludlow A D, Boyd M M, Ye J, Peik E, and Schmidt P O 2015 *Rev. Mod. Phys.* **87**, 637
[24] Del'Haye P, Arcizet O, Schliesser A, Holzwarth R, and Kippenberg T J 2008 *Phys. Rev. Lett.* **101**
[25] Kippenberg T J, Holzwarth R, and Diddams S A 2011 *Science* **332**, 555
[26] Cagnac B, Grynberg G, and Biraben F 1973 *J. Phys.* **34**, 845
[27] Poulin M, Latrasse C, Cyr N, and Tetu M 1997 *Phot. Tech. Lett.* **9**, 1631
[28] Edwards C S, Barwood G P, Margolis H S, Gill P, and Rowley W R C 2005 *Metrologia* **42**, 464
[29] Burke J, in *SPIE Photonics West* (San Francisco, 2016).
[30] Felder R 2005 *Metrologia* **42**, 323
[31] Hummon M T et al. 2018 *Optica* **5**, 443
[32] Vanier J and Audoin C 2005 *Metrologia* **42**, S31

[33] Essen L and Parry V I 1955 *Nature* **176**, 280
[34] Li C, Chai X, Wei B, Yang J, Daruwalla A, Ayazi F, and Raman C 2019 *Nat. Commun.* **10**, 1831
[35] Douahi A, Nieradko L, Beugnot J C, Dziuban J, Maillote H, Guerandel S, Moraja M, Gorecki C, and Giordano V 2007 *Elec. Lett.* **43**, 279
[36] Thomas J E, Hemmer P R, Ezekiel S, Leiby C C, Picard R H, and Willis C R 1982 *Phys. Rev. Lett.* **48**, 867
[37] Hemmer P R, Shahriar M S, Lamelarivera H, Smith S P, Bernacki B E, and Ezekiel S 1993 *J. Opt. Soc. Am. B* **10**, 1326
[38] Olson J, Fox R W, Fortier T M, Sheerin T F, Brown R C, Leopardi H, Stoner R E, Oates C W, and Ludlow A D 2019 *Phys. Rev. Lett.* **123**, 073202
[39] Woetzel S, Talkenberg F, Scholtes T, Ijsselsteijn R, Schultze V, and Meyer H G 2013 *Surf. Coat. Tech.* **221**, 158

Journal of Physics: Conference Series **2889** (2024) 012016 doi:10.1088/1742-6596/2889/1/012016

Compact clocks & cavities for space and ground applications

P Gill, S Mulholland, H Klein, G Huang, I Hill, B Allen, M Haji, J Stacey, A Spampinato, P Tsoulos, G Hockley, N Vincent, M Knapp, M Trigatzis and G Barwood

National Physical Laboratory, Hampton Road, Teddington, UK

E-mail: patrick.gill@npl.co.uk

Abstract. Position, navigation and timing capability worldwide is predominantly underpinned by global navigation satellite systems (GNSS), which provide the underlying capability for critical national infrastructures, science, industry, commerce and society. Recently, the vulnerability of such systems to interference, whether intentional or not, instrument failure or space weather, has been recognised, pointing to the need for alternative independent navigation and timing solutions. In particular, the availability of low size, weight and power (low-SWaP) traceable portable clocks with extended holdover times for use both in space and on ground is seen as paramount. We report on progress in the development of compact cold ion microwave and optical clocks at NPL that point to capability to out-perform existing commercial clock systems and identify future space applications that could benefit from this technology.

1. Introduction

High accuracy timing and positioning provided by global navigation satellite systems (GNSS) are now increasingly essential for economies in respect of transport, logistics, just-in-time manufacturing, retail, high-speed high-volume data transfer, communications and high frequency trading. They also play a major role in critical national infrastructure (CNI) such as emergency services, defence and security, synchronisation of telecommunication networks and power grids due to the GNSS potential vulnerability to spoofing and jamming, space weather and space clock failure. In the latter respect, there have been some temporary GNSS outages due to clock problems on some constellations [1]. Satellite clocks with upgraded designs providing more accuracy and robustness would also help reduce "fly-by" range errors of ~ 10 m to ~ 1 m with well-calibrated receivers in fixed locations. However, even with improved clock accuracy, other uncertainties such as satellite orbit drift, and ionospheric and tropospheric contributions also provide limitations. Next generation clocks, both on satellite and ground segment, offer improved holdover times and faster signal re-acquisition in the event of signal interruption.

Current microwave GNSS satellite clocks and commercial ground-segment clocks can provide holdover for short periods (e.g., hours to a few days, dependent on the allowable integrated error build-up) in situations where GNSS disciplining is lost. NPL are developing compact laser-cooled ion holdover clocks, both microwave and optical, that could out-perform industry-standard commercial Cs clocks, and challenge commercial maser capability but with smaller size, weight and power (SWaP). NPL systems include a portable laser-cooled trapped $^{171}Yb^+$ multi-ion microwave system [2], as well as a portable laser-cooled single $^{88}Sr^+$ ion optical clock [3-6] that employs a multi-wavelength dual-axis cubic cavity-stabilised clock control unit [7-9] suitable for space or ground deployment. The single $^{88}Sr^+$ ion physics package and cubic cavity sub-system are being developed under European Space Agency (ESA) studies for future resilient ground or space segment portable clocks with long holdover times. Further, the use of lasers stabilised to high stability reference cavities in low or medium Earth orbits (LEO or MEO) offer GNSS constellation augmentations by reducing the significant level of satellite clock updating from ground, as proposed in the Kepler idea [10], as well as offering space clocks and ultra-stable lasers for Earth observation and space science. Such space-based ultra-stable lasers in the first instance, followed by full optical clock systems in a later epoch, can feed directly into Earth

Journal of Physics: Conference Series **2889** (2024) 012016 doi:10.1088/1742-6596/2889/1/012016

observation (e.g. next generation gravity missions (NGGM) for climate change studies) and space-based fundamental physics (e.g. gravitational wave observation via laser interferometer space antenna (LISA)). These applications are briefly discussed at the end of this paper.

Cold atom and ion optical clocks in some National Metrology Institutes (NMIs) and other research laboratories have achieved uncertainties at the 10^{-18} fractional frequency level under well-controlled lab conditions, representing a x100 improvement over state-of-the-art microwave Cs fountain clocks. These results point the way towards the potential of future portable high accuracy optical timing references with extended holdover times and resilience, recognising that low-SWaP remote clock operation in remote and/or mobile situations may well see some trade-off in performance specification against environmental conditions, whether in space or on ground.

2. Cold ion microwave clock

Currently, the laser-cooled Cs fountain clock is the primary standard for realising the SI second, with a fractional frequency uncertainty in the range of ~ 1 – 2 x 10^{-16}, but with a large SWaP footprint coupled with the need for good control of e.g. tilt from vertical, and it is not suited to operating at full capability on moving platforms or in remote locations. As a result, there is interest in alternative microwave clocks that have better stability & accuracy than e.g. 5071 thermal beam Cs clocks and can compete with maser systems but with much lower SWaP and cost. To this end, we are addressing the development of a rack-mountable laser-cooled multi-ion $^{171}Yb^{+}$ 12.6 GHz microwave clock [2] capable of mounting within a 3U or 4U Eurorack arrangement. The 12.6 GHz clock transition frequency operates between the $^{171}Yb^{+}$ ion two hyperfine ground states with magnetically insensitive Zeeman levels to first order, as shown in figure 1. The ions are created via photo-ionisation of a weak thermal ytterbium atomic beam. Photo-ionisation is achieved by exciting Yb neutral atoms to the $^{1}P_{1}$ resonance level via 399 nm light, followed by 369 nm light to promote the atoms to the ion ground state. The resulting ions are trapped by means of a linear 4-rod ion trap and laser-cooled with 369 nm laser light on the Yb^{+} $^{2}S_{1/2} – {}^{2}P_{1/2}$ resonance transition. The ions can decay to the $^{2}D_{3/2}$ level with a branching ratio of 1:200, thereby leaking from the cooling cycle. In order to recover those ions back into the cooling cycle, repump light at 935 nm is used. Occasional ion loss into the long-lived low-level $^{2}F_{7/2}$ metastable state can also be recovered by driving the ions with light at 760 nm (see figure 1), though this has not been found necessary so far with this multi-ion system. Early results with a 6U system demonstrated a 12.6 GHz clock transition central Zeeman component linewidth of ~0.2 Hz and short term stability of 3.6 x10^{-12} $\tau^{-1/2}$, reaching 7 x10^{-14} at 1500 s [2].

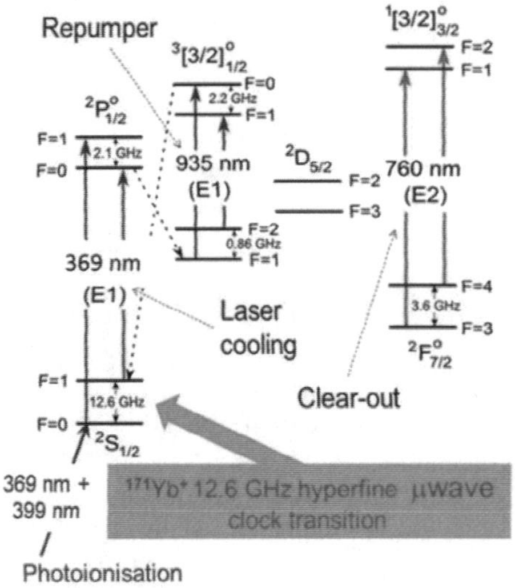

Figure 1: $^{171}Yb^{+}$ ion energy level diagram

Recently, a new lower-SWaP trap physics chamber with dimensions 110 mm x 60 mm x 60 mm has been developed, demonstrating cold Yb^+ ion crystals comprising more than 10,000 ions at mK temperatures. The ions demonstrate laser-cooled storage lifetimes of a few weeks per single load and capable of auto-enabled fast reload when necessary. The linear trap comprises rods of 6 mm diameter and an ion axis-rod spacing of 2.6 mm and is driven symmetrically at 1.4 MHz. Ultra-high-vacuum (UHV) pressures < 10^{-10} mbar are maintained by means of a small sputter ion pump external to the trap chamber and internal non-evaporable getters. Initial operation was demonstrated by monitoring 369 nm cooling fluorescence as a probe microwave frequency was ramped across the 12.6 GHz clock transition frequency. Rabi profiles (figure 2a) of the clock transition with linewidths of ~ 20 Hz full width at half-maximum (FWHM) observed on the central Zeeman component without any magnetic shielding, limited by coherence time due to a magnetic field gradient resulting from the temporary use of a permanent magnet to split the ground state Zeeman components. Currently, custom mu-metal shields and magnetic field coils are being installed to allow nulling of the Earth's field and dispensing with the permanent magnet, which will allow the recovery of sub-Hz linewidths previously observed. In addition to the Rabi profiles, preliminary Ramsey fringes (again with no shielding and a permanent magnet) with 5 ms $\pi/2$ pulses and a 15 msec dark time are shown in figure 2b.

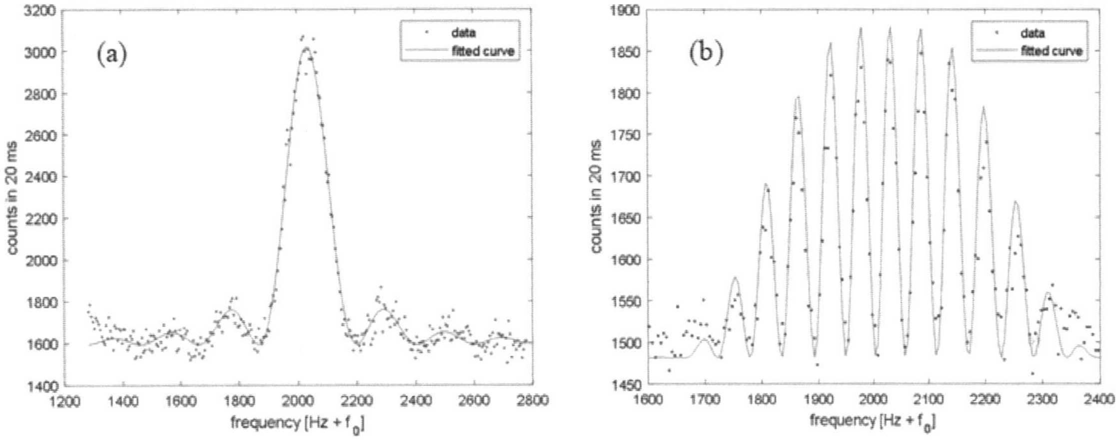

Figure 2: 12.6 GHz clock transition preliminary Rabi (a) and Ramsey (b) profiles; Rabi data points are single shot, measurements, Ramsey data are averages of 4 single shot measurements.

3. Cold ion optical clock

The improved high accuracy of state-of-the-art cold atom and ion optical clocks over Cs fountain clocks is due in part to the much higher frequency (~ hundreds of THz) of optical clocks, combined with the use of a weak narrow linewidth "forbidden" clock transition. Here, Doppler and collisional broadening effects are minimised by laser cooling and operation under UHV. Natural linewidths for these weak transitions in the Hz to mHz range are typical. By use of a very high finesse optical cavity to pre-stabilise a laser to similarly narrow linewidths, the laser can then be frequency-stabilised to the cold ion clock transition, by monitoring a reduction in the strong single ion cooling fluorescence when the clock laser is resonant with the clock transition. As a result, the international frequency standards community is preparing for an optical redefinition of the SI second within the next decade [11].

One of the optical systems under investigation at NPL is the laser-cooled $^{88}Sr^+$ single strontium ion optical clock. This has a relatively simple energy level scheme, with accessible clock, cooling and auxiliary transitions in the visible & near infra-red where DFB lasers are readily available. The strontium ion quadrupole clock transition at 674 nm is already one of the radiations internationally accepted as an optical secondary representation of the second. The partial term scheme of the $^{88}Sr^+$ ion is shown in

Journal of Physics: Conference Series **2889** (2024) 012016 doi:10.1088/1742-6596/2889/1/012016

Figure 3. We have already demonstrated frequency agreement between two $^{88}Sr^{+}$ single ion clocks at a level of a few x10^{-17} [3] using an early single ion trap design. More recently, we have developed an enhanced end-cap trap design [4] that makes use of improved materials and geometry, providing reduced sensitivities to environmental temperatures, important for minimising black body radiation frequency shifts, and low phonon heating rates. Current work is underway to increase robustness and autonomy of the strontium ion trap physics package to a Technology Readiness Level of 5/6 by making use of finite element analysis simulations. The elements of this design have already been incorporated into other trap systems elsewhere. It is anticipated that this single ion trap design together with all necessary diode lasers for cooling, repumping and probing the clock transition should be possible within a 3U Eurorack system, including the necessary reference cavity sub-system for pre-stabilising all the lasers.

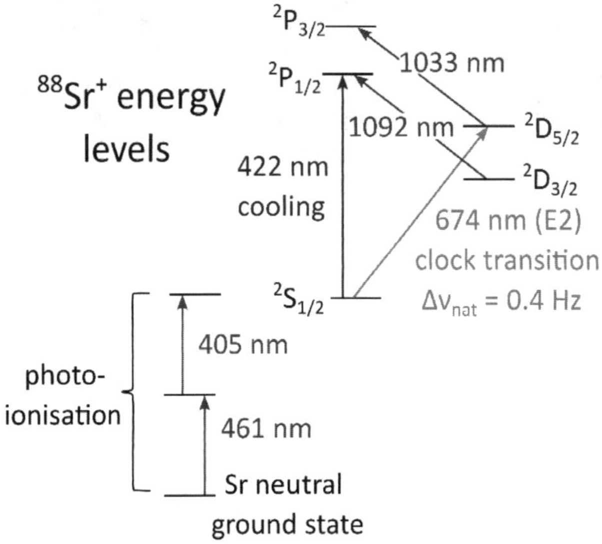

Figure 3: $^{88}Sr^{+}$ ion energy level partial scheme

The $^{88}Sr^{+}$ ion energy level scheme is shown in Figure 3. A single $^{88}Sr^{+}$ ion is loaded into an end-cap trap via photo-ionisation of a weak Sr atomic beam, by means of 461 nm exciting the atom to the $^1P_1$ atomic resonance level, with photo-ionisation to the ion continuum via a 405 nm broadband laser. Laser cooling takes place at 422 nm with light currently generated by frequency doubling an 844 nm extended cavity diode laser (ECDL) or directly with a 422 nm ECDL. The ion decays to the $^2D_{3/2}$ metastable level with a branching ratio of 1:13 and is recovered to the cooling cycle by repumping with a 1092 nm DFB diode. The $^2S_{1/2} - {}^2D_{5/2}$ quadrupole clock transition at 445 THz has a natural linewidth of 0.4 Hz and is driven with the 674 nm cavity-stabilised diode and detected via the level of quantum jumps in the cold ion 422 nm fluorescence. The ion is then recovered to its ground state using a 1033 nm distributed feedback laser (DFB) to clear out the $^2D_{5/2}$ metastable level, allowing the next cooling, state preparation, clock probe and detection sequence to follow.

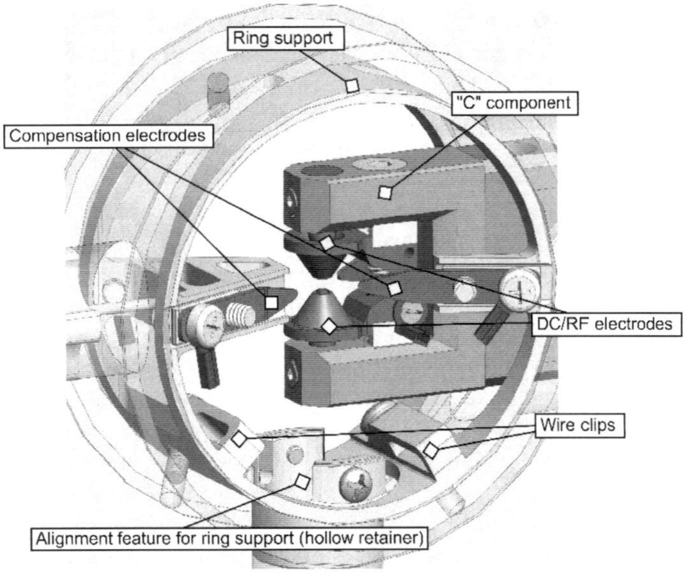

Figure 4: End-cap ion trap assembly

Figure 4 shows the end-cap ion trap assembly. The RF electrodes have 1 mm diameter and 1 mm separation, with a specific drive frequency of 14.4 MHz, where the RF motional Doppler shifts can be offset against electric field Stark shifts of opposite sign [12]. Whilst based on the single ion trap design given in [4], the design is being adapted for future space operation by focusing on robustness, ability to survive vibration and shock during spacecraft launch and satellite deployment. This includes the maintenance of functional performance in micro-gravity within the various thermal and radiation environments encountered in orbit. Finite element analysis (FEA) had already been carried out on the early trap design to assess the temperature increase in the immediate electrode structure due to dielectric loss in the small amounts of insulators within the trap, resulting in ~ 0.14 K temperature rise in the vicinity of the ion [5]. Recently, FEA of the clock system physics package has been undertaken to assess the trap design's ability to cope with vibration and shock levels encountered during launch and thermal vacuum operation during orbit. In this latter respect, it is recognised that significant thermal variations encountered during the satellite's orbit are partially mitigated by means of an internal payload platform together with the vacuum package barrier, both of which would be subject to separate nested thermal servocontrol.

Comprehensive FEA used COMSOL software to examine the structural integrity of the ion trap vacuum package under quasi-static acceleration, sinusoidal and random vibration and mechanical shock. These different categories cover the range of mechanical environments encountered during launch, stage separation, solar panel set-up and in-orbit clock operation. In this instance, a Falcon 9 rocket launch was assumed [6]. As regards target orbit, this will depend on the mission choice, ranging from e.g. an in-orbit demonstrator in low Earth orbit (LEO) through to Lagrange point mission locations, where the particular radiation and thermal vacuum environments in each case will need to be considered.

Expected axial and lateral quasi-static accelerations relevant to a Falcon 9 launch are estimated from the analysis as 7.5 *g* and 2.5 *g*. These FEA simulations [6] have shown that the predicted stress and deflection values for the ion trap and its mounting assembly are safely within the limits for fracture or non-reversible deformation. For sinusoidal vibration in the range 5 Hz - 100 Hz, the maximum calculated stress was found to be on the RF feedthrough rod, which was below that calculated for the quasi-static case. For random vibration in an applied frequency spectrum of 20 Hz – 2 kHz, maximum rms accelerations of ~ 3 g and 2 g were derived for the vacuum chamber and ion trap electrodes respectively. Taking a 3σ for the zero-to-peak value again results in a peak acceleration below that determined for the quasi-static case. High g shock values incident for millisecond times due to stage separation and pyrotechnic bolt firing for opening solar panels suggest a worst case for the DC electrodes and the C-support rod, but again the simulations did not indicate fracture or material deformation.

Thermal analyses on the ion trap design were carried out to simulate (a) operational conditions in respect of trap component heating due to an applied 300 V RF potential, (b) the time-dependent operation of a resistively-heated custom strontium mini-oven for generating a weak Sr atomic beam needed to load a single ion into the trap, and (c) the time necessary for the trap structure temperature to reach equilibrium during loading. The resulting temperature increase of the trap due to RF heating during normal operation indicated a maximum increase of 0.2 °C above ambient, with associated black-body radiation shift at the 1.5 x10^{-18} fractional frequency level. The Sr mini-oven Ta tube demonstrated reaching an on-demand peak operating temperature of 300 °C after 50 s from turn-on, generating sufficient flux of Sr atoms to load the trap, followed by a few seconds 1/e decay time back to ambient temperatures. These simulations demonstrate that occasional loss of the Sr^+ ion can be mitigated by automatic reload of a new ion within 2 minutes, where the clock software would also apply an operational frequency data acquisition block for the same period. Finally, simulations were carried to demonstrate that the combined ion trap and vacuum chamber design was capable of adhering to the thermal vacuum space qualification requirements. ESA qualification requires non-operational temperature cycling of payload equipment over the temperature range of -50 °C to +70 °C with a 20 °C per minute rate change and 4-hour dwell times. The simulations will be compared with data from platinum resistance thermometers at various locations on the vacuum chamber and on the trap electrodes.

Whilst these simulations underpin the robustness of the ion trap design, the trap hardware will of course need to undergo full environmental testing in due course in order to achieve space qualification for a full mission. Prior to and after this, full functional and operational performance of the ion trap physics package within the laboratory will be undertaken, in order to demonstrate reproducible $^{88}Sr^+$ optical clock system frequency stability and uncertainty.

4. Cubic cavity optical reference systems

In order for the optical clock to provide timing and frequency outputs across the visible and microwave/rf outputs at e.g. 10 MHz and 1 pulse per second, additional sub-systems such as optical cavities and optical frequency combs are needed. In particular, we have developed compact portable low-SWaP optical reference cavities based on our cubic cavity design [7] for stabilising not only the clock laser but also all the necessary cooling and auxiliary lasers needed to run either the strontium ion or the strontium neutral optical lattice clock [8]. The cubic cavity design makes use of a 50 mm spacer-cube of ultra-low-expansion (ULE) glass with zero expansivity close to room temperature and held in a novel tetrahedral mounting arrangement within a temperature-controlled chamber evacuated by a small sputter ion pump. The internal cavity spacer volume is also evacuated by the ion pump by means of one of three perpendicular bores, each centred on opposite pairs of cube faces and the cube mid-point. The clock laser cavity-axis mirrors are optically contacted to opposite faces of the cube, concentric with one of the bores. Cavity finesses in the range of 200,000 are achieved with very high reflectivity mirror coatings. The design is both vibration-insensitive and holding force-insensitive, providing leading passive acceleration sensitivities in all three orthogonal directions, with axial fractional frequency sensitivity of 2.45 x10^{-11} per g [7]. The single-axis cubic cavity has already been subject to vibration / shock and irradiation testing, rendering it suitable for both space and mobile ground deployment [9].

As mentioned above, we have also developed a dual-axis cubic cavity system with cavities organised across two of the orthogonal cube directions. As per the single-axis case, Pound-Drever-Hall (PDH) pre-stabilisation of the clock laser is achieved by servo control to a cavity with the narrow-band very-high-finesse mirror coatings, resulting in a laser linewidth and short-term stability floor at the Hz-level. Typical ULE cavity isothermal drift rates close to the zero expansivity temperature have been measured in the region of ~ 20 mHz·s^{-1}. The cooling and auxiliary lasers are stabilised to a cavity on one of the orthogonal cube axes with a wideband mirror coating of variable cavity finesses in the range 1,000 to 10,000, where the individual wavelength PDH cavity signals are recovered via different PDH modulation frequencies. This dual-axis cubic cavity is an efficient arrangement providing an opto-electronic clock control unit with reduced footprint and avoiding the need for multiple cavities. One such dual-axis cubic cavity with 1064 nm and 1397 nm cavities (the latter the sub-harmonic of the Sr neutral clock frequency) has already been subject to structural testing under both vibration and shock loads. For the sinusoidal case, vibration levels were tested to 20 *g* accelerations for the worst low frequency case (5 - 20 Hz) with frequency responses monitored to 140 Hz. Rms random acceleration levels up to 19 *g* were applied. Shock loads targeting 450 *g* for 0.2 ms demonstrated a response spectrum ranging from 4 *g* at 10 Hz to 780 *g* at 6 kHz, with no observed change in the sinusoidal system dynamic frequency response monitored. Together with the frequency noise power spectral density measured for a 1064 nm laser stabilized to the dual axis cubic cavity, these data show good correspondence with the NGGM target performance.

5. Ongoing development and applications

The development of autonomous robust and resilient atomic clocks and ultra-stable lasers for both space and ground applications is ongoing and benchmarked by technology readiness levels and also, in the space segment, space qualification requirements. The cubic cavity multi-wavelength clock control unit for space application, currently at a technology readiness level TRL4, is being taken forward towards TRL6 within an ESA NavISP programme. It is anticipated that this activity can integrate with the ESA trapped ion space optical clock development discussed above, focused on a possible in-orbit demonstrator in a few years' time. Additionally, there is potential for space optical clocks and/or cubic-cavity stabilized laser systems to address the Kepler idea by the location of a few units in LEO and able to optically communicate with each other and MEO GNSS satellites, thus allowing more accurate satellite orbit data with a much-reduced need for multiple update ground locations.

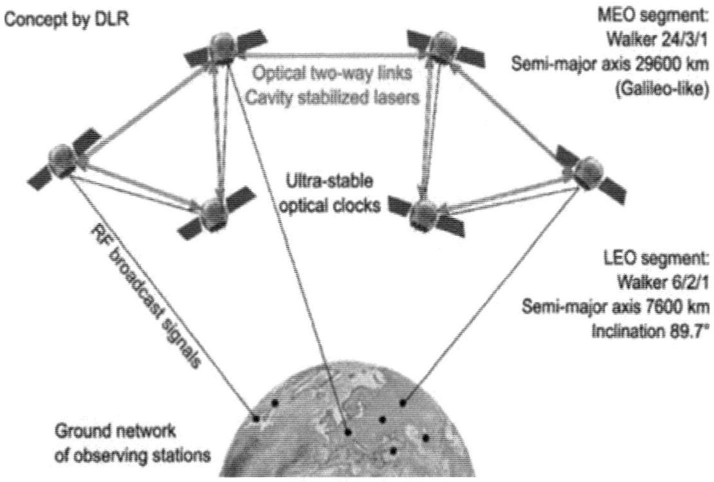

Figure 5: German Space Agency (dlr) concept of the Kepler navigation idea [9], C Guenther / dlr copyright.

One of the earliest mission opportunities for a cubic cavity-based ultra-stable laser interferometer is the ESA Next Generation Gravity Mission (NGGM), part of the NASA/ESA joint Mass Change &

Geophysics International Constellations (MAGIC) scheduled for the late 2020s. This will build on previous campaigns focused on mapping the Earth's gravity field by means of a precision 1064 nm interferometer system, in this case, between a leading and trailing spacecraft pair separated in orbit by ~ 200 km and capable of sensing the differential movement between the spacecraft as they sample different ground terrain during the spacecraft orbit [13]. This data can be used to provide up-to-date geoid mapping with improved resolution, important for climate change studies of ocean circulation and global ice mass coverage. The NGGM/MAGIC collaboration will involve two separate satellites pairs traversing different LEO orbits relative to each other in the "Bender" configuration. The ESA pair is focused on an ultra-stable laser interferometer servoed to NPL cubic cavity systems in the current ESA Phase B study.

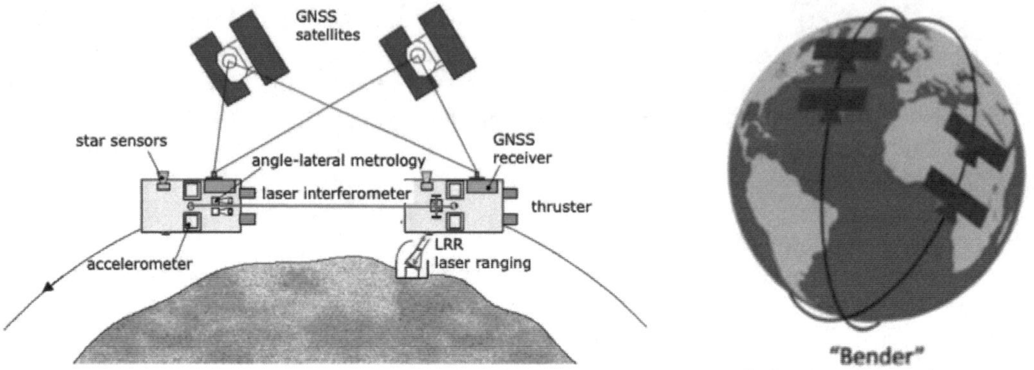

Figure 6: Measurement principle of NGGM (left) and "Bender" constellation (right), left diagram reproduced from [14] and right from [13]

Looking beyond NGGM, the NASA/ESA Laser Interferometer Space Antenna (LISA) for gravitational wave detection from space is provisionally scheduled for launch in the 2030s. This will involve three spacecraft in a two-arm or joined-up three-arm interferometric arrangement trailing about 1 million km behind Earth at the L1 Lagrange point, with spacecraft interferometer arm separations of 2.5 million km. There is potential for the use of cubic cavities either in the spacecraft laser stabilization units or in the ground support system [15]. Follow-on missions for the 2040s for both NGGM and LISA and other fundamental physics missions focused on searches for dark matter and the unification of quantum mechanics and gravity, would have opportunities to make use of high precision optical clocks, assuming that these have reached the requisite level of performance and space qualification by that time.

Finally, we point to the potential for "master" optical clocks in space [16]. Leading optical clocks are now able to detect gravitational redshifts (GRS) at the 10^{-18} fractional frequency level per cm height difference of clocks at ground level. This needs to be accounted for in global international clock comparisons that will maintain the 10^{-18} uncertainty throughout the comparison, placing strong demand on knowledge of the height difference to sub-cm precision between the remote clocks. One way of circumventing this would be to locate a few state-of-the-art globally-distributed optical clocks in geostationary orbits, where the GRS is much reduced over that on the Earth's surface and direct line-of-sight optical communication between master clock satellites is available. These master clocks could then provide reference frequencies to lower orbit satellite clocks and ground clocks, with appropriate GRS adjustment. Whilst this scenario is not imminent, perhaps one can point to it gaining traction by the 11th or 12th Frequency Standards & Metrology Symposia.

Acknowledgements
The Authors would like to acknowledge funding from the UK Government Department of Science, Industry and Technology for the Yb^+ microwave clock, and the European Space Agency in the TEC

(engineering) and Navigation Directorates under the GSTP and NAVISP programs respectively, together with the UK Space Agency, for Sr^+ ion optical clock and cubic cavity funding.

References

[1] www.esa.int/Applications/Navigation/Galileo_clock_anomalies_under_investigation
[2] Mulholland S *et al* 2019, Laser-cooled ytterbium-ion frequency standard, *App. Phys.* B **125** 198
[3] Barwood G P *et al* 2014 Agreement between two $^{88}Sr^+$ optical clocks to 4 parts in 10^{17}, *Phys. Rev.* A **89** 050501(R)
[4] Nisbet-Jones P B R *et al* 2016, A Single-Ion Trap with Minimized Ion-Environment Interactions, *App. Phys.* B **122** 57
[5] Dolezal M *et al* 2015 Analysis of thermal radiation in ion traps for optical frequency standards, *Metrologia* **52** 842
[6] Spampinato A *et al* 2024, An ion trap design for a space-deployable strontium-ion optical clock, *Proc. R. Soc.* A **480** 20230593
[7] Webster S and Gill P 2011, Force-insensitive optical cavity, *Opt. Lett.* **36** 3572
[8] Hill I R *et al* 2021, Dual-axis cubic cavity for drift-compensated multi-wavelength laser stabilisation, *Opt Express* **29** 36758
[9] Cole G D *et al* 2024, Towards space-deployable laser stabilization systems based on vibration-insensitive cubic cavities with crystalline coatings, *Opt. Express* **32** 5380
[10] Guenther C 2018, Kepler – satellite navigation without clocks and ground infrastructure, *Proc 31st meeting Inst. of Nav. ION GNSS+*, 849
[11] Dimarcq N *et al* 2024, Roadmap towards the redefinition of the second, *Metrologia* **61** 012001
[12] Dubé P *et al* 2014, High-accuracy measurement of the differential scalar polarizability of a $^{88}Sr^+$ clock using the time-dilation effect, *Phys Rev Lett* **112** 173002
[13] Haagmans R *et al* 2020, ESA's next-generation gravity mission concepts, *Scienze Fisiche e Naturali* **31** (Suppl 1) S15-S25
[14] Silvestrin P *et al* 2015, Satellite formations and constellations for synergetic missions: a paradigm for international cooperation in earth observation, *Proc. 66th Int. Astronautical Congress* (Jerusalem)
[15] Stacey J *et al* 2023, Laser frequency stabilisation for the LISA mission using a cubic cavity, *Proc SPIE – Int Conf Space Optics 2022*, **12777** 127777F-1
[16] Kleppner D 2006, Time too good to be true, *Physics Today* **59** 10

Progress on Optical Clock Technology for Operational Timescales

S. Peil[1], W. Tobias[2], J. Whalen[2], B. Hemingway[1], T. G. Akin[1]

[1] Precise Time Department, United States Naval Observatory, Washington, DC 20392, USA
[2] Computational Physics Inc., Springfield, VA 22151, USA

E-mail: steven.e.peil.civ@us.navy.mil

Abstract. While optical clock technology has advanced rapidly in recent years, incorporating the technology into operational timescales has progressed more slowly. The highest accuracy frequency standards for groundbreaking measurements do not easily translate to critical timing where continuous, uninterrupted operation over many months and years is required. For example, intermittent steering of a hydrogen maser with an optical standard fails to harness all of the dramatic improvements possible with optical technology. Here we present progress on development and integration of optical-clock technology for operational timescales. An optical oscillator steered to an atomic fountain comprises a hybrid clock with optical-level stability at short times and a reliable long-term reference, and obviates the need for a steered maser. Atomic-beam optical clocks are being developed to support 24/7 operations at a level that improves upon the performance of the U.S. Naval Observatory's rubidium fountains. An optical-lattice is being developed as a gold-standard frequency reference, complementing the role of the atomic beam clocks.

1. Introduction - Operational Timescales

Coordinated Universal Time (UTC) is the internationally agreed upon timescale derived from hundreds of atomic clocks maintained across the globe. Frequency standards are used to calibrate the timescale, imposing a fundamental time interval tied to the SI definition of the second and ensuring long-term stability through accurate evaluation of systematic perturbations. Clocks that operate continuously serve as a "flywheel", maintaining a timing output in between absolute frequency calibrations. Each metrology institute or timing lab maintains a number of clocks that contribute to UTC and also generate a local realization of UTC.

This general architecture for UTC and local timescales has been utilized since inception of atomic time a half-century ago. Perhaps the most significant evolution has been the contributing clock technology; thermal beam primary standards have been replaced by laser cooled atomic fountains [1], and commercial cesium clocks are now weighted significantly less than hydrogen masers. In the past decade, rubidium fountains at the U.S. Naval Observatory (USNO) have contributed significant weight to UTC as continuous clocks (as opposed to secondary frequency standards), the first cold-atom clocks to contribute in this way [2].

The recent advent of optical-clock technology has led to the contribution of strontium (Sr) and ytterbium (Yb) optical lattices to UTC as (secondary) frequency standards, with anticipation that other optical-lattice or trapped-ion clocks will contribute in the near future. On the other hand, the complexity of these systems prevents continuous, uninterrupted operation, and besides frequency calibration of a microwave clock (usually a maser), optical clocks have seen limited use in operational timescales. There are currently no optical clocks contributing to UTC as flywheels, and only vapor-cell clocks [3, 4] show significant promise as continuous clocks. The limited use of optical clocks in timescales forces the continued use of microwave clocks and associated technology, limiting the timescale performance and capabilities over what is achievable with optical technology. We will discuss several efforts underway at USNO to more fully incorporate optical-clock technology into an operational timescale.

2. Optical Oscillators

The use of optical clocks for timing has almost exclusively consisted of an optical-lattice frequency standard intermittently measuring and steering a hydrogen maser [5, 6] or ensemble of masers [7]. This architecture suffers from the lack of a suitable optical flywheel oscillator, relying instead on a proven microwave clock for continuity. Short-term frequency stability is limited by the maser to $\sim 10^{-13}$ at 1 s, and the maser impacts longer-term performance to a degree that depends on the uptime of the lattice.

Optical oscillators, consisting of a cavity-stabilized laser and a frequency-comb divider, have matured significantly in recent years and are becoming much more reliable. Fiber frequency combs routinely stay mode-locked for months on end, and the optical phase locks stabilizing the comb and the CW laser are even more robust. The radio-frequency signals generated maintain the frequency stability of the optical signal to a very high degree. While not yet reliable enough to replace masers in critical operations, optical oscillators are poised to find wider use in timing applications in the near future.

Working toward an experimental "maserless" timescale, we employ an optical oscillator constructed from telecom-wavelength components. A 1542 nm diode laser is stabilized to a high-finesse cavity made from a ULE spacer, silica mirror substrates and crystalline mirror coatings, resulting in a Hz-level linewidth. A fiber frequency comb centered at 1550 nm and with a 250 MHz repetition rate is phase-locked to the narrow laser and generates RF signals at 10 GHz and 200 MHz[1] via two different opto-electronic pathways. The system has proven to be robust, maintaining locks and producing an uninterrupted RF output for up to 6 months in a poorly regulated lab.

We further divide the 200 MHz signal to 5 MHz, the native frequency for many timing systems, and input the signal to our clock measurement system, which measures the scores of atomic clocks that contribute to the USNO timescale. While maintaining optical-clock-level stability when dividing to 5 MHz is not realistic [8], the 1 s instability is still well below that of a maser, and, for applications that can support a 200 MHz or higher frequency output, the impact of frequency division is negligible.

The optical oscillator exhibits an average frequency drift of 3 kHz/day, with significant nonlinear variations strongly correlated with temperature over shorter times. Since optical frequency standards are not reliable enough for continuous operation, we steer the optical oscillator with a rubidium fountain. The USNO fountains have demonstrated reliable performance over 11 years of continuous operation and contribution to UTC, and an optical oscillator disciplined to an atomic fountain can form an operational clock combining the best of both systems.

The USNO rubidium fountains typically exhibit a white-frequency noise level of 1.5 to 2×10^{-13} when a quartz crystal drives the microwave frequency chain. By instead using the 200 MHz signal generated from our optical oscillator, fountain performance can be improved, approaching more closely the quantum-projection noise limit that is as low as 5×10^{-14} in some of our systems. We average the frequency difference between the optical oscillator and fountain

[1] The 4th harmonic of the repetition rate at 1 GHz is divided to generate 200 MHz.

and feedback to the frequency-comb repetition rate to produce the steered output.

This hybrid clock can be expected to exhibit an Allan deviation curve like that illustrated in Figure 1. Short-term stability governed by the local oscillator can be as low as 10^{-15} for the 10 GHz output, though degradation to 10^{-14} is expected when dividing to 5 MHz. The optical oscillator's frequency drift intersects the fountain's integration at about 40 s, after which the hybrid clock integrates as a fountain, shown with a white-frequency noise level of 6×10^{-14}. The frequency stability of the hybrid clock should be better than 10^{-14} for all averaging times, which would be a significant improvement over the short-term performance achievable with a maser-based timescale.

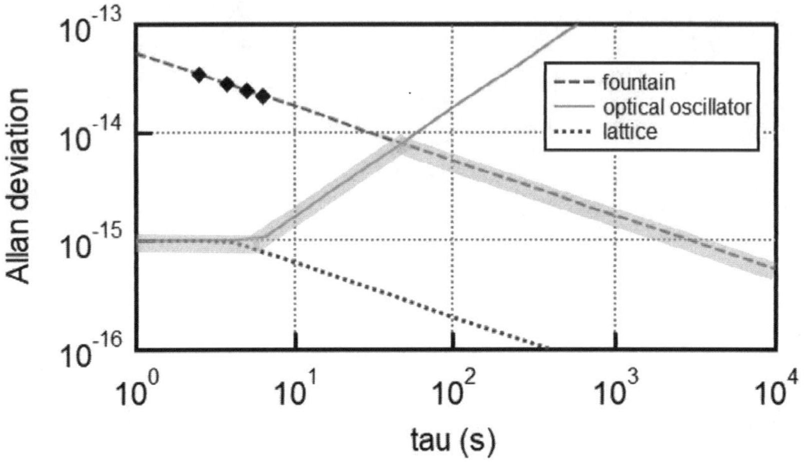

Figure 1. Modeled behavior of hybrid optical-microwave clock. Solid and dashed curves labeled in graph legend show modeled Allan deviation for USNO rubidium fountain using optical oscillator for microwave generation (red, dashed), 10 GHz signal derived from optical oscillator (blue, solid), and lattice clock with 2×10^{-15} white-frequency noise level (black, dotted). The thick grey band illustrates the expected performance of the composite clock, which should exhibit stability better than 10^{-14} for all averaging times. The data points (diamonds) are measured in-loop fountain performance when running with the (unsteered) optical oscillator, validating the displayed white-frequency noise level.

A first test of this prototype hybrid clock was enabled by a portable Yb lattice developed at NIST, which made its maiden journey to USNO. The 10 GHz signal was measured against the 10 GHz output from the Yb lattice, which is expected to have comparable performance at 1 s but superior performance for all longer averaging times. Multiple runs gave frequency records indicating 1 s instability of our hybrid clock at or below 5×10^{-15}, and instability reaching 10^{-14} before integrating as white-frequency noise. There was no convenient, comparable reference to use to measure the stability of the 5 MHz signal.

This hybrid-clock demonstration shows the potential for implementing timescales without masers and associated steering synthesizers, harnessing the capabilities of optical technology for the timescale's front-end.

3. Operational Optical Clocks

The USNO rubidium fountains have been an integral part of the USNO timescale and a significant contributor to UTC for more than a decade. Over that period, new timing requirements have emerged, and new applications, such as relativistic geodesy [9], are becoming a reality. Clocks that can perform significantly better than USNO's rubidium fountains, with 1 s stability of order 10^{-14} and a floor below 10^{-16}, are expected to be needed to meet near future requirements.

With frequencies more than 10^4 times higher than microwave transitions, optical clocks offer substantial headroom for improving upon a microwave standard. The natural or realized linewidth of the clock transition can be relaxed compared to the 1 Hz atomic fountain linewidth while still providing a significant improvement in stability. The ${}^1S_0 - {}^3P_1$ intercombination transition at 657 nm in neutral calcium has been of interest as an optical clock candidate off and on for decades [10]. With a 400 Hz natural linewidth it is well suited for interrogation in an atomic beam, which offers a good compromise between the complexity of trapped atom systems and the non-ideal environment of vapor cells.

3.1. Thermal

The most robust architecture that we expect will meet our performance goals is Ramsey-Bordé spectroscopy on a thermal atomic beam. We have studied this system extensively using a laser geometry that gives a fringe width (equivalent to the full width at half maximum) of 7.5 kHz, culminating in a stability budget that lays out the regulation requirements to reach our frequency stability floor [11]. New integrated optical assemblies will allow us to operate at a fringe width of 2.5 kHz, narrow enough to easily satisfy our short-term stability goals.

Excitation of the clock transition requires a cavity-stabilized laser, and we are investigating the tradeoffs of using a 657 nm laser and cavity versus an IR laser and cavity, with stability transferred to 657 nm via frequency comb. Detection using the $4s4p\,{}^3P_1 - 4p^2\,{}^3P_0$ cycling transition at 431 nm provides an atom-background-free signal that enhances SNR at the expense of an additional laser at a challenging wavelength [12].

The open question regarding a thermal-beam optical clock is the long-term frequency instability due to residual Doppler shifts driven by oven temperature instability (second-order Doppler) and by changes in optical alignment with respect to the atomic beam (first-order Doppler). We are investigating using the technique of reversing the direction of laser-beam propagation (or "k−reversal") to compensate for frequency changes due to changing optical alignment [13]. Figure 2 shows Ramsey-Bordé fringes obtained from each direction of laser propagation in one of our atomic-beam systems. Tracking the measured frequency for each direction should allow us to suppress the impact of alignment instabilities.

3.2. Slowed

Adding a stage of laser cooling to produce a slow calcium beam is a promising approach to significantly reducing residual Doppler instabilities. We are investigating the tradeoffs involved in Ramsey-Bordé spectroscopy using a cooled calcium beam.

Figure 3 shows Ramsey-Bordé fringes obtained with a 70 m/s calcium beam. The cooled beam is generated using a longitudinal Zeeman slower and two 2D MOT stages for transverse cooling and beam steering. Some of the 423 nm cooling light is used for detection. The 2 kHz-wide fringes show higher coherence than in the thermal beam, with fringes extending throughout both recoil components.

Challenges in moving to a cooled beam include not just the additional complexity of the apparatus but also lower atom flux, impact of excited state (lifetime of 420 μs) decay, and narrower Fourier spectrum of the clock excitation (fewer transverse velocity classes addressed). Additional concerns include sensitivity of the cooled beam properties to the frequency and intensity of cooling light and potential light shifts from stray cooling light.

Journal of Physics: Conference Series **2889** (2024) 012017 doi:10.1088/1742-6596/2889/1/012017

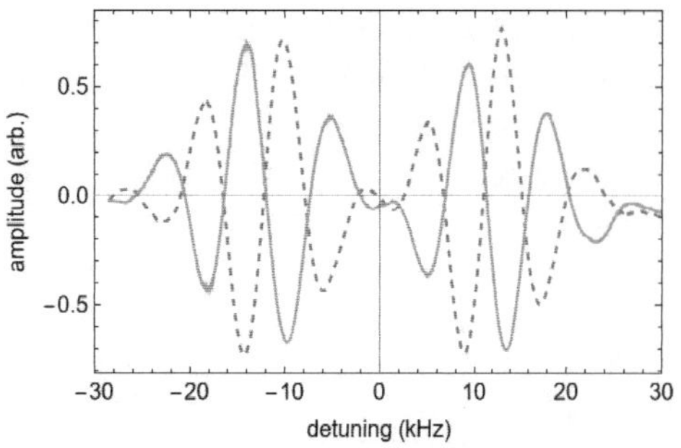

Figure 2. Plot of Ramsey-Bordé signal for each direction of laser propagation. The 2.5 kHz fringe width enables clear resolution of the two recoil resonances separated by 22 kHz. The horizontal "detuning" axis is roughly centered between these two resonances.

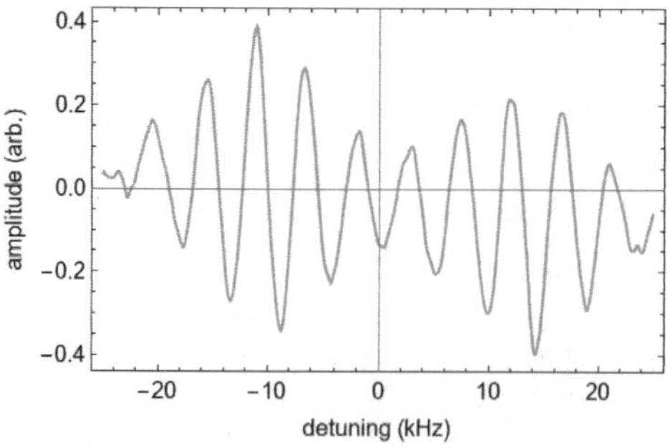

Figure 3. Ramsey-Bordé fringes measured in a slowed atomic beam. In a cold beam, the lifetime of the $^3P_1$ state reduces the contrast of the higher-frequency recoil resonance.

4. Optical Frequency Standard

The gold standard for an atomic frequency reference is an optical-lattice clock. The remarkable performance of such a system is rivaled only by its complexity - in addition to the frequency comb and high-finesse cavity needed for most optical clocks, a strontium (Sr) lattice clock also requires 7 CW lasers, some at challenging wavelengths and some at high powers. Optical-lattice clocks have yet to demonstrate the continuous reliability required for timing operations, and their application to timescales to date have failed to harness their full potential.

We are developing a Sr optical-lattice clock to improve timing capabilities at USNO. We add an additional CW laser to our architecture, a 1542 nm telecom laser that is stabilized to a high-finesse cavity and provides stability for the frequency comb and, in turn, all of the other CW lasers. The addition of this laser provides the system with a robust flywheel based on telecom-wavelength components if one of the other 7 lasers, or some component of the atomic physics system, fails.

The conventional application of a Sr lattice to timing is to measure or steer the frequency of a hydrogen maser. As discussed, this significantly limits the potential advantages of incorporating an optical clock into a timescale. We are investigating the possibility of steering the optical oscillator to a rubidium fountain and creating a hybrid clock, as discussed in Section 2, with stability always better than 10^{-14}. During periods when the Sr lattice is operational, the frequency stability will integrate along the lattice curve shown in Fig. 1, bringing the average instability for the composite clock system lower in proportion to the lattice uptime. Additionally, we may choose to report the Sr lattice to the BIPM by calibrating the frequency of the rubidium fountain, rather than the frequency of a maser. This could be beneficial depending on the averaging time required for the report.

Acknowledgments

This work is supported by Department of Navy Research, Development, Test and Evaluation funds provided for USNO's Clock Development program. We are grateful to the NIST portable Yb lattice team, A. Ludlow, T. Bothwell and W. Brand.

References

[1] Wynands R and Weyers S 2005 *Metrologia* **42** S64

[2] Peil S, Hanssen J, Swanson T B, Taylor J and Ekstrom C R 2016 *Journal of Physics: Conference Series* **723** 012004
[3] Roslund J D, Cingöz A, Lunden W D, Partridge G B, Kowligy A S, Roller F, Sheredy D B, Skulason G E, Song J P, Abo-Shaeer J R and Boyd M M 2023 Optical clocks at sea (*Preprint* 2308.12457)
[4] Martin K W, Phelps G, Lemke N D, Bigelow M S, Stuhl B, Wojcik M, Holt M, Coddington I, Bishop M W and Burke J H 2018 *Phys. Rev. Appl.* **9**(1) 014019
[5] Grebing C, Al-Masoudi A, Dörscher S, Häfner S, Gerginov V, Weyers S, Lipphardt B, Riehle F, Sterr U and Lisdat C 2016 *Optica* **3** 563–569
[6] Hachisu H, Nakagawa F, Hanado Y and Ido T 2018 *Scientific Reports* **8**(1) 4243
[7] Yao J, Sherman J A, Fortier T, Leopardi H, Parker T, McGrew W, Zhang X, Nicolodi D, Fasano R, Schäffer S, Beloy K, Savory J, Romisch S, Oates C, Diddams S, Ludlow A and Levine J 2019 *Phys. Rev. Appl.* **12**(4) 044069
[8] Hati A, Nelson C W, Barnes C, Lirette D, Fortier T, Quinlan F, Desalvo J A, Ludlow A, Diddams S A and Howe D A 2013 *IEEE Transactions on Ultrasonics, Ferroelectrics, and Frequency Control* **60** 1796–1803
[9] Mehlstäubler T E, Grosche G, Lisdat C, Schmidt P O and Denker H 2018 *Reports on Progress in Physics* **81** 064401
[10] Olson J, Fox R W, Fortier T M, Sheerin T F, Brown R C, Leopardi H, Stoner R E, Oates C W and Ludlow A D 2019 *Phys. Rev. Lett.* **123**(7) 073202
[11] Hemingway B, Akin T G, Taylor J and Peil S 2020 *2020 Joint Conference of the IEEE International Frequency Control Symposium and International Symposium on Applications of Ferroelectrics (IFCS-ISAF)* pp 1–4
[12] Taylor J, Hemingway B, Hanssen J, Swanson T B and Peil S 2018 *J. Opt. Soc. Am.* B **35** 1557–1562
[13] Ito N, Ishikawa J and Morinaga A 1994 *Optics Communications* **109** 414–421 ISSN 0030-4018

Contributions of the optical lattice clock NICT-Sr1 to TAI calibration and UTC(NICT) steering

H Hachisu, N Nemitz, M Tønnes, H Ito, N Ohtsubo, Y Miyauchi, M Morikawa, K Matsubara, T Gotoh and T Ido

Space-Time Standards Laboratory, NICT, Koganei, Tokyo 184-8795, Japan

E-mail: hachisu@nict.go.jp

Abstract. Our efforts with the optical lattice clock NICT-Sr1 are now focused on contribution to both the international and the local timescale. Since 2018, NICT-Sr1 has been recognized as a secondary frequency standard and frequently contributed to BIPM's monthly calibration of the rate of the international atomic time TAI. This fulfils one of the criteria that the Consultative Committee for Time and Frequency has set for the redefinition of the SI second. Another, ancillary condition is addressed by NICT-Sr1's contributions to the generation of Japan Standard Time, which began in 2021 and enabled a reduction of the deviation from the international timescale to typically 4 ns over more than two years.

1. Introduction

A variety of optical clocks [1-3] have published budgeted uncertainties approaching or surpassing 2×10^{-18}, and accurate frequency comparisons through optical carrier transfer [4-6] rather than microwave links are now routinely conducted. The Consultative Committee for Time and Frequency (CCTF) of the International Committee for Weights and Measures (CIPM) is considering a redefinition of the unit of the second of the International System of Units (SI) at the General Conference of Weights and Measures (CGPM) in 2030, moving from the current definition based on a caesium hyperfine transition to one based on one or more optical clock transitions. CCTF is urging [7] all institutes operating optical atomic clocks to demonstrate their suitability to maintain and improve the infrastructure of international time such as the international atomic time TAI. One of the mandatory criteria, "regular contributions of optical frequency standards to TAI (as secondary representations of the second)" [7] already seems to be well on track for a redefinition, with a growing number of optical clocks contributing to TAI calibration [8]. The other condition relevant to timescales is "regular contributions of optical frequency standards to UTC(k)" [7]. While this is an ancillary condition, it has recently gained significant attention [9-14].

In this article we summarize the contributions of the optical lattice clock NICT-Sr1 [15] to international time and the generation of the local timescale. There it helps fulfil NICT's responsibility to generate and distribute reference time and frequency signals to provide nationwide access to Japan Standard Time JST based on UTC(NICT) adjusted to the Coordinated Universal Time UTC to the limits of available technology.

2. Applications and achievements of NICT-Sr1

A strontium optical lattice clock as a next-generation frequency standard developed at NICT has been operational since 2011, and has participated in demonstrating many kinds of frequency comparisons since then. Frequency comparisons between distant optical clocks are already being applied for relativistic geodesy [16] and the search for variation of fundamental constants [17]. Comparisons are also essential in metrology to measure frequency ratios to determine or redefine the SI second.

In addition to internal comparisons of NICT-Sr1 to a calcium ion clock [18] and an indium ion clock [19], using frequency combs to measure ratios between different atomic species, we also demonstrated an early remote frequency comparison of NICT-Sr1 to a strontium lattice clock at the University of Tokyo using an optical fibre link [20]. Optical carrier transfer over fibre currently offers the best performance for remote comparisons, but wireless radio-frequency comparisons are still needed for intercontinental and other long-baseline comparisons. Comparisons to a strontium lattice clock and an ytterbium ion clock at Physikalisch-Technische Bundesanstalt PTB [21] using two-way satellite frequency transfer by carrier phase (TWCP) constituted the first intercontinental optical clock comparisons. The same technique has been applied in comparisons to an ytterbium lattice clock at the Korea Research Institute of Standards and Science KRISS [22]. In a different approach, we demonstrated an intercontinental comparison to an ytterbium lattice clock at the National Metrology Institute of Italy INRiM using Very Long Baseline Interferometry VLBI [23]. We have also contributed to the international hunt for dark matter [24] using optical clocks. NICT performed multiple absolute frequency measurements using remote references such as TAI [25] through intermittent measurements of a local time reference [15, 26, 27]. Our reported frequency ratios and absolute frequency measurements contributed to the updates of the frequency recommended by CIPM for the ^{87}Sr clock transition as a secondary representation of the second SRS since 2012 [7, 28]. After the Working Group on Primary and Secondary Frequency Standards assessed NICT's evaluation of the rate of TAI, the optical lattice clock NICT-Sr1 has been officially recognized as a secondary frequency standard in 2018, and its contributions have been accepted for calibration of TAI. In 2016, we first demonstrated the generation of a real-signal timescale steered by an optical clock [29] and since August 2021, NICT-Sr1 has been contributing to JST, which is generated and distributed by NICT [30].

3. Contribution of NICT-Sr1 to TAI calibration

The primary (PFS) and secondary frequency standards (SFS) evaluate the rate of TAI using a flywheel oscillator, such as a hydrogen maser (HM) or the local timescale UTC(k), generated in the institute where the standards are operated. While originally intended for microwave caesium PFS and rubidium SFS, this arrangement also enables optical clocks to contribute to the TAI calibration performed monthly by BIPM. NICT-Sr1 began contributing in 2018 [31], when it was recognized as an SFS, following the adoption of the first optical SFS at LNE-SYRTE.

In such contributions, any uncertainty in the link between the frequency standard and the local timescale or microwave flywheel oscillator participating in the generation of the Free Atomic Time EAL is described in BIPM's Circular T by statistical and systematic uncertainties respectively labelled $u_{\mathrm{A/Lab}}$ and $u_{\mathrm{B/Lab}}$ to distinguish them from the uncertainty $u_{\mathrm{l/TAI}}$ in the long-distance link to TAI [32], a convention we will follow here. In our evaluation the Type A uncertainty $u_{\mathrm{A/Lab}}$ represents the linear trend estimation of the HM ($u_{\mathrm{l/trend}}$) and the uncertainty due to the stochastic noise of the HM during unobserved intervals ($u_{\mathrm{l/stoch}}$). The Type B uncertainty $u_{\mathrm{B/Lab}}$ is the uncertainty due to the frequency comparison between microwave and optical signals including distribution of the microwave signals.

At NICT, we intermittently measure the frequency of the HM chosen as a flywheel oscillator with respect to NICT-Sr1 for several hours with the help of a frequency comb. Such a measurement is usually scheduled once a week. In determining the mean frequency of the HM over a target interval of typically a month, the intermittent mode of operation [15, 33] enables an uncertainty of low 10^{-16}, in a good match to the uncertainty of the GNSS link. The GNSS comparisons determine the time difference of the local reference timescale UTC(NICT) relative to UTC, which is identical in frequency to TAI. The time differences between UTC(NICT) and the HMs in the JST ensemble are recorded every hour.

Table 1 shows an overview of a recent TAI evaluation [34]. The reported value is the mean fractional frequency deviation $\overline{y(\mathrm{UTC(NICT)})}$ determined by NICT-Sr1. The GNSS comparison then allows BIPM to complete the frequency chain to TAI. $u_{\mathrm{A/Sr}}$ is the Type A uncertainty of NICT-Sr1 as an optical standard. It represents the statistical uncertainty determined by interleaved measurements. u_{B} is the Type B uncertainty of NICT-Sr1, including the uncertainty of the gravitational redshift. The clock operated for three to six hours only once per week, resulting here in a monthly uptime ratio of 4.4%. This leaves time for maintenance and other laboratory work to be performed on non-operating days, avoiding unplanned interruptions that would otherwise increase the satellite link uncertainty, calculated as $u_{\mathrm{l/TAI}} = (\sqrt{2}\sigma_{\mathrm{A/link}}/T_0)\,(T_0/T)^{0.9}$ [35] for the duration T of the reporting interval. Here $T_0 = 5$ d is the duration of the Circular T evaluation interval, and $\sigma_{\mathrm{A/link}} = 0.3$ ns is the statistical timing uncertainty of UTC – UTC(NICT) obtained over this interval.

The instability of the HM flywheel oscillator is sufficiently low that extending the evaluation interval to the full length of the reporting interval, typically $T = 30$ d, usually provides the smallest overall uncertainty. The calculation of the statistical measurement and dead time uncertainties counted towards $u_{\mathrm{A/Lab}}$ in Table 1 is detailed in reference [27]. The residual dead time uncertainty can be decreased even more by including additional HMs to form an ensemble average.

The evaluation employs the recommended value for ^{87}Sr as secondary representation of the second as approved by the CCTF in March 2021, with relative uncertainty $u_{\mathrm{SRS}} = 1.9 \times 10^{-16}$.

Table 1. TAI evaluation by NICT-Sr1 from MJD59669 to 59699. The notation of the top row follows that of Circular T, and all uncertainties are given in units of 10^{-16}.

Period of Estimation	$\overline{y(\mathrm{UTC(NICT)})}$	u_{A}	u_{B}	$u_{\mathrm{A/Lab}}$	$u_{\mathrm{B/Lab}}$	$u_{\mathrm{l/TAI}}$	u	u_{SRS}
MJD59669 – 59699	+2.27	0.21	0.67	1.51	0.21	2.0	2.6	1.9
Contribution	Uncertainty							
$u_{\mathrm{A/Sr}}$	0.21	✓						
u_{B}	0.67		✓					
$u_{\mathrm{l/trend}}$	1.17			✓				
$u_{\mathrm{l/stoch}}$	0.96			✓				
Optical-microwave comparison including microwave transfer	0.21				✓			
Local uncertainty, excl. $u_{\mathrm{l/TAI}}$ and u_{SRS}	**1.68**							
$u_{\mathrm{l/TAI}}$ of satellite link	2.0					✓		
Total uncertainty, excluding u_{SRS}	**2.6**						✓	
u_{SRS} for ^{87}Sr	1.9							✓

4. Integration of NICT-Sr1 into locally generated timescales

4.1. Demonstration of a real-signal timescale based on an optical clock

Optical clocks are rapidly advancing from laboratory demonstrations to reliable measurement instruments. Their performance promises all-optical timescales of unprecedented short- and long-term stability, but so far this has mostly been demonstrated in post-processed calculations [13]. They still face obstacles to the long-term uninterrupted operation that is crucial for such a system.

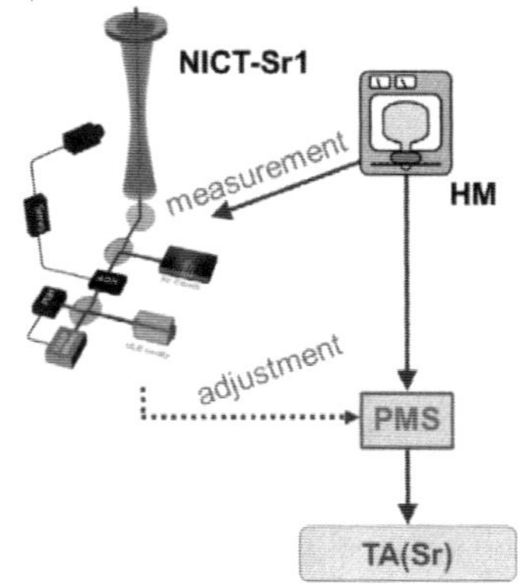

Figure 1. Configuration used to generate the optically steered timescale TA(Sr). Using a frequency comb, a HM source oscillator was measured with respect to NICT-Sr1 for several hours, multiple times per week. A phase micro stepper (PMS) compensates the frequency offset and drift of the HM.

At the 8th Symposium on Frequency Standards and Metrology held in Potsdam 2015, we presented a post-processed timescale based on NICT-Sr1 to demonstrate the feasibility of an optically steered timescale [9]. Over five months in 2016, we subsequently generated the first real-signal timescale based on an optical clock TA(Sr) [30], independent from UTC(NICT). This employed a HM as a source oscillator, which was measured with respect to NICT-Sr1 more than once a week at that time, using a frequency comb to down-convert the optical signal. The frequency offset and drift of the HM was then

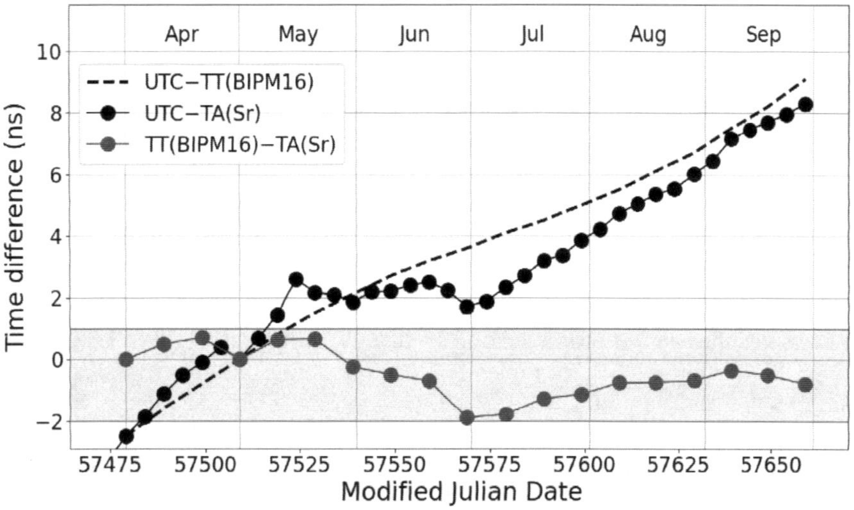

Figure 2. Time difference between UTC and TA(Sr) (black points) agrees with the difference of UTC from the post-processed timescale TT(BIPM16) (dashed line). Red points show the difference between TT(BIPM16) and TA(Sr), which remains within the shaded 3 ns wide band.

Journal of Physics: Conference Series **2889** (2024) 012018 doi:10.1088/1742-6596/2889/1/012018

compensated by a phase micro stepper (PMS) (figure 1) by extrapolating the measurements of the previous 25 days to update the frequency compensation of the PMS every four hours.

Figure 2 shows the result of the optically steered timescale, which over five months deviated by less than 2 ns from TT(BIPM2016), the post-processed timescale calculated by BIPM using all TAI evaluation data from PFS and SFS up to December 2016. This heralded the possibility of using TA(Sr) to improve UTC(NICT), which at this time had a typical deviation from UTC of ±20 ns.

4.2. Contribution to the nationally-distributed timescale UTC(NICT) over two years

Before the inclusion of NICT-Sr1, UTC(NICT) was generated by adjusting a HM source oscillator based on a calculated free ensemble timescale derived from 18 commercial caesium atomic clocks [36]. For redundancy, NICT generates three separately adjusted candidate timescales, each with its own HM source oscillator. One of the timescales is selected as UTC(NICT), from which JST is generated by adding +9 hours. Additional corrections to UTC(NICT) were manually included only when the time difference from UTC was expected to exceed 20 ns.

We have resumed the independent generation of TA(Sr) in July 2021, using the same technical implementation demonstrated in 2016. We now schedule one day of operation for NICT-Sr1 per calendar week, each aiming to obtain three to seven hours of measurements. Each operation yields a HM frequency value that is stored in a database, where it is used in an automatic hourly evaluation that models the evolution of the HM frequency in terms of a linear drift over an interval of typically 35 days. The required frequency correction is then predicted and applied to the HM signal. For a continuous evaluation, it is necessary to limit the separation between measurements (figure 3).

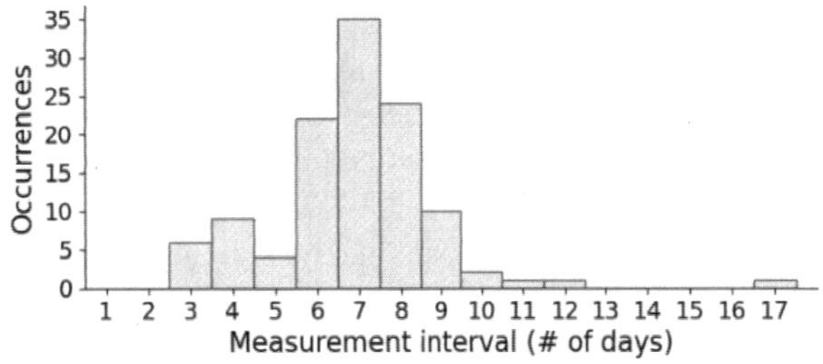

Figure 3. A histogram of the intervals separating two successive measurements of NICT-Sr1 for 115 measurements from 2021 to 2023. The operating schedule targets a one-week interval.

NICT-Sr1 began contributing to the JST generation in August 2021 [30]. To maintain robustness, the optical clock has been incorporated by a loose coupling of UTC(NICT) to TA(Sr) (figure 4). This was realized by adjusting the frequency of UTC(NICT) twice a week to remove the frequency difference y[UTC(NICT) − TA(Sr)], determined from hourly time differences collected over 24 hours. An additional, smaller frequency correction was then added to zero the current time difference between UTC(NICT) and TA(Sr) over a period of 60 days.

The twice-weekly steering of the conventionally generated UTC(NICT) was maintained over two years, and has reduced the peak-to-peak variation of the time difference between UTC and UTC(NICT) from approximately 40 ns since 2017 to typically 5 ns after August 2021, as shown in figure 5. There have been two exceptions: The first, labelled A in the figure, was a timing step of approximately 6 ns that occurred when UTC(NICT) was switched to another of the three redundant candidate implementations. The second, labelled B, resulted from the undetected failure of a mechanical shutter in NICT-Sr1, which exposed the atoms to leaked laser light for the 689 nm second cooling stage during interrogation. The resulting frequency shift was discovered based on the observed timing drift of TA(Sr) relative to the rapid realization UTCr and was initially compensated by manual frequency adjustment. Once the cause had been identified, interleaved measurements of NICT-Sr1 confirmed a systematic frequency shift of several 10^{-15}, consistent with the applied correction. The loose coupling in steering

UTC(NICT) provided an opportunity to correct for the erroneous measurements and avoid the accumulation of a larger time difference until the error could be identified and corrected. Including these events, the peak-to-peak variation of the deviation of UTC(NICT) from UTC has been 12 ns.

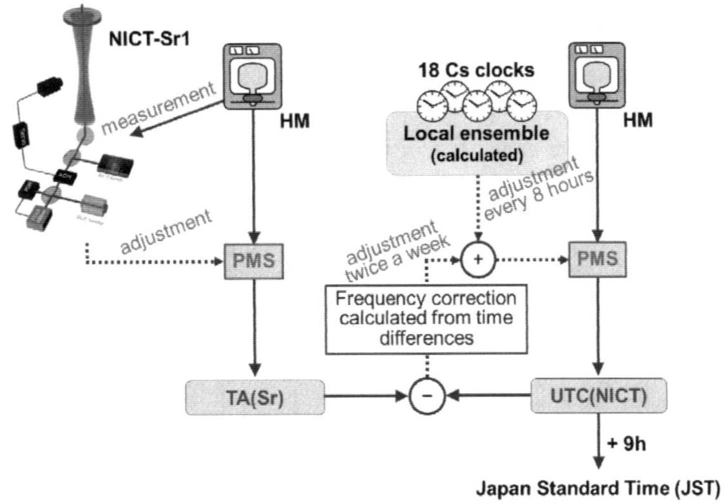

Figure 4. Configuration of the JST generation based on NICT-Sr1. UTC(NICT) is initially steered by an ensemble of 18 commercial caesium atomic clocks and then adjusted to TA(Sr) twice a week.

At the present time, plans for integrating optical clocks with existing radio-frequency timescale systems, with a HM providing the needed robustness, are in progress world-wide [9-12, 14, 29]. At NICT, the automatic steering of UTC(NICT) has been momentarily suspended since August 2023 to evaluate adjustments to the calculated ensemble timescale that place the largest weight on drift-compensated HM contributions. TA(Sr) continues to be available as a reference for manual corrections.

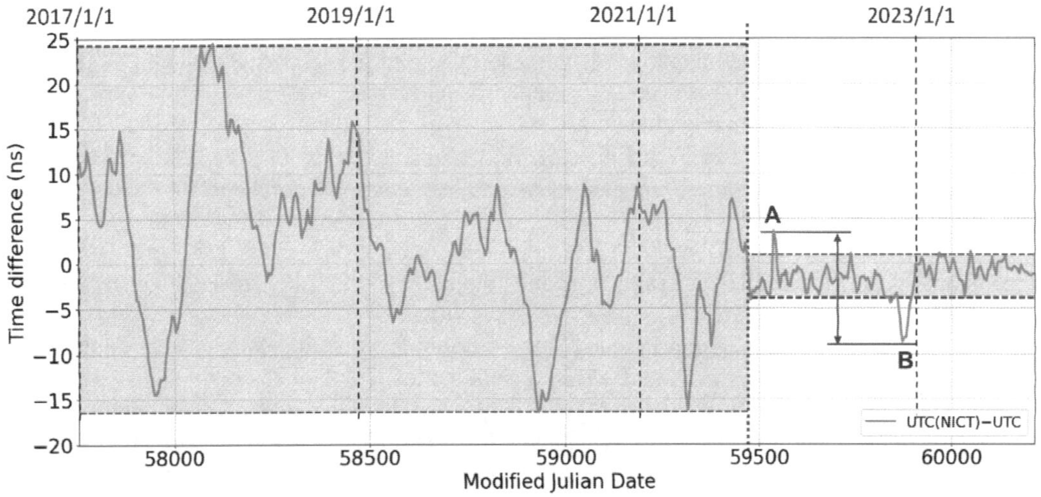

Figure 5. Time difference between UTC and UTC(NICT) since 2017. The use of the optically steered timescale TA(Sr) as reference for UTC(NICT) started in August 2021, indicated by the red dotted line.

5. Conclusion

NICT-Sr1 has officially contributed as a SFS to BIPM's monthly calibration of the rate of TAI since 2018. The number of optical SFS is growing lately [8] and the weight of the optical clock contributions has increased greatly following CCTF 2022. This builds important experience with maintaining and improving international time based on optical clocks.

NICT-Sr1 has also been contributing to the generation of JST by providing a stable, independent reference signal in the form of the timescale TA(Sr) since 2021. To our knowledge, this was the first time that an optical clock has been incorporated in the generation of a nationwide disseminated timescale. The improved steering reduced the deviations of UTC(NICT) from UTC by a factor of three to four. Further reduction of the deviation is expected by improving the reliability of the optical clock and its measurement systems.

The optical clock improves the resilience of UTC(NICT) by providing a precise local reference in the case of an interruption of the international time links that rely on the global navigational satellite networks. The deliberately loose coupling to TA(Sr) avoids introducing a new single-point-of-failure and thus maintains the robustness of the timescale.

We believe that these contributions to the international and the nationally-distributed timescale help demonstrate the suitability of optical clocks as the basis for a redefined SI second, as set out in the mandatory criteria and ancillary conditions of the CCTF roadmap.

References

[1] McGrew W F *et al.* 2018 Atomic clock performance enabling geodesy below the centimetre level *Nature* **564** 87–90
[2] Brewer S M, Chen J-S, Hankin A, Clements E R, Chou C W, Wineland D J, Hume D B and Leibrandt D R 2019 An $^{27}Al^{+}$ quantum-logic clock with systematic uncertainty below 10^{-18} *Phys. Rev. Lett.* **123** 033201
[3] Bothwell T, Kedar D, Oelker E, Robinson J M, Bromley S L, Tew W L, Ye J and Kennedy C J 2019 JILA SrI optical lattice clock with uncertainty of 2.0×10^{-18} *Metrologia* **56** 065004
[4] Lisdat C *et al.* 2016 A clock network for geodesy and fundamental science *Nat. Comm.* **7** 12443
[5] Takano T, Takamoto M, Ushijima I, Ohmae N, Akatsuka T, Yamaguchi A, Kuroishi Y, Munekane H, Miyahara B and Katori H 2016 Geopotential measurements with synchronously linked optical lattice clocks *Nat. Photon.* **10** 662–6
[6] Beloy K *et al.* 2021 Frequency ratio measurements at 18-digit accuracy using an optical clock network *Nature* **591** 564–9
[7] Dimarcq N *et al.* 2024 Roadmap towards the redefinition of the second *Metrologia* **61** 012001
[8] Petit G 2021 Record number of frequency standards contribute to International Atomic Time *BIPM news article* available at https://www.bipm.org/en/-/2021-12-21-record-tai
[9] Ido T, Hachisu H, Nakagawa F and Hanado Y 2016 Rapid evaluation of time scale using an optical clock *in Proceedings, 8th Symposium on Frequency Standards and Metrology J. Phys.: Conf. Ser.* **723** 012041
[10] Grebing C, Al-Masoudi A, Dörscher S, Häfner S, Gerginov V, Weyers S, Lipphardt B, Riehle F, Sterr U and Lisdat C 2016 Realization of a timescale with an accurate optical lattice clock *Optica* **3** 563
[11] Yao J, Parker T E, Ashby N and Levine J 2017 Incorporating an optical clock into a time scale *IEEE Trans. Ultrason. Ferroelectr. Freq. Control.* **65** 127
[12] Zhu L, Lin Y, Wang Y, Jia Z, Wang Q, Li Y, Yang T and Fang Z 2022 Preliminary study of generating a local time scale with NIM ^{87}Sr optical lattice clock *Metrologia* **59** 055007
[13] Milner W R *et al.* 2019 Demonstration of a timescale based on a stable optical carrier *Phys. Rev. Lett.* **123** 173201
[14] ROCIT: Robust optical clocks for international timescales at https://empir.npl.co.uk/rocit/
[15] Hachisu H, Petit G, Nakagawa F, Hanado Y and Ido T 2017 SI-traceable measurement of an optical frequency at the low 10^{-16} level without a local primary standard *Opt. Express* **8** 8511

[16] Takamoto M, Ushijima I, Ohmae N, Yahagi T, Kokado K, Shinkai H and Katori H 2020 Test of general relativity by a pair of transportable optical lattice clocks *Nature Photonics* **14** 411–5
[17] Roberts B M, Delva P, Al-Masoudi A, Amy-Klein A, Bærentsen C, Baynham C F A, Benkler E, Bilicki S, Bize S and Bowden W 2020 Search for transient variations of the fine structure constant and dark matter using fiber-linked optical atomic clocks *New J. Phys.* **22** 093010
[18] Kensuke M, Hachisu H, Li Y, Nagano S, Locke C, Nogami A, Kajita M, Hayasaka K, Ido T and Hosokawa M 2012 Direct comparison of a Ca^+ single-ion clock against a Sr lattice clock to verify the absolute frequency measurement *Opt. Express* **20** 22034–41
[19] Ohtsubo N, Li Y, Nemitz N, Hachisu H, Matsubara K, Ido T and Hayasaka K 2020 Frequency ratio of an $^{115}In^+$ ion clock and a ^{87}Sr optical lattice clock *Opt. Lett.* **45** 5950–3
[20] Yamaguchi A, Fujieda M, Kumagai M, Hachisu H, Nagano S, Li Y, Ido T, Takano T, Takamoto M and Katori H 2011 Direct comparison of distant optical lattice clocks at the 10^{-16} uncertainty *App. Phys. Exp.* **4** 082203
[21] Hachisu H *et al.* 2014 Direct comparison of optical lattice clocks with an intercontinental baseline of 9000 km *Opt. Lett.* **39** 4072–5
[22] Fujieda M *et al.* 2018 Advanced satellite-based frequency transfer at the 10^{-16} Level *IEEE TUFFC* **65** 973–6
[23] Pizzocaro M *et al.* 2020 Intercontinental comparison of optical atomic clocks through very long baseline interferometry *Nature Physics* **17** 223–7
[24] Wcisło P *et al.* 2018 New bounds on dark matter coupling from a global network of optical atomic clocks *Sci. Adv.* **4** eaau4869
[25] Yamaguchi A, Shiga N, Nagano S, Li Y, Ishijima H, Hachisu H, Kumagai M and Ido T 2012 Stability transfer between two clock lasers operating at different wavelengths for absolute frequency measurement of clock transition in ^{87}Sr *Appl. Phys. Express* **5** 022701
[26] Hachisu H, Petit G and Ido T 2017 Absolute frequency measurement with uncertainty below 1×10^{-15} using International Atomic Time *Appl. Phys.* B **123** 34
[27] Nemitz N, Gotoh T, Nakagawa F, Ito H, Hanado Y, Ido T and Hachisu H 2021 Absolute frequency of ^{87}Sr at 1.8×10^{-16} uncertainty by reference to remote primary frequency standards *Metrologia* **58** 025006
[28] Riehle F, Gill P, Arias F and Robertsson L 2018 The CIPM list of recommended frequency standard values: guidelines and procedures *Metrologia* **55** 188
[29] Hachisu H, Nakagawa F, Hanado Y and Ido T 2018 Months-long real-time generation of a time scale based on an optical clock *Sci. Rep.* **8** 4243
[30] Hachisu H, Ito H, Nemitz N, Ohtsubo N, Miyauchi Y, Morikawa M, Matsubara K and Ido T 2023 UTC(NICT) referenced to a timescale based on the optical clock NICT-Sr1 *Proc. 2023 Joint Conference of the European Frequency and Time Forum and IEEE International Frequency Control* (EFTF/IFCS) DOI: 10.1109/EFTF/IFCS57587.2023.10272119
[31] BIPM Time Department database *Primary and Secondary Frequency Standards participating to TAI, since September 2003* at https://webtai.bipm.org/database/show_psfs.html
[32] Bureau International des Poids et Mesures Circular T available at https://www.bipm.org/en/time-ftp/circular-t
[33] Hachisu H and Ido T 2015 Intermittent optical frequency measurements to reduce the dead time uncertainty of frequency link *Japan. J. Appl. Phys.* **54** 112401
[34] BIPM Time Department database *PFSF reports of evaluation* at https://webtai.bipm.org/database/psfs_access.html?psfs=nict-sr1&yymm=all
[35] Panfilo G and Parker T E 2010 A theoretical and experimental analysis of frequency transfer uncertainty, including frequency transfer into TAI *Metrologia* **47** 552–60
[36] Hanado Y, Imamura K, Kotake N, Nakagawa F, Shimizu Y, Tabuchi R, Takahashi Y, Hosokawa M and Morikawa T 2008 The new generation system of Japan Standard Time at NICT *Int. J. Navig. Obs.* **2008** 841672 DOI: 10.1155/2008/841672

Journal of Physics: Conference Series **2889** (2024) 012019 doi:10.1088/1742-6596/2889/1/012019

Hydrogen maser flywheels for optical clocks

N Nemitz, H Hachisu, H Ito, N Ohtsubo, Y Miyauchi, M Morikawa, K Matsubara and T Ido

Space-Time Standards Laboratory, NICT, Tokyo, Japan

Nils.Nemitz@nict.go.jp

Abstract. We consider the statistical and systematic uncertainties that arise from using a hydrogen maser as a flywheel oscillator for an intermittently operating optical clock. Diurnal delay variations observed in the distribution and measurement of the radiofrequency signals are the most significant effects for a clock that repeatedly operates at a similar time of day. In typical operation, we find an overall systematic uncertainty contribution $u_{\mathrm{B/Lab}} = 2.0\times10^{-17}$ for the link between the clock and the reference timescale. Data obtained over more than two years of weekly characterization measurements confirms that the real-world behaviour of the flywheel hydrogen maser is in excellent statistical agreement with our stochastic model.

1. Introduction

The performance of the most accurate optical frequency standards [1–5] promises the generation of timescales that remain synchronized at the nanosecond level over decades. The practical realization, however, relies on flywheel oscillators to provide a continuous output signal between measurements of the frequency standards, whether it is during the preparation time between atomic interrogations [6], or during intentional and unintentional breaks in operation. Lasers stabilized to optical resonators with spacers machined from single-crystal silicon are the most promising candidates to realize an entirely optical timescale system [7], but even their capacity for frequency hold-over quickly degrades once calibrations by an optical frequency standard are interrupted for more than a day.

For the foreseeable future, active hydrogen masers will remain the backbone of physically realized timescales, because they provide frequencies that can be predicted for multiple weeks into the future and operate reliably enough to cover years without loss of signal phase continuity. This function as flywheel makes a hydrogen maser (HM) the logical interface for the integration of optical frequency standards into real-world timekeeping systems, a major topic of interest in the context of the planned adoption of an SI second based on one or more optical atomic transitions [8].

2. Intermittent hydrogen maser evaluation

We structure the typical measurement schedule of NICT's strontium optical lattice clock NICT-Sr1 [9] around the HM performance. In this scheme of intermittent operation [10], NICT-Sr1 now measures the frequency of the reference HM (Anritsu SD1T03C) for approximately $\tau = 20\,000$ s (or 0.2 days), suppressing white frequency noise (WFN) contributions to a level close to the flicker frequency noise (FFN) floor, which we observe at approximately 2×10^{-16} (figure 1) in terms of the fractional Hadamard deviation (HDev). This is inherently insensitive to the HM's continuous frequency drift and allows characterization of slower noise processes that do not converge for the more commonly used Allan deviation (ADev). For convenience, we will occasionally plot this HDev model as a guideline

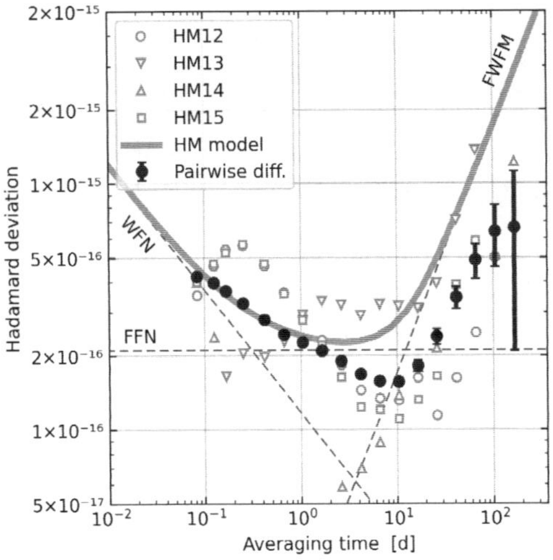

Figure 1. HM instability as overlapping HDev. Black circles and errorbars are an average over the variances of pairwise differences, adjusted to represent a single HM. For this more recent data, instabilities tend to fall below the previously established model shown by the wide green line. Dashed lines indicate noise contribtions considered in this article. The evaluation of individual stabilities (open symbols) by a three-corner-hat method is less robust. The results are shown only for illustration, and not used in the modeling.

together with the ADev of investigated uncertainty contributions. As commonly normalized, the HDev shares the ADev's response to WFN, and its response to FFN differs by only −11% [11].

After calibration by the clock measurements, the HM provides a predictable frequency reference with stable drift until a slowly varying noise contribution best described as flicker-walk frequency modulation (FWFM) becomes dominant for $\tau > 10$ d. To maintain the calibration in the presence of this effect, we repeat the clock measurement typically once per week. Variations of this scheme have enabled uncertainties below 2×10^{-16} for an absolute frequency determination of NICT-Sr1 [11] and for calibrating International Atomic Time TAI [12,13]. The uncertainty budget for these measurements needs to account not only for the clock itself, but also for the distribution and measurement of the radiofrequency (rf) reference signals and the optical-to-rf comparison itself. In the Circular T of the International Bureau of Weights and Measures (BIPM), the uncertainties associated with this lab-side frequency link are $u_{\mathrm{A/Lab}}$ and $u_{\mathrm{B/Lab}}$. The first is random noise with no expected correlation over repeated measurements. The second is a systematic uncertainty that describes the possibility for a frequency error which persists over many or all measurements. The following sections provide an updated evaluation of $u_{\mathrm{B/Lab}}$ as it applies in the intermittent operation of NICT-Sr1.

We distinguish two separate types of error: A *continuous* error will affect all data equally, or at least with similar likeliness. A *synchronous* error, on the other hand, arises when the measurements repeatedly sample the same section of a cyclical disturbance.

3. Sources of synchronous error

3.1. Diurnal variation of signal distribution delay

NICT-Sr1 usually operates between 12:00 and 20:00 (local time), creating a sensitivity to effects with a daily cycle. The first effect to consider is the connection between the reference HM at the Japan Standard Time (JST) facility in an adjacent building and the frequency comb laboratory where the rf-to-optical comparison takes place (figure 2). The HM frequency is transmitted as a 100 MHz signal over an almost 50 m long coaxial cable (Andrew FSJ1-50A). The daily temperature cycle can be expected to create a change $\Delta\tau_{\mathrm{d}}$ in the propagation delay that then results in a diurnal variation in the fractional frequency deviation $\Delta y(t) = -\frac{d}{dt}\Delta\tau_{\mathrm{d}}(t)$. To evaluate the distribution system, we added a separate connection to the JST facility that then loops back to measure the delay variation using a phase comparator in the comb laboratory. The loop also includes a distribution amplifier identical to the one in the main distribution path.

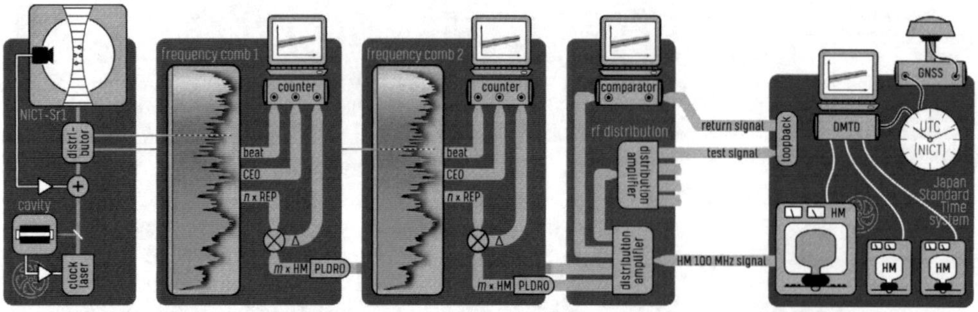

Figure 2. Overview of the lab-side frequency link between NICT-Sr1 and UTC(NICT) and its uncertainty evaluation. The laser stabilized by the clock is sent to two frequency combs by a fibre-noise cancelled distribution system. The repetition rate of each comb is stabilized based on a 100 MHz signal from the flywheel HM operated as part of the Japan Standard Time system in a building adjacent to the clock laboratory. The delay variation of the distribution is estimated by a phase comparator monitoring a separate loopback path that also includes a distribution amplifier identical to the one supplying signals to the frequency combs. Each comb system includes a phase-locked dielectric resonator oscillator (PLDRO) to upconvert the 100 MHz signal to 9.3 GHz (8.2 GHz), used to mix down a harmonic of the comb's repetition rate close to 9.25 GHz (8.2 GHz). The resulting difference signal is stabilized by acting on the comb repetition rate and counted on a multi-channel deadtime-free frequency counter that also records the stabilized carrier-envelope offset (CEO) frequency and the beat signal with the clock-stabilized laser. The flywheel HM is continuously compared to other HMs and to the UTC(NICT) signal that is the local reference for remote comparisons, including the international link contributing to UTC.

From October 2021 to November 2023, we have collected more than 950 000 one-minute measurement intervals of the phase comparator, accounting for a total of 665 days (figure 3), with gaps typically arising from unintended restarts of the computer. For short interruptions, the phase evolution is continued from the last stored value, for interruptions of more than one day, it is reset to zero. The phase measurements are evaluated by calculating a Lomb-Scargle periodogram after removing slow variations using a high-pass filter with $f_{\text{cut}} = 1/(10\text{ d})$. We then determine a sinusoidal test function

$$\Delta\tau_{\text{t}}(t) = a_{\text{dist}} \sin\frac{2\pi\, t}{1\text{ d}} \tag{1}$$

that matches the integrated power[1] of the observed diurnal component when sampled and processed identically. Its amplitude is $a_{\text{dist}} = 0.18$ ps, which agrees with the perceived daily delay variation. The corresponding fractional frequency variation is

$$\Delta y_{\text{dist}}(t) = \frac{2\pi}{1\text{ d}} a_{\text{dist}} \cos\frac{2\pi\, t}{1\text{ d}}\ , \tag{2}$$

and we take the maximal frequency excursion $\Delta\hat{y}_{\text{dist}} = 2\pi\, a_{\text{dist}}/(1\text{ d})$ to bound the uncertainty due to diurnal variations in the HM signal distribution as $u_{\text{dist}} = 1.33 \times 10^{-17}$. This applies to any short measurement, but for extended durations $T > (1\text{ d})/\pi$ (or 27 502 s), the peak-to-peak variation of $\Delta\tau_{\text{t}}(t)$ further limits the frequency excursion to $\Delta\hat{y}_{\text{dist}}^{\text{ext}} = 2\, a_{\text{dist}}/T$. We do not scale the results for the doubled length of the loopback connection since the distribution amplifier and other non-duplicated components in the comb laboratory are expected to contribute significantly to the delay variation.

[1] We consider a frequency bandwidth corresponding to ±0.5 d phase variation over one year.

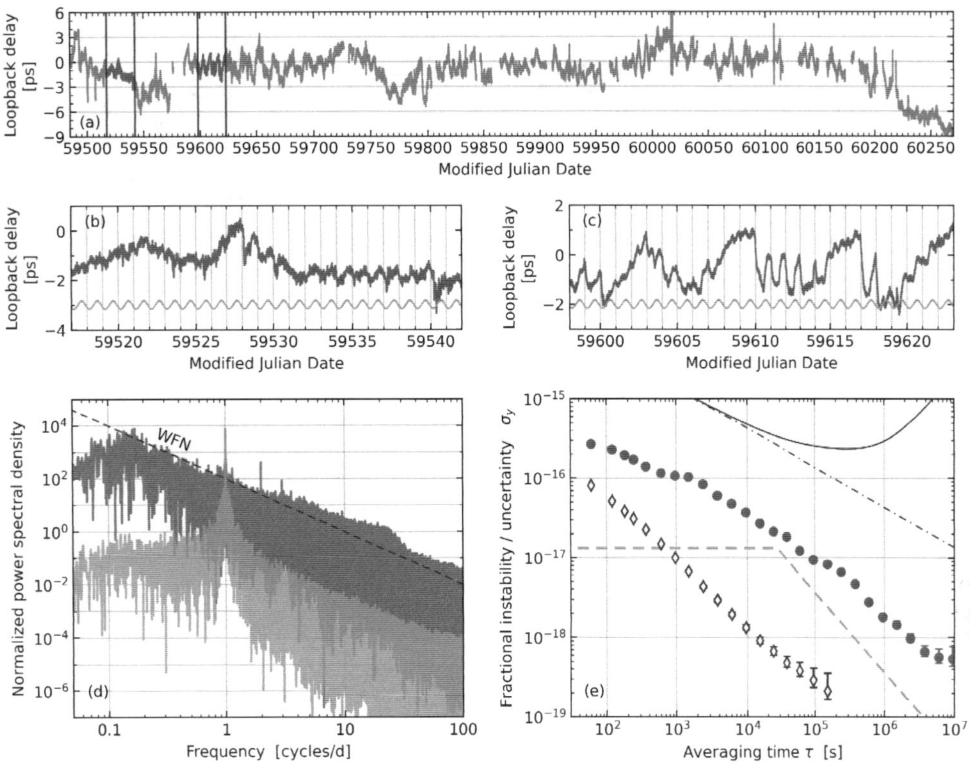

Figure 3. (a) Loopback delay fluctuations from October 2021 to November 2023. The phase is reset to zero for interruptions > 1 d. (b, c) Expanded plots of the continuous segments highlighted in (a). The orange curve (offset for clarity) represents the amplitude of the diurnal component extracted from the periodogram. (d) Lomb-Scargle periodogram of the pre-filtered data (blue). The sharply peaked orange distribution is the periodogram of a diurnal sinusoidal variation with 0.18 ps amplitude after identical sampling and filtering. The dashed line indicates the expected slope for WFN. (e) Allan deviation of the loopback measurement (full circles) and a null measurement of the phase comparator (open diamonds). The orange dashed line shows the modelled systematic uncertainty contribution due to diurnal delay variations, which decreases for continuous measurements with $T > 27\,502$ s. Black lines represent the HDev of the HM instability: The solid line shows the full HM model, the dot-dashed line is the WFN contribution.

3.2. Diurnal variation of time difference measurements

We perform a similar test for a diurnal variation in the measurements of the dual-mixer time-difference (DMTD) measurement system [14] central to the JST system. At any time, the source of the UTC(NICT) signal is one of three candidate signals A to C, which should thus appear identical to the DMTD system. The changes of the selected candidate signal split the full span of 3365 days into 11 sub-intervals. The residual difference in the DMTD measurements is analysed separately for each sub-interval. Fixed and slowly varying measured time offsets are again removed by a high-pass filter with $f_{\mathrm{cut}} = 1/(10\ \mathrm{d})$ before a Lomb-Scargle periodogram is calculated to find a power-matched sinusoidal diurnal time variation. The evaluation of one example sub-interval is shown in figure 4. Over these of 632 days, candidate signal C serves as the reference, and the matched diurnal test function has an amplitude of 0.26 ps.

An average weighted by sub-interval duration yields $a_{\mathrm{DMTD}} = 0.19$ ps as the amplitude of diurnal variations in the timing measurements between HMs and UTC(NICT). This corresponds to an

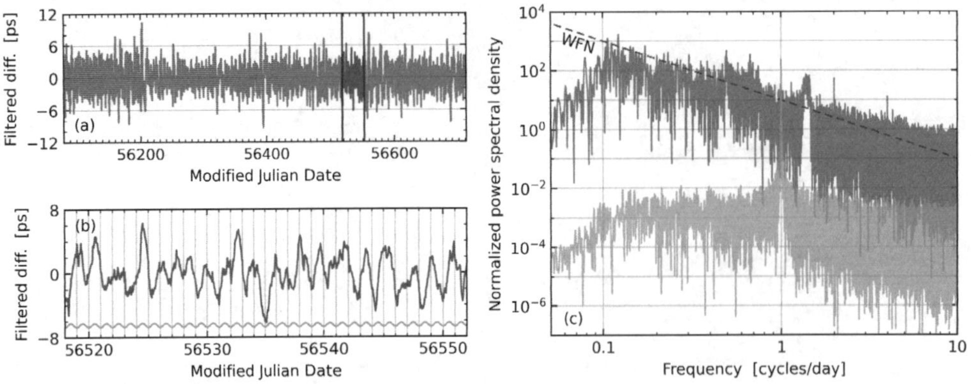

Figure 4. (a) One of the 11 sub-intervals used to investigate diurnal variation of DMTD measurements, showing the filtered time difference between UTC(NICT) and the presently selected candidate signal C. (b) Expansion of the segment highlighted in (a): While a periodic variation is visible, it does not follow a daily cycle. The orange curve (offset for clarity) represents the extracted diurnal component. (c) Lomb-Scargle periodogram of the filtered data (violet). The sharply peaked orange distribution is the periodogram of a diurnal sinusoidal variation with 0.26 ps amplitude, filtered and sampled identically. Dashed line is the slope expected for WFN.

uncertainty $u_{\mathrm{DMTD-d}} = 1.35 \times 10^{-17}$ for short frequency measurements, and $u^{\mathrm{ext}}_{\mathrm{DMTD-d}} = 3.72 \times 10^{-13}$ s / T for extended durations $T > 1$ d/π. Where the reference HM acts as a flywheel, comparisons against UTC(NICT) are usually performed over such extended durations, leading to significantly reduced uncertainties. The full uncertainty $u_{\mathrm{DMTD-d}}$ mainly applies where time-resolved measurements from additional HMs are used to form an ensemble average and suppress the effects of the reference HM's stochastic deviations from a linear drift [11].

4. Sources of continuous error

4.1. Frequency evaluation errors

We also investigate the DMTD system for continuous frequency errors. For this, the DMTD measurements of the 5 MHz input signals are compared to measurements of the associated 1-PPS signals taken by a time interval counter (TIC). Potential sources of error such as slips in counting the DMTD input signals, or numerical errors in the data evaluation, are not expected to equally affect both systems.

We use the records of the JST system over 4110 days to evaluate the difference $\Delta y_{\mathrm{m}} = y_{\mathrm{DMTD}} - y_{\mathrm{TIC}}$. After removing[2] disturbances from HM maintenance, wiring or frequency changes, and technical problems with the DMTD system, we obtain 413 058 hourly intervals in 77 continuous segments across measurements of 8 HMs, sufficient to overcome the phase noise contribution of $\sigma_y(\tau) \approx 10^{-10}/(\tau/\mathrm{s})$ present in the TIC measurements (figure 5c). Over the full set of data, the weighted mean is $\overline{\Delta y_{\mathrm{m}}} = -2.2 \times 10^{-18}$ with an uncertainty of $\sigma = 2.2 \times 10^{-18}$ (adjusted for overscatter with $\chi^2 = 1.1$). From this we set a constraint to a systematic, continuous error in the measurements of the DMTD system as $u_{\mathrm{DMTD-c}} = (\overline{\Delta y}_{\mathrm{m}}^2 + \sigma^2)^{1/2} = 3.1 \times 10^{-18}$.

An alternative interpretation of a discrepancy would be a timebase error in the delay measurements of one of the two systems. This manifests as a frequency error $\Delta y_{\mathrm{m}} = my$, proportional to the measured frequency deviation. This deviation differs between segments due to the varying and drifting HM frequencies. We fit the results of the individual segments to find $m = -0.2(5.1) \times 10^{-6}$. In recent

[2] We inspect and select the data used in TAI or clock frequency measurements to exclude the same problems.

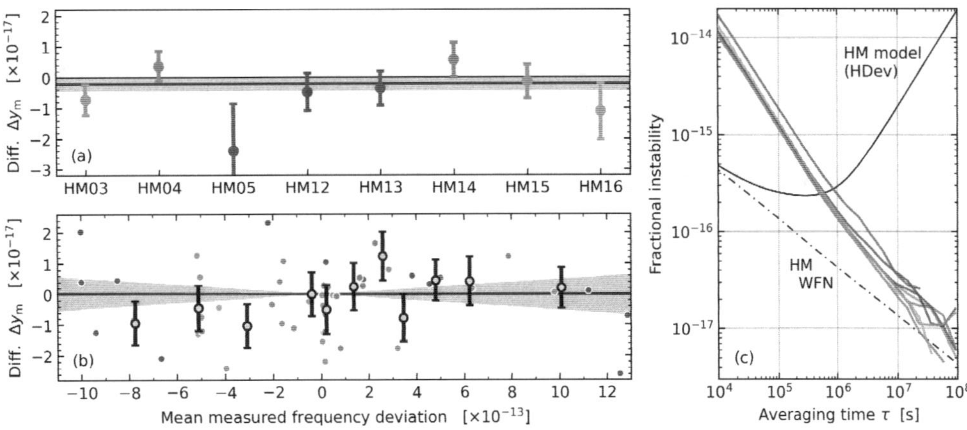

Figure 5. (a) The overall difference between frequency measurements by DMTD and TIC systems is $-2.2(2.2) \times 10^{-18}$ (blue confidence band) and the results for individual HM channels are likewise consistent with zero. (b) Plotting the difference over the measured frequency tests for a timebase error. The blue confidence band shows the slope of $m = -0.2(5.1) \times 10^{-6}$ for a fit to the means of individual segments indicated by full circles coloured according to (a). More points fall outside the plotted range. To visualize the absence of a slope, all such points are sorted by frequency and averaged in 11 bins chosen to represent similar amounts of measurement time. The binned results are shown by black open circles with error bars. (c) Lines coloured according to (a) show the ADev of the concatenated differences for individual HMs. The short-term instability results from the TIC measurements. Black lines show the HDev HM model and its WFN component (dot-dashed) for reference.

operation, HM frequency adjustments to $|y| \leq 4 \times 10^{-13}$ constrain the resulting uncertainty to less than $u_{\mathrm{DMTD-c}}$, such that there is no need for an increase. As neither evaluation shows statistically significant evidence for a measurement error, $u_{\mathrm{DMTD-c}}$ may reduce further with future evaluation.

4.2. *Errors in optical to radiofrequency comparison*

The final link in the chain illustrated in figure 2 is the frequency comb system that accepts the 100 MHz signal from the HM and upconverts it to the optical domain, where it is compared with the stabilized 429 THz laser output provided by NICT-Sr1. To test for errors, we have implemented two parallel systems with different configurations.

For frequency comb 1, the 100 MHz signal is upconverted to 9.3 GHz by a phase-locked dielectric resonator oscillator (PLDRO). Its output signal is used to downmix the 37th harmonic of the comb's repetition rate $f_{r1} \approx 250$ MHz. The resulting 50 MHz difference signal is then used to both measure and stabilize the comb's repetition rate. The frequency comb itself is installed on a temperature-stabilized optical breadboard and the clock laser frequency is measured directly as a beat signal with the frequency-doubled output of a comb branch with a suitably tailored spectrum.

Frequency comb 2 is constructed by a different manufacturer and operates at the laboratory temperature. A separately distributed 100 MHz signal is upconverted to 8.2 GHz by a PLDRO, where it is used to downmix the 82nd harmonic of the comb's repetition rate, which has been set to $f_{r2} \approx$ 99.995 MHz. The approximately 400 kHz difference signal is used to measure and stabilize the repetition rate. A transfer laser at 215 THz is phase-locked to the clock laser using a waveguide PPLN frequency doubler to create a beat signal. The beat of this laser with the frequency comb is then counted to determine the clock laser frequency.

Both combs operate continuously to avoid thermalization effects, and all frequency measurements, including the $f - 2f$ signal used to stabilize the carrier-envelope offset, are recorded by deadtime-free

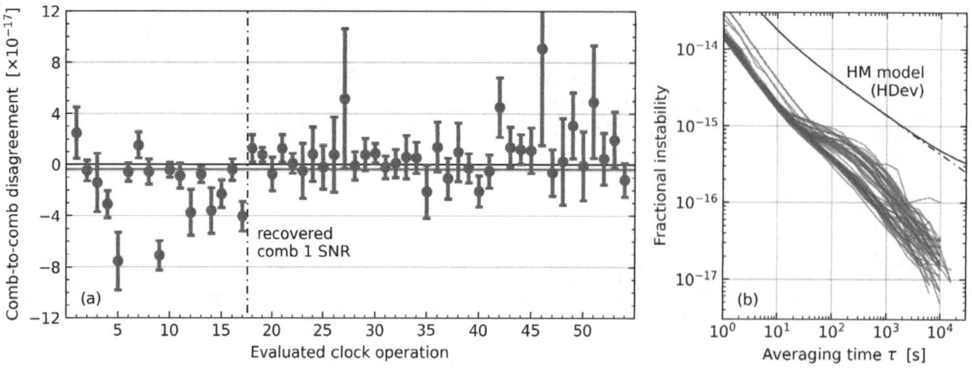

Figure 6. (a) Difference between HM frequency measurements of combs 1 and 2. After adjustment of comb 1 (dot-dashed line), individual differences are statistically consistent with zero. The blue line is the 1σ confidence band $\overline{\Delta y}_{\mathrm{cc}} = -3.8(1.6) \times 10^{-18}$. (b) ADevs calculated over individual time-resolved data sets (coloured lines) confirm suppression of common-mode HM noise. Black lines show the HM model HDev for reference.

phase-averaging multichannel counters. The measurements are synchronized to within a millisecond by using the UTC(NICT) PPS signal as a starting trigger. The combs were operated simultaneously throughout a 2021 to 2022 TAI calibration campaign, and the results are shown in figure 6. Initial measurements included significant disagreement with $\Delta y_{\mathrm{cc}} = y_{\mathrm{c2}} - y_{\mathrm{c1}}$ of up to -8×10^{-17}. After adjusting frequency comb 1 to recover lost signal-to-noise ratio in the beat signal, later measurements agree within statistical expectations. Over the 54 measurements in the campaign, we find a weighted mean of $\overline{\Delta y}_{\mathrm{cc}} = -3.8 \times 10^{-18}$ with an uncertainty of $\sigma = 1.6 \times 10^{-18}$. From this we set a constraint to the expected persistent error in the measurements of either comb as $u_{\mathrm{comb}} = (\overline{\Delta y}_{\mathrm{cc}}^2 + \sigma^2)^{1/2} = 4.1 \times 10^{-18}$, which includes the contributions from the PLDROs and the laser distribution system. If we were to exclude the first 17 measurements, we would find $\overline{\Delta y}_{\mathrm{cc}}^{\mathrm{rec}} = +3.7 \times 10^{-18}$ instead, with no significant impact on the uncertainty evaluation.

Calculating the ADev over the time-resolved Δy_{cc} for each measurement shows a common-mode suppression of the HM instability contribution by at least one order of magnitude. The noise introduced by the optical-to-rf comparison will thus remain insignificant until measurements with close to 100% uptime seek to exactly characterize the complete HM signal over an extended period.

5. Systematic uncertainty of the lab-side frequency link

These four components, summarized in table 1, characterize the uncertainty in the link between the frequency standard NICT-Sr1 and the local timescale reference UTC(NICT) from systematic effects, as listed in the Circular T. The value of $u_{\mathrm{B/Lab}} = 2.0 \times 10^{-17}$ applies for working conditions in a TAI calibration measurement such as the one described in references [12,13], the first to reach the evaluation uncertainty of $< 2 \times 10^{-16}$ (table 2) targeted for optical clock contributions in the roadmap [8] to the redefinition of the SI second prepared by the Consultative Committee for Time and Frequency.

6. Confirming deadtime and interpolation uncertainties

The largest uncertainty contribution in table 2 is $u_{\mathrm{l/Tai}}$ for the GNSS link to TAI. Managing this uncertainty relies on the flywheel HM to connect intermittent measurements into a long uninterrupted period. The HM's frequency noise and instability then determine the statistical contribution $u_{\mathrm{A/Lab}}$ of the link from NICT-Sr1 to the UTC(NICT) signal that serves as reference in the GNSS connection.

Our previously presented stochastic model of the HM behaviour [11] can now be tested using continuous HM characterization data obtained in NICT-Sr1's typically weekly operating schedule,

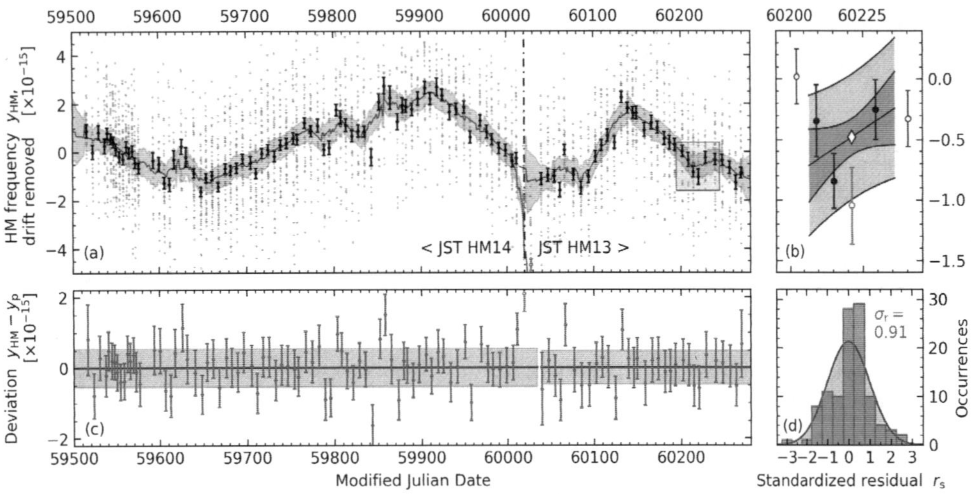

Figure 7. (a) The 110 measurements of JST HM14 and JST HM13 between Oct 29, 2021, and Nov 21, 2023, provide a continuous evaluation of HM frequencies shown by the blue 1σ confidence band with $< 5 \times 10^{-16}$ uncertainty in regular operation. For each HM, a constant linear drift has been removed to better show stochastic variations. Light blue points represent partially binned data to visualize HM WFN. The final three measurements of HM14 (red open circles, only two are visible) were excluded from the statistical analysis due to progressive failure of the HM. (b) Interpolation and prediction. The displayed region corresponds to the rectangle marked in (a). The result of the selected measurement (central red open circle) is predicted (open diamond) by fitting all other datapoints inside an optimized interval, here $D = 29$ d. Prediction uncertainties are obtained from the fit (dark-shaded inner confidence band) and combined (outer confidence band) with the calculated FFN and FWFM contributions. (c) For each measurement, we determine its deviation Δ from the prediction and a combined uncertainty u_Δ indicating expected variance. The blue shaded regions show the standard deviation of Δ for JST HM14 (5.4×10^{-16}) and JST HM13 (4.8×10^{-16}). (d) The distribution of the standardized residuals $r_s = \Delta/u_\Delta$ (histogram) agrees with the expectation of an unbiased normal distribution of unity width (blue bell curve).

consisting of 75 points for JST HM14 (October 29, 2021, to March 8, 2023) and 32 points for JST HM13 (April 5 to November 21, 2023). The final measurements of HM14 exhibited abnormal behaviour due to a depleted hydrogen supply and have been excluded from the evaluation shown in figure 7.

Individual measurements yield a fractional HM frequency deviation $y_{\mathrm{HM},i}$ with statistical uncertainty σ_i for a barycentre time t_i. A successful model will predict each measurement $y_{\mathrm{HM},i}$ from the set of other measurements $y_{\mathrm{HM},j\neq i}$ with a well-characterized precision. We determine the prediction $y_{\mathrm{p}}(t_i)$ from a linear fit of all data within a duration D centred on t_i, except for the data contributing to $y_{\mathrm{HM},i}$ itself. The fit provides σ_{p} for those uncertainties in determining mean frequency and drift that result from WFN in the $y_{\mathrm{HM},j\neq i}$ data. We also calculate the effects of stochastic variation in long-term HM behaviour as

$$u_{\mathrm{stoc}}^2 = \int_0^\infty S_y(f)|G_\Delta(f)|^2\,df\,, \tag{3}$$

as used in reference [11] and described in detail in the supplement to reference [15]. Here

$$S_y(f) = \underbrace{4.0 \times 10^{-32}(f/\mathrm{Hz})^{-1}}_{\mathrm{FFN}} + \underbrace{3.0 \times 10^{-45}(f/\mathrm{Hz})^{-3}}_{\mathrm{FWFM}} \tag{4}$$

is the modelled power spectral density of the HM's FFN and FWFM. $G_\Delta(f)$ is the Fourier transform of a differential weighting function $g_\Delta(t) = g_i(t) - g_{j\neq i}(t)$. The normalized distributions g_i and $g_{j\neq i}$

describe the separate clock operating intervals used to determine either $y_{\mathrm{HM},i}$ or the $y_{\mathrm{HM},j\neq i}$ used in the linear fit. We evaluate different durations D to find the smallest overall prediction uncertainty u_p according to $u_\mathrm{p}^2 = \sigma_\mathrm{p}^2 + u_\mathrm{stoc}^2$. For the example shown in figure 7(b), this yields $D = 29$ d and $u_\mathrm{p} = 5.3 \times 10^{-16}$ (from $\sigma_\mathrm{p} = 1.9 \times 10^{-16}$ and $u_\mathrm{stoc} = 5.0 \times 10^{-16}$).

The WFN from $y_{\mathrm{HM},j\neq i}$ is included in σ_p, but the WFN in the measurement of $y_{\mathrm{HM},i}$ still needs to be accounted for. The deviation from the prediction, $\Delta = y_{\mathrm{HM},i} - y_\mathrm{p}(t_i)$, is then expected to show a variance $u_\Delta^2 = \sigma_i^2 + \sigma_\mathrm{p}^2 + u_\mathrm{stoc}^2$. To test the model, we introduce the standardized residual $r_\mathrm{s} = \Delta/u_\Delta$. Over the total number of 107 predictions, we find a mean value of $\bar{r}_\mathrm{s} = 0.05$ with a standard deviation $\sigma_\mathrm{r} = 0.91$, in excellent agreement with the expectation of an unbiased, unity-width normal distribution. The standard deviation over all Δ is 5.4×10^{-16} for JST HM14 and 4.8×10^{-16} for JST HM13.

We use the same procedure to create a daily record of interpolated HM frequencies, representing the expected result for a full-day measurement by DMTD or satellite link. The uncertainties of the interpolation are $< 5 \times 10^{-16}$ except for individual days during larger-than usual gaps between measurements.

Table 1. Contributions to the systematic lab-side link uncertainty $u_\mathrm{B/Lab}$ between NICT-Sr1 and UTC(NICT) under usual working conditions.

Contribution		Uncert. $[10^{-17}]$
Signal distribution, diurnal	u_dist	1.33
Frequency eval., diurnal	$u_\mathrm{DMTD-d}$	1.35
Frequency eval., continuous	$u_\mathrm{DMTD-c}$	0.31
Optical-to-rf comparison	u_comb	0.41
Total	$\boldsymbol{u_\mathrm{B/Lab}}$	**1.96**

Table 2. Uncertainties for the TAI calibration reported in Circular T 408 (Dec 2022). NICT-Sr1 operated for 7.2% of a 40-day interval.

Contribution		Uncertainty $[10^{-15}]$
Clock statistics	u_A	0.01
Clock systematics	u_B	0.07
Link statistics	$u_\mathrm{A/Lab}$	0.08
Link systematics	$u_\mathrm{B/Lab}$ [a]	0.02
GNSS link to TAI	$u_\mathrm{l/Tai}$	0.15
Total	$\boldsymbol{u_\mathrm{tot}}$	**0.19**

[a] A preliminary value of $u_\mathrm{B/Lab} = 2.3 \times 10^{-17}$ was used in the report.

7. Summary

We have evaluated the link of our optical lattice clock NICT-Sr1 to the rf timescale signal UTC(NICT) with a systematic uncertainty $u_\mathrm{B/Lab} < 2 \times 10^{-17}$. This link makes it possible to use one of the HMs in the ensemble operated as part of the Japan Standard Time system as a flywheel for the optical clock, allowing it to contribute to national [16] and international time [13] with statistical uncertainties down to $u_\mathrm{A/Lab} < 10^{-16}$, even with less than 10% overall duty fraction (tables 1 and 2). Due to the excellent long-term stability, the flywheel HM can maintain its frequency characterization to $< 10^{-15}$ fractional uncertainty even if the clock is not (or cannot be) operated for a week or more. This opens a smooth migration path from rf frequency standards, through the initial integration of intermittently operating optical standards, to robust optical systems that operate with little or no downtime.

Even before the technology is ready for the leap to all-optical timescales [7], the small magnitudes of the excess statistical and systematic uncertainties support notable improvements in stability and timekeeping from the application of optical standards and the introduction of an optical SI second.

References

[1] Huntemann N, Sanner C, Lipphardt B, Tamm C and Peik E 2016 Single-ion atomic clock with 3×10^{-18} systematic uncertainty *Phys. Rev. Lett.* **116** 063001 DOI: 10.1103/PhysRevLett.116.063001

[2] McGrew W F *et al.* 2018 Atomic clock performance enabling geodesy below the centimetre level *Nature* **564** pp 87–90
DOI: 10.1038/s41586-018-0738-2

[3] Brewer S M, Chen J-S, Hankin A M, Clements E R, Chou C W, Wineland D J, Hume D B and Leibrandt D R 2019 $^{27}Al^{+}$ quantum-logic clock with a systematic uncertainty below 10^{-18} *Phys. Rev. Lett.* **123** 033201
DOI: 10.1103/PhysRevLett.123.033201

[4] Oelker E *et al.* 2019 Demonstration of 4.8×10^{-17} stability at 1s for two independent optical clocks *Nature Photonics* **13** pp 714–719
DOI: 10.1038/s41566-019-0493-4

[5] Huang Y, Zhang B, Zeng M, Hao Y, Zhang H, Guan H, Chen Z, Wang M and Gao K 2022 Liquid-nitrogen-cooled Ca^{+} optical clock with systematic uncertainty of 3×10^{-18} *Phys. Rev. Applied* **17** 034041
DOI: 10.1103/PhysRevApplied.17.034041

[6] Dick G J 1987 Local oscillator induced instabilities in trapped ion frequency standards *in Proc. of 19th Precise Time and Time Interval Meeting, Redondo Beach* pp 133–147 (Washington, DC: U.S. Naval Observatory)

[7] Milner W R *et al.* 2019 Demonstration of a timescale based on a stable optical carrier *Phys. Rev. Lett.* **123** 173201
DOI 10.1103/PhysRevLett.123.173201

[8] Dimarcq N *et al.* 2023 Roadmap towards the redefinition of the second *accepted for publication in Metrologia*
DOI 10.1088/1681-7575/ad17d2

[9] Hachisu H, Petit G, Nakagawa F, Hanado Y and Ido T 2017 SI-traceable measurement of an optical frequency at the low 10^{-16} level without a local primary standard *Opt. Express* **25** pp 8511–23
DOI: 10.1364/OE.25.008511

[10] Hachisu H and Ido T 2015 Intermittent optical frequency measurements to reduce the dead time uncertainty of frequency link *Jpn. J. Appl. Phys.* **54** 112401
DOI: 10.7567/JJAP.54.112401

[11] Nemitz N, Gotoh T, Nakagawa F, Ito H, Hanado Y, Ido T and Hachisu H 2021 Absolute frequency of ^{87}Sr at 1.8×10^{-16} uncertainty by reference to remote primary frequency standards *Metrologia* **58** 025006
DOI: 10.1088/1681-7575/abc232

[12] Hachisu H *et al.* 2022 Evaluation of y(UTC(NICT)) with respect to NICT-Sr1 for the period MJD 59539 to 59579 *available online from BIPM at* www.bipm.org/en/time-ftp/data as https://webtai.bipm.org/ftp/pub/tai/data/PSFS_reports/nict-sr1_59539-59579.pdf

[13] BIPM 2022 Circular T 408 *available online at* www.bipm.org/en/time-ftp/circular-t

[14] Hanado Y, Imamura K, Kotake N, Nakagawa F, Shimizu Y, Tabuchi R, Takahashi Y, Hosokawa M and Morikawa T 2008 The new generation system of Japan Standard Time at NICT *Int. J. Navig. Obs.* **2008** 841672
DOI: 10.1155/2008/841672

[15] Grebing C, Al-Masoudi A, Dörscher S, Häfner S, Gerginov V, Weyers S, Lipphardt B, Riehle F, Sterr U and Lisdat C 2016 Realization of a timescale with an accurate optical lattice clock *Optica* **3** pp 563-9
DOI: 10.1364/OPTICA.3.000563

[16] Hachisu H, Ito H, Nemitz N, Ohtsubo N, Miyauchi Y, Morikawa M, Matsubara K and Ido T 2023 UTC(NICT) referenced to a timescale based on the optical clock NICT-Sr1 *in Proc. EFTF/IFCS*, Toyama, Japan 2023
DOI: 10.1109/EFTF/IFCS57587.2023.10272119

Journal of Physics: Conference Series **2889** (2024) 012020 doi:10.1088/1742-6596/2889/1/012020

Robust design and performance of NPL Cs fountain clocks

K Szymaniec[1], R J Hendricks[1], J Whale[1], A Wilson[1], S Walby[1,2], M Knapp[1], C J Foot[2]

[1]National Physical Laboratory, Teddington, UK
[2]Department of Physics, University of Oxford, Oxford, UK

email: krzysztof.szymaniec@npl.co.uk

Abstract
We report on developments in the atomic fountain systems being built and operated at NPL. An improved generation of Cs fountains has been developed, with units being constructed for use by both NPL and commercial customers. These systems combine world-class stability and accuracy with increased reliability and can run for long periods of time without maintenance. Here we describe how the NPL fountains work, and present performance data for the latest systems. We describe some of the applications of these fountains, both in time scale implementation and fundamental science. We also present an overview of a miniature atomic fountain that is being developed, which will help make fountain technology accessible to a wider range of sectors.

1. Introduction

Atomic fountain clocks have become ubiquitous in maintaining the most stable national time scales within a small margin of UTC and, as primary or secondary frequency standards, have the dominant role in the computation of the TAI time scale. A number of national measurement institutes and other labs have pursued their development, reaching accuracy close to only one part in 10^{16} (for recent evaluation reports see, for example, [1][2][3]). After three decades of effort and many improvements the designs of fountain clocks have reached a high level of maturity and recent work has been aiming towards increased robustness and reliability.

At NPL we have adopted a distinctive design approach offering the highest accuracy and best short-term stability, yet with relatively simple construction and operation. With the confidence gained from several years of operation of NPL-CsF2, the approved primary frequency standard for the UK [4][5], we have designed an upgraded version of the fountain, which was subsequently offered and delivered to customers under the rigour of commercial contracts. Devices have been constructed and sent to other national measurement institutes, large scientific facilities, and potential future customers include providers of telecommunication and financial services.

In this presentation we review the key features of the NPL design (section 2) and the performance of selected realisations of the clocks (section 3). We also refer to a technique for further accuracy improvement by limiting a leading cause of systematic bias, the distributed cavity phase effect. Implementation of the highly reliable fountain clocks to steer a local time scale is covered in section 4. Finally, in section 5, we present the concept of *mini-fountain*, a compact version of the clock, which aims to achieve comparable performance to the NPL fountains but with much reduced size of the physics package and other subsystems including the optical bench.

Journal of Physics: Conference Series **2889** (2024) 012020 doi:10.1088/1742-6596/2889/1/012020

2. Features of the NPL Cs fountain design

The general features of fountain clocks are replicated in the majority of working devices [6][7]. Laser cooled atoms are collected and launched vertically in a cloud with a temperature of a few microkelvin. On both ascent and descent, they pass through a microwave cavity that is resonant with the clock transition between the hyperfine sublevels of the ground state atoms. The two interactions with microwaves in the cavity separated by the field-free period constitute a Ramsey interrogation scheme, with a typical Ramsey fringe width of about 1 Hz. Laser induced fluorescence signals from a two-zone time-of-flight detection region enable direct inference of normalised probability of the clock transition.

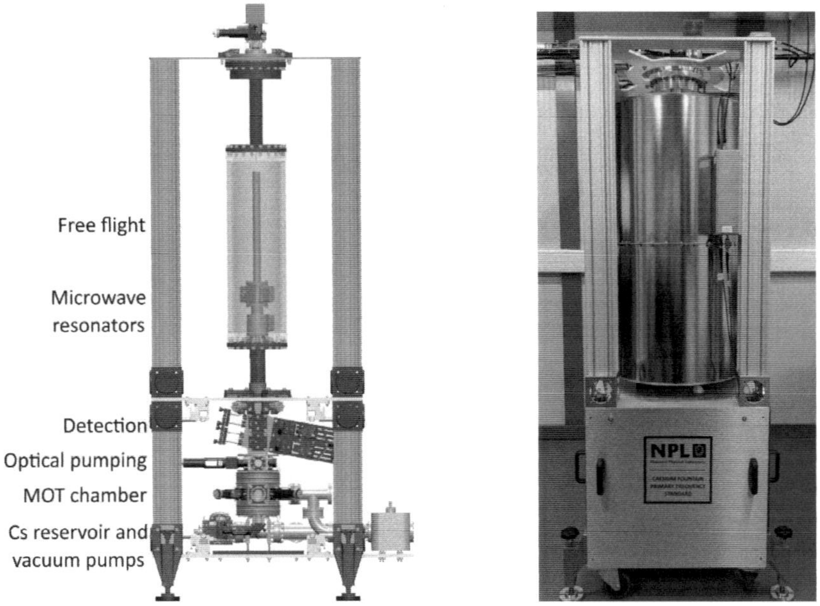

Figure 1. View of the 'full size' physics package of the improved NPL Cs fountain design. *Left:* drawing showing the layout and main functional regions (C-field coil and magnetic shields are not shown). *Right:* photograph of an assembled unit.

The variations between different fountain designs generally relate to how the cold atoms are produced and prepared in a suitable state for the interrogation. Furthermore, various approaches are used to control systematic effects such as cold collisions, blackbody radiation, microwave leakage or distributed cavity phase effects. The specific features of the NPL design are summarised below and more details can be found in [8]. See also fig. 1 for the illustration of the physics package.

- The cold caesium atoms are initially collected and cooled in a single-stage vapour-loaded magneto-optical trap (MOT), using an easy to align (0,0,1) optical configuration. The vapour is produced by a temperature controlled (cooled) reservoir of caesium metal.
- After the MOT fields are switched off the initial atom cloud size is tuned with a short and variable period of expansion before the subsequent phase of molasses cooling.
- The population of the initial clock state (F=4, m_F=0) is boosted by an optical pumping (OP) pulse [9] applied in an additional light access area between the MOT and detection chambers. The pulse is derived from the cooling laser and tuned to the (F=4 $\rightarrow$ F'=4) component of the D2 transition. A pulsed magnetic field from a pair of small Helmholtz coils defines the quantisation axis for the linear π polarisation of the OP light. This technique increases the detection signal approximately five-fold, which is an optimum between a higher spin polarisation and heating due to random scattering of photons resulting in truncation of the cloud.
- The microwave cavity assembly consists of two cavities, with the lower one used for the usual clock state purification, which is necessary due to the limitations of the OP method used. The

upper one is the Ramsey cavity specially designed to minimise the distributed cavity phase effect [10]. The atoms' flight above the Ramsey cavity is entirely enclosed in a narrow tube that forms a waveguide below cut-off and prevents their exposure to any leaked resonant microwave field.

- The chamber housing the cavities assembly and the flight tube is built with a double wall water jacket, which is used to control the temperature of this area. This helps to finely tune the Ramsey cavity to the atomic (clock) resonance as well as to improve the uniformity of the temperature of the surroundings and the resulting blackbody radiation.
- The detection chamber is designed to maximise the solid angle of the fluorescence collection with retroreflectors nearly doubling the time-of-flight signal. In addition, the laser beams inducing the fluorescence are tilted by 10° from the horizontal to reduce resonant scattering by the thermal background by means of the Doppler effect [11]. The intensity of these beams is stable to 0.1%

The complete system, including the physics package, can be transported fully assembled by road or air. Subsystems, such as control electronics, microwave generation and optical bench are all mounted in just two full height 19-inch racks. The optics (including lasers) are built in a modular way and entirely mounted in one of the racks without need for an optical table. They incorporate a distributed Bragg reflector (DBR) type diode laser for cooling and detection and a distributed feedback diode (DFB) as repump laser.

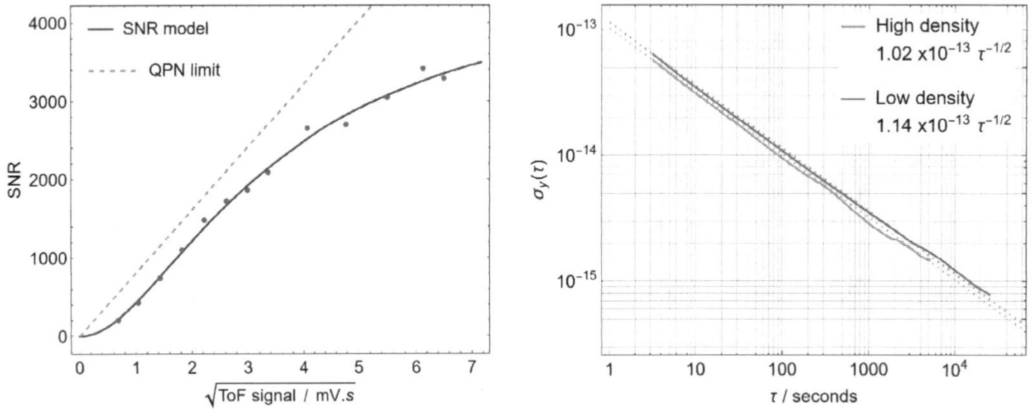

Figure 2. *Left:* Measurements of fountain signal-to-noise ratio (SNR) as a function of the square root of the atom density, showing SNR values in excess of 3000. The fitted SNR model enables us to estimate the effect of quantum projection noise (QPN) and shows that this, rather than technical noise, is dominant except at very high and low atom number. *Right:* Measurements of the short-term stability at high and low atom density when operating with a quartz local oscillator. The very similar stability for different densities indicates that the measurement is limited by the local oscillator noise.

3. Performance

The combination of launching from a MOT and using the optical pumping technique enables the NPL fountains to operate with a large number of detected atoms. Together with optimised detection, which reduced technical noise, this leads to an excellent signal-to-noise ratio (SNR) that is largely dominated by the quantum projection noise due to the atoms being detected in a superposition of the clock states (fig. 2). Using a low phase noise local oscillator (LO) based, for example, on a cryo-cooled sapphire crystal or on a femtosecond frequency comb locked to an ultra-stable laser, the corresponding short-term stability $\sigma_y(\tau)$ of the fountain would be in the range of 3×10^{-14} at $\tau = 1$ s integration time. When used with an LO based on a room temperature quartz crystal, which has worse phase noise, we observe that the stability is degraded to $\sigma_y(\tau)$ of 1.0×10^{-13} (1 s). For maximum accuracy the atom number (density) must be alternated between high and low, and the measured frequency is extrapolated to zero density. This further degrades the effective short-term stability by up to a factor 3, depending on the LO properties [12].

For accuracy evaluations with low uncertainty as well as for the day-to-day operation of the fountain clock, both a good SNR and a high uptime are required. To improve the latter, we have identified key failure cases, such as, the laser frequency stabilisation, the system control software, temperature control of the flight tube and caesium reservoir, and durability of mechanical shutters on the optical bench. Upon addressing the operational robustness in these cases, the NPL built fountains have demonstrated operation over several months without any unintentional stoppages. NPL-CsF2 has routinely reached uptime of 99%, as measured monthly when not limited by the availability of the microwave signal being measured.

An illustrative uncertainty budget for a fountain clock of the NPL design is given in table 1. The type B values presented are the typical or expected values taken from previous evaluations and measurements, leading to the total uncertainty u_B of below 2×10^{-16}. This is similar to the best fountains currently approved for reporting of data to TAI [1].

Systematic effect	uncertainty /10^{-16}
2nd order Zeeman	0.4
Blackbody radiation	0.7
Cold collisions	0.3
Background gas	0.3
Microwave leakage	0.6
Distributed cavity phase	1.1
Microwave lensing	0.3
Other	0.4
Total u_B	**<2**

Table 1. Values of type-B uncertainties compiled from previously published evaluations [refs NPL, NRC, AOS]. The values shown present an illustrative budget of the accuracy currently achievable by the NPL-built Cs fountain primary frequency standards. Implementation of our new DCP evaluation technique will further improve the accuracy of the system.

Historically, the frequency shift due to cold collisions had been one of the limiting factors in the accuracy of Cs fountain primary frequency standards. Here, this shift is virtually eliminated by tuning the effective collisional energy and adjusting the clock states ratio between the Ramsey interactions [13]. This is achieved despite loading the atoms in the MOT, launching them in a small high-density cloud and implementing the OP stage which largely preserves the high density. As described above, the residual collisional shift must still be corrected for by measuring at both high and low atom densities [12].

In the budget presented in table 1 the largest uncertainty is associated with the distributed cavity phase (DCP) effect, which is essentially a Doppler shift caused by travelling wave components of the microwave field in the cavity. The established treatment follows [14] where these components are approximated as the lowest order terms of an azimuthal series, $m = 0, 1, 2$. It can be shown that the $m = 1$ and $m = 2$ terms may lead to a negligible frequency shift if the atomic trajectories precisely retrace the cavity crossing positions on ascent and descent. Using conventional approaches to optimising the atomic trajectories, the frequency uncertainty introduced by the DCP effect can be limited to about 1×10^{-16}. Following on an earlier work [15] we have developed a technique to precisely determine the crossing positions of the centre of mass of the ensemble of atoms that contribute to the detection time-of-flight signal. The technique uses the approximately linear variation of the transverse components of the cavity field, and we were able to detect and centre the crossing positions to better than 50 μm, which can limit the DCP shift and its uncertainty to less than 1×10^{-17} [16].

4. Applications

The main motivation for the development of atomic fountains was their potential use as primary or secondary frequency standards and, indeed, for the last two decades they have been the dominant source of accuracy for the international time scale UTC [17]. Another related application is as a frequency reference for ultra-high precision spectroscopy measurements, an example being the ALPHA experiment on antihydrogen [18] or measurements of frequency ratios between the clock transitions in Cs and candidate transitions for optical clocks [19][20], the latter leading to constraints on fundamental constants variations, especially on the electron-to-proton mass ratio [21].

While for PFS operation intermittent availability of a fountain is adequate, with their significantly improved reliability they can now operate quasi-continuously, and it is possible to utilise them as a resource enhancing the stability of local UTC time scales. Over the last 5 years, we have used the high availability of the NPL-CsF2 fountain and provided regular and frequent steering corrections for the UTC(NPL) master clock (a hydrogen maser). The fountain's long-term stability is significantly better than that of any maser hence it is sufficient to use a relatively simple algorithm for the steering process where the frequency of the reference clock is acted upon (not the phase). It consists of two components, a frequency feedback loop, and a phase lock loop [22][23]. The first component comes from a direct measurement of the reference maser frequency with an added term due to its anticipated linear drift. The drift is also determined by the fountain, but over a longer period of 2 months. For the phase lock component, the high uptime of the fountain is even more important as it effectively acts as a holdover clock. This term additionally includes a correction for an accumulated time offset between UTC and UTC(NPL) due to imperfect steering during the preceding latency period. Fig. 3 shows the exemplary performance of UTC(NPL) prediction over a typical BIPM reporting period with an added latency.

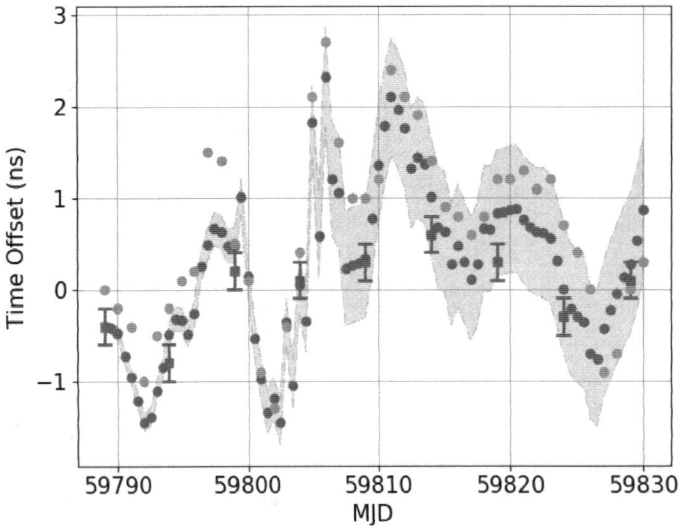

Figure 3. Time offset of UTC(NPL) from UTC as estimated in real time by Cs fountain NPL-CsF2 (blue points). The method's capability is illustrated by a data set when a less stable maser was used as the reference for UTC(NPL). The first point is obtained from the preceding month's Circular T, and the shaded region shows the uncertainty of this estimate as the time progresses. Also plotted are subsequent UTCr (orange circles) and UTC data (red squares), which arrive with a latency of up to 10 and 45 days, respectively.

Operational time scales typically use hydrogen masers as flywheel oscillators to cover any gaps in the frequency standards' availability. To reduce the phase noise on the maser reference signal and the corresponding degradation of the fountain short-term stability a clean-up oscillator is implemented (see above sec 3 and fig. 2). We have recently acquired a cryo-cooled sapphire oscillator (CSO) and have investigated its application not only as a clean-up but also as a flywheel, that is, its stability in long term. We have observed substantial CSO linear frequency drift but noticed that it is highly predictable and can be efficiently removed in a feed-forward loop. With such correction in place, it should be possible to operate a time scale with a CSO as reference, which could be disciplined directly by a Cs fountain (without a need for a maser in the system). The flicker floor of a corrected CSO would be at the level of 10^{-15}, fractionally, such that the typical gaps in fountain operation of a few hours or a day could be tolerated.

5. Mini-fountain concept

In the preceding sections we have described the optimised design and applications of a 'full size' caesium fountain clock, which achieves the best stability, accuracy, and reliability. The mini fountain concept (*mini-f*) described in this section is driven by the needs of industrial users and service providers for a time scale realisation which would be stable to several nanoseconds over up to 6 weeks, the latency period of UTC feedback. The key design requirement is to reduce as much as practical the size, weight and power consumption (SWaP), complexity, and cost, while achieving similar performance to a fountain clock operating with a quartz-based local oscillator, i.e. short-term stability of 3×10^{-13} (1 s), and in the long-term better than 1×10^{-15}. The ability to independently characterise the accuracy (systematic frequency shifts), as in primary or secondary frequency standards, should be possible but is not essential.

Fountain technology appears well suited to meet these requirements, provided appropriate design choices are made. For example, while reducing the overall size of the physics package, we would leave unchanged the actual fountain height (distance from the cavity centre to the atomic trajectory apex) and hence the clock linewidth. The Ramsey cavity would be sealed as part of the vacuum vessel and the magnetic shielding would only have two layers. Importantly, there would be no separate detection chamber or state selection cavity. A simple antenna emitter would be used for the state selection and the unselected $m_F \neq 0$ population would be removed by a pulse from the downwards-going cooling beams.

The final clock states' populations will be detected by the same beams used for the initial trapping and cooling. A new normalised detection method will be implemented as shown in fig. 4a. First, as the descending cloud reaches the MOT region, a pulse from all six cooling beams induces fluorescence only from the population of the 'upper' clock state, N_u, which decays as the atoms are off-resonantly pumped to the 'lower' clock state, which only couples to the repump, with the initial population N_l. Next, a short repump pulse puts all the atoms in the upper state. This is followed by another cooling beam pulse, which induces a second fluorescence signal proportional to $N_{tot} = N_l + N_u$. Alternatively, the repump pulse can be combined with the second cooling pulse, as would be necessitated by the use of an EOM for generation of repumping light. A substantial background fluorescence, resulting from the atomic vapour present in the MOT chamber, is measured and subtracted after one more cooling pulse, delayed so that the cold atoms do not contribute. The clock transition probability is directly obtained as $P = N_u/N_{tot}$.

We have tested the new detection scheme in a simple cold atom source demonstrator and in our experimental fountain setup NPL-CsF3. The scheme was shown to work robustly, and preliminary results are shown in fig. 4b. The high-atom-number limit to the SNR was found to be around 800, limited by laser noise. However, at an atom number corresponding to a typical MOT loading time ($N_{at}\approx 0.5$ a.u.), the SNR is limited by technical noise to only around 100. This noise can be reduced with a purpose-designed cooling/detection chamber, where the fluorescence collection solid angle is optimised, such that we anticipate a detection signal-to-noise ration of around 500. The tests were performed on Cs, however, in the actual mini-f clock we plan to use ^{87}Rb to increase the clock-state population without the need for optical pumping, as well as to reduce effects arising from the cold-collision frequency shift.

Performing the detection in the trapping chamber will also simplify the optical bench as no separate detection beams will be needed. Furthermore, for compactness, we will use all fibre-coupled optics, especially where alignment is critical. Only one stabilised narrowband laser will be necessary for cooling and the repump light will be generated by an electro-optic modulator sideband; a tapered amplifier will boost the light intensity to the level necessary for efficient trapping and cooling.

The mini-f physics package is shown in fig. 4c and has been designed in-house. It will have a volume of 25 litres, which is about 30 times smaller than that of the 'full-size' fountain. The design also includes a microwave cavity with a bent waveguide to feed the cavity symmetrically from a single cable source (based on a design by [24]). The waveguide will be on the airside, while the cavity will be under vacuum and sealed by dielectric viewports added over the coupling holes between the waveguide and the cavity. The remaining subsystems such as the optics, microwave synthesizer and control electronics will utilise commercial-off-the-shelf solutions to accelerate the development.

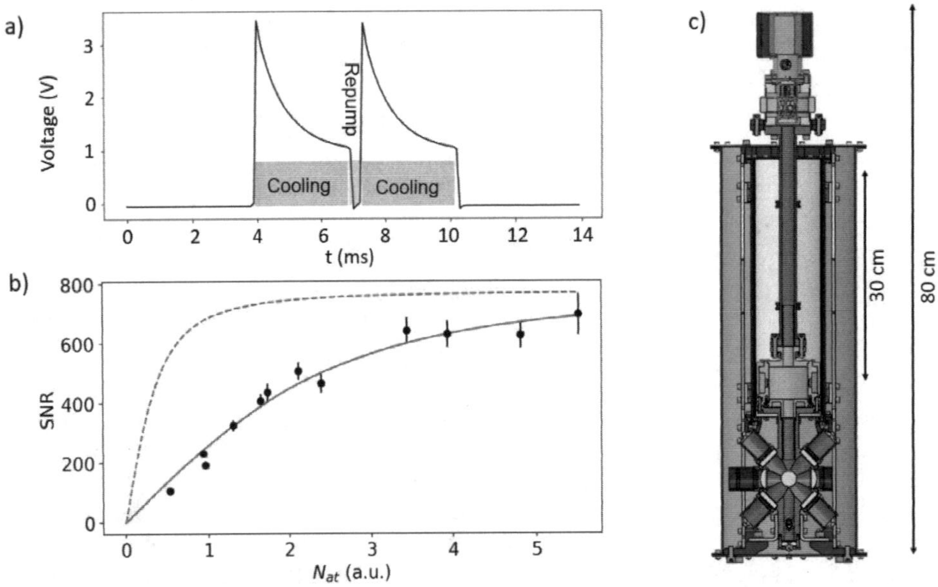

Figure 4. a) Detector output voltage as a function of time for the detection pulse sequence. b) Signal-to-noise ratio vs atom number measured in experiments in CsF3 (data points, solid curve), and expected for improved fluorescence collection in a custom-designed cooling/detection chamber (dashed curve). c) Diagram of the mini-f prototype physics package.

6. Conclusions

We have reviewed the development and performance of the Cs fountain primary frequency standards built by NPL, including several units delivered under commercial contracts. Despite a simple design, the fountains are capable of achieving the best performance in terms of short-term stability and type-B uncertainty as compared to values reported by other groups. Furthermore, the uncertainty of DCP, one of the largest systematic effects, typically evaluated at 1×10^{-16} following conventional methods, can be reduced 10-fold with a technique requiring only a minor modification to the physics package.

We note that further improvements of the accuracy of fountain clocks by reducing uncertainties of other systematic effects would be increasingly difficult to achieve. This is, on one hand, due to a limited ability to model some of these effects (e.g. blackbody radiation shift) and, on the other hand, because there is no clear path to further improve the short-term stability and reduce the averaging time needed to reach a given level of the frequency measurement resolution.

This implies that development effort is now better spent on improving the reliability of fountain clocks. The high uptime, above 99%, that we have demonstrated allows us to use the atomic fountain as a holdover clock for periods between Circular T publications or longer. It should also be possible to use a CSO as a flywheel oscillator without a need for a hydrogen maser.

Improved reliability and simplicity of operation makes it possible to run fountain clocks outside of laboratory environments. NPL is making use of this feature to build a more resilient time scale with infrastructure distributed across multiple locations [25]. More robust atomic fountains also open up applications in a wider range of sectors, such as telecoms and financial services. This is even more true of the mini-fountain, which promises to bring the benefits of compactness, simplicity, and affordability.

Acknowledgments

We are grateful to Kurt Gibble for stimulating discussions and to Kathryn Kathry, Filip Ozimek and Peter Lovelock for technical contributions at various stages of this work. We also acknowledge Peter Whibberley for critical reading of the manuscript.

References

[1] Weyers S, Gerginov V, Kazda M, Rahm J, Lipphardt B, Dobrev G and Gibble K 2018 *Metrologia* **55** 789

[2] Beattie S, Jian B, Alcock J, Gertsvolf M, Hendricks R, Szymaniec K and Gibble K 2020 *Metrologia* **57** 035010

[3] Wang X-L *et al.* 2023 *Metrologia* **60** 065012

[4] Li R, Gibble K and Szymaniec K 2011 *Metrologia* **48** 283

[5] Szymaniec K, Lea S N, Gibble K, Park S E, Liu K and Głowacki P 2016 *J. Phys.: Conf. Ser.* **723** 012003

[6] Wynands R and Weyers S 2005 *Metrologia* **42** S64

[7] Guéna J *et al* 2012 *IEEE Trans. UFFC* **59** 391

[8] Hendricks R J, Ozimek F, Szymaniec K, Nagórny B, Dunst P, Nawrocki J, Beattie S, Jian B and Gibble K 2019 *IEEE Trans. UFFC* **66** 624

[9] Szymaniec K, Noh H-R, Park S E, Takamizawa A 2013 *Appl. Phys. B* **111** 527

[10] Gibble K, Lea S N and Szymaniec K 2012 *Proc. Conference on Precision Electromagnetic Measurement (Washington DC, USA)* 700

[11] Laurent Ph *et al.* 2006 *Appl. Phys. B* **84** 683

[12] Szymaniec K, Park S E 2011 *IEEE Trans. Meas. Instrum.* **60** 2475KS+SEP IEEE 2012

[13] Szymaniec K, Chałupczak W, Tiesinga E, Williams C J, Weyers S and Wynands R 2007 *Phys. Rev. Lett.* **98** 153002

[14] Li R and Gibble K 2010 *Metrologia* **47** 534

[15] Nemitz N, Gerginov V, Wynands R and Weyers S 2012 *Metrologia* **49** 468

[16] Burrows K, Hendricks R J, Szymaniec K, Gibble K, Beattie S and Jian B 2020 *Metrologia* **57** 065003

[17] BIPM Time Department Database *https://webtai.bipm.org/database/show_psfs.html*

[18] Madsen N 2021 *Europhys. News* **52** 18

[19] Godun R M, Nisbet-Jones P B R, Jones J M, King S A, Johnson L A M, Margolis H S, Szymaniec K, Lea S N, Bongs K and Gill P 2014 *Phys. Rev. Lett.* **113** 210801

[20] Lange R, Huntemann N, Rahm J M, Sanner C, Shao H, Lipphardt B, Tamm Ch, Weyers S and Peik E 2021 *Phys. Rev. Lett.* **126** 011102

Journal of Physics: Conference Series **2889** (2024) 012020 doi:10.1088/1742-6596/2889/1/012020

[21] Barontini G *et al.* 2022 *Eur. Phys. Journal Quantum Tech.* **9** 12
[22] Rovera G D, Bize S, Chupin B, Guéna J, Laurent Ph, Rosenbusch P, Uhrich P and Abgrall M 2016 *Metrologia* **53** S81
[23] Galleani L, Signorile G, Formichella V and Sesia I 2020 *Metrologia* **57** 065015
[24] Schröder R, Hübner U and Griebsch D 2002 *IEEE Trans. UFFC* **49** 383
[25] Jones D J M *et al.* 2024 (submitted to this issue)

Journal of Physics: Conference Series **2889** (2024) 012021 doi:10.1088/1742-6596/2889/1/012021

Frequency combs for precision synthesis and characterization of optical atomic standards

Tara Fortier as a representative of the BACON collaboration
Time and Frequency Division, National Institute of Standards and Technology, 325 Broadway MS 847, Boulder CO 80305 USA

E-mail: tara.fortier@nist.gov

February 2024

Abstract. Over the past 20 years, optical frequency combs with atomic clocks, have been a powerful and enabling technology in the context of time and frequency measurement. Impressively, optical atomic clocks have yielded a 100 million-fold improvement in uncertainty in the past 30 years. These improvements are fueling a push toward redefinition of the SI second to optical atomic references, as well as application of atomic clocks to tests of fundamental physics and as relativistic gravitational sensors. Unfortunately, the long times needed to average down clock quantum projection noise and laser noise to reach a measurement stability at and beyond the 10^{-18} level limit the feasibility of next-generation applications. Here I describe the measurement advances in clock comparisons enabled by optical frequency combs and how differential measurement can improve the measurement stability. This paper will also include a discussion of optical frequency combs and their application to precision time/frequency metrology.

1. Introduction

Atomic clocks began their development in the mid 20th century and quickly became a mainstay of modern-day life; serving as the definition of the SI second since 1967 and being the time reference for global positioning system (GPS) starting in the early 1980's. To date, the best microwave atomic clocks based on a ground state hyperfine transition in ^{133}Cs, have demonstrated repeatability and measurement accuracy near 1 part in 10^{16} [1]. The latter requires impressive control, measurement and knowledge of how stray fields impact the 9.2 GHz carrier at the microhertz level. While there still exists some opportunity to improve the accuracy of existing microwave atomic standards, the development of clocks based on electronic optical transitions enables a natural route to much higher performance [2, 3, 4]. This is possible given that atomic clock accuracy and resolution scale inversely with the transition frequency. After more than two decades of development, optical atomic clocks have been evaluated to operate with uncertainties near, or even slightly below 10^{-18} [5, 6, 7] and with short-term instabilities approaching 10^{-16} at a 1-s averaging time, Fig.4. These gains in performance have provided a

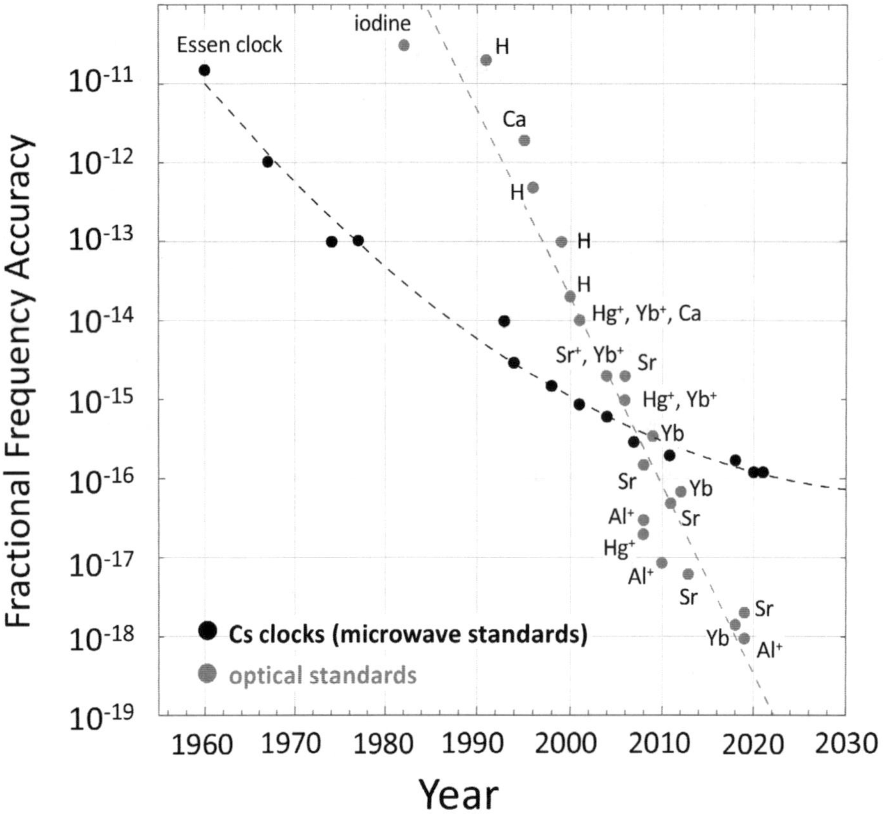

Figure 1: Historical progress of optical and microwave atomic references [4]. Optical clocks leverage their large transition frequencies to push past accuracy limitations experienced by microwave clocks near the 10^{-16} level. Since their first experimental demonstration in the 1980's, optical references have improved in performance with an average slope of 2.5 orders of magnitude per decade, or by a factor of 10 million since 1990! While their progress has yet to cease, optical clocks will encounter serious uncertainty limitations due to gravitational red shift as they near 10^{-19}.

compelling case for the international community to adopt a new definition of the SI second to an optical atomic transition with a fast-approaching target date of 2030 [8, 9, 10].

2. Frequency combs for clock synthesis and time/frequency applications

While optical atomic clocks have exceeded microwave clocks in performance, they are not without their challenges. Unlike microwave clocks, whose electronic carrier can be easily counted and conditioned using standard radio frequency electronics, optical carriers are far too fast to count using standard electronic counters and impossible to use as reference oscillators without the development of novel optical synthesis and

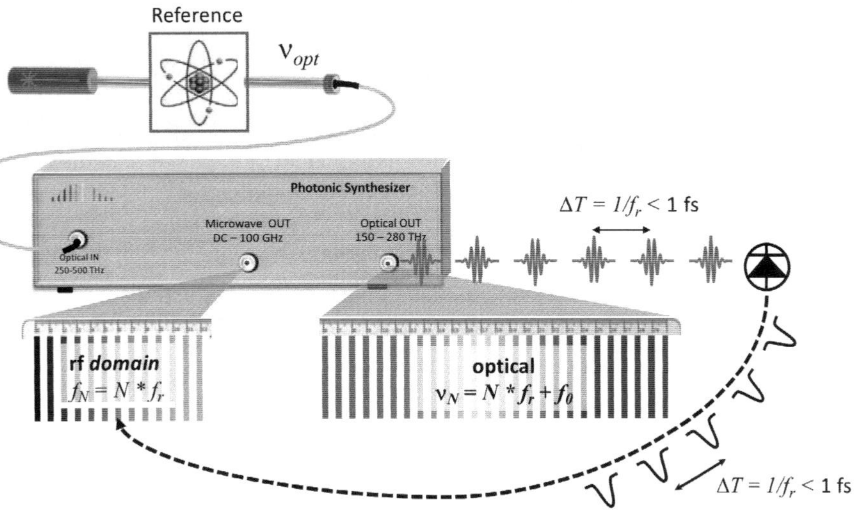

Figure 2: The frequency comb as a pulsed optical and microwave synthesizer. The frequency comb is based on a pulsed laser system. It outputs optical pulses with pulse-to-pulse period, T_r, and a corresponding optical spectrum of harmonically related modes with spacing, f_r frequency offset, f_0, and optical mode numbers N. Here the mode number, $N \approx 300{,}000$, multiply f_r into the optical domain. The repetition rate, f_r, is the lowest order resonant mode of the laser cavity with a frequency that is usually ¡ 1 GHz. Photodetection of the optical pulse train derives an electronic pulse train with a corresponding rf spectrum represented by harmonics ($N = 1, 2, 3$, etc.) of f_r. The frequency comb can be stabilized to an optical reference using a self-referencing technique [11, 12], transferring the reference stability to the pulse-to-pulse timing, which in turn rigidly sets the frequency of the optical and rf modes of the comb spectrum with a deterministic frequency relationship to the reference frequency, ν_{opt}. In this way, the optical frequency comb takes a single frequency optical reference and broadly synthesizes optical and microwave modes that carry the references' stability.

conditioning techniques.

The frequency comb, developed at the turn of the century, helped fill this technological gap and simplified the measurement of, and vastly increased the application space of optical references [13, 14, 15]. Based on modelocked femtosecond lasers, optical frequency combs act like pulsed optical synthesizers with phase-coherent and harmonically related optical and microwave modes [11, 12, 16], see Fig. 2.

Since the early 2000's, frequency combs have been developed based on numerous fiber [17, 18, 19] and free-space laser platforms [20, 21, 22, 23, 24, 25, 26] enabling direct coverage of wavelengths spanning the visible to the mid IR [27, 28, 22, 23], see Fig. 3. When coupled with wavelength conversion techniques, such as difference frequency generation, second harmonic generation and high harmonic generation, frequency combs provide a convenient means to transfer the frequency and phase information

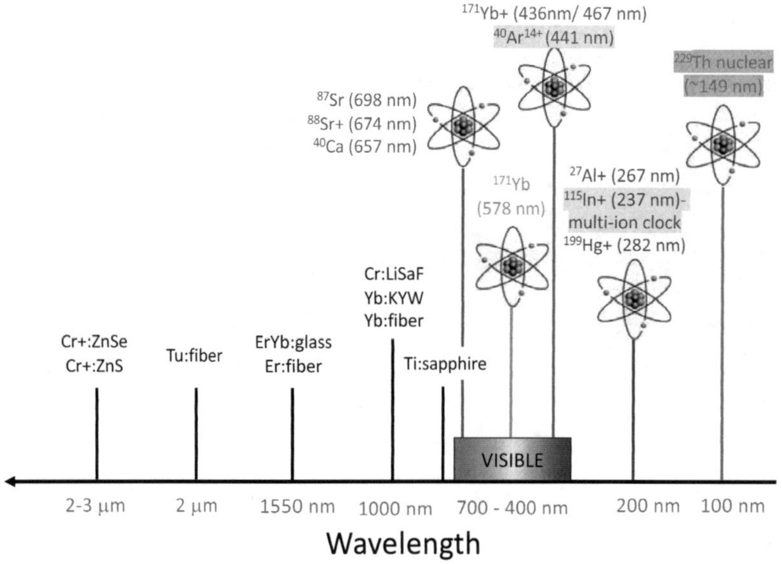

Figure 3: Frequency comb laser gain medium versus center wavelength. Also shown are a selection of optical atomic clock species versus transition wavelength. Highlighted are some of the newest species of optical clocks based on highly charged or multiple trapped ions (yellow), as well as a proposed frequency reference based on a nuclear transition in ^{229}Th (blue). For a list of recommended values of standard frequencies, see https://www.bipm.org/en/publications/mises-en-pratique/standard-frequencies.

of atomic references across much of the electromagnetic spectrum. Finally, a plethora of applications have been enabled by combining the stability and uncertainty of optical references with the synthesis abilities of optical frequency combs. These applications take advantage of the fs-level optical and microwave pulse timing stability, the optical and microwave frequency stability, and high phase coherence possible with optical frequency combs [14, 15].

2.1. Comb operationg and frequency division of optical standards

Frequency combs were first developed to divide optical frequencies to the radio frequency domain to derive electronic signals from optical references. The derived signals are used to characterize the optical transitions against the ^{133}Cs primary standard [29] and to demonstrate continuity between the current and future definition of SI second to optical standards [8]. Moreover, the generation of microwave timing signals from optical atomic references will be central in future optical timescales used to define the SI second to optical atomic standards.

In the time domain, the act of modelocking creates a massively amplitude modulated optical carrier via the coherent addition of the 100,000 to 10^6 harmonically related longitudinal optical modes in the laser cavity. The resulting optical field manifests as short bursts of light, representing just a few cycles of radiation for each round trip in the cavity. The resulting low duty cycle pulse train (pulse width 10 fs – 1 ps, and pulse-to-pulse period of 0.1 - 10 ns) acts like an optical metronome with a pulse repetition rate, f_r of 100 MHz to 10's of GHz, depending on the free spectral range of the laser cavity.

As seen in Fig. 2, direct photodetection of this optical pulse train with high linearity and high-power diodes [30] can permit the derivation of an electronic pulse train with a short-term timing stability and uncertainty matching that of the optical atomic reference, $\Delta T_r/T_r = \Delta\nu/\nu_{opt}$ [31]. Spectral decomposition of this pulsed electronic signal yields harmonics of the frequency comb laser pulse repetition rate that can be counted using an electronic counter referenced to a ^{133}Cs primary standard. The latter measurements enable evaluation of the divided optical standard against the definition of the Hertz. The low residual instabilities and inaccuracies in optical-to-microwave conversion have been achieved despite distortion of the recovered electronic signals due to saturation and nonlinear effects experienced at high photo-carrier concentration per pulse in the photodiode active region [32, 33, 30]. The technique of optical frequency division, beyond the synthesis of microwave timing signals from optical references, also enables the generation of harmonically related and exceptionally pure microwave tones for use as local oscillators in systems that require high stability and low phase noise [34, 33, 31].

2.2. Clock characterization beyond the definition of the Hertz

When performing an absolute optical frequency measurement against a ^{133}Cs primary standard, the measurement is presently limited by the realization of the Hz, fractionally near 10^{-16}. To evaluate optical atomic clocks beyond this limit requires a comparison against a reference with similar or lower uncertainty, namely another optical atomic clock. Optical frequency combs facilitate such measurements using their broad harmonic optical spectrum to cover bandwidths of 100's of THz in the optical domain, connecting optical references operating at disparate frequencies. As seen in Fig. 3, while optical frequency combs usually operate with center frequencies lower than that of many of the current optical references, nonlinear optical techniques can convert clock probe laser frequencies from the infrared to the visible to connect the frequency comb spectra to optical atomic clock transitions. For developing ^{229}Th nuclear atomic clocks [35], high harmonic generation via semi-ionization and recombination of electrons within a noble gas of Xenon or Argon using amplified pulses from the frequency comb can be used to coherently extend comb sources from the infrared and visible to frequencies near 2 PHz [36, 37].

Optical atomic clock comparisons are performed by using the comb as an optical frequency ruler. By measuring the comb mode spacing, f_r the number of modes separating two clocks, $M - N$, and by measuring the difference frequencies between the atomic references and the nearest optical mode of the comb, a relative frequency

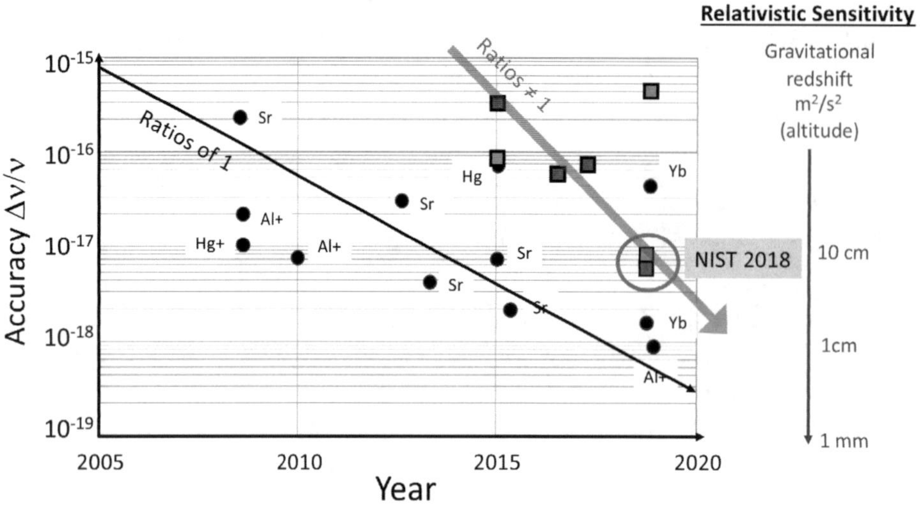

Figure 4: Selected results for clock self-evaluations, frequency ratios = 1, black circles, and inter-species clock comparisons, frequency ratios ≠ 1, colored markers. Also shown is the relativistic sensitivity to earth's gravitational potential in relative altitude difference.

measurement can be performed. Here, the common mode noise in the measurement of f_r(primary standard limitations in the definition of the Hertz) cancel when taking the ratio of two optical frequencies. It is important to note that the frequency comb yields negligible errors in the comparison of clocks. More specifically, the additive instability of the frequency comb in optical synthesis has been demonstrated at levels near 10^{-18} at 1-s averaging time, and with additive fractional frequency errors below 10^{-20} [38, 39, 40].

3. Next generation applications of optical atomic clocks

The highest performance atomic clocks currently report fractional frequency systematic evaluations near 10^{-18} [5, 6, 7]. These "self" evaluations are often assessed by measuring ratios of 1 between two atomic clocks of the same species, and in the same laboratory. These evaluations predict the uncertainty of a clock based on the level of control achieved over stray fields, the measurement error of the classical field sensors, and the theoretical and experimental assessment of how stray fields impact the clock transition frequency. As seen in Fig. 4, historically, the comparison of dissimilar species has yielded ratio results (ratios ≠ 1) with worse measurement uncertainty than self-evaluations (ratios of 1). The observed disagreement is likely the result of common mode noise shared between like systems, or unassessed "dark" systematics for a particular clock [41].

Despite, the disagreement in uncertainty, optical atomic clock comparisons have achieved remarkable performance. At the current uncertainty levels, optical clocks can discern the effects of time dilation due to changes in altitude relative to

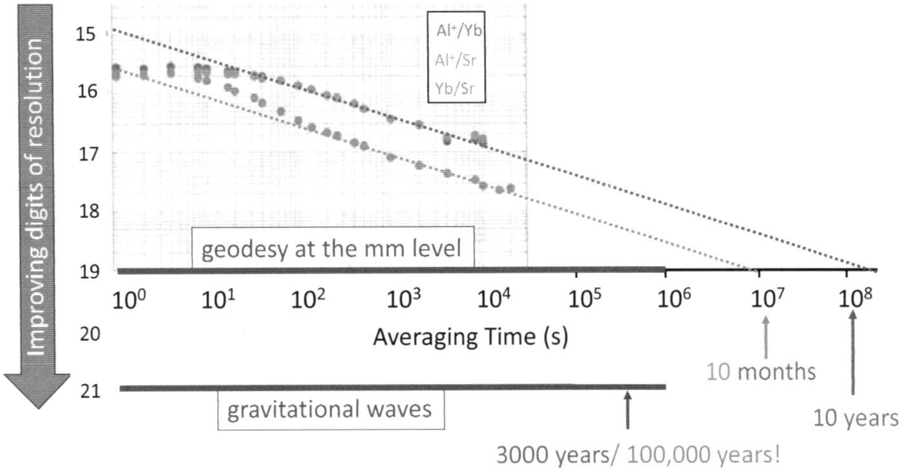

Figure 5: Allan deviation of the 2018 measurement instability for the ratios of the $^{27}Al^{+}$, ^{171}Yb and ^{87}Sr clocks at NIST [41]. Illustrated here are the averaging times needed to reach target resolutions for associated clock applications. Because optical atomic clocks exhibit Gaussian noise, the instability, or frequency resolution, improve as the square root of the averaging time. As seen in the blue, yellow and green data, a factor of 10 improvement in 1-s frequency instability results in a factor of 100 reduction in measurement time for resolution fractional frequency resolutions targets.

the earth's gravitational potential at the cm-level [42, 43]. Because traditional measurements of the gravitational potential are currently limited near the mm-level (see https://geodesy.noaa.gov/library/pdfs/NOAA_TM_NOS_NGS_0077.pdf for the 2018 NIST geodesic survey results), earth-based optical clock comparisons will experience an uncertainty limit near 10^{-19}.

3.1. Clock comparisons for tests of fundamental physics

Beyond their sensitivity to relativistic geodesy [44], the electronic and nuclear spectra of atoms and ions reflect the value of our fundamental constants of nature and the physics internal to the atom. Shifts in the clocks transition frequencies reflect how external fields couple to these transitions. While clock comparisons have yet to reveal new physics beyond the standard model, they have provided constraints on the time variation of the fine structure constant at the 10^{-18} level and the proton-to-electron mass ratio below 10^{-17} [45]. Clocks have also been used to constrain violations of Local Position Invariance [45], and the coupling strength of ultralight dark matter to "normal" matter [41].

4. Differential laser spectroscopy of atomic clocks for improved measurement resolution

While much of the discussion regarding optical atomic clock is on improving their uncertainty, clock instabilities dictate the total measurement time needed to attain a target fractional frequency resolution for a specific application. In Fig. 5, we see that the traditional method (asynchronous frequency measurement) for comparing different optical clocks at NIST, would require between 3 weeks to 10 months of continuous averaging to reach a fractional measurement resolution of 10^{-19}!

4.1. Technical laser noise and its limitations to short-term clock instability

To address the measurement limitations in atomic clock comparisons, differential measurements can be used to reduce, or eliminate, noise from various technical sources. For instance, measurements at JILA have demonstrated differential frequency measurements of ^{87}Sr atoms within the same atom cloud. While this measurement achieved the highest instability measurement of the relative local geopotential [46], because the atoms shared the same environment and same laser, the measurement was highly common mode. While impressive, this level of common mode measurement is not be possible for field tests of geodesy [43], or proposed gravitational wave detection over long baselines using remote clocks [47].

One of the primary limits to optical clock stability is the technical noise of the cavity stabilized probe laser. Laser noise is aliased onto the atomic measurement for clocks operating at probe duty cycles less than 100 % [48]. Additionally, drift of the optical reference cavity, used to pre-stabilize the probe laser, results in phase walk off between the probe laser and the atomic transition oscillations. The latter limits the maximum probe times, or free evolution time of the atoms, T_p, decreasing the measurement resolution for the atomic system.

To address this noise limitation, a common probe laser can be shared between atom clocks of the same species. Using this technique, two ^{27}Al$^+$ clocks at NIST were compared, yielding close to a factor of 10 reduction in measurement time over two ^{27}Al$^+$ clocks probed asynchronously with independent lasers [49]. This technique can be extended to dissimilar clocks by using an optical frequency comb to phase coherently transfer the probe laser frequency from one clock transition to another [50].

4.2. Differential spectroscopy of the $^{27}Al^+$ and Yb^{171} optical clocks

As seen in Fig. 6, the ^{27}Al$^+$ and 171^{171}Yb lattice clocks are synchronously compared at NIST by sharing the ^{171}Yb probe laser with ^{27}Al$^+$ using an Er/^{171}Yb:glass laser frequency comb [51]. Because the quantum projection noise, and hence the stability of the ^{27}Al$^+$ clock, is significantly limited by state detection from a single ion, the predominant route to higher stability is achieved via longer probe times. Using the ^{27}Al$^+$ cavity stabilized laser, atom-laser decoherence limited the Al$^+$ probe time to 150

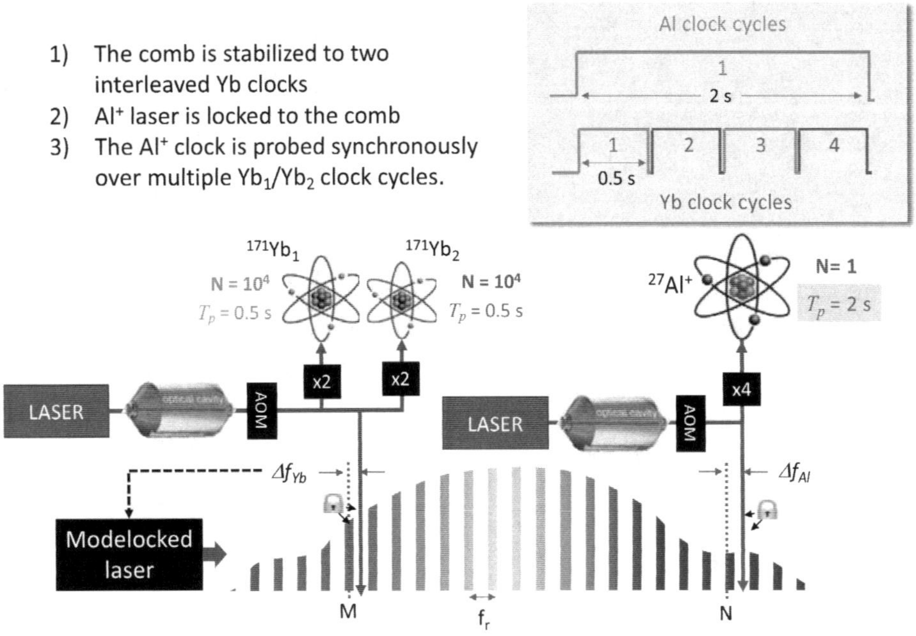

Figure 6: Experimental set up for realization of differential spectroscopy between two ^{171}Yb lattice clocks, run in a zero-dead time configuration, and a single ion ^{27}Al$^+$ clock. In this measurement, the Yb clock laser was coherently transferred to the ^{27}Al$^+$ clock transition frequency using an optical frequency comb. The clocks were synchronously probed, with 4 Yb clock cycles combined to increase the Al$^+$ probe cycle from 0.15 s to close to 2 s.

ms. Lower drift in the Yb cavity stabilized laser enabled a maximum atomic probe time for ^{171}Yb of close to 0.5 s. While the latter represents a factor 3 improvement in probe time over spectroscopy using the cavity stabilized Al$^+$ optical cavity, the upper state lifetime of ^{27}Al$^+$ is 20.6 s [52].

To increase the ^{27}Al$^+$ probe time beyond 0.5 s, two ^{171}Yb clock clocks can be run in a zero-dead time configuration. By interleaving the operation of two ^{171}Yb clocks, the phase walk-off of their cavity stabilized laser can be corrected every 0.5 s. Using this configuration, multiple cycles of the two ^{171}Yb clocks can be combined to increase the ^{27}Al$^+$ probe time to arbitrarily long times. However, the lowest observed 1-s measurement instability between the ^{27}Al$^+$ and ^{171}Yb clocks was achieved for an Al$^+$ probe time of 1.7 s. One possible limiting noise source is phase walk-off between the ^{171}Yb and ^{27}Al$^+$ atom clouds induced by noise in the optical network, comprised of 10's of meters of fiber and free-space links, connecting the two atomic systems. While the stability of the links was established between cavity stabilized lasers as averaging down as 10^{-17} $t^{-1/2}$, the cancellation of laser noise between systems requires low total integrated phase noise, not just high frequency stability.

Journal of Physics: Conference Series **2889** (2024) 012021 doi:10.1088/1742-6596/2889/1/012021

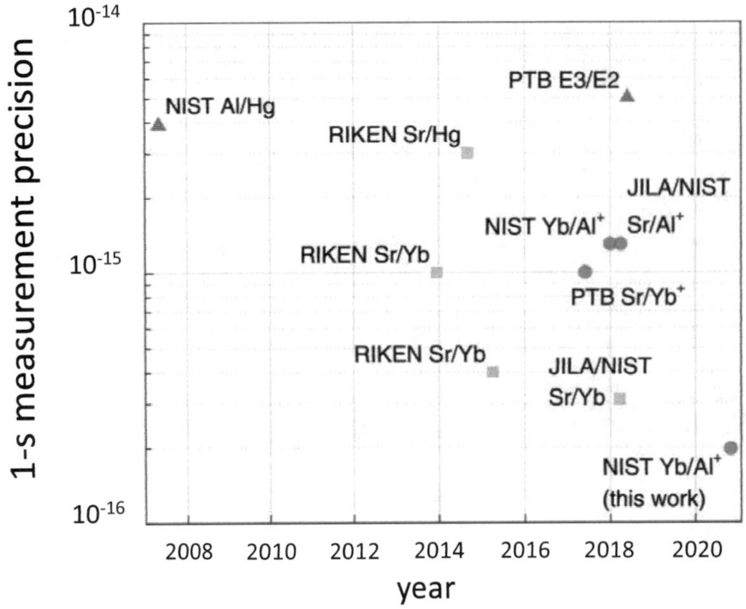

Figure 7: Selected 1-s measurement precision from NIST's differential spectroscopy results as compared with previous inter-species optical atomic clock comparisons. Image credit: M. Kim *et al.* Nature Photonics **19** (2021) [50].

4.3. Results

Finally, in this measurement the clocks were run synchronously with close to 100% probe duty cycle, minimizing the effect of laser noise aliasing in the comparison of the two clock systems. The combination of increased $^{27}Al^+$ probe time (0.15 to 1.7 s) and the minimization of laser noise aliasing enabled nearly a factor of 10 improvement in the relative measurement instability of the $^{27}Al^+$ and ^{171}Yb clock near 2×10^{-16} [50], see Fig. 7. Aside from enabling the lowest instability inter-species clock comparison to date, differential spectroscopy permits a 100 $\times$ speed up in total measurement time improving the feasibility of next generation atomic clock applications.

5. Conclusion and future outlook: the next 7 years.....?

Based on the current trends in optical atomic clock performance, Fig.4, one might expect researchers to report uncertainty results in 2030 near or below 10^{-19}. At this point, local clock comparisons performed over relatively short baselines will experience limitations due to the current uncertainty of traditional geodesic measurement techniques. Comparisons performed over continental differences will suffer from accumulated errors due to laser leveling and GPS uncertainty, currently at the 10^{-16} level.

The lack of access to dark fiber networks constitutes a significant barrier for the remote comparison of optical atomic clocks within North America and between North

America and the rest of the world. Transportable optical atomic clocks [53, ?, 54] will hopefully provide a means to connect continents and circumvent gravitational uncertainties due to earth's geoid. Additionally, ground to satellite free-space optical time/frequency transfer [55, 56, 43] will provide a means to disseminate signals at the level with an uncertainty better than 10^{-18}, which is not currently possible using GPS-based microwave time transfer methods.

Differential spectroscopy of clocks over longer fiber-based of free-space baselines can provide a means for improved measurement stability to ensure reasonable measurement times for upcoming applications in relativistic geodesy [43] and clock comparisons for gravitational wave detection. Such a measurement will encounter potential challenges in maintaining phase coherence $\Delta\phi_{rms} < \pi/2$ over the total measurement time for common mode rejection of laser technical noise over long links.

Waxing more speculatively, at the next Frequency Symposium will we see the beginnings of high-accuracy clock comparisons performed over quantum networks [57]? Finally, I look forward to seeing how well-established time and frequency techniques will be applied to problems in quantum technology.

6. Acknowledgements

I would like to acknowledge all the various contributors to the Boulder Atomic Clock Network collaboration (BACON) from the NIST timescale group, the precision photonic synthesis group, the Al^+, Yb and Sr clock groups. I would like to specifically acknowledge the work of May E. Kim, William F. McGrew, Nicholas V. Nardelli, Ethan R. Clements, Youssef S. Hassan, Xiaogang Zhang, Jose L. Valencia, Holly Leopardi, David B. Hume, Andrew D. Ludlow and David R. Leibrandt on the differential spectroscopy work. I would like to also thank Nick Nardelli who helped with Figures 1 and 5. Thank you also to Esther Baumann, Ladan Arissian and William McGeehee for the careful reading of this manuscript.

References

[1] Sortais Y, Bize S, Abgrall M, Zhang S, Nicolas C, Mandache C, Lemonde P, Laurent P, Santarelli G, Dimarcq N, Petit P, Clairon A, Mann A, Luiten A, Chang S and Salomon C 2001 *Physica Scripta* **2001** 50 URL `https://dx.doi.org/10.1238/Physica.Topical.095a00050`

[2] Ludlow A D, Boyd M M, Ye J, Peik E and Schmidt P O **87** 637–701

[3] Peik E 2023 *Photonic Quantum Technologies: Science and Applications* **1** 333–348

[4] Poli N, Oates C, Gill P and Tino G 2013 *La Rivista del Nuovo Cimento* **36** 555–624

[5] Bothwell T, Kedar D, Oelker E, Robinson J M, Bromley S L, Tew W L, Ye J and Kennedy C J 2019 *Metrologia* **56** 065004

[6] McGrew W F, Zhang X, Fasano R J, Schäffer S A, Beloy K, Nicolodi D, Brown R C, Hinkley N, Milani G, Schioppo M, Yoon T H and Ludlow A D 2018 *Nature* **564** 87–90

[7] Brewer S M, Chen J S, Hankin A M, Clements E R, Chou C W, Wineland D J, Hume D B and Leibrandt D R 2019 *Phys. Rev. Lett.* **123** 033201

[8] Riehle F 2015 *Comptes Rendus Physique* **16** 506–515

[9] Dimarcq N, Gertsvolf M, Mileti G, Bize S, Oates C, Peik E, Calonico D, Ido T, Tavella P, Meynadier F *et al.* 2023 *Metrologia*
[10] Lodewyck J 2019 *Metrologia* **56** 055009
[11] Jones D, Diddams S, Ranka J, Stentz A, Windeler R, Hall J and Cundiff S 2000 *Science* **288** 635–639
[12] Telle H R, Steinmeyer G, Dunlop A E, Stenger J, Sutter D H and Keller U 1999 *Appl. Phys. B* **69** 327–332
[13] Newbury N R 2011 *Nature photonics* **5** 186–188
[14] Fortier T and Baumann E 2019 *Communications Physics* **2** 1–16
[15] Diddams S A, Vahala K and Udem T 2020 *Science* **369** eaay3676
[16] Picqué N and Hänsch T W 2019 *Nature Photonics* **13** 146–157
[17] Ruehl A, Marcinkevicius A, Fermann M E and Hartl I **35** 3015–3017
[18] Newbury N R and Swann W C **24** 1756–1770
[19] Kim J and Song Y 2016 *Advances in Optics and Photonics* **8** 465–540
[20] Fortier T M, Bartels A and Diddams S A 2006 *Opt. Lett.* **31** 1011–1013
[21] Nardelli N V and Fortier T M 2023 Optical and microwave synthesis with an er/yb: Glass frequency comb for precision metrology of atomic clocks *Frontiers in Optics + Laser Science 2023 (FiO, LS)* (Optica Publishing Group) p LM1F.1 URL `https://opg.optica.org/abstract.cfm?URI=LS-2023-LM1F.1`
[22] Vasilyev S, Smolski V, Peppers J, Moskalev I, Mirov M, Barnakov Y, Mirov S and Gapontsev V 2019 *Optics express* **27** 35079–35087
[23] Vicentini E, Gambetta A, Canella F, Coluccelli N, Laporta P and Galzerano G 2023 *Optics Express* **31** 21482–21492
[24] Schilt S and Südmeyer T 2015 *Applied Sciences* **5** 787–816
[25] Holzwarth R, Zimmermann M, Udem T, Hänsch T, Russbüldt P, Gäbel K, Poprawe R, Knight J, Wadsworth W and Russell P S J 2001 *Optics Letters* **26** 1376–1378
[26] Mitaki M, Sugiyama K and Kitano M 2018 *Applied Optics* **57** 5150–5160
[27] Schliesser A, Picqué N and Hänsch T W 2012 *Nature photonics* **6** 440–449
[28] Jiang J, Mohr C, Bethge J, Fermann M and Hartl I Fully stabilized, self-referenced thulium fiber frequency comb *CLEO/Europe and EQEC 2011 Conference Digest* p PDB_1
[29] Ye J, Hall J L and Diddams S A 2000 *Opt. Lett.* **25** 1675–1677
[30] Fortier T M, Quinlan F, Hati A, Nelson C, Taylor J A, Fu Y, Campbell J and Diddams S A 2013 *Opt. Lett.* **38** 1712–1714
[31] Nakamura T, Davila-Rodriguez J, Leopardi H, Sherman J A, Fortier T M, Xie X, Campbell J C, McGrew W F, Zhang X, Hassan Y S *et al.* 2020 *Science* **368** 889–892
[32] Millo J, Boudot R, Lours M, Bourgeois P Y, Luiten A N, Coq Y L, Kersalé Y and Santarelli G 2009 *Opt. Lett.* **34** 3707–3709
[33] Xie X, Bouchand R, Nicolodi D, Giunta M, Hänsel W, Lezius M, Joshi A, Datta S, Alexandre C, Lours M, Tremblin P A, Santarelli G, Holzwarth R and Coq Y L 2017 *Nat. Photon.* **11** 44–47
[34] Fortier T M, Kirchner M S, Quinlan F, J Taylor J C B, T Rosenband N L, Ludlow A, Jiang Y, Oates C W and Diddams S A 2011 *Nat. Photon.* **5** 425–429
[35] Peik E and Tamm C 2003 *Europhysics Letters* **61** 181
[36] Gohle C, Udem T, Herrmann M, Rauschenberger J, Holzwarth R, Schuessler H A, Krausz F and Hänsch T W **436** 234–237
[37] Pupeza I, Zhang C, Högner M and Ye J 2021 *Nature Photonics* **15** 175–186
[38] Leopardi H, Davila-Rodriguez J, Quinlan F, Olson J, Sherman J A, Diddams S A and Fortier T M 2017 *Optica* **4** 879–885
[39] Giunta M, Hänsel W, Fischer M, Lezius M, Udem T and Holzwarth R 2020 *Nature Photonics* **14** 44–49
[40] Shi H, Jiang Y, Yao Y, Li B, Wang C, Yu H and Ma L 2023 *APL Photonics* **8**
[41] Collaboration B A C O N B 2021 *Nature* **591** 564–569

[42] Chou C W, Hume D B, Koelemeij J C J, Wineland D J and Rosenband T **104** 070802
[43] Takano T, Takamoto M, Ushijima I, Ohmae N, Akatsuka T, Yamaguchi A, Kuroishi Y, Munekane H, Miyahara B and Katori H 2016 *Nature Photonics* **10** 662–666
[44] Mehlstäubler T E, Grosche G, Lisdat C, Schmidt P O and Denker H **81** 064401
[45] Lange R, Huntemann N, Rahm J, Sanner C, Shao H, Lipphardt B, Tamm C, Weyers S and Peik E 2021 *Physical Review Letters* **126** 011102
[46] Bothwell T, Kennedy C J, Aeppli A, Kedar D, Robinson J M, Oelker E, Staron A and Ye J 2022 *Nature* **602** 420–424
[47] Kolkowitz S, Pikovski I, Langellier N, Lukin M D, Walsworth R L and Ye J 2016 *Physical Review D* **94** 124043
[48] Santarelli G, Audoin C, Makdissi A, Laurent P, Dick G J and Clairon A 1998 *IEEE transactions on ultrasonics, ferroelectrics, and frequency control* **45** 887–894
[49] Clements E R, Kim M E, Cui K, Hankin A M, Brewer S M, Valencia J, Chen J S, Chou C W, Leibrandt D R and Hume D B 2020 *Physical review letters* **125** 243602
[50] Kim M E, McGrew W F, Nardelli N V, Clements E R, Hassan Y S, Zhang X, Valencia J L, Leopardi H, Hume D B, Fortier T M *et al.* 2023 *Nature Physics* **19** 25–29
[51] Nardelli N V, Leopardi H, Schibli T R and Fortier T M 2023 *Laser & Photonics Reviews* **17** 2200650
[52] Rosenband T, Schmidt P O, Hume D B, Itano W M, Fortier T M, Stalnaker J E, Kim K, Diddams S A, Koelemeij J, Bergquist J *et al.* 2007 *Physical review letters* **98** 220801
[53] Grotti J, Koller S, Vogt S, Häfner S, Sterr U, Lisdat C, Denker H, Voigt C, Timmen L, Rolland A *et al.* 2018 *Nature Physics* **14** 437–441
[54] Zeng M, Huang Y, Zhang B, Hao Y, Ma Z, Hu R, Zhang H, Chen Z, Wang M, Guan H *et al.* 2023 *Physical Review Applied* **19** 064004
[55] Giorgetta F R, Swann W C, Sinclair L C, Baumann E, Coddington I and Newbury N R 2013 *Nature Photonics* **7** 434–438
[56] Kang H J, Yang J, Chun B J, Jang H, Kim B S, Kim Y J and Kim S W 2019 *Nature communications* **10** 4438
[57] Nichol B, Srinivas R, Nadlinger D, Drmota P, Main D, Araneda G, Ballance C and Lucas D 2022 *Nature* **609** 689–694

Robust Optical Clocks for International Timescales (ROCIT)

H S Margolis[1], R M Godun[1], N Huntemann[2], R Le Targat[3], M Pizzocaro[4], M Zawada[5], M. Abgrall[3], D Akamatsu[6], H Álvarez Martínez[3,7], A Amy-Klein[8], M Andia[3], E Benkler[2], N M Bhatt[2], S Bilicki[5], S Bize[3], M Bober[5], D Calonico[4], V Cambier[3], E Cantin[8], C Chardonnet[8], M Cifuentes Marín[3], C Clivati[4], S Condio[4,9], E. A. Curtis[1], A Czubla[10], M Doležal[11], S Dörscher[2], P Dunst[12], C-H Feng[1], M Filzinger[2], R Folman[13], T Fordell[14], V Formichella[4], Y Foucault[3], L Galleani[9], I Goti[4], D Groswasser[13], M Gruszczyński[10], C Guo[3], K J Hanhijärvi[14], H N Hausser[2], I R Hill[1], K Hosaka[6], M Y H Johnson[1], J Keller[2], J Klose[2], T Kobayashi[6], S Koke[2], D Kovačić[5,15], P Křen[11], A Kuhl[2], A Ledziński[5], D Lemański[12], F Levi[4], T Lindvall[14], C Lisdat[2], H Liu[2], J Lodewyck[3], O Lopez[8], L Lorini[3], M Lours[3], P Mašika[11], M Mazouth-Laurol[3], T E Mehlstaubler[2,16], W Moreno[3], P Morzyński[5], M Narożnik[5], J Nawrocki[12], A Nishiyama[6], P Nogaś[12], T Nordmann[2], A O Parsons[1], B Pointard[3], P E Pottie[3], M Risaro[4], B I Robertson[1], J Romero González[7], M Schioppo[1], I Sesia[4], H Shang[3], G Signorile[4], K Stahl[2], M Steinel[2], U Sterr[2], J Suárez Ramírez[7], A Tofful[1], M Tønnes[3], A Tran[1], J Tunesi[1], A E Wallin[14], T Waterholter[2], M Zarei[5], C Zyskind[3]

[1] National Physical Laboratory (NPL), Teddington, Middlesex TW11 0LW, UK
[2] Physikalisch-Technische Bundesanstalt (PTB), 38116 Braunschweig, Germany
[3] LNE-SYRTE, Observatoire de Paris (OP), Université PSL, CNRS, Sorbonne Université, 75014 Paris, France
[4] Istituto Nazionale di Ricerca Metrologica (INRIM), 10135 Torino, Italy
[5] Institute of Physics, Faculty of Physics, Astronomy and Informatics, Nicolaus Copernicus University, PL-87-100 Toruń, Poland
[6] National Metrology Institute of Japan (NMIJ), National Institute of Advanced Industrial Science and Technology (AIST), 1-1-1 Umezono, Tsukuba, Ibaraki, 305-8563, Japan
[7] Real Instituto y Observatorio de la Armada, 11100 San Fernando (Cádiz), Spain
[8] Laboratoire de Physique des Lasers (LPL), Université Sorbonne Paris Nord, CNRS, Villetaneuse, France
[9] Politecnico di Torino, Corso Duca degli Abruzzi 24, 10129 Torino, Italy
[10] Central Office of Measures (GUM), 00-139 Warsaw, Poland
[11] Czech Metrology Institute (CMI), Okružní 31, 63800 Brno, Czech Republic
[12] Space Research Center Polish Academy of Sciences, Warsaw, Poland
[13] Ben-Gurion University of the Negev, Beer-Sheva 84105, Israel
[14] VTT Technical Research Centre of Finland Ltd, National Metrology Institute VTT MIKES, PO Box 1000, FI-02044 VTT, Finland
[15] Institute of Physics, Bijenička cesta 46, 10000 Zagreb, Croatia
[16] Leibniz Universität Hannover, 30167 Hannover, Germany

helen.margolis@npl.co.uk

Abstract. The recently concluded collaborative European project "Robust optical clocks for international timescales" (ROCIT) tackled some of the key challenges on the roadmap towards a redefinition of the SI second. This paper gives an overview of progress made on improving the robustness and automation of optical clocks and verifying their uncertainty budgets through coordinated international comparison campaigns. It also presents work on the incorporation of optical clocks into time scales, covering both their use to steer local physical time scales and their use for evaluations of hydrogen masers contributing data for the computation of International Atomic Time (TAI). The overall objective of the project was to bring European optical clocks to the stage where they could be operated routinely as secondary frequency standards, regularly contributing to TAI.

1. Introduction

Due to the unprecedented stability and accuracy achievable with optical atomic clocks, which now surpass the performance of the best caesium microwave primary frequency standards by up to two orders of magnitude, an optical redefinition of the second is anticipated. The Consultative Committee for Time and Frequency (CCTF) has prepared a detailed roadmap [1], which sets out the key challenges that remain to be addressed before the redefinition can be implemented, along with a set of criteria that enables an assessment to be made of readiness for the change. These criteria are separated into mandatory criteria that must be met before changing the definition, and ancillary conditions that should be well advanced by the time of the redefinition but that do not necessarily need to be completely achieved.

The overall aim of the "Robust optical clocks for international timescales" (ROCIT) project was to bring European optical clocks to the stage where they can contribute regularly to International Atomic Time (TAI) as secondary frequency standards. A key enabler for the use of optical clocks in time scales is improvement to their robustness and automation, making this the initial area of focus for the project consortium (section 2) even though high reliability of optical frequency standards is one of the ancillary conditions in the roadmap towards redefinition. Significant effort was also devoted to validation of the uncertainty budgets of the optical clocks through a programme of international comparisons (section 3), which also addressed the mandatory criterion of continuity with the present definition of the second. Section 4 focuses on progress towards the incorporation of optical clocks into UTC(k) timescales, the physical realisations of UTC maintained by national timing laboratories, while section 5 covers optical clock contributions to TAI. Conclusions and future prospects are presented in section 6.

2. Robustness and automation of optical clocks

Within the ROCIT project, advances were made in the robustness and automation of both trapped ion optical clocks and neutral atom optical lattice clocks, with the focus on the systems that were known to require most frequent user intervention.

Developments included new approaches for automatic long-term control of laser systems and for automated adjustment of optical setups. For example, methods were developed for automatically relocking lasers to external stabilisation cavities, external enhancement cavities used for second-harmonic generation and enhancement cavities for optical lattices. Systems were also developed for automatic laser beam alignment into optical fibres and alignment of objectives for ion imaging.

Another area of work, the priority of which was emphasized by the COVID-19 pandemic, was the development of systems for remote monitoring and control. It is now possible to diagnose, and in many cases rectify, many common problems without entering the laboratory, making running extended measurement campaigns far less arduous than in the past. Some of the software developed for this purpose has been published as open source on GitHub, for example tools for remote monitoring and locking lasers using a commercial wavemeter [2] and for almost real-time monitoring of the blackbody radiation shift in an optical lattice clock [3]. Routines to recover automatically from certain erroneous clock states were also developed, although for most systems some error conditions still require human intervention. In most such cases, an alert system has been implemented to inform operators of the error condition, supporting the minimisation of downtime.

A range of software was also developed to schedule and perform key operational checks, and to determine through interleaved measurements the magnitude of the environmental parameters that cause frequency shifts. For example, routines developed for ion-based clocks included ion loading, minimisation of residual electric fields via measurement of ion micromotion, and minimisation of magnetic fields via measurement of the Zeeman components of the clock transition.

Based on these developments, LNE-SYRTE, NPL and PTB were all able to demonstrate unattended optical clock operation with uptimes exceeding 80 % over 2 weeks for at least one of their optical clocks. In all cases the remaining downtime was dominated by a few longer interruptions to operation, and further engineering improvements are envisaged to address their causes.

3. International consistency of optical clocks

The mandatory criteria set out in the roadmap towards the redefinition of the second include the validation of optical clock uncertainties through the comparison of systems developed independently in different laboratories around the world, with the numerical target for comparison uncertainties set at 5 parts in 10^{18}. With this objective in mind, the ROCIT project included an extensive programme of optical clock comparisons, including both local and remote measurements (Figure 1). This programme involved two new optical clocks that were completed and operated for the first time as part of the project: the $^{88}Sr^+$ optical clock at VTT MIKES [4] and the $^{115}In^+$ optical clock at Leibniz University Hannover/PTB. In addition, the ^{88}Sr optical lattice clock in Toruń joined international comparisons for the first time as part of the project and data was shared from the new $^{88}Sr^+$ optical clock at PTB [5].

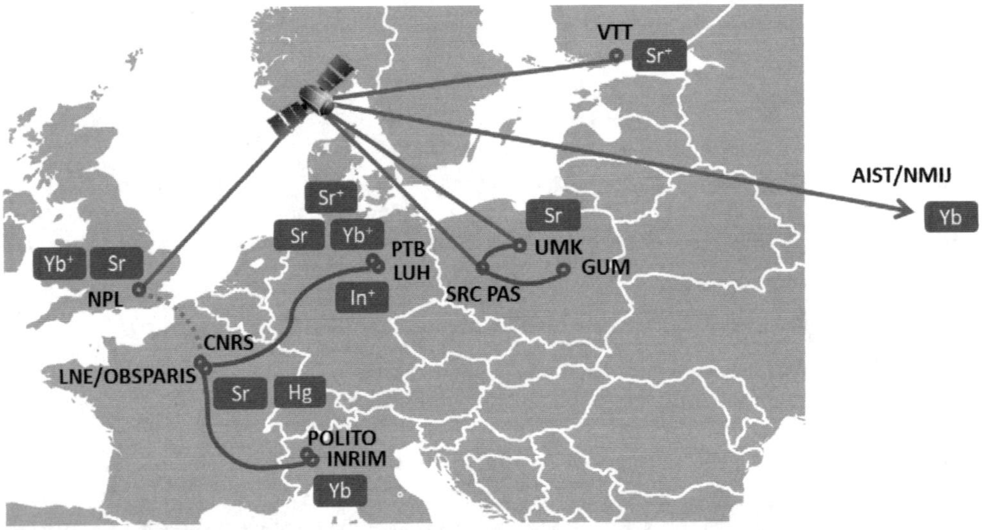

Figure 1. Optical clocks (red boxes) and laboratories involved in the ROCIT optical clock comparison programme, which included both local and remote comparisons. NMIJ joined the programme as a collaborator with the European consortium. The red lines indicate optical fibre links between laboratories (the link between NPL and LNE/OBSPARIS is dotted because it could not be used in the March 2022 campaign due to an unplanned interruption). The blue lines indicate the laboratories that could only be linked by GNSS time and frequency transfer.

3.1. Remote clock comparisons

Currently the only way to link optical clocks in different locations and reach comparison uncertainties that are comparable to the estimated uncertainties of the clocks themselves is by transmission of an optical carrier signal over a specially configured, phase-coherent optical fibre link. Where suitable infrastructure exists, this is therefore the technology of choice for remote optical clock comparisons.

During the ROCIT project, a new optical fibre link was completed between Paris and Turin [6], to extend the reach of the pre-existing European optical clock network linking LNE-SYRTE, NPL and PTB [7, 8] and allow the ^{171}Yb optical lattice clock at INRIM to join international comparisons by fibre link for the first time. Comparisons using this network benefit from the European quasi-geoid model computed as part of an earlier European project, "International Timescales with Optical Clocks" (ITOC) [9], which incorporated new gravity measurements carried out at and around the NMI sites [10, 11]. That prior work unified the description of the geopotential for European NMIs and the uncertainty in the knowledge of the local geopotential at the four NMIs linked by the network meets the numerical target in the roadmap towards definition.

Two coordinated programmes of frequency comparisons were carried out during the project. The first, planned for March 2020, was badly disrupted by the COVID-19 pandemic. The second, performed in March 2022, was far more successful, involving eleven optical clocks in seven different countries, including one from outside Europe (the ^{171}Yb optical lattice clock at NMIJ in Japan). To maximise the number of laboratories that could participate, satellite-based techniques were used in addition to the optical fibre network.

A detailed description of this campaign and the results obtained are being prepared for publication elsewhere, so only an outline is provided here. In total 27 frequency ratios between remote optical clocks were measured, with the comparisons via fibre link reaching uncertainties in the low parts in 10^{17} range while the comparisons via satellite link were limited to the low parts in 10^{16} range. When compared with the 2021 CIPM recommended frequency values for secondary representations of the second, two groups of outliers were observed. The first group of outliers involves the $^{115}In^+$ optical clock, but in this case the offset is well within the uncertainty in the recommended frequency value, which is determined from earlier less accurate measurements of other $^{115}In^+$ optical clocks. We thus expect that the new measurements will have a strong influence on the next update to this recommended frequency value. The other group of outliers involves the new $^{88}Sr^+$ trapped ion optical clock at VTT, and in this case the offset is statistically significant. However, the ratios involving the VTT clock agree with recent results from other $^{88}Sr^+$ optical clocks [5, 12], in particular the only optical ratio involving $^{88}Sr^+$ that has so far been published [5]. It therefore seems likely that the problem lies with the recommended frequency value rather than the VTT clock.

3.2. Local clock comparisons

The remote comparisons were supplemented by local optical frequency comparisons. Local optical frequency ratio measurements provide a way to perform consistency checks without the need for a physical link between locations, if the same ratio is measured independently in different laboratories. Such measurements enabled the European optical clocks to be compared with those on other continents, even though no highly stable and accurate links are yet available over such long distances. Traceability to the SI second was ensured by including caesium primary frequency standards in the comparison programme. As identified in the roadmap towards redefinition, absolute frequency measurements of the optical standards at the limit set by the caesium primary standards are essential to avoid any discontinuity occurring at the time of the redefinition.

During the project, 51 local clock comparisons were performed between clocks on the same site. These included 33 absolute frequency measurements, 26 of which were made using local caesium fountain primary frequency standards and seven of which were performed using TAI as the reference. In most cases, the uncertainties in the absolute frequencies were dominated by the reference to the SI second. The local clock comparisons also included 15 optical frequency ratio measurements, several of which had never been determined directly before, and three frequency ratio measurements involving the rubidium fountain secondary frequency standard at LNE-SYRTE. Most of the local optical frequency ratio measurements achieved uncertainties in the low parts in 10^{17} range. Some local frequency comparison results have already been published [5, 13–17], whilst others are still being finalised for publication.

These local measurements, together with the remote frequency comparison results, will greatly augment the body of high-accuracy clock comparison data available worldwide, with uncertainties being reduced by up to an order of magnitude in some cases. Correlation coefficients between the different measurements were also computed, to feed into future updates to the CIPM recommended values of standard frequencies (section 5.1).

4. Optical clocks in UTC(k) time scales

4.1. *Real-time data validation and on-the-fly correction of systematic frequency shifts*

Essential prerequisites for using optical clocks to steer UTC(k) time scales are the ability to validate the clock data in real time and to evaluate and correct for systematic frequency shifts at an appropriate level on the fly.

Our work on real-time data validation considered the whole metrological connection between the optical reference and the hydrogen maser flywheel used to generate the time scale, including the frequency combs and any local optical fibre links as well as the optical clocks themselves. Techniques used included cycle slip detection in frequency comb measurements by redundant counting, algorithms for detecting sudden unexpected changes in frequency measurements, and the setting of a threshold level for validation of atom number in optical lattice clocks.

Experimental control systems were extended to perform on-the-fly evaluation and correction of systematic frequency shifts by developing routines to measure the magnitude of residual frequency shifting parameters and to store the data acquired in a database for subsequent use by shift calculation routines. For example, residual magnetic fields are monitored via the Zeeman splitting of atomic states, while the blackbody radiation shift is derived from accurate monitoring of the thermal environment using temperature sensors. Residual electric fields in trapped ion clocks are monitored by measuring the ion micromotion.

These developments enabled LNE-SYRTE, NPL and PTB each to demonstrate on-the-fly evaluation and correction of systematic frequency shifts in at least one of their optical clocks. These advances are beneficial not only for real-time steering of UTC(k) time scales, but also for faster analysis of data from clock comparison campaigns (section 3) and for low latency contributions of optical clock data to TAI (section 5.2).

4.2. *Steering algorithm development*

While microwave clocks used to steer UTC(k) time scales typically operate with an uptime close to 100 %, most optical clocks do not yet achieve this level of reliability, even though they have better stability and accuracy. This means that steering algorithms used with optical clocks must be able to handle lower data availability.

Two independent steering algorithms were designed to adapt to these conditions. The first addressed the effect of gaps in the optical clock data and was adapted to be able to handle low uptimes and long periods of data unavailability [18]. Using simulated optical clock data, it was demonstrated that in the ideal scenario of continuous optical clock operation, the fluctuations of the time scale can be maintained within a few hundred picoseconds even if the steering corrections are applied only once per day. With the optical clock operation reduced to one 24-hour period per week, the fluctuations remain within several nanoseconds, close to the best conventionally steered time scales available today.

A second algorithm was tailored to adapt to possible lags between data acquisition and correction, as well as to unexpected changes of the frequency drift of the flywheel. Tests of this algorithm showed that, with typical fluctuations in maser flywheel drift, time scale fluctuations could be maintained at

Journal of Physics: Conference Series **2889** (2024) 012022 doi:10.1088/1742-6596/2889/1/012022

the nanosecond level with a lag of one day in data availability, but that higher latency can rapidly lead to excursions of several tens of nanoseconds.

4.3. Prototype UTC(k) steering using optical clocks

The automated procedures for real-time data validation and on-the-fly correction of systematic frequency shifts (section 4.1), together with the lessons learned from the algorithm testing (section 4.2) were exploited to realise experimental prototypes of optically steered time scales, which we refer to as UTCx(*k*).

During the March 2022 international clock comparison campaign, two prototype time scales UTCx(NPL) and UTCx(OP) were independently and simultaneously demonstrated at NPL and LNE-SYRTE, in each case using data from two local optical clocks that achieved high uptimes. The approach was similar to that used to steer the operational UTC(NPL) and UTC(OP) time scales using caesium primary frequency standards, with a frequency offset generator connected to a hydrogen maser flywheel being steered using data from the local optical clocks. The difference lies in the interval between steers being applied. For example, at NPL caesium-fountain-based steers are applied three times a week, while the interval between optical clock steers was reduced to one hour.

Despite a few data validation issues and technical issues with the frequency offset generators used for steering of the hydrogen maser flywheels, both UTCx(*k*) time scales remained less than 2 ns away from UTC over the 30 days of the test. The two UTCx(*k*) time scales were also compared directly via a satellite-based time transfer link, with their offset found to remain smaller than the offset between the corresponding operational UTC(*k*) time scales over the same period.

A more detailed description of this experiment is being prepared for separate publication.

5. Optical clocks in TAI

Incorporation of optical clocks into the global time scale as secondary representations of the second is achieved by comparing the optical clocks with hydrogen masers that are used in the computation of TAI and submission of this frequency comparison data to the BIPM for inclusion in the monthly bulletin Circular T. While the definition of the SI second is still based on the ground state hyperfine transition in ^{133}Cs, optical clocks may only contribute to TAI with the uncertainty of the recommended frequency value for the secondary representation of the second.

5.1. Optimised frequency values for secondary representations of the second

The recommended values of these standard frequencies are computed by the CCL-CCTF Frequency Standards Working Group (WGFS) [19], based on the worldwide body of clock comparison data. Since 2015, this has been done using least-squares analysis methods developed within the aforementioned ITOC project [20].

The most recent update to the list, approved by the CCTF in March 2021, differed in one crucial respect from previous adjustments performed in 2015 and 2017 [21]. For the first time, detailed consideration was given to correlations between the individual frequency measurements to ensure that the recommended frequency values derived from the data are unbiased and that their uncertainties are not underestimated. This change was influenced by work performed within the ROCIT project, in particular our guidelines on the evaluation and reporting of correlation coefficients [22]. Correlations arise from various sources, with the most significant being correlations from the use of the same primary or secondary frequency standard to access the SI second and correlations between different measurements performed within the same coordinated comparison campaign.

The new recommended values for standard frequencies were used for the first time for the calculation of TAI in Circular T no. 412 (April 2022). From that point, optical clocks have been able to steer TAI with lower uncertainty, as the recommended frequency values for the two types currently contributing (strontium and ytterbium optical lattice clocks) both now have uncertainties at the limit set by caesium fountain primary frequency standards. The update thus resulted in a significant increase in the weight of optical clocks in TAI (Figure 2).

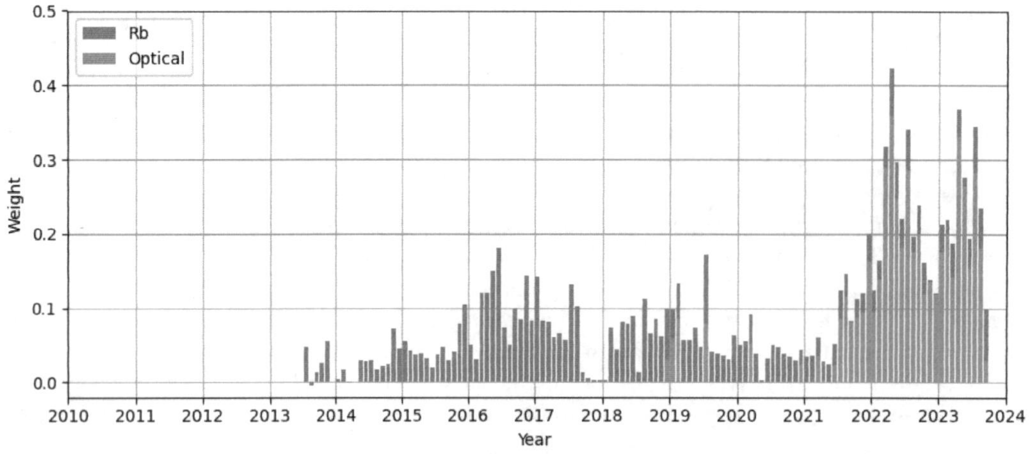

Figure 2. Graphical representation of the weights of secondary representations of the second in TAI, separated into the contributions from Rb and optical clocks, based on data published at https://www.bipm.org/en/time-ftp/other-products ("Fractional frequency of EAL from primary and secondary frequency standards").

5.2. Optical clock contributions to TAI

During the ROCIT project, twelve optical clocks within the consortium were used to evaluate the frequency of hydrogen masers contributing to the computation of TAI, with measurement periods aligned with the 5-day reporting periods of Circular T. A total of 55 periods ranging from 5 to 35 days were recorded during the project, representing a total cumulative data acquisition period exceeding 1000 days.

Five contributions from the SYRTE optical lattice clock SYRTE-Sr2 were used for TAI steering in 2021 and 2022, with one further contribution made in early 2023.

The INRIM ^{171}Yb optical lattice clock IT-Yb1 received approval from the CCTF Working Group on Primary and Secondary Frequency Standards (WGPSFS) in November 2019 and the submitted data was used for TAI steering that month. From February 2022 to May 2023 it contributed monthly to Circular T, a total of 16 contributions, and had the highest weight of any clock in TAI in July 2022.

The first data from NPL's ^{87}Sr optical lattice clock NPL-Sr1 was submitted to the BIPM in December 2022 and following approval by the WGPSFS was used for TAI steering in March 2023. By the time of the 9th symposium NPL-Sr1 had contributed three evaluations carried out within the month of TAI evaluation.

Data collected from other clocks will be submitted to the BIPM pending updated uncertainty budgets, peer-reviewed publications, and approval by the WGPSFS.

6. Conclusions

The work performed within the ROCIT project has significantly advanced the status of optical clocks within Europe towards meeting some of the key milestones in the international roadmap towards a redefinition of the SI second. The robustness and automation of optical clocks has been significantly improved and the coordinated clock comparison campaigns have greatly expanded the worldwide body of frequency comparison data necessary to validate the uncertainty budgets of these clocks. Methods have been developed for real-time data validation and on-the-fly correction of systematic frequency shifts, paving the way for initial demonstrations of two prototype optically steered UTC(k) time scales and a direct comparison between them.

At the time of the 8th Symposium on Frequency Standards and Metrology in 2015, no optical clocks at all had contributed to TAI. Eight years later, at the 9th Symposium, eight optical clocks in seven different laboratories had been approved by the WGPSFS and had contributed a total of 67 evaluations carried out within the month of TAI evaluation [23]. Of these eight optical clocks, four (SYRTE-Sr2, SYRTE-SrB, IT-Yb1 and NPL-Sr1) are operated by members of the ROCIT consortium, and between them they account for almost 40 % of the "on-time" contributions from optical clocks. This represents significant progress towards the mandatory criterion of regular contributions of optical frequency standards to TAI as secondary representations of the second. However further work is still required to reach the numerical target of at least three state-of-the-art calibrations of TAI each month for at least one year, with uncertainty at or below the 2×10^{-16} level (neglecting the recommended uncertainty of the secondary representation of the second), from a set of at least five optical frequency standards.

Acknowledgements

The ROCIT project (18SIB05) received funding from the European Metrology Programme for Innovation and Research (EMPIR), which is co-financed by the Participating States and from the European Union's Horizon 2020 research and innovation programme.

References

[1] Dimarcq N *et al.* 2024 Roadmap towards the redefinition of the second *Metrologia* **61** 012001

[2] Keller J, Wang D, Kiethe J and Fürst H A 2021 https://github.com/QUEST-EQM/wavemeter

[3] Pizzocaro M 2021 https://github.com/INRIM/BBRMon

[4] Lindvall T, Hanhijärvi K J, Fordell T and Wallin A E 2022 High-accuracy determination of Paul-trap stability parameters for electric-quadrupole-shift prediction *J. Appl. Phys.* **132** 124401.

[5] Steinel M, Shao H, Filzinger M, Lipphardt B, Brinkmann M, Didier A, Mehlstäubler T E, Lindvall T, Peik E and Huntemann N 2023 Evaluation of a $^{88}Sr^+$ optical clock with a direct measurement of the blackbody radiation shift and determination of the clock frequency *Phys. Rev. Lett.* **131** 083002

[6] Clivati C *et al.* 2022 Coherent optical-fiber link across Italy and France *Phys. Rev. Appl.* **18** 054009

[7] Delva P *et al.* 2017 Test of special relativity using a fiber network of optical clocks *Phys. Rev. Lett.* **118** 221102

[8] Schioppo M *et al.* 2022 Comparing ultrastable lasers at 7×10^{-17} fractional frequency instability through a 2220 km optical fibre network *Nat. Commun.* **13** 212

[9] Margolis H S *et al.* 2013 International timescales with optical clocks (ITOC) Proc. 2013 Joint European Frequency and Time Forum & International Frequency Control Symposium (EFTF/IFCS) p 908

[10] Denker H, Timmen L, Voigt C, Weyers S, Peik E, Margolis H S, Delva P, Wolf P and Petit G 2018 Geodetic methods to determine the relativistic redshift at the leve of 10^{-18} in the context of international timescales: a review and practical results *J. Geodesy* **92** 487–516

[11] Riedel F *et al.* 2020 Direct comparisons of European primary and secondary frequency standards via satellite techniques *Metrologia* **57** 045005

[12] Jian B, Bernard J, Gertsvolf M and Dubé P 2022 Improved absolute frequency measurement of the strontium ion clock using a GPS link to the SI second *Metrologia* **60** 015007

[13] Pizzocaro M, Bregolin F, Barbieri P, Rauf B, Levi F and Calonico D 2020 Absolute frequency measurement of the $^1S_0$ – $^1P_0$ transition of ^{171}Yb with a link to International Atomic Time *Metrologia* **57** 035007

[14] Schwarz R, Dörscher S, Al-Masoudi A, Benkler E, Legero T, Sterr U, Weyers S, Rahm J, Lipphardt B and Lisdat C 2020 Long-term measurement of the ^{87}Sr clock frequency at the limit of primary Cs clocks *Phys. Rev. Research* **2** 033242

[15] Dörscher S, Huntemann N, Schwarz R, Lange R, Benkler E, Lipphardt B, Sterr U, Peik E and Lisdat C 2021 Optical frequency ratio of a $^{171}Yb^{+}$ single-ion clock and a ^{87}Sr lattice clock *Metrologia* **58** 015005

[16] Lange R, Huntemann N, Rahn J M, Sanner C, Shao H, Lipphardt B, Tamm C, Weyers S and Peik E 2021 Improved limits for violations of local position invariance from atomic clock comparisons *Phys. Rev. Lett.* **126** 011102

[17] Goti I, Condio S, Clivati C, Risaro M, Gozzelino M, Costanzo G A, Levi F, Calonico D and Pizzocaro M 2023 Absolute frequency measurement of a Yb optical clock at the limit of the Cs fountain *Metrologia* **60** 035002

[18] Formichella V, Galleani L, Signorile G and Sesia I 2022 Robustness tests for an optical time scale *Metrologia* **59** 015002

[19] Riehle F, Gill P, Arias F and Robertsson L 2018 The CIPM list of recommended frequency standard values: guidelines and procedures *Metrologia* **55** 188–200

[20] Margolis H S and Gill P 2015 Least-squares analysis of clock frequency comparison data to deduce optimized frequency and frequency ratio values *Metrologia* **52** 628–634

[21] Margolis H S, Panfilo G, Petit G, Oates C, Ido T and Bize S 2024 The CIPM list "Recommended values of standard frequencies": 2021 update *Preprint* arXiv:2401.14537 [physics.atom-ph]

[22] Margolis H and Pizzocaro M 2020 Guidelines on the evaluation and reporting of correlation coefficients between frequency ratio measurements (http://empir.npl.co.uk/rocit/project-outputs/)

[23] BIPM 2024 Primary and secondary frequency standards participating to TAI https://webtai.bipm.org/database/show_psfs.html

Journal of Physics: Conference Series **2889** (2024) 012023 doi:10.1088/1742-6596/2889/1/012023

Incorporation of hydrogen masers into a free ensemble timescale aimed for optically steered UTC(*k*)

T Ido, H Ito, H Hachisu, N Nemitz, N Ohtsubo, Y Miyauchi, M Morikawa, M Tønnes and K Matsubara

Space-Time Standards Laboratory, NICT, Tokyo, Japan

ido@nict.go.jp

Abstract. The combination of a single hydrogen maser and an intermittently operated optical clock can generate a highly accurate timescale, but it is difficult to address a sudden deterioration of the hydrogen maser. This issue can be resolved by the use of an ensemble. By estimating the frequency drift of multiple individual hydrogen masers, it is possible to incorporate them as the main contributors to an ensemble time scale that benefits from their long-term stability. Such a timescale is in operation at UTC(NICT) since August 2023. To determine the required adjustments for frequency offset and drift of the free-running timescale, we refer to a separate timescale generated from a hydrogen maser steered according to intermittent operation of an optical clock. In combination, this realizes a robust and accurate timescale.

1. Introduction

Hydrogen masers (HMs) have a short-term stability that is an order of magnitude better than commercial caesium atomic clocks, and some also outperform the caesium clocks with long-term instabilities at the 10^{-16} level. While they lack the frequency accuracy to generate a timescale on their own, the addition of a separate frequency or time reference enables HMs to play a fundamental role as flywheel oscillators [1].

This realization motivated BIPM to change the weighting algorithm for the generation of the free atomic timescale EAL (Échelle Atomique Libre) in 2011, moving the dominant weight from the caesium clocks to the HMs. This change led to an improved stability of the international atomic time TAI. The new strategy was supported by caesium atomic fountain clocks entering reliable operation at multiple institutes as frequency references, and by GNSS and TWSTFT links connecting the fountains to EAL with characterized uncertainties. A few years later, institutes began improving the stability of their local implementations of UTC, UTC(*k*), by direct reference to their caesium atomic fountain clocks [2, 3]. Similarly stable operation of optical clocks is now becoming a reality, and timescales will again benefit from their inclusion.

At NICT, the optical lattice clock maintains operation in an intermittent scheme, and now acts as a reliable optical standard that makes the existing ensemble of commercial Cs clocks unnecessary as frequency references. Since the ensemble timescale underlying UTC(NICT) then needs to provide stability rather than frequency accuracy, it was decided to revise its calculation to better utilize the HMs operated at NICT. This paper summarizes the background leading up to this and the preliminary results.

Journal of Physics: Conference Series 2889 (2024) 012023 doi:10.1088/1742-6596/2889/1/012023

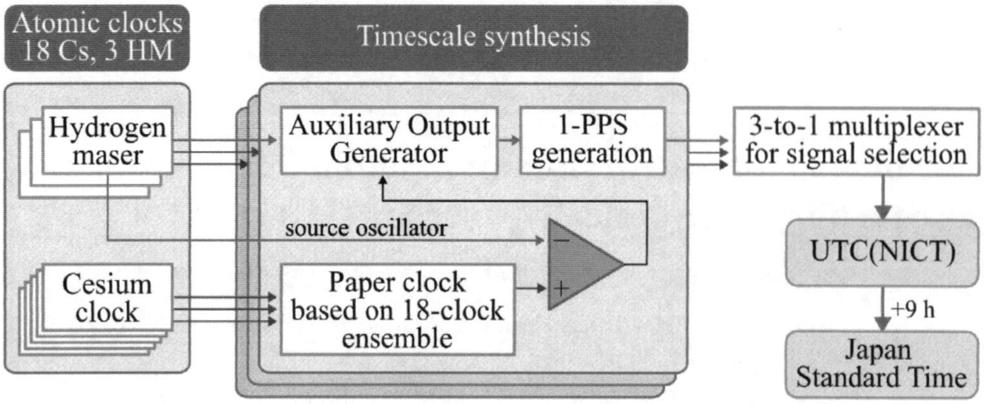

Figure 1. Schematic diagram of the Japan Standard Time system used in the synthesis of the redundant candidate signals for UTC(NICT), before the change of the ensemble timescale calculation. PPS: Pulse per second

2. Generation of UTC(NICT)

The current UTC(NICT) system started operation in 2006. It consists of three redundant subsystems which each use a HM as the source oscillator that is then adjusted for time and frequency through an Auxiliary Output Generator (AOG) to realize the frequency and phase of the ensemble timescale. The ensemble timescale otherwise only exists as a calculated "paper clock" timescale. It was originally determined from a pool of commercial caesium atomic clocks (figure 1).

Accurate measurement of the time difference between the Cs atomic clocks, hydrogen masers, and AOG outputs is necessary for both the paper clock and actual signal generation. A 24-channel DMTD (Dual-Mixer Time-Difference) system was developed specifically for this. Comparing the measurements of three DMTD implementations and a time interval counter ensures accuracy and reliability. The instability of the DMTD system reaches 10^{-16} at 10^4 seconds [4], sufficient for the generation of UTC(NICT).

Multiple HMs with flicker floors in the drift-insensitive Hadamard deviation below 5×10^{-16} have been secured as source oscillators (figure 2). Particularly, several Anritsu-made HMs purchased in the mid-2010s achieved instability at the 10^{-16} level up to 10 days or more.

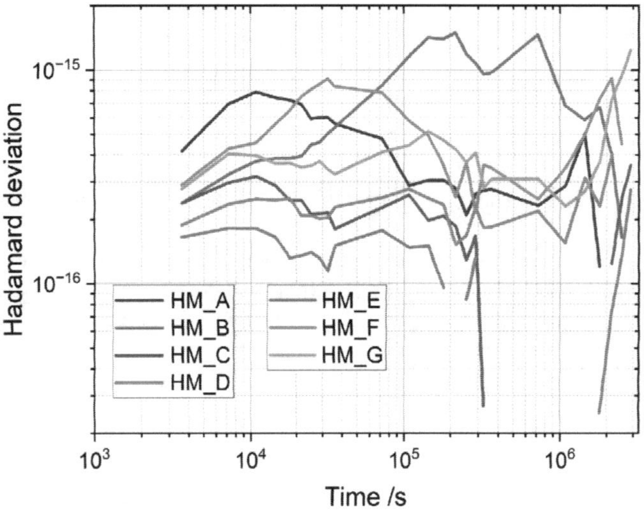

Figure 2. Instabilities of several HMs operated as part of the JST system. Individual Hadamard deviations have been calculated by a three-corner-hat method.

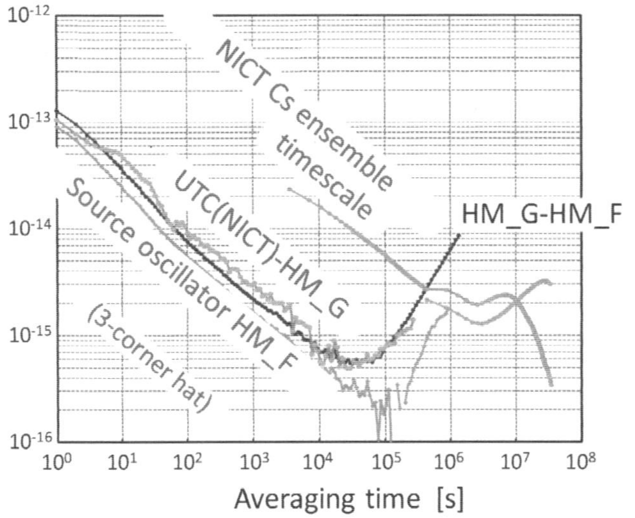

Figure 3. Instability contributions in the generation of UTC(NICT). The stability of the source oscillator HM_F (green) is estimated by a 3-corner hat method. Its short-term stability is transferred to UTC(NICT), as shown by comparison to HM_G (blue). For averaging time of 5 days and more, the steering to the calculated caesium ensemble timescale (magenta) takes effect.

The ensemble timescale was initially calculated using a pool of typically 18 caesium atomic clocks, employing a least-squares method similar to the algorithm for EAL [5]: The frequency value of each clock is estimated with reference to the most recent data of the ensemble, and the ensemble timescale is in turn adjusted by the weighted sum of the time difference between the prior estimate based on the obtained frequency values and the posterior measurement. For this calculation scheme, the Allan deviations shown in figure 3 indicates an instability of $1–2\times10^{-15}$ for UTC(NICT) over an averaging time of one month. The excellent long-term stability of multiple HMs is one of the strengths of the Japan Standard Time system, but steering them with the ensemble of less stable caesium atomic clocks can easily negate their superior characteristics.

3. Strategies for incorporating hydrogen masers into the ensemble

Early ensemble algorithms characterized the contributing atomic clocks solely by a frequency offset taken into account when determining the prior estimate. As described above, the differences between prior estimate and the posterior measurement were then used to iteratively renew the ensemble timescale. However, HMs show a significant deterministic frequency drift that is not captured by this model. Incorporating their excellent stability into the ensemble timescale requires accounting for this drift in the determination of the prior estimate. This exemplifies the importance of being aware of the order of characteristics of atomic clocks: Generalized higher-order variances have recently been proposed [6] as an indicator for these.

Another crucial point is setting the weights used to determine the ensemble mean. With the subtraction of the individual clocks' drift, the drift-insensitive Hadamard deviation is an effective indicator of stability. Although weights are commonly assigned inversely proportional to the *variance*, UTC(NICT) has employed weights based on the Allan *deviation* since 2006, and the new algorithm likewise assigns relative weights as the inverse of the Hadamard deviation. Our investigations have not shown a distinguishable difference in performance for weights based on the Hadamard variance. In an alternative approach, the EAL algorithm defines a new "predictability" parameter [7].

4. Utilization of Hydrogen Masers in UTC(NICT)

NICT started the calculation of the ensemble timescales RTAHM1 and RTAHM2 in June 2022, where five HMs are typically included in the ensemble. RTAHM1 includes a first order detrend corresponding to a frequency offset, whereas RTAHM2 also accounts for frequency drifts of the included HMs through a 2nd order detrending of the phase measurements.

The algorithmically simpler optical timescale TA(Sr) implemented through a single HM steered by intermittent frequency calibration [8] provides a reference for the evaluation of the ensemble timescale. Figures 4 and 5 confirm that the new calculated timescale achieves an instability at the mid-10^{-16} level, now with the high reliability of an ensemble of multiple HMs. In response to these results, the method to generate UTC(NICT) was revised in August 2023 to use RTAHM2 as the target timescale. Over the months following the change, the reduced timing fluctuation of UTC(NICT) relative to UTC has already become apparent as shown in figure 6.

5. Conclusion and outlook

The advancement of optical clocks is remarkable but generating a local UTC(k) timescale entirely in the optical domain will require an optical flywheel oscillator [9] steered using multiple optical clocks to maintain continuous operation. Commercial availability of accurate optical clocks will make this significantly more feasible but is still a future prospect.

Until then, forming an intermediate HM-based ensemble timescale as described here is an effective method to increase the performance of a hybrid system using optical clocks as the reference to periodically adjust frequency offset and frequency drift offset. While the system implemented at NICT is currently aligned to the monitored optical time scale TA(Sr) only through manual adjustments, future

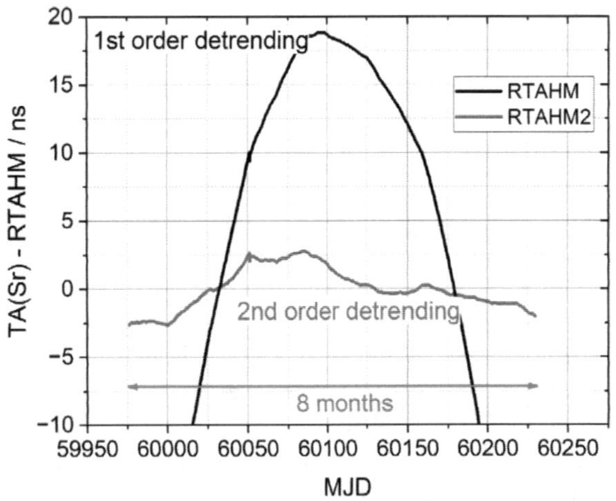

Figure 4. Record of the HM ensemble timescales RTAHM and RTAHM2 with respect to TA(Sr).

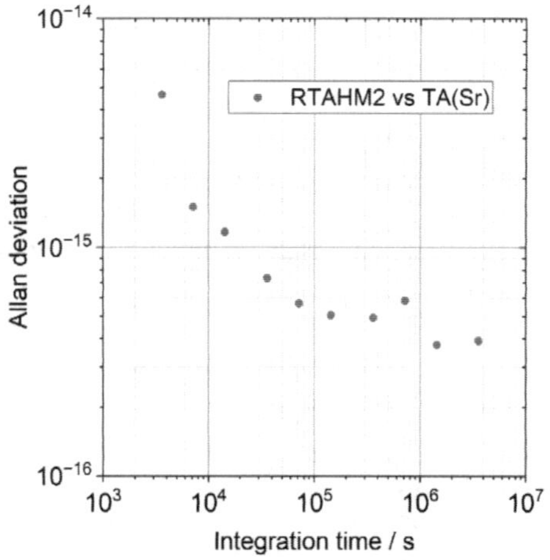

Figure 5. Allan deviation off RTAHM2 calculated with respect to TA(Sr).

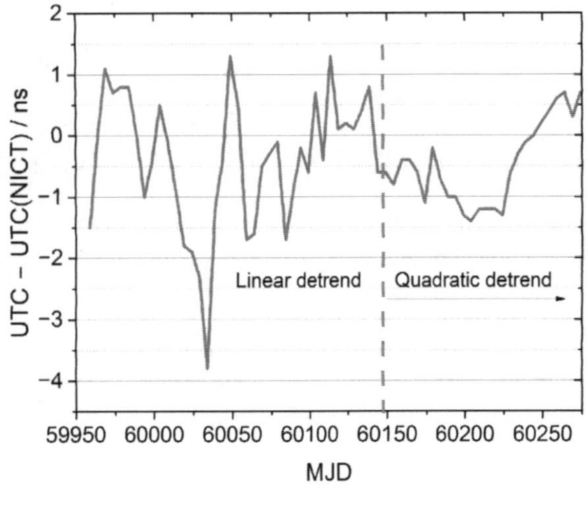

Figure 6. Improvement in the stability of UTC(NICT) after the adoption of RTAHM2 as ensemble timescale with quadratic detrending.

developments target a systematic method of control. A Kalman filter may enable combining information on the residual frequency error provided by local optical clock measurements with time differences provided by the international time links.

Optically referenced local timescales may soon begin outperforming the stability provided by the GNSS links used in the calculation of UTC itself, which may create a need to prioritize either the accuracy of the timescale interval or the international agreement, or alternatively maintain two timescales to fulfil both demands.

References

[1] Parker T 2002 Time scales using hydrogen masers *Tutorial at 34th Annual Precise Time and Time Interval Systems and Applications Meeting*

[2] Bauch A *et al.* 2012 Generation of UTC(PTB) as a fountain-clock based time scale *Metrologia* **49** 180

[3] Guena J *et al.* 2012 Progress in atomic fountains at LNE-SYRTE *IEEE Trans. Ultrason. Ferroelectr. Freq. Control* **59** 391

[4] Nakagawa F *et al.* 2005 Development of multichannel dual-mixer time-difference system to generate UTC (NICT) *IEEE Trans. Instr. Meas.* **54** 829

[5] Panfilo G and Arias F 2010 Algorithms for international atomic time *IEEE Trans. Instr. Meas.* **57** 140

[6] Ishizaki T *et al.* 2024 Higher-order Allan variance for atomic clocks of arbitrary order: mathematical foundation *Metrologia* **61** 015003

[7] Panfilo G, Harmegnies A and Tisserand L 2014 A new weighting procedure for UTC *Metrologia* **51** 285

[8] Hachisu H *et al.* 2018 Months-long real-time generation of a time scale based on an optical clock *Scientific Reports* **8** 4243

[9] Milner W R *et al.* 2019 Demonstration of a timescale based on a stable optical carrier *Phys. Rev. Lett.* **123** 173201

Journal of Physics: Conference Series **2889** (2024) 012024 doi:10.1088/1742-6596/2889/1/012024

Optically steered time scale generation at OP and NPL and remote comparisons

M Abgrall[1], B Chupin[1], P Uhrich[1], L Lorini[1], R Le Targat[1], B Pointard[1], J Lodewyck[1], J Tunesi[2], D B A Tran[2], A O Parsons[2], A Tofful[2], B I Robertson[2], M Y H Johnson[2], C-H Feng[2], M Schioppo[2], E A Curtis[2], I R Hill[2], R M Godun[2] and H S Margolis[2]

[1] LNE-SYRTE, Observatoire de Paris, Université PSL, CNRS, Sorbonne Université, 61 avenue de l'Observatoire, 75014 Paris, France
[2] National Physical Laboratory, Hampton Road, Teddington TW11 0LW, United Kingdom
Email: Michel.abgrall@obspm.fr

Abstract. In this paper, we present real-time optically steered time scales generated simultaneously at the Observatoire de Paris (OP) and the National Physical Laboratory (NPL). Throughout one month, experimental UTCx(*k*) time scales were generated concurrently in the two laboratories alongside the local operational UTC(*k*) time scales. The UTCx(*k*) time scales were based on hydrogen masers whose frequencies were calibrated by local optical frequency standards using optical frequency combs, with algorithms being developed for outlier filtering and frequency steering estimations in the two laboratories. The performance of the experimental time scales was assessed through local comparisons against the corresponding UTC(*k*) time scales, and also through remote comparisons performed both via Coordinated Universal Time (UTC) and by using the GPS Precise Point Positioning (PPP) technique. The two optically steered time scales remained within 4 ns of each other, outperforming the corresponding UTC(*k*) over the same period. To our knowledge, this is the first reported comparison of two independent "optical time scale" prototypes, and the results demonstrate the capability of optical clocks to produce operational time scales.

1. Introduction

Optical frequency standards (OFS) under development in many institutions worldwide have demonstrated impressive progress in terms of accuracy and stability [1–5], surpassing the performance of atomic fountain microwave frequency standards by two orders of magnitude. In the context of the anticipated redefinition of the second in the International System of units (SI) [6], several OFS currently considered as secondary representations of the SI second are already contributing to the steering of International Atomic Time (TAI), calculated monthly by the Bureau International des Poids et Mesures (BIPM). Local time scales maintained by National Metrology Institutes will also benefit from the accuracy and stability of OFS, which are expected to enable the time offset of these UTC(*k*) time scales (where *k* refers to the institute) from Coordinated Universal Time (UTC) to be maintained in the 100 ps range or lower [7–10].

Journal of Physics: Conference Series **2889** (2024) 012024 | doi:10.1088/1742-6596/2889/1/012024

In this paper, we present real-time optically steered time scales generated simultaneously at the Observatoire de Paris (OP) and the National Physical Laboratory (NPL). In this experiment, performed during the Robust Optical Clocks for International Timescales (ROCIT) project [11], independent experimental time scales UTCx(*k*) were generated concurrently for one month in the two laboratories in parallel to the local operational UTC(*k*) time scales. The experimental UTCx(*k*) time scales were based on hydrogen masers whose frequency was calibrated by the local OFS (SYRTE-SrB and SYRTE-Sr2 optical lattice clocks [12] at OP, NPL-Sr1 strontium optical lattice clock [13] and NPL-E3Yb+3 ytterbium ion optical clock [14] at NPL) via optical frequency combs. From these frequency calibrations, steering corrections were updated on an hourly basis via frequency offset generators (FOGs), to compensate for the real time maser frequency fluctuations more frequently than in the local UTC(*k*) time scales, where steering parameters are typically updated with a periodicity of one or a few days. After describing the experimental measurement chains, we present the algorithms that were implemented for outlier filtering and frequency steering estimations. We then analyse the performance of the experimental time scales based firstly on local comparison against the local UTC(*k*) and secondly on remote comparisons performed via UTC and using the GPS Precise Point Positioning (PPP) technique, before outlining strategies for improvement. Further details will be available in a publication in preparation.

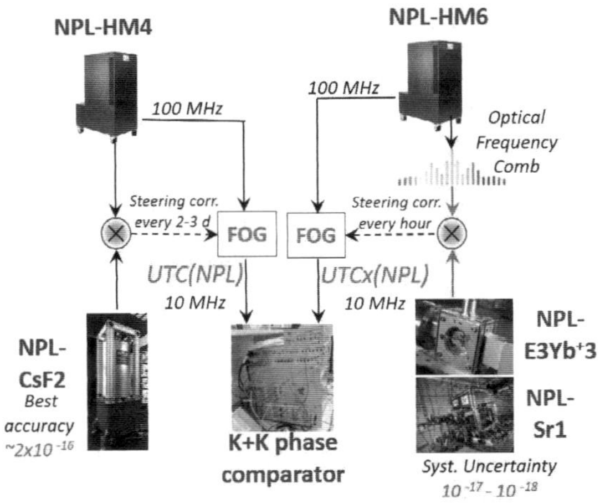

Figure 1. Experimental setup implemented at NPL

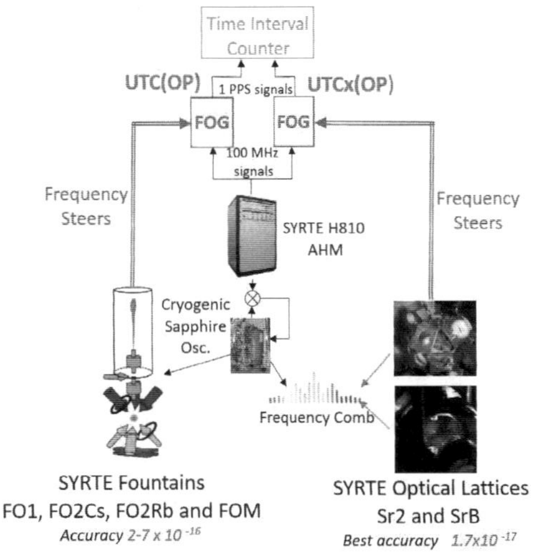

Figure 2. Experimental setup implemented at OP

2. NPL optically steered time scale: UTCx(NPL)

At NPL an optical lattice clock based on the $5s^2\ {}^1S_0$ to $5s5p\ {}^3P_0$ transition in ^{87}Sr (labelled as NPL-Sr1) [13] and a trapped ion optical clock based on the ${}^2S_{1/2} \rightarrow {}^2F_{7/2}$ electric octupole transition in $^{171}Yb^+$ (labelled as NPL-E3Yb+3) [14, 15] were operated during an international clock comparison campaign in March 2022. During this period, both optical clocks were used to steer a local microwave flywheel oscillator (active hydrogen maser NPL-HM6) in order to realise a prototype optical time scale. Figure 1 shows a simplified scheme of the experimental setup of UTCx(NPL) where the output frequency from the maser is up-converted to the optical domain using a fibre-based optical frequency comb (OFC) and then compared to the frequencies of the two optical clocks.

The frequency of the optical clocks compared against NPL-HM6 is measured by the OFC with a sampling time of 1 s and validated in hour-long chunks. The optical clock data is also validated in real time, with on-the-fly correction of the systematic frequency shifts. From the validated comparison data, we derive the fractional frequency difference (FFD) y[X – HM6] between the optical clocks and NPL-HM6 where X is either NPL-Sr1 or NPL-E3Yb+3. The computed FFD is used to steer the flywheel by means of a FOG. The steering parameter is updated every hour according to the following scenarios:

- If both optical clocks provide data for at least 25% of the hour, the averaged FFD y[X-HM6] values from each clock are combined via a weighted mean, with weights proportional to the relative proportion of available data.
- If only one optical clock is available within the hour, its averaged FFD is used.
- If neither clock provides sufficient data, then past steering parameters and the frequency drift of NPL-HM6 are used to predict the steering parameter.

An assessment of the stability of the prototype UTCx(NPL) time scale is enabled through a comparison with the local UTC(NPL) time scale. A simplified scheme of the generation of UTC(NPL) is also shown in Figure 1; it is based on another active hydrogen maser (NPL-HM4) whose frequency is steered by the NPL caesium fountain primary frequency standard (NPL-CsF2) [16,17]. A phase comparator is used to measure the phase difference between UTCx(NPL) and UTC(NPL).

During the one month experiment, the offsets of UTCx(NPL) from UTC, computed using the local measurement UTCx(NPL) – UTC(NPL) and data from Circular T, remained within ±0.64 ns. This is significantly better than the operational UTC(NPL) time scale, which differed from UTC by as much as 2.7 ns over the same period. One limiting factor to the operation of UTCx(NPL) was the appearance of a few phase jumps in the resulting time scale. These glitches were identified as being due to two main problems: firstly the FOG would occasionally add an instantaneous phase jump when a frequency steer was applied, and secondly invalid data occasionally passed through our filters. Both types of glitches were rapidly apparent from the comparison between UTCx(NPL) and UTC(NPL), making it possible to correct for these erroneous phase offsets in-situ.

3. OP optically steered timescale: UTCx(OP)

At OP, we operated an experimental time scale UTCx(OP) based on combining data from the two strontium optical lattice clocks SYRTE-Sr2 and SYRTE-SrB [12] during the measurement campaign performed in March 2022 and compared it to our operational local time scale UTC(OP). The experimental setup is shown in Figure 2. The frequency reference of the laboratory is based on the active hydrogen maser 1400810 filtered by a cryogenic sapphire oscillator (CSO) presenting an excellent short term stability. The resulting hybrid oscillator provides a signal at 11.98 GHz that is compared to the two optical clocks via an OFC, and also feeds the microwave synthesizer of the OP caesium and rubidium atomic fountains (FO1, FO2Cs, FO2Rb and FOM) [18, 19].

UTC(OP) is generated using a frequency offset generator (FOG) developed by SKK Electronics [20] to steer a 100 MHz signal from the hydrogen maser. More details on the generation of UTC(OP) can be found in [21]. During this experiment, UTCx(OP) was generated using a second FOG to steer the same reference maser. The steering algorithm uses as input the 1 s sampled maser – SrX data (SrX = SYRTE-Sr2 and SYRTE-SrB, respectively), corrected for the strontium systematic frequency shifts and made available in real time. There is no additional steering towards UTC.

The steering correction is calculated every hour. After outlier filtering of the maser – SrX data, we compute a "pack" of data averaged over periods of $0.1d$ for each OFS. We also compute extrapolated values using the available pack data of both OFS during the previous $5d$ with respect to the time of the computation correction. The steering is updated to the FOG every hour according to the following approach: if the last pack is older than $0.4d$, we take the last extrapolated value for the new steering, otherwise, we take the last valid pack data. If the difference between the previous steering and the new one is larger than 2×10^{-14}, the steering is considered as an outlier and thus not updated.

The experimental time scale UTCx(OP) was compared to UTC(OP) using a time interval counter and to UTC using Circular T data. Both differences UTCx(OP) – UTC(OP) and UTCx(OP) – UTC remained below 2 ns during the entire duration of the campaign, while UTC(OP) was no more than 0.5 ns away from UTC over the period. As at NPL, the time scale prototype at OP was affected by phase jumps due to bad outlier rejection. This led to incorrect steering in four instances during the month.

If we ignore the anomalous steering events, the time scale showed its potential to remain within a few hundred ps of UTC over the one-month period. In order to demonstrate this, we performed simulations with the same algorithm using maser – SrX data after a manual filtering of the outliers. The simulations were performed using Sr2 and SrB independently for comparison purposes. The accumulated phase of the simulated time scale UTCxSrB with respect to UTC remained below 600 ps over the 30*d* period. It was a bit larger (1.2 ns) for UTCxSr2 due to the effect of missing data close to the end of the period.

4. Distant comparison of the independent timescale prototypes

To compare the experimental time scales, we combined the local UTCx(*k*) – UTC(*k*) comparisons with UTC(NPL) – UTC(OP) comparisons performed both via UTC and using the GPS Precise Point Positioning (PPP) technique. These comparisons showed that the time scale prototypes remained within 4 ns of each other, with the observed fluctuations dominated by the anomalous events on the two time scales. The different comparison techniques agree to within a few hundred ps. The comparisons of the *a posteriori* corrected time scales resulted in differences lower than 2 ns. This is an improvement by a factor of two with respect to the comparison between the local time scales UTC(NPL) and UTC(OP).

5. Conclusion

In the context of a future redefinition of the second [6], this paper presents one of the first attempts to generate real-time optically steered time scales. NPL and OP demonstrated the first out-of-loop comparison of two independent time scales steered exclusively with optical clock data that exhibited a lower difference than for that of the corresponding operational UTC(*k*) time scales over the same period. The results of the experiment were affected by a few phase jumps due to data validation issues and unexpected behaviour of the FOGs used for steering. Correction of these issues in post processing showed a factor of two improvement in time deviation over the period with respect to the local UTC(*k*) comparison.

This result shows the potential of using OFS for the generation of highly accurate time scales. Further work is ongoing at NPL and OP, as well as in other institutions, towards increasing the reliability of operation of the optical clocks and measurement chains, improving the steering algorithms and testing optically-steered time scale prototypes over much longer periods.

Acknowledgements

The ROCIT project (18SIB05) received funding from the European Metrology Programme for Innovation and Research (EMPIR), which is co-financed by the Participating States and from the European Union's Horizon 2020 research and innovation programme.

References

[1] Ushijima I, Takamoto M, Das M, Ohkubo T and Katori H 2015 Cryogenic optical lattice clocks *Nature Photonics* **9** 185-189
[2] Huntemann N, Sanner C, Lipphardt B, Tamm Chr and Peik E 2016 Single-ion atomic clock with 3×10^{-18} systematic uncertainty *Phys. Rev. Lett.* **116** 063001
[3] McGrew W *et al.* 2018 Atomic clock performance enabling geodesy below the centimetre level *Nature* **564** 87-90
[4] Bothwell T, Kedar D, Oelker E, Robinson J M, Bromley S L, Tew W L, Ye J and Kennedy C J 2019 JILA SrI optical lattice clock with uncertainty of 2.0×10^{-18} *Metrologia* **56** 065004
[5] S. Brewer, Chen J S, Hankin A M, Clements E R, Chou C W, Wineland D J, Hume D B and Leibrandt D R 2019 $^{27}Al^{+}$ quantum-logic clock with a systematic uncertainty below 10^{-18} *Phys. Rev. Lett.* **123** 033201
[6] Dimarcq N *et al* 2024 Roadmap towards the redefinition of the second *Metrologia* **61** 012001
[7] Formichella V, Galleani L, Signorile G and Sesia I 2021 Robustness tests for an optical time scale *Metrologia* **59** 015002

Journal of Physics: Conference Series **2889** (2024) 012024 doi:10.1088/1742-6596/2889/1/012024

[8] Grebing C, Al-Masoudi A, Dörscher S, Häfner S, Gerginov V, Weyers S, Lipphardt B, Riehle F, Sterr U and Lisdat C 2016 Realization of a timescale with an accurate optical lattice clock *Optica* **3** 563–569

[9] Hachisu H, Nakagawa F, Hanado Y and Ido T 2018 Months-long real-time generation of a time scale based on an optical clock *Scientific Reports* **8** 4243

[10] Yao J *et al.* 2019 Optical-clock-based time scale *Phys. Rev. App.* **12** 044069

[11] Margolis H S *et al.* 2024 Robust optical clocks for international timescales (ROCIT) *in this proceedings*

[12] Lodewyck J *et al.* 2016 Optical to microwave clock frequency ratios with a nearly continuous strontium optical lattice clock *Metrologia* **53** 1123

[13] Hobson R, Bowden W, Vianello A, Silva A, Baynham C F A, Margolis H S, Baird P E G, Gill P and Hill I R 2020 A strontium optical lattice clock with 1×10^{-17} uncertainty and measurement of its absolute frequency *Metrologia* **57** 065026

[14] Nisbet-Jones P B R, King S A, Jones J M, Godun R M, Baynham C F A, Bongs K, Doležal M, Balling P and Gill P 2016 A single-ion trap with minimized ion-environment interactions *Applied Physics B* **122** 57

[15] Godun R, Nisbet-Jones P, Jones J, King S, Johnson L, Margolis H, Szymaniec K, Lea S, Bongs K and Gill P 2014 Frequency ratio of two optical clock transitions in $^{171}Yb^{+}$ and constraints on the time variation of fundamental constants *Phys. Rev. Lett.* **113** 210801

[16] Szymaniec K, Park S E, Marra M and Chalupczak W 2010 First accuracy evaluation of the NPL-CsF2 primary frequency standard *Metrologia* **47** 363

[17] Li R, Gibble K and Szymaniec K 2011 Improved accuracy of the NPL-CsF2 primary frequency standard: evaluation of distributed cavity phase and microwave lensing frequency shifts *Metrologia* **48** 283

[18] Guéna J *et al.* 2012 Progress in atomic fountains at LNE-SYRTE *Trans. Ultrason. Ferroelectr. Freq. Control* **59** 391

[19] Guéna J, Abgrall M, Clairon A and Bize S 2014 Contributing to TAI with a secondary representation of the SI second *Metrologia* **51** 108

[20] Rovera G D, Abgrall M and Siccardi M 2012 Characterization of an auxiliary offset generator for steering of H Masers *Proc. 2012 EFTF* 14–15

[21] Rovera G D, Bize S, Chupin B, Guéna J, Laurent P, Rosenbusch P, Uhrich P and Abgrall M 2016 UTC(OP) based on LNE-SYRTE atomic fountain primary frequency standards *Metrologia* **53** S81

Journal of Physics: Conference Series **2889** (2024) 012025 doi:10.1088/1742-6596/2889/1/012025

A new resilient time and frequency infrastructure for UTC(NPL)

D J M Jones, M Aikomo, S Ashford, A Ashkhasi, J A Davis, B Devine, B Eglin, S B Everett, R Foot, D Galbraith, R J Hendricks, K Khatry, C Langham, R Lewis, J Newton-Griffiths, H Owen, S L Shemar, A Sheppard, C Smyth, K Szymaniec, S Walby, D F M Weston, J Whale, P B Whibberley, A Wilson, C Wilson and H S Margolis

National Physical Laboratory, Teddington, Middlesex, TW11 0LW, UK

Email: douglas.jones@npl.co.uk

Abstract. The Resilient Enhanced Time Scale Infrastructure (RETSI) is an atomic clock network distributed geographically in secure locations, that is being developed as part of work in the National Timing Centre programme. It will become the source of UTC(NPL), the physical realisation of Coordinated Universal Time (UTC) maintained by the National Physical Laboratory (NPL) in the United Kingdom (UK). Dissemination points called User Access Nodes will link to RETSI via Orchestration Nodes, allowing users to access time and / or frequency references traceable to UTC(NPL). This will help to improve security and resilience in national time and frequency generation and distribution, enabling the UK to reduce its dependence on timing signals from Global Navigation Satellite Systems.

1 Introduction

The UK's critical national infrastructure (CNI) is heavily reliant on timing signals from Global Navigation Satellite Systems (GNSS), meaning that the use of GNSS could act as a single point of failure for many of the systems used by CNI. This risk was identified in the Blackett review [1], a study into the critical dependencies on GNSS in the UK, and was further highlighted in the 2023 update to the UK's national risk register [2].

The National Timing Centre (NTC) programme is enabling the UK to reduce its reliance on GNSS. A major part of this programme involves designing and implementing a new Resilient Enhanced Time Scale Infrastructure (RETSI), which will replace the UK's current UTC(NPL) infrastructure. Here we define the resilience of a time scale as its ability to withstand, adapt to and recover from challenges, and continue to generate its time and frequency signals. This is the key design objective for RETSI, which will have a distributed architecture at secure sites located across the UK. RETSI will form the generative core of the UK's timing system and produce GNSS-independent and UTC-traceable time and frequency signals that can be disseminated to UK CNI sectors.

In the following we present the design features of RETSI and then discuss its future applications as part of the wider NTC programme.

2 Design of the resilient national time and frequency infrastructure

RETSI is designed to be a mesh of four linked time scale laboratories (Core Nodes), that are separated geographically within the UK and located at secure sites. These Core Nodes comprise atomic clocks and other time and frequency distribution hardware, that generate several time scale realisations. At any given time, one of these realisations at one of the Core Nodes is identified as UTC(NPL), with the associated site called the 'primary site'. The other Core Nodes in the network (at secondary sites) generate backups to UTC(NPL). The multi-site design and geographic separation of the Core Nodes improve the resilience of the overall system as it keeps them isolated from challenges local to a given site [3][4]. The high-level overview of the node structure of RETSI is shown in figure 1.

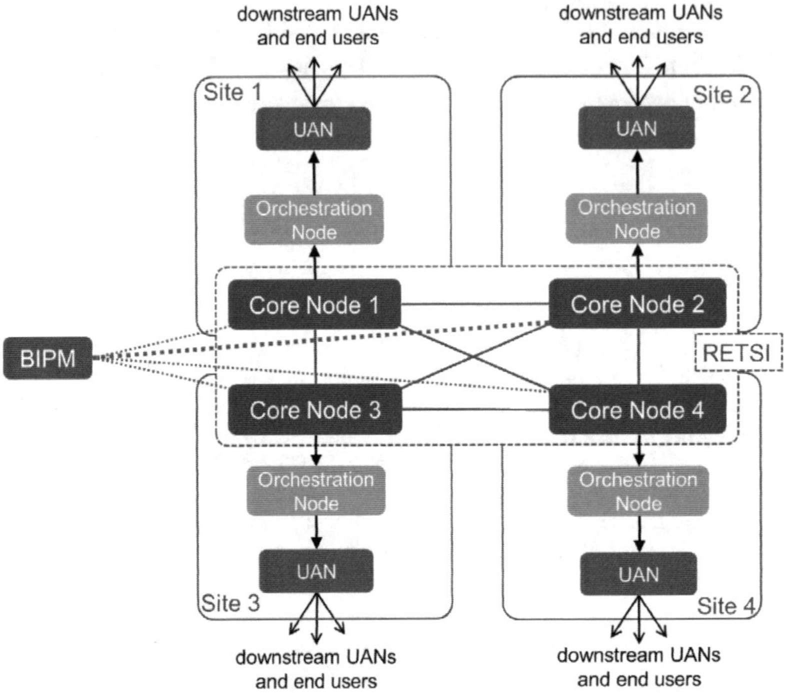

Figure 1. Concept for the design of RETSI, showing the four geographically distributed and interconnected sites, with their time scale laboratories (Core Nodes) and the interfaces to the BIPM. Solid lines indicate that both, physical time and frequency signals and data, are transferred, while dotted lines are used to indicate that only data is transferred. At any time, there is only one Core Node that generates UTC(NPL) and interfaces with the BIPM. In the example shown in this figure, this is Core Node 2 (indicated by the thicker dotted line to the BIPM), but the Core Node that generates UTC(NPL) may be changed from time to time. The Orchestration Nodes interface to RETSI and distribute its timing signals out to users via tier 1 User Access Nodes (UAN) located at RETSI sites. From the tier 1 UANs, timing signals may be distributed to end users either directly or via lower tier UANs beyond the RETSI sites.

Each Core Node produces four time scale realisations, each denoted UNR. These UNRs are outputs of frequency offset generators (FOG) that are fed from independent active hydrogen masers (figure 2). One of the UNRs at each site is designated as the reference time scale realisation for that site and is denoted UN_HUB. One of these UN_HUBs is selected to be the physical realisation of UTC(NPL). The unique aspect in the design of RETSI, compared to other national time scale infrastructures, is that the UN_HUB that defines the physical realisation of UTC(NPL) can be swiftly changed to a different physical realisation at a different site, in a process called "site selection". This means the UTC(NPL) time point can be relocated within the time scale network. Locally, at a given site, the UN_HUB can be switched to a different UNR at that site.

The Core Nodes are connected to each other by a network of time and frequency transfer links for comparisons of the outputs of UN_HUBs. Primarily, RETSI will use time and frequency transfer over optical fibres running between sites to facilitate such comparisons with low noise [5]. Additionally, Two-Way Satellite Time and Frequency Transfer (TWSTFT) links will be operated between sites. Operating these different modalities of time and frequency transfer increases resilience, since they have distinct failure modes, meaning they are unlikely to fail at the same time, and are both completely independent of GNSS. RETSI will continue to use GNSS time transfer methods, for cross-checking the measurements of the other transfer methods and calibration of link delays. However, RETSI is not operationally reliant on these GNSS links, hence a disruption of GNSS will not disrupt the RETSI output. RETSI will maintain traceability to UTC via the international TWSTFT links (along with calibrated GNSS measurements) between UTC(NPL) at the primary site and other UTC(k) labs.

A diagram in figure 2 shows the Core Node hardware including the independent maser clocks and FOGs forming the UNR chains, the time scale switches, and distribution and measurement hardware.

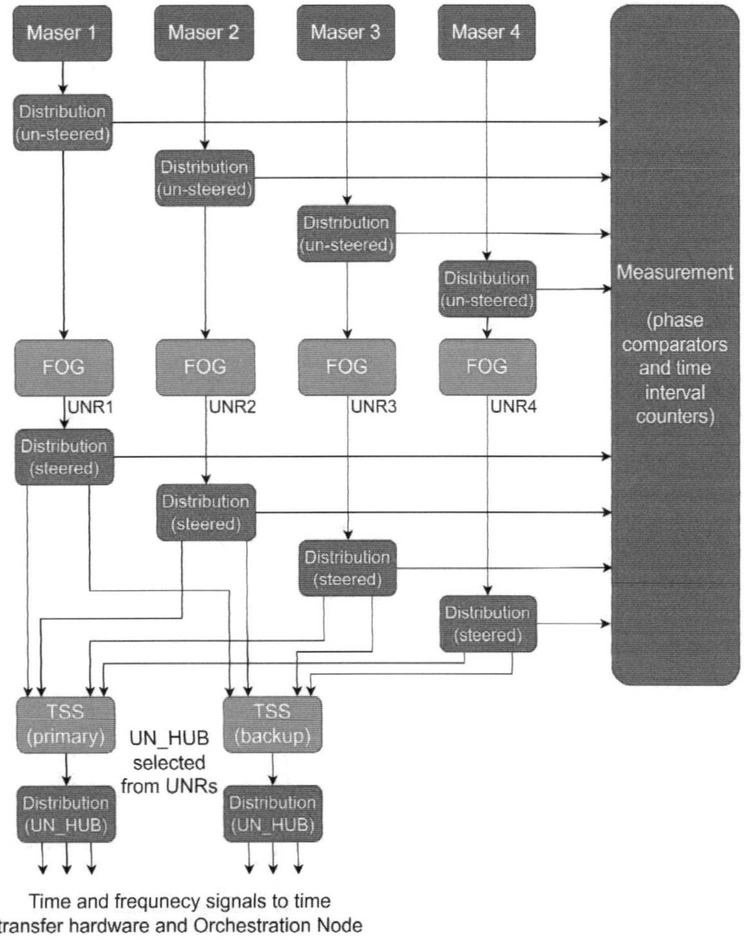

Figure 2. The concept for the generation, distribution and measurement of time and frequency signals in a single RETSI Core Node. The four maser chains allow for the generation of four independent time scale realisations (UNRs). Steers to the UNRs are applied with frequency offset generators (FOG). Time scale switches (TSS) facilitate the selection of which UNR is designated as the site reference time scale (UN_HUB). Arrows in the diagram represent the direction of time and frequency signal distribution.

Four free-running (un-steered) hydrogen masers produce 1 PPS, 10 MHz and 100 MHz signals that are multiplied in distribution amplifiers, and then measured with reference to the site's UN_HUB. Other signal copies are sent to FOGs that apply the frequency steers calculated in the RETSI steering algorithm to generate the UNR. The FOG UNR outputs are then, distributed for measurement against the site's UN_HUB. The time scale switches (TSS) implement the selection of UN_HUB from one of the UNRs. The TSS and the distribution amplifiers for UN_HUB output signals are duplicated in a separate hardware chain in order to eliminate a potential single point of failure. A timecode is also generated at each site from the local UN_HUB to provide adequate synchronisation for devices within RETSI that require it.

The UN_HUBs will be aligned to UTC(NPL) with a residual time offset of less than 3 ns, so that time steps are minimised when the UTC(NPL) realisation is changed in a site selection procedure. Similarly, the UNRs at a given site will be kept aligned to the local UN_HUB within 1 ns. The time offset of each UNR from the local UN_HUB is measured using a combination of time interval counters (for 1 PPS signals) and phase comparators (for standard 10 MHz and 100 MHz signals).

A steering algorithm will be implemented to keep all the UNRs, UN_HUBs and UTC(NPL) aligned with each other, and, importantly, to keep UTC(NPL) within a small offset from UTC (target: <5 ns; aspiration: <1 ns). The algorithm builds on the approaches of [6] and [7] and comprises both a frequency lock and a phase lock part. A stable frequency reference is derived from an ensemble of caesium (Cs) fountain primary frequency standards operated at different Core Nodes. In future, this ensemble could be expanded to include optical clocks. The frequency lock part of the algorithm steers every UNR independently using information from all available Cs fountains, including remotely located ones. The phase lock part combines time offset measurements made in real time with Cs fountain data and post-processed data on deviation from UTC published in Circular T by the BIPM. As the latter is only available with long latency, up to 40 days, the use of ultra-stable frequency standards to discipline UNRs is essential. Each month, RETSI will also submit the estimates of the average frequency of UTC(NPL) made by its primary frequency standards to the BIPM, and thereby contribute to the evaluation and steering of the UTC scale interval.

The status and performance of the Core Node output signals and time transfer functionality is continually assessed in RETSI in case of needing to change to a different physical realisation of UTC(NPL). Additionally, the primary site may need to be changed i.e., a new site selected as primary, for example when:

- the primary site is unresponsive,
- links from the primary site to other sites have failed,
- international TWSTFT links from the primary site to other NMIs have failed,
- major site maintenance is planned.

The design and implementation of RETSI respond to key needs for accurate, traceable, and highly available time reference signals across CNI sectors. As a result, its operational software adheres to the highest standards of security and reliability, to ensure its resilience to withstand, adapt and recover. The software is run on dedicated servers, and the outputs of the software are archived and stored securely. The computed outputs of the software are available to operators on a monitoring dashboard. Should problems occur during the operation, operators will be notified with monitoring alarms [8]. Hardware and software performance metrics and logs will be monitored and feed the alarms. A hierarchy of alarms, warnings and notifications will be established for the information that needs to be delivered to operators.

With the resilience of the entire system and security of the Core Nodes in mind, the distribution of RETSI output signals will be facilitated by an interface called an Orchestration Node (ON). The ON interfaces will be arranged on each site for the provision of RETSI timing signals to User Access Nodes (UAN), the nodes accessible by the users of RETSI (figure 3). ONs provide two services:

- Timing Signal Service: provision of physical signals (1 PPS and 10 MHz) from the co-located and remote UN_HUBs, 10 MHz outputs from co-located free-running masers, and timecode distribution.
- Status Signal Service: provision of information to RETSI users, that is updated every minute, indicating whether the distributed time and frequency signals are within specification. Additionally, information about pending leap seconds will be broadcast by this service. The status signal gives users confidence in the timing signals they are receiving, a unique feature and advantage of RETSI over GNSS services.

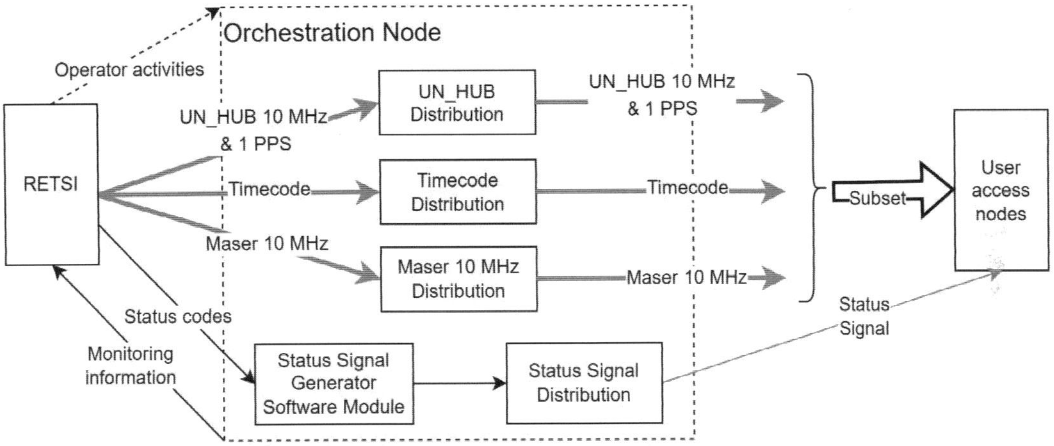

Figure 3. Orchestration Node high level design, showing interfaces with RETSI and User Access Nodes, and interfaces to the Orchestration Node. The dashed line represents the key operator management and configuration of the Orchestration Node. The subset of signals provided to an individual User Access Node is decided by its users and managed by RETSI operators.

3 Outlook

Implementation of RETSI and the Orchestration Nodes is underway. In the short-term, equipment is being installed site by site with links being formed between them. One of the RETSI Cs fountains is now operational, with the systematic effects being evaluated. Another Cs fountain is under construction. Once a robust programme of testing has been completed, a measurement campaign will be undertaken to relocate the UTC(NPL) timepoint from the legacy infrastructure at Teddington into the RETSI mesh.

The UK CNI sectors, which are dependent on time traceable to UTC, have expressed the need for dissemination of UTC(NPL) in resilient timing infrastructure. Precision time and frequency signals will be disseminated out of the RETSI time scale laboratory network via multiple UANs, which will supply industrial, academic and other specific users. UANs that interface directly to the ONs are termed 'tier 1 UANs'. Lower tier UANs can stem out from the tier 1 nodes, thus distributing the precision time generated by RETSI to broader classes of users, especially those with critical applications for timing signals. This architecture will eventually reduce dependence on GNSS and strengthen the resilience of UK CNI sectors. Furthermore, distribution of these signals to users will drive innovation in the UK's time and frequency sector and associated supply chain. The RETSI design allows for accommodation of the anticipated developments in optical frequency standards and the expected future redefinition of the SI second [9]. This work is strongly aligned with the UK's Position Navigation and Timing policy framework to provide resilient, terrestrial, sovereign, and high-quality timing for the UK [10].

Acknowledgements

The authors acknowledge the funding from the UK government Department for Science, Innovation and Technology (DSIT) through the National Timing Centre Programme.

References

[1] Government Office for Science 2018 Satellite-derived time and position: a study of critical dependencies *HM Government*
[2] Cabinet Office 2023 National risk register 2023 *HM Government* chapter 4 pp 91– 92
[3] Hutsell S T and Koppang P A 2000 USNO alternate master clock steering *Proceedings of the 32nd Annual Precise Time and Time Interval Systems and Applications Meeting* pp 289–300
[4] Cosart L, Helmut I and Zampetti G 2022 cnPRTC — coherent network primary reference time clock: a geographically distributed resilient timescale for telecommunications *IEEE Communications Magazine* **61** pp 28–32
[5] English E L, Ashkhasi A and Whibberley P 2021 Distributing time by fibre *Contemporary Physics* **62** pp 24–38
[6] Rovera G D, Bize S, Chupin B, Guéna J, Laurent P, Rosenbusch P, Uhrich P and Abgrall M 2016 UTC(OP) based on LNE-SYRTE atomic fountain primary frequency standards *Metrologia* **53** S 81–88
[7] Galleani L, Signorile G, Formichella V and Sesia I 2020 Generating a real-time time scale making full use of the available frequency standards *Metrologia* **57** 065015
[8] Kaczmarek J, Miczulski W, Kozioł M and Czubla A 2013 Integrated system for monitoring and control of the national time and frequency standard *IEEE Transactions on Instrumentation and Measurement* **62** pp 2828–2838
[9] Dimarcq N *et al*. 2024 Roadmap towards the redefinition of the second *Metrologia* **61** 012001
[10] Department for Science Innovation and Technology 2023 Critical services to be better protected from satellite data disruptions through new position, navigation and timing framework *HM Government*

Journal of Physics: Conference Series **2889** (2024) 012026 doi:10.1088/1742-6596/2889/1/012026

A definition of the SI second based on several optical transitions

Jérôme Lodewyck

LNE–SYRTE, Observatoire de Paris, Université PSL, CNRS, Sorbonne Université, 61 avenue de l'Observatoire, 75014 Paris, France

E-mail: jerome.lodewyck@obspm.fr

Abstract. A new definition of the SI second based on optical transitions is expected to be adopted within the next 10 years. Several options for this redefinition are currently under consideration. Among them, a definition based on several transitions would take advantage of the variety of high performance optical frequency standards. In this paper, we review practical aspects such a definition entails, and propose a detailed analysis of its strengths and weaknesses.

1. Introduction

The rapid progress in optical frequency standards over the last two decades led the Consultative Committee for Time and Frequency (CCTF) to craft a roadmap [1] for the future redefinition of the SI (International System of Units) second based on optical clock transitions, aimed at replacing the current definition based on the caesium atom. Three options are under consideration:

- *Option 1:* a single atomic transition i replaces to the Cs hyperfine transition. Its frequency would be the new defining constant for the second, and its numerical value would be set to:

$$\nu_i = N \text{ Hz} \tag{1}$$

  where the exact numerical constant N is the defining value.

- *Option 2:* the definition is based on the weighted geometric mean of the frequencies of several transitions. This average, having the dimension of a frequency, would be the new defining constant for the second, and its numerical value would be set to:

$$\prod_i \nu_i^{w_i} = N \text{ Hz} \tag{2}$$

  where the exact normalisation constant N and the weights w_i (summing to one) are the defining values.

- *Option 3:* the second is defined by fixing the numerical value of a fundamental constant. For instance, the electron mass m_e would be the new defining constant, and its numerical value would be set to:

$$m_e = M \text{ kg} \tag{3}$$

  where the exact numerical constant M is the defining value.

Journal of Physics: Conference Series **2889** (2024) 012026 doi:10.1088/1742-6596/2889/1/012026

Option 1 is a natural continuation of the current definition of the second, simply replacing caesium with another atomic species. Option 3 is an intellectually appealing extension of the 2018 definitions of the other base units of the SI, with the caveat that the electron mass is a property of a specific particle, unlike the other constants used in the SI, which are more universally rooted in the fundamental equations of physics. However, because the realisation uncertainty of such constants is many orders of magnitude larger than the uncertainty of the best frequency standards, this options is deemed unpracticable.

Option 2 has been introduced in [2]. The motivation for this proposition is two-fold: First, the large number of optical frequency standards with equivalent performances makes it difficult to reach a consensus and to single out one of them as a new Primary Frequency Standard (PFS). Second, in order to cope with the diversity of frequency standards, the Consultative Committee for Length (CCL) and CCTF Working Group on Frequency Standards (WGFS) has put in place a procedure that consists in collecting all measured frequency ratios between these various primary and secondary frequency standards in order to derive, by a least square procedure [3], a consistent set of best estimates for these frequency ratios. Although the output of this procedure is formally a set of "recommended frequencies for Secondary Representations of the Second (SRS)" [4], hence linked to the Cs transition, the genuine mathematical output of the fit is actually a set of frequency ratios (given by the ratio of recommended frequencies) together with their covariance matrix (with uncertainties that can be smaller than the uncertainties on recommended frequencies). It is thus completely independent of the definition of the second, and, more generally, does not distinguish a particular transition. In light of this *de facto* procedure, option 1, as a continuation of the current definition of the second, arguably maintains an artificial, arbitrary, and to some extent unnecessary bias in the definition of the second. Therefore, a more "decentralized" definition, based on the fit output would be a more elegant choice, practically taking advantage of the many frequency ratio measurements currently available. Such an idea was already put forward by P. Gill at the 8th Symposium on Frequency Standards and Metrology [5]:

> [...] a definition comprising a "universal" value determined from the matrix of individual best values for different species.

In [2] we explicitely proposed a physical quantity with the dimension of a frequency, namely the geometric mean (2), derived from the "matrix", that could effectively give corpse to this "universal" value. The discussion in [5] continues with:

> However, in this scenario, the definition would not be based on a fundamental constant or a "constant of nature" (as in the current definition), but on an aggregation of values for different systems, with little physical significance. Further, this value would change every time individual values were upgraded, and change significantly with the introduction of a new species into the value matrix.

This notice illustrates the questions and concerns raised by a definition that does not rely on a single atomic transition. In this paper, we aim at clarifying practical aspects of option 2 (section 2), and at proposing a strengths and weaknesses analysis answering potential concerns about option 2, such as its physical significance as cited above (section 3). Finally, we discuss the possibility and the need for regular and smooths updates of the defining values [2] (section 4).

2. Practical implementation of option 2

2.1. Realizing the unit

For a definition to be practical, it must be possible to practically realize it. In option 2, the definition of the SI second (2) could in principle be directly realised by building a complete set of clocks, each realizing one of the transitions entering the definition, and by combining

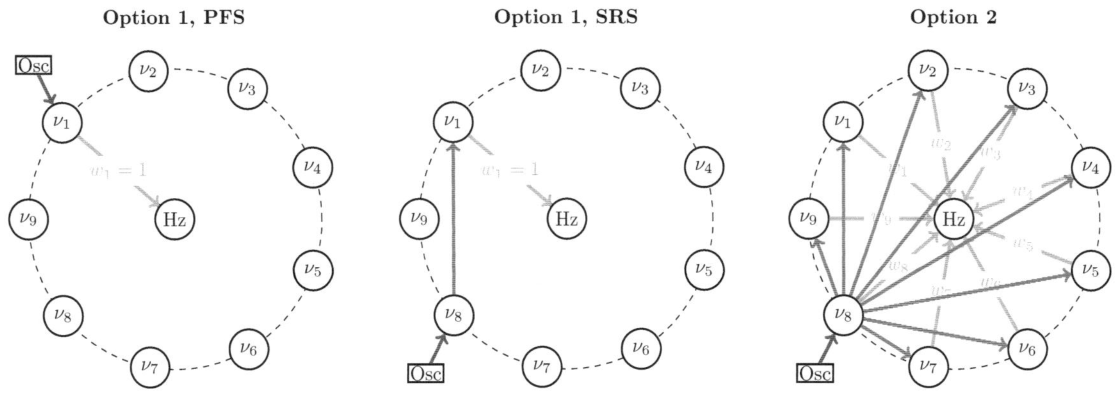

Figure 1. Realizing the second in options 1 and 2. Option 1 can be seen as a specific case of option 2 for which a single transition (here labelled 1) has a non-zero weight: $w_i = \delta_{i,1}$. Measuring the frequency of an oscillator in the SI can be done either by directly comparing this oscillator to a Primary Frequency Standard (left figure), or to a Secondary Representation of the Second (SRS) (middle figure) knowing the recommended frequency for this SRS (orange arrow). In option 2, the oscillator can be equivalently compared to any frequency standard entering the definition, using its recommended frequency. It is connected to the unit through the global fit of frequency ratios between the various frequency standards.

their output. More precisely, measuring the frequency ν_{osc} of an oscillator[1] in Hz would involve measuring its frequency ratio $\rho_{\text{osc},i}$ against all clocks i and then deduce, knowing the exact numerical constants N and w_i:

$$\nu_{\text{osc}} = N \prod_i \rho_{\text{osc},i}^{w_i} \text{ Hz}. \tag{4}$$

In practice, building a full set of clocks realizing all the transitions composing the unit can be involving. But even if only a single clock i_0 is available, it is still possible to determine ν_{osc} in Hz, by relying on otherwise known frequency ratios $\rho_{i,j}$ between the clock transitions:

$$\nu_{\text{osc}} = \rho_{\text{osc},i_0} N_{i_0} \text{ Hz} \quad \text{with} \quad N_{i_0} = N \prod_i \rho_{i_0,i}^{w_i}. \tag{5}$$

The numerical constant N_{i_0} is independent of the oscillator being measured, and can be viewed as the recommended frequency for transition i_0, the equivalent of the recommended frequency for SRS in the current definition of the second (see figure 1).

The output of the global fit of frequency ratios by the WGFS can be used to compute the recommended frequencies N_i, and they can be regularly published and updated by the CIPM, in the same way as the recommended frequencies of SRS are currently published. The covariance matrix output by the global fit can also be used to provide uncertainties for these recommended frequencies (see equation (A.5) of reference [2]), which add up to the realisation uncertainty of the frequency standard used to realized the unit. This added realisation uncertainty fundamentally comes from our limited knowledge of frequency ratios.

Equations (4) and (5) correspond to the extreme cases for which all transitions and a single transition, respectively, are used to realise the unit. They can be easily extended to any intermediate cases, and feature an added realisation uncertainty that decreases as more clocks are implemented.

[1] For instance, the local pivot oscillator used to steer the International Atomic Time (TAI)

2.2. *Choice of defining values*

The unit is defined by its defining values N and w_i. In this section, we detail the procedure to determine these values, set at the time of the redefinition.

2.2.1. Weights In order to reach a consensus on the implementation of option 2, it is preferable to choose the weights w_i of the various species composing the unit following systematic and quantitative rules. In [2], it was shown that, in the special case of a unit composed of two transitions, the ratio w_2/w_1 should be strictly larger than the inverse ratio between the clock uncertainties u_1/u_2, so that the realisation uncertainty of the unit with the best clock would not be significantly hindered by the uncertainty of the other clock. From this result, it sounds reasonable to choose weights proportional to the inverse *squared* uncertainty $1/u_i^2$. More generally, it can be proven by a least square calculation that, for an arbitrary number of clock transitions, this choice actually minimizes the overall added realisation uncertainty.

The choice of weights proposed above requires that the clock uncertainty u_i is a well-defined quantity. However, the uncertainty u_i is different from realisation to realisation, and is thus not intrinsic to the clock transition itself. Therefore, a method to estimate a typical uncertainty u_i for the transition i must be established, using the set of reported realisation uncertainties. Because the ultimate aim of a unit is to effectively be able to build frequency standards capable of operational measurements, it seems reasonable to select reported uncertainties of clocks that were involved in actual high precision frequency ratio measurements. In addition, it is preferable that a transition that is connected to other transitions only by a single frequency ratio measurement should not have a too large weight in the unit, to ensure redundancy in the evaluation of frequency ratios. From these considerations, a heuristic method was proposed in [2]: the weight w_i is obtained by averaging the inverse squared relative uncertainties of the two best frequency ratios involving the transition i. Although this method, and variations thereof, was found to provide reasonable and consistent weights when applied to the set of frequency ratios published before the 2017 CCTF, it lacks rigorous foundations, and also prevents option 2 from eventually converging to option 1: if a single transition stands out, its weight will be at most identical to the weight of the second best transition. To resolve these issues, we developed a rigorous method to set the weights, in the spirit of the three-cornered hat method: u_i is obtained as the output of a least squares adjustment that aims at matching the relative uncertainty on the frequency ratio $\rho_{i,j}$ with the quadratic sum of u_i and u_j. This method takes as input the covariance matrix of the outcome of the global fit of frequency ratios: by construction, it only relies on uncertainties associated with frequency ratio measurements. Furthermore, it provides a fair estimate of the uncertainty of a single clock transition, provided a realisation of this transition have been compared to at least two other transitions.

An independent publication in preparation will aim at detailing the results of this section.

2.2.2. Normalization constant The normalisation constant N is a fixed, exactly known constant, that is chosen to ensure the continuity of the unit across the redefinition. In option 2, as well as in option 1, this choice is driven by the set of frequency ratio measurements available at the time of the redefinition: N is computed from the set of recommended frequencies derived by fitting the set of frequency ratios. Quantitatively, the normalisation constant is chosen such that the recommended frequency for the caesium clock transition at the moment of the redefinition is identical to the previously fixed value. The uncertainty on this recommended frequency yields the typical, unavailable mismatch between the new and the old definition. Such a mismatch, arising from our limited knowledge of frequency ratios is unavailable, and on the same order, for both options 1 and 2.

2.3. After the redefinition

After the redefinition, the various optical frequency standards will likely be further improved, and new and more precise frequency ratios between theses standards will be measured. We can thus legitimately ask whether these improvements may *change* the definition of the second, or worse, be *incompatible* with the definition. After all, would it be possible that updated measured frequencies for the various transitions composing the unit would eventually not geometrically average any-more to the fixed value chosen for the definition? This concern can be answered with two properties of option 2:

(i) *The weights w_i and the normalisation constant N univocally define the unit*, without the need to specify a set of measured frequency ratios. Indeed, if we were to specify the duration of a second to an extra-terrestrial civilization, the sole transmission of w_i and N would be sufficient. From these values, the alien civilization would be able to realize the second, by simply building clocks implementing each species composing the unit and applying equation (4), possibly with a lower realisation uncertainty that we can currently attain ourselves. The limited knowledge of frequency ratios only limits our ability to *realize* the otherwise completely specified definition of the second, and acquiring more precise values for these ratios can only help reducing the realisation uncertainty, in the same way as acquiring a more precise value of atomic polarisability or collision cross-section can help reducing the realisation uncertainty of the current definition SI second based on caesium. Said otherwise, the knowledge of physical quantities (frequency ratios for Option 2, polarisability, cross-sections,... for Options 1 and 2) is necessary to connect a physical realisation of a frequency standard, to the actual, idealized, definition of the unit; and uncertainties on these quantities constitute the realisation uncertainty of the perfectly defined unit.

(ii) *The normalisation constant N is unconstrained by frequency ratio measurements.* Assuming new clock comparisons yield refined measured frequency ratios after the redefinition, with improved uncertainties, these new measurements will allow us to derive improved recommended frequencies for all the transitions composing the unit. But while the recommended frequencies do change with improved frequency ratios measurements, their weighted geometric mean stays invariably exactly equal to N, regardless of the actual values of measured frequency ratios.
In order to prove this fact, we note $\bar{\rho}_{i,j}$ the set of best estimates for frequency ratios as output by the least squared global fit of frequency ratio measurements. By construction, they satisfy the relations $\bar{\rho}_{i,i} = 1$ and $\bar{\rho}_{i,j}\bar{\rho}_{j,k} = \bar{\rho}_{i,k}$, which are constraints built in the fit. The updated recommended frequencies would then read $N_i = N \prod_k \bar{\rho}_{i,k}$, from where it can be shown that $\prod_i N_i^{w_i} = N$, independently of the actual values of $\bar{\rho}_{i,j}$:

$$\prod_i N_i^{w_i} = \prod_i N^{w_i} \prod_k \bar{\rho}_{i,k}^{w_i w_k} = N \frac{\prod_{i,k} \bar{\rho}_{i,i_0}^{w_i w_k}}{\prod_{i,k} \bar{\rho}_{k,i_0}^{w_k w_i}} = N \frac{\prod_i \bar{\rho}_{i,i_0}^{w_i}}{\prod_k \bar{\rho}_{k,i_0}^{w_k}} = N, \tag{6}$$

where we mathematically used a specific, arbitrary transition i_0 as a pivot.

3. Comparison between options 1 and 2

Choosing between the various options for the redefinition of the second requires to compare the advantages and drawbacks of each option. In this section, we propose a detailed analysis the strengths and weakness of option 2 that have been identified in the roadmap [1], focusing on those for which option 2 differs from option 1.

3.1. Strengths

Option 2 brings several advantages, as compared to option 1:

- *Reaching a consensus.* There are currently many possible technologies (ion clocks and optical lattice clocks) and atomic species for the redefinition of the SI second, with comparable performances in the low 10^{-18}. With option 1, a unique atomic species is singled out, introducing a somehow arbitrary distinction between the possible candidates. Reaching a consensus on a single species can potentially be a difficult task, and may delay the redefinition of the second. Option 2 solves this choice problem by attributing weights to the different candidates, objectively based on their demonstrated performances in measuring high accuracy frequency ratios. Option 2 would thus lead to an easier and earlier consensus.
- *Promoting diversity.* While the vast variety of optical frequency standard is a challenge for the redefinition of the second, it is a wealth for the scientific community, because it encourages innovation, and large data sets of interspecies comparisons are helpful to improve the robustness of optical frequency standards, and to test fundamental physics. Option 1 may lead to a strong focus of resources on a single species, which will tend to lessen the applications of optical frequency standards. On the contrary, option 2 not only does not strongly bias the definition on a single species dubbed "primary frequency standard", but also specifically encourages the production of new and more precise frequency ratio measurements, in order to decrease the realisation uncertainty. Said otherwise, the best way to improve the realisation of the SI second with option 1 is to build several PFS of the same kind, while in option 2, it is to build several different frequency standards and to compare them.
- *Readily and simply implementable.* All the tools to implement option 2 are already in place: the choice of defining values and the publication of recommended frequencies directly rely on the global fit of frequency ratios, for which the WGFS has now a decade-long experience. Steering the International Atomic Time (TAI) will be done in the same was as it is currently done with SRS.
- *Mitigate risks.* Despite the best effort of the time and frequency community in implementing frequency standards, it remains possible that a particular systematic bias has been overlooked, or wrongly estimated for a specific clock transition. Such a mistake would have a different impact on the realisation of the second in option 1 and 2: on the one hand, it would only impact option 1 if it affects the transition chosen as the definition, whereas it would impact option 2 if it affects any transition entering the definition. On the other hand, the error would be reduced in option 2 because the transition only partly contributes to the definition. Likewise, if, at the redefinition, a recommended frequency happens to have an underestimated uncertainty, the step at the redefinition would be more probable with option 2, but smaller. In summary, option 1 favours high risk low probably events, whereas option 2 favours lower risk but higher probability events. A more far-fetched application of option 2 is the possibility to choose the weights of transitions in order to make the unit insensitive to a putative variation of α.

3.2. Weaknesses
Nevertheless, option 2 is a rather disruptive way to define a unit, and hence triggers questions and concerns. Here, we give a list of possible drawbacks of option 2, together with arguments mitigating them.

- *Match with the spirit of the SI.* With the introduction of a fixed value for the speed of light in 1983, and a fixed value for other physical constants in 2018, the spirit of the SI system is to base units on the laws of physics. Option 2 contrasts with this trend, with a definition that is more driven by technological readiness than fundamental considerations. However, it can be argued that the aim of a system of unit should first to be practical, and then, among all practical options, the most elegant and fundamental one can be selected.

For instance, it was only possible to redefine the kilogram by fixing the Planck constant after the uncertainty of practical realisations could allow for it. This is the reason why option 3, although more elegant than other options, is not considered as a viable option for the redefinition. In contrast, equation (1) for option 1, or equation (2) for option 2, essentially consist in fixing the numerical value of a function of fundamental constants, solution of the physical equations describing the state of one or more many-electron atoms. Although fixing such a function amounts in effect to fixing a combination of fundamental constants, this function is currently unknown at the level of precision of the best realisation of the second, even for an atom as simple as hydrogen. Consequently whether this function involves a single or several atomic species is essentially insignificant from the point of view of fundamental constants of physics.

- *Physical interpretation of the normalisation constant.* Even though we argued above that option 2 is not less fundamental than option 1, it remains that in option 1, the normalisation constant N has a physical meaning: it is the frequency, in Hz, of the specific clock transition used as a definition. On the contrary, the normalisation constant N of option 2 does not directly map to the frequency of a particular physical quantity in Hz (beyond the hypothetical physical realisation of the geometric mean (2)). This criticism is certainly valid, but is arguably inessential, because the actual value of the normalisation constant N in both options is accidental (chosen for continuity with historical definitions of the second); and furthermore, the recommended frequencies derived from the defining values of the unit have a physical meaning, and are used in practice to realize the second with option 2.
- *Realization uncertainty.* Both options 1 and 2 necessarily yield a realisation uncertainty, given that actual frequency standards always deviate to some extent from the idealized unit. For instance, atoms interact with the ambient black-body radiation, and correcting for this effect involves the imperfect knowledge of the atomic polarisability over the black-body spectrum. As a consequence, the so-called "Primary Frequency Standards" (PFS) of the current definition and of option 1 are not genuine PFS because they do not strictly implement the definition of the second (non-interacting atoms at rest and at 0 K). From this point of view, the fact that the notion of PFS disappears in option 2 has not much physical significance: in practice, all frequency standards must be corrected to match the definition, and the uncertainty on this correction has to be estimated regardless of physical origin.

  A significant difference still remains between options 1 and 2: while the realisation uncertainty solely arise from characteristics of a single atomic species in option 1, it also depends on the knowledge of frequency ratios in option 2. As a consequence, option 2 suffers from an additional realisation uncertainty as compared to option 1. However, three observations mitigate this drawback.

  First, frequency ratios, as measured physical quantities are often more reliable than other physical quantities such as atomic polarisability. Indeed, the latter are usually measured with high precision on dedicated experiments [6, 7], and seldom reproduced in different laboratories nor at different times. On the contrary, frequency ratios are usually repeatedly measured, and their consistency is cross-checked by the fitting procedure performed by the WGFS.

  Second, the added realisation uncertainty of option 2 tends to be negligible as the number of species entering the definition increases [2]. This averaging effects comes from the multiplicity of connections between a specific transition and the definition, as illustrated in figure 1.

  Third, although the realisation uncertainty in option 2 is a little increase for the transition that would be chosen as the PFS of option 1, it is more significantly decreased for every transition acting as SRS of option 1. As a result, in a situation where SRS are widely used, as it has been the case for contributions to TAI for the last couple of years, the overall realisation uncertainty could actually be lower with option 2, as compared to option 1.

- *Fit industrial need.* In can be argued that industries developing frequency standards could be less confused if a single species is put forward as PFS (option 1), rather than being offered the choice to select one of the several transitions composing the unit of option 2. However, the past experience does not support this argument: driven by technological need, other atoms than Cs have been widely considered in the industry, such as Rb for microwave clocks ranging from small chip scale clocks to large higher performance atomic clocks [8]. In the last few years, collaborations between academic laboratories and industries led to the demonstration of industrial-grade optical clocks [9, 10].
- *Compatibility with national legislations.* Because the legal time of most countries is based on UTC, itself based on the SI second, it is necessary that the new definition of the second is compatible with national legislations. While this is most certainly true for option 1, it is less obvious for option 2, especially if allowing for the dynamic updates of defining values, as described in section 4. For option 2, the compatibility with legislations must therefore be anticipated and verified with national delegates.
- *Understanding by the general public.* Before 2018, the second was defined as "the duration of 9 192 631 770 periods of the radiation corresponding to the transition between the two hyperfine levels of the ground state of the caesium 133 atom". Although the wording of the definition is now different, this physical interpretation can still be used to explain to the general public how the second is defined, and such an explanation remains valid in the case of option 1. In contrast, option 2 cannot be explained in these terms, as the expression (2) of the unit is mathematically more involving. However, it is probable that the general public can understand that, given that many different high performing atomic clocks are available, the second is defined from an average (mentioning the weighted geometric mean in unnecessary) of these frequency standards, and that it is practically realizable with any clock, because we have very good measurements of the frequency ratios between these clocks. Indeed, the idea of averaging can be found behind many concepts in other domains, such as the computation of inflation, that the non-specialist public can easily grasp. It can also be argued that such an explanation does not seem more difficult to understand than how fixing the Planck constant can define the kg, especially when it comes to explaining the mise en pratique.

The purpose of this section was to focus on the specific advantages and drawbacks of option 2. Making an enlightened choice for the new definition of the second will rather require an analysis of both options on an equal footing. This may prove to be difficult, as different strengths and weaknesses may be hard to weight against each other.

Another strategy could be to start from a list of *a priori* requirements and specifications the new definition should or could meet (*e.g.* ability to reach a consensus, adapted to applications, durability, opportunity to build realisations, uncertainty of realisations, acceptability, ...), and then to rate options against these criteria.

4. Evolution of the unit

After the SI second had been defined using the caesium hyperfine transition in 1967, caesium beam clocks, and then caesium fountain clocks have remained the best frequency standards for the following 40 years, only being surpassed by optical frequency standards in the late 2000s.

Ideally, such a situation would be met after the SI second has been redefined, either with a single transition having all the weight (option 1) or with a set of transitions (option 2), this choice remaining relevant throughout the 21st century. However, the field of optical frequency metrology is currently evolving fast, and there is a significant probability that the choice settled in the next few years will become obsolete within a couple of decades. Consequently, before the second is redefined, this eventuality has to be considered, and a clear strategy to face it has to be elaborated. Several possibilities can be considered:

(0): The chosen optical second can remain unchanged for more than half a century, disregarding the eventual progress of new frequency standards. Frequency ratios between these standards would be measured with an uncertainty lower than the uncertainty of the best realisations of the SI second.

(1): A new redefinition is adopted a decade or two after the initial redefinition with optical frequency standard(s).

(2): The new definition allows for a regular update of the defining values by the CIPM (International Committee for Weights and Measures), following predefined rules.

All three options above have disadvantages: (0) means that we are satisfied with a SI second disconnected from the uncertainty of the best clocks, similarly to the current situation with the Cs-based definition. Therefore it would question the motivation to redefine the second based on optical frequency standards at all. (1) means that we may be in a continuous redefinition process, as the interval between redefinitions matches the time required to reach a consensus on a new defintiion. (2) implies that the update process becomes partly independent of the CGPM, but more fundamentally that more frequent steps are introduced in the definition of the second, with their associated uncertainty.

It is important to note that these considerations about how the unit may evolve in the near future are essentially independent of whether option 1 or option 2 is chosen: we may exhaustively call these various possibilities options 1.0, 1.1, 1.2, 2.0, 2.1, and 2.2. However, option 2.2 has a specific behaviour when compared to 1.2. Indeed, when changing the definition of the second, a discrepancy is introduced between the old and the new definition, due to our imperfect knowledge of frequency ratios. Quantitatively, the variance of the step in the definition is given, in relative units, by [2]:

$$v_m = \sum_{k,l} (w'_k - w_k)(w'_l - w_l)\Sigma'_{k,l}, \tag{7}$$

where the double sum runs over all transitions composing the unit, w'_i and w_i are the new and old weights of transition i respectively, and $\Sigma'_{k,l}$ is the output covariance matrix of the global fit of frequency ratios, expressed in relative units[2]. In the specific case of option 1, where changing from transition i to transition j, the expression (7) reduced as expected to:

$$v_m = \left(\frac{\delta\rho_{i,j}}{\rho_{i,j}}\right)^2 \quad \text{for option 1}, \tag{8}$$

i.e. the typical step in the unit is given by the relative uncertainty of the frequency ratio between the new and the old definition.

After repeatedly changing the definition, the steps v_m accumulate, resulting in a random walk of the unit. From equation (7), we can see that option 2.2 has a clear advantage over options 1.1, 1.2 and 2.1: a regular but smooth update of the weights yields smaller steps in the redefinition, mitigating the risk of divergence of the unit. In this process, the definition should be changed only if v_m, which is is an objective and quantitative criterion which can be calculated from the global fit of frequency ratios, significantly reduces from one redefinition to the next. The main drawback of option 2.2 is that it constitutes a more radical change from the traditional way the units are defined.

[2] Using notations of [2] appendix A, we define $\Sigma'_{k,l} = \Sigma_{k,l}/\bar{\nu}_k\bar{\nu}_l$, such that the relative uncertainty on the frequency ratios $\rho_{k,l}$ is given by $(\delta\rho_{k,l}/\rho_{k,l})^2 = \Sigma'_{k,k} + \Sigma'_{l,l} - 2\Sigma'_{k,l}$

Options 1.1, 1.2, 2.1, and 2.2, imply that the unit will change frequently, which could be a concern if the second eventually becomes incompatible with an anterior definition, after several steps have accumulated. This is all the more a concern that all other units of the SI (except the mole) are derived from the second, although in practice these other units are realized with uncertainties several orders of magnitude larger than for the second. In fact, it can be shown that the random walk resulting from the successive definition is actually convergent if the series of v_m is convergent. In this case, the well-defined asymptotic value of the definition can be seen as the actual, immovable definition of the second; And at any time, the definition is an approximation of this asymptotic value, with a discrepancy that is guaranteed to be smaller than the uncertainty of the best realizations of the time.

5. Conclusion

In this paper, we discussed the possibility to redefine the second using the weighted geometric mean of several clock transitions, as a generalisation of the definition based a single species promoted as a primary frequency standard. This proposition, more able to reach a consensus given the diversity of optical frequency standards under development, is inspired by, and makes use of the decade-long work of the WGFS in collecting frequency ratio measurements and in compiling them into a set of consistent best estimates. From this point of view, option 2 aims at taking advantage, in the definition, of this effort: A proper choice of weights allows to realize the second with a vast variety of frequency standards, while minimizing the overall realisation uncertainty.

Because it deviates more from the current definition of the second, option 2 legitimately raises concerns and questions, such as its relation to fundamental constants, the physical signification of the normalisation constant N, or the physical origin of its realisation uncertainty. We have proposed answers to these questions, starting from an introspection about how these concepts are dealt with in the current definition or in option 1, and highlighting the many similarities between the options.

How the new definition may evolve in the next decades is a concern that should be addressed, regardless of whether option 1 or 2 is selected. If optical frequency standards continue to be improved, all foreseeable scenarios have drawbacks. However, assuming that the definition will be eventually be updated in order to follow the progress of frequency standards, option 2 offers a smoother transition with a reduced gap, owing to its ability to continuously tune the weights of different species, as opposed to option 1 for which the weights discretely change from one species to another. Therefore, we could take advantage of this property, and allow for regular updates of the unit, based on quantitative and predefined criteria.

Acknowledgments

We acknowledge funding from the French "Agence Nationale de la Recherche" (ANR), project RYCARD (ANR-22-CE47-0009, 2022)

References

[1] Dimarcq N, Gertsvolf M, Mileti G, Bize S, Oates C W, Peik E, Calonico D, Ido T, Tavella P, Meynadier F, Petit G, Panfilo G, Bartholomew J, Defraigne P, Donley E A, Hedekvist P O, Sesia I, Wouters M, Dubé P, Fang F, Levi F, Lodewyck J, Margolis H S, Newell D, Slyusarev S, Weyers S, Uzan J P, Yasuda M, Yu D H, Rieck C, Schnatz H, Hanado Y, Fujieda M, Pottie P E, Hanssen J, Malimon A and Ashby N 2024 *Metrologia* **61** 012001 URL `https://dx.doi.org/10.1088/1681-7575/ad17d2`

[2] Lodewyck J 2019 *Metrologia* **56** 055009

[3] Margolis H S and Gill P 2015 *Metrologia* **52** 628–634

[4] Riehle F, Gill P, Arias F and Robertsson L 2018 *Metrologia* **55** 188 URL `https://dx.doi.org/10.1088/1681-7575/aaa302`

[5] Gill P 2016 *Journal of Physics: Conference Series* **723** 012053 URL `https://dx.doi.org/10.1088/1742-6596/723/1/012053`

[6] Sherman J A, Lemke N D, Hinkley N, Pizzocaro M, Fox R W, Ludlow A D and Oates C W 2012 *Phys. Rev. Lett.* **108**(15) 153002 URL `https://link.aps.org/doi/10.1103/PhysRevLett.108.153002`

[7] Middelmann T, Falke S, Lisdat C and Sterr U 2012 *Phys. Rev. Lett.* **109**(26) 263004 URL `https://link.aps.org/doi/10.1103/PhysRevLett.109.263004`

[8] Marlow B L S and Scherer D R 2021 *IEEE Transactions on Ultrasonics, Ferroelectrics, and Frequency Control* **68** 2007–2022

[9] Takamoto M, Ushijima I, Ohmae N, Yahagi T, Kokado K, Shinkai H and Katori H 2020 *Nature Photonics* **14** 411–415

[10] Stuhler J, Abdel Hafiz M, Arar B, Bawamia A, Bergner K, Biethahn M, Brakhane S, Didier A, Fortágh J, Halder M, Holzwarth R, Huntemann N, Johanning M, Jördens R, Kaenders W, Karlewski F, Kienle F, Krutzik M, Lessing M, Mehlstäubler T, Meschede D, Peik E, Peters A, Schmidt P, Siebeneich H, Tamm C, Vogt E, Wicht A, Wunderlich C and Yu J 2021 *Measurement: Sensors* **18** 100264 ISSN 2665-9174 URL `https://www.sciencedirect.com/science/article/pii/S2665917421002270`

Journal of Physics: Conference Series **2889** (2024) 012027 doi:10.1088/1742-6596/2889/1/012027

Proposal for a computable optical Clock

Omer Amit[1], Vitaly Wirthl[1], Derya Taray[1], Vincent Weis[1], Soumyadeep Sarkar[1], Min–Cheol Kim[1], Mustafa Waqar Syed[1], Jorge Moreno[1], Florian Egli[1], Johannes Weitenberg[1,2], Muhammad Thariq[1], Theodor Strobl[1], Malte Bieringer[1], Linus Hein[1], Savely Karshenboim[1], Randolf Pohl[3], Dylan C. Yost[4], Akira Ozawa[1], Theodor W. Hänsch[1,5], and Thomas Udem[1,5]

[1]Max–Planck Institute für Quantenoptik, Garching, Germany
[2]Fraunhofer–Institut für Lasertechnik ILT, Aachen, Germany
[3]Johannes Gutenberg–Universität, Mainz, Germany
[4]Colorado State University, Fort Collins, Colorado, USA
[5]Ludwig–Maximilians–Universität, München, Germany

E-mail: thomas.udem@mpq.mpg.de

Abstract. With the recent update of the SI system, all but one of the units are now based on defining the values of some fundamental constants. This development began in 1983 when the speed of light was assigned an exact fixed value. The advantage of this method is that it separates the definition from the realization, allowing new realizations to be introduced as technology advances without further redefinition. In addition, it allows unit realizations that are adapted to the scale of their intended use. Because of these advantages, we expect that one day also the last remaining object in the current SI system, the caesium atom, will also disappear. The purpose of this proposal is to outline possible paths for realizations of a future SI second based on the definition of the value of the Rydberg constant. Hydrogen and hydrogen–like systems would be the obvious candidates. The emphasis here is on the development of optical clock systems that circumvent difficulties associated with the short wavelength lasers otherwise required for cooling and driving the clock transition. The proposed clock systems based on atomic hydrogen and hydrogen–like He^+, should be no more complex than current optical lattice clocks.

1 International System of Units SI

In the old days units have been defined by referring to an object such as the king's foot, a platinum–iridium rod or the Cs atom. Among objects, the latter is the better choice than an artifact, because it comes with a large number of identical copies and hence cannot be lost or degrades over time. The even better way is to assign exact and fixed values to physical constants with dimensions. Such a unit might be more difficult to explain to high school students, but is more powerful to use as we explain in this section.

The speed of light c was the first constant that received a promotion to become a unit in 1983 [1], even though theoreticians used natural scale units in which $c = 1$ for a long time. This definition does neither fix the meter nor the second, but their ratio. It can be used to realize the meter in at least two different ways. One can measure the travel time of an optical pulse or use an interferometer to realize a

known multiple of the wavelength $\lambda = c\nu$ using the optical frequency ν. In both cases one needs a clock to either measure the travel time or ν. Obviously, the travel time realization works for large distances, say for Lunar ranging [2], while the interferometer is good for lab scale applications. The unit realization can be adapted to the scale.
With the latest reform of the SI [3], the Planck constant received a fixed value among others. This fixes the "mass of the Cs photon" $h\Delta\nu_{\mathrm{Cs}}/c^2$, because c and $\Delta\nu_{\mathrm{Cs}}$, the Cs ground state hyperfine splitting, already have defined values. Therefore

$$\frac{1\ \mathrm{kg}}{h\Delta\nu_{\mathrm{Cs}}/c^2} = \frac{917097121160018}{621541050725904 75} 10^{42} = 1.4755213997352709160650259536221241563890597\ldots \times 10^{40} \tag{1}$$

Cs photons make one kg. This is much more accurate than the notoriously imprecise [4] platinum—iridium cylinder served as the international prototype of the kilogram. However, this is just the definition. For the realization one would certainly not generate and store this many photons. Instead, either a Kibble balance or a silicon sphere is used to realize a macroscopic mass reference [5]. Again, more than one realization is possible using the same definition. A third one would be to use the mass of a single Cs photon. This would indeed use the full accuracy of (1), but operates 40 orders of magnitude below the "human" scale of kg, m and s. Such an extremely large dynamic range is unheard of for any standard that is based on an object. In fact the scale problem has prevented the redefinition of the kg in terms of a naturally occurring object for long time. The electron mass m_e for example, is 30 orders of magnitude below the human scale. Very accurate mass comparisons between microscopic particles can be performed in Penning traps [6], but the extremely large range prevents these comparisons between the microscopic and the human scale. For the speed of light the situation is different as it takes only 8 orders of magnitude to connect to the human scale. This is why the definition of the value of c came first.

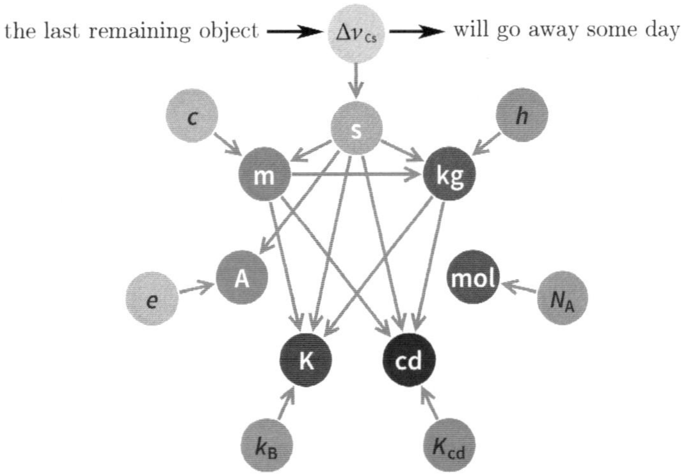

Figure 1: The current state of the SI system of units: The inner ring lists the base units and the outer the underlying definitions. All but one are now based on fixing values of dimensional fundamental constants. The only one that is based on a physical object is the SI second based on the ground state Cs hyperfine splitting $\Delta\nu_{\mathrm{Cs}}$. While one is free to pick any suitable method to realise any of the other units, the SI second can only be realized using the Cs atom. Source: Emilio Pisanty for Wikipedia. Creative Commons Attribution–Share Alike 4.0 International license [7].

Interestingly, one can have a unit system completely based on defined constants which might be irritating at first sight as it seems to be decoupled from any real object. However, only the definition of the units would have no reference to the physical world, but the realization certainly has. All but one of the units are now based on defined constants (see fig. 1). The only remaining object is the Cs atom that is used to define and realize the SI second by fixing the value of $\Delta\nu_{\mathrm{Cs}}$. A natural way to proceed is to remove the last object from the SI by defining the value of the Rydberg constant R_∞. For hydrogen–like systems, the energies of the bound S–states are given by [8]:

$$E_n = hcR_\infty \left(-\frac{1}{n^2} + f(\alpha, m_e/m_N, ...) + \frac{C_{\rm NS}}{n^3} \langle r_N^2 \rangle \right) \tag{2}$$

with the Schrödinger term $-1/n^2$, many relativistic and QED corrections $f(\alpha, m_e/m_N, ...)$, that depend on other constants such as the nuclear mass m_N, and the finite nuclear size term with the squared nuclear charge radius $\langle r_N^2 \rangle$ and the constant $C_{\rm NS}$ known from theory. The Rydberg constant

$$R_\infty = \frac{m_e \alpha^2 c}{2h} \tag{3}$$

links the natural energy scale of atomic systems and the SI unit system. At the same time it connects the mass of the electron m_e and the fine structure constant α. The energy and hence energy differences of any atomic or molecular system can be written in the form (2), i.e. as a product of the Rydberg constant and a dimensionless theoretical expression. Fixing the value of the Rydberg constant fixes in theory all atomic and molecular frequencies. While this is in principle true also for $\Delta\nu_{\rm Cs}$, the theoretical expression for the hydrogen–like systems is of course the most precise and hence recommend themselves for the realization.

Strictly speaking the theory part of (2) becomes fully computable only if all dimensionless constants, such as α and m_e/m_N, become computable. At this time it is not known if and how that will be possible, but since these constants are dimensionless, it is at least conceivable. Dimensionless constants can not be removed by promoting them to become a unit. One could also imagine redefining the second by fixing another constant like the electron mass for example. A realization would then be possible by precisely measuring the frequency of the 511 keV annihilation photon. According to (3), fixing the value of the Rydberg constant also fixes the mass of the electron. How is this possible? Is the kg not already fixed with h? No, it is fixed by h and c *and* $\Delta\nu_{\rm Cs}$. R_∞ would take over the role of $\Delta\nu_{\rm Cs}$ if it gets a defined value.

2 Hydrogen Clock

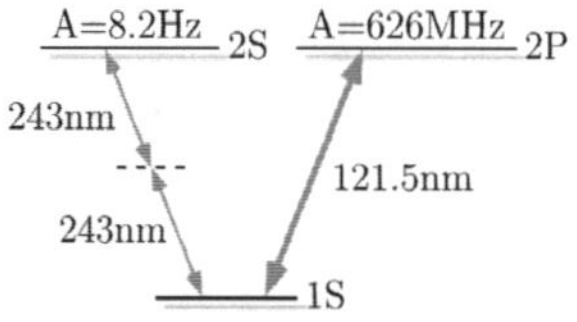

Atomic hydrogen has an ideal level scheme for an optical clock. Since there is no 1P state, the 2S state can neither decay nor be excited with a single photon dipole transition, at least not in a field–free environment. This has three important advantages: The lifetime of the 2S is very long leading to natural line width of $A/2\pi = 1.3$ Hz. This is a good value for an optical clock. Because the 1S–2S clock transition at 2 466 THz is excited with two photons, the required laser operates at 243 nm rather than 121.5 nm (Lyman–α). Moreover, the two–photon excitation can be arranged such that it is free of the Doppler effect in first order. This eliminates a set of systematic effects. It also implies that only moderately low temperatures and no strong confinement (Lamb–Dicke regime) of the trapped atoms is required. The magic wavelength [9] for the 1S–2S transition is at a convenient wavelength of 515 nm [10]. High power narrow band lasers are readily available by frequency doubled Yb–based lasers. The 1S–2P cooling transition is strong (line width of $A/2\pi = 100$ MHz). However, the corresponding Lyman–α wavelength is very difficult to obtain with current laser technology. The demonstrated continuous wave sources [11] do not reach the power level to operate a MOT or a Zeeman slower. In addition, because of the low mass and large photon momentum, cooling would be rather inefficient with the Doppler and recoil limit at high temperature of 2.39 mK and 1.29 mK respectively. So laser cooling is not a good option. A number of proposals have been published to circumvent this problem [12, 13, 14, 15, 16]. This one adds another one to the list with a proposed trap loading scheme that avoids laser cooling. In contrast to laser cooling, the proposed method works better the lower the atomic mass and the shorter transition wavelengths are.

It should be mentioned that atomic hydrogen has been trapped before using superconducting magnets [17]. Antihydrogen is generated inside such a trap on a regular basis at CERN [18]. Even cooling with a pulsed Lyman–α laser has been achieved in both cases [19, 20]. With a typical repetition rate of ~10 Hz these lasers are easier to build [21, 22]. Very long cooling times of 15 min (hydrogen) and several hours (antihydrogen) are required. This rules out laser cooling as a means for loading a trap. Most other atoms can be loaded into a trap when laser cooling is available. In contrast, atomic hydrogen can be laser cooled (so far) only when it is already trapped. Magnetic traps are large and

cumbersome and their fields on the order of a few Teslas gives rise to a Zeeman shifts on the order of several tens of GHz. Therefore, we do not consider this approach as an option for a clock. Instead, our setup may not be more complex than a common optical lattice atomic clock.
By utilizing trapped atomic hydrogen, one can achieve greater precision and potentially develop a hydrogen optical atomic clock. This clock, based on the most basic of all atoms, has been referred to as a "computable clock" [23]. Its transition frequency can be linked directly to fundamental constants such as the Rydberg constant. Defining the value of the Rydberg constant would make it possible to establish a new SI second. A similar proposal of a hydrogen optical lattice clock has recently been published by J.P. Scott [24] and coworkers, albeit without a method of loading, which is the focus of this work.

2.1 System overview
The figure below sketches a possible geometry for a hydrogen clock, albeit the optimum design still has to be determined. An atomic hydrogen beam is obtained from a cryogenic nozzle. In contrast to our existing 1S–2S spectrometer [25], no laser beam needs to go trough the nozzle to be collinear with the atomic beam. This gives us more freedom in the design and we expect to be able to use that freedom to obtain a significantly larger flux of slow atomic hydrogen. The atomic beam obtained with this nozzle emerges from a 5 K cold surface of a permanent magnet after being deposited there for thermalization. Hydrogen atoms are guided to the nozzle from a gas discharge that breaks up the molecules. Teflon tubing prevents the atoms from recombining into molecules.

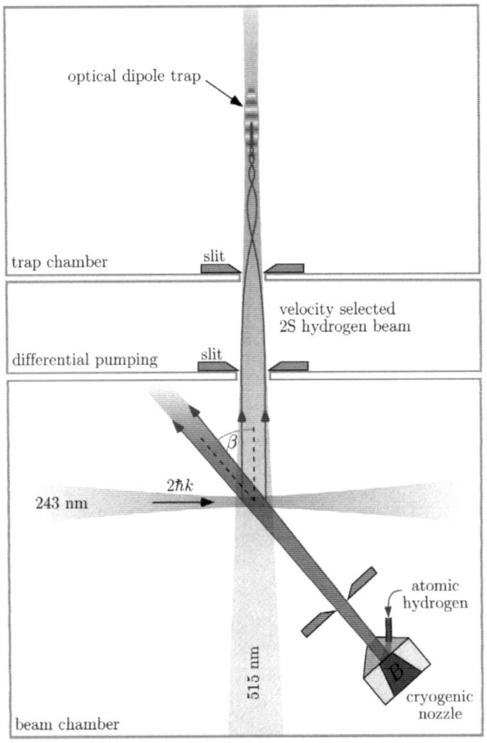

Upon leaving the nozzle, the high–field–seeking atoms are slowed down, so that their number in the slow tail of the Maxwell–Boltzmann distribution is enhanced. Some of the slow atoms are later removed again by collisions with the faster atoms within the beam [26].
The atomic beam then crosses a 243 nm laser beam that drives the 1S–2S two–photon transition. The laser light is tuned to the Doppler–sensitive component absorbing two photons from either direction which gives them a two–photon recoil of $v_r = 3.26$ m/s. The atoms are thereby deflected with an angle $\beta = \arcsin(v_r/v_i)$ that dependent on their initial velocity v_i. The undeflected atoms can be employed for normalization of the line shape, which had previously turned out to be essential for high–quality data in our 1S–3S experiment [27]. A range of velocities can be selected using a long and narrow tube or a series of slits or skimmers as shown in the sketch. This or a similar geometry also acts as the separator for differential pumping between the beam chamber with high pressure and the ultra high vacuum (UHV) trap chamber required for stable trapping. At this time we do not know what the lowest velocities are that are still present in the beam because of the above mentioned deviations from the Maxwell–Boltzmann distribution. From our previous experiments we have good reasons to assume that $v_i = 5$ m/s can still be selected, which corresponds to a deflection angle of $\beta = 41°$. The deflected atoms are then funneled (curved lines) by the 515 nm trapping laser dipole force towards the trapping region, where they are predominately "quenched" to the 1S state. This happens because the trapping laser mixes the metastable 2S state with the short–lived P states. The decay gives rise to a large photon recoil in a random direction and changes the velocity by as much as 4 m/s, bringing some of the 5 m/s atoms almost to rest. Without this dissipation of kinetic energy, atoms that fly into the trap would fly out on the other side. With this process we get the dissipation exactly where we want it and not at any other position outside the trap where it would be useless. A fraction of the atoms receive the proper recoil to let them stay in the trap such that a steady state number of trapped atoms develops from the equilibrium of the trapping rate and loss rate. The loss rate is dominated by collisions of trapped atoms with the background gas rather than from the incoming atomic beam.

The energy of a single 515 nm photon is not sufficient to ionize the 2S state. However, at high intensities two–photon ionization is no longer negligible [28]. In this case the quenching rate for trapping competes with the two–photon ionization rate. The former is proportional to the intensity while the latter is proportional to the intensity squared. Limiting the trapping laser intensity allows to obtain a useful trapping rate while keeping the two–photon ionization rate sufficiently suppressed. Obviously, low–velocity atoms are better to load into a shallow trap.
Once the trap is filled, the atomic beam can be turned off for example by switching off the 243 nm laser. An evaporative cooling step might be added if the temperature and therefore the second–order Doppler shift turns out to be too large. Note that the 1S–2S clock transition will be driven as a Doppler–free two–photon transition that does not require a very low temperature. The 1S–2S clock transition can then be driven with the same 243 nm laser that also provides the atomic beam deflection. In the simplest case, the signal for a successful 1S–2S transition would originate from the subsequent ionization that is possible with a third 243 nm photon. Ions can be detected with a quantum efficiency close to unity with a very low background. To improve the duty cycle a non–destructive detection scheme such as the dispersive detection used by J. Lodewyck and coworkers could be implemented [29]. The whole setup may even be operated continuously without interrupting the loading process.
Our estimations indicate that a reasonable loading rate should be possible with a Watt–level deflection laser and a kW–level trapping laser. Both lasers might be resonantly enhanced in a standing wave or running wave ring cavity. Details of these estimations will be presented in a subsequent publication.

2.2 Expected systematics

In this section we are giving a brief list of expected systematic uncertainties of the proposed hydrogen clock. As the clock transition is free of the first order Doppler effect, the second order Doppler shift $-3k_BT/2mc^2$ dominates. Approximating the temperature by the trap depth of say 0.1 mK, this shift will be around 10^{-17} in relative units. It will give rise to a line asymmetry [24] that should be straight forward to model at the percent level. Evaporative cooling can further reduce this shift to well below 10^{-18}.
The AC–Stark shift is intrinsically large in two–photon transitions, because they are driven off–resonantly and require large laser intensities. It is of the order of the Rabi–frequency, which can be chosen at the Hz level, adapted to the natural line width, for trapped atoms. The relative shift would be 5×10^{-17}. Again there is room for improvements by an even lower probe laser intensity, by modeling and by extrapolating to zero probe laser intensity. Another option is to cancel the negative AC–Stark shift with the positive hyperpolarization shift by stabilizing the probe and trapping laser intensity ratio.
The Landé g–factors of the 1S and 2S levels are identical such that the Zeeman shift vanishes for the $M = \pm 1$ components. The two–photon S$\rightarrow$S selection rule is $\Delta F = \Delta M = 0$ for any polarization. Relativistic corrections to the g–factors lead to a splitting between the $M = \pm 1$ components of 37 Hz/Gauss. Magnetic fields can be shielded and the splitting does not necessarily lead to a line shift. Therefore, the Zeeman shift can be significantly lower than any medium–term accuracy goal.
The DC–Stark shift of the 1S–2S clock transition is quadratic in the electric field strength E and given by $+0.36\ E^2$ Hz m^2/V^2. It should be possible to shield stray fields to better than 1 V/m [25] and hence passively reduce the DC–Stark shift below the Hz level. To further improve this shift, one can make use of the quadratic dependence on the external field. The vertex of the parabola describing the observed clock transition frequency as a function of an applied electric field gives both the unperturbed clock transition frequency and the stray electric field in the direction of the applied field.
The pressure shift of the 1S–2S clock transition has been determined experimentally at large pressures p to $-4.14(36)\ p$ MHz/mbar with a H_2 molecular background gas [30]. A comparable value has been been obtained as $-3.8(0.8) \times 10^{-16} \rho$ Hz m^3 at lower pressure in an environment of cryogenic spin polarized atomic hydrogen with a density ρ [31]. We use the former value to estimate a pressure shift in the mHz regime due to the molecular background gas. Assuming a density of $\rho \approx 10^{16}$ of trapped spin polarized atoms, we use the latter number to estimate a Hz level pressure shift due the trapped atoms. The pressure shift may be reduced by a larger trapping volume or traded for statistics by reducing the number of trapped atoms.
The AC–Stark shift due to the blackbody radiation of the environment has been calculated in [32] to 1.036 Hz at $T = 300$ K for the hyperfine centroid of the 1S–2S clock transition. Knowing the temperature of the environment within 1 K, reduces the uncertainty of the blackbody shift of the centroid to 14 mHz.
This slope of the AC–Stark shift due to the trapping laser as a function of its wavelengths is very similar to other lattice clocks [10, 33]. However, as the hydrogen clock is expected to operate at larger trap laser intensities, the slope is also more critical. At a trap depth of 0.1 mK, the trapping laser

Journal of Physics: Conference Series **2889** (2024) 012027 doi:10.1088/1742-6596/2889/1/012027

frequency has to be known with 6 orders of magnitude lower accuracy than the accuracy goal of the clock. This should not pose a principle problem with an optical atomic clock at hand. However, the spectral purity of the trapping laser is certainly an issue in this regard.
In lattice clocks, the important mechanisms that perturb the concept of the magic wavelength are vector and tensor polarizations and the hyperpolarizability [34]. Both, the associated vector and the tensor Stark shifts vanish when using the $F = 0$ atoms. Thanks to the $\Delta F = 0$ selection rule for two–photon transitions, this also applies to the excited state.
The hyperpolarizability cannot be eliminated by picking the proper hyperfine level. It is due to the 2nd order AC–Stark shift (4th order perturbation) that scales with the square of the trapping laser intensity I_t and is given for the 1S [35] and 2S level [36]. The hyperpolarizability of the 2S level dominates and reaches ≈ 100 Hz for a trap depth of 0.1 mK. Using the wavelength dependence, a magic wavelength can be found if there are only linear or only quadratic terms of the AC–Stark shift. With both contributions this is not exactly possible. This shift seems to require more attention than the others. The mitigation strategies might include extrapolating to zero intensity. Thanks to the quadratic dependence, a reduction of laser power by one order of magnitude reduces the 2nd order AC–Stark by two orders of magnitude. By employing the $F = 1$ atoms, the hyperpolarizability can be reduced by trapping at an "operational magic wavelength", where frequency shifts are insensitive to changes in trap depth [37]. There may also exists a "magic ellipticity" that cancels the hyperpolarizability shift [33].

3 He^+ Clock

Another candidate for realizing a new SI second based on a defined value of the Rydberg constant is the hydrogen–like He^+ ion. In contrast to atomic hydrogen, He^+ is easily stored in an ion trap and sympathetically cooled, for example with laser cooled Be^+ ions. Again, laser cooling on the 1S–2P transition at 30.4 nm is neither available nor required. The theoretical expression for its energy levels is of the same form as (2). We have been working towards high–precision spectroscopy of the 1S–2S transition at 60.8 nm [38, 39] to test the predictions of Quantum Electrodynamics. In contrast to hydrogen, the values for the nuclear charge radius obtained from electron scattering [40] and muonic He^+ [41] agree.
Currently, the main challenge is the very short wavelength of the 1S–2S clock transition. Narrow band laser radiation in this extreme ultraviolet (XUV) region is thus far only available in form of a frequency comb via high harmonic generation (HHG). Direct frequency comb spectroscopy seems to be the only option for now. To the best of our knowledge, no high resolution spectroscopy has been performed in this spectral region yet. With a two–photon transition, the modes of the comb add pairwise such that one obtains the transition rate corresponding to the power of all modes combined while the laser line width is given by the width of a single mode [42]. We have used this method for the hydrogen 1S–3S transition in the UV region and obtained a line width limited by the natural line width [27] using an atomic beam. Two–photon direct frequency comb spectroscopy is expected to work even better with trapped ions as several systematic effects are absent. Ion trap clocks are the only real competitors to the optical lattice clocks. The Al^+ ion clock using an ultraviolet clock transition at 267 nm now reaches a fractional uncertainty of 10^{-18} [43]. With this proposal we are addressing both of these very successful strategies, the lattice clock and the ion clock scheme. As the quantum limited Allan deviation of a clock scales with the inverse of the transition frequency, XUV optical clocks are the next logical step. For the same reason microwave clocks have been superseded by optical clocks. In this respect the He^+ clock goes even beyond the proposed ^{229}Th nuclear clock [44].
For our existing XUV comb system [39], we employ an ultra–stable high–power infrared optical frequency comb [45] that is further power enhanced in a resonator that contains the gas target for HHG. An XUV frequency comb at 60.8 nm is generated inside the resonator with ~ 10 μW of power extracted from the resonator. The high–power infrared driving laser and the intracavity HHG is a fairly large and complex system that is not suited to be operated continuously as a clock. Moreover, the required mechanical stability of a large system is difficult to maintain even with several active path length stabilizations, as the phase instability scales with the inverse of the wavelength.
Here we propose a new compact setup that would use much lower average laser power. The HHG, XUV waveguides and an ion trap would all be integrated in a single device that may be only be ~ 10 cm in size. As all path lengths are short and common mode, we expect the device to be much more stable to operate and introduce negligible phase noise to the XUV frequency comb. We also want to avoid any kind of XUV optics because these are available only with large losses. To achieve this, we plan to use a single pass solid target for HHG without an enhancement resonator. While the reduced laser power and the solid state HHG reduce the XUV efficiency, we plan to compensate for that by reducing the repetition rate of the infrared driving laser. Fewer pulses per second with the same time averaged power

give rise to a larger pulse energy and hence boost the very non–linear HHG. This principle has been used for decades with chirped pulse amplification with pulse repetition rates below the kHz regime. While the pulse energy is the main quantity in most time domain applications, in the frequency domain the main question would be whether or not the dense frequency comb would be washed out by a little bit of excess noise. As it seems, this question has not been addressed in the literature until recently when we have been able to demonstrate a clear mode structure for a repetition rates as low as 40 kHz [47].

3.1 System overview

The schematics of the proposed setup is shown at the left hand side of the figure below. The entrance of an XUV waveguide is covered with a solid–state target for HHG. The best material has yet to be identified and has been investigated intensively in the last ten years. In 2011 it was found that nonlinear frequency conversion up to the 25th order showed non–perturbative behaviour when a ZnO crystal is illuminated with short pulses at mid–infrared wavelength [48]. Only XUV light generated close to the end surface of the medium contributes because of reabsorption. However, this process can be driven with a modest intensity of 10^{11} to 10^{13} W/cm^2 and still reach XUV wavelengths. Besides ZnO, SiO_2, sapphire and fluoride crystals are possible candidates. Also, engineered micro–structured media have been demonstrated to improve the harmonic yield and could also be used for our application [49].
Before coupling the XUV light into a waveguide, the high power driving laser light has to be removed to prevent damage on the subsequent structure and avoid ionization and Stark shifting the He^+ ion. For this purpose a small void structure is formed right after the HHG target as shown at the right hand side of the figure. This prevents illumination of the waveguide substrate with high intensity. A few holes are added towards this void, such that the volume can be evacuated to low pressures. This prevents plasma generation and reabsorption even with possible outgassing enhanced by laser heating. The IR beam after HHG is then lead to the bulk medium of SiO_2 with highly reflective coating so that a significant fraction of the IR radiation is back reflected. The IR beam in the XUV waveguide is further suppressed because the IR beam has a larger divergence angle than the XUV beam. The residual IR beam is then eliminated by a thin Al filter which is standing free in the subsequent waveguide.

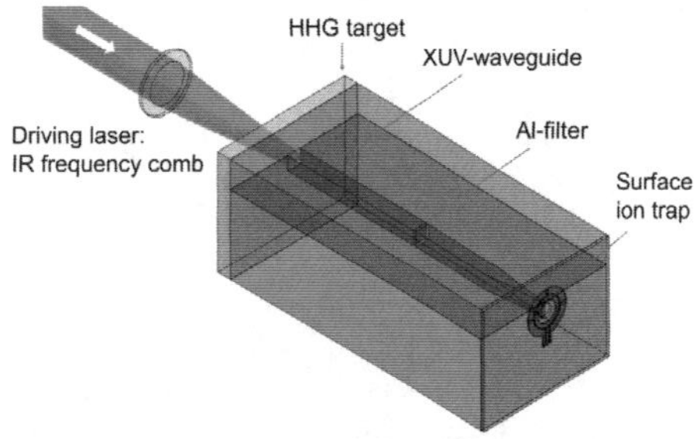

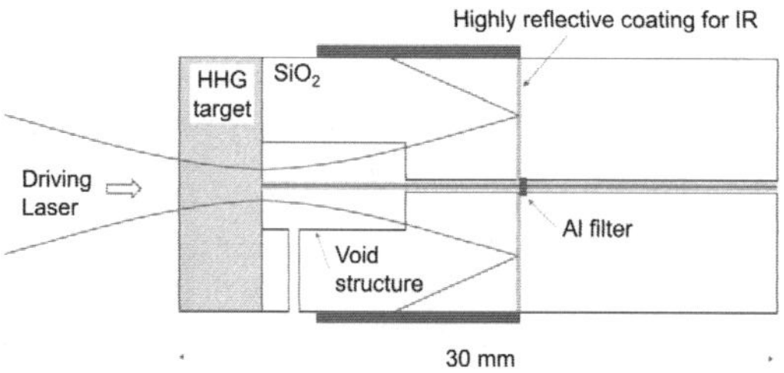

To form a hollow core waveguide that supports XUV radiation, we employ a refractive index smaller than unity that often goes along with strong absorption, namely a large imaginary part of the refractive index. For all eigenmodes of the waveguide, the optical field penetrates more or less into the cladding, otherwise the field cannot be affected by the cladding. Therefore, a low loss due to the cladding material is another important design criterion. However, since the propagation length planned in this project is short (less than 30 mm), some loss can be tolerated. Currently we are investigating SiO_2 and Si as the most promising candidates for the cladding material. These materials are most commonly used in microfabrication using lithography and etching processes. This enables precise fabrication of the waveguide structure with a nm–scale precision and low surface roughness. We have already modeled the spatial modes by numerical simulations. We have also tested samples of waveguides and confirmed that the transmission agrees with the simulations [50].
In a future step the end of the waveguide will be fitted to a spectroscopy target, for example, a surface ion trap. We are also investigating the possibility to focus the XUV light by tapering the end of the waveguide. A part of the waveguide could be coated with gold and electrically biased with respect to the ion trap so that photoelectrons do not interfere with stable ion trapping. In the end we hope to be able to fabricate these components on a single chip that would fit into a small vacuum chamber evacuated by a small getter pump without employing any differential pumping. This could lead to a compact, transportable and reliable XUV optical clock that would be a realization of a new SI second based on a defined value of the Rydberg constant.

3.2 Expected systematics

Obviously, all the laser trap induced shifts as discussed in section 2.2 are replaced by ion trap induced shifts. The DC–Stark effect scales as $\propto 1/Z^6$ and is reduced relative to atomic hydrogen to $+5.6\ E^2$ mHz $\mathrm{m}^2/\mathrm{V}^2$ [38]. This is still about 5 and 4 orders of magnitudes larger than the DC–Stark shift of the Al^+ [51] and the $^{88}\mathrm{Sr}^+$ [52] ion standards respectively. The cause of this large sensitivity is the Lamb shift, i.e. a small splitting between the 2S and the 2P levels which occurs in the denominator of the 2nd order perturbation theory describing the 2S DC–Stark shift. Hence, the DC–Stark shift could be an important limitation when micromotion causes the stored ion to sample the non–zero trapping fields around the potential minimum. The obvious mitigation is to minimize the micromotion. Micromotion also leads to a second order Doppler shift that has the opposite sign in the case of the He^+ 1S–2S transition. Operating the ion trap at an RF frequency around 12 MHz, these two shifts cancel [52]. For all other ions, the accuracy of this cancellation depends on how accurately the DC–Stark shift is known theoretically. For hydrogen–like systems this is not a limitation. The remaining trap related systematics are not expected to be much different from other high performance ion clocks [43]. The blackbody shift does not simply scale with some power of Z but has been computed for the 1S–2S in He^+ to 1.2×10^{-17} in relative units at $T = 300$ K [53]. This is only a factor 5 larger than the blackbody shift of the Al^+ clock that has one of the lowest sensitivities [54].
The Zeeman shift does not depend directly on Z, but the small residual splitting as discussed in section 2.2 is $\propto Z^2$. With $^4\mathrm{He}^+$ there is no hyperfine structure and only the small line splitting of 4×37 Hz/Gauss. The splitting scales the same way as the transition frequency.
The leading uncertainties are presumably the AC–Stark shift caused by the clock laser as well as the residual infrared of the driving laser and the line broadening due to ionization of the excited state by the clock laser. The latter is intrinsic to the hydrogen–like systems because one laser photon resonant to the two–photon 1S–2S transition carries the energy of $(3/8)Z^2R_\infty$ while it requires only $(1/4)Z^2R_\infty$ to ionize the 2S state. To reduce this effect the Rabi frequency needs to be low as the excitation rate scales with the square of the laser intensity while ionization rate scales with the third power. This also reduces the AC–Stark shift, which is on the order of the Rabi frequency. We expect that the natural line width of 83 Hz can be reached eventually with a transition frequency of almost 10^{16} Hz.

4 Discussion and Outlook

Even when a simple and compact apparatus for a hydrogen based clock becomes available, there are two main objections to use them for time keeping. The first one concerns the accuracy: With 92 atoms and even more ions to choose from, it seems very unlikely that hydrogen or a hydrogen–like ion will turn out to have the best properties among all of them. However, as the gravitational redshift within the trapped atom clouds is now setting the limit [55], it appears that at this level, a ground based clock no longer measures time but elevation. Reaching such a level of accuracy with a hydrogen based clock, might lead to a situation where accuracy is no longer the only figure of merit for a clock.
Another unsolved problem is the lack of a precise values of the nuclear charge radii that enter the theory (2) and higher order nuclear terms that are included in $f(\alpha, m_e/m_N, ...)$. Possible solutions are advancements in lattice QCD techniques, which in principle would allow their calculation [56]. Measurements of the charge radii may be obtained from other transitions in atomic hydrogen and muonic hydrogen [58, 41]. Effectively this means combining different transition frequencies such that the nuclear charge radius term in (2) cancels out. Improvements in electron scattering data might also be employed [59], but there is no underlying precise theory to such a cancellation. Pure leptonic systems such as positronium could also solve this problem, but the short lifetime would be challenging for the accuracy [57].
The main limitation of hydrogen–like systems are the possibility of ionization of the excited state by the clock laser and the large DC–Stark effect for He^+ in an ion trap. A new definition of the SI second also requires progress of the underlying theory. However, even without a new SI second or clock applications in mind, the proposed experiments are useful to improve hydrogen spectroscopic data. These are important for fundamental research to test QED and to find more evidence for physics beyond the standard model. To measure the 1S—2S transition frequency in He^+ is particularly important in this regard because in this case, the fine structure constant in (2) is replaced by $Z\alpha$ in many of the terms. In the higher order QED terms, the fine structure appears in high orders. Hence He^+ is a more sensitive probe for higher order QED corrections.

5 Acknowledgements
This project has received funding from the European Research Council (ERC) under the European Union's Horizon 2020 research and innovation program (Grant Agreement No. 742247). We also received funding from MCQST (EXC-2111/Projekt-ID: 390814868) and from the DFG (Projektnummer: 534953889). The project/research is part of the Munich Quantum Valley, which is supported by the Bavarian state government with funds from the Hightech Agenda Bayern Plus. T.W. Hänsch acknowledges support from the Max-Planck Foundation.

References
[1] "Mise en pratique for the definition of the meter in the SI", BIPM SI Brochure, 9th edition – Appendix 2 (2019).

[2] P.L. Bender *et al.*, Science 182, 229 (1973).

[3] D.B. Newell *et al.*, Metrologia 55, L13 (2018).

[4] G. Girard, Metrologia 31, 317 (1994).

[5] "Mise en pratique for the definition of the kilogram in the SI", BIPM SI Brochure, 9th edition – Appendix 2 (2019).

[6] J. Dilling, K. Blaum, M. Brodeur, and S. Eliseev, Annu. Rev. Nucl. Part. Sci. 68, 45 (2018).

[7] https://creativecommons.org/licenses/by-sa/4.0/

[8] E. Tiesinga, P.J. Mohr, D.B. Newell, and B.N. Taylor, Rev. Mod. Phys. 93, 025010 (2021).

[9] H. Katori, Proceedings of the 6th Symposium on Frequency Standards and Metrology, University of St Andrews, Fife, Scotland 9–14 September 2001.

[10] C.M. Adhikari, A. Kawasaki, and U.D. Jentschura, Phys. Rev. A 94, 032510 (2016).

[11] D. Kolbe, M. Scheid, and J. Walz, Phys. Rev. Lett. 109, 063901 (2012).

[12] R. deCarvalho, N. Brahms, B. Newman, J.M. Doyle, D. Kleppner, and T. Greytak, Can. J. Phys. 83: 293–300 (2005).

[13] I.C. Lane, Phys. Rev. A 92, 022511 (2015).

[14] R. Côté, M.J. Jamieson, Z-C. Yan, N. Geum, G.-H. Jeung, and A. Dalgarno, Phys. Rev. Lett. 84, 2806 (2000).

[15] S.F. Vázquez-Carson, Q. Sun, J. Dai, D. Mitra, and T. Zelevinsky, New J. Phys. 24 083006 (2022).

[16] S.A. Jones, New J. Phys. 24 023016 (2022) .

[17] H.F. Hess, G.P. Kochanski, J.M. Doyle, N. Masuhara, D. Kleppner, and T.J. Greytak, Phys. Rev. Lett. 59, 672 (1987).

[18] The ALPHA collaboration, Nature 557, 71 (2018).

[19] I.D. Setija, H.G.C. Werij, O.J. Luiten, M.W. Reynolds, T.W. Hijmans, and J.T.M. Walraven, Phys. Rev. Lett. 70, 2257 (1993).

[20] The ALPHA collaboration, Nature 592, 35 (2021).

[21] J.M. Michan, G. Polovy, K.W. Madison, M.C. Fujiwara, and T. Momose, Hyp. Int. 235, 29 (2015).

[22] G. Gabrielse *et al.*, Opt. Lett. 43, 2905 (2018).

[23] J.L. Flowers, H.A. Klein, D.J.E. Knight, and H.S. Margolis, NPL Report CBTLM 11, (2001).

[24] J.P. Scott, R.M. Potvliege, D. Carty, and M.P.A. Jones, Metrologia 61, 025001 (2024).

[25] C.G. Parthey *et al.*, Phys. Rev. Lett. 107, 203001 (2011).

[26] I. Estermann, O.C. Simpson, and O. Stern, Phys. Rev. 71, 238 (1947).

[27] A. Grinin, A. Matveev, D.C. Yost, L. Maisenbacher, V. Wirthl, R. Pohl, T.W. Hänsch, and Th. Udem, Science 370, 6520 (2020).

[28] W. Zernik, Phys. Rev. 135, 51 (1964).

[29] J. Lodewyck, P.G. Westergaard, and P. Lemonde, Phys. Rev. A 79, 061401(R) (2009).

[30] D.H. McIntyre, R.G. Beausoleil, C.J. Foot, E.A. Hildum, B. Couillaud, and T.W. Hänsch, Phys. Rev. A 39, 4591 (1989).

[31] T.C. Killian, D.G. Fried, L. Willmann, D. Landhuis, S.C. Moss, T.J. Greytak, and D. Kleppner, Phys. Rev. Lett. 81, 3807 (1998).

[32] J.W. Farley, and W.H. Wing, Phys. Rev. A 23, 2397 (1981).

[33] H. Katori, V.D. Ovsiannikov, S.I. Marmo, and V.G. Pal'chikov, Phys. Rev. A 91, 052503 (2015).

[34] A. Derevianko, and H. Katori, Rev. Mod. Phys. 83, 331 (2011).

[35] D.P. Shelton, Phys. Rev. A 36, 3032 (1987).

[36] N.L. Manakov, S.I. Marmo, and E.A. Pronin, J. Exp. & Theo. Phys. 98, 254 (2004).

[37] . R.C. Brown *et al.*, Phys. Rev. Lett. 119, 253001 (2017).

[38] M. Herrmann, *et al.*, Phys. Rev. A 79, 052505 (2009).

[39] J. Moreno *et al.*, Eur. Phys. J. D 77, 67 (2023).

[40] I. Sick, Phys. Rev. C 77, 041302 (2008).

[41] J.J. Krauth *et al.*, Nature 589, 527 (2021).

[42] Ye.F. Baklanov, and V.P. Chebotayev, Appl. Phys. 12, 97 (1977).

[43] S.M. Brewer *et al.*, Phys. Rev. Lett. 123, 033201 (2019).

[44] K. Beeks, T. Sikorsky, T. Schumm, J. Thielking, M.V. Okhapkin, and E. Peik, Nat. Rev. Phys. 3, 238 (2021).

[45] F. Schmid, J. Moreno, J. Weitenberg, P. Russbüldt, T.W. Hänsch, Th. Udem, and A. Ozawa, APL Photon. 9, 026105 (2024).

[46] J. Limpert, F. Roser, T. Schreiber, and A. Tünnermann, IEEE J. Sel. Top. Quantum Electron. 12, 233 (2006).

[47] F. Canella *et al.*, Optica 11, 1 (2024).

[48] S. Ghimire *et al.*, Nat. Phys. 7, 138 (2011).

[49] H. Liu, *et al.*, Commun. Phys. 3, 1 (2020).

[50] T. Strobl, L. Hein, T.W. Hänsch, Th. Udem, T. Sukegawa, M. Koyama, and A. Ozawa, CLEO/Europe 2023, paper ck_7_3.

[51] M.S. Safronova, M.G. Kozlov, and C.W. Clark, Phys. Rev. Lett. 107, 143006 (2011).

[52] P. Dubé, A.A. Madej, Z. Zhou, and J.E. Bernard, Phys. Rev. A 87, 023806 (2013).

[53] U.D Jentschura, and M. Haas, Phys. Rev. A 78, 042504 (2008).

[54] M.S. Safronova, M.G. Kozlov, and C.W. Clark, IEEE T. Ultrason. Ferr. 59, 439 (2012).

[55] T. Bothwell, C.J. Kennedy, A. Aeppli, D. Kedar, J.M. Robinson, E. Oelker, A. Staron, and J. Ye, Nature 602, 420 (2022).

[56] The PACS collaboration, Hyp. Int. 239, 29 (2018).

[57] L. Gurung, T.J. Babij, S.D. Hogan, and D.B. Cassidy, Phys. Rev. Lett. 125, 073002 (2022).

[58] A. Antognini, *et al.*, Science 339, 417 (2013).

[59] W. Xiong, *et al.*, Nature 575, 147 (2019).

Towards practical quantum secure time transfer

E. Villaseñor[1], K. Shimizu[2], R. Chandrasekara[1], W. Y. S. Lau[4] and S. Gensemer[3]

[1] CSIRO Manufacturing, Research Way, Clayton, VIC 3168, Australia
[2] CSIRO Manufacturing, Gate 13, Kintore Ave Adelaide, SA 5000, Australia
[3] CSIRO Manufacturing, Bradfield Road, West Lindfield, NSW 2070, Australia
[4] CSIRO Manufacturing, Entrance 4, Waite Road, Urrbrae, SA 5064, Australia

E-mail: eduardo.villasenoralvarez@csiro.au

Abstract. Accurate clock synchronization and time transfer are vital for positioning, navigation, and timing systems, however, current systems are susceptible to attacks, such as spoofing. Quantum secure time transfer (QSTT) can address this threat by providing a method to validate the authenticity of the time signals used. In this work, we analyze the feasibility of free-space QSTT between two mobile transceivers. Using various techniques, we model the free-space channel to estimate the maximum ranges achievable while guaranteeing a defined level of security. The results indicate that QSTT is feasible for daytime operations up to a distance of 14 km, while nighttime operations can be achieved up to 30 km. This study can be used to design the next pathfinder experiments towards the development of practical QSTT.

1. Introduction

Time transfer mechanisms ensure that different devices and systems operate on the same time reference, enabling them to coordinate their actions effectively. In particular, accurate time transfer is essential for precise positioning calculations in navigation systems. However, time transfer can be susceptible to man-in-the-middle attacks, where an attacker intercepts and alters time synchronization messages. Proper security measures can help mitigate these attacks, such as using secure communication channels, digital signatures, or additional authentication mechanisms [1]. Nonetheless, classical security measures leave the possibility of spoofing attacks whereby an attacker can delay a copy of the intercepted signal in time while remaining undetected, compromising the ability to synchronize clocks. Using polarization-encoded single photons as the time signal can guarantee the time signals' validity based on the techniques developed to perform quantum key distribution (QKD) [2].

This protocol is known as quantum secure time transfer (QSTT). In particular, using single photons is advantageous since it allows for precise time synchronization compared to other time signals [3]. Several studies have used single photons to synchronize clocks remotely via fiber and the free space optical (FSO) channels [4, 5, 6, 7]. On the other hand, a set of proof-of-concept experiments of free-space QKD have been done over free-space for horizontal [8], mobile airborne [9], and satellite-to-ground [10] channels. These experiments, in addition to the demonstration presented in [11], strongly suggest that free-space QSTT is not only viable but practical.

A careful study of QSTT for FSO channels is missing from the literature; this is the contribution of this paper. In this work, we use various models and measurements to determine

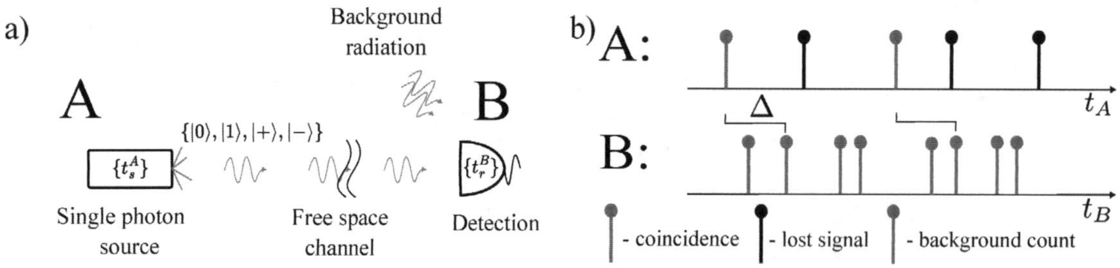

Figure 1. a) Alice transmits single photons to Bob via the FSO channel. Alice keeps timestamps for each photon created, while Bob keeps timestamps for every photon received. b) In the timestamps kept by Alice and Bob, true coincidences (green) have a time difference Δ in their local clocks. Due to the free-space channel, many photons sent by Alice will not be detected by Bob (black). Alternatively, Bob's data will contain many instances of background radiation (red).

the environmental conditions and system requirements for a practical QSTT system. The results indicate that QSTT can be effectively implemented for a free-space horizontal channel during the daytime for distances up to 14 km, with much larger distances, up to 30 km, achievable for nighttime operations.

2. Protocol

Here, we briefly review the protocol to perform two-way QSTT, following ref [4]. We consider two parties, Alice and Bob, to be separated by an unknown distance, L, between them. Both parties carry precise local clocks that are unsynchronized, i.e., at any given time,

$$t^A = t^B + \delta_{AB}, \tag{1}$$

with t^A and t^B the local times are reported by each party, and δ_{AB} is unknown. The objective of two-way QSTT is for the two parties to find L and δ_{AB} with high accuracy while guaranteeing security. In particular, the goal is to calculate

$$\begin{aligned} \Delta_1 &= t_r^B - t_s^A; \\ \Delta_2 &= t_r^A - t_s^B, \end{aligned} \tag{2}$$

where $t_s^{A(B)}$ and $t_r^{A(B)}$ are the times at which the signals were sent and received, respectively, as recorded by each party. From these values, it follows that:

$$\begin{aligned} \delta_{AB} &= \frac{1}{2}(\Delta_1 - \Delta_2); \\ L &= \frac{c}{2}(\Delta_1 + \Delta_2), \end{aligned} \tag{3}$$

with c the speed of light.

Thus, there are two components to the protocol: the first is finding the values of $\Delta_{1(2)}$, and the second is measuring the quantum bit error rate (QBER) of the signal to guarantee security. Given that the process is symmetric, we discuss specifically finding Δ_1, referred to as Δ, from now on. Finding Δ_2 follows the same steps with reversed transmitter and receiver roles.

To estimate Δ, the process begins with the preparation of single photons in one of the four polarization states following the BB84 protocol [12]. Subsequently, these photons are transmitted

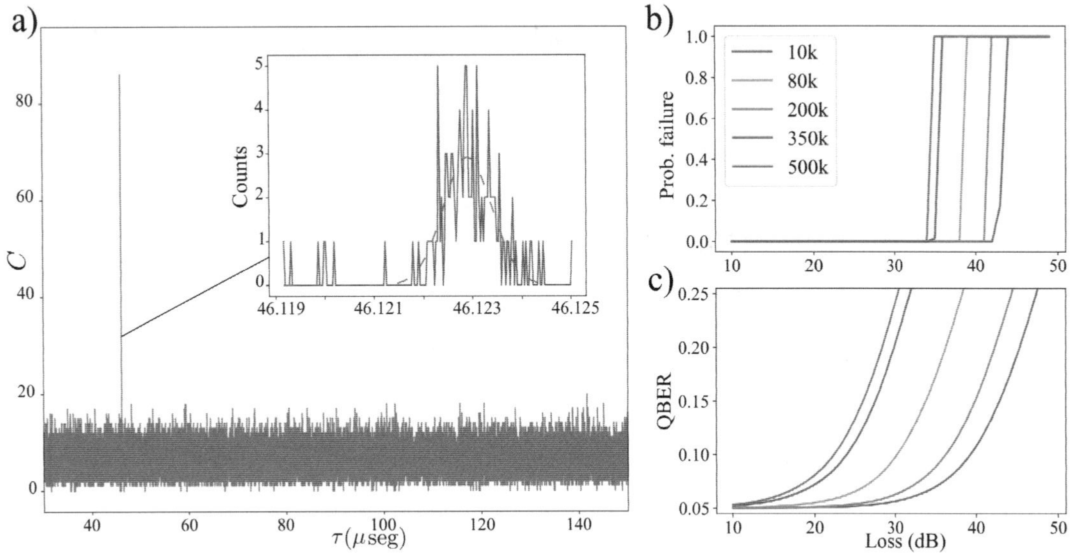

Figure 2. a) Example of computation of Δ. First, a rough computation of Δ is performed by computing the cross-correlation function with a coarse grid. (inset) After identifying the maximum, a fine cross-correlation allows us to identify Δ with a precision corresponding to σ_j. b) The probability of failing to identify Δ for different values of background counts per second. c) The QBER obtained as a function of loss.

from Alice to Bob through the FSO channel, as illustrated in Fig. 1. For each photon sent, Alice records a timestamp, denoted as t_s^A. Upon reception, Bob measures the polarization of each photon by randomly selecting the polarization orientation, typically referred to as the X and Z bases. The measurement results are stored, and Bob creates a detection timestamp, marked as t_r^B. The next crucial step involves one of the parties transmitting the timing and polarization data to determine Δ. This transmission occurs via an authenticated classical channel.

2.1. Processing the data

With one party in possession of all the data, finding Δ corresponds to simply finding the maximum value of the cross-correlation function between the timestamp data of Alice and Bob. The time correlation can be efficiently computed using the discrete Fourier transform as [4],

$$C_k = \mathcal{F}^{-1}\left[\mathcal{F}^*[a_k]\mathcal{F}[b_k]\right]. \tag{4}$$

where a_k, and b_k correspond to Alice and Bob's discrete timing data arrays, defined as,

$$a_k = \sum_i \delta_{k,\lfloor (t_i/\delta t)\mathrm{mod}(N)\rfloor}; \qquad k = 0, 1, ..., N-1, \tag{5}$$

where $\delta_{k,j}$ is the Kronecker delta, $\lfloor\cdot\rfloor$ is the floor function, t_i iterates over the timestamps in Alice's data, mod$(\cdot)$ is the modulus, δt is the time resolution of each discrete bin in the array, and N the total number of bins. An analogous array b_k is computed using Bob's data.

Assuming δ_t has been correctly selected, Δ corresponds to the maximum value of C_k, as shown in Fig. 2a. However, due to the presence of random noise (as a result of background radiation), the protocol is only successful when the maximum is considerably larger than the noise baseline of the cross-correlation function. We can compute the probability of failure by finding the probability of noise having a higher cross-correlation than the signal. See that in Alice and Bob's discrete timing arrays, the probability of bins a_k and $b_{k+\tau}$ for a given k, to both have the value of 1 due to a background count on Bob's array, is given by

$$p_b = r_s r_b \delta t^2. \tag{6}$$

where r_s is the rate at which the signal is created, and r_b is the rate of background detection events on Bob's side. The noise will be distributed following a Poisson distribution, with mean and variance $\mu_b = \sigma_b^2 = p_b N$. Assuming that $C_k - \mu_b$ equals the total number of signal received, N_s^t, then the desired failure probability is given by [4],

$$\begin{aligned} \text{Prob}(X > N_s) &= \frac{N}{2}\text{erfc}\left(\frac{N_s^t}{\sqrt{2\sigma_b^2}}\right) \\ &= \frac{N}{2}\text{erfc}\left(\sqrt{\frac{r_s T \eta^2}{2 r_b \delta t}}\right), \end{aligned} \tag{7}$$

where T is the protocol time, and η is the total optical loss.

2.2. Security of time transfer

The security of QSTT heavily relies on measuring the QBER to verify the integrity of signal photons against tampering. In the context of time transfer operations, an attack is defined as any attempt to manipulate the measured value of Δ. During such an attack, the adversary intercepts signal photons, conducting measurements to extract sufficient information for recreating a plausible signal at a later chosen time. Specifically for time transfer, the actual attack — attempting to manipulate Δ — can only occur in the moments between Alice sending a photon and Bob receiving and measuring it. This constraint contrasts with QKD, where more sophisticated attacks involve entangling a quantum system with the transmitted signals. QKD typically involve the adversary awaiting Alice and Bob's announcement of their respective bases before executing measurements, thereby gaining additional information about the cryptographic key. Consequently, during time transfer, an attacker is restricted to what is known as "intercept and resend" attacks only (see [2] for a definition of possible attacks on QKD). Intercept and resend attacks incur a QBER of 25%, a security limit adopted for QSTT, implying that if the measured QBER exceeds this limit, the protocol is considered insecure[1].

During the protocol, upon successful identification of Δ, Alice and Bob can pair their information about the prepared and measured quantum states to verify the QBER, thus validating the protocol's security. The QBER measured will have two main contributions. First, the QBER contribution attributed to device imperfections when creating and measuring the quantum states, such as those that slightly deviate from the BB84 states. Second, the error induced by background counts that coincide with the time window around Δ for the data points in Alice. By chance, a small fraction of background counts will land on the time correlation window matching the signal photons and, as such, cannot be filtered out. We can quantify how many background counts will be incorrectly identified as true coincidences after sifting as,

$$N_b = T r_s r_b \delta_t / 2. \tag{8}$$

[1] It's worth noting that this discussion assumes the attacker does not possess a quantum memory device. A quantum memory, capable of increasing the time of flight of photons at will, could modify Δ without interfering with the polarization quantum states. Although difficult to implement in practice, such a device could compromise the proposed protocol. In such cases, additional security conditions, similar to those used in classical time transfer [1], must be imposed to ensure the protocol's security.

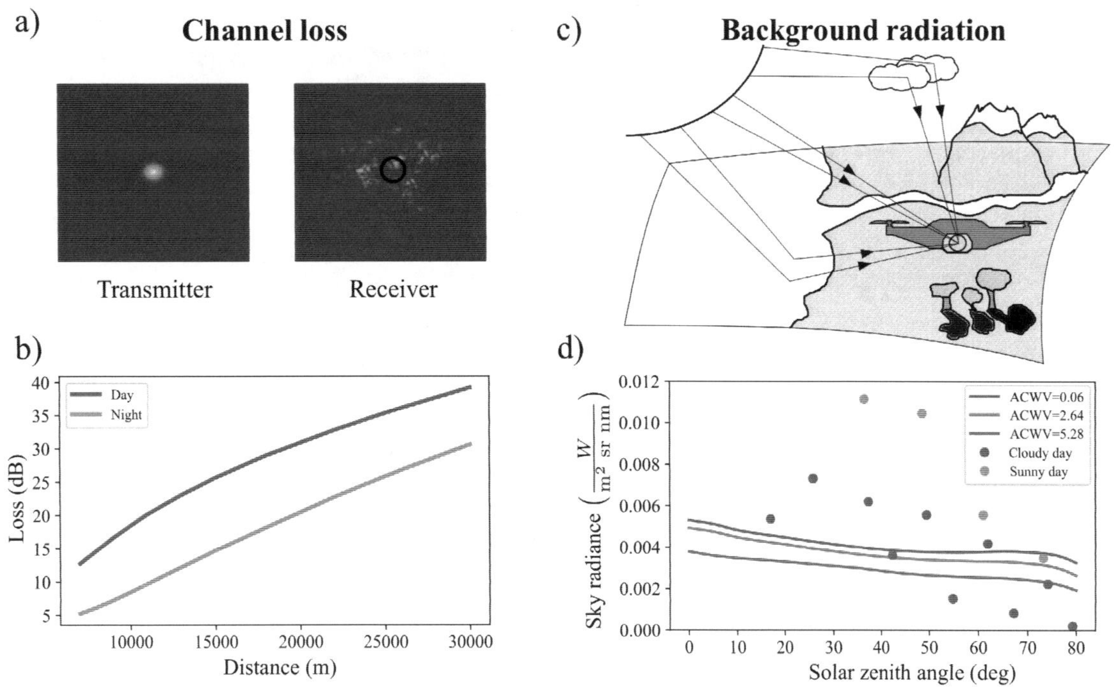

Figure 3. a) An example of the beam at the transmitter and receiver planes; the circle marks the area of the aperture used to collect the light. b) Using the parameters specified in Tab. 1, we compute the total loss of the optical channel, η_{ch}. c) The measurement system will unavoidably detect radiation from the environment. d) Sky radiance, as a function of the sun's zenith angle, during daytime. The solid lines represent the values obtained from the MODTRAN model for different values of atmospheric column water vapor (ACWV), and the dotted points represent values measured by a photodiode for different weather conditions.

Then, the total QBER measured during the protocol is,

$$\begin{aligned} \mathrm{QBER} &= \frac{e_d N_s + \frac{1}{2} N_b}{N_s + N_b} \\ &= \frac{e_d \eta + \frac{1}{2} r_b \delta_t}{\eta + r_b \delta_t}, \end{aligned} \tag{9}$$

with e_d is the QBER associated with all the device imperfections, and $N_s = N_s^t/2$ and N_b are the numbers of true signal and background photon coincidences after basis sifting.

3. Free-space single-photon quantum communications

Characterization of the free-space quantum channel depends on a series of parameters that define the communications setup. We are considering a symmetric configuration where Alice and Bob use the same telescopic aperture of radius a, to send and receive single photons of the wavelength λ. Due to the FSO channel loss, η_{ch}, and the detection efficiency, η_d, only a small fraction of photons are detected by the receiving party. The relevant parameters of the system are summarized in Tab. 1 for the setup considered here. Crucially, the numbers that must be calculated to characterize the free-space channel are the channel loss and the number of counts induced by background radiation.

The loss of the FSO channel is presented in Fig. 3b for both day and night conditions. We use the numerical methods presented in [13, 14] to determine the loss. These numerical methods allow us to estimate the optical loss induced by the beam deformations and wandering induced by the atmosphere combined with the geometric losses determined by the telescope sizes and the diffraction of the beam as it travels from Alice to Bob[2]. The main parameter defining the strength of the atmospheric effects on any traversing optical beam is the refractive index structure, C_n^2. Furthermore, the results presented in Fig. 3b include the extra loss caused by the scattering of the light by the atmosphere for the chosen wavelength (see Tab. 1). Finally, adding the loss of Bob's detection system, the total optical loss (in dB) is $\eta = \eta_{ch} + \eta_d$.

Computing the number of background counts that will be detected depends uniquely on the radiance of the sky around the wavelength of the signal, $L_{e,\Omega,\lambda}$, as well as the properties of the optics used by the receiver to collect the signal. The sky radiance was estimated using a MODTRAN model and taking direct measurements using a photodiode. In Fig. 3d, the results are presented; we see good agreement between the MODTRAN model and the measured values for a cloudy day. For a sunny day, the values measured are considerably higher in the ranges, indicating that daytime operations, if possible, are restricted to weather and factors such as the sun's position relative to transceivers. To simplify the following calculations, we assume a daytime radiance of 0.004 $\frac{W}{\mathrm{m^2\ sr\ nm}}$. Night-time radiance is taken to be 20X less, 0.0002 $\frac{W}{\mathrm{m^2\ sr\ nm}}$.

Spectral and spatial filters can reduce the amount of background radiation detected. We assume that advanced adaptive optics systems can be implemented at the receiver to compensate for the atmospheric turbulence on the incoming rays, such that a spatial filter can be used to reduce the field of view, f_v, [15]. Thus, the rate at which background counts will be detected per second can be computed as,

$$r_b = L_{e,\Omega,\lambda}\Omega\pi a^2\delta\lambda\eta_d\frac{\lambda}{hc}, \tag{10}$$

where $\Omega = 4\pi\sin^2(f_v/2)$ is the solid angle corresponding to the field of view, h is Plank's constant, and the remaining parameters are defined as in Tab. 1.

To generate the single photons, we consider that Alice first creates pairs of photons locally using spontaneous down conversion (SPDC) processes in a Type-0 periodically poled $\mathrm{KTiOPO_4}$ crystal in a Sagnac loop configuration as in [16]. The result of such configuration is the creation of polarization-entangled photon pairs at a rate r_{SPDC}. Both photons are highly correlated in time for each pair, allowing Alice to measure one locally, normally called the idler photon. The idler photon is measured along one of the polarization axes chosen passively using a beam splitter; since the pair is entangled, this has the effect of preparing the remaining photon, referred to as the signal photon, in one of the four polarization states. Since Alice's detection system is lossy, the generation rate of signal photons is $r_s = \eta_d r_{\mathrm{SPDC}}$. Afterward, the signal photon is transmitted to Bob via the FSO channel to perform QSTT.

The total time jitter comes from the measurement devices used to detect the single photons, and the main contribution to the jitter comes from the single-photon avalanche photodiodes used to detect the photons. We consider each party to have a time jitter $\sigma_{j_i} = 500$ ps on their respective measurements, a value corresponding to current commercially available detectors. Thus, the total jitter of the system is $\sigma_j = \sqrt{2}\sigma_{j_i}$.

4. Results

Using Eq. 9, we calculate the QBER that will be measured after transmission according to the models presented in the previous section. To reduce the number of background counts

[2] For more information, we refer the reader to the work [14]. Unless specified otherwise, all the corresponding parameters of the simulations are set to be as specified in [14].

Parameter	Description	Value	Source
λ	Wavelength of the single photons	810 nm	Design
a	Aperture radius of telescope	75 mm	Design
f_v	Field of view	24 μrad	Technological requirement
r_{SPDC}	Entangled pair generation rate	20 MHz	[16]
σ_j	Total system jitter	754 ps	Hardware properties
e_d	Device imperfections QBER	0.05	Hardware properties
$L_{e,\Omega,\lambda}$	Daytime sky radiance	$0.004 \frac{W}{\mathrm{m}^2\ \mathrm{sr\ nm}}$	Measured & MODTRAN
$\delta\lambda$	Spectral width at detection	2.3 nm	[16]
η_d	Single photon detection loss	5 dB	Hardware properties
η_s	Scattering loss under high visibility	0.45 dB/km	[17]
$C_n^{2(day)}$	Daytime refractive index structure	$1.52\mathrm{x}10^{-14}\mathrm{m}^{-\frac{2}{3}}$	[18]
$C_n^{2(night)}$	Nighttime refractive index structure	$1.25\mathrm{x}10^{-15}\mathrm{m}^{-\frac{2}{3}}$	[18]

Table 1. Parameters used to model the FSO quantum channel.

erroneously identified as signals, we implement a time filter by setting $\delta_t = 2\sigma_j$. This time filter increases the signal-to-noise ratio as the background counts follow a uniform distribution while the signal is normally distributed, although, at the cost of reduced number of coincidences detected. We consider that time transfer updates are made every .1 s, using the aggregate timing data of all the coincidences detected during that integration time. The measured polarization data is accumulated in blocks of 1 s integration time to measure the QBER. The block is declared secure if the measured QBER is four standard deviations below 0.25. After the QBER is measured, the polarization data is discarded to prevent a sudden attack from going undetected as a consequence of averaging large data sets. New data is used to validate the security of the next block.

The results are shown in Fig. 4; we see that for daytime operations, QSTT is possible up to distances of 14 km, while for nighttime operations, the maximum operation range is vastly larger, up to 30 km. For further distances during nighttime operation, the high optical loss makes it impossible to distinguish signal from noise in the autocorrelation function. Following the central limit theorem, we report a precision of $2\sigma_j/\sqrt{N_c}$ using N_c successful coincidence detections, with a 95% confidence of the true mean being within the reported precision. The precision obtained is in the order of tens of picoseconds, vastly superior to the one of the current GPS systems.

5. Conclusions

Free-space quantum communications using single photons are attractive as they can dynamically deploy QKD or QSTT protocols on the field without needing pre-existing infrastructure, such as optical fiber. In this work, we presented an overview of the QSTT protocol and studied its feasibility over the horizontal free space channel. Here, we show that QSTT can be implemented over long ranges of 30 km during nighttime. During the day, QSTT could be implemented over 14 km, albeit conditioned to weather conditions and the availability of optical stabilization technology such as adaptive optics. The results motivate further experiments of free-space quantum communications during the daytime.

6. Acknowledgements

We thank Alberto Peruzzo for the valuable discussion that helped improve this work. We acknowledge the support of the Quantum Photonics Laboratory and the ARC Centre of Excellence for Quantum Computation and Communication Technology at RMIT University.

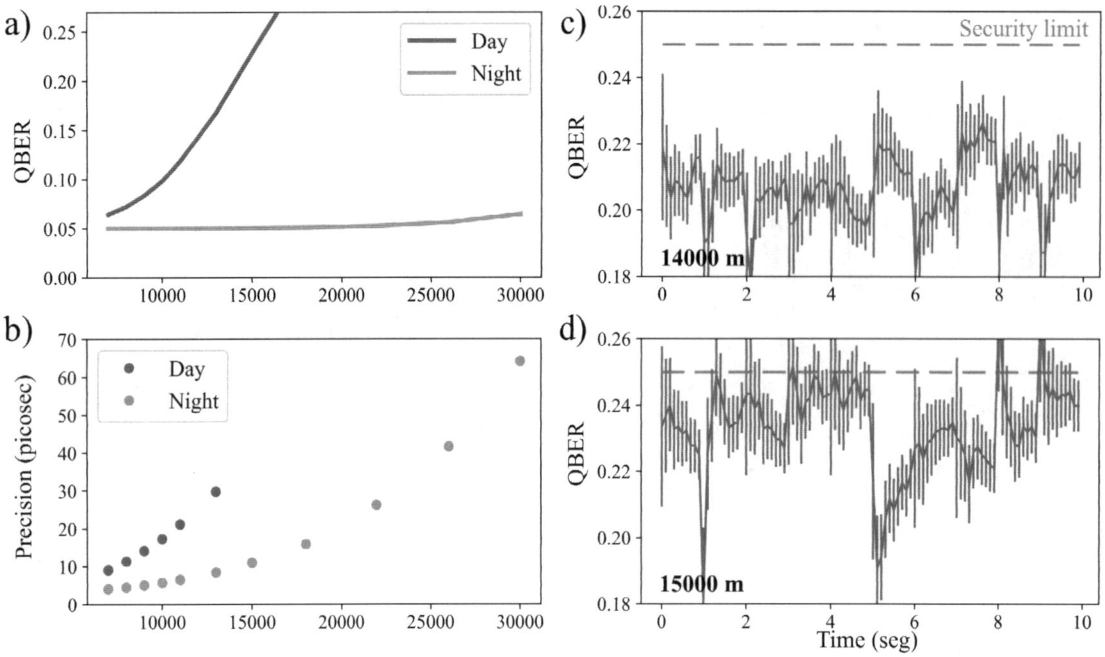

Figure 4. a) QBER as a function of range computed using the loss and background rate obtained from the models. b) When time transfer can be done securely, Δ can estimated from the number of coincidences detected at a definitive precision. c) Time transfer is secure when the measured QBER is sufficiently far from the security limit. d) As the distance increases, the background radiation increases the QBER closer to the security limit, and thus, time transfer can not be deemed secure.

References

[1] Narula L and Humphreys T E 2018 *IEEE Journal of Selected Topics in Signal Processing* **12** 749–762 ISSN 1932-4553 URL `https://dx.doi.org/10.1109/jstsp.2018.2835772`

[2] Pirandola S, Andersen U L, Banchi L, Berta M, Bunandar D, Colbeck R, Englund D, Gehring T, Lupo C, Ottaviani C, Pereira J L, Razavi M, Shaari J S, Tomamichel M, Usenko V C, Vallone G, Villoresi P and Wallden P 2020 *Adv. Opt. Photon.* **12** 1012–1236 URL `https://opg.optica.org/aop/abstract.cfm?URI=aop-12-4-1012`

[3] Spiess C and Steinlechner F 2022 Clock synchronization methods with single photon sources *Quantum 2.0 Conference and Exhibition* (Optica Publishing Group) p QTu3B.2 URL `https://opg.optica.org/abstract.cfm?URI=QUANTUM-2022-QTu3B.2`

[4] Ho C, Lamas-Linares A and Kurtsiefer C 2009 *New Journal of Physics* **11** 045011 ISSN 1367-2630 URL `https://dx.doi.org/10.1088/1367-2630/11/4/045011`

[5] Lafler R and Lanning R N 2023 *Phys. Rev. Appl.* **20**(2) 024064 URL `https://link.aps.org/doi/10.1103/PhysRevApplied.20.024064`

[6] Lee J, Shen L, Utama A N and Kurtsiefer C 2022 *Optics Express* **30** 18530 ISSN 1094-4087 URL `https://dx.doi.org/10.1364/OE.455542`

[7] Spiess C, Töpfer S, Sharma S, Kržič A, Cabrejo-Ponce M, Chandrashekara U, Döll N L, Rieländer D and Steinlechner F 2023 *Phys. Rev. Appl.* **19**(5) 054082 URL `https://link.aps.org/doi/10.1103/PhysRevApplied.19.054082`

[8] Schmitt-Manderbach T, Weier H, Fürst M, Ursin R, Tiefenbacher F, Scheidl T, Perdigues J, Sodnik Z, Kurtsiefer C, Rarity J G, Zeilinger A and Weinfurter H 2007 *Phys. Rev. Lett.* **98**(1) 010504 URL `https://link.aps.org/doi/10.1103/PhysRevLett.98.010504`

[9] Pugh C J, Kaiser S, Bourgoin J P, Jin J, Sultana N, Agne S, Anisimova E, Makarov V, Choi E, Higgins B L and Jennewein T 2017 *Quantum Science and Technology* **2** 024009 URL `https://dx.doi.org/10.1088/2058-9565/aa701f`

[10] Lu C Y, Cao Y, Peng C Z and Pan J W 2022 *Reviews of Modern Physics* **94** 035001 rMP URL https://link.aps.org/doi/10.1103/RevModPhys.94.035001

[11] Dai H, Shen Q, Wang C Z, Li S L, Liu W Y, Cai W Q, Liao S K, Ren J G, Yin J, Chen Y A, Zhang Q, Xu F, Peng C Z and Pan J W 2020 *Nature Physics* **16** 848–852 ISSN 1745-2473 URL https://dx.doi.org/10.1038/s41567-020-0892-y

[12] Bennett C H and Brassard G 2014 *Theoretical Computer Science* **560** 7–11 ISSN 0304-3975 URL https://www.sciencedirect.com/science/article/pii/S0304397514004241

[13] Mudge K A, Silva K K M B D, Clare B A, Grant K J and Nener B D 2011 Scintillation index of the free space optical channel: Phase screen modelling and experimental results *2011 International Conference on Space Optical Systems and Applications (ICSOS)* pp 403–409

[14] Villaseñor E, Malaney R, Mudge K A and Grant K J 2021 Atmospheric effects on satellite-to-ground quantum key distribution using coherent states *GLOBECOM 2020 - 2020 IEEE Global Communications Conference* pp 1–6 ISBN 2576-6813

[15] Hampson K M, Turcotte R, Miller D T, Kurokawa K, Males J R, Ji N and Booth M J 2021 *Nature Reviews Methods Primers* **1** 68 ISSN 2662-8449 URL https://doi.org/10.1038/s43586-021-00066-7

[16] Steinlechner F, Gilaberte M, Jofre M, Scheidl T, Torres J P, Pruneri V and Ursin R 2014 *J. Opt. Soc. Am. B* **31** 2068–2076 URL https://opg.optica.org/josab/abstract.cfm?URI=josab-31-9-2068

[17] Kim I I, McArthur B and Korevaar E J 2001 Comparison of laser beam propagation at 785 nm and 1550 nm in fog and haze for optical wireless communications *Optical Wireless Communications III* vol 4214 ed Korevaar E J International Society for Optics and Photonics (SPIE) pp 26 – 37 URL https://doi.org/10.1117/12.417512

[18] Dix-Matthews B P, Karpathakis S F E and Schediwy S W 2023 *Opt. Lett.* **48** 5519–5522 URL https://opg.optica.org/ol/abstract.cfm?URI=ol-48-21-5519

Field-deployable interferometric fibre link terminals with very low temperature sensitivity

A Kuhl and J Kronjäger

Physikalisch-Technische Bundesanstalt, National Metrology Institute, Bundesallee 100, 38116 Braunschweig, Germany

E-mail: alexander.kuhl@ptb.de, jochen.kronjaeger@ptb.de

Abstract. We present a novel setup for optical frequency transfer through fibre which is free from uncompensated paths and, as a result, extremely insensitive to the environment. Our test setup involving a 75 km spooled fibre link achieves an instability approaching 10^{-21} at 10^4 s of averaging, even with one of the terminals temperature-cycled over 15 K. The temperature coefficient of the phase error is estimated at approximately 2 as/K, without any active temperature stabilization or thermal shielding. Fibre link terminals using the topology presented here are suitable for field deployment, for example in geodetic applications together with transportable optical clocks.

1. Introduction

Long distance optical frequency dissemination via interferometric fibre links (IFLs) achieves fractional instabilities and uncertainties of 10^{-19} and below [1-3]. IFLs enable a wide range of high precision applications, such as comparison of atomic clocks [4], chronometric levelling [5-7], tests of fundamental physics [8] or searches for Dark Matter [9]. However, conventional IFL topologies, including those implemented in commercial equipment [10], generally involve reference arms and other uncompensated paths that need to be passively stable. This can be achieved, for example, by keeping these paths as short as possible, balancing path lengths and placing them in a temperature-stabilized environment. Nevertheless, the long-term instability of practical IFLs is often limited by residual sensitivity to environmental conditions [11].

In this work, we present a novel optical setup for IFLs with an inherently very low temperature sensitivity, based on standard fibre and components, in a topology free of uncompensated optical paths. The terminals are designed to be deployed outside of well-controlled labs, for example for chronometric levelling with transportable optical clocks.

2. Setup of field-deployable, temperature-insensitive fibre link terminals

2.1. Concept

We consider a situation where two optical frequency standards, e.g., optical clocks or cavity-stabilized lasers, are to be compared across a typical inter-city distance, using a long-haul fibre link with typically around 15 dB attenuation. Generally, optical frequency combs will be used at either end to translate the optical frequency standard to a wavelength in the telecom C-band (1530 – 1565 nm) for transmission through the fibre link. Based on this scenario, each of the two terminals which are linked by the long-haul optical fibre also needs to have a local branch, linking it to the local optical frequency comb (or to the local optical frequency standard directly, in case their wavelengths match sufficiently well). In addition, one of the terminals has a third branch connecting to a local auxiliary cavity-stabilized laser, which is used to provide a low-phase-noise reference to suppress delayed self-heterodyne noise in the long-haul link.

The local branches are inherently part of the comparison chain and need to be phase-stabilized along with the long-haul link to eliminate frequency shifts due to the Doppler effect. The key idea of our setup is to implement separate IFLs with a single common reference plane in each terminal for both the local and the long-haul link.

We also modified the setup of the IFL reference arm to reduce optical losses by using a circulator and a partially reflecting mirror (Figure 1). Only a small reflected part (in our case 5%) is required to obtain sufficient optical power from the reference arm, so that almost all light coming back from the link to be stabilized is available for the beat signal.

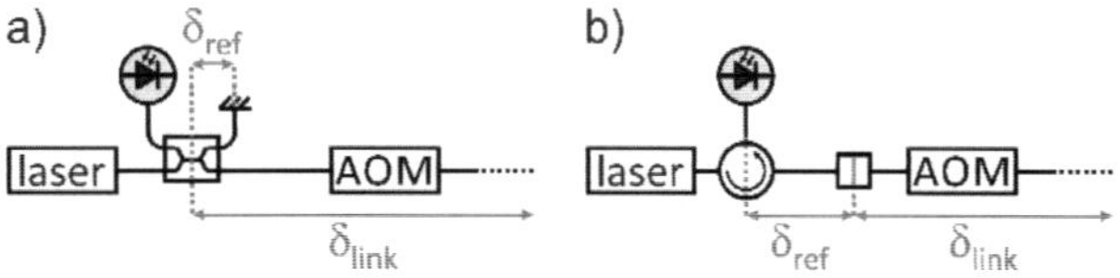

Figure 1. a) Standard setup of the reference arm from an IFL. b) Alternative setup using a circulator and a partially reflecting mirror.

2.2. Long IFL and local links

The setup of our two terminals is shown in Figure 2. At the local terminal A, a transfer laser (laser 1) operating at around 194.4 THz is phase locked to a cavity-stabilized laser via a beat note from photodiode PD4. At the remote terminal B, another transfer laser is phase locked to the light received at RP2 from laser 1 through long link 1, using the beat signal from PD2, with a frequency offset f_{2L}. With ν_x the optical frequency of laser 1 and laser 2, respectively, defined at the corresponding RPx, f_{AOMx} the frequency shift of AOM1 and AOM2, respectively, and δ_{1x} the frequency noise of the link section 1A and 1B, respectively, we have:

$$f_{2L} = \nu_2 - \nu_1 - f_{AOM1} - f_{AOM2} - \delta_{1A} - \delta_{1B}$$

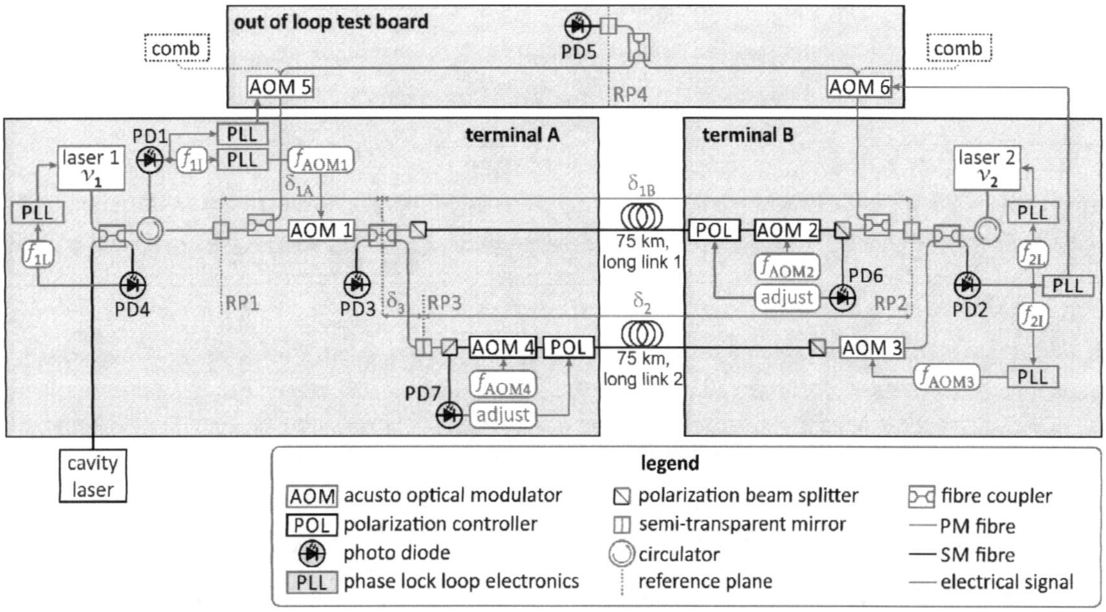

Figure 2. Setup of the two link terminals in a test configuration with 75 km spooled fibre for the long links. The two local links are connected to form an out-of-loop beat note for characterisation.

Journal of Physics: Conference Series **2889** (2024) 012029 doi:10.1088/1742-6596/2889/1/012029

AOM2 is operated at a fixed frequency to shift away Rayleigh scattered light, which would otherwise interfere with laser 2 light from the reference arm at terminal B, causing phase noise in the beat note for the phase lock.

Part of the light from laser 2 is transmitted back through long link 1. The beat note on PD1 with laser 1 light reflected from RP1 is used to adjust the frequency of AOM1, so that the beat frequency $f_{1\mathrm{I}}$ is stabilized:

$$f_{1\mathrm{I}} = \nu_2 - \nu_1 + f_{\mathrm{AOM1}} + f_{\mathrm{AOM2}} + \delta_{1\mathrm{A}} + \delta_{1\mathrm{B}}$$

As a result, the IFL through long link 1 is stabilized:

$$\nu_2 = \nu_1 + \frac{f_{1\mathrm{I}}}{2} + \frac{f_{2\mathrm{L}}}{2}$$

In addition to the long IFL, each terminal has a short IFL, in the following called local link, which is stabilized by feedback to an AOM based on beat signals from PD1 and PD2, respectively. The local links simply use a partially reflecting mirror to produce a return signal. Due to the short length of the fibre, neither a phase-locked laser nor an additional frequency-shifting AOM are required at the end points. In Figure 2, which shows a test configuration of the setup, the partially reflecting mirror at RP4 provides the return signal for both local links. In a real-world application, where the two terminals are not co-located, separate partially reflecting mirrors would be used.

At terminal A, RP1 is the common reference plane for the local link and the long IFL. At terminal B, RP2 is the common reference plane for the local link and the laser phase lock. Effectively, both terminals together act to match the phase of the optical signal at the end points of the two local links, with no uncompensated paths in between.

2.3. Integrity check

In a real remote comparison scenario, it is desirable to have an independent way of verifying the proper operation of the long-haul link. Therefore, terminal B provides an additional IFL back to terminal A on a second fibre (long link 2), which is stabilized by tuning the frequency of AOM3 to keep the in-loop beat $f_{2\mathrm{I}}$ at PD2 constant. To minimize the optical losses of long link 2, the reflection at the end should be as high as possible, and therefore using RP1 is not ideal. Instead, we use another reference plane RP3 with a higher reflecting ratio (in our case 90%), thus avoiding the need for optical amplification or another phase-locked laser. At RP3, we get

$$\nu_2 + \frac{f_{2\mathrm{I}}}{2} = \nu_1 + \frac{f_{1\mathrm{I}}}{2} + \frac{f_{2\mathrm{I}}}{2} + \frac{f_{2\mathrm{L}}}{2}$$

However, we now need to relate the phase at RP1 and RP3 to obtain an end-to-end integrity check. Here we use the idea of extracting a signal along a stabilized link (also known as eavesdropping) [12]. This method also reduces the optical power of unwanted signals on long link 2 and suppresses unwanted beats, especially on PD2. From laser 2 light transmitted through long link 2, together with laser 1 light reflected from RP1, we get the beat B_{PD1} at PD1:

$$B_{\mathrm{PD1}} = f_{1\mathrm{I}} + \frac{f_{2\mathrm{I}}}{2} - f_{AOM2} - \delta_{1B} + \delta_3$$

Here, δ_3 is the frequency noise of the short fibre section between the coupler for PD3 and the reference plane RP3. We also get the beat between forward and backward propagating light on long link 1 at PD3:

$$B_{\mathrm{PD3}} = f_{2\mathrm{L}} + 2f_{\mathrm{AOM2}} + 2\delta_{1\mathrm{B}} - 2\delta_3$$

With the correct combination of these beats, we finally obtain an integrity check C:

$$C = B_{\mathrm{PD1}} + \frac{B_{\mathrm{PD3}}}{2} = f_{1\mathrm{I}} + \frac{f_{2\mathrm{I}}}{2} + \frac{f_{2\mathrm{L}}}{2}$$

The signal generated in this way encompasses all major sources of cycle slips and thus provides an integrity check for both long link operations and for the lock of laser 2.

2.4. Polarization management

The core parts of the link terminals are constructed with polarization-maintaining components. Long-haul fibre links generally utilize non-polarization maintaining, single mode fibres (SMF). Therefore, both long links employ an automatic polarization controller at the end, which adjusts the polarization by minimizes the optical power at the output of a polarizing beam splitter (PBS), detected by PD6 and PD7. In addition, there are PBS at the beginning of the long links which remove mismatched polarization components that would be subject to large phase shifts in the PM fibres.

3. Results

To verify the performance of the design, we connect terminal A and B via 75 km SMF spools to simulate a typical long-haul link. We connect both local links via a coupler to the same partially reflecting mirror (RP4, see Figure 2, out of loop test board). The beat note at PD5 provides an out-of-loop measurement of the residual phase error of the entire setup. To test sensitivity to temperature, we put the optics, the lock electronics, and the laser of terminal B into a climate chamber and measure the performance under stable and under cycling temperature condition (see Figure 3). The relevant beat frequencies are counted with dead time-free totalizing counters (type K&K FXE) in phase averaging mode with 1 s gate time (Λ_{1s}). The integrity check C is calculated in post processing. The FXE counter was operated outside of the climate chamber. Because the electronics generated a considerable amount of heat, which the climate chamber was not originally designed to cope with, fans were placed inside

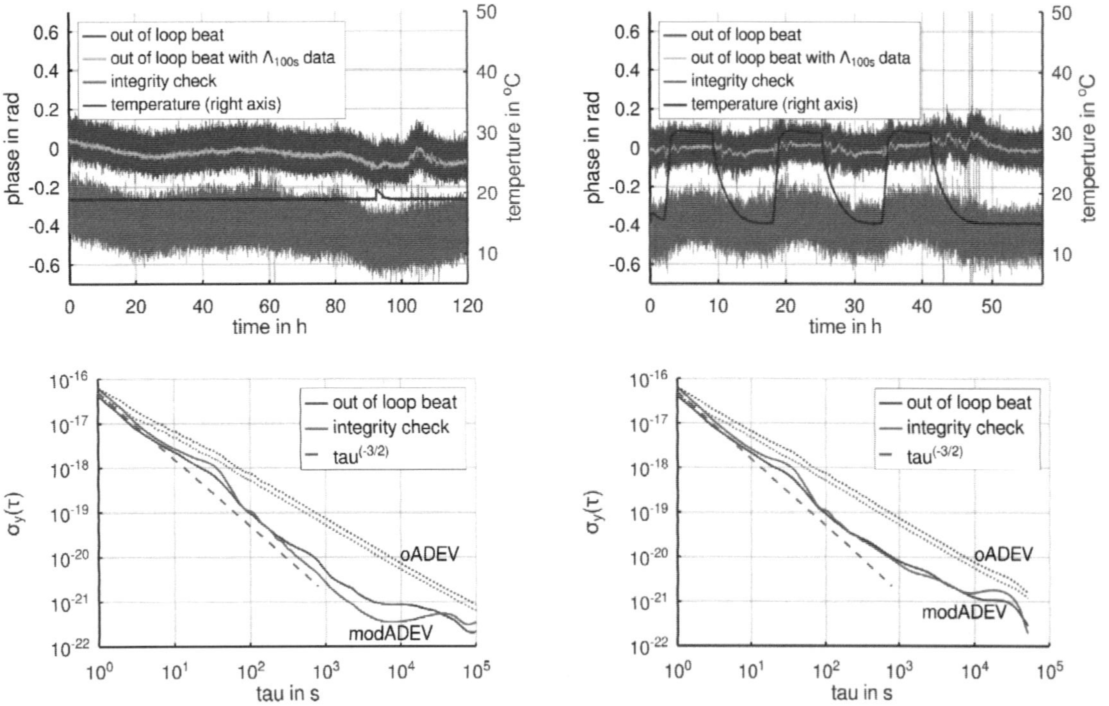

Figure 3. Top: Phase and temperature as a function of time. **Left** with a nominally constant temperature (cycle slips corrected: out of loop 0, integrity check 1); **Right** with temperature cycling (cycle slips corrected: out of loop 2, integrity check 12). **Bottom:** Corresponding overlapping Allan deviation (oADEV) and modified Allan deviation (modADEV).

the climate chamber to help distribute the heat more evenly. Temperature steps were limited to 15 K due to the constraints of the climate chamber and to protect the electronics.

The setup achieves a relative instability (modified Allan deviation) below 10^{-20} at an averaging time of 10^3 seconds and approaching 10^{-21} at 10^4 seconds. We observe long-term phase changes over 100 h below 0.25 rad, and the highest phase drift over 1 h was below 0.1 rad. From the phase steps associated with 15 K temperature steps, see Figure 4 (Left), we estimate the temperature coefficient to be $\partial\phi/\partial T = 2.1 \pm 0.6$ mrad/K, which corresponds to $(2\pi\nu_0)^{-1}\, \partial\phi/\partial T = (1.7 \pm 0.5) \times 10^{-3}$ fs/K with $\nu_0 = 194.4$ THz. While even lower thermal sensitivity has been claimed for PLC-based laser repeater stations [13], the demonstrated performance is comparable and is likely to be limited by other factors [14].

As the modified Allan deviation levels off at around 10^4 s, we take this value as the statistical uncertainty of the link. Figure 4 (Right) shows the average frequency for every block of 10^4 seconds, calculated from the difference of consecutive averages of the phase (Λ_{10000s}), with error bars given by the modified Allan deviation at 10000 s. The effect of temperature cycling can clearly be seen, even though the number of samples outside $0 \pm \mathrm{mod}\sigma_y(10000\ s)$ (grey bar in Figure 4) is still roughly consistent with a zero-mean Gaussian distribution. Comparing the temperature-stable and temperature-cycled cases (Figure 3), it is also clear that other long-term noise processes of almost the same magnitude superimpose the temperature step effect.

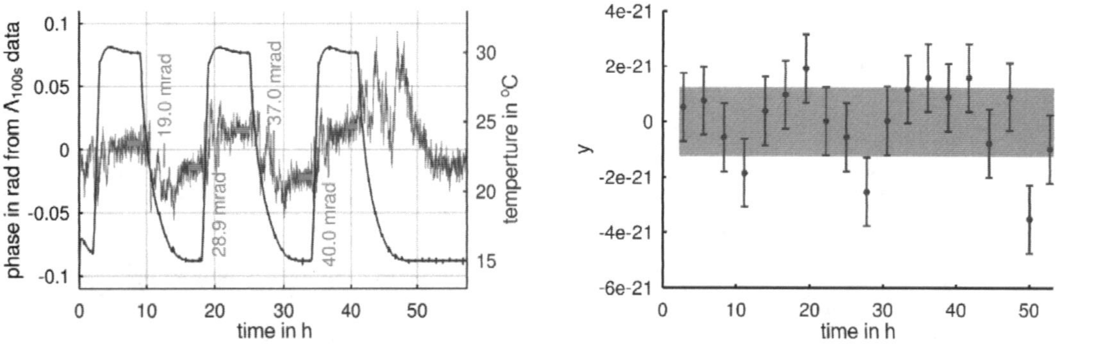

Figure 4. Left: Estimation of the temperature coefficient from phase averages just before a temperature step, where the temperature has stabilised from the previous step. The numbers indicate the difference between consecutive averages. The section of data averaged for each temperature step is 2 h, indicated by the yellow bars. The last temperature drop has been omitted due to strong disturbances, the reason for which is unknown. **Right:** Λ_{10000s}-averaged fractional frequency deviation for the temperature-cycled case shown left, with error bars corresponding to the modified Allan deviation at 10000 s. The grey bar indicates zero within one error bar, i.e., $0 \pm \mathrm{mod}\sigma_y(10000\ s)$.

We observe that all cycle slips in the out-of-loop beat are also detected in the integrity check. However, the integrity check also indicates false positive cycle slip detections. This can happen if one of the cleanup oscillators for the detection of the B_{PD1}, B_{PD3} beats, or for the long link 2 stabilization, experience a cycle slip.

In general, care must be taken when selecting the frequencies of the AOMs and the offset of the lasers (Table 2), to ensure that the required beats do not coincide with each other or with other (unwanted) beats. For example, in our setup, on PD1 only the beats generated with laser 1 at 22 MHz and 70 MHz for the stabilization from the long and the local link, respectively, and 104 MHz for the integrity check are relevant. In reality, we observe many more beats (see Figure 5), of which we have identified 21 as being generated by the 7 optical signals listed in Table 2.

The optical frequency $\nu_2 = \nu_1 + f_{1l}/2 + f_{2L}/2$ delivered at terminal B depends on the radiofrequency references at both locations A and B through the offsets $f_{1l}/2$ and $f_{2L}/2$, respectively. Consequently,

Table 2. Nominal frequencies of AOMs and phase lock offset frequencies used in our setup.

Frequency shift	AOM1	AOM2	AOM3	AOM4	AOM5	AOM6	laser 1 offset f_{1L}	laser 2 offset f_{2L}
Nominal value [MHz]	-73	+47	-72	+36	+35	-39.5	+160	+30

Table 2. Each pair of signals produces a beat on PD1.

Frequency	Signal
ν_1	Laser 1
$\nu_1 - 22$ MHz	Back transmitted signal from long link 1
$\nu_1 + 70$ MHz	Back reflected signal from local link 1
$\nu_1 - 146$ MHz	Rayleigh scattering from long link 1
$\nu_1 - 104$ MHz	Laser 2 via long link 2
$\nu_1 - 160$ MHz	Cavity-stabilized laser
$\nu_1 - 0.5$ MHz	Crossover signal from local link 2 via out of loop test board setup*

* Present only in test setup

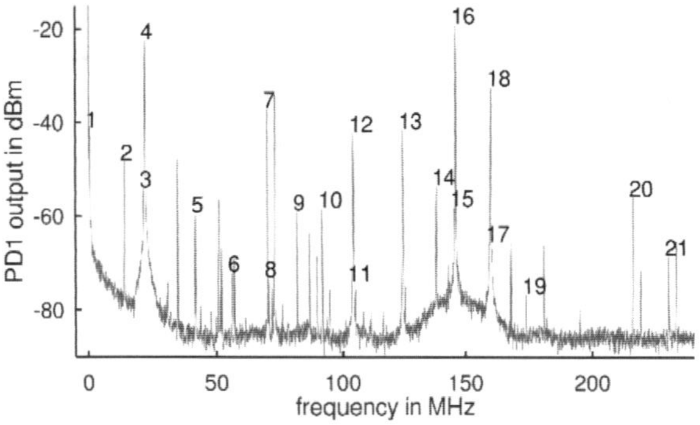

Figure 5. Measured spectrum of the output from PD1. Of these beats, 21 can be identified with the signal combinations from Table 2, but only 3 are required for system operation.

the instability and inaccuracy of the references will have an impact on ν_2[1]. However, in terms of fractional frequency deviation, any error in the references at A and B will be suppressed by a factor $2\nu_2/f_{1l}$ and $2\nu_2/f_{2L}$, respectively. To maintain the performance shown in Figure 3, for example, references with a modified Allan deviation below 10^{-10} at 1 s, averaging down to 10^{-15} at 10^4 s and with an accuracy of the same order of magnitude, would be sufficient. Time and frequency transfer techniques like ELSTAB [15] readily support this level of instability and accuracy and can be deployed alongside optical frequency transfer on the same fiber, using wavelength multiplexing.

This is actually not a concern in the specific situation discussed in this paper, where the purpose is to compare two optical clocks $\nu_{\text{clock A}}$ and $\nu_{\text{clock B}}$ at locations A and B, respectively, and optical frequency combs can be assumed to be available at both locations. Through a convenient definition of a virtual optical transfer frequency, the dependencies on the local reference at A and B can be separated,

$$\nu_{\text{virtual}} := \nu_1 + \frac{f_{1l}}{2} = \nu_2 - \frac{f_{2L}}{2},$$

where the equality is guaranteed by the fiber link. Thanks to the frequency combs, the frequency ratio at both locations, and thus the frequency ratio of the optical clocks, can be determined independently of the local references:

$$\frac{\nu_{\text{clock A}}}{\nu_{\text{clock B}}} = \frac{\nu_{\text{clock A}}}{\nu_{\text{virtual}}} \times \frac{\nu_{\text{virtual}}}{\nu_{\text{clock B}}} = \frac{\nu_{\text{clock A}}}{\nu_1 + \frac{f_{1l}}{2}} \times \frac{\nu_2 - \frac{f_{2L}}{2}}{\nu_{\text{clock B}}}$$

[1] In the results presented here, this impact is not visible, as the same RF reference was used for both terminals and for the frequency counter.

4. Conclusion

We have demonstrated fibre link terminals that achieve optical frequency transfer with a relative uncertainty below 10^{-20}, despite strong fluctuations in the ambient temperature and without shielding the optics. The temperature coefficient of the remote terminal is below 10 as/K, equivalent to the thermal sensitivity of about a quarter of a millimetre of SMF28 fibre [16], and two orders of magnitude below the commercial off-the-shelf solution [10]. Cycle slips in any of the feedback loops involved in the long link can be reliably detected at the local terminal, using an integrity check signal based on a second, independently stabilized fibre link.

Due to their very low environmental sensitivity, the terminals are suitable for deployment outside of the lab, for example in data centres or in improvised settings, such as for chronometric levelling with transportable optical clocks.

Acknowledgments

Funded by the Deutsche Forschungsgemeinschaft (DFG, German Research Foundation) – Project-ID 434617780 – SFB 1464 and under Germany's Excellence Strategy – EXC-2123 QuantumFrontiers – 390837967.

References

[1] Koke S, Benkler E, Kuhl A and Grosche G 2021 *New J. Phys.* **23** 093024
[2] Cantin E, Tønnes M, Targat R L, Amy-Klein A, Lopez O and Pottie P-E 2021 *New J. Phys.* **23** 053027
[3] Tønnes M B K, Schuller F, Cantin E, Lopez O, Targat R L, Amy-Klein A and Pottie P-É 2022 *Metrologia* **59** 065004
[4] Lisdat C et al. 2016 *Nat Commun* **7** 12443
[5] Takano T, Takamoto M, Ushijima I, Ohmae N, Akatsuka T, Yamaguchi A, Kuroishi Y, Munekane H, Miyahara B and Katori H 2016 *Nature Photon* **10** 662–6
[6] Mehlstäubler T E, Grosche G, Lisdat C, Schmidt P O and Denker H 2018 *Rep. Prog. Phys.* **81** 064401
[7] Grotti J et al. 2023 *Phys. Rev. Appl.* **21** L061001
[8] Delva P et al. 2017 *Phys. Rev. Lett.* **118** 221102
[9] Roberts B M et al. 2020 *New J. Phys* **22** 093010
[10] Guillou-Camargo F et al. 2018 *Appl. Opt.* **57** 7203–10
[11] Jürss T, Grosche G and Koke S 2023 *Photon. Res.* **11** 1113–24
[12] Grosche G 2014 *Opt. Lett.* **39** 2545
[13] Akatsuka Ta et al. 2020 *Opt. Express* **28** 9186–97
[14] Xu D, Lopez O, Amy-Klein A and Pottie P-E 2021 *Opt. Express* **29** 17476–90
[15] Krehlik P, Sliwczynski L, Buczek L, Kolodziej J and Lipinski M 2016 *IEEE Trans. Ultrason., Ferroelect., Freq. Contr.* **63** 993-1004
[16] Stefani F, Lopez O, Bercy A, Lee W-K, Chardonnet C, Santarelli G, Pottie P-E and Amy-Klein A 2015 *J. Opt. Soc. Am.* B **32** 787

Journal of Physics: Conference Series **2889** (2024) 012030 doi:10.1088/1742-6596/2889/1/012030

Synergetic Repetition Frequency Locking of an Optical Frequency Comb

Ziyi Jin[1], Dongrui Yu[2], Ziyang Chen[2], Hong Guo[2]

[1]State Key Laboratory of Information Photonics and Optical Communications, Beijing University of Posts and Telecommunications, Beijing 100876, China

[2]State Key Laboratory of Advanced Optical Communication Systems and Networks, School of Electronics, and Center for Quantum Information Technology, Peking University, Beijing 100871, China

E-mail: chenziyang@pku.edu.cn

Abstract. This paper proposes a method to achieve long-term locking of the repetition frequency of the optical frequency comb under severe temperature fluctuations by cooperatively adjusting the cavity length of the resonator using piezoelectric ceramic, temperature control module, and delay line module. This method enhances the practicability and usability of optical frequency comb at a low cost.

1. Introduction

Fiber-based optical frequency combs (OFCs) have achieved rapid development in recent years [1, 2, 3, 4, 5]and have been widely applied in fields such as geodesy, fundamental physics, medical diagnosis, etc [6, 7, 8, 9, 10, 11]. OFCs are different from ordinary lasers and require locking of the carrier-envelope offset frequency and the repetition frequency f_r. Common methods for locking f_r include: 1) using piezoelectric ceramic (PZT) [12]; 2) using electro-optic modulators (EOM) [13]; 3) using magneto-optic modulators (MOM) [14]. EOM can provide a high feedback bandwidth to effectively suppress high-frequency noise, but it also introduces additional dispersion and suffers from mechanical resonance due to the piezoelectric effect [15]. MOM can offer a higher feedback bandwidth than EOM, making the locking of f_r more stable, but it has a complex structure, high cost, and is difficult to be widely used [14]. PZT is a more mainstream scheme compared with the above two schemes. Although it has a smaller feedback bandwidth, it directly adjusts the cavity length, is less dependent on polarization and wavelength, and has the advantages of small size, fast response speed, no heat generation, good controllability, low cost, and simple structure, etc [16, 17].

However, the driving voltage range of ordinary PZT is limited, generally positive voltage and less than 100 volts. Under severe temperature fluctuations in the environment, even with temperature control on the OFC, the residual impact can make the f_r of the OFC drift out of the PZT control range, resulting in the inability to lock for a long time. In addition, the voltage-rise response curve of PZT is not linear in the range of zero to 100 volts. Large-scale changes in this range will cause PID feedback instability and affect the quality of signal loading. To achieve long-term locking of the f_r of OFC under severe temperature fluctuations and improve its practicability and usability, this paper proposes an OFC f_r locking scheme based on delay line, temperature control, and PZT. Using computer equipment, according to the PZT control voltage and the output voltage of the temperature control (TEC) module, we dynamically adjust the temperature control setting value and delay line setting value to make them work cooperatively to achieve long-term f_r locking of OFC.

2. Method and Principle

The cooperative feedback control system consists of three sub-servo stabilization systems: the PZT control system, the TEC system, and the delay line control system, as shown in Figure 1. PZT has the smallest adjustable range, the fastest adjustment speed, and the highest precision. TEC can only realize the slow adjustment of the cavity length. The delay line has a large adjustable range and fast adjustment speed, but its precision is slightly lower. The working principle of this system is as follows: The reference signal and the OFC signal are phase-detected by a fiber-loop optical-microwave phase detector (FLOM-PD). The error signal is obtained by filtering the FLOM-PD output signal by using a low-pass filter, which reflects the frequency difference between the f_r and the reference signal. Then, the error signal is fed into the PZT control system for feedback control. The output of the PID control circuit in the PZT control system is amplified by a high-voltage driver module (HDM), which controls the PZT in the OFC to adjust the laser cavity length, thus achieving dynamic adjustment of the OFC f_r. While the PZT control system is working, the TEC system constantly monitors the driving voltage of PZT. Suppose the driving voltage of PZT exceeds the preset range. In that case, the TEC system will operate and adjust the fiber temperature in the resonant cavity according to the deviation value of the driving voltage of PZT, thereby tuning the cavity length and maintaining the driving voltage of PZT within the preset range. Compared with the PZT control system and the TEC system, the delay line control system is usually in standby mode, and it continuously monitors both the driving voltage of the PZT and the output voltage of the TEC system. There may be two scenarios in the experiment: 1) The difference between the preset temperature of TEC and the actual fiber temperature is too significant, and TEC runs at total capacity; 2) The environmental temperature changes too sharply, the deviation of the driving voltage of PZT from the preset range is large. When these scenarios occur, the delay line control system will function and directly adjust the OFC's cavity length to ensure that the PZT driving voltage is within the set range while preventing the temperature control system from overloading. The principle is also illustrated in Figure 2.

It is worth noting that the tunable range of the fiber delay line is extensive, which often can adjust the comb f_r by several MHz, but the minimum step size of the delay line is generally at the 100 Hz level, which is far less precise than the PZT and TEC for the laser cavity length. The high-frequency adjustment will affect the f_r locking, which is why the delay line control system is generally in standby mode, with a shallow working frequency, and it only starts when the PZT control system and the TEC system are both unable to perform feedback control properly.

3. Experiments

The system diagram of the experiment is shown in Figure 3. We use homemade OFCs with a f_r of 100 MHz. The OFCs have a central wavelength of approximately 1560 nm, a spectral width of around 10 nm, and an output optical power of about 100 mW. In the experiment, the employed delay line is a reflective optical path variable fiber-type electrically controlled delay line [18]. It incorporates low-loss corner cube mirrors, high-precision one-dimensional linear displacement stages, and high-performance servo motor closed-loop control technology, enabling high-speed positioning and rapid scanning. This delay line can be directly adjusted in length using computer equipment, with an adjustable range of 0-39999 μm and a minimum step size of 3 μm. The corresponding adjustable range for repetition frequency is approximately 0-1.27 MHz, with a minimum step size of about 95 Hz. The computer device involved in the cooperative

Journal of Physics: Conference Series **2889** (2024) 012030 doi:10.1088/1742-6596/2889/1/012030

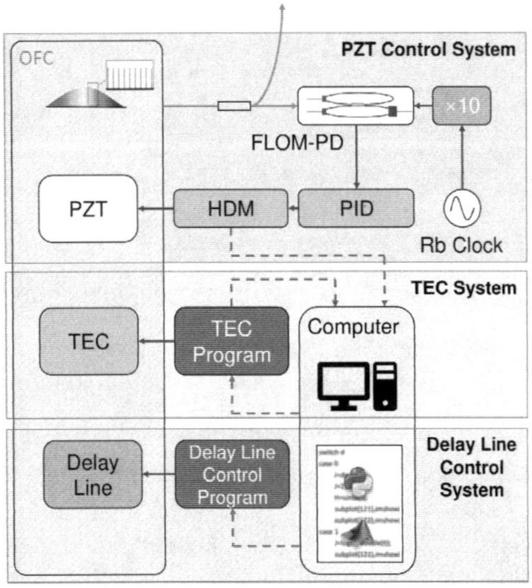

Figure 1: Schematic diagram of the locking system of the OFC with cooperative feedback control of PZT, TEC, and delay line. OFC: optical frequency comb; PZT: piezoelectric ceramic transducer; TEC: temperature control; PID: proportional integral derivative control system; HDM: high-voltage driver module.

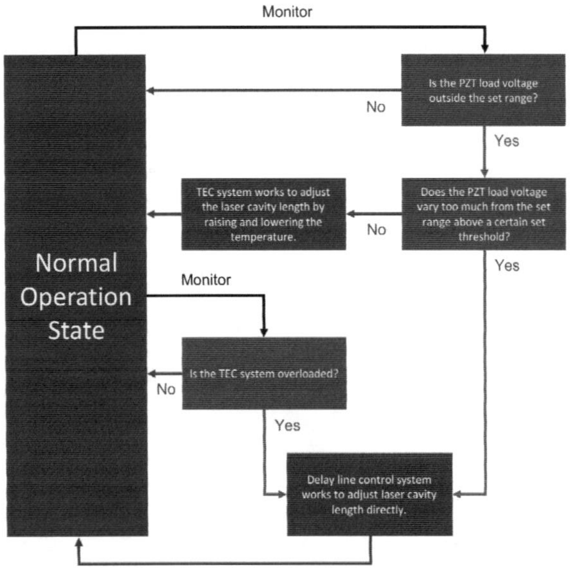

Figure 2: Cooperative feedback control system working principle diagram

feedback control was an ordinary laptop. There were two versions of the complete program, which had the same functionality, but one was written in Matlab and the other in Python. The program used in this experiment was the Python version. Due to the experimental conditions, we only locked the f_r of the OFCs in the experiment. We turned off the air conditioners in the laboratory and performed a long-term f_r locking test on the OFCs under the condition of diurnal

Journal of Physics: Conference Series **2889** (2024) 012030 doi:10.1088/1742-6596/2889/1/012030

temperature difference exceeding 5°C and severe environmental temperature fluctuation. One OFC used the cooperative feedback control system for f_r locking, while the other did not. The test method was to detect the output signal of the OFCs by a photodetector and use a phase noise analyzer to obtain the phase drift between them and the output of the Rb clock, and then calculate their Allan variances using Stable32.

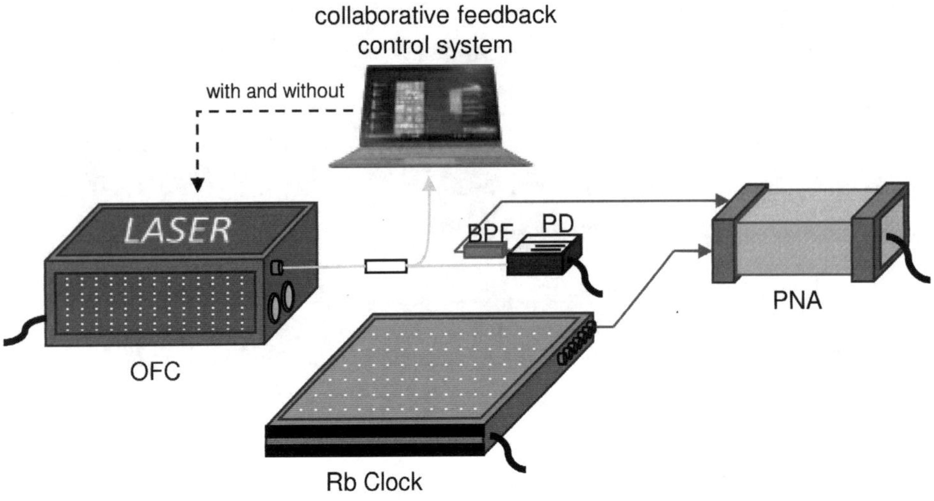

Figure 3: Schematic diagram of the cooperative feedback system test. OFC: optical frequency comb; BPF: band-pass filters with a center frequency of 100 MHz and bandwidth of 10 MHz; PD: photoelectric detector; PNA: phase noise analyzer.

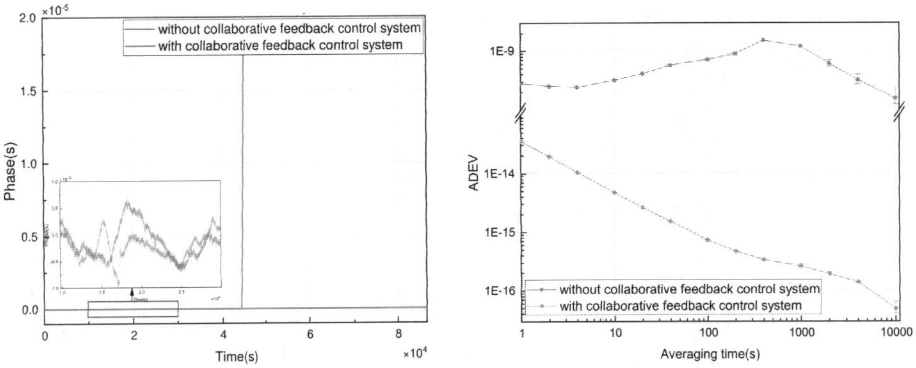

(a) (b)

Figure 4: Result of the cooperative feedback control system. (a) The phase drift; (b) Allan deviations. The red line: without cooperative feedback control system; the blue line: with cooperative feedback control system.

4. Results and Discussion

The experimental results are shown in Figure 4. It can be seen that at approximately 45,000 s, the OFC without the cooperative feedback control system lost its f_r locking, and the phase started to fluctuate violently. At the same time, the corresponding frequency instability also reached the order of 10^{-9}. We analyzed that this result was because around 45,000 s, the night came, and the outdoor temperature dropped sharply. The laboratory temperature decreased accordingly when the air conditioner was not turned on. The weak performance of the temperature control system made it difficult to ensure the stability of the fiber temperature, which caused the f_r to drift out of the PZT controllable range, resulting in the f_r unlocking. However, the OFC that used the cooperative feedback control system for f_r locking, because of the participation of the delay line in the overall dynamic adjustment, was almost unaffected by the temperature drop in terms of its f_r locking.

The experimental results show that by using the cooperative feedback system, the OFC with low PZT and temperature control performance can enable long-term stable locking of its f_r under drastic environmental temperature changes, enhancing the usability and practicality of the OFC with minimal cost.

5. Conclusion

5.1. Future work

The computer equipment used to achieve cooperative feedback control in the current experiment is a standard laptop. Still, in fact, the system has low requirements for the performance of computer equipment. Devices can replace it with smaller sizes and lower cost, such as microcontrollers, which can also meet the demand. In the future, we envision integrating and modularizing the cooperative feedback control system to make it easier to use and lower cost.

Meanwhile, in principle, the cooperative feedback control system can be used not only on our homemade OFCs but also on other OFCs that have two or more tunable OFC f_r systems. We will improve the versatility of the cooperative feedback control system in the future to be used on more OFCs.

5.2. Summary

We used the cooperative feedback control system to achieve stable f_r locking of OFC for more than 24 hours in a laboratory environment with temperature change exceeding 5°C. The research results show that the cooperative feedback system makes it possible to use OFC under severe temperature changes.

In this work, we do not need to replace or upgrade the devices of OFC itself, nor do we need high-performance computer equipment to achieve long-term repetition frequency locking of OFC in an environment with severe temperature change, which has strong practicability and reduces the cost. Based on this technology, it is expected to lower the threshold for using OFC in the future and provide technical support for the rational commercialization of OFC.

6. Acknowledgment

The research is supported by National Natural Science Foundation of China (Grants Nos. 62201012, and 61531003) and National Hi-Tech Research and Development (863) Program.

References

[1] Newbury N R and Swann W C 2007 *J. Opt. Soc. Am. B* **24** 1756–1770
[2] Schibli T, Hartl I, Yost D, Martin M, Marcinkevičius A, Fermann M and Ye J 2008 *Nat. Photonics* **2** 355–359
[3] Giorgetta F R, Swann W C, Sinclair L C, Baumann E, Coddington I and Newbury N R 2013 *Nat. Photonics* **7** 434–438
[4] Fortier T and Baumann E 2019 *Commun. Phys.* **2** 153

[5] Chang L, Liu S and Bowers J E 2022 *Nat. Photonics* **16** 95–108
[6] Minoshima K and Matsumoto H 2000 *Appl. Opt.* **39** 5512–5517
[7] Yamada K, Onae A, Hong F L, Inaba H, Matsumoto H, Nakajima Y, Ito F and Shimizu T 2008 *J. Mol. Spectrosc.* **249** 95–99
[8] Wang C and Sahay P 2009 *Sens.* **9** 8230–8262
[9] Diddams S A 2010 *J. Opt. Soc. Am. B* **27** B51–B62
[10] Shen Q, Guan J Y, Ren J G, Zeng T, Hou L, Li M, Cao Y, Han J J, Lian M Z, Chen Y W *et al.* 2022 *Nature* **610** 661–666
[11] Caldwell E D, Deschenes J D, Ellis J, Swann W C, Stuhl B K, Bergeron H, Newbury N R and Sinclair L C 2023 *Nature* **618** 721–726
[12] Denisov V I, Ivanenko A V, Nyushkov B N and Pivtsov V S 2008 *Quantum Elec.* **38** 801
[13] Hudson D D, Holman K W, Jones R J, Cundiff S T, Ye J and Jones D J 2005 *Opt. Lett.* **30** 2948–2950
[14] Nakamura T, Tani S, Ito I and Kobayashi Y 2017 *Opt. Express* **25** 4994–5000
[15] Nyushkov B N, Pivtsov V S, Koliada N A, Kaplun A B and Meshalkin A B 2015 *Quantum Elec.s* **45** 486
[16] Briles T C, Yost D C, Cingöz A, Ye J and Schibli T R 2010 *Opt. Express* **18** 9739–9746
[17] Ma Y, Xu B, Ishii H, Meng F, Nakajima Y, Matsushima I, Schibli T R, Zhang Z and Minoshima K 2018 *Opt. Lett.* **43** 4136–4139
[18] Zhou L, Wang X, Lu L and Chen J 2018 *CHIN OPT LETT* **16** 101301

REFIMEVE frequency and time network and applications

E Cantin[1], O Lopez[1], C Chardonnet[1], A Amy-Klein[1], M Rabault[2], G Coget[2], P Rosenbusch[2], V Ménoret[2], N Quintin[3], M Tønnes[4], B Pointard[4], M Mazouth-Laurol[4], H Alvarez-Martinez[4,5], C B Lim[4], P Tuckey[4], M Abgrall[4], R Le Targat[4], P-E Pottie[4]

[1] Laboratoire de Physique des Lasers, Université Sorbonne Paris Nord, CNRS, 99 avenue Jean-Baptiste Clément, 93430 Villetaneuse, France
[2] Exail SAS - Quantum Systems Division, Institut d'Optique d'Aquitaine, rue François Mitterrand, 33400 Talence, France
[3] RENATER, 23-25 rue Daviel, 75013 Paris, France
[4] LNE-SYRTE, Observatoire de Paris, Université PSL, CNRS, Sorbonne Université, 61 avenue de l'observatoire, 75014 Paris, France
[5] Real Instituto y Observatorio de la Armada, Cecilio Pujazón s/n, E-11110 San Fernando, Spain

email: amy@univ-paris13.fr

Abstract. We present the development and current status of the REFIMEVE national research infrastructure, which disseminates time and frequency reference signals generated at LNE-SYRTE to research labs all over France. Using mainly the optical fiber backbone of RENATER, the French National Research and Education Network, REFIMEVE currently provides an optical frequency reference to around 15 labs, and to connection points to Germany, the UK, Italy and Switzerland. The robustness of the various types of equipment specifically developed for the transfer of ultrastable frequencies and the efficient supervision system have enabled us to demonstrate uptimes of over 90%. We show the latest development we have carried out to extend the network to CERN in Switzerland, with a branching link showing residual frequency fluctuations of 3×10^{-15} for 1 s measurement time, and a few 10^{-19} at long term. We discuss the issue of data processing for long periods of time and how to mitigate the effects of missing data to achieve reliable link uncertainty estimates. Thanks to these performances, the REFIMEVE network has been used for the precise comparison of primary and optical clocks in Europe, and the REFIMEVE optical signal is currently being exploited in France by user laboratories for photonics, laser stabilisation or control and atomic and molecular high-resolution spectroscopy.

1. Introduction

In order to compare two distant clocks, one possibility is to physically transport a third clock between them. This method is still very powerful today, but may be limited because mobile clocks usually have lower performance than static ones [1,2]. The alternative of transmitting the clock information encoded on a carrier signal of some sort is very attractive for its practical simplicity, and its ability to be broadcast to multiple users. Therefore, since the first transatlantic radio-broadcast by Marconi in 1902, radio-emission based clock comparisons using GNSS or telecommunications satellites have become the most used techniques for high-performance applications such as the realization of International atomic time (TAI) and Universal coordinated time (UTC) [3,4]. However, the satellite-based comparison methods are not sufficient for comparing optical clocks because their contribution to the measurement uncertainty is greater than that of the clocks (for all currently practicable measurement durations). For the past two decades optical fiber links have been explored as a very promising means to achieve the goal of ultra-precise clock comparisons, with almost no degradation of the clock performances.

The principle of fiber links is quite simple, and was described in the late 70s and 80s (see for example [5]). Today the basic principles are the same, but for the highest-stability links the implementation has progressed significantly: an ultrastable laser operated in the C-band generates the optical reference signal, which feeds into a single mode fiber link, which carries it to the far end of the link. At the far end, part of the signal is injected back into the same fiber, causing it to undergo a complete round trip. A highly imbalanced Michelson interferometer is used to detect the phase variation of the light that experiences the round-trip in the fiber, compared to the light that experiences a round-trip in the short reference arm [6]. In such links it is crucial that the light propagates in both directions in the same fiber, since this allows the propagation noise to be very well measured and compensated.

To evaluate the performance of the frequency transfer to the far end of the link, a second bidirectional fiber link is commonly used to transfer the transmitted reference signal back to the initial point. One can then measure the optical phase difference between the re-transferred optical frequency reference and the ultrastable laser signal feeding the first link. This so-called end-to-end measurement enables us to obtain an upper value of the residual frequency fluctuations of the link and the uncertainty of the frequency transfer [7].

Fiber links have numerous advantages. First of all, guided propagation allows low-loss transmission, about 0.2 to 0.3 dB/km. In addition, noise is relatively low, and even more importantly has excellent statistical properties: the noise averages very well toward zero from 1 s to years of measurement time. The main advantage is the excellent reciprocity of paths for signals travelling forward and backward in the same fiber. This allows a coherent regime to be achieved, where relative frequency stability scales down as the inverse of the integration time, or better, which is faster than with satellite-based methods. Fiber links also have some limitations. The fundamental limit set by the propagation delay in the fiber, limits both the noise rejection amplitude and the bandwidth of the servo-loop that corrects the propagation phase noise [8]. More technical limits are related to the uncontrolled polarisation of the light in single-mode fiber and to the instrumental noise floor of the interferometers [8]. The main practical disadvantage of coherent fiber links is that they are limited to point-to-point or point-to-a-few dissemination [9]. There is also a very attractive prospect of using fiber links in uni-directional configurations, but this topic is still little explored so far and is the subject of debate [8,10,11].

The range of fiber links has been extended up to thousands of km; they can be concatenated together and joined to form large networks [12,13]. Fiber links have reached a high technology readiness level [14]. Currently the main challenge for fiber links lies in accessing existing fiber infrastructure designed for high rate data transfer, or alternatively in creating dedicated fiber infrastructure. In terms of performance, fiber links outperform current satellite-based methods by several orders of magnitude. To date, free-space optical transfer techniques are the only alternative for achieving better than 10^{-18} resolution [15,16]. The technological breakthrough constituted by fiber links opens many new scientific opportunities: clock comparisons [13,17], chronometric geodesy [1,18], tests of special relativity [19], the search for dark matter [20], atomic and molecular spectroscopy [21–24], seismology [25,26] and VLBI [27].

2. Current status of REFIMEVE

The REFIMEVE network is a national research infrastructure designed to disseminate ultrastable time and frequency signals from LNE-SYRTE to academic labs, research institutes and companies throughout France, with connections to other European National metrological Institutes. It totalizes currently 3800 km of pairs of fiber links, as shown on Figure 1, where links in operation are displayed with green solid lines and future links with green dotted lines. It has 4 European connections, to the fiber links of UK, Germany and Italy, and to CERN. The ultrastable signals generated at LNE-SYRTE are distributed through the fiber network of RENATER, the French National Research and Education

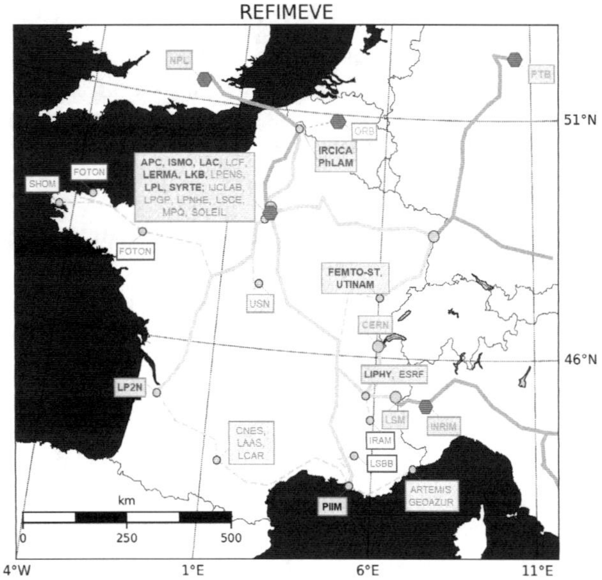

Figure1: current status of the REFIMEVE research infrastructure. Continuous green lines: operational REFIMEVE links, dotted green lines: future REFIMEVE links, blue lines: European links to which REFIMEVE is connected, blue dotted line: future European link. Currently-connected labs are indicated in green, labs which will be connected in the future are shown in grey.

Network (NREN). The ultrastable signals are multiplexed onto the same fibers which carry RENATER's digital data. Two signals are currently disseminated by REFIMEVE, an ultrastable optical frequency reference and a White Rabbit time and RF frequency reference, the latter being disseminated only in the Paris region at present.

The optical frequency reference is generated from an ultrastable laser that is actively de-drifted versus either a cryogenic oscillator or an H maser [28]. It has a fixed frequency, known to within a range of 1 Hz when controlled to the cryogenic oscillator, with a stability below 10^{-15} at 1 s averaging time and a routine uncertainty of a few 10^{-15}. As already discussed, for the non-degraded dissemination of this signal with an optical link, the key issue is the use of bidirectional light propagation within a single fiber. This is unusual for telecommunication networks, which are based on the unidirectional propagation of light using two fibers, one for each propagation direction. Thus, every (unidirectional) piece of equipment of the telecommunication network has to be bypassed using Optical Add-and-Drop Multiplexers (OADM). The second fiber of the telecommunication pair can be used for the "backwards" bidirectional optical frequency reference transfer link, allowing end-to-end measurement as discussed above.

The REFIMEVE optical frequency signal is amplified using low-noise bidirectional amplifiers, which are part of the technology development and knowledge transfer carried out within REFIMEVE. To build a network on a national scale, we use Repeater Laser Stations (RLS) and Multi-branch Laser Stations (MLS) for signal regeneration and fiber noise compensation, using a cascaded approach that provides better propagation noise rejection compared with using the longest possible fiber spans. These devices are commercially available, following their knowledge transfer to Exail [14]. The RLS are placed every 300 to 500 km on the fiber network and include a laser, which is controlled to copy the optical phase of the incoming signal and distribute it to the next optical link. In addition, a user output enables the ultrastable signal to be distributed to a local user or used to characterize the metrological performance of the transported signal. The MLS are based on the same design as the RLS but have several phase-coherent outputs, allowing signals to be distributed to multiple links to build a star network configuration [9].

As a national research infrastructure, we have to achieve the highest possible uptime and control the proper frequency transfer to users, but with very limited human resources; we therefore aim to reduce human intervention in the operation of the network as much as possible. To that aim, we have developed automated servo loops and a supervision system. This allows remote operation of the network for parameter tuning and optimization and gives us the ability to monitor all links in real time. In addition, as the majority of REFIMEVE uses RENATER's active telecommunications network, this supervision enables RENATER's Network Operating Centre to control the lasers and equipment in the metrology network and thus to have the possibility of switching them off. This is a prerequisite for sharing a network with a telecommunications operator. Finally, the supervision system includes an automated data processing and management system and the possibility to raise alarms to users.

3. Performance of REFIMEVE

3.1. Robustness of the REFIMEVE network

Figure 2 shows an example of the operation of 4 national links and 1 regional link of REFIMEVE during 73 days in late 2022. Each graph displays the residual frequency fluctuations measured using the forward and backward links, with an averaging time of 1 s. The link lengths range from 2×50 km to 2×900 km and exhibit residual fluctuations which increase with this length and range from a few 10^{-15} to a few 10^{-14}. Note that the Lyon-Marseille link (red trace) has a quite high noise level despite not being very long. It is in fact a branching link on the main Paris-Modane link (as depicted in Figure 3 a)) and therefore it is affected by some of the residual noise from this main link. Thanks to the remote supervision and control of the network equipment, as well as the maintenance of the fiber network by RENATER, the uptime of the links is quite high, ranging from 73% to 91%. Each interruption has been analyzed in order to improve the operation. For instance, the long interruption on the Paris-Modane link (green trace) was due to unusual noise, undoubtedly attributed to work along the fiber route. Some interruptions occurred on all 4 or 5 links and are attributed to a failure of one of the master MLS of the network at SYRTE or in the main datacenter in Paris. Finally, the main interruptions on the Lyon-Marseille link (red trace) (showing a short and a long operating period) and on the Paris-Orsay link (purple trace) are due to a failure of the end-to-end instability characterization, but the signal was still disseminated to the remote labs and used there.

These traces show the robustness of the REFIMEVE network, which we are continuing to improve with the aim of achieving an uptime of over 90% on all the links, and over 99% during dedicated periods.

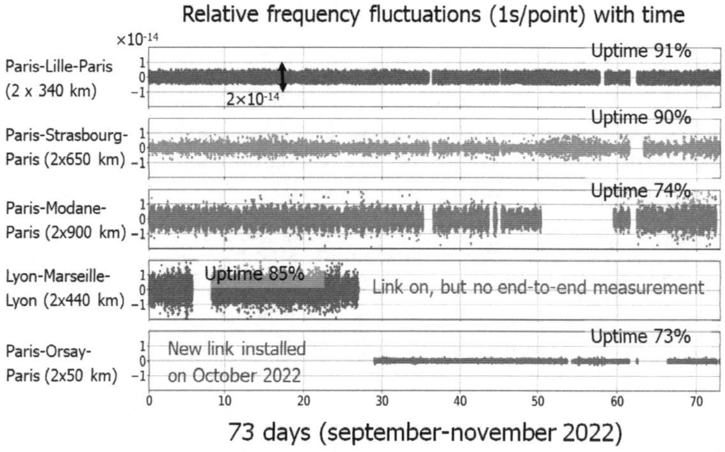

Figure 2: relative frequency fluctuations of 5 links versus time during 73 days in 2022, with their uptimes. Averaging time: 1s/point.

3.2. Example of the SYRTE-CERN link

Here we focus on the latest link implemented on the REFIMEVE network, connecting SYRTE (Paris) to CERN (Geneva, Switzerland). This is a branching link of 200 km along the long, cascaded link of 1000 km between Paris and Italy, as shown in Figure 3 a). The residual instability of the branch link to CERN and back (2×200 km) is displayed in Figure 3 b) for a data set of 5 days with an uptime of 97 %, showing that the link is quite robust. The stability is 3×10^{-15} at 1 second averaging time, it reaches the 10^{-18} level in less than 2 hours, and 3×10^{-19} at 10^5 seconds.

This residual instability is higher than that of other links of similar length in the REFIMEVE network. The noise is in fact increased by the residual noise of the previous segments of the main link, this residual noise being quite high since it is increasing with the length of the link. This stability could be improved in the future with a better signal regeneration in Grenoble, using for instance an ultrastable laser as a regeneration laser instead of a low-noise laser diode. The uncertainty in the transferred frequency is estimated from the mean relative frequency and the long-term instability to be $(1.4 \pm 4.2) \times 10^{-19}$.

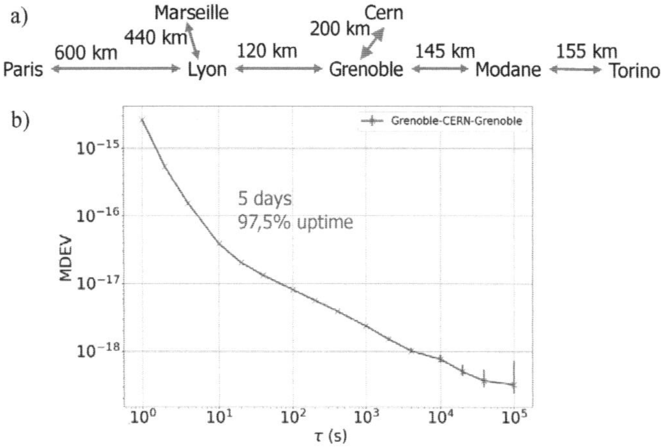

Figure 3: a) schematic of the link from Paris (SYRTE) to Geneva (CERN); b) residual instability (modified Allan deviation) of the link from Grenoble to CERN and back.

3.3. Noise study over several years

We took advantage of the fact that a REFIMEVE link of 700 km, connecting SYRTE to the German border, has been operating since 2015, to study the noise statistics over a very long period [29,30]. For this study we face the issue of missing data, since the link did not operate continuously throughout the period, due to fiber replacement or other interruptions. A basic approach to data processing consists in concatenating the data, but in this case the temporal behaviour is modified, especially for long interruptions. Another possibility is to analyse the link during a long operational period without interruption, in order to estimate the noise parameters. This then enables us to simulate the noise and replace missing data with simulated noise. These various processing methods were used and the mean relative frequency offsets of the processed data are displayed on Figure 4, together with the corresponding modified Allan deviations. The lowest uncertainty is obtained when missing data are replaced with simulated data. In this case the measured frequency offset is compatible with the one obtained using concatenated data. Similarly, when replacing missing data by simulated data, we obtain a stability which is compatible with the short-term stability, and which is lower than that obtained with concatenation. Finally, this study gives a long-term stability below 10^{-20} at 10^7 s, whatever the processing method used, and in the low 10^{-21} for an averaging time of one year, when missing data are replaced with simulated noise. As far as we know, this is the longest optical link data set ever studied, containing 786 million points after filtering of outliers, giving us the possibility to compare the processing methods up to a third of a year of averaging time.

4. Applications of REFIMEVE

Thanks to its connection to 3 neighbouring European links, REFIMEVE enables European clocks to be compared with the best possible uncertainty, which is clearly below 10^{-18} as recommended by the BIPM [31]. The first European clock comparisons by optical fiber link occurred in 2015, comparing the Sr optical clocks of SYRTE and PTB through a 1400-km link [17]. The clocks frequencies agreed, with an uncertainty of $(4.7\pm5)\times10^{-17}$ after taking into account the gravity potential correction. The comparison uncertainty was not limited by the link residual fluctuations, which contributed only 2×10^{-19}. This seminal clock comparison was followed by numerous comparison campaigns, mainly in the frame of the EURAMET projects OFTEN and ROCIT (see for instance [13]). These campaigns are based on the

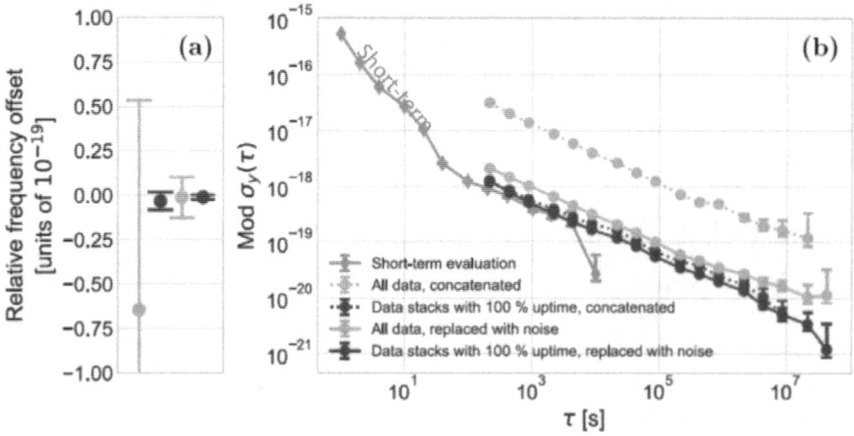

Figure 4: (a): Relative frequency offsets and (b): modified Allan deviations (MDEV) of the Paris-Strasbourg link (2x700 km) over 5 years, calculated in 4 different ways: from all data stacks (green), only data stacks with 100 % uptime (blue), all data stacks where missing data is replaced with simulated noise (light brown), and only data stacks with 100 % uptime where missing data is replaced with simulated noise (purple) (from [29]).

star network between SYRTE and NPL, PTB and INRIM and on the technology of the RLS developed as part of REFIMEVE, using for instance two-way frequency comparison of the RLS user outputs. These comparison campaigns are essential for T&F metrology and the future redefinition of the second [31]. In addition, they enable fundamental test of physics to be carried out such as the search for dark matter [20] or tests of special relativity [19]. For these comparisons, it is very important that the residual instability and uncertainty of optical links are lower than those of the clocks, and this is the case, with uncertainty in the low 10^{-19} range for all REFIMEVE links.

Beyond clock comparisons, REFIMEVE has enabled partner laboratories to carry out a wide range of precision measurements, from high-resolution molecular spectroscopy [32–34] to photonics and noise characterization (see for instance [35]). In addition, the network itself can be exploited as a giant and distributed noise sensor, both for sensing seismic noise [25,30] or tiny Earth rotation fluctuations through the Sagnac effect [36]. Figure 5 gives an example of the detection of an earthquake of magnitude

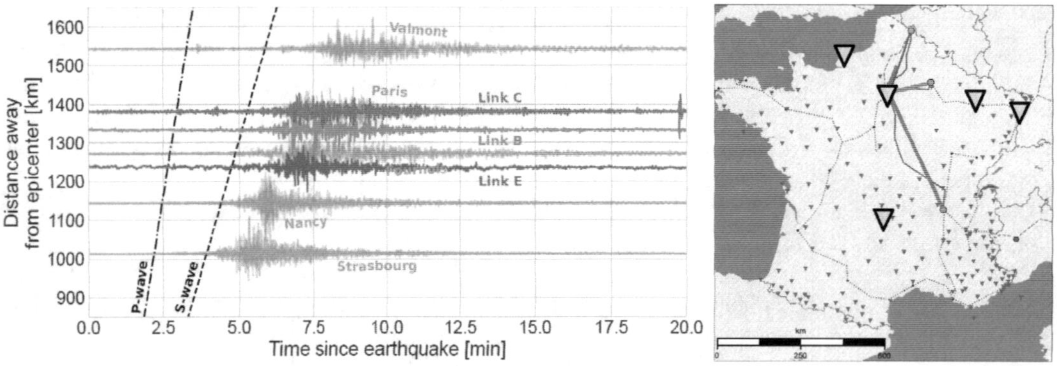

Figure 5: example of earthquake sensing with 3 links of the REFIMEVE network (green, orange and purple signals) and comparison with 5 standard seismometers (pale-coloured signals). The links and the seismometer positions are shown on the map on the right (from [29]).

5.7 in Bosnia using three links of the REFIMEVE network. The detected signals compare very well with those of low-noise broadband seismometers (such as the Trillium Compact), with similar or lower sensitivity for a 0.5 Hz bandwidth in the best case.

5. Conclusion

To conclude, the REFIMEVE research infrastructure currently connects 15 labs throughout France, and also provides cross-connections to 4 European countries, enabling clock comparisons with the UK, Germany and Italy. The network will soon be extended to the south and west of France and new European connections are planned, for instance with the Royal Observatory of Belgium (BOOSTED project).

Current developments also involve upgrading the network, which will be equipped with MLS at each node of the network, to simplify multi-user dissemination. Ultrastable lasers could be also set up in key network nodes to improve signal regeneration.

The service provided by REFIMEVE is currently being extended to RF reference and time transfer, using White Rabbit technology. Currently the entire Paris region already benefits from a White Rabbit signal, and this coverage will soon be extended to a radio-observatory in central France.

Beyond fundamental metrology, REFIMEVE has a wide range of applications. Many of them concern tests of fundamental physics such as the search for variations of the fundamental constants or for dark matter, and high-precision spectroscopy. In the future, we anticipate extending the applications to sea-level monitoring with chronometric levelling, earthquake sensing, and synchronization for astrophysics and high-energy physics experiments.

Funding

The authors have received financial support from Agence Nationale de la Recherche ANR T-REFIMEVE (ANR-21-ESRE-0029), REFIMEVE+ (ANR-11-EQPX-0039), Labex First-TF (ANR-10-LABX-48-01), Idex PSL (ANR-10-IDEX-0001-02 PSL), LNE, Conseil Régional Bourgogne-Franche-Comté, Conseil Régional Ile de-France (DIM SIRTEQ), the European Metrology Programme for Innovation and Research (EMPIR) in project 15SIB05 (OFTEN) and 18SIB05 (ROCIT). EMPIR is jointly funded by the EMPIR participating countries within EURAMET and the European Union.

Acknowledgements

We acknowledge the strong support of RENATER, and especially Emilie Camisard and Laurent Gydé. We are grateful to Samir Kassi and co-workers (LIPHY) for their commitment in hosting the equipment for the Grenoble-CERN link.

References

[1] Grotti J *et al.* 2018 Geodesy and metrology with a transportable optical clock *Nature Physics* **14** 437–41
[2] Takamoto M, Ushijima I, Ohmae N, Yahagi T, Kokado K, Shinkai H and Katori H 2020 Test of general relativity by a pair of transportable optical lattice clocks *Nature Photonics* 1–5
[3] Panfilo G and Arias F 2019 The Coordinated Universal Time (UTC) *Metrologia* **56** 042001
[4] Jiang Z, Zhang V, Parker T E, Petit G, Huang Y-J, Piester D and Achkar J 2019 Improving two-way satellite time and frequency transfer with redundant links for UTC generation *Metrologia* **56** 025005
[5] Primas L, Lutes G and Sydnor R 1988 Fiber optic frequency transfer link *Proceedings of the 42nd Annual Frequency Control Symposium, 1988.* Proceedings of the 42nd Annual Frequency Control Symposium, 1988. pp 478–84
[6] Ma L-S, Jungner P, Ye J and Hall J L 1994 Delivering the same optical frequency at two places: accurate cancellation of phase noise introduced by an optical fiber or other time-varying path *Optics Letters* **19** 1777

[7] Lopez O, Haboucha A, Kéfélian F, Jiang H, Chanteau B, Roncin V, Chardonnet C, Amy-Klein A and Santarelli G 2010 Cascaded multiplexed optical link on a telecommunication network for frequency dissemination *Opt. Express, OE* **18** 16849–57

[8] Williams P A, Swann W C and Newbury N R 2008 High-stability transfer of an optical frequency over long fiber-optic links *Journal of the Optical Society of America B* **25** 1284

[9] Cantin E, Tønnes M, Targat R L, Amy-Klein A, Lopez O and Pottie P-E 2021 An accurate and robust metrological network for coherent optical frequency dissemination *New J. Phys.* **23** 053027

[10] Bercy A, Stefani F, Lopez O, Chardonnet C, Pottie P-E and Amy-Klein A 2014 Two-way optical frequency comparisons at 5×10^{-21} relative stability over 100-km telecommunication network fibers *Physical Review A* **90** 061802(R)

[11] Xu D, Lopez O, Amy A and Pottie P-E 2021 Polarization scramblers to solve practical limitations of frequency transfer *J. Lightwave Technol.* **39** 3106

[12] Raupach S M F, Koczwara A and Grosche G 2015 Brillouin amplification supports 1×10^{-20} uncertainty in optical frequency transfer over 1400 km of underground fiber *Physical Review A* **92**

[13] Clivati C *et al.* 2022 Coherent Optical-Fiber Link Across Italy and France *Phys. Rev. Appl.* **18** 054009

[14] Guillou-Camargo F *et al.* 2018 First industrial-grade coherent fiber link for optical frequency standard dissemination *Appl. Opt., AO* **57** 7203–10

[15] Dix-Matthews B P, Schediwy S W, Gozzard D R, Savalle E, Esnault F-X, Lévèque T, Gravestock C, D'Mello D, Karpathakis S, Tobar M and Wolf P 2021 Point-to-point stabilized optical frequency transfer with active optics *Nat Commun* **12** 515

[16] Caldwell E D, Deschenes J-D, Ellis J, Swann W C, Stuhl B K, Bergeron H, Newbury N R and Sinclair L C 2023 Quantum-limited optical time transfer for future geosynchronous links *Nature* **618** 721–6

[17] Lisdat C *et al.* 2016 A clock network for geodesy and fundamental science *Nature Communications* **7** 12443

[18] Takano T, Takamoto M, Ushijima I, Ohmae N, Akatsuka T, Yamaguchi A, Kuroishi Y, Munekane H, Miyahara B and Katori H 2016 Geopotential measurements with synchronously linked optical lattice clocks *Nature Photon* **10** 662–6

[19] Delva P *et al.* 2017 Test of Special Relativity Using a Fiber Network of Optical Clocks *Physical Review Letters* **118** 221102

[20] Roberts B M *et al.* 2020 Search for transient variations of the fine structure constant and dark matter using fiber-linked optical atomic clocks *New Journal of Physics* **22** 093010

[21] Friebe J, Riedmann M, Wübbena T, Pape A, Kelkar H, Ertmer W, Terra O, Sterr U, Weyers S, Grosche G, Schnatz H and Rasel E M 2011 Remote frequency measurement of the1S0->3P1transition in laser-cooled 24Mg *New J. Phys.* **13** 125010

[22] Matveev A *et al.* 2013 Precision Measurement of the Hydrogen 1 S − 2 S Frequency via a 920-km Fiber Link *Physical Review Letters* **110** 230801

[23] Argence B, Chanteau B, Lopez O, Nicolodi D, Abgrall M, Chardonnet C, Daussy C, Darquié B, Le Coq Y and Amy-Klein A 2015 Quantum cascade laser frequency stabilization at the sub-Hz level *Nature Photonics* **9** 456–60

[24] Clivati C *et al.* 2016 Measuring absolute frequencies beyond the GPS limit via long-haul optical frequency dissemination *Optics Express* **24** 11865

[25] Marra G, Clivati C, Luckett R, Tampellini A, Kronjäger J, Wright L, Mura A, Levi F, Robinson S, Xuereb A, Baptie B and Calonico D 2018 Ultrastable laser interferometry for earthquake detection with terrestrial and submarine cables *Science* **361** 486–90

[26] Marra G *et al.* 2022 Optical interferometry–based array of seafloor environmental sensors using a transoceanic submarine cable *Science* **376** 874–9

[27] Clivati C *et al.* 2017 A VLBI experiment using a remote atomic clock via a coherent fibre link *Scientific Reports* **7** 40992

Journal of Physics: Conference Series **2889** (2024) 012031
doi:10.1088/1742-6596/2889/1/012031

[28] Guena J *et al.* 2012 Progress in atomic fountains at LNE-SYRTE *IEEE Transactions on Ultrasonics, Ferroelectrics, and Frequency Control* **59** 391–409

[29] Tønnes M B K, Schuller F, Cantin E, Lopez O, Le Targat R, Amy-Klein A and Pottie P-É 2022 Coherent fiber links operated for years: effect of missing data *Metrologia* **59** 065004

[30] Tønnes M B K 2022 Understanding and exploiting metrological fiber networks for Sagnac sensing, frequency dissemination and geosensing, *PhD thesis, Université PSL*

[31] Dimarcq N *et al.* 2023 Roadmap towards the redefinition of the second *Metrologia* **61** 012001

[32] Santagata R *et al.* 2019 High-precision methanol spectroscopy with a widely tunable SI-traceable frequency-comb-based mid-infrared QCL *Optica* **6** 411–23

[33] Votava O, Kassi S, Campargue A and Romanini D 2022 Comb coherence-transfer and cavity ring-down saturation spectroscopy around 1.65 μm: kHz-accurate frequencies of transitions in the $2\nu_3$ band of $^{12}CH_4$ *Phys. Chem. Chem. Phys.* **24** 4157-4173

[34] Leuliet M, Mbardi A, Argence B, Karr J-P and Hilico L 2023 SI-referenced formic acid (HCOOH) spectroscopy at the sub ppt level, submitted to publication

[35] Manamanni K, Steshchenko T, Wiotte F, Le Targat R, Abgrall M, Lopez O, Cantin E, Pottie P-É, Amy-Klein A, Roncin V and Du-Burck F 2022 Limitations due to residual interference in a fiber-based optical frequency reference at 1.55 μm *J. Opt. Soc. Am. B* **39** 438

[36] Clivati C, Calonico D, Costanzo G A, Mura A, Pizzocaro M and Levi F 2013 Large-area fiber-optic gyroscope on a multiplexed fiber network *Optics Letters* **38** 1092

Journal of Physics: Conference Series **2889** (2024) 012032 | doi:10.1088/1742-6596/2889/1/012032

Progress on the optically detected magnetic-state-selected cesium beam clock

Yuanhao Li, Sifei Chen, Chen Liu, Lifeng Fan, Chaojie Li, Yanhui Wang

Institute of Quantum Electronics, School of Electronics, Peking University, Beijing, China

E-mail: wangyanhui@pku.edu.cn

Abstract. The optically detected magnetic-state-selected cesium beam clock is a new type of compact cesium clock developed by Peking University. The principle and performance of our clock are introduced in the article as well as our recent researches on beam optics, signal-to-noise ratio, spectrum linewidth and light shift.

1. Introduction

Commercial compact cesium clocks are widely used in the fields of communication, timekeeping, navigation and so on. We have developed a new type of compact cesium clock: the optically detected magnetic-state-selected cesium clock (OMCC). In OMCC, the cesium atoms in $6S_{1/2}$ $|F = 3\rangle$ are selected by a non-uniform magnetic field and the state detection is realized by using a laser that resonates with D_2 line $|F = 4\rangle - |F' = 5\rangle$. The idea was proposed by Wang's team from Peking University in 2009, and after years of development, the 4U size prototype with stability of 4.0×10^{-13}@100 s, 4.5×10^{-14}@10000 s and 2.2×10^{-14}@1 d is realized [1, 2].

In this article, we introduce the principle and performance of OMCC, and an improvement version of OMCC, Reversed OMCC, which using laser that resonates with D_2 line $|F = 3\rangle - |F' = 2\rangle$. Reversed OMCC has a better signal-to-noise ratio and a narrower linewidth. The recent research on beam optics and two new methods on light shift suppressing are also introduced.

2. Principle and Performance of OMCC

Fig. 1 shows the structure of OMCC. A cesium oven with a collimator is used to generate atomic beam. The magnets generate a non-uniform magnetic field to select the atoms. Atoms of different energy levels own different paths in the non-uniform magnetic field because of the different effective magnetic moments. After passing through the magnets, the atoms in the $|F = 3\rangle$ state enter the microwave cavity. The microwave cavity is a Ramsey cavity, narrowing the microwave spectrum linewidth. The length of the microwave interaction zone and the drift zone is $l = 0.01$ m, $L = 0.16$ m. A weak magnetic field (usually called C-field) distributed inside the microwave cavity is used to split the magnetic sublevels. In the detection zone, a laser resonates with the 4-5 cycling transition line (see in Fig. 2) interacts with the atoms and a photo diode (PD) is used to detect atomic fluorescence. The fluorescence magnitude reflects the density of atoms in $|F = 4\rangle$, so that the microwave frequency can be locked by the PD signal.

Journal of Physics: Conference Series **2889** (2024) 012032 doi:10.1088/1742-6596/2889/1/012032

Distributed feedback (DFB) diode is used to generate laser (Fig. 1b) as DFB can avoid mode-hop over a wide current range, which is important for a compact clock and ensures reliability for long-term operation. A liquid crystal variable retarder (LCVR) is used to lock the light power. The detecting light spot diameter is about 5 mm. The beam fluorescence spectroscopy is adopted to stabilize the laser frequency [3], e.g. the atomic fluorescence is used to lock the microwave frequency and laser frequency as well.

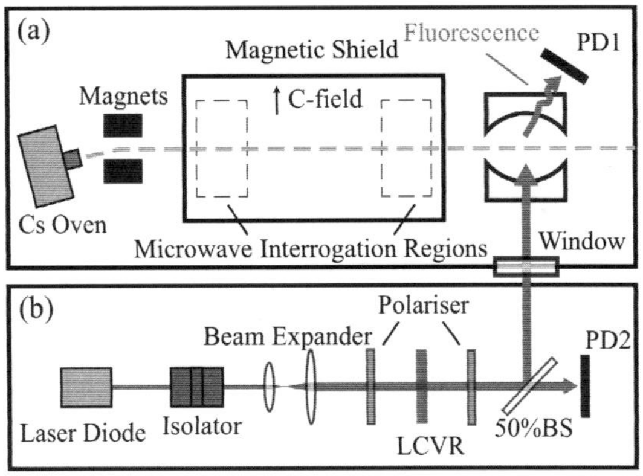

Figure 1. The structure of OMCC. (a) beam tube; (b) optical package; PD: photo diode, LCVR: liquid crystal variable retarder, BS: beam splitter.

Compared with the traditional Cs clock, OMCC avoids the complex detection system which contains the second inhomogeneous magnetic field, the ionizer and the electron multiplier [4]. Compared with the optical pumping beam clock, OMCC's optical system is simpler, because there is no pumping light [5]. It also has a narrower and lower velocity distribution resulting in a narrower linewidth.

We optimize parameters such as detecting light power, cesium oven temperature, positions of the cesium oven and beam collimator, microwave modulation frequency, microwave modulation depth, lock loop gains and so on, which improve our frequency stability gradually. Fig. 4 shows the frequency stabilty of OMCC. The best OMCC compact prototype reachs 4.0×10^{-13}@100 s, 4.5×10^{-14}@10000 s and 2.2×10^{-14}@1 d, and the SWaP (size, weight and power) is 456 mm×561 mm×177 mm (4U), $\leq$ 40 kg, $\leq$ 50 W.

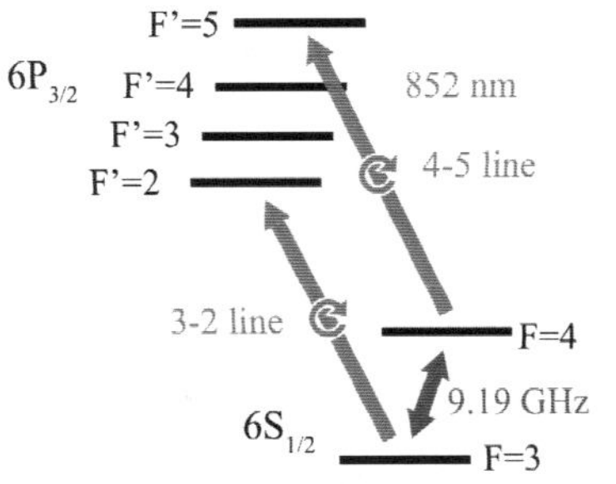

Figure 2. ^{133}Cs D_2 line enengy levels. Two cycling transition lines are labeled.

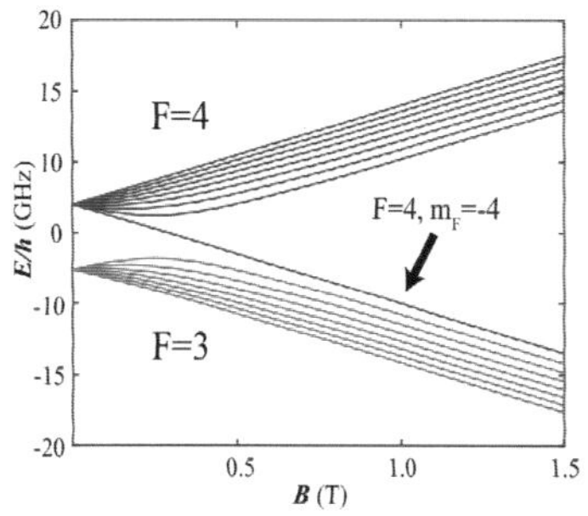

Figure 3. ^{133}Cs hyperfine structure in an external magnetic field.

However, further improvement of the frequency stability becomes more difficult due to a drawback. Fig.3 shows ^{133}Cs energy levels shift in the static magnetic field. In the magnetic state selection, when $|F=3\rangle$ atoms are selected, atoms in $|F=4, m_F=-4\rangle$ experience the same force as atoms in $|F=3\rangle$. This means that $|F=4, m_F=-4\rangle$ will follow $|F=3\rangle$ into the microwave cavity and reach the detection zone. These atoms do not interact with the microwave field but contribute to the detection noise. This limits the improvement of signal-to-noise ratio (SNR) in OMCC. Therefore, we adjusted the energy levels used for state selection and detection. Atoms in $|F=4\rangle$ state is selected and laser that resonates with D_2 3-2 line is used to detect atoms. We call the new scheme "Reversed OMCC", which will be described in detail in the next section.

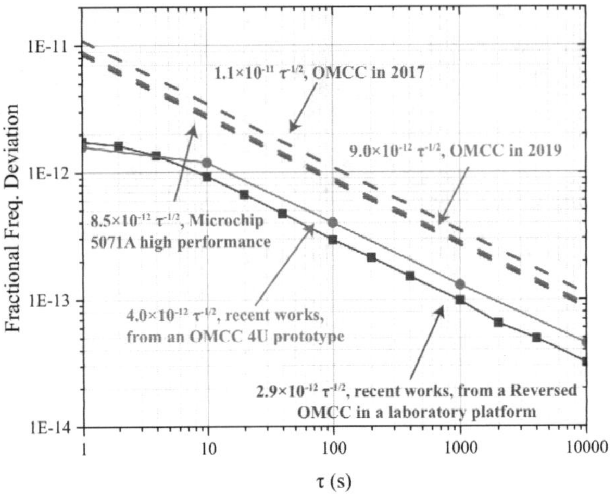

Figure 4. Allan deviation of OMCC in year 2017 [6], year 2019 [7], 5071A high performance [8], recent OMCC compact prototype and Reversed OMCC in a laboratory platform [9].

3. Improvement: Reversed OMCC

Compared with OMCC, Reversed OMCC reverses the orientation of the state-selection magnets and let atoms in $|F=4\rangle$ go into the microwave cavity. Laser that resonates with D_2 3-2 line ($|F=3\rangle-|F'=2\rangle$) is used to detect atoms in $|F=3\rangle$. This scheme avoid the influence of atoms in $|F=4, m_F=-4\rangle$ thus has higher SNR. Also in the reversed scheme most of the detectable atoms pass through the state-selection zone close to the concave pole, i.e., the weak-field region. The gradient is smaller as well as more uniform, which means that the atoms need to travel longer to reach a similar deflection angle in OMCC scheme. This brings the narrower and slower velocity distribution [9].

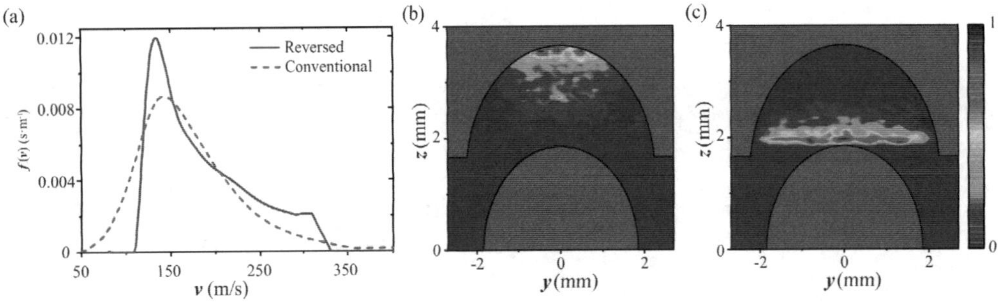

Figure 5. (a) Velocity distributions of atoms (b) & (c) Densities of the detectable atoms when entering the state-selecting zone of the reversed (b) and conventional (c) schemes.

Fig. 6 shows OMCC and Reversed OMCC's microwave spectrum. Considering these two advantages mentioned above, the Reversed OMCC has better SNR as well as a narrower linewidth, therefore gives better short-term stability. Fig. 4 shows the frequency stability of Reversed OMCC, which reaches $2.9\times10^{-12}\tau^{-1/2}$. However, when σ light is used for detection, we cannot see the atomic fluorescence due to the ground-state coherence effects [10]. Fig. 7 shows the 3-2 line energy level structure in weak magnetic field. OMCC does not have this problem because 4-5 line is insensitive to ground-state coherence effects. Thus in Reversed OMCC, an extra static magnetic field (about 3 G) is needed in the optical detection region to remove the ground-state coherence.

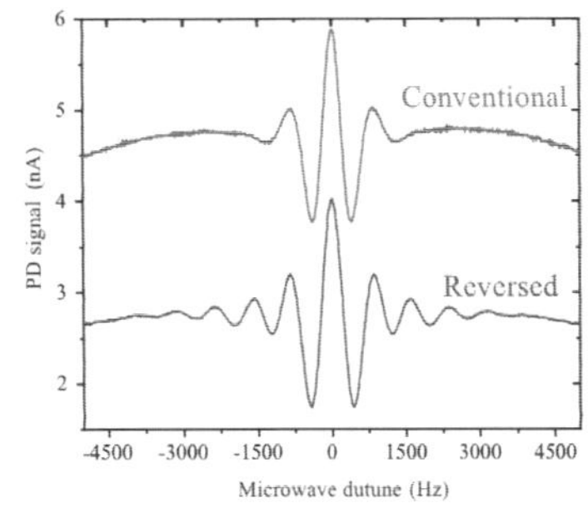

Figure 6. Conventional OMCC and Reversed OMCC's Ramsey fringes.

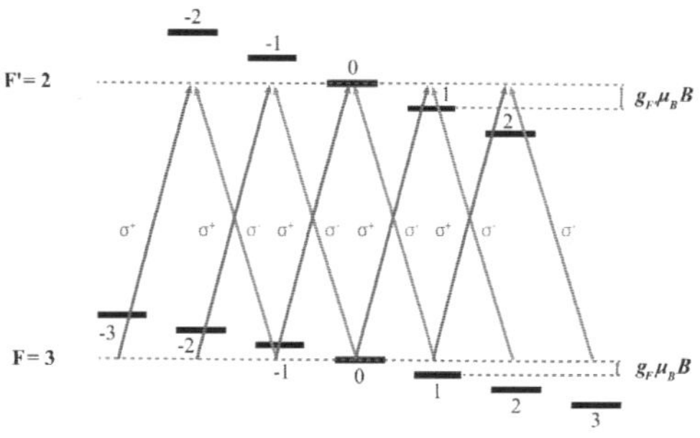

Figure 7. 3-2 line energy level structure in weak magnetic field.

4. Research on Beam Optics

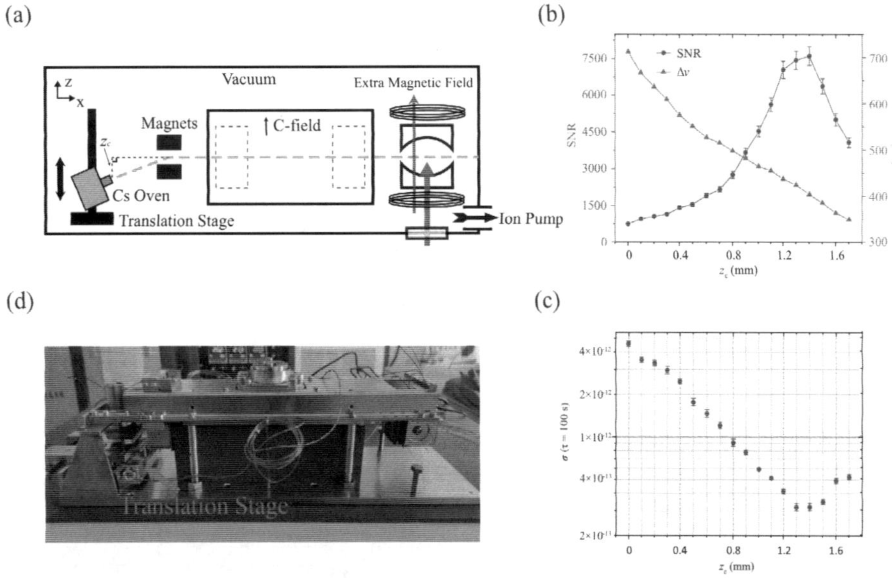

Figure 8. (a) Schematic of the experiment setup of beam optics; (b) Experimentally measured linewidth of the Ramsey fringes (red) and SNR at the modulation frequency with 1 Hz bandwidth (blue) at different position z_c; (c) The Allan deviation at an averaging time of 100 s of different position z_c; (d) Photographs of the experimental setup.

Beam optics is important in the system where magnetic state selection is applied. The position and angle of the atomic oven and collimator determine the energy level distribution and density of atoms interacting with microwave fields and detecting light. We use Monte-Carlo method for simulations [11] and a electronic translation stage (a translation stage with a stepping motor) that can operate in vacuum environment for experiments. The translation stage provides 1-dimension movement of the atomic oven and the collimator (connected together), i.e., scanning z_c in Fig. 8a. An ion pump maintains vacuum of the experimental system. In this experiment, the size of the microwave cavity, detection zone, etc. are consistent with the compact prototype version to ensure that the experimental data can be applied to the optimization of the compact prototype. In the Reversed OMCC scheme with the translation stage, we finally achieved $2.9 \times 10^{-12}\tau^{-1/2}$ at $z_c = 1.3$ mm, which is the source of the data given in Fig. 4.

5. Research on Light Shift

Light detection brings light shift, i.e. AC-Stark shift [12]. When atoms interact with microwave fields, there will be stray light passing through the exit hole of the microwave cavity. The stray light will influence the atoms energy levels and cause shift of the clock's output frequency. The stray light contains both the fluorescence released by the atoms and the diffused light generated by reflections of the windows and spherical mirrors in the detection zone.

Here we introduce two methods to suppress light shift in OMCC: multi-frequency light detection and pulsed light detection. Both methods have been experimentally validated, but both also deteriorate short-term frequency stability.

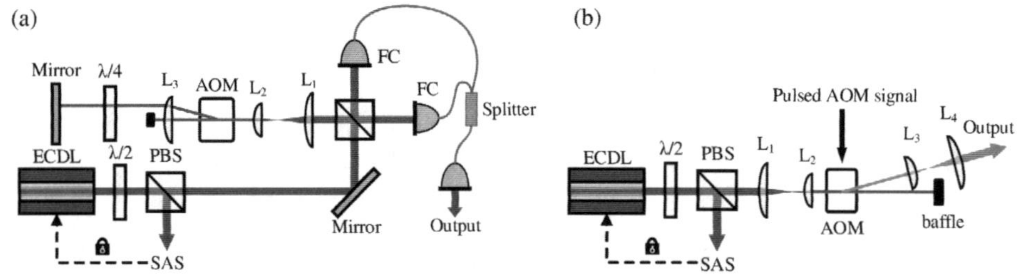

Figure 9. (a) Optical system setup of multi-frequency light detection; (b) Optical system setup of pulsed light detection; FC: fiber coupler, PBS: polarization beam splitter, ECDL: external-cavity diode laser, SAS: saturated absorption spectroscopy, AOM: acousto-optic modulator, $L_{1,2,3,4}$: flat convex lens.

5.1. Multi-frequency Light Detection

The light shift ω_{LS} is mainly related to the optical frequency ω_l and the detecting light power I. Define two light shift coefficients:

$$\alpha = \frac{\partial \omega_{LS}}{\partial I}/\omega_0, \quad \beta = \frac{\partial \omega_{LS}}{\partial \omega_l}/\omega_0,$$

where $\omega_0 = 2\pi \cdot 9192631770$ Hz.

After considering atomic velocity distribution, the light resonating with 4-5 line cannot satisfy either $\alpha = 0$ or $\beta = 0$. If a second light with a specific detuning frequency and a specific light power ratio with the detecting light is introduced addtionally, the combined α and β can be both zero [13, 14]. Fig. 9a shows the experiment setup. Fig 10 and Fig 11 show the theoretical calculations and experimental results. Here the light power ratio $r_m = I_{\text{detuned}}/I = 1.0$, and the detuned frequency $\nu_m = \omega_{\text{detuned}} - \omega_l = -2\pi \cdot 560$ MHz. α coefficient is reduced from $(1.25 \pm 0.03) \times 10^{-12}$/mW to $(0.87 \pm 0.85) \times 10^{-13}$/mW. The short-term stability of the clock deteriorated by a factor of 1.4.

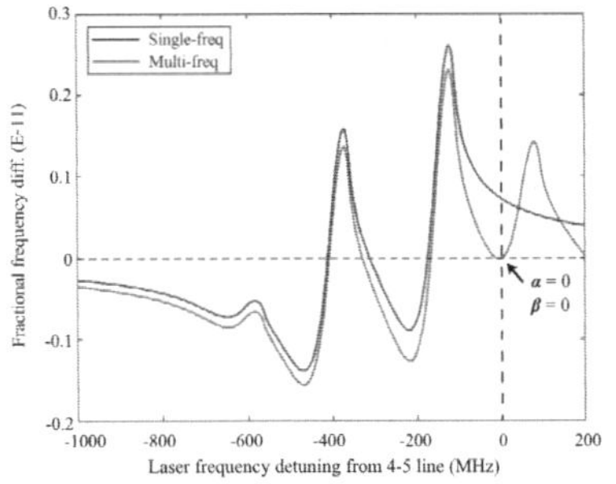

Figure 10. Theoretical calculations of multi-frequency light detection.

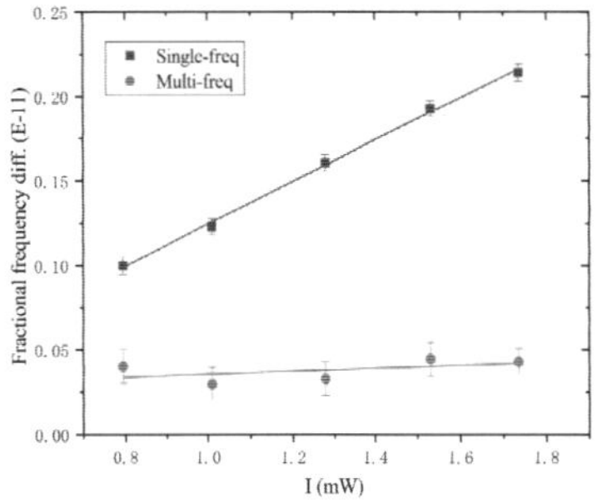

Figure 11. Experimental results of multi-frequency light detection

5.2. Pulsed Light Detection

Using high-frequency pulsed light (kHz) to detect atoms can also suppress light shift in thermal beam clock [15]. The key of this method is the separation of atoms interacting with microwaves field and light at the same time, which is similar to the POP clock (pulsed optically pumping clock) [16]. But the situation is different when it comes to beam clocks for the atoms are constantly moving instead of kept in a cell.

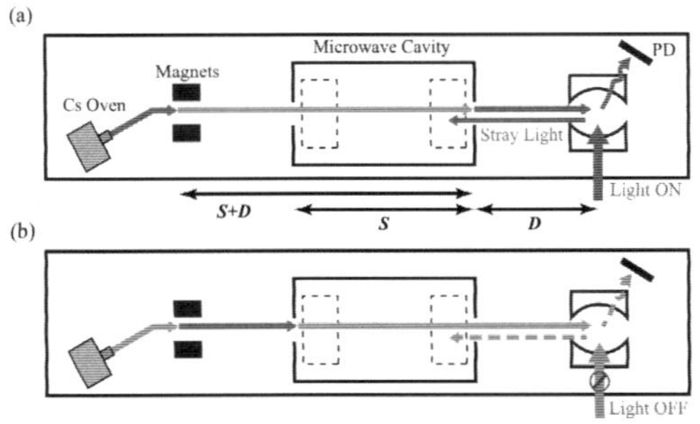

Figure 12. Process of pulsed detection. The green part represents the atoms which irradiated by light and will not be detected; the blue part represents the atoms which are not irradiated by light.

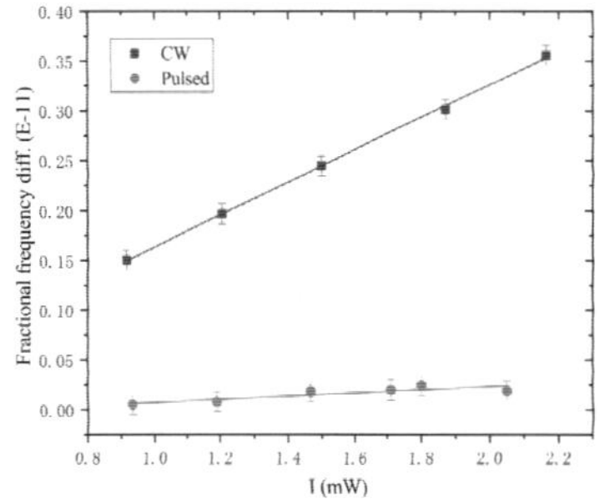

Figure 13. Experimental results of pulsed light detection. The pulsed frequency is 640 Hz, and the duty cycle is 30%.

Fig. 12 shows the process of pulsed light detection. Assume that atomic velocity $v = u$ is a const. Considering a period of $T = (S + 2D)/u$, where D is the distance from the microwave cavity to the detection zone and $S = L + 2l$ is the length of the whole Ramsey cavity. We turn on the light (i.e., the moment shown in Fig.12a) only for the time D/u and detect part of atoms of length D. At this time the atoms behind (length $S + D$) will be irradiated when passing through the microwave cavity and therefore need to be abandoned. We turn the light off when the irradiated atoms reach the detection zone (i.e., the moment shown in Fig.12b).

During this period of time, not only the irradiated atoms are abandoned, but also the atoms of length D in the next cycle are not irradiated when passing through the microwave cavity. Repeat (a)-(b)-(a)-(b)-(a)-..., thus only the atoms that do not interact with microwave fields and light at the same time will be detected.

Since actual atoms have velocity distributions, such a mechanism cannot simply suppress LS to zero. Also, as fewer atoms are detected per unit time, the signal-to-noise ration decreases, thus causing a degradation of the short-term stability. Fig.13 shows the experimental results. The pulsed frequency is 640 Hz, and the duty cycle is 30%. α coefficient is reduced from $(1.63 \pm 0.03) \times 10^{-12}$/mW to $(1.59 \pm 0.45) \times 10^{-13}$/mW. The short-term stability of the clock deteriorated by a factor of 1.7.

6. Conclusion

The article introduces our recent progress on the optically detected magnetic-state-selected cesium beam clock (OMCC). The 4U compact prototype achieves $4.0 \times 10^{-12}\tau^{-1/2}$, but further optimization of stability is difficult due to the stray atoms in $|\mathrm{F} = 4, \mathrm{m_F} = -4\rangle$. The new scheme, Reversed OMCC, uses 3-2 line for detection and has achieved $2.9\times10^{-12}\tau^{-1/2}$ in the experimental platform. In the researches on beam optics, we use Monte-Carlo method for simulations and an electron translational stage that can be operated in vacuum to perform experimental validation. Two methods to suppress the light shift in OMCC are introduced. In the multi-frequency light detection method, the light shift coefficient is reduced from $(1.25 \pm 0.03) \times 10^{-12}$/mW to $(0.87 \pm 0.85) \times 10^{-13}$/mW, and the short-term stability deteriorates by 1.4 times. In the pulsed light detection method, the light shift coefficient is reduced from $(1.63 \pm 0.03) \times 10^{-12}$/mW to $(1.59 \pm 0.45) \times 10^{-13}$/mW, and the short-term stability deteriorates by 1.7 times.

References

[1] Liu C, Zhou S and Wang Y 2015 Noise investigation on optical detection in a cesium beam clock with magnetic state selection
[2] Liu C, Chen S, Chen Z, Li L, Xu S, Li Y, Wang J, Wang Y, Hou S, Zhang J, Dong R and Jiang X 2021 *Rev Sci Instrum* **92** 073302
[3] Gerginov V and Tanner C E 2003 *Optics Communications* **222** 17–28
[4] Chen J, Wang J, Ma P, Guo L, Tu J, Yang W, Cui J, Ma Y and Cheng D 2018 *Journal of Time and Frequency* **41** 190–193
[5] He X, Fang S, Yuan Z, Xie W, Chen N, Xiong Z, Wang Q, Qi X H and Chen X Z 2021 *Appl Opt* **60** 10761–10765
[6] Liu C, Wang S, Chen Z, Wang Y, Jia J, Sun Y and Zhang B 2017 A caesium atomic beam microwave clock detected by distributed feedback laser diodes
[7] Liu C, Wang S, Chen Z, Wang Y and Hou S 2019 *Measurement Science and Technology* **30** 075044
[8] Kalliomaki K, Mansten T and Mannermaa J 2004 *2004 18th European Frequency and Time Forum (EFTF 2004)* pp 409–410 ISBN 0537-9989
[9] Chen S, Liu C, Fan L, Li C, Li Y, Liu C and Wang Y 2023 *Review of Scientific Instruments* **94** 043304
[10] Théobald G, Dimarcq N, Giordano V and Cérez P 1989 *Optics Communications* **71** 256–262
[11] Chen S, Liu C, Xu S, Li Y, Wang J, Wang Y, Liu Y and Jiao W 2022 *Journal of Applied Physics* **131** 114401
[12] Jun J W, Lee H S, Kwon T Y and Minogin V G 2001 *Metrologia* **38** 221–227
[13] Affolderbach C, Andreeva C, Cartaleva S, Karaulanov T, Mileti G and Slavov D 2005 *Applied Physics B* **80** 841–848
[14] Xu S, Chen S, Liu C, Li Y, Wang J, Wang Y, Zhang J, Dong R and Jiang X 2021 *Europhysics Letters* **136** 23002
[15] Li Y, Xu S, Chen S, Liu C, Wang J and Wang Y 2022 *Optics Express* **30** 43271–43280
[16] Kang S, Gharavipour M, Affolderbach C, Gruet F and Mileti G 2015 *Journal of Applied Physics* **117**

Progress in the miniature trapped ion optical clock development with 16 cc sealed trap tube

Thai M. Hoang[1], Wei Zhang[1], Thanh Le[1], Sean Mulholland[1,2], Mehdi Langlois[1], Yuna Park[1], Andrey B. Matsko[1], John Prestage[1], and Nan Yu[1]

[1]Jet Propulsion Laboratory, California Institute of Technology, Pasadena, CA 91109, USA
[2] Current address: National Physical Laboratory, Teddington, Middlesex, UK

E-mail: nan.yu@jpl.nasa.gov

Abstract. Optical clocks have well surpassed the frequency stability performance of microwave clocks. They have been mainly limited to tabletop setups and large rack-mount systems. Miniaturizing optical clocks will make the new technology and capability more ubiquitously applied to applications from navigation to fundamental science. Here we aim to develop a miniature space optical clock (mSOC) with the size of a compact microwave clock and capable of achieving a short-term stability of $10^{-14}\tau^{-1/2}$ and noise floor of 10^{-16}. In this paper, we will describe the mSOC concept, discuss the experimental setup, and present recent results. We have demonstrated trapping single ions and characterized the ion lifetime in a 16-cc trap tube. The trap tube is completely sealed off without any active pump. We also constructed a clock laser referenced to a small ultra-low expansion reference cavity at a low 10^{-14} level and observed the clock transition.

1. Introduction

At $3 \times 10^{-13}\tau^{-1/2}$ and a stability floor at 10^{-15} level, the Deep Space Atomic Clock (DSAC) is the state-of-the-art microwave clock of its size, which is designed for deep space applications [1]. To reach beyond the frequency stability of DSAC, one will have to take the new approach of the optical clock where the clock ticking rate is at hundreds of THz rather than tens of GHz. The high optical frequency enables the clock stability and accuracy to significantly exceed what microwave clocks can achieve, with accuracies beyond 10^{-17} [2]. The challenge is being able to take advantage of the optical clock performance with a small enough size, weight, and power to be deployable in deep space platforms. The miniature space optical clock (mSOC) development program focuses on reducing the size and power of an optical clock while still outperforming the stability of any microwave clocks of a similar size by an order of magnitude in all time scales. In particular, our objective is to develop and demonstrate an mSOC concept with a short-term stability of $10^{-14}\tau^{-1/2}$ and flicker noise floor of 10^{-16}.

2. Technical Approach

From a system engineering perspective, a singly trapped ion atomic clock offers a practical path to minimize SWaP. The core of mSOC is a miniature ultra-high vacuum tube in which a single ytterbium ion is confined in an RF trap and laser-cooled as the atomic reference. The single

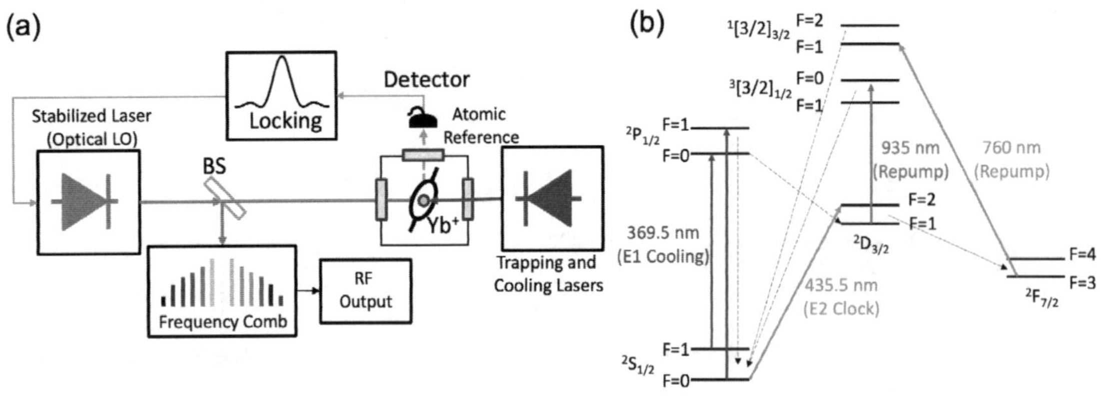

Figure 1. (a) Simplified clock concept. A single ^{171}Yb$^+$ is trapped and laser-cooled inside a small vacuum trap tube. A stabilized laser is used to probe the 435.5-nm electric quadrupole transition between the ground $|^2S_{1/2}, F=0\rangle$ and $|^2D_{3/2}, F=2\rangle$ states. The stabilized laser is tuned and locked to the clock transition based on the signal from the trapped ions. (b) Relevant energy levels of ^{171}Yb$^+$ show the cooling, repump, and clock transitions.

trapped ion is well isolated from environmental perturbations, suitable for a high-performance clock. Operation of single laser-cooled ions generally requires very little laser power to operate. In the case of a single ytterbium ion in mSOC, the cooling and clock laser beams need about tens μW of power each. Less laser optical powers lead to lower overall electrical power consumption including that for thermal management. Laser systems for the ytterbium ion are commercially available. Miniature and chip-based lasers are also being developed through various programs elsewhere. For the miniature trap tube, the necessary ultra-high vacuum condition can be maintained with only a built-in passive getter pump. The environmental controls, such as thermal isolation and magnetic shield, can be implemented more easily for a small vacuum package. The clock laser will be stabilized to a cm-sized reference cavity. In the current plan, the mSOC will use a commercially available small optical frequency comb to down-convert the stabilized optical frequency to the standard RF output. A more compact microcomb can be implemented when it becomes available in the future.

mSOC is developed as an operational clock where we are mostly interested in the stability performance rather than emphasizing on accuracy. Based on the mSOC targeted long-term stability objectives, the 435.5-nm electric quadrupole clock transition is chosen because it requires only μW of laser power to drive. With a clock laser probe time of 10 ms (20% duty cycle and 50% excitation rate), the short-term stability can be $7 \times 10^{-15}\tau^{-1/2}$. Key factors affecting the long-term instability include the second-order Zeeman, second-order Doppler, blackbody radiation, and quadrupole shifts. To achieve the frequency stability floor below the 10^{-16}-level, for the second-order Zeeman shift, the magnetic field needs to be stabilized within 0.03 μT for a bias field of 3 μT. The ion temperature should be kept under 20 mK during the probe for the second-order Doppler shift. The vacuum trap tube temperature should be stabilized to within 5 K for the blackbody radiation shift. The quadrupole shift may be the most challenging, caused by the presence of largely unknown patch electric potentials across the ion trap electrodes. The shift is a function of the angle between the electric field gradient and the bias field. For the electric field gradient estimated at about 1 V/mm^2 at the ion location for the trap we use, the bias magnetic field direction needs to be reproduced within 2% uncertainty. Since the ytterbium vapor will coat the molybdenum surface of the ion trap electrodes during each ion loading, the patch potential and thus the quadruple shift can change over time. We expect that after a period

Journal of Physics: Conference Series 2889 (2024) 012033 doi:10.1088/1742-6596/2889/1/012033

of ion-loading operation, the trap surfaces will be fully coated with Yb and the associated patch potentials will reach a sufficiently steady state.

In this paper, we will focus on two key components, the small vacuum trap tube and the compact ultra-low expansion (ULE) reference cavity. We have demonstrated single ytterbium ion trapping and observed the clock transition in a 16-cc trap tube that is completely sealed off and standalone without any active pump. We will also describe the construction and characterization of a compact ULE high-finesse cavity.

3. Experimental Concept

A simplified mSOC concept is illustrated in Fig. 1(a). Currently, we use $^{171}Yb^{+}$, but other ion species are possible. Relevant energy levels of $^{171}Yb^{+}$ for the cooling, repump, and electric quadrupole clock transitions are shown in Fig. 1(b). The ion is laser-cooled to the Doppler limit of mK by the 369.5-nm laser through the $|^2S_{1/2}, F = 1\rangle \rightarrow |^2P_{1/2}, F = 0\rangle$ electric dipole transition with a natural linewidth of 19.6 MHz [3, 4, 5, 6, 7, 8, 9]. The 369.5-nm cooling laser is typically red-detuned at about 10 MHz. The 369.5-nm cooling laser can non-resonantly excite the ion to the $|^2P_{1/2}, F = 1\rangle$ state which then spontaneously decay to the $|^2S_{1/2}, F = 0\rangle$ state. This effectively takes the ion out of the $|^2S_{1/2}, F = 1\rangle \rightarrow |^2P_{1/2}, F = 0\rangle$ cooling transition. To keep the cycling transition, a 14.7-GHz sideband (about 0.5%) is added to the 369.5-nm cooling laser beam to depopulate the $|^2S_{1/2}, F = 0\rangle$ state. A strong magnetic field is also applied to avoid optical-pumping the ion into the $|^2S_{1/2}, F = 1\rangle$ dark states during the cooling cycle[10]. For the clock loop operation, the stabilized 435.5-nm laser locked to the electric quadrupole transition of the $^{171}Yb^{+}$ ion. The 435.5-nm clock laser is generated by a frequency-doubled and stabilized 871-nm laser system. The 871-nm laser system consists of a commercial 871-nm laser stabilized to a compact ULE high-finesse cavity (see below). A frequency comb down-converts the stabilized optical frequency to an RF output for characterizations and measurements.

In our proof-of-concept tabletop setup, the cooling, ionization, and repump lasers were all referenced to a wavemeter. We generally operated with 40 μW of 369 nm, over 100 μW of 399 nm, over 1 mW of 760 nm, and over 1 mW of 935 nm. We were able to deliver about 12 μW of 435.5-nm laser at the ion trap, generated by about 10 mW of 871-nm laser before the frequency doubler. All the laser beams were focused onto the trap through a 100 mm lens. The diameters of the 369 nm, 399 nm, 435 nm, 760 nm, and 935 nm laser beams were about 40 μm, 150 μm, 80 μm, 190 μm, and 150 μm at the ion location, respectively. Ytterbium ions are generated using the isotope selective two-photon-ionization method with 399-nm and 369.5-nm ionization lasers. The ion fluorescence signal was collected and detected by a CCD camera for imaging and a photomultiplier tube (PMT) for photon counting at the same time. While $^{171}Yb^{+}$ ion is being used for the clock operation, we often used $^{172}Yb^{+}$ ion for characterization and diagnosis. Operating with $^{172}Yb^{+}$ ions is simpler since it does not require a high magnetic field and microwave sideband for laser cooling.

4. Results

4.1. Miniature vacuum trap tube

We have designed and constructed two generations of miniature vacuum Yb^{+} trap tubes. This development leverages our experience in developing small vacuum trap tubes for mercury ion [11, 12, 13, 14]. The vacuum trap tubes consist of a ring ion trap, sapphire optical windows, Yb oven appendage, non-evaporative getter (NEG), copper pinch-off tube, and corresponding electrical feedthroughs as illustrated in Fig. 2(a). The Yb oven appendages contain a natural-abundance ytterbium metal source and can be heated to over 300 C using an external heater. The ring ion trap consists of a molybdenum ring (about 1 mm in diameter) and two notional end caps. The end caps are also used for DC compensation along the end cap direction. The ion trap is housed inside a 16-cc miniature titanium cubic chamber. The trap vacuum tube components

Journal of Physics: Conference Series **2889** (2024) 012033 doi:10.1088/1742-6596/2889/1/012033

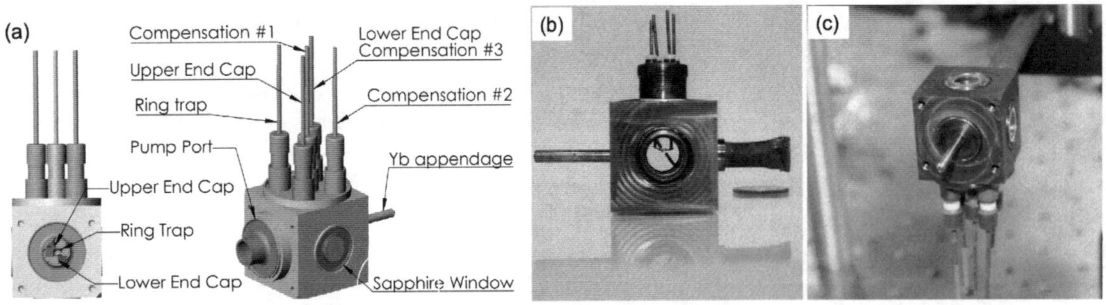

Figure 2. (a) Computer-aided design of the vacuum trap tube viewing from different angles. Relevant components are labeled. (b) First-generation trap tube with one end cap being the only DC compensation electrode. (c) Second-generation trap tube with three DC compensation electrodes in three directions as illustrated in Figure (a).

were welded together with electron-beam welding without nuts and bolts to minimize the volume footprint. The trap tube was first baked to 400 ^{o}C to achieve a base pressure of a nano torr or beyond, and the passive NEG inside was activated at the same time. After the preparation process, the copper tube pump port was pinched off and the trap tube was completely separated from the pumping station. The NEG pump can maintain the vacuum integrity in the trap tube for many years with zero power consumption. A pinched-off first-generation trap tube is shown Fig. 2(b). This tube does not have additional electric field compensation ability and therefore, large micromotion efforts were unsuppressed. The excess micromotion will not only induce a large second-order Doppler shift in the clock transition but may also limit the current ion storage lifetime of days in the trap. For this reason, a second-generation trap tube with all-axis compensation capability has been designed and constructed, as shown in Fig. 2(c). The second generation trap tube was baked and sealed at a base pressure of about 5×10^{-10} torr. We want to emphasize that all of the data in this paper were taken with the first-generation trap tube.

4.2. Ion temperature

For the ion trap, we applied about 300 V peak-to-peak at 6.5 MHz to the ring electrode and end caps. A DC voltage can be applied between the two end caps to compensate for the stray DC electric field as needed. However, without the ring direction DC compensations in the first generation trap tube, we observed relatively large micromotions and the ion temperature higher than the Doppler cooling limit $T_D = 2\hbar\Gamma/3k_B = 0.63$ mK. The ion temperature can normally be measured simply by scanning the 369.5-nm cooling laser frequency through the $|^2S_{1/2}, F = 1\rangle \rightarrow |^2P_{1/2}, F = 0\rangle$ transition [15, 16]. In this method, the ion fluorescence would quickly drop off once the cooling laser frequency is tuned to the blue-detuned side. The blue-detuned cooling laser would heat the ions instead of cooling them. The resulting asymmetric fluorescence signal on the red-detuned side could be fitted to a Voigt profile to extract the Lorentzian and Gaussian linewidth (full width at half maximum) components.

We measured the ion temperature via ^{171}Yb^{+} ions by scanning the 14.7-GHz sideband of the 365.5-nm cooling laser as shown in Fig. 3(a). The 14.7-GHz sideband only occasionally excites the ions along the $|^2S_{1/2}, F = 0\rangle \rightarrow |^2P_{1/2}, F = 1\rangle$ transition and repumps the ions back to the $|^2S_{1/2}, F = 1\rangle \rightarrow |^2P_{1/2}, F = 0\rangle$ cycling cooling transition. Therefore, the laser cooling effect is dominated by the red-detuned main cooling carrier even if its sideband is on the blue-detuned side. In this way, we were able to obtain a symmetric resonance spectral profile. For each sideband frequency step, we collected the count for 1.99 seconds, reset the sideband frequency

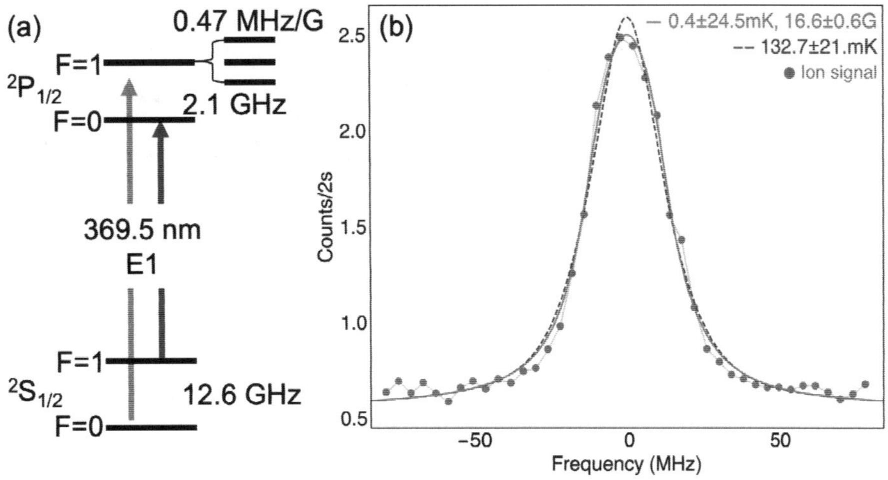

Figure 3. (a) The main cooling transition transition (blue arrow) and the 14.7-GHz sideband of the cooling transition (blue arrow). (b) Temperature of several trapped $^{171}Yb^{+}$ ions. The data are fitted to a single Voigt function (dashed blue curve) and triple Voigt function (red curve) to obtain the ion temperature.

back to the resonant value for 30 seconds for cooling, and moved to the next frequency step. We can fit the full spectrum data to a Voigt profile to extract the Gaussian linewidth. Since the power broadening linewidth is negligible, the Lorentzian linewidth of the Voigt profile effectively is equal to the natural linewidth of the $|^2S_{1/2}, F = 0\rangle \rightarrow |^2P_{1/2}, F = 1\rangle$ transition, about 19.6 MHz [9]. Fig. 3 shows a typical spectrum of several $^{171}Yb^{+}$ ions. A direct fitting of the data to a Voigt profile would yield a temperature of 133 ± 21 mK. However, since we had to apply a strong magnetic field to avoid optical pumping in the laser cooling, the Zeeman sub-levels of the $|^2P_{1/2}, F = 1\rangle$ state will be separated by several MHz. This Zeeman splitting significantly broadened the $|^2S_{1/2}, F = 0\rangle \rightarrow |^2P_{1/2}, F = 1\rangle$ spectrum, and the Gaussian linewidth would appear larger than its actual value. As the $|^2S_{1/2}, F = 0\rangle \rightarrow |^2P_{1/2}, F = 1\rangle$ spectrum contains 3 separate Zeeman peaks, we fitted the data to a triple Voigt profile (a sum of three functions) to extract the Gaussian linewidth. The triple Voigt profile yielded an ion temperature of 0.4 ± 25 mK, setting an upper-temperature limit of 25 mK. The separation of the three Voigt profiles indicated a magnetic field of 16.6 G.

4.3. Compact ULE cavity

We constructed a compact Fabry-Perot cavity made of ULE glass as a reference for the clock laser, as shown in Fig. 4(a). The total cavity volume is 23 cc. The spacer is 9.3 mm thick with a 50.8 mm outer diameter and 13 mm inner diameter. The spacer has three evenly distributed vacuum vent holes of 2 mm diameter. The mirror substrates (25.4 mm diameter and 6.35 mm thickness) are made of ULE glass with a radius of curvature of 1 meter. The substrates with a high reflective coating at 871 nm were optically contacted onto the spacer to form a resonator. The ring-down measurement yields the cavity finesse of 180000, which can support a cavity linewidth of 88 kHz for laser frequency stabilization. To minimize the temperature-induced frequency noise, the cavity was housed inside a two-layer thermal shield as shown in Fig. 4(b). The shield (110 mm in diameter and 60 mm in height) consists of two nested, cup-shaped enclosures made of copper and a ULE ring on which three Teflon rods sit. The contact area between each part is minimized to provide low thermal conductivity. The time constant of the

Journal of Physics: Conference Series **2889** (2024) 012033
doi:10.1088/1742-6596/2889/1/012033

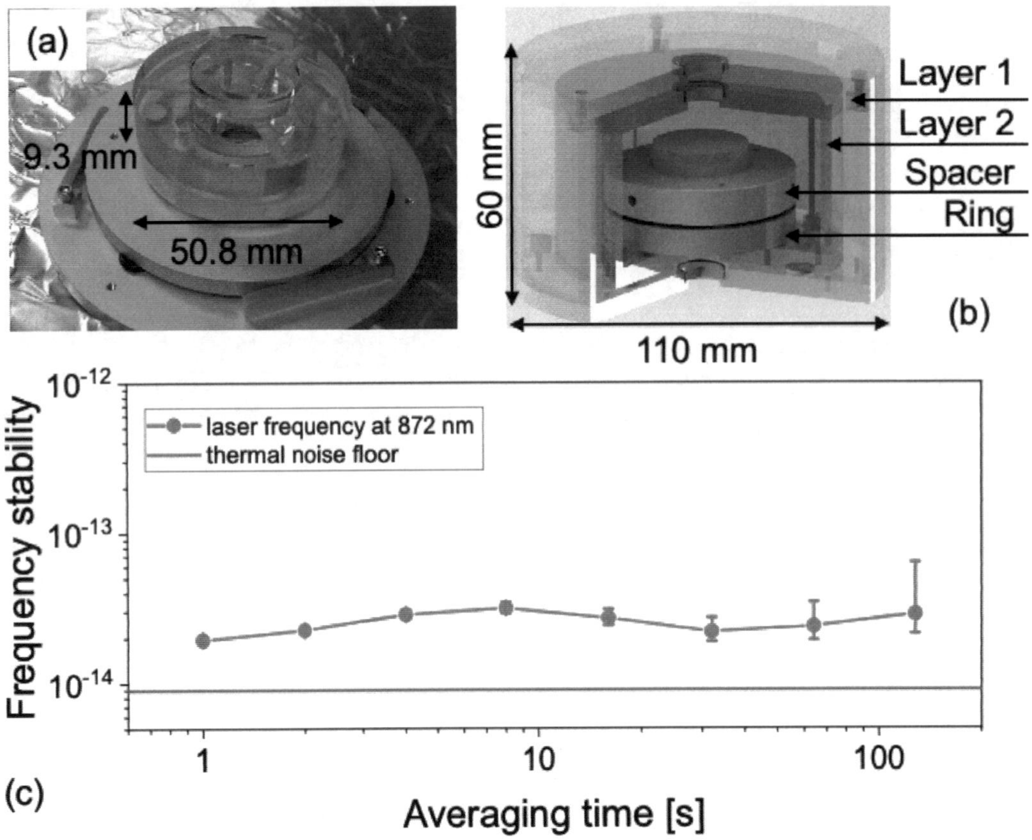

Figure 4. (a) Photograph of the compact Fabry-Perot cavity, with 9.3 mm in length and 50.8 mm in diameter. (b) Computer-aided design model of the cavity with two layers of thermal shield. (c) Measured frequency stability of the stabilized laser (red). For averaging time between 1 s to 100 s, the laser stability is 2×10^{-14}, limited by the cavity thermal noise (gray) at 9×10^{-15}. The lines connect the markers to guide the eyes.

thermal shield was measured to be 10 hours, which is long enough to support thermal noise limit stability.

To evaluate the fractional frequency stability of the cavity, the 871-nm clock laser was locked to the cavity by the Pound-Drever-Hall scheme [17]. The stabilized 871-nm laser was heterodyned with an Er-fiber frequency comb that is frequency-locked to a commercial cavity-stabilized reference laser of a Hz-level stability at 1550 nm. To compare the frequency at 871 nm, the comb spectrum around 1740 nm was extracted by a bandpass filter and frequency-doubled to 871 nm. The corresponding heterodyned beat note was used to calculate the Allan deviation of the 871 nm cavity stabilized laser. Fig. 4(c) shows the measured frequency stability. The linear drift was measured to be 6.5 Hz/s. The frequency stability (red line) was calculated to be 2×10^{-14} after removing the linear drift. The cavity thermal noise contribution (grey line) is 9×10^{-15}, dominated by the Brownian noise from the substrate (6.8×10^{-15}) and coating (6×10^{-15})[18].

To suppress the vibration-induced frequency noise, we employed the vibration-insensitive design [19]. Three counter-bored holes are evenly distributed on the spacer at about 20.95 mm from the spacer center. Three short Teflon rods were inserted into the holes to vertically support the cavity. According to the finite-element analysis, the vibration sensitivity of the cavity is minimized by adjusting the offset between the support surface of the Teflon rod and the top surface of the spacer. The offset was chosen to be 4.75 mm. To verify the vibration sensitivity, we also constructed an identical cavity with a high-reflective coating at 1550nm. The 1550-nm cavity was mounted on an active vibration isolation (AVI) table which can be excited by an external modulation. The beat note with the commercial reference laser at 1550 nm was recorded by a frequency counter when the AVI table was modulated. A three-axis accelerometer was used to calibrate the arbitrary vibration noise. The vibration sensitivity along vertical and horizontal directions were derived to be 7.8×10^{-10}/g and 5×10^{-10}/g, respectively.

4.4. Electric quadrupole clock transition

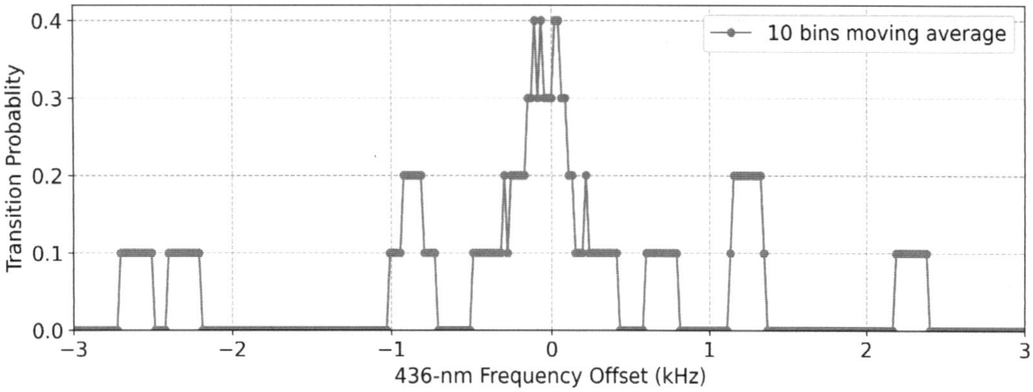

Figure 5. Electric quadrupole clock transition. The zero frequency offset point was put at about the center of the resonance peak. Each data point is a moving-average of 10 data points.

With the demonstrated robust single ion trapping in the first generation trap tube and the frequency stabilized 435 nm clock laser, we further demonstrated the observation of $^{171}Yb^{+}$ electric quadruple clock transition of the trapped single ion. The experiments demonstrated how the atomic reference system, lasers, cavity, control software, and frequency comb work together. For these experiments, the Yb oven was operated at the lowest temperature (about 260°C) to increase the chance of loading no more than one ion at a time. We turned off the Yb oven temperature right after an ion signal appeared (a few minutes). The ion fluorescence was about 1000 counts/s, while the background count was about 3000 counts/s. In the current setup, the fluorescence collected is equally split into both the CCD camera and the PMT. Only the PMT signal was used for clock transition detection while the CCD camera was used for convenience and helpful for diagnosis. The 435-nm laser was locked at a 7 GHz offset from the resonance of the compact ULE cavity by adding a sideband to the laser using an EOM. This offset bridges the possible frequency difference between the ion clock transition and the nearby reference cavity mode (the cavity-free spectral range is 10 GHz). The clock frequency scan is accomplished by an AOM. It enables us to scan the 435-nm clock laser in sub-hertz resolution across the electric quadrupole clock transition at 688,358,979,309,307 Hz [7, 20]. With the help of a frequency comb referenced to a hydrogen maser, we were able to set the 435.5-nm laser to within 1 kHz of the clock transition.

In a typical clock transition spectroscopy scan, each frequency step included 156 ms of laser cooling, 20 ms of state preparation (14.7-GHz sideband off), 10 ms of clock interrogation (435.5-nm laser on), and 4 ms of detection. The full power of 12 μW of 435.5-nm laser power was used for the clock transition. A transition occurrence is determined when the ion signal is below a certain threshold. This simply means that if the ion makes the $|^2S_{1/2}, F = 0\rangle \rightarrow |^2D_{3/2}, F = 2\rangle$ transition, the ion will be in the dark state and does not absorb or emit photons. Other conditions include that the ion signal is above a certain threshold before the clock interrogation, and the ion signal is below a certain threshold after the state preparation. A typical obtained spectrum is shown in Fig. 5. The frequency step was 20 Hz. The maximum average transition probability is expected 0.5 (or 50%) at the resonance for the continuous excitation. We observed a maximum transition probability of 0.4 (or 40%). This discrepancy is likely due to the strict threshold level we set to reduce the transition noise which also reduces valid transition events (transition probability).

5. Summary

We have demonstrated the feasibility of laser-cooling and trapping of single ions for an optical clock in a completely sealed standalone 16-cc trap tube. We also constructed and characterized a 10-mm long compact ULE reference cavity used for stabilizing the clock laser. We were able to observe the $^{171}Yb^+$ electric quadrupole transition with the clock laser. These results represent the first step in developing critical components for achieving a low-SWaP optical clock. In the next stage of development, we will use the second-generation trap tube with all-axis compensation electrodes to reduce and eliminate the excess micromotion which will allow us to achieve the natural linewidth-limited ion clock transition spectroscopy, demonstrate the clock operation, and validate its frequency stability.

6. Acknowledgments

We would like to acknowledge William Diener for the reference signals from the frequency standards test laboratory and Dr. Sheng-wey Chiow for the helpful discussions. The research was carried out at the Jet Propulsion Laboratory, California Institute of Technology, under a contract with the National Aeronautics and Space Administration. Sean Mulholland's research was supported by an appointment to the NASA Postdoctoral Program at the NASA Jet Propulsion Laboratory, administered by Universities Space Research Association under contract with NASA.

References

[1] Burt, E. *et al.* Demonstration of a trapped-ion atomic clock in space. *Nature* **595**, 43–47 (2021).
[2] Dörscher, S. *et al.* Optical frequency ratio of a $^{171}Yb^+$ single-ion clock and a ^{87}Sr lattice clock. *Metrologia* **58**, 015005 (2021).
[3] Wineland, D. J. & Itano, W. M. Laser cooling of atoms. *Phys. Rev. A* **20**, 1521 (1979).
[4] Bell, A. *et al.* Laser cooling of trapped ytterbium ions using a four-level optical-excitation scheme. *Phys. Rev. A* **44**, R20 (1991).
[5] Itano, W. M., Bergquist, J. C., Bollinger, J. J. & Wineland, D. J. Cooling methods in ion traps. *Phys. Scr.* **1995**, 106 (1995).
[6] Leibfried, D., Blatt, R., Monroe, C. & Wineland, D. Quantum dynamics of single trapped ions. *Reviews of Modern Physics* **75**, 281 (2003).
[7] Tamm, C., Engelke, D. & Bühner, V. Spectroscopy of the electric-quadrupole transition $^2S_{1/2}(F = 0) -^2 D_{3/2}(F = 2)$ in trapped $^{171}Yb^+$. *Phys. Rev. A* **61**, 053405 (2000).
[8] Tamm, C., Weyers, S., Lipphardt, B. & Peik, E. Stray-field-induced quadrupole shift and absolute frequency of the 688-THz ^{171}Yb+ single-ion optical frequency standard. *Phys. Rev. A* **80**, 043403 (2009).
[9] Olmschenk, S. *et al.* Measurement of the lifetime of the $6p^2P^0_{1/2}$ o level of Yb^+. *Phys. Rev. A* **80**, 022502 (2009).
[10] Berkeland, D. & Boshier, M. Destabilization of dark states and optical spectroscopy in zeeman-degenerate atomic systems. *Phys. Rev. A* **65**, 033413 (2002).
[11] Hoang, T. M. *et al.* Performance of micro mercury trapped ion clock. In *2019 Joint Conference of the IEEE International Frequency Control Symposium and European Frequency and Time Forum (EFTF/IFC)*, 1–2 (2019).

[12] Hoang, T. M. *et al.* Demonstration of a mercury trapped ion clock prototype. In *2021 Joint Conference of the European Frequency and Time Forum and IEEE International Frequency Control Symposium (EFTF/IFCS)*, 1–4 (2021).

[13] Hoang, T. M. *et al.* Integrated physics package of micromercury trapped ion clock with 10^{-14}-level frequency stability. *Appl. Phys. Lett.* **119**, 044001 (2021).

[14] Hoang, T. M. *et al.* Micro mercury trapped ion clock prototypes with 10^{-14} frequency stability in 1-liter packages. *Sci. Rep.* **13**, 10629 (2023).

[15] Fisk, P., Lawn, M. & Coles, C. Laser cooling of $^{171}Yb^{+}$ ions in a linear paul trap. *Appl. Phys. B* **57**, 287–291 (1993).

[16] Kielpinski, D., Cetina, M., Cox, J. & Kaertner, F. X. Laser cooling of trapped ytterbium ions with an ultraviolet diode laser. *Opt. Lett.* **31**, 757–759 (2006).

[17] Drever, R. W. *et al.* Laser phase and frequency stabilization using an optical resonator. *Appl. Phys. B* **31**, 97–105 (1983).

[18] Numata, K., Kemery, A. & Camp, J. Thermal-noise limit in the frequency stabilization of lasers with rigid cavities. *Phys. Rev. Lett.* **93**, 250602 (2004).

[19] Chen, L. *et al.* Vibration-induced elastic deformation of fabry-perot cavities. *Phys. Rev. A* **74**, 053801 (2006).

[20] Tamm, C. *et al.* Cs-based optical frequency measurement using cross-linked optical and microwave oscillators. *Phys. Rev. A* **89**, 023820 (2014).

Journal of Physics: Conference Series **2889** (2024) 012034 doi:10.1088/1742-6596/2889/1/012034

A compact cold atom cavity clock

A. Bregazzi[1], E. Batori[2], B. Lewis[1], C. Affolderbach[2], G. Mileti[2], P. Griffin[1] and E. Riis[1]

[1] Department of Physics, University of Strathclyde, G4 0NG, Glasgow, UK
[2] University of Neuchâtel, Institute of Physics, Laboratoire Temps-Fréquence, Avenue de Bellevaux 51, 2000 Neuchâtel, Switzerland

e.riis@strath.ac.uk

Abstract. A sample of laser cooled atoms are created inside an additively manufactured loop-gap microwave cavity using a grating magneto-optic trap requiring only a single laser cooling beam. Using a Ramsey excitation scheme with free evolution times of up to 20 ms and based on the ^{87}Rb ground-state clock transition, we demonstrate a short-term stability of $1.5\times10^{-11}\ \tau^{-1/2}$, averaging down to 2×10^{-12} after 100 s. The short-term stability limit is found to be dominated by the signal to noise ratio of the Ramsey fringes while for longer timescales the limitation is magnetic field noise due to the experiment being magnetically unshielded. Potential improvements to the setup and its operation point to a viable route forward for a miniaturised atomic microwave clock.

1. Introduction

Laser cooling and trapping of atoms and ions have brought substantial advantages to the field of atomic clocks. Most notably this is exemplified by the laser cooled atomic fountain [1,2], which currently provides the basis for the realisation of the SI second using a ground-state microwave transition in atomic Cs. Equally, a future redefinition is expected to take advantage of optical transitions in atoms cooled and trapped in optical fields, again taking advantage of a long interrogation of atoms in a pristine environment [3]. The drive towards improved clock stability, however, has led to ever increasing experimental complexity, size and laser performance requirements. This is contrasted by parallel development of optical-microwave double resonance clocks using hot vapour cells with applications expected in future portable, compact and space-grade clocks [4-7]. While impressive performance has been achieved, these systems are ultimately limited by coupling with fluctuating environmental parameters [4,8], vapour-cell ageing [9] and high dephasing rates. Vapour cell clocks based on coherent population trapping (CPT), where the chip-scale atomic clock [10] (CSAC) has achieved 10^{-10} level performance with low power consumption and an impressive small form factor, share these limitations.

In an effort to limit the impact of the operating environment and extend the possible interrogation time, solutions have been presented for compact cold atom microwave clocks based on spherical [11,12] and cylindrical cavities [13,14] as well as more recently on a loop-gap-resonator [15]. Common to these schemes and the one presented in this paper is that they all take advantage in the same way as the atomic fountain of the low perturbation environment cold atoms experience in free fall. However, they all have interrogation times in the 10s of ms region rather than the ~1 s flight time in a fountain [2]. While this does increase the observed atomic linewidth it vastly reduces the size of the apparatus and offers the potential for a reduced cycle time and hence increased data rate.

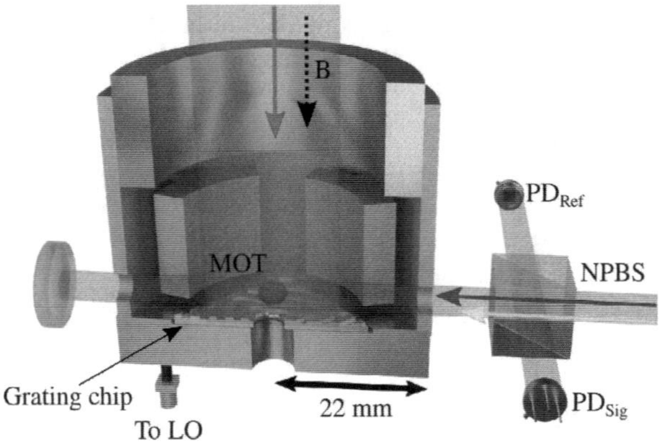

Figure 1. Simplified experimental setup showing cold atoms trapped inside loop-gap resonator. A single beam of laser cooling light (red arrow) impinges on the micro-fabricated grating at the bottom of the cavity forming the basis for the compact GMOT. State preparation and normalised atomic read-out (green/blue arrows) are obtained with a weak retroreflected beam from the side. The dashed black arrow indicates the direction of the bias magnetic field.

In this work we present a demonstration of an optically pumped cold-atom ^{87}Rb clock operating in the conventional dual-pulse Ramsey scheme and utilising an additively manufactured loop-gap resonator [16] and a grating magneto-optical trap [17] (GMOT). The relatively complex cavity structure would be difficult to realise with traditional machining techniques and is enabled using additive manufacturing, while the GMOT architecture reduces the complexity of the optical delivery system required to trap and cool an atomic cloud.

2. Experimental setup

The setup for the laser cooling and atomic clock part of the experiment is shown in Figure 1. The use of the GMOT simplifies the atomic trapping and cooling process as only a single beam of light is required while providing similar atom numbers and temperatures to conventional laser cooling schemes. In the present setup the GMOT provides the additional advantage that it lends itself to implementation inside the bore of a loop-gap resonator without the need to create further holes for cooling beams. The grating is placed at one end of the resonator with the bore size chosen at 20 mm to match the cooling beam diameter. This setup is typically demonstrated to trap $> 3\times10^6$ atoms and cool them to below 10 μK [18]. The MOT coils are wound coaxially with and external to the resonator and the whole setup mounted inside a vacuum system maintained at a pressure of $\sim10^{-8}$ mbar and with resistively heated Rb metal dispensers providing a partial Rb pressure of $\sim10^{-9}$ mbar.

The internal electrode structure of the loop-gap resonator shown in Figure 1 allows for the realisation of sub-wavelength cavities and hence is the basis for a compact physics package. The control required in the manufacturing process to ensure repeatability and accuracy of the electrode structure is afforded by additive manufacturing. The material chosen for this resonator was $AlSi_{10}Mg$ and was demonstrated to lead to insignificant additional outgassing at the operational vacuum pressure [16].

The loop-gap resonator operates in a TE_{011}-like mode at a resonance frequency of 6.8347 GHz with the nearest cavity modes more than 1 GHz away. The design, simulation and testing of the loop-gap resonator including assessment of the field alignment and homogeneity are described in detail elsewhere [16]. Due to the presence of the grating at the bottom of the resonator the Q-factor is rather low at 360. However, this is seen as an advantage as it minimises the cavity-pulling effect.

The atomic states are prepared and read out optically and hence require additional laser beam access. This is done through two diagonally opposite 4 mm holes cut in the resonator with no significant observable effect on the microwave field. After trapping and cooling the atoms are optically pumped into the dark $F = 2$, $m_F = 0$ ground state while in free fall. This step relies crucially on the alignment of the linear polarisation of the light with the ~100 mG bias magnetic field along the resonator axis. This state preparation is then followed by two microwave $\pi/2$ pulses separated by a Ramsey time T_R of up to 20 ms, limited by the expansion and fall of the atomic cloud. The 200 μs long $\pi/2$ pulses have a power level of ~0.04 mW, demonstrating that the low cavity-Q is not a problem. Following this the F=2 and total populations are determined by observing the absorption of a probe beam aligned through the holes in the cavity. Full details of this process including the relevant laser frequencies present at the different stages are provided elsewhere [19]. The experimental cycle is repeated at a rate of up to 7 Hz taking advantage of recapturing 80-90% of the atoms before they leave the GMOT optical overlap region [20].

The correct $\pi/2$ pulse parameters are obtained by observing the Rabi oscillations as the pulse area is increased through several times π [16]. This additionally acts a sensitive diagnostic on the microwave field homogeneity of the loop-gap resonator. As the cold atoms probe a volume of the resonator set by the size of the atom cloud, they experience the inhomogeneity of the microwave field amplitude and hence a variation of the Rabi frequency across the cloud. Experimentally this is observed as a dampening of the Rabi oscillation with a 1/e pulse area of ~10π. In terms of field inhomogeneity this corresponds to a 6.3% across the volume sampled by the atoms [16].

3. Results

Figure 2 shows an example of this normalised signal for a Ramsey time of 10 ms corresponding to a fringe linewidth of 50 Hz. The SNR is measure to ~110 at the steepest points on either side of the central fringe. This is expected to translate into an Allan deviation at 1 s of an oscillator stabilised to the atomic transition of 8.95×10^{-12} for a 10 ms Ramsey time [19].

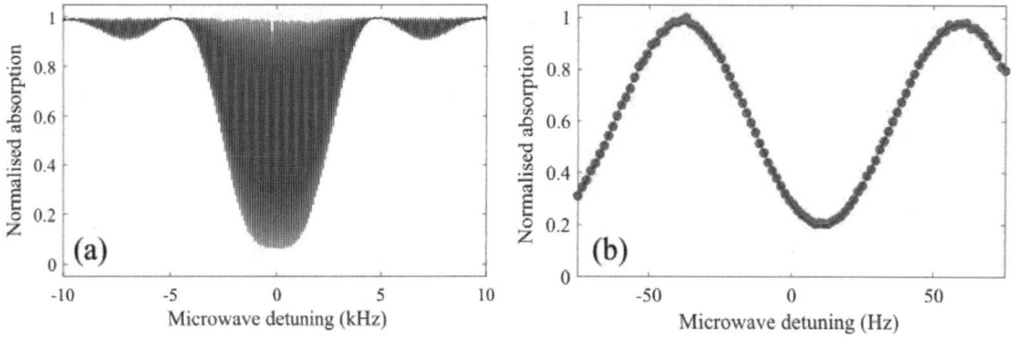

Figure 2. (a) Typical recorded Ramsey signal taken with a 10 ms free evolution time. (b) Closer look at the central Ramsey fringe obtained at 10 ms evolution time. Red points indicate measured normalised probe absorption while the black curve shows a sinusoidal fit to the data. The 100 mG bias field causes the observed fringe offset from 0 Hz detuning due to the second-order Zeeman effect.

To assess the stability of our system we lock the local oscillator frequency to the atomic signal. An error signal, which is used to steer the local oscillator through its internal reference, is obtained by jumping the microwave detuning between the steepest slopes on either side of the central fringe and comparing the recorded signal levels. The overlapping Allan deviation of the resulting atomic clock when compared to an oven-controlled quartz crystal oscillator disciplined to GPS is shown in Figure 3. The observed clock stability averages down with a $1.49\times10^{-11}\ \tau^{-1/2}$ dependence out to 10 s, where it starts deviating from the $\tau^{-1/2}$ dependence. The main reason for this is that the experiment is currently operating without magnetic shielding and as a result of that, magnetic field fluctuations will directly influence the clock stability through the second-order Zeeman effect. This has been confirmed by operating the clock on a first-order magnetic field sensitive transition ($|F = 1, m_F = 1\rangle \rightarrow |F' = 2, m_{F}' = 1\rangle$). The observed linear shift at 1.4 MHz/G on this transition can be converted for an equivalent field fluctuation to the expected effect on the second-order sensitive (575 Hz/G^2) clock transition. This is shown in the black curve in Figure 3 indicating that magnetic field fluctuations are a limiting factor for the stability in the 10s of seconds region.

At the 100s of seconds averaging time the clock stability is limited by the GPS disciplined quartz local oscillator. The Allan deviation of this oscillator is shown in the yellow curve in Figure 3. Together these two effects explain the observed levelling off of the Allan deviation of the cold atom microwave clock at longer averaging times. However, in both cases there is a clear path to substantially reducing the effects through the inclusion of magnetic shielding and by the use of a lower-noise local oscillator.

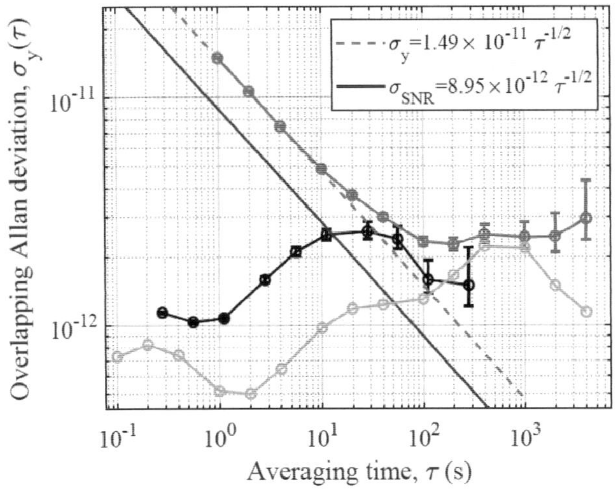

Figure 3. Red points: Overlapping Allan deviation of local oscillator when locked to the atomic signal. Red dashed line: $\tau^{-1/2}$ dependence of the measured 1 s stability. Blue solid line: stability expected from measured fringe SNR. Black points: stability limit set by observed magnetic field fluctuations through second-order Zeeman shift. Yellow points: GPS disciplined reference oscillator stability measured against two commercial Cs beam clocks (three-corner hat). Error bars represent 1σ confidence bound.

The stability budget at τ = 1 s is summarised in Table 1. The dominating identified source is the SNR of the observed absorption signal. At a value of ~110 this is about an order of magnitude worse than the quantum projection limit of the square root of the atom number. The primary route to approaching this limit is an optimisation of the detection process through the suppression of probe laser intensity and frequency noise.

The next most significant limitation to the 1 s clock stability is electronic noise on the signal used to steer the local oscillator. In the current setup this is not a major concern, but improvements to the SNR and shielded operation are expected to readily enable operation of the clock below the 10^{-12} level of this noise source and this would hence require further optimisation.

Table 1. The various identified noise sources and their contributions to the Allan deviation at 1 s integration time.

Noise source	σ contribution (τ = 1s)
SNR	8.95×10^{-12}
Electronic noise	1.14×10^{-12}
Dick Effect	6.12×10^{-13}
Zeeman Effect	1.07×10^{-12}
QPN	4.90×10^{-13}
Total $\left(\sqrt{\sigma_{SNR}^2+\sigma_{Elec}^2+\sigma_{Dick}^2}\right)$	9.04×10^{-12}
Measured	1.49×10^{-11}

Currently the Dick effect [21], which is a manifestation of local oscillator phase noise, is not a limiting factor but can be reduced further when solving the GPS disciplined reference oscillator stability problem evident in the data.

Overall, Table 1 demonstrates a reasonable agreement between the observed stability obtained at a 10 ms Ramsey time and the theoretical expectations of the identified main noise components based on separate analysis of their individual contributions.

4. Conclusion

We have implemented an additively manufactured loop-gap resonator in a compact microwave clock based on ^{87}Rb atoms cooled in a grating magneto-optic trap. In the present implementation the experimentally optimised clock stability is measured as $1.5\times10^{-11}\,\tau^{-1/2}$, averaging down to 2×10^{-12} after 100 s. The short-term stability is primarily limited by the SNR of the Ramsey signal. In the medium-term the clock stability is limited by the second-order Zeeman shift due to magnetic field noise while a further limit is due to the reference oscillator itself. Measures to reduce these limiting effects are outlined above. Additionally, it would be possible to extend the Ramsey time by an order of magnitude through implementing a low launch-height grating-chip fountain [22]. In a CPT clock experiment signals were observed from atoms returning after 100 ms. In addition to eliminating the issue of the atoms dropping out of view of the probe beam this would also have the advantage of the two Ramsey pulses probing the atoms in approximately the same location of the cavity. To gain the full advantage of this approach it may be necessary to lower the atomic temperature further, e.g. through the use of grey molasses, which has also been demonstrated at for a grating chip [23].

The loop-gap resonator inherently allows for a compact setup and the physics package is amenable to further miniaturisation and integration. The additive manufacturing process not only ensures a scalable manufacturing process, it has also recently been demonstrated to be compatibility with the requirements for vacuum chambers operating at UHV [24] pressure levels. This could potentially lead to the cavity body itself doubling up as a UHV chamber, enabling the entire physics package to be manufactured as a single component, thereby reducing the size and complexity of the system. Furthermore, passively pumped vacuum chambers have been shown to maintain the pressure required for atom trapping for extended periods of time [25,26]. The implementation of this would not only lead to a reduced power consumption by negating the continuous use of an ion pump but also a further reduction in size and the elimination of a permanent magnet. With a corresponding further integration of the lasers and optical manipulation requirements [27] there would appear to be a genuine opportunity to realise an integrated microwave clock benefitting from many of the cold-atom advantages of the atomic fountain, but trading off a small factor of stability for a vast reduction of experimental complexity and not least size, weight and power.

References

[1] Kasevich MA, Riis E, Chu S and DeVoe RG 1989 *Phys. Rev. Lett* **63** 612
[2] Bize S 2019 *Comptes Rendus Physique* **20** 153
[3] Dimarcq N *et al.* 2023 *Metrologia* in press https://doi.org/10.1088/1681-7575/ad17d2
[4] Almat N, Gharavipour M, Moreno W, Affolderbach C and Mileti G 2019 *Joint Conf. of the IEEE Intern'l Freq. Control Symposium & European Frequency and Time Forum* 1
[5] Batori E, Almat N, Affolderbach C and Mileti G 2020 *Adv. Space Res.* **68** 4723
[6] Huang M, Little A and Camparo J 2022 *Joint Conference of the European Frequency and Time Forum & IEEE International Frequency Control Symposium* 1
[7] Batori E, Affolderbach C, Pellaton M, Gruet F, Violetti M, Su Y, Skrivervik AK and Mileti G 2022 *Phys. Rev. Appl.* **18** 054039
[8] Moreno W, Pellaton M, Affolderbach C and Mileti G 2018 IEEE *Trans. Ultrason. Ferroelectr. Freq. Control* **65** 1500
[9] Abdullah S, Affolderbach C, Gruet F and Mileti G 2015 *Appl. Phys. Lett.* **106** 163505
[10] Knappe S, Shah V, Schwindt PDD, Hollberg L, Kitching J, Liew LA and Moreland J 2004 *Appl. Phys. Lett.* ***85*** 1460
[11] Esnault FX, Holleville D, Rossetto N, Guerandel S and Dimarcq N 2010 *Phys. Rev.* A **83** 033436
[12] Esnault F, Rossetto N, Holleville D, Delporte J and Dimarcq N 2011 *Adv. Space Res.* **47** 854
[13] Müller ST, Magalhães DV, Alves RF and Bagnato VS 2011 *J. Opt. Soc. Am.* B **28** 2592
[14] Liu P, Meng Y, Wan J, Wang X, Wang Y, Xiao L, Cheng H and Liu L 2015 *Phys. Rev.* A **92** 062101

[15] Lee S, Choi GW, Hong HG, Kwon TY, Lee SB, Heo MS and Park SE 2021 *Appl. Phys. Lett.* **119** 064002
[16] Batori E, Bregazzi A, Lewis B, Griffin PF, Riis E, Mileti G and Affolderbach C 2023 *Journal of Applied Physics* **133** 224401
[17] Nshii CC, Vangeleyn M, Cotter JP, Griffin PF, Hinds EA, Ironside CN, See P, Sinclair AG, Riis E and Arnold AS 2013 *Nature Nanotechnology* **8** 321
[18] McGilligan JP, Griffin PF, Elvin R, Ingleby SJ, Riis E and Arnold AS 2017 *Scientific Reports* **7** 384
[19] Bregazzi A, Batori E, Lewis B, Affolderbach C, Mileti G, Riis E and Griffin PF 2024 *Scientific Reports* **14** 931
[20] Hoth GW, Elvin R, Wright M, Lewis B, Arnold AS, Griffin PF and Riis E 2019 in *Proc. SPIE 10934, Optical, Opto-Atomic, and Entanglement-Enhanced Precision Metrology* eds. S. M. Shahriar and J. Scheuer (SPIE) 109342E
[21] Santarelli G, Audoin C, Makdissi A, Laurent P, Dick GJ and Clairon A 1998 *IEEE Transactions on Ultrason. Ferroelectr. Freq. Control* **45** 887
[22] Lewis B, Elvin R, Arnold AS, Riis E and Griffin PF 2022 *Appl. Phys. Lett.* **121** 164001
[23] Barker DS, Norrgard EB, Klimov NN, Fedchak JA, Scherschligt J and Eckel S 2022 *Opt. Express* **30** 9959
[24] Cooper N *et al.* 2021 *Addit. Manuf.* **40** 101898
[25] Little BJ, Hoth GW, Christensen J, Walker C, DeSmet DJ, Biedermann GW, Lee J and Schwindt PDD 2021 *AVS Quantum Sci.* **3** 035001
[26] Burrow OS, Osborn PF, Boughton E, Mirando F, Burt DP, Griffin PF, Arnold AS and Riis E 2021 *Appl. Phys. Lett.* **119** 124002
[27] Lee J *et al.* 2022 *Nature Comm.* **13** 5131

Acknowledgements
The authors would like to thank R. Elvin and J. P. McGilligan for useful conversations. A.B. was supported by a Ph.D. studentship from the Defence Science and Technology Laboratory (dstl). E. B., C.A. and G.M. acknowledge funding from the European space Agency (ESA, Contract No. 4000131046) and the Swiss Space Office (Swiss Confederation). We gratefully acknowledge funding from the UK Engineering and Physical Sciences Research Council through the International Network for Microfabrication of Atomic Quantum Sensors (EPSRC EP/W026929/1).

Active optical clock: seventeen years of progress and next steps

Jia Zhang[1], Tiantian Shi[1,2,3*], Jianxiang Miao[1], Xiaolei Guan[1], Xun Gao[1], Jingbiao Chen[1,4]

1. State Key Laboratory of Advanced Optical Communication Systems and Networks, Institute of Quantum Electronics, School of Electronics, Peking University, Beijing 100871, China
2. School of Integrated Circuits, Peking University, Beijing 100871, China
3. National Key Laboratory of Advanced Micro and Nano Manufacture Technology, Beijing 100871, China
4. Hefei National Laboratory, Hefei 230088, China

E-mail: `tts@pku.edu.cn`

Abstract. The concept of the active optical clock (AOC) was proposed seventeen years ago by prof. Jingbiao Chen. After early calculations and experiments that demonstrated the advantages of "cavity-pulling suppression" and "narrower linewidth", the development of AOCs soon became an active competition involving many research groups that proposed schemes based on different atomic species, and different energy-level structures. In this paper, after a brief background introduction, we will mainly focus on the most recent experimental and theoretical progress of AOCs in Peking University. In addition, we give an outlook on AOCs based on cold atoms that can be operated continuously, and report a novel AOC scheme based on diffuse laser cooling that can be operated continuously for the first time, which is proposed and performed by Peking University. The AOC has great promise for applications in quantum metrology, as it can overcome the thermal noise of the reference cavity that limits conventional lasers, making AOC a promising candidate for future atomic optical clocks.

1. Introduction

Current state-of-the-art optical atomic clocks based on lattice-trapped atoms and single-ion systems have reached fractional frequency uncertainty under the level of 10^{-18} [1, 2, 3]. Unprecedented improvements of the clocks can significantly broaden a wide range of applications such as testing of general relativity [3, 4], searching for dark matter [5], detecting gravity waves [6], precision physical constant measurement [7], and studing of atomic interactions [8]. However, the above clocks typically operate in a passive mode, whose local laser works in the good-cavity limit with the cavity bandwidth being far narrower than the spectral width of the gain profile. Their performance is mainly limited by the local lasers that mostly utilize Pound-Drever-Hall (PDH) for frequency stabilization. Now, the spectral linewidth of passsive optical clock is approaching fundamental limitations by the thermal Brownian motion of the cavity mirror [9, 10], which is one of the obstacles for further precision improvements of the atomic optical clocks.

To solve this problem, Prof. Jingbiao Chen proposed active optical clock (AOC) [11, 12] seventeen years ago. Recent theoretical and experimental studies showed that it is advantageous to utilize AOCs working in the bad-cavity regime [13, 14, 15, 16], where the cavity field decays much faster than the bare atomic transition (cavity-mode linewidth wider than atomic gain linewidth), analogous to active hydrogen maser operation in the microwave domain [17]. In a laser operating in the bad-cavity region, the frequency of the emitted light depends on the transition frequency of the atomic gain medium rather than the optical cavity, allowing suppression of the cavity noise and reduction of the cavity-pulling effect.

Ever since the AOC was proposed [11, 12], research groups across the globe conduct research based on a variety of systems, including thermal atomic vapor cells [18, 19], Raman systems [20], atoms trapped in optical lattice [14, 21, 22, 23], atomic beam [15, 24], and Faraday atomic filter types [25]. Also, there are some preliminary considerations summarized and presented in the 7th [26] and 8th [27] Symposium on Frequency Standards and Metrology. Currently, international research groups such as Peking University, JILA, NIST, University of Colorado, University of Amsterdam, Vienna University of Technology, Aarhus University, University of Innsbruck, Leibniz University Hannover and others have carried out research on AOC. Recently, JILA has observed and characterized a 698 nm AOC superradiant laser based on optical-lattice-trapped ^{87}Sr atoms with a linewidth on the order of 10 Hz, a fractional Allan deviation of 6.7×10^{-16} at 1 s of averaging, and accuracy of 4×10^{-15} [13]. The University of Copenhagen used an atomic beam scheme to realize pulsed AOC superradiant laser output with a maximum output power close to 1 μW [15]. The University of Hamburg [14] and Aarhus University [28] have also realized AOC superradiant output based on different quantum systems.

In this paper, we will mainly focus on the most recent experimental and theoretical progress of AOCs in Peking University, including: 1) A continuous wave (cw) 1470 nm AOC [18] with a linewidth of 53 Hz realized based on the dual-wavelength (DW) good-bad-cavity technique. 2) A theoretical calculation of AOC based on optical-lattice-trapping of Cs atoms [21]. 3) An inhibited laser [19], which expands the working regime of the classical AOC from the resonant cavity condition to the far off-resonance condition. 4) Exact expressions for the FWHM and the finesse of Fabry-Pérot (FP) cavity Airy distribution [29], which solves the problem of inapplicability under ultra-low reflectivity conditions in the conventional formula. 5) An extremely bad-cavity laser [30] with a cavity finesse close to the limit of 2. In addition, we discuss the future development of AOCs, focusing on the realization of cold-atom-based AOCs that can be operated continuously. In particular, we introduce the recently proposed and experimentally performed AOC based on diffuse laser cooling that can operate continuously.

2. Recent works on active optical clock in Peking University

2.1. Dual-wavelength good-bad-cavity active optical clock

Although a four-level AOC can effectively suppress the cavity-pulling effect and theoretically achieve an AOC signal with mHz linewidth, if the cavity length of the AOC resonant cavity is not stabilized, the clock laser is limited by the frequency drift caused by the residual cavity-pulling effect, ultimately leading to the broadening of the laser linewidth. In order to solve this problem, a DW good-bad-cavity scheme [18] was proposed based on the traditional four-level AOC.

The key of this scheme is that two different wavelength lasers must be co-cavity and output at the same time. The two lasers work in the good-cavity region and the bad-cavity region respectively, called the good-cavity laser and bad-cavity laser, and the bad-cavity laser is used as the clock laser. Compared with the good-cavity laser stabilized using PDH technique, the bad-cavity laser has the advantage of cavity-pulling suppression. To measure its performance, two DW-AOCs were built. The cavity lengths of the two DW-AOCs are stabilized by phase locking of the good-cavity laser, thus eliminating the influence of the residual cavity-pulling effect on the AOC. In principle, if only the influence of the cavity-pulling effect on the AOC is considered, the bad-cavity laser linewidth will be further narrowed on the basis of the good-cavity laser, and the narrowing factor depends on the bad-cavity factor.

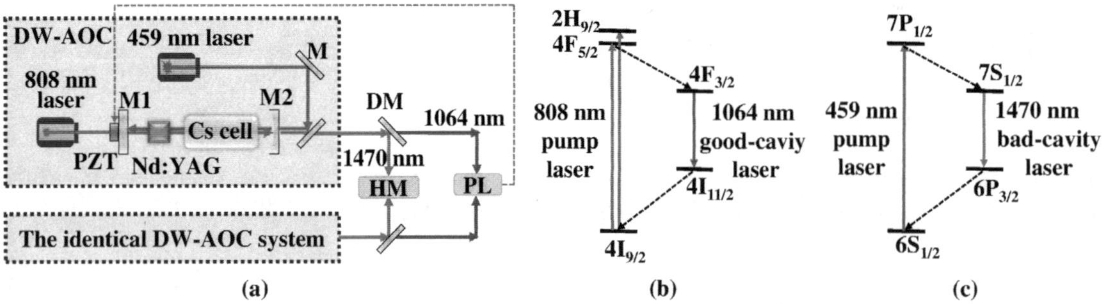

Figure 1. (a) Experimental scheme. M1 and M2 are the cavity mirror; HM, 1470 nm heterodyne module; PL, phase locking system; DM, dichroic mirror. **(b)** The energy level of good-cavity laser. **(c)** The energy level of Cs four-level AOC (bad-cavity laser).

To stabilize the cavity length, we initially proposed a scheme of He-Ne 633 nm and Cs 1359 nm DW-AOC [27]. However, this experiment failed because the cavity length was too long and the atomic gain was not large enough. Later, we improved the scheme. Experimentally, the Nd:YAG crystal and the Cs vapor cell share a common optical resonance cavity. The 808 nm semiconductor laser pumps the Nd:YAG crystal to output a 1064 nm laser, and the frequency-stabilized 459 nm semiconductor laser pumps the Cs atoms in the vapor cell to output a 1470 nm laser, thus realizing the simultaneous co-cavity output of the DW laser signals. By selecting appropriate parameters, the 1064 nm laser operates in the good-cavity mode and the 1470 nm laser operates in the bad-cavity mode. The phase locking of the two 1064 nm good-cavity lasers is used to realize the high-precision tracking of the cavity lengths of the two DW-AOCs, so as to eliminate the influence of the common-mode cavity-length noise on the linewidth of the 1470 nm laser beating frequency, and to verify the performance of the 1470 nm laser after the cavity-length locking. By optimizing the phase-locking process, the tracking accuracy of the two DW-AOCs' cavities is 3×10^{-16} for an averaging time of 1 s, and 1×10^{-17} for an averaging time of 1000 s. The phase locking has a fourfold narrowing effect on the beating frequency of the 1470 nm laser, narrowing the linewidth from 223 Hz down to 53 Hz.

2.2. Optical-lattice-based Cs active optical clock

In order to solve the problem of Doppler broadening within a four-level AOC using thermal atoms, Peking University, Beijing National Laboratory for Condensed Matter Physics of the Chinese Academy of Sciences, and the Physical Research Laboratory of India collaborated in a experimental scheme based on optical lattice [21]. They calculated in detail the magic wavelengths corresponding to the $|7S_{1/2}; F = 4, M_F = 0\rangle \rightarrow |6P_{3/2}; F = 3, M_F = 0\rangle$ clock transition in Cs atoms, which have a strong binding effect on multiple excited state atoms in the four-level scheme. This work predicts an AOC power up to 24 μW, linewidth on the order of mHz, and a frequency uncertainty prediction of 2×10^{-15}.

In this work, four suitable hyperfine energy levels are used to trap cold Cs atoms in a one-dimensional optical lattice, as a gain medium. A 455 nm laser is used as the pump light, which can populate Cs atoms from the ground state to the excited state. Meanwhile, an 894 nm repump light is used for Sisyphus cooling, and the magic wavelength 1181 nm laser corresponding to the clock transition $|7S_{1/2}; F = 4, M_F = 0\rangle \rightarrow |6P_{3/2}; F = 3, M_F = 0\rangle$ is theoretically calculated as the lattice light, so as to reduce the light shift. Finally, a cw AOC lasing at 1470 nm can be achieved. In addition, the clock frequency due to the blackbody radiation shift, the light shifts casued by pump and repump lasers, and the second-order Zeeman shifts are evaluated. By

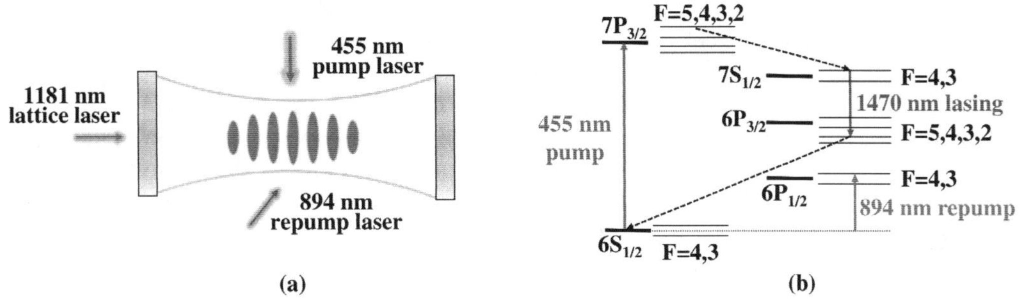

Figure 2. **(a)** Experimental scheme. **(b)** The energy level of optical-lattice-based Cs AOC.

simulation of real experimental conditions, the predicted frequency uncertainty is in the order of 10^{-15} and the linewidth is a few mHz. The use of cold atoms trapped in the optical lattice as a gain medium can overcome the inherent thermal effects in previous thermal-atom systems, thus significantly enhancing the short-term stability.

2.3. Anti-resonant FP cavity with ultralow finesse

FP cavity is a very important device in the fields of laser frequency stabilization and atomic clocks. However, traditional research has mostly focused on the aspect of resonant cavities with high reflectivity. Existing theories and experiments have fully analyzed the properties of resonant cavities with high reflectivity, such as cavity finesse, cavity-mode linewidth of Airy function, and cavity enhancement effect. However, the above characterization is limited to resonant high-Q-factor conditions only, neglecting analysis under low-reflectivity conditions. We reexamined the properties of the FP cavity by expanding its operating conditions from high to low reflectivity and from resonant to anti-resonant regions, and obtained some surprising properties.

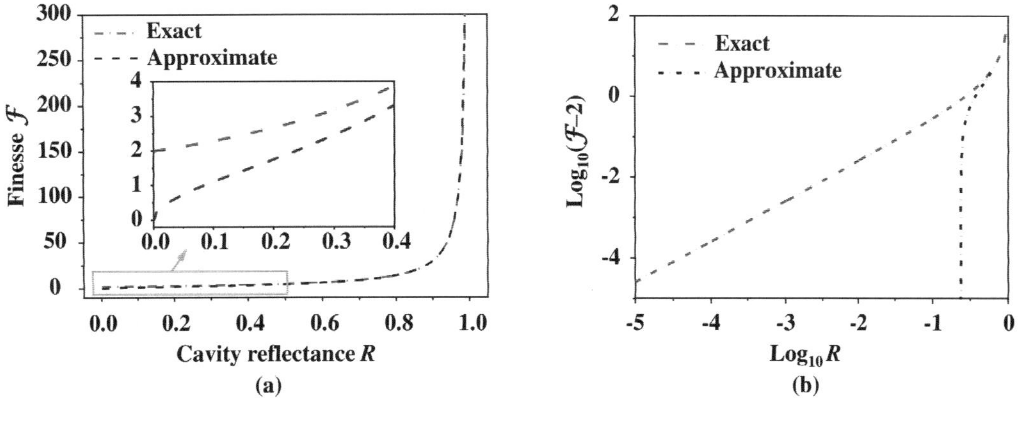

Figure 3. **(a)** Finesse $\mathcal{F}$ with the change of cavity-mirror reflectivity $R = r_1 r_2$. The red line represents the exact finesse expression, $\mathcal{F} = \frac{\pi}{\arccos(\frac{2R}{1+R^2})}$. The black line represents the approximate finesse expression $\mathcal{F} = \frac{\pi\sqrt{R}}{1-R}$, which can be usually seen in textbooks. **(b)** The finesse in logarithmic axes gives a more intuitive view of the difference between the exact expression and the approximate expression.

In this work, we derive exact expressions for the cavity finesse and full width at half maximum (FWHM) from the expressions for the Airy function, including in particular the low finesse case [29], which is important for the study of AOC. Theoretically, it is obtained that the FWHM is close to half of the free spectral range, rather than infinity [31], when the cavity reflectivity tends to zero. And in this case, the finesse is close to the limit 2, rather than 0, as shown in Fig. 3(a). Here, we define the logarithmic finesse as $\log_{10}(\mathcal{F}-2)$. The use of logarithmic axes gives a more intuitive representation of the difference between the exact expression and the approximate one. In addition to this, we expand the cavity from the resonant region to the anti-resonant region. Theoretically it is obtained that it is easier to realize anti-resonant lasing at low finesse, which is also experimentally verified. Under the anti-resonant case, we obtained that the cavity-length thermal noise can be suppressed by a factor of $(\frac{1+R}{1-R})^2$, where R represents the cavity reflectivity.

2.4. An inhibited laser

In order to overcome the cavity-length thermal noise problem of the ultrastable optical cavity that exists in the frequency stabilization of traditional passive optical clocks, the concept of AOC was proposed. However, whether it is the traditional passive optical clock or AOC, the clock laser works near the resonant region of the cavity mode, and the research on the laser working in the completely anti-resonant region of the cavity lacks the corresponding theoretical analysis and experimental verification. In the completely anti-resonant region, the cavity length is equal to an odd multiple of the laser quarter-wavelength (as shown in Fig. 4(a)), and we refer to this special laser as the inhibited laser [19], that is, anti-resonant laser.

Based on the bad-cavity laser system, a detailed study of the working mode of anti-resonant lasers has been carried out, which extends the concept and research scope of AOCs. The present work is based on the weak feedback of low-finesse FP cavity combined with a Cs four-level quantum system, which outputs bad-cavity lasers by stimulated radiation. Further, the stimulated radiation signal operating in the bad-cavity region is extended from the cavity resonance to anti-resonant region where the atomic transition frequency is located at the center frequency of two neighboring cavity modes. Experimentally, the optical properties of the inhibited laser, such as power, linewidth and cavity-pulling suppression effect, are investigated in this work, which proves that the inhibited laser can effectively suppress the influence of cavity-length thermal noise on the laser frequency, and is expected to realize a higher-performance active optical frequency standard.

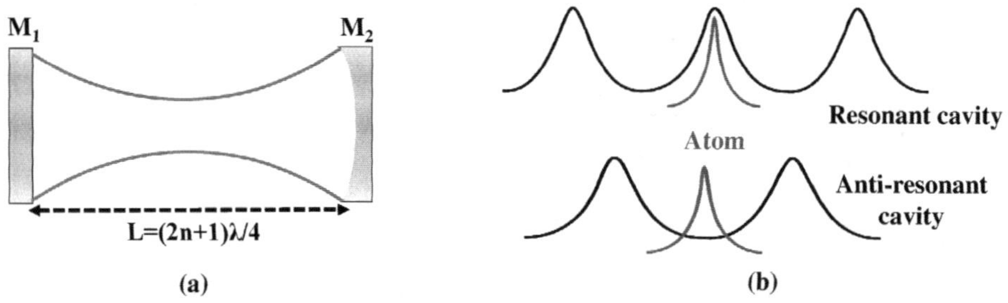

Figure 4. (a) The anti-resonant cavity, its cavity length equals to an odd multiple of the laser quarter-wavelength. **(b)** For resonant cavity, the atomic gain line is exactly at the center of cavity mode. On the contrary, the inhibited laser is realized in an anti-resonant cavity, where the atomic gain line is exactly at the center of two adjacent cavity modes.

When the pump intensity and atomic number density are low, the anti-resonant laser has better robustness to changes in pumping laser's intensity and vapor cell's temperature, which is expected to narrow the laser linewidth. As a result, the linewidth of anti-resonant laser is not broader than the resonant laser. Moreover, the cavity-pulling suppression effect of the inhibited laser is further enhanced by a factor of $(\frac{2\mathcal{F}}{\pi})^2$ compared with the resonant AOC, where $\mathcal{F}$ is the cavity finesse. This work is expected to further improve the performance of AOCs and provide new research ideas for quantum metrology, cavity QED, and laser physics.

2.5. An extremely bad-cavity laser

Conventional resonant cavities used for PDH frequency stabilization show an increasing trend of finesse over time, as shown in Fig. 5. However, for cavities used in AOCs, it shows a decreasing trend. This is because to ensure the realization of a bad cavity, the cavity-mode linewidth needs to be wider than the atomic gain linewidth. Moreover, when other conditions are the same, stronger cavity-pulling suppression effect can be obtained as the cavity finesse decreases. Therefore, we expand the cavity to the extremely low reflectivity case, so that the usual textbook finesse formula $\mathcal{F} = \frac{\pi\sqrt{R}}{1-R}$ is no longer applicable. In the case of extremely low cavity reflectivity, the textbook expression for finesse is close to 0, but the exact finesse expression $\mathcal{F} = \frac{\pi}{\arccos(\frac{2R}{1+R^2})}$ is close to 2 [29].

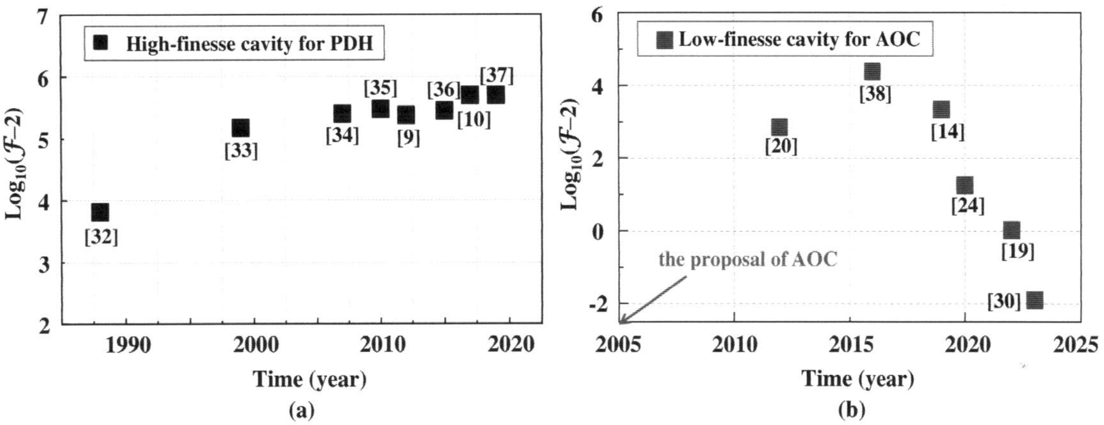

Figure 5. **(a)** Trends in logarithmic finesse of high-finesse cavities for PDH over time [32, 33, 34, 35, 9, 36, 10, 37]. In order to reflect the trend more intuitively, logarithmic coordinates are used here. **(b)** Trends in logarithmic finesse of low-finesse optical cavities used for AOCs over time [20, 38, 14, 24, 19, 30].

Experimentally, the cavity finesse close to its limit of 2. Therefore, we call the laser produced under such conditions an extremely bad-cavity laser [30]. In this case, the feedback of the resonant cavity is finely tuned, and a stable 1470 nm AOC signal output is still achieved, and the power could be maintained at a few tens of microwatts. By building two identical systems for frequency beating, the cavity-pulling coefficient is measured. The influence of cavity-length thermal noise on the clock laser is suppressed to nearly 70 times using this extremely low-finesse cavity. In addition, the linewidth is measured, and similar to the higher finesse case. Finally, the factors affecting the linewidth: the power fluctuations of 459 nm pumping laser and the temperature fluctuations of thermal vapor cell are analyzed, and it can be found that the temperature fluctuations of the thermal atoms are the most important factor affecting the linewidth. That is to say, for the thermal-atom AOC, its performance is mostly limited by the thermal-atom gain medium, including collisional broadening, Doppler broadening, and so on. Therefore, to solve this problem, it is necessary to adopt cold atoms as the gain medium.

3. The future development of active optical clock

3.1. Trends in international development

The AOC uses the stimulated radiation signal of atoms directly as an optical frequency standard and operates in the deep bad-cavity region. The gain linewidth is much narrower than the cavity-mode linewidth, so the laser frequency depends mainly on the quantum transition rather than the cavity-mode center frequency. Currently, one of the major directions in the development of AOCs is cw operation.

In order to realize cw AOC superradiant laser, Kazakov et al. proposed a "hybrid" AOC scheme [39, 40, 41], in which the atoms are prepared outside the cavity and sent into the cavity to complete the excitation process by means of a "guided beam" or "optical conveyor belt". For European Union, the iqClock project led by the University of Amsterdam [41, 42, 43], uses a moving optical lattice to transport cold Sr atomic beams continuously as a gain medium for AOC. Its goal is to achieve a output linewidth of 100 mHz within 5 years, and a linewidth on the order of mHz within 10 years. JILA has proposed a cw AOC scheme using alkaline earth metal atomic beams as the gain medium, which can theoretically achieve linewidths on the order of mHz. In addition, JILA [44] and Katori's group in Japan [45] also carry out experiments with moving optical lattice based on Sr atom, so as to transport atoms continuously and realize cw AOCs. FEMTO-ST [23] in France uses an optical conveyor belt to transport Yb atoms to a ultra-stable cavity in order to realize a cw AOC superradiant laser. In the near future, we believe that continuously operating AOC superradiant lasers based on moving optical lattices will be realized.

The research in our group focuses on thermal atom four-level AOCs, because one great advantage lies in the possibility of achieving continuous output. Previously, a linewidth of 53 Hz was realized in the thermal-atom system by suppressing the remaining cavity-pulling noise with a DW good-bad-cavity system [18], but the final performance is still limited by the thermal atoms. On the theoretical side, the magic wavelength of the 1470 nm transition of the Cs atom [21] was calculated with the population inversion. But in general, it seems that continuously operated AOC based on cold atom have not been realized until now. Here, we propose for the first time an AOC based on diffuse laser cooling that allows continuous operation.

3.2. Active optical clock based on diffuse laser cooled atoms

Typically, cold-atom AOCs are more likely to operate in a pulsed mode due to the preparation cycle of the cold atoms. To achieve a cw output cold-atom AOC, it is necessary to achieve cw laser cooling or to maintain phase coherence between the output pulses. In order to realize a cold-atom AOC that can be operated continuously, we adopt the scheme of a four-level structure combined with diffuse laser cooling, with a view to realizing continuous cooling, repumping, pumping, and suppressing the technical noise factor in the thermal-atom system with the help of cold-atom. The use of a four-level scheme, where the pumping laser and clock laser energy levels are separated from each other, not only reduces the effect of the light shift of the pumping laser, but also allows continuous pumping and thus realize a cw cold-atom AOC. The use of diffuse cooling allows for the preparation cold atom cloud with large volume. In a diffuse light field, which can also be described as an isotropic homogeneous light field, the velocity of atoms correspond to red-detuning of the cooling laser, which is matched automatically by the incident angle, so that diffuse laser cooling allows the capture of atoms over a wide range of velocities.

In the experiment (photographs of the experimental system as shown in Fig. 6), we utilize two 780 nm lasers as a cooling and repumping light, respectively. After appropriate frequency shifting, cooling light is bundled with repumping light and coupled into multimode optical fiber beamsplitters, and then enter into the one-meter-long atomic gas chamber, so that we can prepare a cold atom cloud up to the order of meter long as the gain medium. After preparing the cold atoms, a 420 nm laser is used as the pumping light to excite the cold atoms. Under

Journal of Physics: Conference Series **2889** (2024) 012035 doi:10.1088/1742-6596/2889/1/012035

the weak feedback of the optical cavity, the stimulated radiation of the 1367 nm AOC signal is formed and output. Preliminary experimental results from this work are now being obtained, and a preliminary 1367 nm fluorescence signal output has been achieved without the cavity. We have made a preliminary probe of its power, and the power obtained by probing from one end of the vacuum gas chamber is up to 0.3 μW. Subsequently, we will continue to promote AOC experiments based on diffuse laser cooling, coupled with cavity to achieve a truly stimulated radiation, and conduct performance tests such as linewidth, cavity-pulling suppression and frequency stability.

(a) (b)

Figure 6. (a) Physical picture of the AOC based on diffuse laser cooling. (b) Gas chambers with lengths of 100 cm and 50 cm, respectively, are used to achieve large-volume atomic clouds. They are illuminated by the diffuse light field as well as the pump light and display beautiful colours.

4. Conclusion

In this work, we focus on several recent progress around AOCs at Peking University. The experimental works include: DW good-bad-cavity AOC, an inhibited laser, an extremely bad-cavity laser, and AOC based on diffuse laser cooling that can be realized in continuous operation. Theoretical works include: optical-lattice-based Cs AOC, anti-resonant FP cavity with ultralow finesse. For further improving the frequency stability of AOCs, it can be combined with PDH frequency stabilization, that is to say, combining active and passive optical clocks aiming at realizing higher-performance optical clocks. Moreover, to reduce the classical technical noise due to the motion of thermal atoms in the thermal-atom system, AOCs based on cold atoms need to be further developed. In addition, the realization of a cw cold-atom AOC is also an issue that deserves in-depth investigation. Among the existing technical solutions, the use of a moving optical lattice to continuously transport atoms into cavity and the AOC based on diffuse laser cooling, which was first proposed and experimented by Peking University, are both very promising solutions. In the near future, we believe that continuously operating AOC based on the above two methods will be realized.

Acknowledgments

This research was funded by the National Natural Science Foundation of China (NSFC) (91436210), Innovation Program for Quantum Science and Technology (2021ZD0303200), China Postdoc-toral Science Foundation (BX2021020), and Wenzhou Major Science & Technology Innovation Key Project (ZG2020046).

References

[1] Bloom B J, Nicholson T L, Williams J R, et al. 2014 *An optical lattice clock with accuracy and stability at the* 10^{-18} *level, Nature* **506** p 71

[2] Brewer S M, Chen J S , Hankin A M, et al. 2019 $^{27}Al^{+}$ *quantum-logic clock with a systematic uncertainty below* 10^{-18}, *Physical Review Letters* **123** p 033201

[3] Takamoto M, Ushijima I, Ohmae N, et al. 2020 *Test of general relativity by a pair of transportable optical lattice clocks, Nature Photonics* **14** p 411

[4] Chou C W, Hume D B, Rosenband T, et al. 2010 *Optical clocks and relativity, Science* **329** p 1630

[5] Wcislo P, Ablewski P, Beloy K, et al. 2018 *New bounds on dark matter coupling from a global network of optical atomic clocks, Science Advances* **4** p eaau4869

[6] Cagnoli L, Gammaitoni J, Hough J, et al. 2000 *Very high Q measurements on a fused silica monolithic pendulum for use in enhanced gravity wave detectors, Physical Review Letters* **85** p 2442

[7] Godun R, Nisbet-Jones P, Jones J, et al. 2014 *Frequency ratio of two optical clock transitions in* $^{171}Yb^{+}$ *and constraints on the time variation of fundamental constants, Physical Review Letters* **113** p 210801

[8] Zhang X, Bishof M., Bromley S L, et al. 2014 *Spectroscopic observation of SU (N)-symmetric interactions in Sr orbital magnetism, Science* **345** p1467

[9] Kessler T, Hagemann C, Grebing C, et al. 2012 *A sub 40-mHz-linewidth laser based on a silicon single-crystal optical cavity, Nature Photonics* **6** p 687

[10] Matei D, Legero T, Hafner S, et al. 2017 *1.5 μm lasers with sub-10 mHz linewidth, Physical Review Letters* **118** p 263202

[11] Chen J and Chen X, 2005 *Optical lattice laser, In: Proceedings of the 2005 IEEE International Frequency Control Symposium (IFCS)*, p 608-610

[12] Chen J, 2009 *Active optical clock, Chinese Science Bulletin* **54** p 348

[13] Norcia M A, Cline J R, Muniz J, et al. 2018 *Frequency measurements of superradiance from the strontium clock transition, Physical Review X* **8** p 021036

[14] Laske T, Winter H, Hemmerich A, 2019 *Pulse delay time statistics in a superradiant laser with calcium atoms, Physical Review Letters* **123** p 103601

[15] Schäffer S A, Tang M, Henriksen M R, et al. 2020 *Lasing on a narrow transition in a cold thermal strontium ensemble, Physical Review A* **101** p 013819

[16] Zhang Y, Shan C, Mølmer K, 2022 *Active frequency measurement on superradiant strontium clock transitions, Physical Review Letters* **128** p 013604

[17] Goldenberg H M, Kleppner D, Ramsey NF, 1960 *Atomic hydrogen maser, Physical Review Letters* **5** p 361

[18] Shi T, Pan D, Chen J, 2019 *Realization of phase locking in good-bad-cavity active optical clock, Optics Express* **27** p 22040

[19] Shi T, Pan D, Chen J, 2022 *An inhibited laser, Communications Physics* **5** p 1

[20] Bohnet J G, Chen Z, Weiner J M, et al. 2012 *A steady-state superradiant laser with less than one intracavity photon, Nature* **484** p 78

[21] Pan D, Arora B, Yu Y, et al. 2020 *Optical-lattice-based Cs active clock with a continual superradiant lasing signal, Physical Review A* **102** p 041101

[22] Norcia M A, Winchester M N, Cline J R, et al. 2016 *Superradiance on the millihertz linewidth strontium clock transition, Science Advances* **2** p e1601231

[23] Ponciano-Ojeda F S, El Badawi J, Hauden M, et al. 2023 *Progress on a* ^{171}Yb*-based active optical atomic clock, Joint Conference of the European Frequency and Time Forum and IEEE International Frequency Control Symposium (EFTF/IFCS)* pp 1-2

[24] Liu H, Jäger S B, Yu X, et al. 2020 *Rugged mHz-linewidth superradiant laser driven by a hot atomic beam, Physical Review Letters* **125** p 253602

[25] Zhuang W, Chen J, 2014 *Active faraday optical frequency standard, Optics letters* **39** p 6339

[26] Chen J, 2009 *Active optical clocks, Proceedings of the 7th Symposium in Frequency Standardsand Metrology* p 525

[27] Pan D, Xue X, Zhang X, et al. 2016 *Different ways to active optical frequency standards, Journal of Physics: Conference Series* **723** p 012028

[28] Debnath K, Zhang Y, Mølmer K, 2018 *Lasing in the superradiant crossover regime, Physical Review A* **98** p 063837

[29] Shi T, Miao J, Zhang J, et al. 2023 *Antiresonant Fabry-Pérot cavity with ultralow finesse, Physical Review A* **107** p 023517

[30] Zhang J, Shi T, Miao J, et al. 2023 *An extremely bad-cavity laser, arXiv preprint* pp 2310.14240

[31] Ismail N, Kores C C, Geskus D, et al. 2016 *Fabry-Pérot resonator: spectral line shapes, generic and related Airy distributions, linewidths, finesses, and performance at low or frequency-dependent reflectivity, Optics Express* **24** p 16366

[32] Salomon C, Hils D, Hall J L, 1988 *Laser stabilization at the millihertz level, Journal of the Optical Society of America B* **5** p 1576-1587
[33] Young B C, Cruz F C, Itano W M, et al. 1999 *Visible lasers with subhertz linewidths, Physical Review Letters* **82** p 3799
[34] Ludlow A D, Huang X, Notcutt M, et al. 2007 *Compact, thermal-noise-limited optical cavity for diode laser stabilization at* 1×10^{-15}, *Optics Letters* **32** p 641-643
[35] Jiang Y, Fang S, Bi Z, et al. 2010 *Nd: YAG lasers at 1064 nm with 1-Hz linewidth, Applied Physics B,* **98** p 61-67
[36] Häfner S, Falke S, Grebing C, et al. 2015 8×10^{-17} *fractional laser frequency instability with a long room-temperature cavity, Optics Letters* **40** p 2112-2115
[37] Robinson J M, Oelker E, Milner W R, et al. 2019 *Crystalline optical cavity at 4 K with thermal-noise-limited instability and ultralow drift, Optica* **6** p 240-243
[38] Norcia M A, Thompson J K, 2016 *Cold-strontium laser in the superradiant crossover regime, Physical Review X* **6** p 011025
[39] Kazakov G A, Schumm T, 2013 *Active optical frequency standard using sequential coupling of atomic ensembles, Physical Review A* **87** P 013821
[40] Kazakov G A, Schumm T, 2014 *Active optical frequency standards using cold atoms: perspectives and challenges. In 2014 European Frequency and Time Forum (EFTF)* pp 411-414
[41] Kazakov G, Dubey S, Famà F, et al. 2021 *Towards an active optical clock using an optical conveyor within a ring cavity, 2021 Joint Conference of the European Frequency and Time Forum and IEEE International Frequency Control Symposium (EFTF/IFCS)* pp 1-5
[42] Chen C C, Bennetts S, Escudero R G, et al. 2019 *Continuous guided strontium beam with high phase-space density, Physical Review Applied* **12** p 044014
[43] Escudero R G, Chen C C, Bennetts S, et al. 2021 *Steady-state magneto-optical trap of fermionic strontium on a narrow-line transition, Physical Review Research* **3** p 033159
[44] Cline J R K, Schäfer V M, Niu Z, et al. *Continuous collective strong coupling between atoms and a high finesse optical cavity, arXiv preprint* pp 2211.00158
[45] Takeuchi R, Chiba H, Okaba S, et al. 2023 *Continuous outcoupling of ultracold strontium atoms combining three different traps, Applied Physics Express* **16** p 042003

Journal of Physics: Conference Series **2889** (2024) 012036
doi:10.1088/1742-6596/2889/1/012036

Progress on cadmium-ion and ytterbium-ion Microwave Frequency Standards at Tsinghua University

S. N. Miao[1], Y. Zheng[1,2], Y. T. Chen[1], W. X. Shi[1], J. W. Zhang[1] and L. J. Wang[1,2]

[1] State Key Laboratory of Precision Measurement Technology and Instruments, Department of Precision Instruments, Tsinghua University, Beijing 100084, China
[2] Department of Physics, Tsinghua University, Beijing 100084, China

E-mail: zhangjw@tsinghua.edu.cn

Abstract. As a candidate for the next generation microwave clock, microwave atomic clocks based on trapped ions are currently studied worldwide. Since 2010, our team at Tsinghua University has been committed to developing a high-performance $^{113}Cd^{+}$ microwave ion clock. In 2021, the laser-cooled cadmium-ion microwave clock constructed in our laboratory demonstrated remarkable performance, with a short-term instability of $4.2 \times 10^{-13}/\sqrt{\tau}$ and a relative frequency uncertainty of 1.8×10^{-14}, improving nearly three orders of magnitude compared to the result from JPL. The 0-0 ground-state hyperfine splitting frequency of $^{113}Cd^{+}$ was determined to be 15199862855.02799(27) Hz, currently the most accurate measurement result for this transition frequency. Apart from that, we introduced the sympathetic cooling technology into the experiment of cadmium-ion microwave clock. Using $^{40}Ca^{+}$ as the coolant ion, we obtained Ramsey patterns with a free evolution time of up to 100 seconds, ultimately improving the short-term instability of the system to $3.48 \times 10^{-13}/\sqrt{\tau}$. Moreover, a microwave frequency standard based on laser-cooled $^{171}Yb^{+}$ ions has also been developed in our laboratory since 2021. The short-term frequency instability was measured to be $8.5 \times 10^{-13}/\sqrt{\tau}$. The ground-state hyperfine splitting frequency of $^{171}Yb^{+}$ was determined to be 12642812118.4674(8) Hz with a fractional frequency uncertainty of 6.33×10^{-14}.

1. Introduction

The atomic clock is an instrument that achieves precise frequency output based on the transition frequency of atoms or ions. The transition frequency between two energy levels of an atom is an inherent property that remains constant in the absence of external disturbances. Depending on the operating frequency, atomic clocks can be divided into two main categories: microwave atomic clocks and optical atomic clocks. Among them, optical atomic clocks surpass microwave atomic clocks in terms of accuracy and stability. However, due to factors such as size and cost, optical atomic clocks are currently mainly applied in fundamental research [1-2]. Microwave atomic clocks have a relatively simple structure, and low-cost, compact-volume microwave atomic clocks can be tailored to different needs. They remain the primary frequency standard used in timekeeping, synchronization, and navigation. Currently, all of the microwave atomic clocks are based on the transition spectra of neutral atoms or ions. Compared to neutral atoms, microwave atomic clocks based on trapped ions have advantages in miniaturization and stability, making them widely researched and considered candidates for the next generation of commercial practical microwave clocks. Commonly used ions in the development of microwave ion clocks include $^{199}Hg^{+}$, $^{171}Yb^{+}$ and $^{113}Cd^{+}$, with corresponding clock transition frequencies of 40.5 GHz, 12.6 GHz and 15.2 GHz, respectively. Larger clock transition frequencies correspond to higher expected performance.

Due to its large ground-state hyperfine splitting frequency, $^{199}Hg^+$ has become a popular ion for microwave ion clock research. Since the 1980s, the National Aeronautics and Space Administration Jet Propulsion Laboratory (NASA JPL) has been investigating a $^{199}Hg^+$ microwave ion clock based on lamp pumping and buffer-gas cooling. The latest generation of their mercury-ion clock, LITS-10, achieves a short-term frequency stability of $4.5 \times 10^{-14}/\sqrt{\tau}$, surpassing 2.4×10^{-16} in the stability in the long term [3]. Their developed Deep Space Atomic Clock (DSAC) combines advantages such as small size (17L, 16kg), low power consumption (17W), and high performance [4]. Its short-term frequency stability on the ground and in space reaches $1.5 \times 10^{-13}/\sqrt{\tau}$ and $7 \times 10^{-13}/\sqrt{\tau}$, respectively. The stability during continuous operation in space for 23 days is 3×10^{-15} [5]. Additionally, the National Institute of Standards and Technology (NIST) has developed a mercury-ion microwave clock based on laser cooling. By trapping seven $^{199}Hg^+$ ions in an ultra-cold vacuum and maintaining an ultra-low temperature of 25 mK, they can achieve a Ramsey free evolution time of up to 100 seconds, with a short-term frequency instability of $3.3 \times 10^{-13}/\sqrt{\tau}$ [6].

The wavelengths of laser beams required for the cooling and detection of $^{171}Yb^+$ ions are 369 nm and 935 nm, making them more easily accessible. Therefore, microwave ion clocks based on $^{171}Yb^+$ ions have become a research hotspot in recent years. In 2014, the National Institute of Information and Communications Technology (NICT) in Japan developed a laser-cooled ytterbium-ion microwave clock with a short-term frequency stability of $2.09 \times 10^{-12}/\sqrt{\tau}$ and a relative frequency uncertainty of 1.1×10^{-14} [7]. The National Physical Laboratory (NPL) in the United Kingdom developed a portable laser-cooled ytterbium-ion microwave clock. The entire microwave clock system is enclosed in a 6U chassis, with a short-term frequency stability of $3.6 \times 10^{-12}/\sqrt{\tau}$ and a frequency shift uncertainty of 8×10^{-12} [8-9].

Compared to $^{199}Hg^+$ and $^{171}Yb^+$, there has been relatively less investigation into $^{113}Cd^+$. The nuclear spin of $^{113}Cd^+$ is 1/2, and its upper level $^2P_{3/2}$ hyperfine splitting frequency is 800 MHz. Only a 214.5 nm laser is needed to cool, pump, and detect the ion, demonstrating the potential for miniaturization. In 1996, Tanaka et al. from Japan employed a buffer-gas cooling scheme and measured the ground-state hyperfine splitting frequency of $^{113}Cd^+$ as 15199862858(2) Hz with a relative frequency uncertainty of 1.3×10^{-10} [10]. In 2006, Jelenković et al. improved the measurement uncertainty to 1.3×10^{-11} via a microwave optical double resonance technique, where the ions were also cooled by buffer gas, and pumped and detected by a discharge lamp. They also estimated the instability of their $^{113}Cd^+$ clock to be $2 \times 10^{-13}/\sqrt{\tau}$, but no experimental data were reported [11].

Since 2010, our team at Tsinghua University has been committed to developing a high-performance cadmium-ion microwave clock. In 2015, we implemented closed-loop locking of the cadmium-ion microwave clock system, achieving a short-term instability of $6.1 \times 10^{-13}/\sqrt{\tau}$ [12]. Apart from that, we measured the ground-state hyperfine splitting frequency of $^{113}Cd^+$ to be 15199862855.0192(10) Hz, with a relative frequency uncertainty of 6.6×10^{-14} [12]. Subsequently, we further optimized the system and introduced the sympathetic cooling technique into the study of the cadmium-ion microwave clock. Additionally, we initiated the study of a lased-cooled ytterbium-ion microwave clock in 2021, and have achieved significant progress in this phase. This paper reports the latest progress made by our team at Tsinghua University in the research on cadmium-ion microwave clocks and ytterbium-ion microwave clocks.

2. Experimental setup and results

2.1. Laser-cooled cadmium-ion microwave clock

The relevant energy level structure of $^{113}Cd^+$ is shown in Figure 1. The ions are cooled and detected by a 214.5-nm red detuned laser via the cycling transition between the states of $|^2S_{1/2}, F = 1, m_F = 1\rangle$ and $|^2P_{3/2}, F = 2, m_F = 2\rangle$. During laser cooling, ions may transition to the energy level of

$|^2P_{3/2}, F = 1\rangle$ because of the relatively small frequency difference between the states of $|^2P_{3/2}, F = 1\rangle$ and $|^2P_{3/2}, F = 2\rangle$. Part of the ions will fall to the energy of $|^2S_{1/2}, F = 0\rangle$ through radiation. Therefore, a cooling laser with circular polarization connecting two Zeeman sublevels of the cycling transition is applied. In addition, microwave radiation of 15.2 GHz is applied to repump the ions from the dark state of $|^2S_{1/2}, F = 0\rangle$. Because the hyperfine splitting frequency of $|^2P_{3/2}\rangle$ is only 800 MHz, the pump laser beam is generated by blue-shifting the cooling laser beam using acousto-optic modulators (AOMs). The frequency difference between the states of $|^2S_{1/2}, F = 0, m_F = 0\rangle$ and $|^2S_{1/2}, F = 1, m_F = 0\rangle$ is the chosen transition frequency, which is approximately 15.2 GHz.

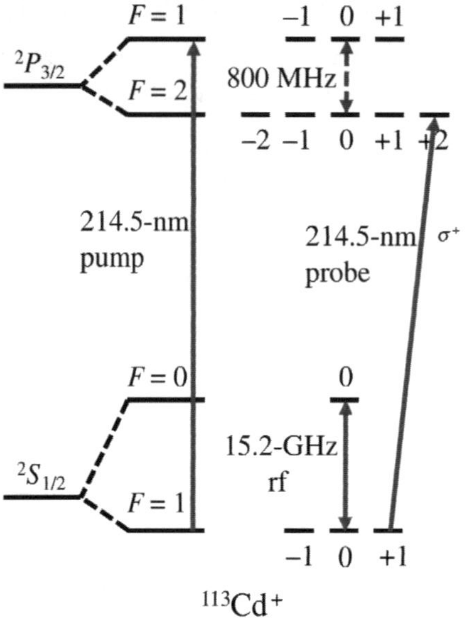

Figure 1. Relevant energy-level structure of $^{113}Cd^+$.

In 2020, we designed and constructed a new cadmium-ion microwave clock system, primarily composed of an ion trap and vacuum system, laser and fluorescence detection system, microwave system, and control system (Figure 2). The core of the entire system is a linear quadrupole ion trap. Details of the ion trap are given in [13-14]; here, a brief description suffices. Each electrode is divided into three segments: the middle segment with length of 40 mm that acts as radio frequency (RF) electrodes for radial confinement of ions, and the two ends of length 20 mm that act as end-cap (EC) electrodes for axial confinement. The radius of each electrode is 7.1 mm. The minimum distance from the nodal line of the trap to the electrode surface is 6.2 mm. The ion trap is horizontally mounted in a stainless-steel vacuum chamber. The pressure in the vacuum chamber is maintained below 1.5×10^{-11} mbar by an ion pump. A high-performance magnetic shield barrel composed of five layers of permalloy and a layer of soft iron around the vacuum chamber is used to minimize stray magnetic fields. In addition, the barrel is motor driven to move up and down. After active shielding, the magnetic field fluctuation in the center of the trap is reduced to less than 0.1 nT. To make the system more compact, three pairs of Helmholtz coils were fixed around the vacuum chamber to generate a static magnetic field to split the Zeeman sublevels and to compensate for any residual magnetic field. Moreover, to reduce electrical noise, three high-precision current sources were used to power the Helmholtz coils.

Figure 2. Schematic diagram of the cadmium-ion microwave clock system.

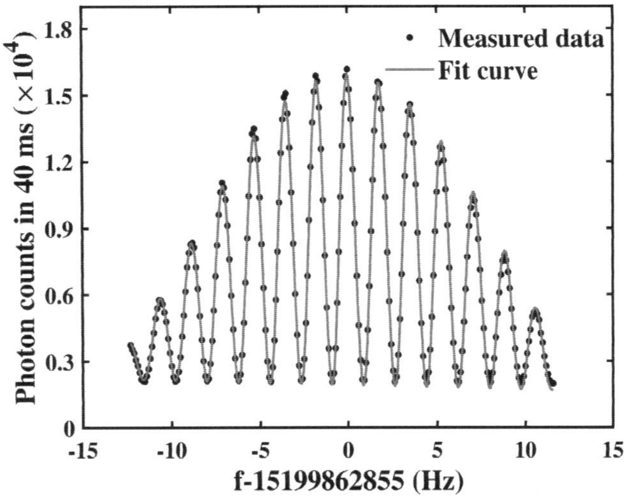

Figure 3. Typical Ramsey fringe pattern obtained with a 500-ms cooling time, a 50-ms pump time, two phase-coherent microwave pulses of 60-ms duration separated by a free evolution time of 500-ms, and a fluorescence signal integration time of 400-ms. The cycle time to collect a point is 1.59 s. The scanning step is 0.1 Hz. The solid line is a nonlinear fit obtained from the experimental data [15].

The Ramsey's method of separated oscillatory fields was employed to acquire the clock transition spectrum. The typical Ramsey fringe is shown in Figure 3. The cooling time was set to 500-ms, with two microwave pulses each lasting 60-ms and spaced 500-ms apart, followed by a fluorescence detection time of 400-ms. In the experiment, the period for acquiring one data point was 1.59 s, and the signal-to-noise ratio (SNR) of the clock transition spectrum was approximately 40. During the closed-loop locking process, the frequency synthesizer (E8257D) was referenced to the local oscillator (LO). The fluorescence intensity at the half-maximum points on both sides of the central peak of the Ramsey pattern for $^{113}Cd^{+}$ was measured, and the frequency deviation between the LO and the ion was calculated. This deviation was fed back to the LO through proportional-integral (PI) control, resulting in the locking of the local oscillator's output frequency to the ion spectrum. Finally, the output frequency of the LO was compared with that of a hydrogen clock to obtain the frequency stability of the cadmium-ion microwave clock, as shown in Figure 4.

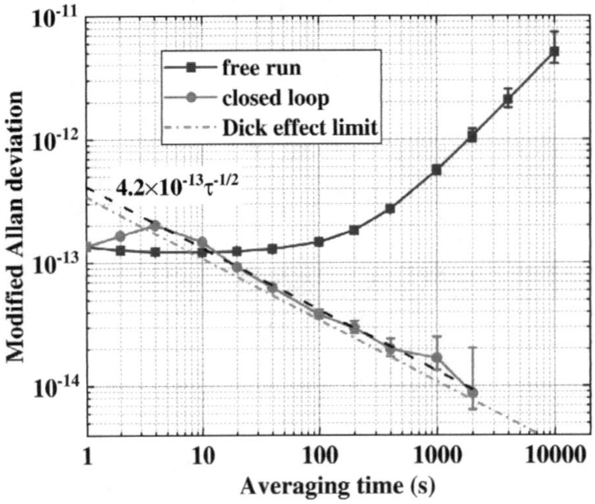

Figure 4. Modified Allan deviations of the $^{113}Cd^{+}$ microwave ion clock. The solid blue line is a fitting of data from the free-running local oscillator, the solid red line is that after closed-loop locking, and the pink dotted line signifies the Dick effect limit. The short-term fractional frequency stability of the clock is estimated to be $4.2 \times 10^{-13}/\sqrt{\tau}$ [15].

The loop time is 3.18 s, which is the time to collect the error data at the half height of the central peak of the Ramsey fringe pattern, and corresponds to the time constant. Allan deviations longer than the time constant stem from the cadmium ions of the clock; those shorter than the time constant stem from the LO itself. After a long duration making comparisons, the fractional frequency stability of the $^{113}Cd^{+}$ microwave ion clock was estimated to be $4.2 \times 10^{-13}/\sqrt{\tau}$ [15], which reduces by nearly half that of the previous one [12]. Under the current experimental parameters with a loop time of 3.18 s, the fractional frequency stability limit due to the Dick effect is $3.43 \times 10^{-13}/\sqrt{\tau}$ [16], which implies that the noise of the LO is already the main factor limiting the short-term stability of the $^{113}Cd^{+}$ microwave ion clock.

Table 1. Estimated systematic frequency shifts and uncertainties of laser-cooled $^{113}Cd^{+}$ microwave ion clock [15].

Shift	Magnitude (10^{-15})	Uncertainty (10^{-15})
Second-order Zeeman shift (SOZS)	105720	3
Second-order Doppler shift (SODS)	-65.9	3.6
BBR Stark shift	-0.1815	<0.1
BBR Zeeman shift	-0.00981	<0.01
Gravitational red shift	4.7	0.1
Pressure shift	0	<0.1
Light shift	0	0
Total	105658.6	4.6

The main systematic frequency shifts and their uncertainties of the $^{113}Cd^+$ microwave ion clock are listed in Table 1. Among them, the second-order Zeeman frequency shift and the second-order Doppler frequency shift are two primary factors limiting the uncertainty of the system. The uncertainty of the second-order Zeeman frequency shift is primarily attributed to fluctuations in the geomagnetic field, which can be minimized by enhancing the performance of the magnetic shielding device. The uncertainty of the second-order Doppler frequency shift arises mainly from the secular motion of trapped ions. The continuous reduction in the number of ions and the heating of ions during the microwave interrogation process also play a role. By employing multipole ion traps to position ions as closely as possible along the axis of the ion trap, the excess micromotion of trapped ions can be reduced. Another approach is to use ions with a mass number greater than that of $^{113}Cd^+$ for sympathetic cooling of $^{113}Cd^+$, ensuring sustained cooling of $^{113}Cd^+$ ions during the microwave interrogation process.

In our experiment, the frequency measurement is referenced against an active hydrogen clock, and that the fractional frequency difference between this active hydrogen clock and coordinated universal time (UTC) was measured by the method of GPS common-view, the difference between UTC and the primary frequency standards (PFSs) was also considered. To obtain the 0-0 ground-state hyperfine transition frequency of $^{113}Cd^+$, the majority of the systematic frequency shifts and corresponding frequency uncertainties have been evaluated carefully. Ultimately, the 0-0 ground-state hyperfine splitting frequency of $^{113}Cd^+$ was determined to be 15199862855.02799(27) Hz, with a relative frequency shift uncertainty of 1.8×10^{-14} [15]. Table 2 lists all the measurement results of the 0-0 ground-state hyperfine splitting frequency for $^{113}Cd^+$. Our measurement result was consistent with previous results. To the best of our knowledge, the measurement accuracy is currently the highest among all laser-cooled $^{113}Cd^+$ microwave atomic clocks.

Table 2. Comparison of different measurement results of $^{113}Cd^+$ hyperfine splitting of the ground state.

Group	Measurement result (Hz)	Uncertainty	Reference
Kansai Advanced Research Center (1996)	15199862858(2)	1.3×10^{-10}	[10]
Jet Propulsion Laboratory (2006)	15199862855.0(2)	1.3×10^{-11}	[11]
Tsinghua University (2012)	15199862854.96(12)	7.9×10^{-12}	[17]
Tsinghua University (2013)	15199862855.0125(87)	5.7×10^{-13}	[18]
Tsinghua University (2015)	15199862855.0287(10)	6.6×10^{-14}	[12]
Tsinghua University (2021)	15199862855.02799(27)	1.8×10^{-14}	[15]

2.2. Sympathetically-cooled cadmium-ion microwave clock

The performance improvement of laser-cooled cadmium-ion microwave clocks is subject to certain limitations: firstly, during the microwave interrogation, effective cooling of cadmium ions cannot be achieved, resulting in an increase in ion temperature; secondly, the cooling process belongs to the system's dead time, which is unavoidable. Thus, we introduced the technology of sympathetic cooling into the cadmium-ion microwave clock. The basic principle of sympathetic cooling relies on the Coulomb interaction between ions, realizing continuous cooling of one type of ion (target ion, sympathetic cooling) by another type of ion (coolant ion, laser cooling). Compared to laser-cooled microwave ion clocks, sympathetically-cooled microwave ion clocks have the following advantages: first, there is no need for the laser cooling process, reducing the dead time of the loop and effectively lowering the Dick effect; second, continuous cooling extends the coherence time of ions, further narrowing the linewidth of clock transition; third, maintaining the low-temperature state of target ions continuously reduces the Doppler frequency shift and its uncertainty of ions.

In 2018, we used $^{24}Mg^+$ as the coolant ion, achieving sympathetic cooling of a large mixed ion crystal containing approximately 3.3×10^5 ions [19]. However, $^{24}Mg^+$ ions easily react with hydrogen gas in the background, generating MgH^+ dark ions, which cannot be used for long-term cooling of $^{113}Cd^+$ ions. In 2019, we changed the coolant to the more stable $^{40}Ca^+$ ion [20]. Figure 5 shows the experimentally obtained $^{40}Ca^+$-$^{113}Cd^+$ two-component ion crystal [21]. Figure 5(a) is an image without a filter; figure 5(b) is a picture taken separately under filters corresponding to the two ions, later synthesized and subjected to simple color correction; figure 5(c) is the result obtained from molecular dynamics (MD) simulation, consistent with the experimental results [22]. In the two-component ion crystal, the lighter $^{40}Ca^+$ forms an ellipsoidal structure at the center, while the heavier $^{113}Cd^+$ surrounds the $^{40}Ca^+$ ellipsoid, forming a hollow cylindrical structure. Under sympathetic cooling conditions, a typical temperature of 80 mK was measured for the $^{113}Cd^+$ ion cloud through 214.5 nm detection light scanning. In addition, optimization was performed on the RF voltage and EC voltage during the experiment.

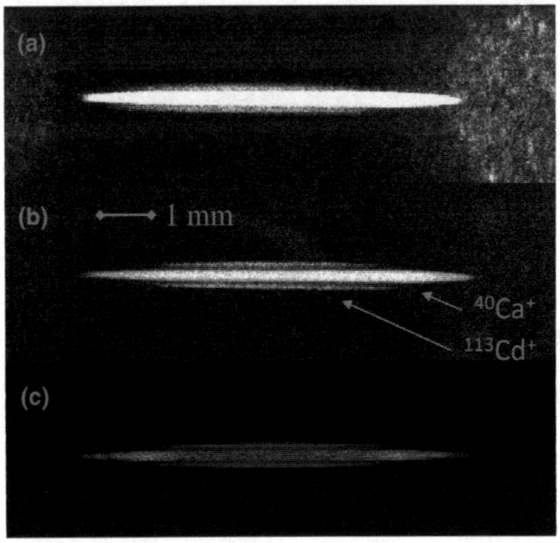

Figure 5. Typical image of $^{40}Ca^+$-$^{113}Cd^+$ two-component ion crystal. In these images, $^{40}Ca^+$ ions appear white and $^{113}Cd^+$ ions appear blue. (a) Image taken by EMCCD without UV filters. (b) Image synthesized from ion images under different wavelength filters, taking into account the chromatic aberration effects. (c) Simulation image of the two-component ion crystal under the same conditions as for (a) and (b) [21].

Using sympathetic cooling technology, we obtained Ramsey fringes with a maximum free evolution time of up to 100 seconds. Considering that the SNR of the signal decreases as the free evolution time increases, a free evolution time of 5 seconds was set during closed-loop locking in the experiment. The loop time was approximately 11 seconds, and the short-term frequency stability of the $^{40}Ca^+$-$^{113}Cd^+$ cadmium-ion microwave clock was measured to be $3.48 \times 10^{-13}/\sqrt{\tau}$ [21]. Compared to the optimal short-term frequency stability of a laser-cooled cadmium-ion microwave clock [15], the sympathetically-cooled cadmium-ion microwave clock demonstrates nearly twice the performance in short-term frequency stability. In comparison to NIST's laser-cooled $^{199}Hg^+$ microwave clock [6], the short-term frequency stability of the sympathetically-cooled $^{113}Cd^+$ microwave clock is similar, despite the clock transition frequency of $^{113}Cd^+$ being only one-third that of $^{199}Hg^+$.

Journal of Physics: Conference Series **2889** (2024) 012036	doi:10.1088/1742-6596/2889/1/012036

Currently, the performance of the sympathetically-cooled cadmium-ion microwave clock is primarily constrained by drawbacks of the $^{40}Ca^+$-$^{113}Cd^+$ sympathetic cooling scheme. In the $^{40}Ca^+$-$^{113}Cd^+$ bi-component ion crystal, the $^{113}Cd^+$ ions, positioned on the outer side of the crystal, experience Coulomb repulsion from the $^{40}Ca^+$ ions, which results in increased excess micromotion of cadmium ions. To overcome the limitations imposed by the second-order Doppler frequency shift and its uncertainty, we have used $^{174}Yb^+$ as the coolant ion. The advantages are as follows: firstly, the mass difference between $^{174}Yb^+$ and $^{113}Cd^+$ is smaller, thereby expecting to improve the efficiency of sympathetic cooling. Secondly, $^{113}Cd^+$ is positioned on the inner side in the formed bi-component ion crystal, resulting in relatively smaller radio-frequency heating experienced by $^{113}Cd^+$. In the future, the short-term frequency stability of the cadmium-ion microwave clock may be improved to $2 \times 10^{-13}/\sqrt{\tau}$ with $^{174}Yb^+$ as the coolant ion.

2.3. *Laser-cooled ytterbium-ion microwave clock*

The partial energy levels of $^{171}Yb^+$ are shown in Figure 6. The ground-state hyperfine splitting frequency is approximately 12.6 GHz. The cycling transition used for laser cooling is $|^2S_{1/2}, F = 1, m_F = 0\rangle \leftrightarrow |^2P_{1/2}, F = 0, m_F = 0\rangle$. The clock signal originates from $|^2S_{1/2}, F = 1, m_F = 0\rangle \leftrightarrow |^2S_{1/2}, F = 0, m_F = 0\rangle$. The laser-cooled ytterbium-ion microwave clock system is illustrated in Figure 7: the entire system, divided into four layers, is based on a movable aluminum frame. The top and bottom layers' plates are used to place laser and electronic devices. The middle-bottom layer houses an ion pump and a titanium pump to maintain ultra-high vacuum. The middle-top layer contains the vacuum chamber and the ion trap for confining ions. Three layers of magnetic shielding surround the coils producing the static magnetic field, the microwave horn, and the vacuum chamber. The parameters of the ion trap used in the system are consistent with those of the cadmium-ion microwave clock [15, 21].

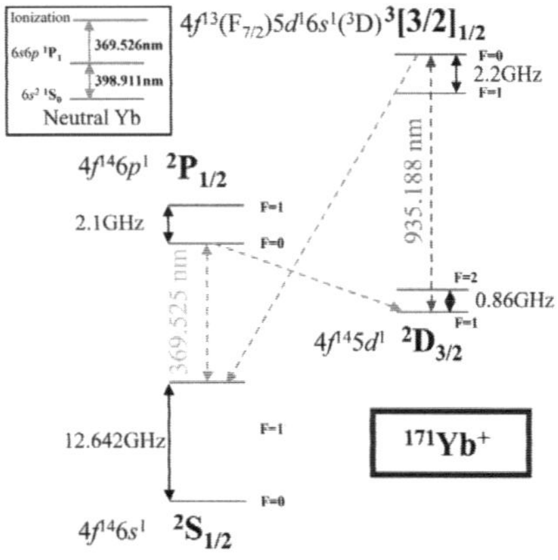

Figure 6. Schematic energy levels of single $^{171}Yb^+$ ion (not to scale) [23].

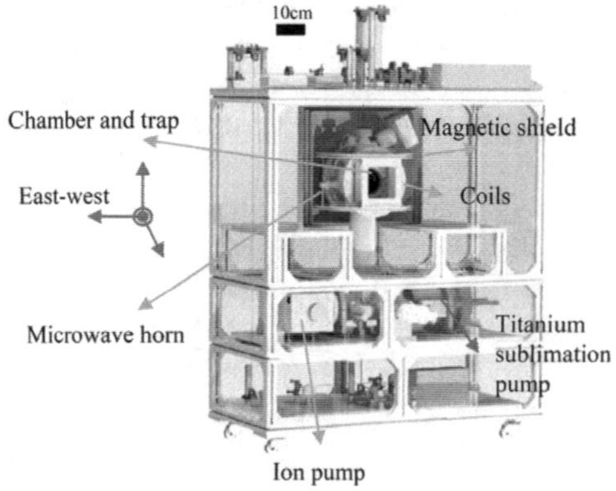

Figure 7. Detailed sketch of the $^{171}Yb^{+}$ ion microwave clock system [23].

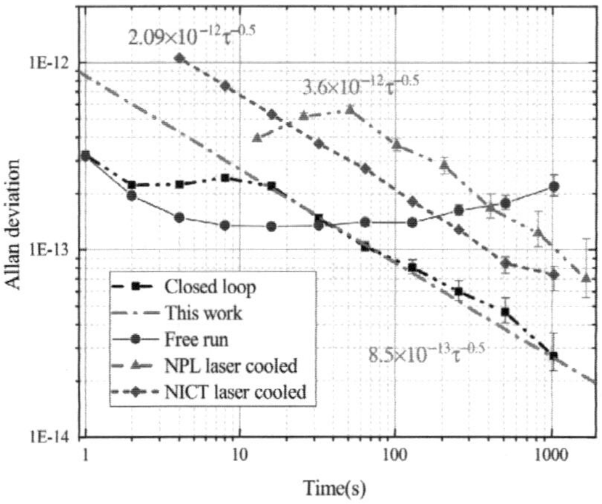

Figure 8. Allan deviations of the $^{171}Yb^{+}$ clock and the free-running OCXO. Other scholars' primary and significant work are also plotted, including results from NICT and NPL. The system was run continuously for 13000s. The frequency reference is generated by a hydrogen maser [23].

During closed-loop locking, a constant temperature crystal oscillator OCXO (Oven Control Crystal Oscillator) was used as the local oscillator, and then the output frequency of the crystal oscillator was compared with the output frequency of an active hydrogen clock using a 51300A phase noise analyzer produced by Microchip. The short-term stability of the system was ultimately determined to be $8.5 \times 10^{-13}/\sqrt{\tau}$ (Figure 8) [23]. Compared to laser-cooled ytterbium-ion microwave clocks built by other teams such as NPL in the United Kingdom [8-9] and NICT in Japan [7], the short-term stability of this system has improved by 2-3 times.

The main systematic frequency shifts and their uncertainties of the laser-cooled ytterbium-ion microwave clock system are listed in Table 3 [23]. Currently, the uncertainty of the second-order Zeeman frequency shift is the main factor limiting the accuracy of the ytterbium-ion microwave clock. In the future, it is possible to improve the magnetic shielding effect by increasing the number of layers of the magnetic shielding device and reducing the gap area of the magnetic shielding device, thereby reducing the second-order Zeeman frequency shift and its uncertainty.

Table 3. Estimated systematic uncertainties of laser-cooled $^{171}Yb^+$ microwave ion clock [23].

Shift type	Uncertainty (10^{-15})
Second-order Doppler shift (SODS)	3
Second-order Zeeman shift (SOZS)	<9
Black body radiation shift	1
Quadratic Stark shift	0.3
Gravitational red shift	0.1
Pressure shift	0
Light shift	0
Total	<9.5

3. Conclusion

As a candidate for the next generation practical microwave clock, microwave atomic clocks based on trapped ions are currently a research focus in the field of atomic clocks worldwide. The $^{113}Cd^+$ ion has attracted extensive attention due to its advantages, such as requiring fewer laser beams and having a large hyperfine splitting frequency. Since 2010, our team at Tsinghua University has been consistently devoted to building a high-performance cadmium-ion microwave clock. Using laser cooling technology, the cadmium-ion microwave clock constructed in our laboratory demonstrated outstanding performance, with a short-term instability of $4.2 \times 10^{-13}/\sqrt{\tau}$ and a relative frequency uncertainty of 1.8×10^{-14}, improving nearly three orders of magnitude compared to the result from JPL. Additionally, the 0-0 ground-state hyperfine splitting frequency of $^{113}Cd^+$ was determined to be 15199862855.02799(27) Hz, currently the most accurate measurement result in the world. To further improve the short-term stability of the cadmium-ion microwave clock, we introduced the sympathetic cooling technology into the experiment. Using $^{40}Ca^+$ as the coolant ion, we obtained Ramsey patterns with a free evolution time of up to 100 seconds, ultimately improving the short-term stability of the cadmium-ion microwave clock to $3.48 \times 10^{-13}/\sqrt{\tau}$, showcasing the advantages of microwave ion clocks based on sympathetic cooling scheme. Furthermore, we conducted research on laser-cooled ytterbium-ion microwave clocks. After closed-loop locking, the short-term stability of the system reached $8.5 \times 10^{-13}/\sqrt{\tau}$. Compared to ytterbium-ion microwave clocks built by other teams, the short-term stability of our system improved by 2-3 times.

Overall, the results obtained by our team at Tsinghua University enrich the studies of microwave ion clocks and advancing the transition of microwave ion clocks towards commercial practical microwave clocks.

Acknowledgments

The authors thank previous team members, Hao-Ran Qin, Nong-Chao Xin, Ji-Ze Han, Ya-Ni Zuo, Kai Miao, Peng-Fei Cheng, Xiao-Lin Sun, Shi-Guang Wang and Zheng-Bo Wang for their outstanding works. This work is supported by National Key Research and Development Program of China (2021YFA1402101), and National Natural Science Foundation of China (12073015).

References

[1] Brewer, S.M., Chen, J.S., Hankin, A.M., Clements, E.R., Chou, C.W., Wineland, D.J., Hume, D.B. and Leibrandt, D.R., 2019. $^{27}Al^+$ quantum-logic clock with a systematic uncertainty below 10^{-18}. *Physical review letters*, **123**(3), p.033201.

[2] Bothwell, T., Kennedy, C.J., Aeppli, A., Kedar, D., Robinson, J.M., Oelker, E., Staron, A. and

Ye, J., 2022. Resolving the gravitational redshift across a millimetre-scale atomic sample. *Nature*, **602**(7897), pp.420-424.

[3] Burt, E.A., Yi, L., Tucker, B., Hamell, R. and Tjoelker, R.L., 2016. JPL ultrastable trapped ion atomic frequency standards. *IEEE transactions on ultrasonics, ferroelectrics, and frequency control*, **63**(7), pp.1013-1021.

[4] Tjoelker, R.L., Prestage, J.D., Burt, E.A., Chen, P., Chong, Y.J., Chung, S.K., Diener, W., Ely, T., Enzer, D.G., Mojaradi, H. and Okino, C., 2016. Mercury ion clock for a NASA technology demonstration mission. *IEEE transactions on ultrasonics, ferroelectrics, and frequency control*, **63**(7), pp.1034-1043.

[5] Burt, E.A., Prestage, J.D., Tjoelker, R.L., Enzer, D.G., Kuang, D., Murphy, D.W., Robison, D.E., Seubert, J.M., Wang, R.T. and Ely, T.A., 2021. Demonstration of a trapped-ion atomic clock in space. *Nature*, **595**(7865), pp.43-47.

[6] Berkeland, D.J., Miller, J.D., Bergquist, J.C., Itano, W.M. and Wineland, D.J., 1998. Laser-cooled mercury ion frequency standard. *Physical Review Letters*, **80**(10), p.2089.

[7] Phoonthong, P., Mizuno, M., Kido, K. and Shiga, N., 2014. Determination of the absolute microwave frequency of laser-cooled $^{171}Yb^+$. *Applied Physics B*, **117**, pp.673-680.

[8] Mulholland, S., Donnellan, S., Barwood, G.P., Gentle, D., Huang, G., Klein, H.A., Patel, P., Walsh, G., Baird, P.E.G. and Gill, P., 2018. A portable microwave clock using laser-cooled trapped $^{171}Yb^+$ ions. *IEEE International Frequency Control Symposium (IFCS)* (pp. 1-2). IEEE.

[9] Mulholland, S., Klein, H.A., Barwood, G.P., Donnellan, S., Gentle, D., Huang, G., Walsh, G., Baird, P.E.G. and Gill, P., 2019. Laser-cooled ytterbium-ion microwave frequency standard. *Applied Physics B*, **125**(11), p.198.

[10] Tanaka, U., Imajo, H., Hayasaka, K., Ohmukai, R., Watanabe, M. and Urabe, S., 1996. Determination of the ground-state hyperfine splitting of trapped $^{113}Cd^+$ ions. *Physical Review A*, **53**(6), p.3982.

[11] Jelenković, B.M., Chung, S., Prestage, J.D. and Maleki, L., 2006. High-resolution microwave-optical double-resonance spectroscopy of hyperfine splitting of trapped $^{113}Cd^+$ ions. *Physical Review A*, **74**(2), p.022505.

[12] Miao, K., Zhang, J.W., Sun, X.L., Wang, S.G., Zhang, A.M., Liang, K. and Wang, L.J., 2015. High accuracy measurement of the ground-state hyperfine splitting in a $^{113}Cd^+$ microwave clock. *Optics letters*, **40**(18), pp.4249-4252.

[13] Zhang, J.W., Wang, S.G., Miao, K., Wang, Z.B. and Wang, L.J., 2014. Toward a transportable microwave frequency standard based on laser-cooled $^{113}Cd^+$ ions. *Applied Physics B*, **114**, pp.183-187.

[14] Wang, Z.B., 2011. Ion storage, cooling and applications in quantum frequency standards. *Scientia Sinica Physica, Mechanica & Astronomica*, **41**(4), p.350.

[15] Miao, S.N., Zhang, J.W., Qin, H.R., Xin, N.C., Han, J.Z. and Wang, L.J., 2021. Precision determination of the ground-state hyperfine splitting of trapped $^{113}Cd^+$ ions. *Optics letters*, **46**(23), pp.5882-5885.

[16] Zhang, J.W., Miao, K. and Wang, L.J., 2015. Dick effect in a microwave frequency standard based on laser-cooled $^{113}Cd^+$ ions. *Chinese Physics Letters*, **32**(1), p.010601.

[17] Zhang, J.W., Wang, Z.B., Wang, S.G., Miao, K., Wang, B. and Wang, L.J., 2012. High-resolution laser microwave double-resonance spectroscopy of hyperfine splitting of trapped $^{113}Cd^+$ and $^{111}Cd^+$ ions. *Physical Review A*, **86**(2), p.022523.

[18] Wang, S.G., Zhang, J.W., Miao, K., Wang, Z.B. and Wang, L.J., 2013. High-accuracy measurement of the $^{113}Cd^+$ ground-state hyperfine splitting at the milli-Hertz level. *Optics express*, **21**(10), pp.12434-12442.

[19] Zuo, Y.N., Han, J.Z., Zhang, J.W. and Wang, L.J., 2019. Direct temperature determination of a sympathetically cooled large $^{113}Cd^+$ ion crystal for a microwave clock. *Applied Physics Letters*, **115**(6).

[20] Han, J.Z., Qin, H.R., Xin, N.C., Yu, Y.M., Dzuba, V.A., Zhang, J.W. and Wang, L.J., 2021. Toward a high-performance transportable microwave frequency standard based on sympathetically cooled $^{113}Cd^{+}$ ions. *Applied Physics Letters*, **118**(10).

[21] Qin, H.R., Miao, S.N., Han, J.Z., Xin, N.C., Chen, Y.T., Zhang, J.W. and Wang, L.J., 2022. High-performance microwave frequency standard based on sympathetically cooled ions. *Physical Review Applied*, **18**(2), p.024023.

[22] Miao, S.N., Qin, H.R., Xin, N.C., Han, J.Z., Chen, Y.T., Zhang, J.W. and Wang, L.J., 2023. Sympathetic cooling of a large $^{113}Cd^{+}$ ion crystal with $^{40}Ca^{+}$ in a linear Paul trap. *Chinese Journal of Physics*, **83**, pp.242-252.

[23] Xin, N.C., Qin, H.R., Miao, S.N., Chen, Y.T., Zheng, Y., Han, J.Z., Zhang, J.W. and Wang, L.J., 2022. Laser-cooled $^{171}Yb^{+}$ microwave frequency standard with a short-term frequency instability of $8.5 \times 10^{-13}/\sqrt{\tau}$. *Optics Express*, **30**(9), pp.14574-14585.

Atom Fountains at NTSC

Jun Ruan[1,2*], Xinliang Wang[1,2], Hui Zhang[1,2], Dandan Liu[1,2], Yong Guan[1,2], Junru Shi[1,2], Yang Bai[1,2,3], Sichen Fan[1,2], Qiang Hao[1,2], Shuai Nie[1,2], Pengyue Lei[1,2,3], Ziyi Zhu[1,2,3], Shougang Zhang[1,2*]

[1]National Time Service Center, Chinese Academy of Sciences, Xi'an 710600, China
[2]Key Laboratory of Time Reference and Applications, Chinese Academy of Sciences, Xi'an 710600, China
[3]University of Chinese Academy of Sciences, Beijing 100049, China

*Corresponding author. E-mail: ruanjun@ntsc.ac.cn, szhang@ntsc.ac.cn

Abstract. Atom fountains as primary frequency standard (PFS) and continuously operated clocks have been developed to improve the local time scale at the National Time Service Center (NTSC), Chinese Academy of Sciences. Cesium fountain NTSC-CsF2 as PFS showed that the type B uncertainty was evaluated at 4.3×10^{-16}. Frequency comparisons were completed between NTSC-CsF2 and the SI second through a GNSS link for 14 months, the results of which were consistent with those of the other fountains within the uncertainty. A cryogenic Cs fountain is being developed to reduce the blackbody radiation shift (BBR) and verify the BBR theory. An Rb fountain clock NTSC-RbF1 was operated to generate local and EAL time scale. The frequency drift was reduced by maintaining a stable vapor pressure. NTSC-RbF2 features a more robust optical system consisting of all fiber-optical components and an ultra-stable OCXO as a local oscillator. An instability time scale below 100 ps is expected based on the Rb atom fountain clocks.

1. Introduction

The stability and accuracy of a thermal beam clock are limited by short Ramsey time and cavity phase difference [1]. An Atom fountain on the way up and down through the same microwave cavity results in a much longer Ramsey time and eliminates the end-to-end cavity phase shift [2]. Two or more orders of magnitude in terms of stability and accuracy have been achieved since the first invention of an atom fountain in 1995 [3-6]. Primary frequency standards and continuously running clocks based on atom fountains play important roles in the generation of UTC. The National Time Service Center (NTSC), Chinese Academy of Sciences, generates and maintains the UTC(NTSC), which connects the BeiDou navigation satellite system time (BDT) to UTC, and broadcasts the time signal to users through the BPM, BPC and BPL systems, networks, and telephones. Cs atomic fountain primary frequency standards (PFS) and Rb fountain clocks were established to produce a more accurate and stable local atomic time UTC(NTSC).

This paper reported the feature of atom fountains including two Cs primary frequency standards and two Rb fountain clocks. The long term frequency stability of the Rb fountain clock is limited by variation in background vapor pressure. The systematic uncertainty of the Cs PFS is evaluated at 4.3×10^{-16}, and the cold collisions and distributed cavity phase (DCP) are the main factors of the accuracy budget. Additionally, some improvements in atom fountains are described also.

2. Cesium fountain primary frequency standards

Cs atomic fountains are used as PFSs to realise definition of the SI second. More than ten Cs atomic PFSs have been used to steer International Atomic Time (TAI) and produce more accurate and stable local atomic times in some laboratories.

2.1. NTSC-CsF2 PFS

NTSC-CsF2 was being operated as a PFS to calibrate a hydrogen maser. A cold atom beam source produced by a 2D MOT was used to increase the loading rate [7]. The 2D-MOT vacuum chamber consists of a rectangular body made of titanium and four rectangular coils to produce a magnetic field of 1 mT/cm with a vacuum of 2×10^{-6} Pa. A small hole with a diameter of 1.5 mm existed to allow the atom beam to pass through the 3D-MOT chamber. The detection zone was aligned at 45° with respect to the Ramsey cavity feeds direction to eliminate the frequency shift of the DCP $m = 2$ term. The Ramsey cavity with a Q value of 12 000 was fed with a double-ended waveguide to evaluate the DCP shift. To prevent microwave leakage, the end caps of the cavity had 7-cm-long cut-off waveguides, and the joints between the end caps and the cavity body, and between the rectangular waveguide and the cavity body were sealed with indium wire. Three calibrated platinum resistors were attached to the flight tube to evaluate the blackbody radiation shift. To detect the magnitude of the microwave leakage, a 20 dBm microwave was fed into the Ramsey cavity before and after assembling the physical package to confirm the microwave leakage in the free drift zone and around the detection zone. An external-cavity diode laser (ECDL) with tapered amplifier (TA) generated a 900 mW laser for cooling, pushing, and detecting atoms. The laser frequency was locked by the saturated absorption spectrum of the Cesium $F = 4 \rightarrow F'$ = 5 hyperfine transition. The frequency during the Sisyphus cooling period was detuned by double-passed through an acousto-optic modulator (AOM). Another ECDL locked to the saturated absorption spectrum of the caesium $F = 3 \rightarrow F' = 4$ hyperfine transition was used for repumping the atoms. The microwave frequency synthesizer's reference was a 100 MHz signal provided by the maser, the output of 9.192 GHz exhibited a residual phase noise of −75 dBc/Hz at 1 Hz and spurious rejection ratio of 53 dB. The microwave signal was divided into two arms through the power divider. One arm included a phase shifter and the other arm included an attenuator for the phase and power adjustments to mitigate the distributed cavity phase shift.

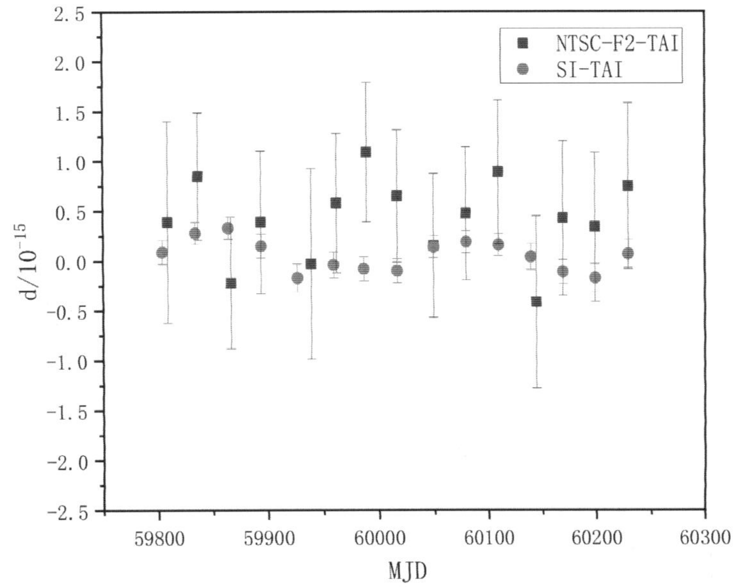

Figure 1. Fractional frequency difference d from August 2022 to October 2023.

The type A uncertainty was calculated from the overlapping Allan deviations of the frequency at high, low, and zero atomic densities. The short-term stability values were $\sigma_y^{(H)} = 2.2 \times 10^{-13}(\tau/s)^{-1/2}$, $\sigma_y^{(L)} = 2.1 \times 10^{-13}(\tau/s)^{-1/2}$, and $\sigma_y^{(Z)} = 6.4 \times 10^{-13}(\tau/s)^{-1/2}$. Type B uncertainty included five physical effects in the correction of the NTSC-CsF2 frequency: second-order Zeeman effect, blackbody radiation, cold atom collisions, gravitational redshift, and microwave leakage. The collisional shift was corrected in real-time via the extrapolation of frequencies running alternately at high and low densities. Both the second-order Zeeman effect and blackbody radiation shift were assessed for each frequency measurement to ascertain their frequency correction. Other effects have fixed frequency corrections and uncertainties. Cold collisions, microwave leakages and DCP frequency shifts are the dominant terms of the uncertainty budget.

The resulting type B uncertainty reaches 4.0×10^{-16}.The type A uncertainty reaches 4.2×10^{-16} in 30 days. Frequency comparison between NTSC-CsF2 and the SI second showed an average frequency difference of $4.2(5.3) \times 10^{-16}$ over 14 months (figure 1) with a reduced chi-square of $\chi^2 = 0.7$. This indicates that NTSC-CsF2 is consistent with that of the other fountains within the uncertainty.

2.2. NTSC-CsF3 PFS

NTSC-CsF3 is being built to operate at cryogenic temperature near 85 K, with a reduced blackbody shift and distributed cavity phase shift. A zero-evaporation liquid nitrogen Dewar device maintains a continuous cryogenic environment in the free flight zone. The cryogenic zone was designed with the sandwich structure of a liquid nitrogen thermostatic chamber, which is fixed by lifting and covered with two layers of cold screens to reduce heat radiation (figure 2). The cryogenic chamber is connected to the condenser through a liquid nitrogen adiabatic tube. Cold nitrogen gas and liquid nitrogen can flow in the same pipe to achieve zero evaporation of the liquid nitrogen in the system.

A cryogenic resonant microwave cavity had been developed [8], with a microwave cavity resonant frequency deviation of 10 kHz from the clock transition frequency and an unloaded quality factor of 25, 000. The leaked microwave power was less than -120 dBm. The effects of temperature, vacuum, humidity and other factors on the resonant frequency were measured experimentally to complete the tuning of the cryogenic microwave resonant cavity. The temperature uncertainties of the temperature sensor and thermometer measurements were less than 0.02 K. The total uncertainty is dominated by temperature distribution which is approximately 0.6 K (figure 3). Consequently, the blackbody radiation frequency shift of the cryogenic Cs atomic fountain clock was evaluated to be 9.05×10^{-17}, and the uncertainty was 2.68×10^{-18}.

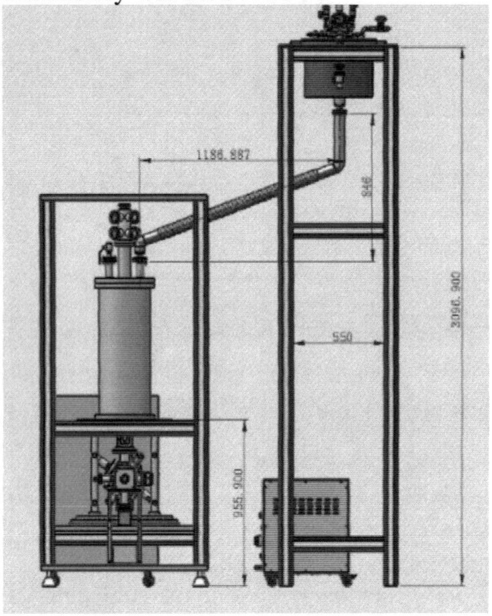

Figure 2. Structure of NTSC-CsF3.

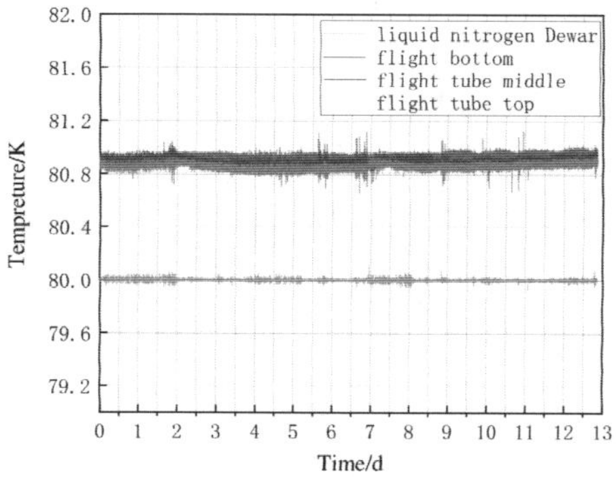

Figure 3. Long term temperature variation at NTSC-CsF3.

Journal of Physics: Conference Series **2889** (2024) 012037 doi:10.1088/1742-6596/2889/1/012037

3. Rb fountain clocks

Compared to Cs fountains, Rb fountains have a lower collision frequency shift [9], which make them inherently better in terms of long term frequency stability. Moreover, the laser source can be built using a frequency-doubling C-band telecom fiber laser at a wavelength of 1560 nm, making it possible to construct a more robust laser system. In this context, the NTSC has developed two ^{87}Rb Fountain Continuous clocks (^{87}Rb RFCC) that operate unattendedly, similar to commercial Cs clocks and hydrogen masers.

3.1. NTSC-RbF1 fountain

Rb atoms were trapped directly from the background vapor into an MOT in the (1, 1, 1) configuration [10]. The MOT chamber had six windows with a diameter of 35 mm for laser cooling and trapping the atoms. Ion pumps and getter pumps were installed at the top and the bottom of the physical package to maintain the vacuum at a pressure of 2×10^{-9} Pa and 6×10^{-9} Pa in the two regions, respectively. Some graphite tubes were placed along the flight tube to absorb the thermal Rb atoms. The inner surfaces of the detection and push zones were painted with colloidal graphite to extinguish the residual stray light. A state-selection cavity with $Q_L = 800$ was coupled with rectangular waveguides via a coupling hole and sealed with sapphire on the midplane of the wall of the cavity. The Ramsey cavity with $Q_L = 15\,800$ is a standard TE_{011} mode cylindrical cavity made of oxygen-free copper. Three layers of magnetic shielding surround the outside of C-field coil, which keeps the spatial fluctuations of the vertical magnetic field in the free flight zone within ±1.5 nT (peak-to-peak) around average value of approximately 113 nT. One magnetic shield layer surrounds the lower detection zone, MOT chamber, state-selection cavity, and detection zone.

The cooling laser source was a commercial 780-nm fiber laser, which was locked by a digital frequency locking and modulation transfer spectroscopy (MTS) to $\nu_{2,3}$ of the ^{87}Rb D_2 line. During the Sisyphus cooling period, the fiber laser's PZT control voltage is ramped for detuning the laser frequency to -120 MHz. The repumping laser source was a commercial 780 nm ECDL, which was locked using saturated absorption spectroscopy (SAS). All components of the optical bench were integrated into a 600 mm× 500 mm× 100 mm aluminium box. These included AOMs, polarization beam splitters (PBSs), 1/2 wave plates, reflectors, lenses, fiber couplers, shutters, MTS, and SAS.

The frequency difference between the maser and fountain was fed back to the DDS of the 6.834 GHz synthesizer providing an interrogating signal to the atoms and an HROG. The standard 5-MHz/1-PPS signal of the ^{87}RFCC were generated by the HROG fed by the maser, which was steered every two fountain cycles to the atomic transition frequency. The fountain exhibits a frequency stability of $1.91 \times 10^{-13}\tau^{-1/2}$, reaching 4.7×10^{-16} at 200 000 s. Since May 2022, the NTSC-RbF1 has provided data(table 1) that are sent to the BIPM to participate as a free running clock for the computation of the Echelle Atomique Libre (EAL).

Table1. Received weight in the EAL timescale

Date	2022			2023											
	10	11	12	1	2	3	4	5	6	7	8	9	10	11	12
Weight	0.300	0.229	0.141	0.120	0.152	0.151	0.184	0.188	0.210	0.206	0.234	0.198	0.229	0.282	0.502

Initially, the clock exhibited a large frequency drift owing to the increasing vapor pressure. Therefore, when the valve between the reservoir and MOT chamber is closed, the frequency drift is reduced because of more stable vapor pressure (figure 4). The frequency drift of a Rb fountain clock may also be affected by the 2nd order Zeeman effect, cold collisions, cavity pulling, microwave leakage, phase transient effects and distribution cavity phase shift.

(1) 2nd order Zeeman effect. The frequency shift is derived from the magnetic field (C field) along the flight of the Rb atoms. The C field may drift owing to the variation in current driver, of the Earth's magnetic field despite the magnetic shielding. However, this frequency shift was checked every month between each run of the clock and was found to be stable within 10^{-16}.

(2) Cold collisions and cavity pulling. These shifts are proportional to the number of cold atom cloud. The number of atoms fluctuated within 15% peak-to-peak. However, the shift was in the low 10^{-16} level and contributed to the floor of the fountain ADEV at the level of 10^{-18}.
(3) Distribution cavity phase. The drift of DCP shift relies on the stability of the cold atom cloud geometry and balance between microwave amplitude and phase at the two opposing feeds. In fact, the shift is of the same order of magnitude as that of cold collisions.

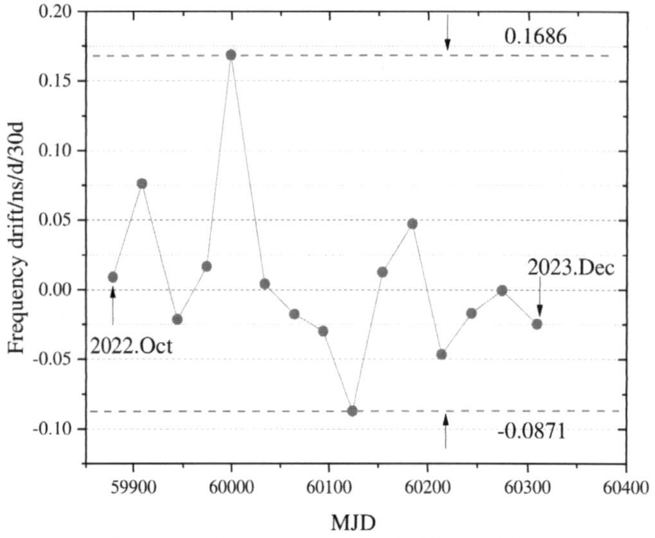

Figure 4. Frequency drift of NTSC-RbF1

3.2. NTSC-RbF2 fountain

NTSC-RbF2 has similar structure to that of NTSC-RbF1 except for the laser setup, which is completely made of fiber-optical components including the fiber lasers, AOMs, shutters and splitter, as shown in figure 5. The cooling and repump lasers were frequency stabilized by modulation transfer spectrum with a high sign-to-noise ratio and a clean background. When the fiber laser was power on, the program

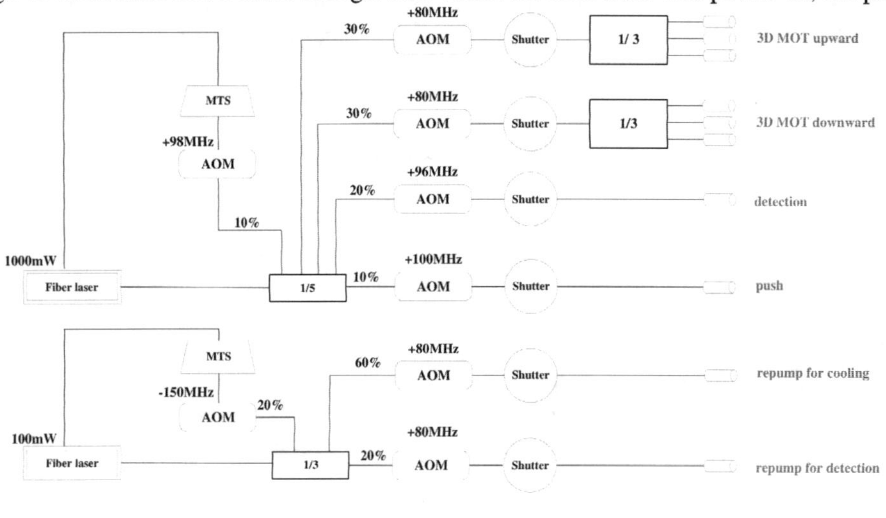

Figure 5. All fiber-optical setup of NTSC-RbF2.

began checking the laser frequency within specific atom resonance level. Subsequently, the frequency was adjusted by digitally setting the diode laser temperature and injection current within the maximum and minimum of the MTS signal. Finally, the laser frequency corrections calculated by the digital PID were output to control diode laser temperature and injection current. Periodically, the program checked the Rb cell fluorescence to detect a possible unlock. In this case, the program opened the loop and restarted locking procedure described above. Several fiber splitters were used to divide the laser for cooling and detecting atoms in fixed proportions. The laser collimators had waveplates and PBSs to tun the laser power to balance the six capture beams. Mechanical shutters were packaged in small aluminium boxes in front of polarization maintain fiber pigtail. A specific fiber acoustic optic modulator is used to shift the laser frequency and control the laser power with a high extinction ratio of >23 dB. An ultra stable oscillator, HSO14, was directly locked to Ramsey fringe. Figure 6 shows the obtained frequency stability at the level of $3 \times 10^{-13}\tau^{-1/2}$. A cryogenic sapphire oscillator is an alternative to improve the stability at 1s down to the 10^{-14} level.

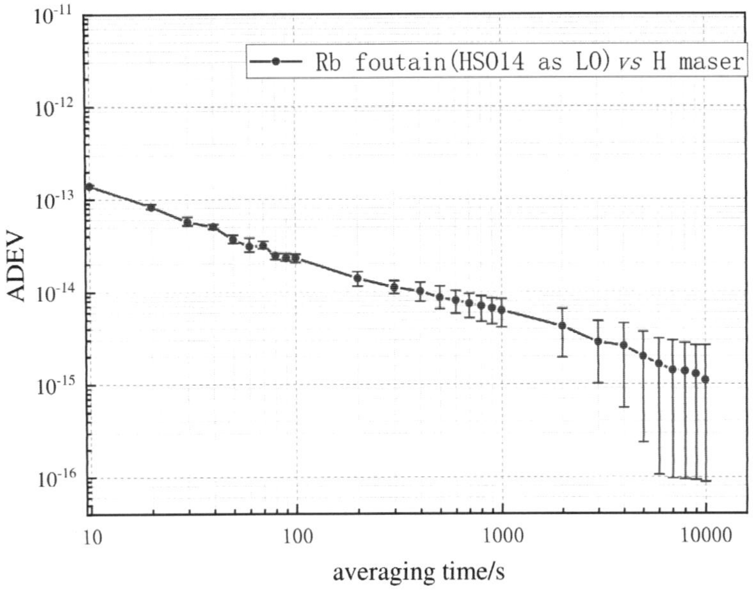

Figure 6. Frequency stability of NTSC-RbF2, HSO14 OCXO as local oscillator.

4. Conclusion and outlook

Primary frequency standards and clocks based on atom fountains were being developed by the NTSC. NTSC-CsF2 shows a systematic uncertainty of 4.3×10^{-16}, dominated by cold collisions and the distributed cavity phase shift. NTSC-CsF3 is design to be cryogenic for reducing the BBR and DCP to improve accuracy, aiming to reach below 2×10^{-16}.

NTSC-RbF1 fountain clock has been in operation since May 2022 and has regularly received weight on the EAL time scale. NTSC-RbF2 fountain clock has being construct, and its optical system, which is based on two industrial 1560 nm fiber laser are expected to be much more robust. This fountain clock uses all fiber-optical components and fits in a single rack, replacing a traditional optical table system. NTSC-RbF2 is expected to operate as a clock in the future, as continuously as Cs beam clocks and hydrogen masers.

Frequency stability of Rb fountain clock is expected to be improved to 2×10^{-14} at 1s and 7×10^{-17} at 1 d when a sapphire cryogenic oscillator is used as the local oscillator [11]. The floor of the frequency stability is expected to be at 10^{-18} level, which would be limited by the long term stability of the cold atom density. When the Rb vapor diffuses across the entire chamber, the effective laser power changes. Consequently, when the cold atom cloud geometry changes slowly, the distributed cavity phase and the cavity pulling shifts may drift. However, time scale below 100 ps instability should be anticipated because of super stable and low drift Rb fountain clocks.

5. Acknowledgements

This research was financially supported by the Natural Science Basic Research Program of Shaanxi Province (Grant No. 2023-JC-YB-078), Key Research and Development Program of Shaanxi (Program No. 2023-YBGY-402), and Large Research Infrastructures Improvement Funds of the Chines Academy of Sciences(DSS-WXGZ-2020-0005).

References

[1] Makdissi A, et al 2001 Metrologia **38** 409
[2] Forman P 1985 Proc. IEEE **73** 1181
[3] Clairon A, et al 1996 Proc. 5th Symp. Frequency Standards and Metrology 45
[4] Guéna J, et al 2012 IEEE Trans. Ultrason. Ferroelectr. Freq. Control **59** 391
[5] Szymaniec K, et al 2010 Metrologia **47** 363
[6] Weyers S, et al 2018 Metrologia **55** 789
[7] Wang X, et al 2023 Metrologia **60** 065012
[8] Yang F, et al 2022 Rev. Sci. Instrum. **93** 044708
[9] Sortais Y, et al 2000 Phys. Rev. Lett. **85** 3117
[10] Zhang H, et al 2022 IEEE Trans. Instrum. Meas. **71** 1008312
[11] Fluhr C, et al 2023 IEEE Trans. Instrum. Meas. **72** 1005408

Journal of Physics: Conference Series **2889** (2024) 012038 doi:10.1088/1742-6596/2889/1/012038

Re-evaluation of the NRC-FCs2 Fountain Clock

Scott Beattie, Bin Jian, and Marina Gertsvolf

National Research Council Canada, 1200 Montreal Rd, Ottawa, Ontario, K1A 0R6, Canada

E-mail: Scott.Beattie@nrc-cnrc.gc.ca

Abstract. The NRC-FCs2 fountain clock was first evaluated in 2020 [1]. Since that time, it has demonstrated both accuracy and robustness and has contributed regularly to the steering of International Atomic Time (TAI) through monthly submissions to the BIPM. It is also currently used to steer UTC(NRC), Canada's official timescale. We have recently undertaken a re-evaluation of several prominent systematic effects in NRC-FCs2. Thus far, this has resulted in a reduction of the overall uncertainty from ~3 x 10^{-16} to 1.7 x 10^{-16}. Here, we will describe the details of the re-evaluation.

1. Introduction

Caesium fountain clocks continue to provide the most accurate realization of the SI second and can reach systematic uncertainties at the level of a few parts in 10^{16} in fractional frequency [2-5]. Atomic clocks in the optical domain have made extremely rapid progress over the last two decades, and can now reach uncertainties two orders of magnitude lower than the best Caesium clocks [6-10]. For this reason, the future redefinition of the second will rely on optical clock transitions.

Despite the superior precision of optical clocks, Caesium fountain clocks maintain a critical importance in their present roles as primary frequency standards. This is both for their current realization of the second, as well as for the successful establishment of a new definition of the second. The roadmap to a new definition of the second requires high precision absolute measurements of optical transition frequencies that will require comparisons with high performance fountain clocks [11].

2. Performance and error budget

The NRC-FCs2 fountain clock has been operating as a standard since 2020, reporting regularly to the BIPM to contribute to the steering of TAI. The typical systematic uncertainty of NRC-FCs2 has been ~ 2-3 x 10^{-16} in fractional frequency.

Figure 1. NRC-FCs2 fountain clock measurements (blue data) plotted with the scale interval, *d*, calculated from the average of all contributing primary and secondary frequency standards (black line) [12].

As shown in Fig. 1, the NRC-FCs2 fountain clock has shown stable performance in agreement with other reporting standards. From October 2019 to February 2024, NRC-FCs2 had an average deviation from the BIPM calculation of the scale interval, *d*, of 0.6 +/- 3.4 x 10^{-16}. Table 1 shows a typical error budget for NRC-FCs2. Here the bold entries denote re-evaluated uncertainties.

Table 1. A typical error budget for the NRC-FCs2 primary frequency standard. This budget is for the data submitted to the BIPM in Sept. 2022. The bold numbers in parentheses represent the recently re-evaluated uncertainties.

Effect	Bias (10^{-16})	Uncertainty (10^{-16})
Quadratic Zeeman shift	724.9	0.2
Blackbody radiation	-162.6	0.7
Gravitational redshift	104.52	0.03
Cold collisions	-	2.25 **(0.2)**
Distributed cavity phase shift	-0.6	1.4
Microwave lensing	0.6	0.2
Microwave leakage	0.1	1.0 **(0.3)**
Synchronous phase transients	-	0.8 **(0.5)**
Total	666.9	3.05 **(1.7)**

Journal of Physics: Conference Series **2889** (2024) 012038 doi:10.1088/1742-6596/2889/1/012038

3. Re-evaluation

The re-evaluation of the NRC-FCs2 fountain clock systematic uncertainties has focused on the four dominant contributions to the error budget: cold collisions, microwave leakage, synchronous phase transients, and distributed cavity phase. These four effects have typically been responsible for over half of the systematic uncertainty.

The evaluation has been facilitated by the availability of ultra-stable microwaves via an optical frequency comb locked to an ultra-stable laser. This has improved the best highest stability achievable with NRC-FCs2 from $\sigma_y(1s) = 1.1 \times 10^{-13}$ to $\sigma_y(1s) = 3 \times 10^{-14}$, resulting in an order of magnitude improvement in averaging time [13].

3.1 Cold Collisions

Cold collisions have typically had the largest single contribution to the error budget. In our standard operation, the atomic density is alternated between high and low by varying the microwave power in the state selection cavity. By extrapolating to zero density, a collisional-shift-corrected frequency measurement can be obtained.

Our frequency uncertainty is largely due to the uncertainty in the ratio of atomic densities, as it is atomic number that is directly measured. Any distortions to the atomic cloud due to the variation in the power of the state selection cavity would result in a frequency bias. We have used absorption imaging to directly measure the relationship between atomic number ratio and atomic density ratio [14]. Using this method, we have been able to confirm that the error in the ratio of atomic densities when approximating it as the ratio of atom numbers is below 1%. This is an order of magnitude lower than the value of 10% we had previously used and results in a nearly proportional reduction in the overall uncertainty due to the collisional shift.

3.2 Microwave leakage

It was previously determined that the NRC-FCs2 fountain clock has microwave leakage below the Ramsey cavity that results in a fractional frequency shift at the level of 10^{-15}. In normal operation, this effect is mitigated by rapidly switching the Ramsey microwave frequency far from resonance immediately after the atom cloud traverses the Ramsey cavity on its descent. This is done by using the frequency shift key function of the microwave synthesizer, which should toggle the frequency without causing any power transients in the synthesizer.

We have recently put more stringent bounds on the effectiveness of this approach by applying amplified microwaves to the system using a microwave horn, as shown in Fig. 2. In this configuration, the Ramsey microwaves are split, and high-power microwaves (31 dB higher power than the microwaves sent to the Ramsey cavity) are directed towards the physics package via a microwave horn.

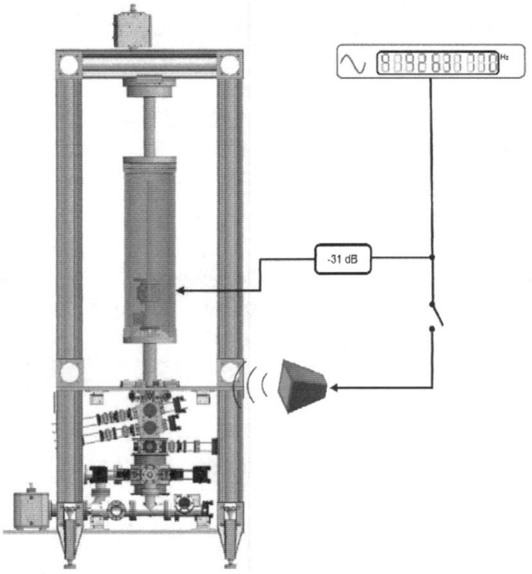

Figure 2. The setup for amplifying the microwave leakage.

Differential measurements were taken with amplified microwaves sent via microwave horn with and without frequency shift keying. As a baseline measurement, data was also taken in our normal operating mode, using the frequency shift keying without amplified microwaves. These measurements allowed more stringent bounds on the efficacy of the synthesizer frequency shift key function as an effective microwave switch. It has been determined that the frequency shift key function reduces the effects of microwave leakage by a factor of over 100. This reduces the associated systematic uncertainty to 3×10^{-17} in fractional frequency.

3.3 Synchronous phase transients

A time-varying perturbation to the Ramsey microwave phase will cause an error in the measured frequency. When these perturbations are synchronized with the experimental cycle, it can result in a frequency bias. Typical causes include elements in the experimental cycle with a high transient power draw, such as amplifiers and optical shutters. In our case, we also consider the effect of any phase transient due to the frequency shift keying applied to the Ramsey microwave frequency. We had previously characterized these contributions, and found that the uncertainty due to synchronous phase transients was 0.8×10^{-16}. Since that evaluation, we have relocated the Ramsey microwave synthesizer to another lab > 30 meters away from any potential source of perturbation which reduces our uncertainty to 0.5×10^{-16}. We expect that a planned re-evaluation of the effect of the frequency shift key function will further reduce this uncertainty, as previous measurements showed no effect from the switching within the measurement uncertainty. Now with access to ultra-stable microwaves it will be possible to characterize the effect to a higher precision.

3.4 Distributed cavity phase shift

Asymmetries in the microwave phase within the Ramsey cavity can lead to a frequency bias when coupled with any horizontal motion of the atomic cloud during its ballistic flight. To quantify and minimize the effect, it is important to ensure that the atoms are launched vertically. For this purpose, we use absorption imaging to measure the horizontal velocity of the atomic cloud after launch.

Initially, we reflect the downwards trapping beam off of a liquid surface below the physics package to align it to be vertical using a beam splitter and a retroreflector. We can then use the downwards beam as a reference to align the upwards trapping beam to be vertical as shown in Fig 3.

Journal of Physics: Conference Series **2889** (2024) 012038 doi:10.1088/1742-6596/2889/1/012038

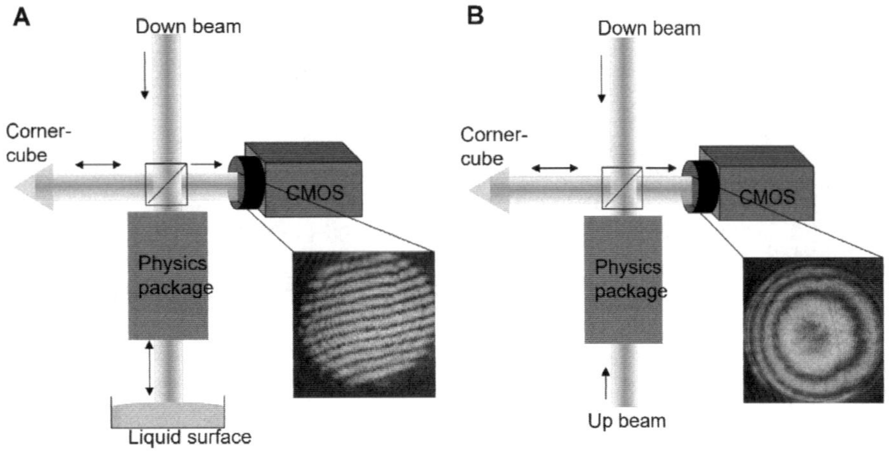

Figure 3. A) Setup for aligning downwards beam to be vertical. The interference pattern is for a ~0.3 mrad misalignment from vertical. B) Setup for aligning the upwards imaging beam verticality via interference with the downward beam.

With a vertically-aligned upwards beam as a reference, it is possible to perform absorption imaging with this beam to track the trajectory of the atoms during their ballistic flight as shown in Fig. 4. Using this technique allows for a more precise alignment of the verticality of the atomic trajectory, and consequently, to reduce the uncertainty due to the distributed cavity phase shift. Preliminary results suggest that the contribution to the error budget from the distributed cavity phase shift will be reduced to the level of 1×10^{-16} in fractional frequency. As these are preliminary results, they are not reflected in Table 1.

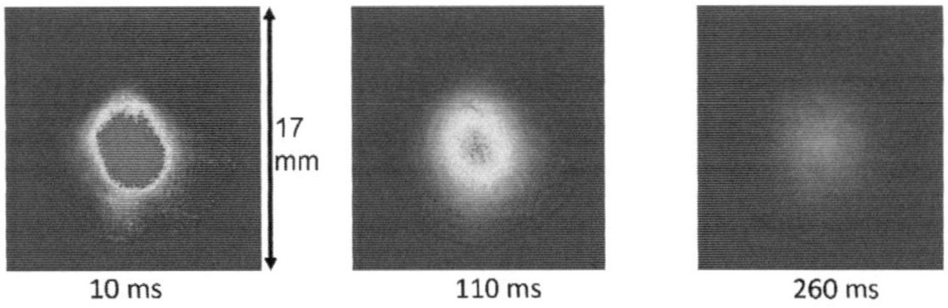

Figure 4. Absorption images with time of flight after launch ranging from 10 to 260 ms.

4. Conclusions

We have recently undertaken a re-evaluation of the systematic effects of the NRC-FCs2 primary frequency standard. Thus far, the re-evaluation has resulted in a significant reduction in three of the 4 dominant contributions to the systematic uncertainty of the fountain clock. We have currently achieved a systematic uncertainty of 1.7×10^{-16} and it is expected that the current evaluation of the distributed cavity phase shift will further reduce the overall systematic uncertainty to $< 1.5 \times 10^{-16}$, an improvement of ~ 2x from our previous typical uncertainty. The improvement will be important for upcoming absolute frequency measurements of the lab-based and portable (under development) $^{88}Sr^+$ optical clocks at the NRC [15,16], which will help in the achievement of the criteria laid out in the roadmap to a redefinition of the SI second.

5. References

[1] Beattie S, Jian B, Alcock J, Gertsvolf M, Hendricks R, Szymaniec K and Gibble K 2020, Metrologia, **57**, 035010

[2] Heavner T P, Donley E A, Levi F, Costanzo G, Parker T E, Shirley J H, Ashby N, Barlow S and Jefferts S 2014 Metrologia **51** 174

[3] Levi F, Calonico D, Calosso C E, Godone A, Micalizio S and Costanzo G A 2014 Metrologia **51** 270

[4] Weyers S, Gerginov V, Kazda M, Rahm J, Lipphardt B, Dobrev G and Gibble K 2018 Metrologia **55** 789

[5] Szymaniec K, Lea S, Gibble K, Park S, Liu K and Głowacki P 2016 NPL J. Phys.: Conf. Ser. **723** 012003

[6] Nicholson T L *et al* 2015 Nat. Commun. **6** 1–8

[7] McGrew W F *et al* 2018 Nature **564** 87–90

[8] Sanner C, Huntemann N, Lange R, Tamm C, Peik E, Safronova M S and Porsev S G **567** 204–8

[9] Brewer S M, Chen J-S, Hankin A M, Clements E R, Chou C-W, Wineland D J, Hume D B and Leibrandt D R 2019 Phys. Rev. Lett. **123** 033201

[10] Takamoto M, Ushijima I, Ohmae N, Yahagi T, Kokado K, Shinkai H and Katori H 2020 Test of general relativity by a pair of transportable optical lattice clocks Nat. Photon. **14** 411–15

[11] Dimarcq N *et al* 2024 Metrologia **61** 012001

[12] BIPM Time Department Database (available at https://webtai.bipm.org/database/d_plot.html)

[13] Marceau C, Beattie S, Pakulski W, and Gertsvolf M, "Frequency comb for 88Sr+ clock frequency comparison with Cs fountain clock at the NRC", 9th Symposium on Frequency Standards and Metrology, 2023, Kingscliff, Australia

[14] Beattie S and Jian B, 2023, Metrologia, **60**, 045004

[15] Jian B, Bernard J, Gertsvolf M and Dubé P 2022 Metrologia **60** 015007

[16] Dubé P, Kato K, Bernard J, and Jian B,"Progress Towards a Transportable and High-Accuracy Sr+ Ion Clock at NRC," *2021 Joint Conference of the European Frequency and Time Forum and IEEE International Frequency Control Symposium (EFTF/IFCS)*, Gainesville, FL, USA, 2021, pp. 1-2, doi: 10.1109/EFTF/IFCS52194.2021.9604288.

Journal of Physics: Conference Series **2889** (2024) 012039 doi:10.1088/1742-6596/2889/1/012039

The development and future of the Faraday laser

Zhiyang Wang[1], Hangbo Shi[1], Zijie Liu[1], Jianxiang Miao[1], Tiantian Shi[2,3*], and Jingbiao Chen[1,4]

1. State Key Laboratory of Advanced Optical Communication Systems and Networks,
Institute of Quantum Electronics, School of Electronics, Peking University, Beijing 100871, China
2. School of Integrated Circuits, Peking University, Beijing 100871, China
3. National Key Laboratory of Advanced Micro and Nano Manufacture Technology, Beijing 100871, China
4. Hefei National Laboratory, Hefei 230088, China

E-mail: tts@pku.edu.cn

Abstract. Faraday lasers, distinguished by their utilization of the Faraday anomalous dispersion optical filter (FADOF), employ atomic transitions for mode selection. This sets them apart from other semiconductor diode lasers, which commonly employ gratings, interference filters, and etalons for frequency selection and initiating laser oscillation. A notable advantage of Faraday lasers is their resistance to noises in driving current and the temperature of the laser diodes. Additionally, the using of FADOF enables the output frequency of the laser to align with atomic transition lines immediately when powering the lasers with driving current above the threshold. This paper presents the advancements in Faraday laser system technology spearheaded by our group, outlining a forward-looking research agenda aimed at continually enhancing the performance of Faraday lasers. This endeavor seeks to establish these lasers as frontrunners in the next generation of laser used in precision spectroscopy, with specific applications in atomic physics, quantum metrology, and quantum precision measurement technology.

1. Introduction

Semiconductor diode lasers operating at a distinct atomic transition frequency with high frequency stability, narrow linewidth, are very important in a lot of research areas, especially in quantum precision measurement, quantum optics, atomic physics [1]. The widely used diode lasers usually employ gratings, interference filters, or etalons to tune the output frequency corresponding with the atomic transition lines, and then achieve frequency locking and linewidth narrowing through the methods such as polarization spectroscopy [2], dichroic atomic vapor laser lock [3], saturated absorption spectroscopy [4] and modulation transfer spectroscopy [5]. The frequency locking of these lasers is also commonly achieved through the use of the Pound-Drever-Hall (PDH) technique [6]. While the frequency of the lasers mentioned above could not stop drifting away from the atomic transition lines with the long-term operation and fluctuations of working parameters. Comparing with tuning the laser frequency repeatedly, the ability of the laser frequency to consistently remain in close proximity to the atomic transition lines during extended operational periods and a large fluctuation in the driving current and working temperature of the laser diode, would greatly enhance the ease and efficiency of initial locking and subsequent re-locking processes after missing frequency lock. Using the FADOF as the intra-cavity frequency selective element instead of the gratings, interference filters, or etalons can provide long term stability [7]. FADOF can limit the laser frequency within the vicinity of the atomic transition frequency

Journal of Physics: Conference Series **2889** (2024) 012039 doi:10.1088/1742-6596/2889/1/012039

for about 1GHz or less. Different from the other lasers using conventional frequency selection elements, whose output frequency is depended on the angle or length of a mechanical reference, the output frequency of the laser is determined on the transmission spectrum of the intra-cavity FADOF, it should result in stronger immunity to the fluctuations of working parameters. While because of the existence of the internal-cavity modes (modes of the laser diode) and the multiple transmission peaks of the FADOF, the frequency of this laser system experiences significant shifts when there are substantial changes in the current and temperature of the laser diode [8]. The laser system with immunity to fluctuations of driving current and working temperature of laser diode[9, 10], was first achieved in 2011 with the creative application of antireflection coated laser diode (ARLD), and then named it the Faraday laser [10]. Since then, we have continuously poured our efforts into the performance improvements and practical applications [11] of the Faraday laser, through the progresses made in the four main components of the Faraday laser system. The initial component is the ARLD, serving as the gain medium. The second component is the FADOF, using atomic transitions acting as the frequency selection mechanism. The third component encompasses the feedback and fine-tuning of the laser. Finally, the fourth component involves the frequency stabilization of the Faraday laser, crucial for meeting the demands of real-world applications.

2. Works on the Faraday laser system

As previously discussed, figure 1 provides a comprehensive overview of the developmental strides made by our group in the Faraday laser system.

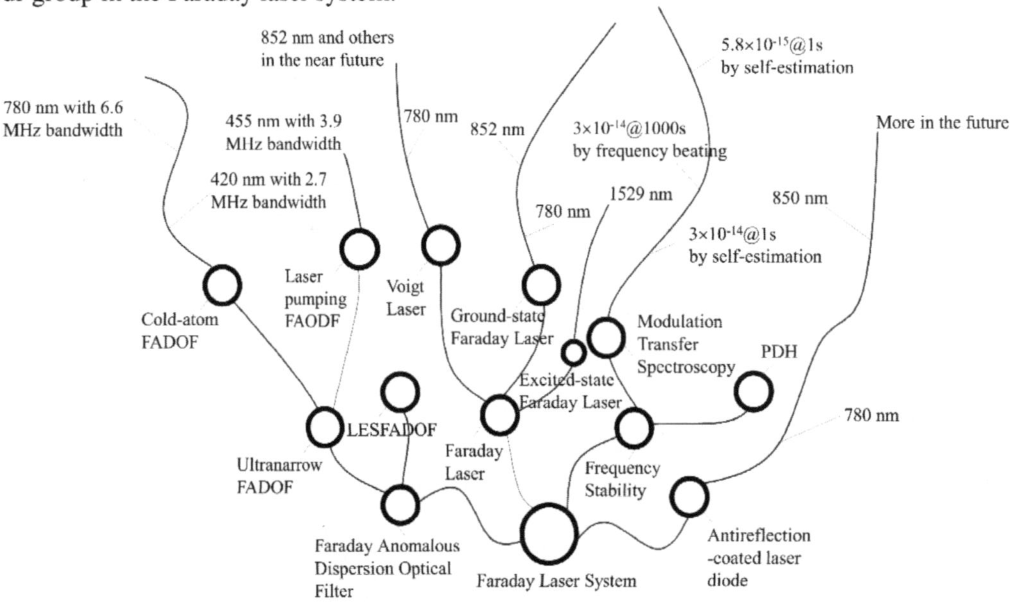

Figure 1. Development venation of the Faraday laser in Peking University.

2.1. The main progress made in FADOF

FADOF, which was first reported in 1956 [12], plays a pivotal role in the Faraday laser for frequency selection. Its schematic representation illustrated in figure 2(a). Additionally, an exemplary transmission spectrum is depicted in figure 2(b). The FADOF mechanism hinges on the Zeeman-induced differential frequency transitions in right and left circularly polarized light, profoundly altering absorption and dispersion characteristics. This dispersion directly influences the refractive indices of the polarization components of light, resulting in the distinct phase shifts for light with different polarization on the same optical path. These shifts culminate in a rotation in the polarization of incident linear light, a phenomenon emblematic of the Faraday anomalous dispersion effect, prominently observed near atomic transition frequencies. Consequently, the FADOF confines the laser frequency within the atomic Doppler broadened line, resulting in the remarkable characteristic that Faraday lasers are impervious to fluctuations in the temperature and current of the laser diodes. This leads to the output wavelength of the Faraday laser aligns closely with the atomic transition lines from the moment the laser is activated.

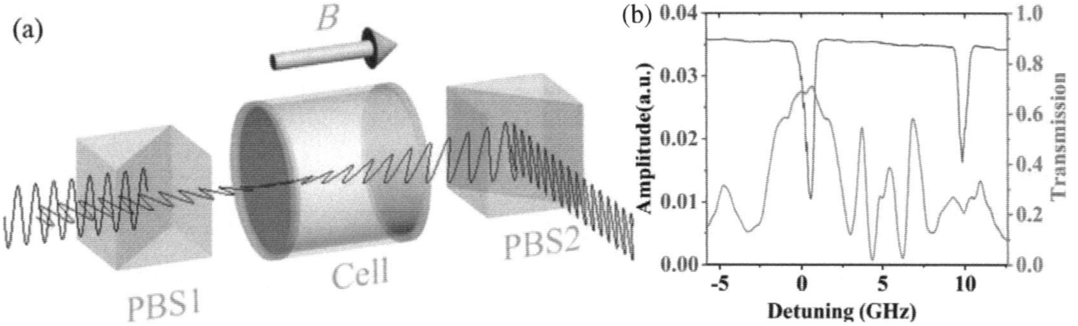

Figure 2. (a) Composition structure of FADOF. (b) Transmission spectrum of FADOF, and the black solid line is the saturated absorption spectroscopy (SAS) of Cs-D2-line, the solid red line is the transmission spectrum.

For the research of FADOF, we achieved ultranarrow bandwidth FADOF [13-15] and the electrodeless discharge lamp-based excited-state FADOF (LESFADOF) [16, 17]. The ultranarrow bandwidth can be achieved through laser-pumping [13] and cold-atoms [14, 15]. The laser-pumping is based on both the Faraday effect and saturated absorption. In this system, a pumping laser with strong intensity is injected to the cell, and the saturation effect is used to pick atomic resonance transitions and their cross-over transitions, preventing them from being overwhelmed by absorption. Under the influence of Faraday effect, the polarization directions of light with frequencies resonant with chosen transitions rotate, and only the light with these frequencies can pass through perpendicular polarizers. As shown in figure 3, the bandwidths of the filter we have constructed [13] are 6.2 MHz for the Cs $6S_{1/2}$, F=4→$7P_{3/2}$, F'=5 transition, 3.7 MHz for the Cs $6S_{1/2}$, F=4→$7P_{3/2}$, F'=4,5 crossover transition, respectively.

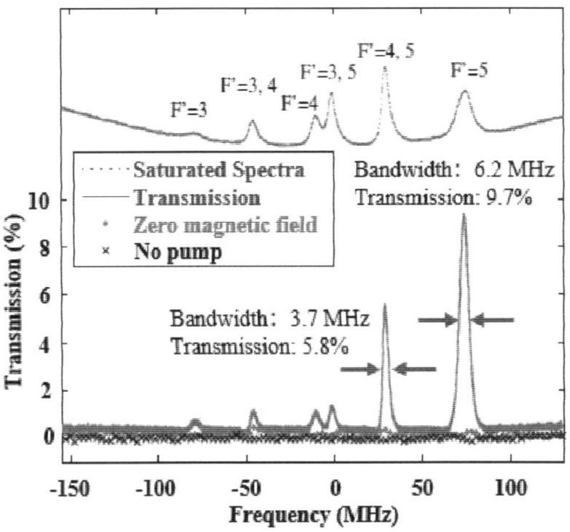

Figure 3. Ultranarrow transmitted spectra of Cs $6S_{1/2}$→$7P_{3/2}$ transition (adapted from [13]).

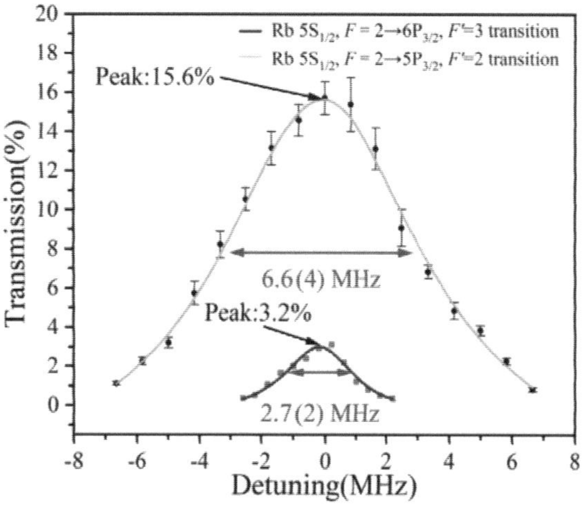

Figure 4. Ultranarrow bandwidth FADOF based on cold atom (adapted from [14] and [15]).

Considering the Doppler broadening induced by the thermal atomic ensemble, the transmitted bandwidth of conventional atomic filter is difficult to be narrowed to the order of natural linewidth. To overcome this, we innovatively combine cold atoms with Faraday anomalous dispersion optical filter. Laser cooling technology can directly slow down atoms, thus effectively overcoming the Doppler effect.

Based on this, an ultranarrow-bandwidth cold-atom FADOF with bandwidth close to the natural linewidth of atomic transition has been realized. The bandwidths are 2.7 MHz (blue solid line in figure. 4) for Rb $5S_{1/2}$, F=2→$6P_{3/2}$, F'=3 transition [14] and 6.6 MHz (green solid line in figure. 4) for Rb $5S_{1/2}$, F=2→$5P_{3/2}$, F'=2 transition [15], respectively.

The LESFADOF is different from other FADOFs that use the ground state, and it can achieve the filter function for the transitions with excited lower states using an electrodeless discharge lamp, such as Rb $5P_{3/2}$→$5D_{5/2}$ 775.9 nm transition [16] and the Rb $5P_{3/2}$→$4D_{5/2}$ 1529 nm transition [17]. Based on the LESFADOF, the laser output frequency correspond to the excited state transition lines can be achieved without the need of an additional frequency-stabilized pump laser to prepare the atoms to excited state.

The stimulated amplified FADOF (SAFADOF) based on population inversion has also been achieved by our group [18]. This filter has the capability to greatly enhance faint laser signals while filtering out interference, resulting in clear signals amidst a strong background. Benefiting from the stimulated emission process, the gain factor can up to 44 dB (25000) when a weak signal of 50 pW within a bandwidth narrow to 13 MHz. To avoid the high temperature needed in the conventional FADOF, the hollow cathode lamp based FADOF is obtained in 2016 [19]. This FADOF operates at the $^{88}Sr(5s^2)^1S_0$→$(5s5p)^1P_1$ transition (461 nm), and the maximum transmittance is 62.5% with a bandwidth of 1.19GHz.

Besides the FADOF, the Voigt anomalous dispersion optical filter (VADOF) has also been realized [20]. The main difference between these two optical filters is the direction of the magnetic field. In the FADOF, the magnetic field and the incident light are parallel with each other, while in the VADOF, they are perpendicular to each other. The VADOF can be used to establish the Voigt laser, which also has immunity to fluctuations in the current and temperature of the laser diode, as well as an output frequency corresponding to atomic transition lines.

Research on FADOF is of great significance to the continuous development of Faraday lasers. Table 1 lists the typical research results of anomalous dispersion optical filter made by our group. In the following sections, we will give more details for the lasers we achieved based on these optical filters.

Table 1. Research results of anomalous dispersion optical filter.

Type of the atom	Transition lines	Cold or Hot atom	Results	Type of the filter
Rb [16]	$5P_{3/2}$→$5D_{5/2}$(775.9 nm)	Hot	Center Transmission: 1.9% Bandwidth: 650 MHz	LESFADOF
Rb [17]	$5P_{3/2}$→$4D_{5/2}$(1529.4 nm) $5P_{3/2}$→$4D_{3/2}$(1529.3 nm)	Hot	Wings Transmission: 21.9%/3% Bandwidth: 300 MHz	LESFADOF
Rb [21]	$5S_{1/2}$→$5P_{3/2}$(780 nm)	Hot	Center Transmission: 30.6% Bandwidth: 1.41 GHz	FADOF
Cs [22]	$6S_{1/2}$, F=4→$7P_{3/2}$, F'=3,4,5(455 nm)	Hot	Wings Transmission: 86% Bandwidth: 1.5 GHz	FADOF
Cs [13]	$6S_{1/2}$, F=4→$7P_{3/2}$, F'=5(455 nm) $6S_{1/2}$, F=3→$7P_{3/2}$, F'=2,3(455 nm)	Hot	Wings Transmission: 9.7%/6.1% Bandwidth: 6.2 MHz/3.9 MHz	FADOF
Cs [23]	$6S_{1/2}$→$7P_{3/2}$(459 nm)	Hot	Wings Transmission: 98% Bandwidth: 1.2 GHz	FADOF

^{88}Sr [19]	$(5s^2)^1S_0 \rightarrow (5s5p)^1P_1$(461 nm)	Hot	Center Transmission: 62.5% Bandwidth: 1.19 GHz	FADOF
Cs [18]	$6P_{3/2}$, F=5→$7S_{1/2}$, F'=4 (1470 nm)		Bandwidth: 13MHz	SAFADOF
^{87}Rb [14]	$5S_{1/2}$, F=2→$6P_{3/2}$, F'=3 (420 nm)	Cold	Center Transmission: 3.2% Bandwidth: 2.7(2) MHz	FADOF
^{87}Rb [15]	$5S_{1/2}$, F=2→$5P_{3/2}$, F'=2 (780 nm)	Cold	Center Transmission: 15.6% Bandwidth: 6.6(4) MHz	FADOF
^{87}Rb [20]	$5S_{1/2}$→$5P_{3/2}$, (780 nm)	Hot	Wings Transmission: ~90% Bandwidth: ~2GHz	VADOF

2.2. The main progress made in Faraday laser and Voigt laser

The challenge of the atomic filter laser stemmed from the competitive interplay between the internal-cavity and external-cavity modes. X. Miao, *et al* creatively built a semiconductor diode laser system with an anti-reflection coated laser diode (ARLD) as the gain medium and a Rb FADOF as the frequency selection element [9]. As figure 5 shows, the frequency output variation of this laser can always be kept within 2 pm (985.6 MHz) when the driving current and temperature change from 55 mA to 142 mA and from 15 °C to 35 °C, respectively. Considering the large range for the change of driving current and temperature, we can say that this laser has immunity to the fluctuations of these two parameters. Z. Tao, *et al* named the laser systems that have the typical characteristic mentioned above as Faraday laser, and this name is widely used [10, 24]. These pioneering works lead to a lot of research interests, contributing to the progress of the whole area.

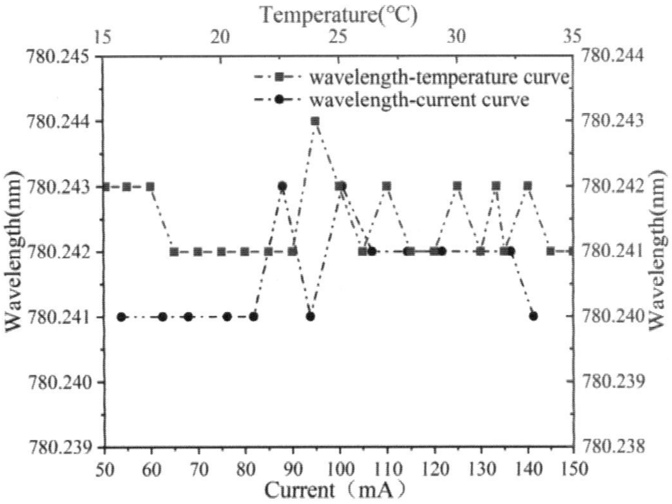

Figure 5. The current-wavelength curve and temperature-wavelength curve of the Faraday laser (adapted from [9]).

In 2014, W. Zhuang, *et al* built an active Faraday optical frequency standard based on the FADOF with ultranarrow bandwidth using Laser-pumping on Cs $6S_{1/2}$, F=4→$6P_{3/2}$, F'=4,5 crossover transition line [25]. An optical heterodyne beat between two comparable independent systems shows that the frequency linewidth reaches 281(23) Hz, which is 1.9×10^4 times smaller than the natural linewidth of the Cs atom.

Based on the LESFADOF, the excited state Faraday laser operating on 1529 nm mainly used for optical communication had also been achieved [26]. The immunity ranges of laser diode are 85 mA to 171 mA and 11 °C to 32°C, with the 24-hour frequency fluctuation below 600 MHz.

Like Faraday lasers based on Rb atoms, the Faraday lasers operating on the Cs-D2 resonance line also have the similar characteristic on immunity to driving current and temperature [27]. Within the driving current range of 60 mA to 130 mA and temperature range of 14 °C to 32 °C, the output frequency of the Faraday laser is immune to the changes of current and temperature. The 48-hour wavelength fluctuation range is always within ± 2 pm, and the most probable linewidth is 17 kHz with Lorentz fitting.

Through changing the parameters that influence the transmission spectrum of FADOF, such as the strength of the magnetic field, the length of the atomic cell and the density of atom in the cell (the temperature of the atomic cell), we can achieve a Faraday laser working stably with the dual-frequency mode. As depicted in figure 6, these two modes match the Cs-D2 resonance lines with F=3 and F=4 ground states, respectively. The most probable linewidth of the beat signal between the two free running modes is 902.95 Hz, which has the potential for providing microwave source when we lock the frequency of both modes [28].

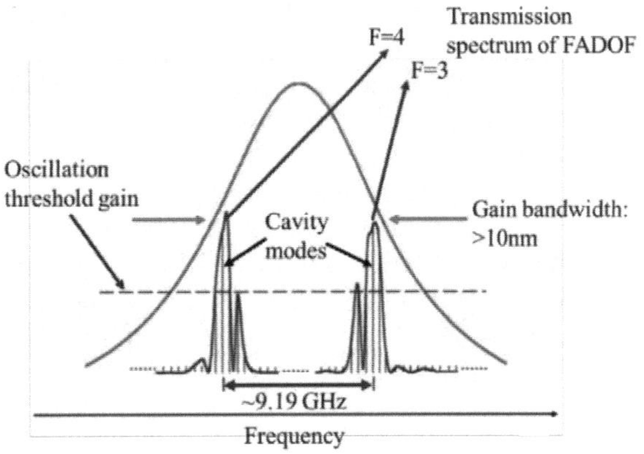

Figure 6. The relationship between the cavity modes, the transmission spectrum of FADOF and the gain bandwidth for the dual-frequency Faraday laser (adapted from [28]).

Recently, the Faraday laser using corner-cube retroflectors [29] and the Voigt laser [20] have also been realized. For the Voigt laser shown in figure 7, it also has great robustness to the fluctuation of the driving current and temperature (± 0.5 pm variation range with changes of driving current from 74 mA to 150 mA and temperature from 12 °C to 30 °C, respectively) of laser diode. During its 48-hour free-running operation, the wavelength fluctuates at a level below ±0.5 pm, illustrating remarkable reproducibility without manual adjustment. For the Faraday laser based on corner cube feedback, laser oscillation can always be established when the angle between the incident light and the optical axis of the corner-cube retroreflector is within ±3° range. Its single mode output is always within the Doppler-broadened Cs atomic transition line while the diode current changes from 55 mA to 155 mA. The corner-cube feedback can be extended to the Voigt lasers for increased robustness.

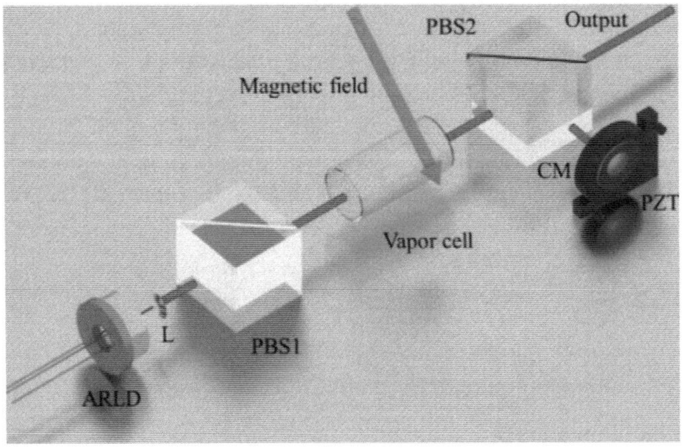

Figure 7. Schematic of the Voigt laser. ARLD: anti-reflection coated laser diode; L: aspheric lens; PBS: polarizing beam splitter; CM: cavity mirror; PZT: piezoelectric ceramic tube (adapted from [20]).

Table 2. Research results of the Faraday laser.

Type of the atom	linewidth	Immunity range	Changing range of output wavelength	Application areas
Natural Rb $5S_{1/2}\rightarrow 5P_{3/2}$ (780 nm) [9]	63.7 kHz	55 mA to 142 mA 15°C to 35°C	within 2 pm (985.6MHz)	Quantum precision measurements
96.5% ^{87}Rb $5S_{1/2}\rightarrow 5P_{3/2}$ (780 nm) [10]	< 221 kHz	55 mA to 142 mA 15°C to 35°C	~1 pm ~2 pm	Quantum precision measurements, Atomic physics
Natural Rb with 2 torr Xe buffer gas $5P_{3/2}\rightarrow 4D_{3/2}$ (1529 nm) [26]	15.5 kHz	85 mA to 171 mA 11°C to 32°C	~500 MHz ~500 MHz	Metrology, microwave photonics and optical communication
Cs $6S_{1/2}\rightarrow 6P_{3/2}$ (852 nm) [27]	17 kHz	60 mA to 130 mA 14°C to 35°C	~6 pm ~4 pm	Quantum precision measurements
Cs $6S_{1/2}(F=3)\rightarrow 6P_{3/2}$ [28] Cs $6S_{1/2}(F=4)\rightarrow 6P_{3/2}$ (852 nm)	30 kHz 33 kHz	With dual-wavelength output at the same time	7.43-8.80 GHz when the temperature of FADOF change from 36°C to 55°C	Atomic physics
^{87}Rb $5S_{1/2}\rightarrow 5P_{3/2}$ (780 nm) [20]	± 0.5 pm for 48-hour operation	73 mA to 150 mA 12°C to 30°C	~1 pm ~0.5 pm	atomic clock, laser cooling

2.3. The main progress made in frequency stabilization of the Faraday laser and Voigt laser

The output frequency of the Faraday laser and the Voigt laser can always be kept within the Doppler broadened atomic transition lines, which means that these semiconductor diode lasers based on the atomic optical filter comes with their own quantum references. The frequency stability measured by beating between two free running Faraday lasers can easily reach the 10^{-10} level. However, in a lot of research areas, such as atomic clocks, laser spectroscopy, atomic physics, the level of the frequency stability can still not meet the requirement. We use modulation transfer spectroscopy (MTS) to stabilize the output frequency of the Faraday laser to an external quantum reference, therefore suppressing the frequency fluctuations. The short-term frequency stability of the Faraday laser by self-estimation can be as low as $5.8\times10^{-15}/\sqrt{\tau}$ level by using the doubler-layer reference vapor cell [30], and when we lock one of the two modes of a Faraday laser operating with dual-frequency mode, the linewidth of the beat-note spectrum can be narrowed to 85 Hz [31]. Compare to the 902.95 Hz free-running linewidth, the linewidth is narrowed by more than 10 times.

Through heterodyne beating between two frequency locked lasers as well as one frequency locked laser and one frequency comb, the short-term frequency stability can reach the 10^{-13}, or even 10^{-14} level. Benefiting from the low cost, robustness and transportability, these lasers can be used as compact optical frequency standards, as well as in many other researches and practical applications.

The PDH technique has also been used for frequency-locking and linewidth-narrowing. We believe that the frequency stability can easily reach a level as low as 10^{-15} level with short term, and combing with the MTS, we can obtain excellent frequency stability both in short-term and long-term. The linewidth now is at the level of tens of hertz, it will be narrow to the hertz level in the near future. Table 3 is a summary of the typical research results of frequency stabilization of the Faraday laser and Voigt laser conducted by our team.

Table 3. Research results of frequency stabilization of the Faraday laser and Voigt laser.

Type of the atom	Transition lines	Results	Descriptions
Cs [31]	$6S_{1/2}F=3\rightarrow6P_{3/2}F'=2$ $6S_{1/2}F=4\rightarrow6P_{3/2}F'=5$	$3\times10^{-14}/\sqrt{\tau}$ with self-estimation 85 Hz linewidth when lock one mode	This laser can work both in single-frequency and dual -frequency mode
Cs [30]	$6S_{1/2}F=4\rightarrow6P_{3/2}F'=5$	$5.8\times10^{-15}/\sqrt{\tau}$ with self-estimation 2.0×10^{-15}@50 s with self-estimation	The double-layer cesium vapor cell was used to provide frequency reference
Cs	$6S_{1/2}F=3\rightarrow6P_{3/2}F'=2$ $6S_{1/2}F=4\rightarrow6P_{3/2}F'=5$	$1.63\times10^{-13}/\sqrt{\tau}$ with heterodyne beating	Can be used for optical frequency standards
^{87}Rb	$5S_{1/2}\rightarrow5P_{3/2}$, (780 nm)	$4\times10^{-13}/\sqrt{\tau}$ with heterodyne beating 9×10^{-14}@100 s with heterodyne beating	Can be used for optical frequency standards

2.4. The main progress made in anti-reflection coated laser diode

For almost all the semiconductor diode lasers, ARLDs are used to provide the essential gain. We have dedicated considerable attention to the development of ARLDs that are specifically engineered to meet our distinct needs. As of now, the Faraday as well as Voigt lasers based on the ARLDs made by us, can also keep the typical characteristic for immunity, the driving current range can be bigger than 120 mA, and the 48-hour wavelength fluctuation range is 4.5 pm when the laser is free running. The short-term frequency stability through heterodyne beating of two identical free running lasers can be as low as 1.2×10^{-10}@1s, and if we suppose the two lasers have the same contribution to the frequency stability, 1.2×10^{-11}@1s and 4.3×10^{-10}@10s can be attained for a single laser. When the laser frequency is locked through MTS, the short-term stability for a single laser acquired through heterodyne beating can also reach the 10^{-13} level. Evidenced by the results mentioned above, we have made some progress in this area, and are no longer dependent on commercial ARLD products.

3. The future of the Faraday and Voigt lasers

To continuously improve the performance and realize the potential to lead the next generation of semiconductor lasers with high frequency stability, the future research plans of our group are described as follows.

First of all, improving the structure of the FADOF and VADOF. Just as figure .2 (a) shows, these two atomic optical filters are composed of two PBSs, the permanent magnet, and the atomic vapor cell. Although the transmittance of the PBS and the atomic vapor cell can be higher than 95%, there is still feedback noise caused by reflection on the surfaces of these components returning to the ARLD. This will limit the frequency stability and other performance of the Faraday laser and the Voigt laser. As figure 8 shows, we intend to utilize an atomic vapor cell with two Brewster windows, combined with the external magnetic field to achieve an atomic optical filter, and then achieve laser oscillation. We believe that this changing will reduce the feedback noise of the lasers, as well as improve mechanical stability.

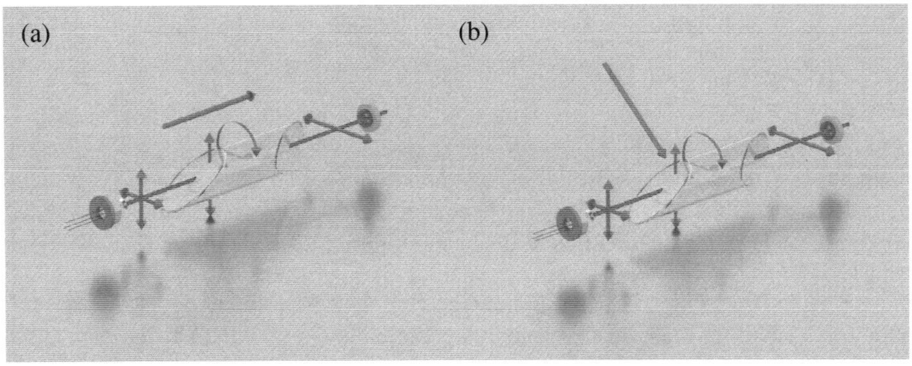

Figure 8. (a) Brewster-structure of Faraday laser. (b) Brewster-structure of Voigt laser.

Second, we try our best to develop the fiber FAODF and VADOF to achieve all-fiber lasers. Compared with the lasers in free space, all-fiber lasers can have higher mechanical stability, reducing the technical noise. Besides, the laser cavity length can be easily extended to the kilometer level or even longer to improve the situation of mode hopping. We think the core technology is the fiber cell, and it can be achieved in the near future by us.

Third, we will further combine the PDH technique with atomic spectroscopy-based stabilization, to obtain excellent frequency stability of the Faraday laser and the Voigt laser, both in short-term and long-term. And the laser can be applied for not only the research focusing on the short-term frequency stability, such as the local oscillator for the optical clock, but also in practical applications, for example the atomic fountain microwave frequency standards, the atomic gravimeter, the microwave frequency standards for deep space, the atomic beam clock and the atomic interferometer, etc.

Fourth, the bad cavity laser using the ultranarrow bandwidth cold-atom FADOF for frequency selection is of interest for research. When the bandwidth of the cavity mode is larger than the gain bandwidth, an active optical frequency standard [32-34] will be established, endowing the laser with the benefits of ultranarrow linewidth and immunity to cavity fluctuation, greatly improving the performance.

Finally, we aim to realize the miniaturized device with all the performances for the revolution of this semiconductor diode laser. We have improved the tunable range of this semiconductor diode laser to 4GHz without mode hopping and reduced the volume of the laser to 85 mm × 50 mm × 56 mm at the same time. We will continuously focus on this, and try to create an on-chip Faraday laser entirely based on chip-scale components.

4. Conclusion

This paper presents the advancements our team have achieved over the past twelve years in several key areas of the Faraday laser system: the development of the atomic optical filter, the laser characterized by its distinctive features, enhancements in frequency stability, and the utilization of ARLD as the gain medium for both Faraday and Voigt lasers with superior frequency stability. Looking forward, we outline research strategies aimed at propelling these technologies to the forefront of the next generation of semiconductor diode lasers. We are confident in our ability to realize these objectives in the near future, and to continuously elevate the performance of these laser systems.

Acknowledgments
This research was funded by the National Natural Science Foundation of China (NSFC) (91436210), Innovation Program for Quantum Science and Technology (2021ZD0303200), China Postdoc-toral Science Foundation (BX2021020), and Wenzhou Major Science & Technology Innovation Key Project (ZG2020046).

References

[1] Wieman C E, Hollberg L. 1991 *Rev. Sci. Instrum.* **62** 1-20.
[2] Wieman C, Hänsch T W. 1976 *Phys. Rev. Lett.* **36** 1170.
[3] Corwin K L, Lu Z T, Hand C F, Epstein R J, and Wieman C E 1998 *Appl. Optics* **37** 3295-98.
[4] Hänsch T W, Schawlow A L. 1975 *Opt. Commun.* **13** 68-9.
[5] Shirley J H. 1982 *Opt. Lett.* **7** 537-9.
[6] Drever R W P, Hall J L, Kowalski F V, Hough J, Ford G M, Munley A J and Ward H 1983 *Appl. Phys.* B **31**: 97-105.
[7] Wanninger P, Valdez E C. 1992 IEEE *Photonics Tech. L.* **4** 94-96.
[8] Ohtsu M, Teramachi Y, Otsuka Y and Osaki A 1986 IEEE *J. Quantum. Elect.* **22** 535-543.
[9] Miao X Y, Yin L F, Zhuang W, Luo B, Dang A H, Chen J B and Guo H 2011 *Rev. Sci. Instrum.* **82** 086106.
[10] Tao Z M, Hong Y L, Luo B, Chen J B and Guo H 2015 *Opt. Lett.* **40** 4348-51.
[11] Zhang J L, Gao G J, Wang B L, Guan X L, Yin L F, Chen J B and Luo B 2022 *Journal of Lightwave Technology* **40** 63-73.
[12] Ohman Y. 1956 *Stockholms Obs. Ann.* **19** 9-11.
[13] Wang Y F, Zhang S N, Wang D Y, Tao Z M, Hong Y L and Chen J B 2012 *Opt. Lett.* **37** 4059-61.
[14] Guan X L, Zhuang W, Shi T T, Miao J X, Zhang J, Chen J B and Luo B 2023 IEEE *Photonic. Tech. L.* **35** 672-5.
[15] Guan X L, Zhuang W, Shi T T, Miao J X, Zhang J, Chen J B and Luo B 2022 Frontiers in Physics **10** 1090483.
[16] Sun Q Q, Zhuang W, Liu Z W and Chen J B 2011 *Opt. Lett.* **36** 4611-13.
[17] Sun Q Q, Hong Y L, Zhuang W, Liu Z W and Chen J B 2012 *Appl. Phys. Lett.* **101** 211102.
[18] Pan D, Shi T T, Luo B, Chen J B and Guo H 2018 *Sci. Rep.* **8** 6567.
[19] Pan D, Xue X B, Shang H S, Luo B Chen J B and Guo H 2016 *Sci. Rep.* **6** 29882.
[20] Liu Z J, Guan X L, Qin X M, Wang Z Y, Shi H B, Zhang J, Miao J X, Shi T T, Dang A H and Chen J B 2023 *Appl. Phys. Lett.* **123** 131103.
[21] Xue X B, Tao Z M, Sun Q Q, Hong Y L, Zhuang W, Luo B, Chen J B and Guo H 2012 *Opt. Lett.* **37** 2274-76.
[22] Wang Y F, Zhang X G, Wang D Y, Tao Z M, Zhuang W and Chen J B 2012 *Opt. Express* **20** 25817-25.
[23] Xue X B, Pan D, Zhang X G, Luo B, Chen J B and Guo H 2015 *Photonics Res.* **3** 275-8.
[24] Keaveney J, Hamlyn W J, Adams C S and Hughes I G 2016 *Rev. Sci. Instrum.* **87** 095111.
[25] Zhuang W, Chen J B. 2014 *Opt. Lett.* **39** 6339-42.
[26] Chang P Y, Peng H F, Zhang S N, Chen Z Y, Luo B, Chen J B and Guo H 2017 *Sci. Rep.* **7** 8995.
[27] Chang P Y, Chen Y L, Shang H S, Guan X L, Guo H, Chen J B and Luo B 2019 *Appl. Phys.* B, **125** 230.
[28] Shi T T, Guan X L, Chang P Y, Miao J X, Pan D, Luo B, Guo H and Chen J B 2020 IEEE *Photonics J.* **12** 1-11.
[29] Wang Z Y, Liu Z J, Miao J X, Shi H B, Qin X M, Guan X L, Gao Z H, Shi T T and Chen J B 2023 *arXiv:* **2309** 04485.
[30] Shi H B, Chang P Y, Wang Z Y, Liu Z J, Shi T T and Chen J B 2022 IEEE *Photonics J.* **14** 1-6.

[31] Chang P Y, Shi H B, Miao J X, Shi T T, Pan D, Luo B, Guo H and Chen J B 2022 *Appl. Phys. Lett.* **120** 141102.
[32] Chen J B and Chen X Z. 2005 *Proc. of the* IEEE *Int. Frequency Control Symp. and Exposition* (Vancouver, BC, Canada) 3 pp.
[33] Chen J B. 2009 *Chinese Sci. Bull.* **54** 348-352.
[34] Shi T T, Pan D, Chen J B. 2019 *Opt. Express* **27** 22040-52.

Validating a lutetium frequency reference.

K J Arnold[1,2], S Bustabad[3], Zhao Qi[1], Qin Qichen[1], Zhiqiang Zhang[1], Zhang Zhao[1] and M D Barrett[1,3]

[1]Center for Quantum Technologies, National University of Singapore, Singapore
[2]Temasek Laboratories, National University of Singapore, Singapore
[3]National Metrology Center, Agency for Science, Technology and Research (A*STAR), Singapore

E-mail: phybmd@nus.edu.sg

Abstract. We review our progress in developing a frequency reference with singly ionized lutetium and give estimates of the levels of inaccuracy we expect to achieve in the near future with both the $^1S_0 \leftrightarrow {}^3D_1$ and $^1S_0 \leftrightarrow {}^3D_2$ transitions. Based on established experimental results, we show that inaccuracies at the low 10^{-19} level are readily achievable for the $^1S_0 \leftrightarrow {}^3D_1$ transition, and the frequency ratio between the two transitions is limited almost entirely by the BBR shift. We argue that the frequency ratio measured within the one apparatus provides a well-defined metric to compare and establish the performance of remotely located systems. For the measurement of an in situ frequency ratio, relativistic shifts drop out and both transitions experience the same electromagnetic environment. Consequently, the uncertainty budget for the ratio is practically identical to the uncertainty budgets for the individual transitions. If the ratios for two or more systems disagree we can be certain at least one of the clock assessments is incorrect. If they agree, subsequent comparisons on one transition would only differ by relativistic effects. Since motional effects are easily assessed and typically small for a heavy ion, only the differential gravitational red-shift will significantly contribute and this can be confirmed by comparison on the second transition.

1. Introduction

Over several years, we have developed the tools and methodologies to utilize singly-ionized lutetium (^{176}Lu$^+$) as a frequency standard of the highest calibre. Recently we have experimentally demonstrated agreement between two independent ^{176}Lu$^+$references at the low 10^{-18} level with clear evidence that mid 10^{-19} is readily attainable [1]. This has been achieved with much less technical effort than needed in other systems, as the clock transition used has the least sensitivity to environmental factors than any other system that we are aware of. We would argue that many systems would be hard pressed to demonstrate the levels of accuracy that lutetium so easily affords.

The relative ease at which a comparison at the 10^{-18} level can be achieved with lutetium should not be under-rated. Comparisons between clock systems claiming this level of accuracy have variations of measured ratios that cannot be explained by uncertainty budgets of the individual systems, which is compelling evidence that large systematic shifts are not being controlled to the levels claimed. Such problems are unlikely to occur for lutetium for the simple fact that systematics are not large enough to cause such variations. Indeed, the total shift and any single contribution to the uncertainty budget given in [1] is, at most, at the level of a few

Journal of Physics: Conference Series **2889** (2024) 012040 doi:10.1088/1742-6596/2889/1/012040

10^{-18}. Moreover, it is self-evident that such low systematics will have substantial room for improvement if current state-of-the-art technologies were employed to take advantage of this potential.

At the level of 10^{-18}, an important question to ask is: how does one know that a clock is truly performing at the claimed level without a direct comparison. Even if a comparison is made via an optical fiber link, the comparison would be influenced by the local gravitational potential, which is the basis for using optical clocks for geodesy [2, 3]. This is a catch-22 situation in which a comparison to determine the differential gravitational potential would require a comparison using knowledge of the gravitational potential to ensure the clocks are sufficiently accurate to make the measurement. Lutetium offers a unique solution to this problem in that it provides two high quality clock transitions within the one atom. Thus a frequency ratio for each individual system can be measured and globally compared. When measured on the same atom, a frequency ratio is independent of any relativistic effects, namely the gravitational redshift and motion. Motional effects are easily assessed and typically small for a heavy ion, so the measured IFR provides independent confirmation on the performance of two remotely located systems: if the ratios for each system disagree, there must be something wrong with one of the assessments; conversely the ratios can only agree if the individual assessments are correct. We neglect the very unlikely scenario that two independently assessed systematics of completely different physical origin are wrong and just so happen to cancel each other out, particularly given the low absolute systematic shifts that lutetium affords. No other atomic system that we are aware of could provide such a validation at least not to the degree that is possible with $^{176}\mathrm{Lu}^+$.

2. Systematics

In this section, we discuss the leading systematics for $^{176}\mathrm{Lu}^+$and the level of assessment that can be made for both $^1S_0 \leftrightarrow {}^3D_1$ and $^1S_0 \leftrightarrow {}^3D_2$ transitions, which we henceforth refer to as the primary and secondary transitions respectively. In both cases we assume that hyperfine averaging (HA) has been used to suppress shifts associated with the electronic angular momentum as first proposed in [4]. For the primary transition it is also assumed hyper-Ramsey spectroscopy is used to suppress the probe-induced ac-Stark shift.

A summary of typical systematics is given in table 1. For the most part, values given in the table are based on measurements that have been carried out in the lab. The two most notable exceptions are the residual quadrupole shift for the secondary transition and the blackbody radiation shift as discussed in sections 2.4 and 2.6 respectively.

2.1. Quadratic Zeeman shifts

For a given transition, quadratic Zeeman shifts are given by

$$\alpha_z \left(B_0^2 + \langle \tilde{B}^2 \rangle \right) + \tilde{\alpha}_z \langle \tilde{B}_\perp^2 \rangle, \tag{1}$$

where B_0 is the applied static field, $\tilde{B}$ is the amplitude of an oscillating field, and $\tilde{B}_\perp$ is the component of $\tilde{B}$ perpendicular the B_0. For lutetium, the most significant oscillating field is from radio-frequency (rf) currents in the electrodes driven by the rf trapping potential. The coefficient α_z arises from magnetic coupling to other fine-structure manifolds and $\tilde{\alpha}_z$ arises from magnetic coupling between hyperfine levels of the fine-structure manifold of interest [5]. For the primary transition, the measured value of α_z is $-4.89264(88)\,\mathrm{Hz/mT^2}$ [1], which gives a fractional frequency shift of $\sim 1.38 \times 10^{-16}$ at our nominal operating field of $B_0 \sim 0.1\,\mathrm{mT}$. Since B_0 is tracked during clock operation at the level of $10\,\mathrm{nT}$, the static term only affects an uncertainty budget at the mid 10^{-20} level.

Table 1. Expected uncertainty budgets for upcoming comparisons. All values are expressed fractionally in units of 10^{-18}. Shifts are typical values measured in our current systems. Uncertainties are based on those reported in [1] along with expected improvements as discussed in the text. Uncertainties are added in quadrature, but for the secondary transition ($^1S_0 \leftrightarrow {}^3D_2$) we omit the BBR shift to better illustrate the extent to which it dominates the uncertainty budget.

	$^3D_1$		$^3D_2$	
Effect	Shift	Unc.	Shift	Unc.
Quad. Zeeman (0.1 mT)	−138	0.04	48.1	0.04
ac-Zeeman (rf)	0.54	0.01	−10.5	0.11
ac-Zeeman (microwave)	−0.10	0.05	0.00	0.00
2nd-order Doppler (thermal)	−0.13	0.06	−0.13	0.06
Micromotion	−0.10	0.10	−0.10	0.10
RQS	0.22	0.02	$\lesssim 0.5$	0.10
Microwave coupling	0.00	0.21	0.00	0.21
BBR (300 ± 5 K)	−1.36	0.16	27.0	1.8
Total		0.30		0.28

The value of $\tilde{\alpha}_z$ is calculated to give a fractional shift of 5.65×10^{-19} for an rms amplitude of $1\,\mu$T, which we take to be accurate at the 1% level[1]. Indeed, calculation of the quadratic Zeeman shifts of individual hyperfine splittings differ only by Clebsch-Gordan factors and agree with measured values to better than 0.5%. As proposed in [5] and demonstrated in [6], the amplitude of $\tilde{B}_\perp$ is measured via an Autler-Townes splitting with an inaccuracy better than 0.1% and different orientations of B_0 can be used to infer $\tilde{B}$ [7]. In our current traps $\tilde{B} \lesssim 1\mu$T, which keeps the contributions to an uncertainty budget below 10^{-20}. The uncertainty value given in the table has been rounded up to avoid a zero entry for a non-zero shift.

For the secondary transition, α_z is yet to be measured, but is calculated to give a fractional shift of 4.8×10^{-17} at our nominal value of B_0. We expect this to be measurable to the same level as the primary transition and to similarly contribute to an uncertainty budget at the low 10^{-20} level. However, the value of $\tilde{\alpha}_z$ is a factor of 20 larger giving a calculated fractional frequency shift of -1.05×10^{-17} for an rms amplitude of $1\,\mu$T for $\tilde{B}$. Consequently, this term will limit the Zeeman uncertainties to 1.1×10^{-19}. In principle, $\tilde{\alpha}_z$ could be assessed more accurately, but the shift can also be suppressed by lowering the trap drive frequency, Ω_{rf}. For a given Ω_{rf} and voltage amplitude U_0, the trap psuedo-potential confinement frequencies, ω_i, scale as U_0/Ω_{rf} whereas the rf currents scale as $U_0\Omega_{\text{rf}}$. Consequently, for fixed ω_i, $\langle \tilde{B}^2 \rangle$ and $\langle \tilde{B}_\perp^2 \rangle$ scale as Ω_{rf}^4.

It is worthwhile commenting on the stability of the $\tilde{B}_\perp$ as comments in the literature are overly pessimistic [8] and misrepresent our own results [6]. The magnitude of $\tilde{B}_\perp$ is directly proportional ω_i and the orientation is fixed by the trap geometry however imperfect that might be. The trap geometry certainly doesn't change, and even a 1% change in ω_i is a substantial change in trap frequencies. It is also a significant change in the orientation of B_0, which influences optical pumping and state preparation in our case. Indeed, in our experiments, the orientation of B_0 is set relative to the linear polarization of the optical pumping beam that prepares the atom in $|^3D_1, F = 7, m_F = 0\rangle$. Maximizing the depumping time out $|^3D_1, F = 7, m_F = 0\rangle$ enables

[1] Magnetic field effects depend mostly on angular momentum considerations and are typically accurately calculated.

Journal of Physics: Conference Series **2889** (2024) 012040 doi:10.1088/1742-6596/2889/1/012040

alignment of B_0 to better than half a degree. One can expect some level of inhomogeneity in the spatial dependence of $\tilde{B}_\perp$, but not significantly over distances bounded by micromotion compensation. Consequently, there is no practical need to remeasure an Autler-Townes splitting on a regular basis as suggested in [8]. Only if there is a deliberate change in the pseudo-potential frequencies ω_i or to the clean-up polarizer in the optical pumping beam line would such a remeasurement be needed. Of course, the situation would be very different if the rf-currents and associated shifts were substantially larger than we see in our current traps.

For the primary transition there is an additional contribution to the ac-Zeeman shift that arises from our interrogation technique, which utilizes microwave transitions between the upper clock states to effect an averaging over all $m = 0$ hyperfine states. The technique, which we refer to as hyperfine-averaged Ramsey spectroscopy (HARS), was first demonstrated in [9], and a more detailed analysis given in [1, Supplemental]. The ac-Zeeman shifts arise from unwanted polarizations in the applied microwave fields that off-resonantly couple to neighbouring Zeeman states. Polarization components can be measured directly by microwave spectroscopy and the most significant contributions to the ac-Zeeman shift arise from coupling to the $F = 7$ states due to the very small g_F-factor. Consequently, this is unlikely to have any significant influence for the secondary transition, for which g_F-factors are more than an order of magnitude larger.

2.2. Micromotion

Considerations of excess micromotion (EMM) are much the same for lutetium as any other ion. For lutetium, we probe micromotion using sideband spectroscopy on the secondary clock transition, specifically the $|{}^1S_0, F = 7, m_F = 0\rangle \leftrightarrow |{}^3D_2, F = 9, m_F = 0\rangle$ transition, which does have benefits. When using sideband spectroscopy, it is important to have good resolution of both the carrier and sideband, and to ensure one remains on resonance with the transition. Exact numbers depend on specific beam orientations and polarizations relative to the applied magnetic field, but generally speaking it is straightforward to achieve a Rabi frequency of $\sim 10\,\mathrm{kHz}$ on the carrier with $< 100\,\mu\mathrm{W}$ of optical power, which is readily available at the clock wavelength of 804 nm. The 10 kHz Rabi frequency is well below any trap frequencies and has an associated ac-Stark shift $\lesssim 10\,\mathrm{Hz}$. This readily allows the detection of micromotion sidebands corresponding to a modulation index below 0.01. For typical trap drive frequencies of $\Omega_{\mathrm{rf}} = 2\pi \times 20\,\mathrm{MHz}$, the associated fractional second-order Doppler shift (SDS) is below 10^{-19}.

The fractional shift from EMM also has an ac-Stark shift arising from the electric fields driving the motion. For the primary transition this contribution is practically negligible owing to the small value of $\Delta\alpha_0(0)$. For the secondary transition, the contribution has opposite sign to the SDS leading to a cancellation as is well known for clock transitions having $\Delta\alpha_0(0) < 0$. This leads to the concept of a "magic rf" at which EMM shifts cancel to leading order. For the secondary transition in $^{176}\mathrm{Lu}^+$, this occurs at approximately $\Omega_{\mathrm{rf}} = 2\pi \times 32\,\mathrm{MHz}$.

Micromotion shifts reported in [1] were specified to bound the largest value observed over a period of several months with a timescale between measurements that was sometimes much longer than the measurement campaign. We are now working towards a more autonomous operation with automated assessment and compensation of micromotion. We consequently expect a significant reduction in this contribution to an uncertainty budget. Preliminary work indicates shifts can be maintained below 10^{-19} with compensation carried out once or twice a day with minimal impact on run time.

2.3. Thermal motion

Thermal motion is much less problematic for heavy ions. The cooling transition in $^{176}\mathrm{Lu}^+$has a linewidth of $\Gamma = 2\pi \times 2.45\,\mathrm{MHz}$, which leads to a fractional SDS at the Doppler cooling limit of -7.7×10^{-20} including the contribution from intrinsic micromotion (IMM) [10]. We typically get within about a factor of two of this limit. As this is a relativistic shift, it is the same for both transitions and cancels for an IFR measurement. In our recent comparison [1], the shift for one trap is substantially larger owing to a significant heating problem of unknown origin. The trap has now been replaced and gives a heating rate consistent with that found in our second trap. We currently use dephasing measurements of the clock transition to infer the thermal state, which is sufficient for any levels of precision we could achieve in the near future.

As shown in [10], there is an additional frequency shift from thermal motion due to sampling of the electric fields over the spatial extent of the external motion. As for EMM this is negligibly small for the primary transition owing to the small value of $\Delta\alpha_0(0)$ and leads to some level of cancellation for the secondary transition owing to the sign of $\Delta\alpha_0(0)$. The uncertainty given in the table for $^3D_2$ is therefore a conservative estimate inferred from measurements on $^3D_1$.

2.4. Residual quadrupole shift

Mixing of fine-structure levels due to the hyperfine interaction brings about modifications to Landé g-factors and quadrupole moments for a given state [11, 12]. These corrections are not cancelled by HA giving rise to a residual quadrupole moment (RQM) for the hyperfine-averaged frequency reference. For lutetium, leading order corrections for both g-factors and quadrupole moments arise from the magnetic dipole moment of the nucleus. Thus, high accuracy g-factors measurements can be used to factor out matrix elements associated with the nuclear magnetic dipole moment for the estimation of the RQM. Using this approach, we have obtained an estimate of $-2.54\times10^{-4}\,ea_0^2$ for the RQM of $^3D_1$ [11], which leads to a shift at low 10^{-19} for typical confinement conditions. A measurement of the shift can then be obtained by scaling the quadrupole shift extracted from clock operation with Ba^+ by the ratio of quadrupole moments. An estimate of the RQM for $^3D_2$ could be achieved in a similar fashion. However, contributions to the $^3D_2$ RQM are primarily from coupling to $^3D_1$ and $^3D_3$. The former can be inferred from the RQM of $^3D_1$ and the latter contribution is reduced by the larger energy splitting. Consequently, we expect the RQM for $^3D_2$ to have a similar magnitude to $^3D_1$ barring any fortuitous cancellation.

A more direct experimental assessment of the RQM could potentially be obtained from differential quadrupole shifts of hyperfine splittings in a multi-ion crystal as used to determine the quadrupole moments of $^3D_1$ and $^3D_2$ in [7]. The quadrupole moments were extracted from measurements of a single hyperfine splitting. To lowest order, ratios of the quadrupole shifts for different hyperfine splittings are completely determined by Clebsch-Gordan coefficients. However, corrections to these ratios from mixing with other fine-structure levels can also be related to the RQM. Indeed it was noted in [7] that the inferred quadrupole moments may well be influenced by this.

2.5. Probe-induced ac-Stark shift

The lifetime of the primary clock transition is estimated to be about 1 week, and this gives rise to a significant probe-induced ac Stark shift, although it is substantially smaller than for the E3 transition in Yb^+ under similar interrogation conditions. For the experiments reported in [1], Ramsey spectroscopy with an interrogation time of 744 ms and optical pulses with an 8 ms π-pulse time, gave a shift of $\sim 10^{-16}$ with an uncertainty at mid 10^{-18} level. Further use of hyper-Ramsey spectroscopy practically eliminates the shift without any need to monitor or control variations in the laser intensity beyond the usual power stabilization. For the secondary transition, the lifetime is substantially reduced and the probe-induced ac-Stark shift is negligible under any reasonable interrogation conditions.

The probe-induced ac-Stark shift is dependent on the interrogation time and π-pulse duration. The ac-Stark shift itself scales as τ_L^{-2}, where τ_L is the π-pulse time, and Ramsey spectroscopy gives a suppression factor that scales as T/τ_L, where T is the Ramsey time. Consequently, the probe-induced ac-Stark shift becomes much less problematic as laser technology allows increasing interrogation times. For example, increasing the interrogation time to 10 s and τ_L to 24 ms, reduces the shift for the primary clock transition to 5×10^{-18} with an uncertainty of 2.5×10^{-19} assuming a modest 5% stability of the laser intensity. At this level, hyper-Ramsey spectroscopy may well be unnecessary.

2.6. Blackbody radiation shift

The blackbody radiation (BBR) shift is characterized by the differential static scalar polarizability, $\Delta\alpha_0(0)$, which has been measured for both transitions [13] using a CO_2 laser. For the primary transition, the shift is just $-1.36(9) \times 10^{-18}$ at 300 K. In our current experiments we do not have the ability to measure temperature but even an assessment of 35(10) °C would give a contribution of just 2.9×10^{-19} to an uncertainty budget. A trap that incorporates a temperature sensor at the trap electrodes is currently under construction, but comparisons under any practical circumstances are unlikely to be influenced by temperature. In contrast, the BBR shift for the secondary transition is about a factor of 20 larger with a fractional shift of $2.70(21) \times 10^{-17}$ at 300 K. Thus, inaccuracy below 10^{-18} on the secondary transition requires a significant improvement in the measured polarizability and a temperature assessment better than ±3 °C.

To improve the assessment of $\Delta\alpha_0(0)$ for the secondary clock transition, we plan to make a comparison measurement using Ba^+ as proposed in [14]. The basic idea is to use the more accurate value of $\Delta\alpha_0(\omega)$ for Ba^+ to better characterize the in situ intensity of the CO_2 laser. Supporting measurements to characterize $\Delta\alpha_0(\omega)$ for Ba^+ have already been carried out [15, 16, 17] that enable an improvement in the fractional inaccuracy of $\Delta\alpha_0(0)$ to 0.5% [18]. This will reduce the uncertainty contribution from $\Delta\alpha_0(0)$ to $< 2 \times 10^{-19}$ for operation at 300 K leaving temperature assessment as the biggest hurdle for the secondary transition.

As $\Delta\alpha_0(\omega) < 0$ for the secondary transition, one could in principle use the approach used for Sr^+ and Ca^+ [19, 20], which makes use of the magic rf for the trap drive at which large EMM shifts are cancelled. However, both EMM shifts and the quadratic Zeeman shift from rf currents in the electrodes vary quadratically as the ion is displaced. With very large amounts of EMM, the contribution from the quadratic Zeeman shift can be expected to be prohibitively large. Moreover, we measure this with Ba^+ for which uncompensated micromotion will lead to a different spatial location than Lu^+ owing to the difference in mass.

2.7. Miscellaneous

Additional shifts that have been considered but outside the categories given above are phase chirp and uncertainties arising from the imperfections in the HARS interrogation technique we have used. Phase chirp has been investigated following the methodology reported in [21] and estimated to be less than 10^{-19}. When using HARS, the most important effects are ac-Zeeman shifts from the microwave fields as discussed in Sect. 2.1, and from the accuracy of the microwave π-pulse times, which is the more significant issue. We estimate the sensitivity to the π-pulse times from simulations, and we expect this to be similar for both transitions.

At the level of $\lesssim 10^{-19}$ there may well be additional shifts, such as collision shifts that appear in some uncertainty budgets, to consider. Without the measurement precision or methodology to investigate further, we consider such considerations to be academic and speculative. Until measurement precision reaches the level of realistic estimates, we omit them.

3. An in situ frequency ratio measurement

On the primary clock transition, inaccuracy below 5×10^{-19} is readily achievable in the near future. Eliminating the large heating rate of one trap, characterizing the quadrupole fields as already done for one trap, and automated micromotion detection and compensation reduces the largest contributions to the uncertainty budgets given in [1] to $\lesssim 10^{-19}$. With a realistically

Journal of Physics: Conference Series **2889** (2024) 012040 doi:10.1088/1742-6596/2889/1/012040

achievable temperature assessment with an inaccuracy of $\pm 5\,^\circ$C, the total uncertainty would be 3.0×10^{-19}, with a significant contribution coming from imperfections in the interrogation technique. The only shift that could realistically affect the transition at that level would be that from the dc applied magnetic field, which is monitored with an interleaved servo. With a 10 s long interrogation via correlation spectroscopy, verification at the level of 5×10^{-19} would be achievable with a few days of integration.

The BBR shift not withstanding, the same argument applies to the secondary transition. Consequently, measurement of an in situ frequency ratio (IFR) would be almost entirely limited by the BBR shift on the secondary transition as is evident from table 1. At a liquid nitrogen temperature of 77 K, the BBR shift is reduced to 10^{-19} and removes this limitation. The measured IFR then becomes an independent figure of merit for clock performance at the level of uncertainty producible by either transition. At room temperature, a 5×10^{-19} uncertainty would require a temperature assessment with an uncertainty $\approx \pm 1.5\,^\circ$C, which is fairly challenging in practice. Arguably, so is the use of cryogenic temperatures, but a measurement of the IFR at 77 K is practically a zero temperature measurement. The corresponding ratio is a property of the atom that can be measured independently by any lab and this can achieved with a given uncertainty if and only if both clock transitions can operate at the same level of performance, which is currently at the level of $\lesssim 5 \times 10^{-19}$.

Once a zero temperature ratio is established, it would allow an IFR measured at room temperature to estimate the BBR shift and hence the effective temperature of the thermal environment seen by the ion. If we assume both the zero-temperature and room-temperature ratios were measured with an inaccuracy of 5×10^{-19}, the contribution from the BBR shift for the room temperature system could be inferred with an inaccuracy of $5\sqrt{2} \times 10^{-19}$ corresponding to a temperature inaccuracy of $\pm 2^\circ$C. Two clocks having been assessed in this manner could then be compared on the primary transition to determine the gravitational red-shift between them, with a small correction due to the temperature. Subsequent comparison on the secondary transition could then be used as a consistency check to confirm the gravitational red-shift and temperature assessment.

Even without the determination of a zero-temperature frequency ratio, use of the secondary transition would still be instrumental in bounding the temperature difference between two clocks. For the secondary transition, the BBR shift is the only significant systematic above mid 10^{-19}, so any difference between two systems would be attributable to temperature. For a given difference in the secondary frequency or primary to secondary ratio, the temperature difference is a monotonically decreasing function of the mean temperature. Taking the temperature of the surrounding as a lower bound for the temperature of the coldest clock, one can then infer bounds on the possible temperature difference between the two systems.

4. Stability Considerations

The level of accuracy attainable with lutetium requires an improvement in stability beyond that achievable with our available laser system. For this reason we have relied upon correlation spectroscopy to extend the achievable interrogation time and validate the quality of a lutetium frequency reference. Nevertheless, there is still a clear need to increase the number of ions beyond the single ion paradigm and proof-of-principle experiments were first demonstrated in [22, 9]. In those experiments we demonstrated the ability to measure and control inhomogeneous shifts to a degree that would allow long (10 s) interrogation times.

For lutetium the most notable source of inhomogeneous broadening is the quadrupole shift induced by the electric fields from neighbouring ions. As demonstrated in [22], this can be mitigated by aligning the magnetic field at angle $\theta_0 = \cos^{-1}(1/\sqrt{3})$ to the ion crystal at which the contribution from neighbouring ions is zero. Although the sensitivity of the suppression is linear in the angle, this should be considered in the context of how well the angle can be set and maintained. As demonstrated in [7], the differential shift between the outmost and innermost ions in a linear ion crystal can be measured using microwave spectroscopy. In that work the field was aligned to maximize the shift and the measured value used to estimate the quadrupole moment. Alternatively, the same method can be used to align the magnetic field to zero the difference.

For the purposes of illustration, consider a 10-ion crystal with an axial confinement of 100 kHz. For angles $\theta = \theta_0 + \delta\theta$, the differential quadrupole shift between the outermost and innermost ions is approximately $7\delta\theta$ Hz when interrogating the $F = 7$ to $F = 8$ microwave transition. It is reasonably straightforward to keep magnetic field variations below 0.1 μT at our operating field of 0.1 mT, which corresponds to $\delta\theta = 10^{-3}$ and reduces the differential quadrupole shift to 7 mHz. Due to the low magnetic field sensitivity of the clock states and practically unlimited lifetime of $^3D_1$, Ramsey interrogation can easily resolve the 7 mHz in a matter of minutes. Moreover, any remaining inhomogeneity would be removed during interrogation when using HARS as demonstrated in [9].

Chip scale ion traps may eventually make it feasible to consider hundreds or even thousands of ions, but near term advances will undoubtedly utilize a more modest number. In any case, it will always be advantageous to extend interrogation times to the limit that technology will allow. Already laser technology permits interrogation times of a few tens of seconds. For the primary transition, a 5-ion lutetium clock with a 10 s interrogation time would provide a projection noise limit of $6.4 \times 10^{-17}/\sqrt{\tau}$, which allows mid 10^{-19} precision in 5 hours. The corresponding instability for the secondary transition is slightly larger ($8.0 \times 10^{-17}/\sqrt{\tau}$) owing to the finite lifetime, but this still enables mid 10^{-19} precision within 7 hours. From the perspective of gaining stability, the importance of lifetime is best illustrated by considering an atom such as In^+. With a lifetime of just 200 ms, the same $6.4 \times 10^{-17}/\sqrt{\tau}$ instability achievable with 5 Lu^+ ions would require over 50 In^+ ions.

Another approach to improving stability of ion-based clocks is the use of differential spectroscopy demonstrated in [23]. In essence, a more stable lattice clock was used to extend the interrogation time of the ion beyond the coherence time of the laser to improve the measurement precision of the frequency ratio. The measured frequency ratio was not given in their report as neither clock was operating with previously claimed levels of performance. Nevertheless the method is of interest for those ions that can provide and maintain superior accuracy. The method equally applies to ensembles of ions and even entangled states. Entanglement could only be advantageous for clock transitions in which lifetime does not become a factor. Given the week-long lifetime of $^3D_1$ and low sensitivity of the primary clock transition to its electromagnetic environment, this approach is ideally suited to lutetium.

5. Discussion

In [24], 37 authors from 21 institutes across 15 different countries outline a roadmap towards the redefinition of the second based on optical frequency standards (OFS). The roadmap lists a set of mandatory criteria to be met for such a redefinition and gives the degree to which the 37 authors consider these criteria to be met. As this is presumably a representation of what is considered to be state-of-the-art, it is of interest to consider our work in the context of these criteria and in comparison to those works considered to have met those criteria. Criteria most relevant to our work are I.1, and I.2, which we give here for convenience.

Mandatory Criteria I.1.b: *At least three frequency evaluations of optical frequency standards based on different reference transitions, either in the same institute or different institutes, have demonstrated evaluated uncertainties $\lesssim 2 \times 10^{-18}$ based on comprehensive, comparable and published accuracy budgets.* This criteria is considered to be more or less fulfilled with one of the three evaluations quoted being that of the NIST Al^+ clock at 9.4×10^{-19} [25]. In that assessment, the researchers observe a 10^{-16} frequency difference when the atom is probed from

Journal of Physics: Conference Series **2889** (2024) 012040 doi:10.1088/1742-6596/2889/1/012040

counter-propagating directions. This is assumed to be a first order Doppler shift of 5×10^{-17} arising from charging and discharging induced by photo-electrons generated by the UV lasers used in the operation of the clock. But no conclusive scientific evidence is given to support this conjecture and no clock comparison has ever been presented that demonstrates cancellation of the shift.

Mandatory Criteria I.2.a: *Unit ratios (frequency comparison between standards with same clock transition): at least three measurements between OFS in different institutes in agreement with an overall uncertainty of the comparison* $\Delta\nu/\nu \lesssim 5\times10^{-18}$ *(either by transportable clocks or advanced links). Applicable to at least one radiation of I.1.* This is closely related to Mandatory Criteria I.1.a with the PTB Yb^+ comparison [26] cited as one of the three comparisons being a "first step in the right direction". We caution that a frequency comparison in and of itself does not validate an uncertainty budget, particularly if every effort is made to make the environmental conditions the same for both systems. If temperatures are made near equal, simulations used to decide absolute temperature and extrapolations to determine the differential scalar polarizability are inconsequential. For the latter, it is readily shown that the accuracy of the extrapolation given in [27] is incorrect by at least a factor of three.

So how does lutetium stack up against this road map and the work of leading metrology institutes? For our comparison reported in [1], the uncertainly for each system is $\leq 0.68\times10^{-18}$, although we openly acknowledge that this does not include the BBR shift as we are unable to measure temperature in either system. However, if we assume the temperature of each system is within the range of $25-65\,°C$, the total uncertainty would still be below the 0.94×10^{-18} claimed by NIST [25]. As assumptions go, we consider this temperature range to be very reasonable. Moreover, a temperature range of $25-110\,°C$ ensures the uncertainty for each system is below the roadmaps 2×10^{-18} requirement. Additionally, given the low systematics of a $^{176}Lu^+$ primary, a deviation by more than 2×10^{-18} would be highly unlikely. With regards to criteria I.2.a, the comparison in [1] has a projection noise limit of 3×10^{-18}, which is below the 5×10^{-18} of criteria I.2.a.

But what of frequency ratios between standards? Frequency ratios with other OFS are only meaningful if carried out with the highest degree of scientific integrity and that starts with each OFS being independently and rigorously assessed. Such an assessment would at least include: (i) a clear set of experimental measurements of atomic properties and environmental factors determining the performance claim, (ii) a same-species comparison with a precision to at least the level of accuracy being claimed, which cannot be better than the measurement precision, and (iii) a stress test on any systematic that is more than an order of magnitude larger than the accuracy claim i.e. a frequency comparison in which the systematic is deliberately shifted to give a predicted shift in the clock frequency. We see a stand-alone uncertainty budget without a same-species comparison as an hypothesis that is to be experimentally tested, which can only be done through a same-species comparison.

Problems that arise for frequency ratios with other OFS should be resolved using impartial, evidence-based investigation. Dismissing inconsistent results with error scaling [28] or a "comprehensive Baysian analysis" [29], or disregarding earlier measurements in conflict with the new [29] has no place in precision metrology. It undermines the purpose of carrying out a frequency comparison in the first place; be it a search for new physics, geodesy, or ensuring the integrity of OFSs around the world. With the widespread acceptance of error scaling practices, any evidence of a variation of fundamental constants or postulated interactions with dark matter will be based on a subjective decision of how or when error scaling should be applied. Equally, it would make "option 2" for the redefinition of the second questionable as the choice of weights for the contributing OFSs would succumb to the same considerations.

We have argued that the frequency ratio between the primary and secondary transitions in $^{176}Lu^+$ is an independent figure of merit that confirms the performance of the individual transitions to the level they have been assessed. Our approach to validating a $^{176}Lu^{+}$ frequency reference will be to establish two liquid nitrogen systems; one at the Center for Quantum Technologies and one at the National Metrology Center in Singapore. The aim is to verify agreement of the IFR for the two systems, and then compare the two systems on the primary to deduce the local gravitational redshift, which can be independently confirmed by conventional technologies and also by comparison on the secondary. Further comparisons with a room temperature system can then confirm the IFR may be used to assess the system temperature. We would then have a complete set of well-defined experimental metrics to characterize and assure the performance of the standard.

Acknowledgments

This research is supported by the Agency for Science, Technology and Research (A*STAR) under Project No. C210917001; the National Research Foundation (NRF), Singapore, under its Quantum Engineering Programme (QEP-P5); the National Research Foundation, Singapore and A*STAR under its Quantum Engineering Programme (NRF2021-QEP2-01-P03) and its CQT Bridging Grant; and the Ministry of Education, Singapore under its Academic Research Fund Tier 2 (MOE-T2EP50120-0014).

[1] Zhang Z, Arnold K J, Kaewuam R and Barrett M D 2023 *Science Advances* **9** eadg1971
[2] Mehlstäubler T E, Grosche G, Lisdat C, Schmidt P O and Denker H 2018 *Reports on Progress in Physics* **81** 064401
[3] McGrew W, Zhang X, Fasano R, Schäffer S, Beloy K, Nicolodi D, Brown R, Hinkley N, Milani G, Schioppo M *et al.* 2018 *Nature* **564** 87–90
[4] Barrett M 2015 *New Journal of Physics* **17** 053024
[5] Gan H C J, Maslennikov G, Tseng K W, Tan T R, Kaewuam R, Arnold K J, Matsukevich D and Barrett M D 2018 *Physical Review A* **98** 032514
[6] Arnold K J, Kaewuam R, Chanu S R, Tan T R, Zhang Z and Barrett M D 2020 *Physical Review Letters* **124** 193001
[7] Kaewuam R, Tan T R, Zhang Z, Arnold K J, Safronova M S and Barrett M D 2020 *Physical Review A* **102** 042819
[8] Beloy K 2023 *Physical review letters* **130** 103201
[9] Kaewuam R, Tan T R, Arnold K J, Chanu S R, Zhang Z and Barrett M D 2020 *Physical review letters* **124** 083202
[10] Berkeland D, Miller J, Bergquist J C, Itano W M and Wineland D J 1998 *Journal of applied physics* **83** 5025–5033
[11] Zhang Z, Arnold K J, Kaewuam R, Safronova M S and Barrett M D 2020 *Physical Review A* **102** 052834
[12] Beloy K, Leibrandt D R and Itano W M 2017 *Physical Review A* **95** 043405
[13] Arnold K J, Kaewuam R, Roy A, Tan T R and Barrett M D 2018 *Nature communications* **9** 1650
[14] Barrett M, Arnold K and Safronova M 2019 *Physical Review A* **100** 043418
[15] Arnold K, Chanu S, Kaewuam R, Tan T, Yeo L, Zhang Z, Safronova M and Barrett M 2019 *Physical Review A* **100** 032503
[16] Zhang Z, Arnold K, Chanu S, Kaewuam R, Safronova M and Barrett M 2020 *Physical Review A* **101** 062515
[17] Woods S L, Hanni M, Lundeen S and Snow E L 2010 *Physical Review A* **82** 012506
[18] Chanu S, Koh V, Arnold K, Kaewuam R, Tan T, Zhang Z, Safronova M and Barrett M 2020 *Physical Review A* **101** 042507
[19] Dubé P, Madej A A, Tibbo M and Bernard J E 2014 *Physical review letters* **112** 173002
[20] Huang Y, Guan H, Zeng M, Tang L and Gao K 2019 *Physical Review A* **99** 011401
[21] Falke S, Misera M, Sterr U and Lisdat C 2012 *Applied Physics B* **107** 301–311
[22] Tan T R, Kaewuam R, Arnold K J, Chanu S R, Zhang Z, Safronova M and Barrett M D 2019 *Phys. Rev. Lett.* **123** 063201
[23] Kim M E, McGrew W F, Nardelli N V, Clements E R, Hassan Y S, Zhang X, Valencia J L, Leopardi H, Hume D B, Fortier T M *et al.* 2023 *Nature Physics* **19** 25–29
[24] Dimarcq N, Gertsvolf M, Mileti G, Bize S, Oates C, Peik E, Calonico D, Ido T, Tavella P, Meynadier F *et al.* 2023 *Metrologia*
[25] Brewer S M, Chen J S, Hankin A M, Clements E R, Chou C, Wineland D J, Hume D B and Leibrandt D R 2019 *Physical review letters* **123** 033201

[26] Sanner C, Huntemann N, Lange R, Tamm C, Peik E, Safronova M S and Porsev S G 2019 *Nature* **567** 204–208
[27] Huntemann N, Sanner C, Lipphardt B, Tamm C and Peik E 2016 *Physical review letters* **116** 063001
[28] Dörscher S, Huntemann N, Schwarz R, Lange R, Benkler E, Lipphardt B, Sterr U, Peik E and Lisdat C 2021 *Metrologia* **58** 015005
[29] Beloy K *et al.* 2021 *Nature* **591** 564–569

Trapping of triply charged thorium-229 for a nuclear clock

A Yamaguchi[1,2,3], Y Shigekawa[4], H Haba[4], M Wada[5], and H Katori[1,2,3]

[1] Quantum Metrology Laboratory, RIKEN, Saitama, Japan
[2] Space-Time Engineering Research Team, RIKEN, Saitama, Japan
[3] Department of Applied Physics, Graduate School of Engineering, The University of Tokyo, Tokyo, Japan
[4] Nishina Center for Accelerator-Based Science, RIKEN, Saitama, Japan
[5] KEK Wako Nuclear Science Center, Saitama, Japan

Email: atsushi.yamaguchi.fv@riken.jp

Abstract. We developed an ion trap system for triply charged thorium-229 ($^{229}Th^{3+}$). We used uranium-233 as the ion source, which supplied $^{229}Th^{3+}$ ions in the nuclear ground state and low-lying isomeric state ($^{229m}Th^{3+}$) at 8.3 eV. Our ion trap system allows the laser spectroscopy of $^{229m}Th^{3+}$ toward the nuclear clock.

1. Introduction

Thorium-229 (^{229}Th) attracts attention owing to its very low-energy transition between the nuclear ground state and the first excited state (called "isomer", ^{229m}Th). The transition energy is approximately 8.3 eV [1–5]. Figure 1 shows the energy of the first nuclear-excited state as a function of atomic number. The ^{229}Th isomer is the lowest-energy nuclear state reported to date. This transition energy is laser-accessible, enabling the high-resolution laser spectroscopy of atomic nuclei. One of its applications is a "nuclear clock" based on the resonance frequency of this nuclear transition [6–9]. Compared with the electronic transitions used in existing atomic clocks, the nuclear transition frequency is more insusceptible to the external environment. Thus, the accuracy of the nuclear clock is expected to outperform that of the current atomic clocks [10]. Taking advantage of its insusceptibility, a solid-state nuclear clock using a ^{229}Th-doped crystal has also been proposed [7,8].

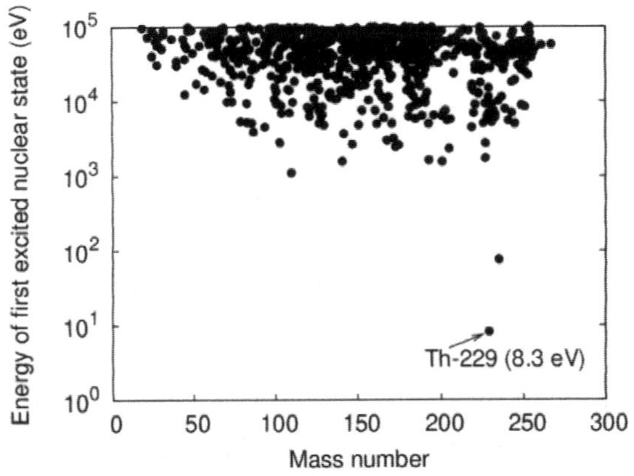

Figure 1. Energy of the first nuclear-excited state is plotted as a function of atomic numbers. ^{229}Th has an exceptionally low-energy nuclear excited state at 8.3 eV.

An ion trap can be an attractive platform for an accurate nuclear clock. The states of ions in a trap can be controlled and read with high fidelity using various laser spectroscopic techniques. For a nuclear clock, the favourable charge state is 3+ because of the closed electronic transitions [6], which are inevitable for laser cooling and laser-induced fluorescence detection of ions [11]. Figure 2 shows the electronic levels of triply charged thorium (Th^{3+}) ions. The $^2F_{5/2}$ state is the ground state, and there are two closed transitions: the $^2F_{5/2}$–$^2D_{3/2}$ transition (1088 nm) and the $^2F_{5/2}$–$^2D_{5/2}$–$^2F_{7/2}$ transition (690 and 984 nm). By combining these two closed transitions, laser cooling of $^{229}Th^{3+}$ ions in the nuclear ground state was successfully demonstrated in 2011 [12].

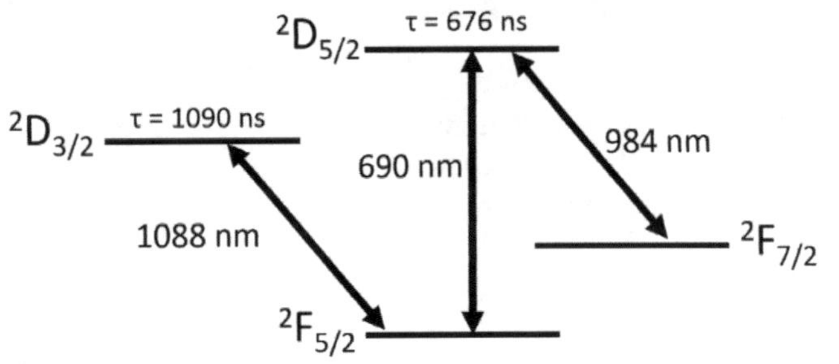

Figure 2. Electronic levels of Th^{3+} ions. The $^2F_{5/2}$ state is the ground state. The 1088-nm and the 690- and 984-nm transitions are the closed transitions. The lifetime of $^2D_{3/2}$ and $^2D_{5/2}$ states were calculated in ref. [15].

Toward a nuclear clock, the detailed properties of ^{229m}Th have been experimentally investigated, including the first experimental confirmation of ^{229m}Th [13], laser spectroscopy of doubly charged ^{229m}Th [14], and precise energy determination of ^{229m}Th [1–5]. These studies improved knowledge of ^{229m}Th significantly. However, no experimental studies have been reported on triply charged ^{229m}Th ($^{229m}Th^{3+}$).

For nuclear clocks based on $^{229}Th^{3+}$, information on the hyperfine structures of the electronic states of $^{229m}Th^{3+}$ is essential for detecting nuclear excitation by an electron-shelving technique using selective fluorescence detection of isomers. The lifetime of $^{229m}Th^{3+}$ is also important because it is related to the natural linewidth of nuclear transition [15]. We developed an ion trap system that allows the trapping and laser spectroscopy of $^{229m}Th^{3+}$.

2. ^{233}U as a source of $^{229}Th^{3+}$ isomer

A straightforward method to obtain ^{229m}Th is the direct laser excitation of ^{229}Th in the nuclear ground state. However, this has not been demonstrated thus far, mainly because of the difficulties in preparing the excitation laser at a vacuum ultraviolet wavelength (149 nm). Instead, three alternative methods have been developed: X-ray pumping, β-decay of actinium-229 (^{229}Ac), and α-decay of uranium-233 (^{233}U). In the X-ray pumping scheme, ^{229}Th nuclei are excited to the second nuclear-excited state at 29.2 keV using X-rays (synchrotron radiation) [16]. Subsequently, the nuclei partially decay to the isomeric state. The latter two methods utilise radioactive decay. The β-decay of ^{229}Ac populates isomer with higher probability (14%–93% [5,17,18]) than the α-decay of ^{233}U (approximately 2% [13,14,19]). However, the half-life of β-decay of ^{229}Ac is only 62.7 min. Thus, measurements should be conducted "online" at nuclear reaction facilities, such as ISOLDE at CERN [5]. On the other hand, the half-life of α-decay of ^{233}U is 1.6×10^5 years. Therefore, once we prepare the ^{233}U ion source, "offline" experiments with ^{229}Th can be performed without beam time constraints. In this study, ^{233}U was selected as the source of the isomer ($^{229m}Th^{3+}$).

We electrodeposited 600 kBq of ^{233}U on the surface of a titanium foil (thickness is 50 μm). The diameter of the deposited area was 90 mm. The average thickness of deposited ^{233}U was estimated to be 25 $\mu g/cm^2$, which was close to the range of recoil ^{229}Th ions in uranium compounds such as U_3O_8. To remove the impurities deposited on the ^{233}U surface, which potentially blocked the recoiled ^{229}Th ions, we heated the source at 600 °C by a heater plate in a vacuum. Figure 3(a) shows the ^{233}U source on the heater plate attached to the vacuum flange. Figure 3(b) shows the temperature and pressure

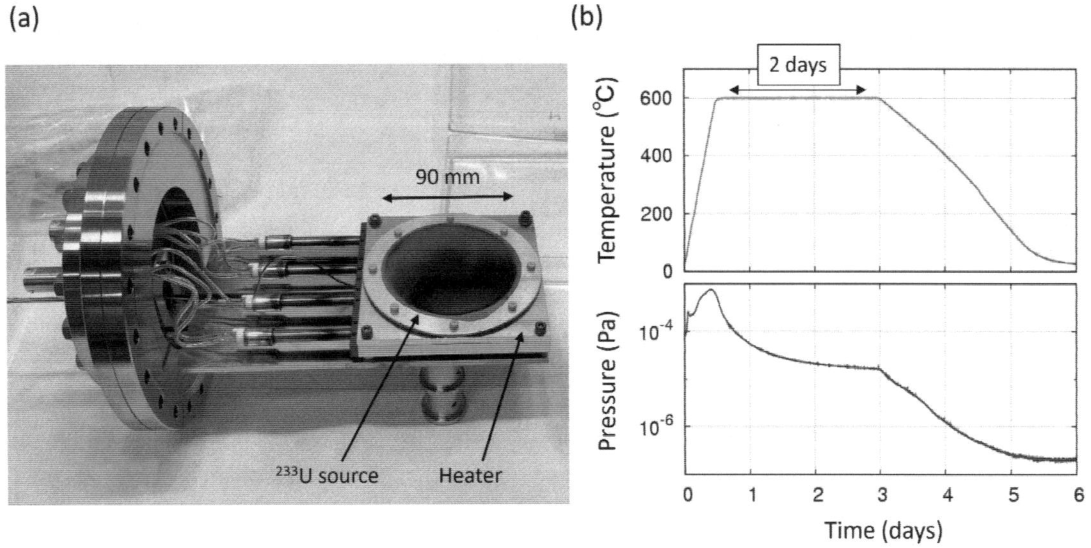

Figure 3. (a) Picture of the apparatus to heat the ^{233}U source up to 600 °C. The ^{233}U source was installed on a heater plate. The heater plate was attached to the vacuum flange. (b) Changes of temperature and vacuum pressure during heating. The ^{233}U source was heated at 600 °C for two days to remove impurities attached on the surface of ^{233}U during the electrodeposition process.

changes during the heating process. We increased the temperature at a rate of 1 °C/minute. After more than two days of heating, the pressure became nearly constant, indicating that the desorption of impurity molecules from the ^{233}U source was complete. We then lowered the temperature at a rate of approximately −0.2 °C/minute. This slow cooling was necessary to prevent wrinkling of the titanium foil caused by its uneven temperature distribution.

3. Extraction and trapping of ^{229}Th^{3+} ions

Figure 4 shows a schematic of the apparatus used for extracting and trapping the recoil ^{229}Th^{3+} ions emitted from the ^{233}U source and trap them. The recoiled ions were cooled using a helium buffer gas. The buffer gas cooling process also aligns the charge state of ^{229}Th ions to 3+ via the following mechanism [20]: Since the α-decay of ^{233}U is an energetic process (energy of the α-particle is 4.9 MeV), ^{229}Th ions are likely to be in the highly charged state just after recoiled from the ^{233}U source. The charge state of ^{229}Th ions was lowered by charge-exchange collisions with helium gas. However, this process stops when ^{229}Th becomes 3+ because the first ionisation energy of helium is larger than the 3rd ionisation energy of Th.

The buffer-gas-cooled ^{229}Th^{3+} ions were extracted using an RF carpet (ion collector) [21], transported using a quadrupole ion guide, and trapped using a linear Paul trap. The incoming rate of

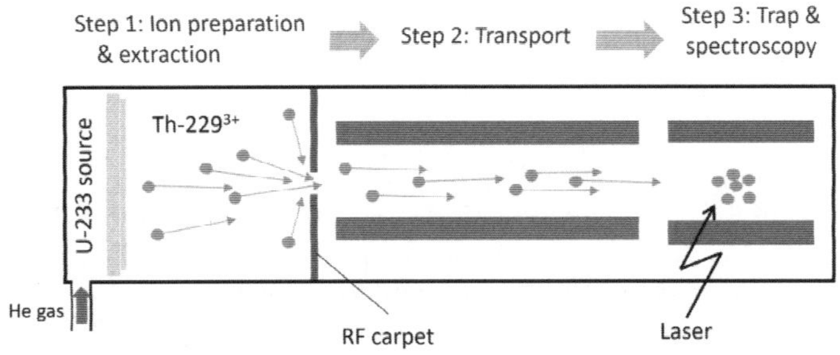

Figure 4. Schematic of ion trap system for ^{229}Th^{3+}. The recoil ^{229}Th^{3+} ions emitted from ^{233}U source were extracted by an RF carpet and transported to the trap.

$^{229}Th^{3+}$ was estimated to be approximately 2000 ions/s. It was estimated that approximately 40% of them were trapped. The trapped ions were monitored via fluorescence at 984 nm, induced by irradiation with 690- and 984-nm lasers simultaneously (see Fig. 2). The emitted photons are captured using a charge-coupled device camera. All the lasers were commercially available diode lasers. The frequency of each laser was stabilised to each reference cavity. The laser frequency was controlled by applying a voltage to the piezoelectric transducer attached to the cavity mirrors.

Figure 5 is a Doppler-broadened spectrum of the $^2F_{5/2}$–$^2D_{5/2}$ (F=5→4, 690 nm) transition of $^{229}Th^{3+}$ in the nuclear ground state. The orange bar indicates the estimated resonance frequency based on the resonance frequency of the $^2F_{5/2}$–$^2D_{5/2}$ transition of ^{232}Th, isotope shift between ^{232}Th and ^{229}Th, and hyperfine constants of ^{229}Th reported in refs. [11,12]. The centre frequency of the observed spectrum is in agreement with the estimated value within the uncertainties. Because 2% of the recoil ^{229}Th from ^{233}U is in the isomeric state, our trap also allows studies of $^{229m}Th^{3+}$.

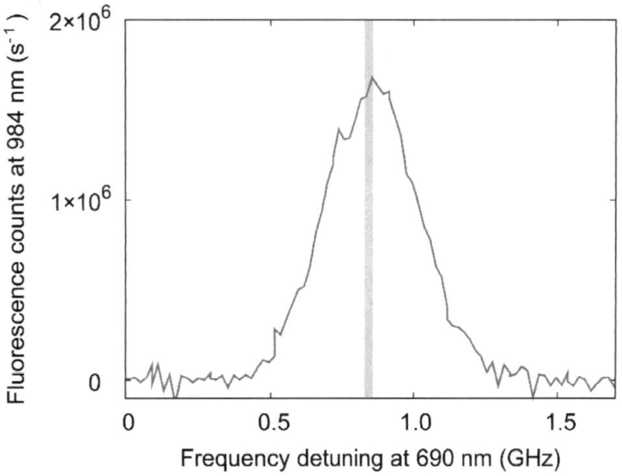

Figure 5. Background subtracted spectrum of the $^2F_{5/2}$–$^2D_{5/2}$ (F=5→4, 690 nm) transition of $^{229}Th^{3+}$ in the nuclear ground state. The orange bar indicates calculated resonance frequency based on results from previous studies. The width of the bar represents its uncertainty.

4. Conclusion

We built an ion trap system for $^{229}Th^{3+}$. As the source of ^{229}Th ions, we used ^{233}U, from which ^{229}Th ions in the nuclear ground and isomeric states could be obtained. Because the half-life of ^{233}U is 1.6×10^5 years, this trap system allows offline laser spectroscopic studies of $^{229}Th^{3+}$ ions.

It has been proposed that the ^{229}Th nuclear clock could be used as a sensor sensitive to variations in the fine structure constant [22]. The sensitivity can be derived from the change in the nuclear charge radius and nuclear electric quadrupole moment [23,24]. These parameters can be experimentally determined using laser spectroscopy of the hyperfine structures of $^{229}Th^{3+}$ in the nuclear ground state and isomer. High-resolution laser spectroscopy using laser-cooled $^{229}Th^{3+}$ ions in both nuclear states allows precise sensitivity determinations. Our trap system, which allows laser cooling of trapped ions, is an optimal platform for nuclear clock development and studies using $^{229m}Th^{3+}$.

References

[1] Beck B R, Becker J A, Beiersdorfer P, Brown G V, Moody K J, Wilhelmy J B, Porter F S, Kilbourne C A, Kelley R L 2007 Energy Splitting of the Ground-State Doublet in the Nucleus ^{229}Th *Phys. Rev. Lett.* **98** 142501

[2] Seiferle B *et al.* 2019 Energy of the ^{229}Th nuclear clock transition *Nature* **573** 243–6

[3] Yamaguchi A *et al.* 2019 Energy of the ^{229}Th Nuclear Clock Isomer Determined by Absolute γ -ray Energy Difference *Phys. Rev. Lett.* **123** 222501

[4] Sikorsky T *et al.* 2020 Measurement of the ^{229}Th Isomer Energy with a Magnetic Microcalorimeter *Phys. Rev. Lett.* **125** 142503

[5] Kraemer S *et al.* 2023 Observation of the radiative decay of the ^{229}Th nuclear clock isomer *Nature* **617** 706–10
[6] Peik E, Tamm C 2003 Nuclear laser spectroscopy of the 3.5 eV transition in Th-229 *Europhys. Lett.* **61** 181–6
[7] Peik E, Okhapkin M 2015 Nuclear clocks based on resonant excitation of γ-transitions *C. R. Physique* **16** 516–23
[8] Peik E, Schumm T, Safronova M S, Pálffy A, Weitenberg J, Thirolf P G 2021 Nuclear clocks for testing fundamental physics *Quantum Sci. Technol.* **6** 034002
[9] Beeks K, Sikorsky T, Schumm T, Thielking J, Okhapkin M V, Peik E 2021 The thorium-229 low-energy isomer and the nuclear clock *Nat. Rev. Phys.* **3** 238–48
[10] Campbell C J, Radnaev A G, Kuzmich A, Dzuba V A, Flambaum V V, Derevianko A 2012 Single-Ion Nuclear Clock for Metrology at the 19th Decimal Place *Phys. Rev. Lett.* **108** 120802
[11] Campbell C J, Steele A V, Churchill L R, DePalatis M V, Naylor D E, Matsukevich D N, Kuzmich A, Chapman M S 2009 Multiply Charged Thorium Crystals for Nuclear Laser Spectroscopy *Phys. Rev. Lett.* **102** 233004
[12] Campbell C J, Radnaev A G, Kuzmich A 2011 Wigner Crystals of ^{229}Th for Optical Excitation of the Nuclear Isomer *Phys. Rev. Lett.* **106** 223001
[13] von der Wense L *et al.* 2016 Direct detection of the ^{229}Th nuclear clock transition *Nature* **533** 47–51
[14] Thielking J, Okhapkin M V, Głowacki P, Meier D M, von der Wense L, Seiferle B, Düllmann C E, Thirolf P G, Peik E 2018 Laser spectroscopic characterization of the nuclear-clock isomer ^{229m}Th *Nature* **556** 321–5
[15] Scharl K *et al.* 2023 Setup for the Ionic Lifetime Measurement of the $^{229m}Th^{3+}$ Nuclear Clock Isomer *Atoms* **11** 108
[16] Masuda T *et al.* 2019 X-ray pumping of the ^{229}Th nuclear clock isomer *Nature* **573** 238–42
[17] Verlinde M *et al.* 2019 Alternative approach to populate and study the ^{229}Th nuclear clock isomer *Phys. Rev. C* **100** 024315
[18] Ruchowska E *et al.* 2006 Nuclear structure of ^{229}Th *Phys. Rev. C* **73** 044326
[19] Barci V, Ardisson G, Barci-Funel G, Weiss B, El Samad O, Sheline R K 2003 Nuclear structure of ^{229}Th from γ-ray spectroscopy study of ^{233}U α-particle decay *Phys. Rev. C* **68** 034329
[20] von der Wense L, Seiferle B, Laatiaoui M, Thirolf P G 2015 Determination of the extraction efficiency for ^{233}U source α-recoil ions from the MLL buffer-gas stopping cell *Eur. Phys. J. A* **51** 29
[21] Wada M *et al.* 2003 Slow RI-beams from projectile fragment separators *Nucl. Instrum. Methods Phys. Res. B: Beam Interact. Mater. At.* **204** 570–81
[22] Flambaum V V 2006 Enhanced Effect of Temporal Variation of the Fine Structure Constant and the Strong Interaction in ^{229}Th *Phys. Rev. Lett.* **97** 092502
[23] Berengut J C, Dzuba V A, Flambaum V V, Porsev S G 2009 Proposed Experimental Method to Determine α Sensitivity of Splitting between Ground and 7.6 eV Isomeric States in ^{229}Th *Phys. Rev. Lett.* **102** 210801
[24] Fadeev P, Berengut J C, Flambaum V V 2020 Sensitivity of ^{229}Th nuclear clock transition to variation of the fine-structure constant *Phys. Rev. A* **102** 052833

Compact and Robust Laser System for a Transportable Strontium Optical Lattice Clock

T Muramatsu[1,2], T Hiroki[1,2], Y Sakai[1,2], Y Hisai[1,2], T Furumiya[1,2], S Uno[1,2], K Tojo[1,2], M Takamoto[2,3] and H Katori[2,3,4]

[1] Technology Research Laboratory, Shimadzu Corporation, Kyoto, Japan
[2] Space-Time Engineering Research Team, RIKEN Center for Advanced Photonics, Saitama, Japan
[3] Quantum Metrology Laboratory, RIKEN, Saitama, Japan
[4] Department of Applied Physics, Graduate School of Engineering, The University of Tokyo, Tokyo, Japan

E-mail: t-mura@shimadzu.co.jp

Abstract. We developed a laser system for a Sr optical lattice clock to down-size and improve robustness and maintainability. The optics were fixed by laser welding to prevent misalignment during transportation and operation. The volume of the laser system was reduced by removing adjustable mechanisms and integrating multi-function electronics. By utilizing the developed laser system, the volume of optical lattice clocks was down to 250 L.

1. Introduction

Optical lattice clocks [1] realizes a fractional frequency uncertainty of 10^{-18} level [2] and is a promising candidate for the redefinition of the SI second. Furthermore, it is expected to be used for geodesy [3] and fundamental physics [4]. Most of optical lattice clocks have been developed as stationary systems in laboratories because of the complexity of the system containing multiple lasers and control electronics. On the other hand, for use outside laboratories, clocks need to be compact and to operate continuously over a long period. In our previous work, a transportable optical lattice clock with the volume of 920 L were developed to test the gravitational redshift at Tokyo Skytree® [5, 6].

One of our goals is to make optical lattice clocks easier to use in various applications. We are downsizing the clock while improving robustness and maintainability. The optics were rigidly assembled by laser welding technique without adjustable mounts. The electronics were highly integrated with multiple functions to control the lasers. To facilitate maintenance, the laser system was divided into modules that can be easily replaced. By utilizing the developed laser system, optical lattice clocks with the volume of 250 L was achieved.

2. Concept of Laser System

2.1. System Configuration

Optical lattice clocks developed by institutes around the world have a variety of configurations. In our study, the laser system was designed for a Sr optical lattice clock. Our laser system and physics package are shown in Fig. 1. We use external cavity diode lasers (ECDL) due to low intrinsic noise and small size. The lasers for Zeeman deceleration and first-stage cooling (461, 496, and 679 nm) are stabilized by a wavelength meter. The lasers for second-stage cooling (689 nm), optical lattice (813 nm), and clock spectroscopy (698 nm) are stabilized to the build-in reference cavity by Pound-Drever-Hall (PDH) technique. The laser of 813 nm is amplified by a tapered amplifier to generate the optical lattice. For users, the system provides a laser output whose frequency is locked to Sr atoms.

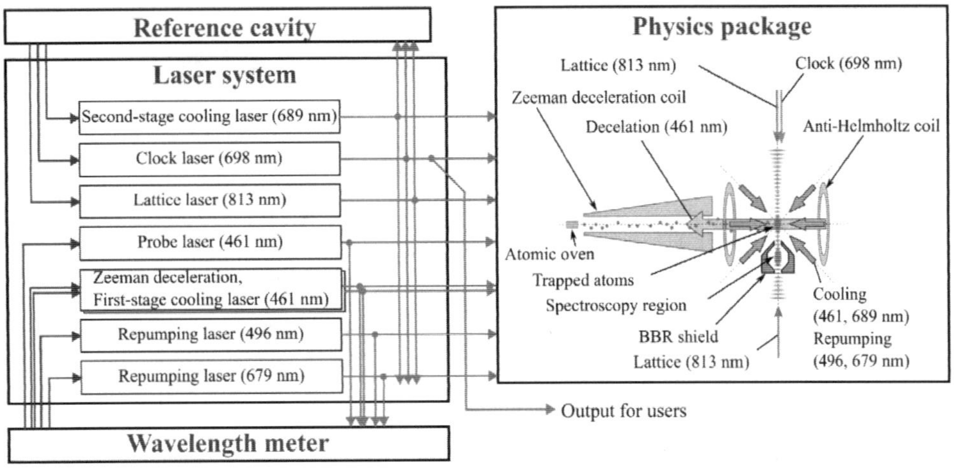

Figure 1. Schematic diagram of the laser system and its roles in a Sr optical lattice clock.

2.2. Maintainability and Flexibility

The concepts of the laser system are shown in Fig. 2. The optics and the electronics are composed of modules divided by function. The module-based system concept allows for high maintainability through easy replacement of the modules and flexible expansion of future configurations. The optics modules and the physics package are connected by optical fibers. Each electronics module is powered and communicates via a backplane board based on the PXI Express® standard. The clock operation is controlled by an embedded controller. Additionally, the electronics module can be controlled by a standard USB interface without a backplane board.

2.3. Down-sizing

The optical components are selected to be particularly compact and are rigidly fixed by the laser welding technique without any adjustment mechanism.

The electronics modules consist of highly integrated multiple functions. The electrical circuits for each function are designed separately. These design results are integrated into circuit boards to realize the multi-function electronics modules.

2.4. Robustness

The laser welding for the optics improves reliability by preventing misalignment during transportation and operation. The optical base plate has built-in heat pipes and connects with a water-cooled plate to ensure high robustness. The heat generated by electrical components is also efficiently exhausted to a water-cooled plate via heat pipes.

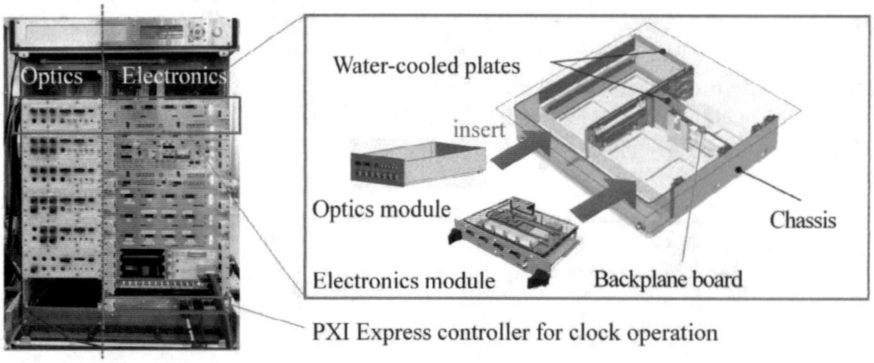

Figure 2. Developed laser system and design concept of the system.

3. Development of Laser System

3.1. Laser System

The developed laser system is shown in Fig. 2. The volume of the laser system is 113 L. The optical lattice clock utilizing our laser system is shown in Fig. 3. The optical lattice clock consists of the physics package, the laser system, the peripheral equipment, and the reference cavity, which are mounted in two 19-inch racks. The volume of the clock is 250 L without racks, approximately one-third the volume of the previously developed clock [5, 6].

Figure 4 shows an example of the developed modules. The entire optical system of the clock are divided into several modules by wavelength. Figure 4(a) shows a module of 496 nm and 679 nm. Basically, an optics module consists of an ECDL, isolators, optical modulators, shutters, lenses for coupling to the output fiber. The ECDL consists of a laser diode, a lens, and a volume Bragg grating (VBG). The laser welding technique is used to assemble the ECDL and the components in the optics module. Since the optics modules have all of the functions, such as frequency and intensity control, to operate the clock, the output fiber is directly connected to the physics package.

The developed electronics modules are summarized in Table 1. While the previously developed electronics consists of many circuit boards interconnected by many wires, the new electronics consist of the modules with multi-channel functions. Fig. 4(b) shows a module of a laser controller for PDH locking, which has all the functions to stabilize two ECDLs by PDH technique.

Figure 3. Sr optical lattice clock utilizing the developed laser system.

Journal of Physics: Conference Series **2889** (2024) 012042 doi:10.1088/1742-6596/2889/1/012042

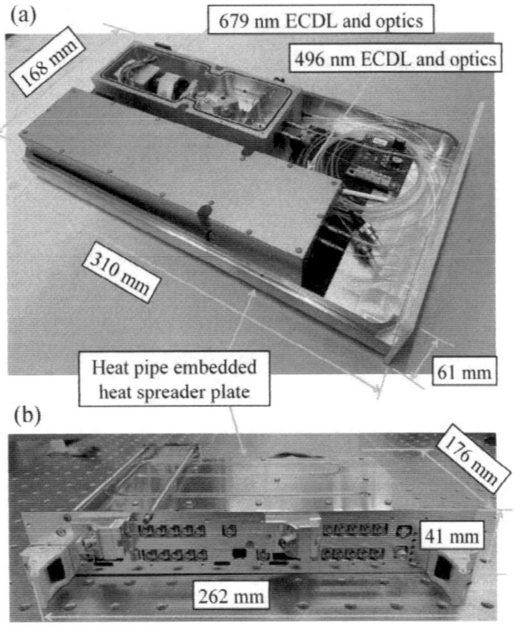

Figure 4. The developed modules. (a) The optics module of 496 and 679 nm. (b) The electronics module for laser locking using PDH technique.

Table 1. Specifications of the electronics modules.

Module type	Integrated functions	Channels
Laser controller	LD driver, TEC controller and PZT driver	4
Laser controller for PDH locking	LD driver, TEC controller, PZT driver and PDH locking circuit	2
RF generator	Direct digital synthesizer (DDS) and RF amplifier	8
Doppler noise canceller (DNC)	DNC circuit with DDS and RF amplifier	4
High power coil driver	Coil driver for Zeeman deceleration and MOT	1
Low power coil driver	Coil driver for magnetic field compensation	4

3.2. Performance Evaluation

The developed system is under evaluation. The already evaluated results are presented here. Figure 5 shows the stability of the clock laser, which requires a particularly important performance in the laser system. The fractional Allan standard deviation for one second averaging time is 1.2×10^{-15}, which is consistent with the thermal noise limit for a 75 mm long ultra-low expansion (ULE) reference cavity. If higher stability is required, the internal clock laser can be phase-locked to a stable external reference laser.

The laser system was built at RIKEN (Wako) and transported by a truck 500 km to Shimadzu Corporation (Kyoto). No significant misalignment of the optics was observed. We are currently evaluating performance as a clock. In the future, we plan to evaluate the robustness such as continuous operation period and effects on environmental temperature and vibrations.

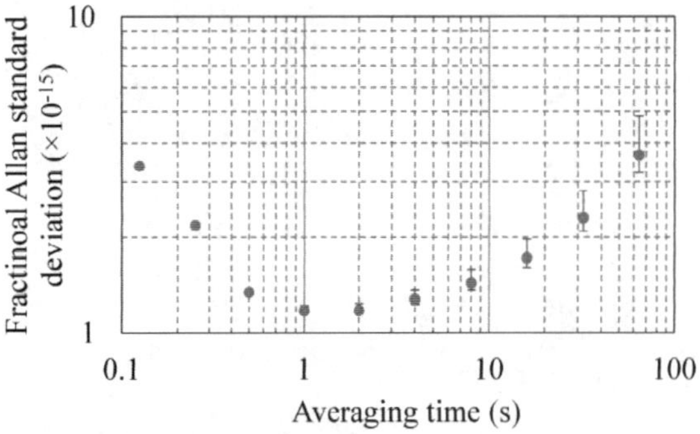

Figure 5. Stability of the clock laser locked to the reference cavity by PDH technique.

4. Conclusion

The laser system for a Sr optical lattice clock with 250 L volume was developed. The optics with laser welding technique and the highly integrated electronics achieved compactness. The clock laser was stabilized to the built-in reference cavity, and the fractional Allan standard deviation is 1.2×10^{-15} for one second averaging time. No re-alignment was required after transporting the system for 500 km. We are currently testing continuous operation to extend operation period to be used the system as an infrastructure for frequency standards in the future.

Acknowledgement

This work was supported by JST-Mirai Program Grant Number JPMJMI18A1, Japan.

References

[1] Katori H, Takamoto M, Pal'chikov V G and Ovsiannikov V D 2003 *Phys. Rev. Lett.* **91** 173005
[2] Bothwell T, Kedar D, Oelker E, Robinson J M, Bromley S L, Tew W L, Ye J and Kennedy C J 2019 *Metrologia* **56** 065004
[3] Tanaka Y and Katori H 2021 *J. Geod.* **95** 93
[4] Derevianko A and Pospelov M 2014 *Nat. Phys.* **10** 933
[5] Takamoto M, Ushijima I, Ohmae N, Yahagi T, Kokado K, Shinkai H and Katori H 2020 *Nat. Photonics* **14** 411
[6] Ohmae N *et al.* 2021 *Adv. Quantum Technol.* **4** 2100015

Journal of Physics: Conference Series **2889** (2024) 012043 doi:10.1088/1742-6596/2889/1/012043

Fringes and Light Shift in CPT-Ramsey Spectroscopy

Atsuo Morinaga[1] **and Shinya Yanagimachi**[2]

[1] R&D Promotion Division, Micromachine Center, Tsukuba, Japan
[2] National Institute of Advanced Industrial Science and Technology, Tsukuba, Japan

E-mail: a_morinaga@mmc.or.jp

Abstract. The CPT-Ramsey fringes are solved based on the time evolution of the Bloch vector and are calculated in the steady-state condition and in the pre-steady state condition. A consistent formula of the light shifts in the CPT-Ramsey fringes is presented due to the detection-time, the free evolution time, and the excitation-time. For a further reduction of the light shift, we propose to use the CPT-Ramsey fringes in the pre-steady state condition.

1. Introduction

The long-term frequency stability of CPT (Coherent population trapping)-based atomic clocks is limited by the light shift which depends on the intensity and detuning frequency of the excited lights [1]. To reduce the light shift, Ramsey's method based on two separated excitation pulses was widely used and it is called CPT-Ramsey [2]. Furthermore, auto balanced Ramsey methods were developed to remove the light shift perfectly, but the complicate spectroscopic process sacrifices the signal-to-noise ratio of fringes [3]. Up to now, the mechanism of CPT-Ramsey was often explained by the idea of the Raman-Ramsey [4] and the cause of the observed light shift has not yet been explained clarify. Unlike a traditional Ramsey resonance [5], the CPT-Ramsey fringes occur in the steady-state condition, where the atomic system is pumped into the dark state. Recently, Micalizio and Godone solved the time evolution of the Bloch vector after the steady-state condition is achieved at the first excitation pulse and derived that there are a detection-time-dependent light shift and a Rabi-pulling induced light shift, for the purpose of investigating the light shift in the CPT maser [6]. We derived a more accurate equation of the CPT-Ramsey fringes not only the steady-state condition but also the pre-steady state condition based on the Bloch vector and presented a consistent formula of the light shift from the CPT resonance to the CPT-Ramsey fringes. Finally, we propose a further subtraction method of the light shift using the CPT-Ramsey in the pre-steady state condition.

2. Solution of derivative equations for density matrix elements

2.1. CPT excitation

We consider a three-level atomic system where two ground hyperfine states of g and e are coupled to a common excited state of i via bichromatic laser fields of ω_1 and ω_2, driven by the two first-order sidebands, as shown in Fig. 1 (a). The system is described by a set of the derivative equations for the density matrix ρ of populations and coherences [7]. Under hypothesis of the two equal Rabi frequencies and in the typical operating conditions of the pumping rate Γ_p and the detuning frequency divided by the relaxation rate δ_0, the set of equations becomes [6,8]

$$\dot{\Delta\rho} + \Gamma'\Delta\rho = 4\Gamma_p\delta_0\rho_{ge}^i \quad (1)$$

$$\dot{\rho_{ge}} + [\Gamma' + i\Delta'_\mu]\rho_{ge} = -\Gamma_p - i\Gamma_p\delta_0\Delta\rho \quad (2)$$

where $\Delta\rho = \rho_{ee} - \rho_{gg}$ is the ground-state population difference and $\rho_{ge} = \rho_{ge}^r + i\rho_{ge}^i$ is the coherence excited between the ground-state sublevels. The superscript *r* and *i* mean the real part and the imaginary part, respectively. Δ_μ is the two photon microwave detuning from the resonance frequency. Δ_{Ls} is the light shift and $\Delta'_\mu = \Delta_\mu - \Delta_{Ls}$. Γ_p is the pumping rate and $\Gamma' = \gamma_c + \Gamma_p$.

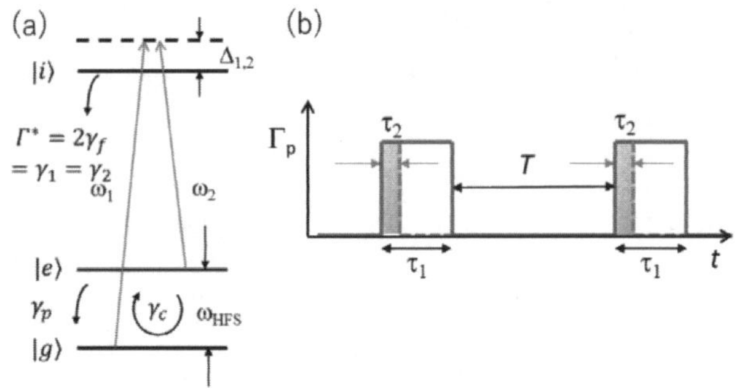

Figure 1. (a) Three-level system. Γ^*, γ_p and γ_c are the relaxation rates for the excited state, ground-state population and coherence, respectively. ω_1 and ω_2 are frequencies of the bichromatic fields. $\Delta_{1,2}$ are the laser detuning and ω_{HFS} is the hyperfine splitting between ground states. (b) Time sequence of the CPT-Ramsey excitation. τ_1 and τ_2 are the durations of the generation pulse and detection pulse. T is the free evolution time.

In Ref. [6], ρ_{ge} was derived from Eq. (2) with $\Delta\rho(t) = 0$, under the assumption that the atomic population is equally distributed among the two ground sublevels and $\Delta\rho(t)$ was solved using Eq. (1). However, at the steady-state condition of $t \gg 1/\Gamma'$, $\Delta\rho(t)$ should be equal to the steady-state value of $\Delta\rho(\infty)$ which is obtained when the all derivative equations become to zero

$$\Delta\rho(\infty) = \frac{4{\Gamma_p}^2\delta_0\Delta'_\mu}{\Gamma'\left(\Gamma'^2+\Delta'^2_\mu+4{\Gamma_p}^2{\delta_0}^2\right)}. \quad (3)$$

Therefore, we suppose the function of $\Delta\rho(t)$ as

$$\Delta\rho(t) = \Delta\rho(\infty)\left(1 - e^{-\Gamma' t}\right). \quad (4)$$

Then, Eq. (2) is solved using the initial conditions of $\rho_{ge}(0) = 0$ and $\Delta\rho(0) = 0$ as a function of the duration *t* of the excitation pulse. The Bloch vector $\mathbf{R}(t)$ is

$$\mathbf{R}(t) = \begin{pmatrix}\rho_{ge}^r(t)\\ \rho_{ge}^i(t)\\ \Delta\rho(t)\end{pmatrix} = \begin{pmatrix} -\frac{\Gamma_p\left(\Gamma'^2+4{\Gamma_p}^2{\delta_0}^2\right)}{\Gamma'\left(\Gamma'^2+{\Delta'_\mu}^2+4{\Gamma_p}^2{\delta_0}^2\right)} + \frac{4{\Gamma_p}^3{\delta_0}^2e^{-\Gamma' t}}{\Gamma'\left(\Gamma'^2+\Delta'^2_\mu+4{\Gamma_p}^2{\delta_0}^2\right)} + \frac{\Gamma_p e^{-\Gamma' t}(\Gamma'\cos\Delta'_\mu t-\Delta'_\mu\sin\Delta'_\mu t)}{\Gamma'^2+\Delta'^2_\mu+4{\Gamma_p}^2{\delta_0}^2} \\ \frac{\Gamma_p\Delta'_\mu}{\Gamma'^2+\Delta'^2_\mu+4{\Gamma_p}^2{\delta_0}^2} - \frac{\Gamma_p e^{-\Gamma' t}(\Gamma'\sin\Delta'_\mu t+\Delta'_\mu\cos\Delta'_\mu t)}{\Gamma'^2+\Delta'^2_\mu+4{\Gamma_p}^2{\delta_0}^2} \\ \frac{4{\Gamma_p}^2\delta_0\Delta'_\mu}{\Gamma'\left(\Gamma'^2+\Delta'^2_\mu+4{\Gamma_p}^2{\delta_0}^2\right)}\left(1 - e^{-\Gamma' t}\right) \end{pmatrix}. \quad (5)$$

In the present paper, we are interested in the real part of the coherence $\rho_{ge}^r(t)$, which determines the profile of CPT resonance as

$$\rho_{ii} = \frac{2\Gamma_p}{\Gamma^*}\left(1 + 2\rho_{ge}^r(t)\right). \quad (6)$$

2.2. CPT Ramsey excitation

Figure 1(b) shows a time sequence of the CPT-Ramsey pulses. τ_1 is the duration of the excitation pulse, *T* is the free evolution time and τ_2 is the time of detection pulse. During the free evolution, the Bloch vector $\boldsymbol{R}(\tau_1)$ at $t = \tau_1$ evolves at an angular frequency Δ_μ according to the free evolution matrix and finally the Bloch vector can be obtained by solving Eq. (2) using $\boldsymbol{R}(\tau_1 + T)$ as initial conditions.

$$\boldsymbol{R}(\tau_1 + T + \tau_2) = \boldsymbol{R}(\tau_2) \cdot \begin{pmatrix} e^{-\gamma_c T}\cos\Delta_\mu T & e^{-\gamma_c T}\sin\Delta_\mu T & 0 \\ -e^{-\gamma_c T}\sin\Delta_\mu T & e^{-\gamma_c T}\cos\Delta_\mu T & 0 \\ 0 & 0 & e^{-\gamma_p T} \end{pmatrix} \cdot \boldsymbol{R}(\tau_1) \tag{7}$$

3. CPT-Ramsey fringes in steady-state

3.1 Steady-state condition

For $\tau_1 \gg 1/\Gamma'$, Eq. (5) reproduces the steady-state behavior. Using $\Delta\rho \equiv \Delta\rho(\infty)$, it turns out to be

$$\mathbf{R}(\tau_1 \gg 1/\Gamma') = \begin{pmatrix} -\dfrac{\Gamma_p(\Gamma'+\delta_0\Delta'_\mu\Delta\rho)}{\Gamma'^2+\Delta'^2_\mu} \\ \dfrac{\Gamma_p(\Delta'_\mu-\Gamma'\delta_0\Delta\rho)}{\Gamma'^2+\Delta'^2_\mu} \\ \Delta\rho = \dfrac{4\Gamma_p{}^2\delta_0\Delta'_\mu}{\Gamma'\left(\Gamma'^2+\Delta'^2_\mu+4\Gamma_p{}^2\delta_0{}^2\right)} \end{pmatrix} = \begin{pmatrix} -\dfrac{\Gamma_p\left(\Gamma'^2+4\Gamma_p{}^2\delta_0{}^2\right)}{\Gamma'\left(\Gamma'^2+\Delta'_\mu{}^2+4\Gamma_p{}^2\delta_0{}^2\right)} \\ \dfrac{\Gamma_p\Delta'_\mu}{\Gamma'^2+\Delta'_\mu{}^2+4\Gamma_p{}^2\delta_0{}^2} \\ \dfrac{4\Gamma_p{}^2\delta_0\Delta'_\mu}{\Gamma'\left(\Gamma'^2+\Delta'^2_\mu+4\Gamma_p{}^2\delta_0{}^2\right)} \end{pmatrix} \tag{8}$$

Then the CPT resonance turns out to be a Lorentzian slightly broadened by δ_0, because $\Delta\rho$ is proportional to Δ'_μ. Here we introduce a parameter $S = \delta_0\Delta\rho/\Gamma'$, instead of $\Delta\rho$, which shows a contribution of antisymmetric Lorentzian function. If *S* is constant not depend on Δ'_μ, the profile of CPT resonance becomes slightly asymmetrical. Practically, we often observe such an asymmetric profile of CPT resonance. From the Eq. (7), the $\rho^r_{ge}(\tau_1 + T + \tau_2)$ becomes

$$\begin{aligned} \rho^r_{ge}(\tau_1 + T + \tau_2) = &-\frac{\Gamma'\Gamma_p}{\Gamma'^2 + \Delta'^2_\mu}\left(1 + S\Delta'_\mu e^{-\gamma_c T}\right) - \frac{\Gamma_p e^{-\Gamma'\tau_2}}{\Gamma'^2 + \Delta'^2_\mu}\left[\{\Gamma'\left(1 + S\Delta'_\mu\right)\cos\Delta_\mu T\right. \\ &-\left(\Delta'_\mu - \Gamma'^2 S\right)\sin\Delta_\mu T\}e^{-\gamma_c T} - \Gamma'\left(1 + S\Delta'_\mu e^{-\gamma_c T}\right)]\cos\Delta'_\mu\tau_2 \\ &+\frac{\Gamma_p e^{-\Gamma'\tau_2}}{\Gamma'^2+\Delta'^2_\mu}\left[\{\Gamma'\left(1 + S\Delta'_\mu\right)\sin\Delta_\mu T + \left(\Delta'_\mu - \Gamma'^2 S\right)\cos\Delta_\mu T\}e^{-\gamma_c T}\right. \\ &\left.-\left(\Delta'_\mu - \Gamma'^2 S e^{-\gamma_c T}\right)\right]\sin\Delta'_\mu\tau_2 \end{aligned} \tag{9}$$

3.2 CPT-Ramsey fringes with S = 0

First, we calculate the CPT-Ramsey fringes with $S = 0$ at the condition of $\gamma_c = 2\pi \times 40$ Hz and $\Gamma_p = 2\pi \times 130$ Hz, which are used throughout the present paper, for $\tau_1 = 1.5$ ms and $\tau_2 = 20$ μs. Figure 2 shows the calculated Ramsey fringes for several free evolution times. They are symmetric Ramsey fringes. The amplitude of Ramsey fringes attenuates with a damping coefficient of $\gamma_c T$.

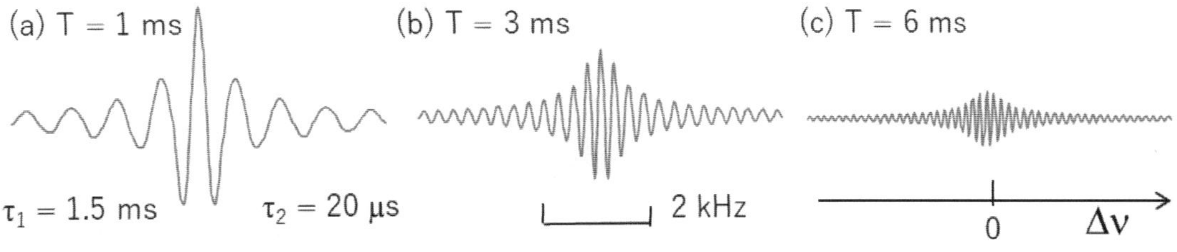

Figure 2 Calculated symmetric steady-state CPT-Ramsey fringes for several free evolution times *T*. $S = 0$, $\gamma_c = 2\pi \times 40$ Hz and $\Gamma_p = 2\pi \times 130$ Hz.

3.3 CPT-Ramsey fringes with S = –0.002

Generally the CPT resonance profile is distorted by the contribution of an antisymmetric Lorentzian function. Figure 3(a) and (b) shows the calculated CPT-Ramsey fringes with S = –0.002 for detection times of 20 and 200 μs, respectively. We observe that the shift of central fringe is –23 Hz for both spectra. In addition to it, we observe that the original profile of CPT resonance is superimposed in the Ramsey fringes measured at τ_2 longer than 100 μs. Thus, a smaller detection-time is preferable to be used as the frequency reference signal.

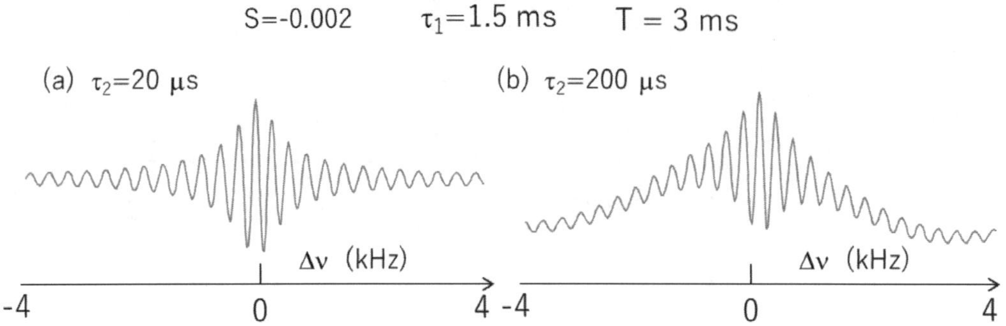

Figure 3 Calculated asymmetric steady-state CPT-Ramsey fringes for detection times. $S = -0.002$, $\gamma_c = 2\pi \times 40$ Hz and $\Gamma_p = 2\pi \times 130$ Hz.

4. CPT-Ramsey fringes in pre-steady state

4.1 Pre-steady state condition

By substituting the $\Delta\rho(\tau_1)$ into Eq. (5), the Bloch vector can be described as

$$\mathbf{R}(\tau_1) = \begin{pmatrix} -\frac{\Gamma_p\Gamma'}{\Gamma'^2+\Delta'^2_\mu+4\Gamma_p^2\delta_0^2} - \frac{\Gamma_p\delta_0}{\Delta'_\mu}\Delta\rho(\tau_1) + \frac{\Gamma_p e^{-\Gamma'\tau_1}(\Gamma'\cos\Delta'_\mu\tau_1 - \Delta'_\mu\sin\Delta'_\mu\tau_1)}{\Gamma'^2+\Delta'^2_\mu+4\Gamma_p^2\delta_0^2} \\ \frac{\Gamma_p\Delta'_\mu}{\Gamma'^2+\Delta'^2_\mu+4\Gamma_p^2\delta_0^2} - \frac{\Gamma_p e^{-\Gamma'\tau_1}(\Gamma'\sin\Delta'_\mu\tau_1 + \Delta'_\mu\cos\Delta'_\mu\tau_1)}{\Gamma'^2+\Delta'^2_\mu+4\Gamma_p^2\delta_0^2} \\ \Delta\rho(\tau_1) = \frac{4\Gamma_p^2\delta_0\Delta'_\mu}{\Gamma'\left(\Gamma'^2+\Delta'^2_\mu+4\Gamma_p^2\delta_0^2\right)}\left(1-e^{-\Gamma'\tau_1}\right) \end{pmatrix} \equiv \begin{pmatrix} \rho^r_{ge}(\tau_1) \\ \rho^i_{ge}(\tau_1) \\ \Delta\rho(\tau_1) \end{pmatrix} \quad (10)$$

Using $\rho^r_{ge}(\tau_1)$, $\rho^i_{ge}(\tau_1)$ and the parameter S, the real part of the coherence $\rho^r_{ge}(\tau_1 + T + \tau_2)$ in the pre-steady state condition turns out to be

$$\begin{aligned} \rho^r_{ge}(\tau_1 + T + \tau_2) = &-\frac{\Gamma_p\Gamma'(1+S\Delta'_\mu e^{-\gamma_c T})}{\Gamma'^2+\Delta'^2_\mu} \\ &+e^{-\Gamma'\tau_2}\left[\{\rho^r_{ge}(\tau_1)\cos\Delta_\mu T + \rho^i_{ge}(\tau_1)\sin\Delta_\mu T\}e^{-\gamma_c T} + \frac{\Gamma_p\Gamma'(1+S\Delta'_\mu e^{-\gamma_c T})}{\Gamma'^2+\Delta'^2_\mu}\right]\cos\Delta'_\mu\tau_2 \\ &-e^{-\Gamma'\tau_2}\left[\{\rho^r_{ge}(\tau_1)\sin\Delta_\mu T - \rho^i_{ge}(\tau_1)\cos\Delta_\mu T\}e^{-\gamma_c T} + \frac{\Gamma_p(\Delta'_\mu - \Gamma'^2 S e^{-\gamma_c T})}{\Gamma'^2+\Delta'^2_\mu}\right]\sin\Delta'_\mu\tau_2 \,. \end{aligned} \quad (11)$$

The calculated CPT-Ramsey fringes with $S = 0$ in the pre-steady state for several τ_1 are shown in Fig. 4. At τ_1 less than 1 ms, the envelope of Ramsey fringes becomes wider than the width of the CPT resonance in the steady-state and it can be approximated by $\mathrm{sinc}^2(\pi\Delta'_\mu\tau_1)$, as shown in a dashed line in Fig. 4. Such a pattern of CPT Ramsey was reported in a previous research [9]. The amplitude of central fringe decreases along a function of $\left(1 - e^{-\Gamma'\tau_1}\right)$, as τ_1 decreases.

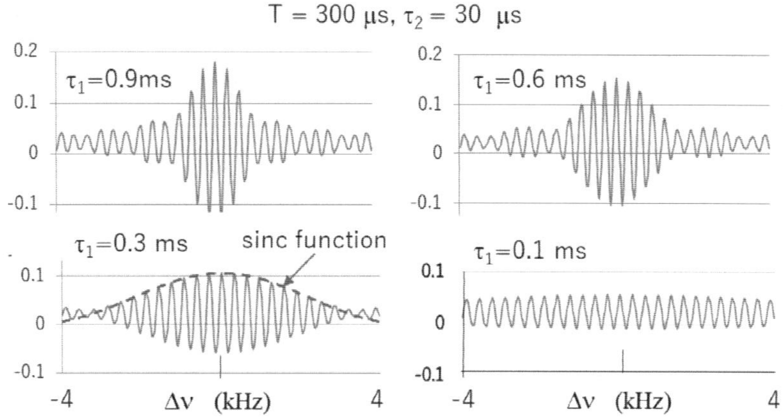

Figure 4 Calculated pre-steady state CPT-Ramsey fringes for several generation time τ_1. $S = 0$, $\gamma_c = 2\pi \times 40$ Hz and $\Gamma_p = 2\pi \times 130$ Hz.

5. Light shift

5.1 Steady-state

As investigated in ref. [6], there are two types of light shift in the CPT-Ramsey fringes. First, in the limit $\Delta'_\mu \ll \Gamma'$ and $\tau_2 \ll T$, the fringe part of Eq. (9) is summarized to

$$\cos\Delta_\mu T \cos\Delta'_\mu \tau_2 - \sin\Delta_\mu T \sin\Delta'_\mu \tau_2 \approx \cos\left(\Delta_\mu - \frac{\Delta_{LS}\tau_2}{T}\right)T. \tag{12}$$

This light shift $\Delta_{DT} = \Delta_{LS}\tau_2/T$ is called the detection-time-dependent light shift and it is reduced to a factor of τ_2/T of that of the CPT resonance.

The second type of light shift is the Rabi-pulling-induced light shift Δ_{RP}. It occurs because that the Rabi envelope is centered at $\Delta'_\mu = 0$, whereas the Ramsey fringes are centered at $\Delta_\mu = 0$. In the limit $\tau_2 \to 0$, Eq. (9) becomes

$$\begin{aligned}\rho_{ge}^r(\tau_1 + T + \tau_2) &\approx -\frac{\Gamma_p\Gamma' e^{-\gamma_c T}}{\Gamma'^2 + \Delta'^2_\mu}\left(\cos\Delta_\mu T - \frac{\Delta'_\mu}{\Gamma'}\sin\Delta_\mu T\right) \\ &= -\frac{\Gamma_p e^{-\gamma_c T}}{\sqrt{\Gamma'^2 + \Delta'^2_\mu}}\cos\left\{\left(\frac{1+\Gamma' T}{\Gamma' T}\right)\left(\Delta_\mu - \frac{\Delta_{LS}}{1+\Gamma' T}\right)T\right\}\end{aligned} \tag{13}$$

It should be noted that the light shift $\Delta_{\mathrm{RP}} = \Delta_{LS}/(1 + \Gamma' T)$ can be reduced from Δ_{LS} at $T = 0$ to $\Delta_{LS}/(\Gamma' T)$ at $T \gg 1/\Gamma'$, although the Ramsey cycle approaches 1/T. Similar idea was discussed in Ramsey-type frequency standards [10]. For $\Gamma'/2\pi = 300$ Hz, Δ_{C-R}/Δ_{LS} decreases as T increases as the red curve shown in Fig. 5 and becomes 0.15 at $T = 3$ ms.

Taking into account both the effects, the light shift of the central Ramsey fringes turns out to be

$$\Delta_{C-R} = \Delta_{DT} + \Delta_{\mathrm{RP}} = \Delta_{LS}\left(\frac{\tau_2}{T} + \frac{1}{1+\Gamma' T}\right). \tag{14}$$

For small Γ_p, the Rabi-pulling shift dominates, while the shift due to the detection-time appears for large Γ_p.

5.2 Pre-steady state

As the envelope of Ramsey fringes in the pre-steady state is a sinc function with a width of $\Delta\nu_{\mathrm{HWHM}} \sim 0.445/\tau_1$, the light shift is

$$\Delta_{C-R} = \Delta_{LS}\left(\frac{\tau_2}{T} + \frac{\tau_1}{\tau_1 + 2.8T}\right). \tag{15}$$

Ratios of Δ_{C-R} to Δ_{LS} for excitation-times τ_1 are shown by green and blue curves in Fig. 5. For example, at τ_1 = 300 μs and T = 3 ms, Δ_{C-R} becomes one-thirtieth smaller than Δ_{LS}.

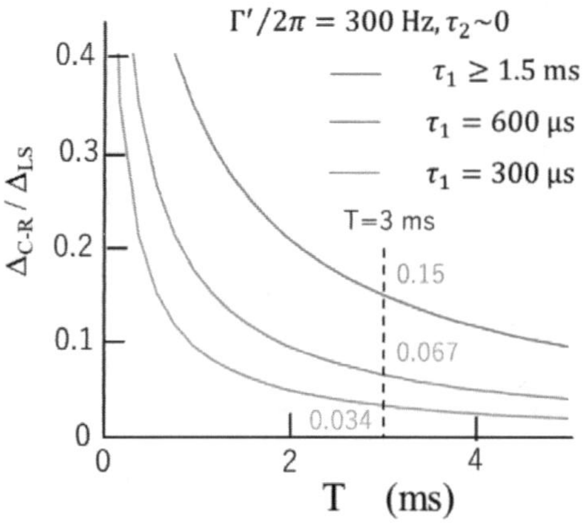

Figure 5 Calculated ratio of Δ_{C-R} to Δ_{LS} versus free evolution time T for several τ_1.

6. Summary

We solved the time derivative equations of density matrix according to a time evolution of Bloch vector and calculated patterns of CPT-Ramsey fringes in the steady-state condition and in the pre-steady state condition. We derived a consistent light shifts formula of CPT-Ramsey depending on the detection-time, the free evolution time, and excitation-time. For further reduction of the light shift, we propose to use the CPT-Ramsey in the pre-steady state condition. The calculation shows the light shift can be reduced to about one-thirtieth of that of the ordinal CPT resonance, at τ_1 = 300 μs and T = 3 ms. We are currently verifying these calculation through experiments. The results will be reported soon.

Acknowledgments
We acknowledge to Drs. S. Kagami, T. Fujisaku, and K. Matsumoto in NEC Corporation for their measurement of Light shift and discussions. This work was supported by Innovative Science and Technology Initiative for Security Grant Number JPJ004596, ATLA, Japan.

References

[1] Yanagimachi S, Harasaka K, Suzuki R, Suzuki M, and Goka S 2020 *Appl. Phys. Lett.* **116** 104102
[2] Castagna N, Boudot R, Guerandel S, Clereq E De, Dimarcq N, and Clairon A 2009 *IEEE Trans. Ultrason. Ferroelectr. Freq. Control.* **56** 246
[3] Hafiz M A, Coget G, Petersen M, Calosso C E, Guerandel S, Clercq E De, and Boudot R 2018 *Appl. Phys. Lett.* **112** 244102
[4] Hemmer P R, Shahriar M S, Natoli V D, and Ezekiel S 1989 *J. Opt. Soc. Am* B **6** 1519
[5] Ramsey N F 1950 *Phys. Rev.* **78** 695
[6] Micalizio S and Godone A 2019 *Phys. Rev. A* **99** 043425
[7] Vanier J 2005 *Appl. Phys.* B **81** 421
[8] Gordone A, Levi F, Micalizio S, and Calosso C 2004 *Phys. Rev. A* **70** 012508
[9] Zanon T, Guerandel S, Clereq E. de, Holleville D, Dimarcq N, and Clairon A 2005 *Phys. Rev. Lett.* **94** 193002
[10] Morinaga A, Murakami M, Nakamura K and Imai H 2016 *Atoms* **4** 23.

^{171}Yb^{+} ion optical clock at NPL: advances in automation and assessment of the trap-induced ac Zeeman shift in an endcap trap geometry

E. A. Curtis, A. Tofful, A. O. Parsons[1], B. I. Robertson[2], and R. M. Godun

National Physical Laboratory, Teddington, UK

E-mail: anne.curtis@npl.co.uk

Abstract. Ensuring robust operation of optical clock systems requires increasingly complex automation of experimental protocols. This includes real-time monitoring of system output for data validation and for high accuracy assessment of systematic frequency shifts. We report here on recent advances in the control system of the ^{171}Yb^{+} ion optical clock at NPL (National Physical Laboratory, UK) and describe how this has led to improvements in systematic shift metrology and reduced the overall uncertainty budget for the optical clock. Additionally we present an assessment of the rf trap drive-induced ac Zeeman effect for our experimental system. These system advances enable us to provide improved contributions to the international efforts in the redefinition of the SI second and other applications such as tests of fundamental physics.

1. Introduction

The accuracy and stability of atom-based optical frequency standards make them ideal metrological tools, with numerous applications in areas requiring precise positioning, navigation and timing. In isolation, the accuracy of an optical clock is difficult to fully assess. Comparing with other clocks via optical frequency comb systems, whether locally or utilising clocks in different laboratories or countries via interconnected fibre networks, can enable a direct exploration of clock performance at the few parts-in-10^{17} level and below [1, 2]. Repeatability of the results year on year gives added confidence in how well the clock system and its long-term behaviour are understood, and is an essential step for redefining the SI second in terms of an optical frequency. The culture of sharing increasingly long-term measurement data has also enabled analysis of clock frequency data on a variety of different timescales, which has great potential for testing fundamental physics theories [3, 4, 5, 6].

The Consultative Committee for Time and Frequency (CCTF) has produced a roadmap towards the redefinition of the SI second in terms of an optical frequency, with a number of conditions required for redefinition [7]. In order to best contribute to this process of redefinition, work is in progress with the Yb^{+} single ion optical clock at NPL (National Physical Laboratory, UK) to (a) improve the uncertainty of the clock system, (b) perform optical frequency ratio measurements between local as well as remotely located frequency standards, and (c) contribute

[1] Present address: Culham Centre for Fusion Energy, Abingdon, UK
[2] Present address: ORCA Computing, LG, 30 Eastbourne Terrace, London, UK

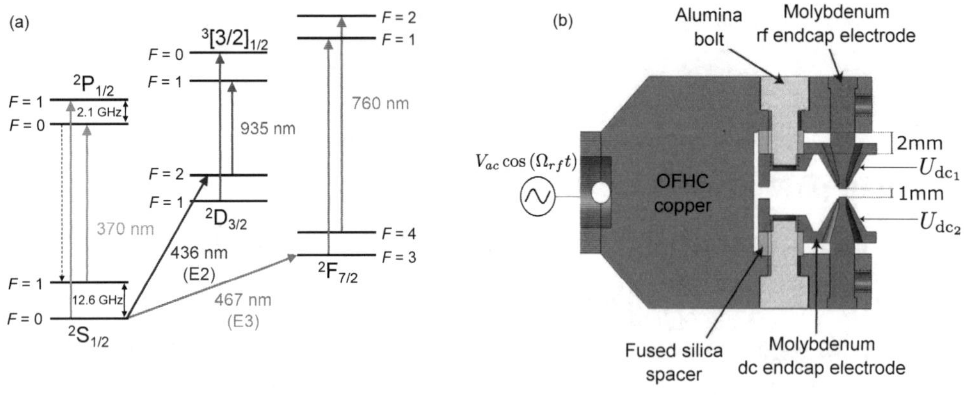

Figure 1. NPL $^{171}Yb^+$ single-ion optical clock: (a) relevant energy level structure and (b) schematic of ion trap hardware [9].

frequency data to the International Bureau of Weights and Measures (BIPM) to begin regular optical frequency contributions to International Atomic Time (TAI).

Advances in automating experimental protocols coordinating clock operation and systematic shift evaluation are essential in our efforts to fulfil the requirements for SI second redefinition and for the implementation of a future optical time scale. As the international metrological community endeavours to perform increasing numbers of simultaneous and extended periods of optical frequency measurement and comparison, the ability to produce valid data with high uptimes becomes a metric for success, in addition to the overall accuracy of the optical clock itself.

Here we describe the $^{171}Yb^+$ clock system at NPL and report on improvements in automation leading to increased robustness of operation during frequency measurement campaigns, extending the utility of the optical clock output data in frequency metrology towards the redefinition of the SI second and in tests of fundamental physics. In addition, we describe the investigation of the rf trap drive-induced ac Zeeman effect in the NPL Yb^+ clock, as part of the systematic shift evaluation for clock operation at the highest accuracy.

2. $^{171}Yb^+$ optical frequency standard at NPL

A recent overview of the experimental apparatus at NPL and methods of clock operation can be found in reference [8]. Here we summarise the relevant details for ion trapping and optical clock operation in the NPL Yb^+ system (see Figure 1). The electronic structure of the $^{171}Yb^+$ ion provides two clock transitions that are used for optical frequency metrology: the ultra-narrow $^2S_{1/2}\rightarrow{}^2F_{7/2}$ electric octupole transition (E3), and the broader $^2S_{1/2}\rightarrow{}^2D_{3/2}$ electric quadrupole transition (E2), which can be additionally utilised as a more sensitive probe of several systematic effects in the clock system. During clock operation, a single $^{171}Yb^+$ ion is held in an endcap trap designed for minimised ion heating rate and symmetric rf delivery to the endcap electrodes [9]. An ion is loaded into the trap by a process of resistively heating a small oven filled with neutral Yb within the vacuum chamber to produce a weak beam of atoms. This atomic beam interacts with overlapping 399-nm and 370-nm light beams at the centre of the trap to produce ions via photoionisation. Once a single ion is successfully loaded, it is continuously cooled using three orthogonal beams of 370-nm light, with additional repumping via excitation using 935-nm light, to mitigate loss due to the ion decaying into a dark $^2D_{3/2}$ state. A Rabi π-pulse is used for clock spectroscopy, with a small ($\sim$3 µT) magnetic field applied to set the quantisation axis for the clock excitation, and the clock laser is detuned to probe on either side of the full-width half-maximum of the Rabi excitation profile. In order to provide an error signal with which to steer the clock laser into resonance, clock excitation is determined by fluorescence collection using a photomultiplier tube (PMT). Using light resonant on the 370-nm cooling transition detects whether clock excitation has occurred, as the ion will only produce a fluorescence signal if it remained in the ground state following the Rabi pulse with the clock laser. By making many such measurements, an excitation fraction can be calculated and compared on either side of line centre, and is used to indicate the direction and magnitude of the frequency correction needed to maintain resonance with the clock transition.

3. Automation of optical clock operation

An ideal control system would allow all parameters to be set initially for clock operation, with protocols for systematic shift analysis scheduled throughout the measurement campaign. Additionally, system operation would be monitored continuously in order to confirm validity of data and also to detect any faults and initiate corrections automatically, without the need for human intervention.

With the goal of achieving a fully automated optical clock system and to improve the overall performance of our clock, we have implemented a computer-controlled FPGA-based infrastructure [10] for experimental control of clock operation and systematic shift evaluation. The strength of using FPGA control protocols is in the scalability and flexibility it brings to experimental control. Within this new control framework, ion loading, minimisation of ion micromotion in the trap, and magnetic field calibration routines are completely automated and can be performed with the push of a button or could be scheduled to run independently at appropriate intervals during a frequency measurement campaign. Additionally, improved monitoring of experimental parameters such as ion fluorescence and laser wavelengths enables more robust operation (e.g., enabling automatic steering of laser frequencies and the implementation of an ion-reloading algorithm). This helps to maximise the uptime of the frequency measurement process, and facilitates complete automation of the validation process, which identifies which frequency data should be included in the final results by reading out the real-time excitation fraction from the associated database. This type of automation combined with monitoring of the clock performance creates a more robust clock system where the systematic shifts can be closely monitored, measured and cancelled, and their respective uncertainties greatly reduced.

3.1. Automation routines for systematic shift evaluation

Recent innovations in automated routines and processes in the Yb^+ optical clock system have been shown to greatly improve the robustness of operation and reduce the overall uncertainty budget for optical frequency measurements. Here we describe in more detail the three most influential improvements to systematic shift evaluation with increased automation tools.

The systematic effects related to the second-order Doppler shift and the rf Stark shift both have components resulting from stray electric fields forcing the displacement of the ion away from the rf node of the ion trap, leading to excess micromotion. Minimising this micromotion is necessary for high-accuracy clock operation, and this can be done by applying a careful selection of dc voltages to the ion trap dc electrodes (see Figure 1). We use a method of rf-photon correlation and an algorithm for finding the voltage values that minimise the ion motion, which requires stepping the electrode voltages through a series of values while monitoring the motion of the ion in each of three orthogonal directions. Our implementation of this protocol is described in detail in reference [11], and has historically required someone to be physically present in the lab to control the beam shutters and rotate waveplates required to probe ion motion with 370-nm light beams in three dimensions. We have now automated this process completely, with motorised, computer-controlled waveplate adjustment and utilising computer-controlled shutters, and the micromotion minimisation routine runs automatically without requiring any human intervention. The experimental configuration is stable enough to require only daily reassessment during a measurement campaign, and the resulting magnitude of the micromotion is stored in a database, from where it could be extracted to calculate the resultant systematic shift and uncertainty values for each day of measurement, but is typically used to provide a collective frequency shift and uncertainty for the entire measurement campaign.

One attribute of the E3 clock transition is its first-order insensitivity to magnetic fields, due to excitation between the $m_F = 0$ levels in the ground and excited states. However, in order to set the quantisation axis for the clock measurement, a small magnetic field (targeting 3 µT) is purposely introduced. This leads to a second-order Zeeman shift that must be characterised throughout the clock measurement campaign. A calibration protocol runs a magnetic field calibration algorithm that performs frequency scans over either of the pairs of $\Delta m_F = 1$ or $\Delta m_F = 2$ Zeeman components of the E2 transition in at least six magnetic field directions, with the computer control of the currents applied to three orthogonal sets of magnetic field coils wound around the ion trap vacuum apparatus. The results are fitted to determine the coil current values necessary to cancel out the background laboratory magnetic field and to produce the quantisation field for the E3 clock measurement. This process has now been fully automated and is performed at least once daily during a measurement campaign.

The upper clock states in Yb^+ have a non-zero electric quadrupole moment, that is, the electric charge distribution is not spherically symmetric in those states. The quadrupole moment is aligned with the quantisation axis, whose direction is determined by the applied magnetic field. This intrinsic property of the ion leads to a frequency shift in the presence of any external dc electric-field gradient, which must be characterised. We have no means of measuring the gradient directly to calculate the shift, but we can make measurements of the frequency shift it induces to probe the gradient's effect. To do this we utilise the higher sensitivity of the E2 transition to this effect and measure its quadrupole-shifted frequency in five magnetic field directions, alternated sequentially. This measurement is interleaved with another measurement in a reference field to remove any contribution of laser frequency drift. The data are fitted to determine the direction and magnitude of the electric-field gradient, which can then be used to calculate the quadrupole shift in any bias magnetic field. It is then possible to scale this result to a shift and uncertainty on the E3 transition. The measurement described above is run once before and after any optical frequency measurement campaign to confirm the stability of the electric-field gradient amplitude and direction over that time period. The entire measurement and computational process for evaluating the shift is automated, enabling the efficient evaluation of this systematic effect, as it requires about 5 days of data acquisition to reach the target uncertainty.

3.2. Fluorescence recovery algorithm

One of the most common failure modes for clock operation is indicated by a loss of fluorescence signal from the ion. This does not necessarily imply that the ion has been lost from the trap, but will result in a failure of the servo of the clock laser to the atomic resonance signal. There are a number of reasons and mechanisms that can cause significant change in the detected output fluorescence, so a fluorescence recovery algorithm was created to step through a check-and-correct programme, exploring each potential problem scenario in turn and then, when the error is identified, take automatic corrective action. A flowchart of the recovery protocol is shown in Figure 2, which contains the following steps. When the fluorescence is determined to have dropped below a designated level, the servo is stopped and a check is made to ensure that all monitored laser wavelengths are correct. If so, the first likely cause of loss of fluorescence is either formation of a molecule or a background gas collision event that excites the ion into a higher motional state. To mitigate either of these scenarios, the ion is exposed to the photoionisation laser to aid in dissociation of molecules, and the cooling laser, far detuned to be more likely to

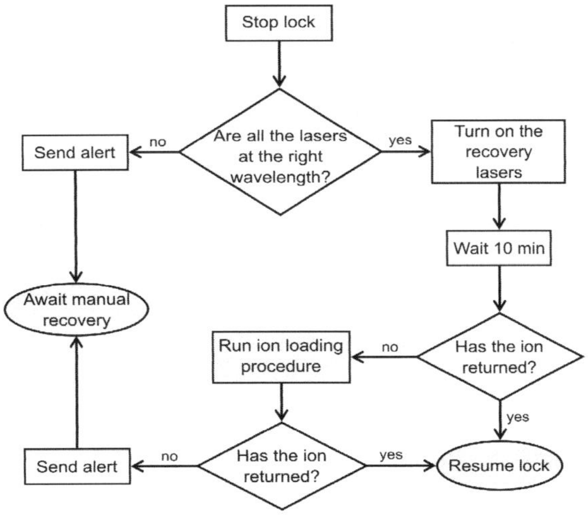

Figure 2. Flowchart schematic for fluorescence recovery algorithm.

be in resonance with a more energetic ion state to aid in reduction of ion motional energy, and both "recovery lasers" are left on for 10 minutes. If the continuously monitored fluorescence signal exceeds a threshold value, the clock servo is reengaged, otherwise the ion is considered lost and an ion-loading protocol is started. The only time human intervention is required is if the wavelength of one of the lasers that is not under computer control is found to have deviated from its nominal value or if an ion is not detected after the loading algorithm completes its cycle. In both these cases an alert is sent to the user to initiate manual corrective action.

3.3. Improved system uptimes and dynamic shift analysis

With the addition of these various monitoring and recovery algorithms, the improvement in uptime of the running Yb^+ clock can be seen in a direct comparison between the uptimes demonstrated during two different month-long measurement campaigns, as shown in Figure 3. The primary criterion for assessing the validity of data to be included as uptime comes from monitoring of the value of the excitation fraction when probing the clock transition on each side of the resonance, and confirming that this value remains within a designated acceptance window. The most significant contribution to improved uptimes came from the automation of the fluorescence recovery algorithm. After its implementation, uptimes exceeded 92% over a 10-day period in a 2023 measurement campaign and enabled a 77% uptime over the entire 30-day measurement campaign. Real-time monitoring is now ubiquitous in the optical clock experimental system, feeding into any number of dashboards interfacing with the database for observation, long-term management, and for initiating correction algorithms. However, there are a few systems that still require physical manual intervention (such as the relocking of the clock lasers to their local ultrastable reference cavities, and other equally rare but possible events), and these will be addressed in the future. A more in-depth discussion of optical clock operation methods, automation and systematics analysis can be found in reference [8], and a detailed description of the finalised systematic error budget for recent absolute frequency measurements of the $^{171}Yb^+$ E3 optical clock transition is in preparation for publication.

On the route to developing an optically-derived time scale, the output of an optical clock system can be utilised to correct the frequency of a microwave-based time scale. If this could be done in real time, this optically-steered time scale could greatly improve local time scales. To this end, efforts have been made to enable dynamic assessment and correction of the frequency

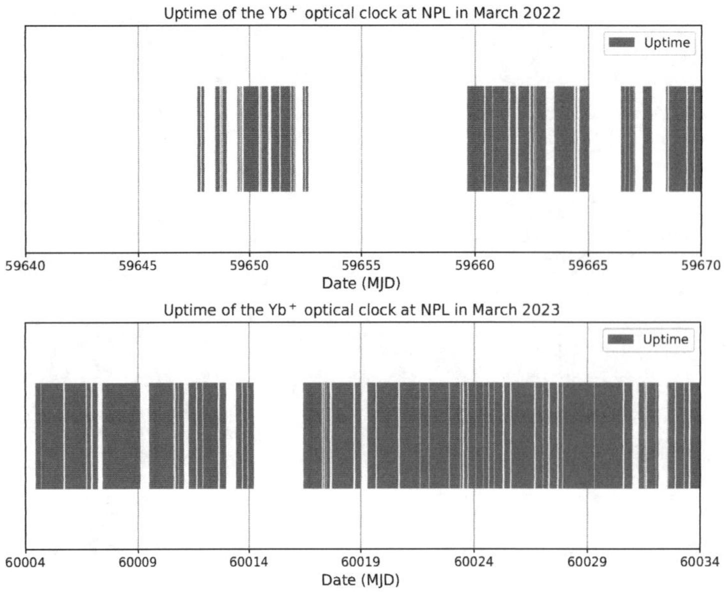

Figure 3. Optical clock system uptimes representing validated data periods during month-long optical frequency measurement campaigns. Comparing campaigns in 2022 (upper) and 2023 (lower) shows an increase in valid data uptimes due to improvements in automation between these two periods.

of the NPL Yb^+ E3 clock output in order to enable its use for steering. A second-by-second calculation of estimated frequency shifts is mapped onto the measurement database, available for hourly updates to steer the time scale. This has been used to explore and test future methods for steering the UTC(NPL) time scale, which are reported separately in this Proceedings.

4. Investigation of rf trap drive-induced ac Zeeman effect

As part of our assessment of the error budget in the NPL Yb^+ E3 clock system, a new investigation of the ac Zeeman effect was required. Previously focus had been placed on assessing the contribution from magnetic field oscillations at low frequencies, such as at 50 Hz from the ac mains powering the experimental systems. Measurements of the linewidth of the excited E2 transition $\Delta m_F \neq 0$ components were used to estimate the size of the effect in the experiment. However, we had not characterised the effect of any ac magnetic field that could be generated at the trap drive frequency, typically on the order of 10 MHz for ion traps. This can occur if there is asymmetry in the rf paths through the ion trap electrodes that create the oscillating trapping field, which could in turn induce an oscillating magnetic field at the same frequency. Care was taken in the rf electrode design of the most recent NPL ion trap [9] and the effect was assumed to be small for this type of endcap trap, but no direct measurement of the effect had been made. Gan *et al.* [12] suggested the use of microwave spectroscopy to investigate this effect directly.

This approach exploits an effect described by Autler and Townes [13], which can be used to observe and quantify any oscillating magnetic field occurring at the trap drive frequency. In the Yb^+ ion energy level structure there is a microwave transition, $^2S_{1/2}(F=0)\rightarrow{}^2S_{1/2}(F=1)$, as shown in Figure 1. If a dc magnetic field is applied, Zeeman splitting of the $F=1$ state will occur. In the presence of a perpendicular ac magnetic field oscillating near resonance with the frequency of the induced Zeeman splitting, state mixing of the Zeeman levels can occur

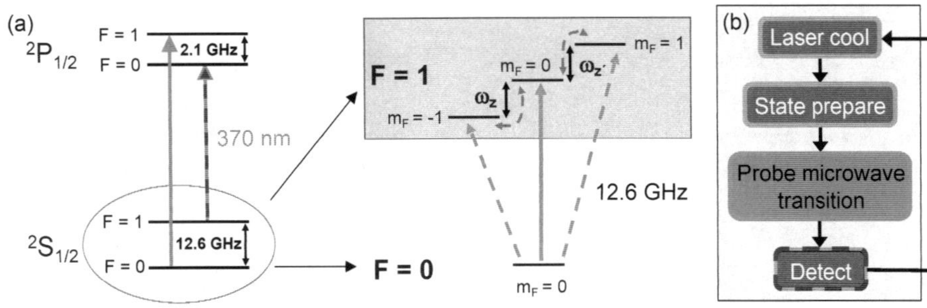

Figure 4. Energy level diagram and experimental protocol for microwave excitation to induce Autler-Townes splitting in the presence of an ac magnetic field. (a) When ω_z is on or near resonance to the rf trap drive frequency, Larmor procession is induced and an Autler-Townes splitting can be observed. (b) Protocol for microwave excitation measurements. The "cool-prepare-probe-detect" cycle is repeated at different microwave frequencies to measure a single excitation profile. Then the process is repeated at different additionally dc magnetic fields in order to map out the Aulter-Townes splitting effect.

(see Figure 4). This produces an "avoided crossing" in the dressed-state picture, which can be experimentally probed and measured by exciting the $(F = 0) \rightarrow (F = 1)$ transition. The observed splitting will have the form $(\delta^2+\Omega^2)^{1/2}$, where $\delta = \omega_z - \Omega_{\rm rf}$ is the difference between the Zeeman shift frequency and the applied rf frequency, and Ω is the coupling strength between the two Zeeman levels. Assuming the rf magnetic field is linearly polarised and its field component perpendicular to the quantisation axis is $B_\perp = b_\perp \cos(\Omega_{\rm rf} t)$, the minimum separation of the split excitation feature is equal to the coupling strength and can be written as

$$\Omega = \frac{g_F \mu_B b_\perp}{\hbar\sqrt{2}}, \tag{1}$$

where g_F is the Landé g-factor and μ_B is the Bohr magneton [12]. The splitting is linearly dependent on $b_\perp$, the amplitude of the ac magnetic field perpendicular to the quantisation axis of the measurement. The quantisation axis is defined by the large applied dc magnetic field required to bring the Zeeman splitting into resonance with the rf-induced ac magnetic field.

The rf trap drive frequency used in our experiment is $\Omega_{\rm rf} = 2\pi\times 13.718$ MHz. The first-order Zeeman shift for the $^2S_{1/2}(F = 1)$ state is 14 kHz/μT, thus requiring a ~1 mT applied dc magnetic field for the Zeeman splitting to match the rf field frequency in order to induce state mixing. The majority of this field magnitude was provided by a fixed source of stacked neodymium magnets placed about 13 cm from the ion trap. The remainder, referred to here as the additional B_y field, was produced with an existing set of field coils wrapped around the ion trap vacuum chamber, providing a field in the laboratory $\hat{y}$ direction. The magnets were initially placed along the same axis so that the sum of the fields would produce the greatest magnitude. A small computer-controlled synthesiser provided the 12.6 GHz source for exciting the microwave transition. A microwave horn was used to direct the output towards the trapped ion and was rotated to adjust the polarisation of the microwaves to excite the $m_F = 0 \rightarrow m_F = 0$ transition or excite to the $m_F = \pm 1$ Zeeman levels, which we used to initially calibrate our permanent magnets' contribution to the applied field.

A control sequence, shown in Figure 4(b), was developed to automate the data collection of frequency scans over the microwave $m_F = 0 \rightarrow m_F = 0$ transition whilst stepping through applied dc magnetic field values so as to map out the excitation frequencies of the Autler-Townes splitting. For each frequency of the microwave probe, initial laser cooling was performed on the

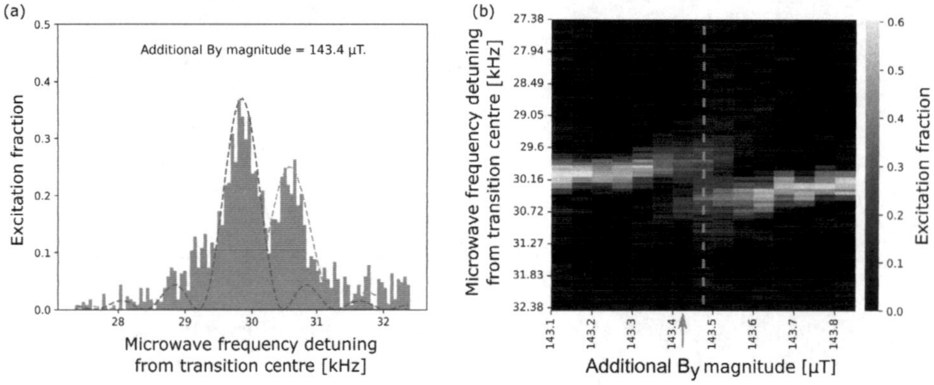

Figure 5. Autler-Townes splitting measurements. (a) A plot of a single microwave scan over the microwave transition frequency at a single applied dc magnetic field value (additional $B_y = 143.4$ µT). Dashed blue and orange lines indicate fits to Rabi excitation features used to identify frequency splitting values. (b) A heatmap comprised of a set of microwave excitation scans, each taken at a different applied dc magnetic field, mapping out the Autler-Townes splitting. An arrow identifies the magnitude of the additional B_y field from (a). The dashed line indicates the approximate position of the minimum distance between excitation peaks, which is used to calculate the magnitude of the ac magnetic field at the rf trap frequency.

ion (370-nm cooling and 935-nm repump lasers engaged), followed by a stage to prepare the ion in the ground $^2S_{1/2}(F = 0)$ state before a microwave Rabi pulse of 1-ms duration was switched on. Excitation was detected using a 370-nm probe resonant with the $^2S_{1/2}(F = 1) \rightarrow {}^2P_{1/2}(F = 0)$ transition. The resulting excitation maps, with magnetic field steps of 0.05 µT, are shown in Figure 5.

The data were analysed to find the magnitude of the frequency separation between measured split excitation peaks. The magnitudes of the frequency splitting for each scan were then plotted versus the applied magnetic field, and fit to a parabola. The minimum of the fit was then used to calculate a value for $b_\perp$, as per Equation 1. This measurement does not require precise knowledge of the absolute value of the dc applied magnetic field, so the uncertainly in the measurement is estimated by considering the stability of the computer-controlled field generation and microwave frequency and the measurement statistics. The measurements give a result of $b_\perp = 58 \pm 14$ nT for our typical operating parameters, where the error is given as one standard deviation. The error is dominated by the statistics of the scatter in the Autler-Townes splitting data results. By increasing the rf power to the ion trap drive, we were able to increase the size of the trap drive-induced ac Zeeman effect, which was shown to increase the measured Autler-Townes splitting as expected.

Measuring the Autler-Townes splitting using a single applied dc magnetic field in the $\hat{y}$ direction will only give information about the magnitude of the trap-induced ac magnetic field perpendicular to that applied field direction. In order to gain more spatial information about the direction and magnitude of the ac magnetic field, an additional set of measurements was made using a second applied magnetic field direction, resulting in a quantisation axis rotated by $37° \pm 5°$ in the $x - y$ plane relative to the initial measurement setup. The $b'_\perp$ field amplitude measured in this configuration was $\sim 40\%$ of the $b_\perp$ field measured with the dc magnetic field in the $\hat{y}$ direction. Using the nominal measured value of $b_\perp$, this allowed us to calculate limits to the size of the trap drive-induced ac magnetic-field in the $\hat{y}$ direction to be in the range 15 nT $\leq |b_y| \leq$ 72 nT.

The relation between the ac magnetic field and the frequency shift of the E3 transition can be written as,

$$\delta\nu_{E3} = \alpha_Z(\mathrm{F}_{7/2}, 3, 0)\left(\langle B_{\parallel}^2\rangle + \frac{5}{8}\langle B_{\perp}^2\rangle\right) - \alpha_Z(\mathrm{S}_{1/2}, 0, 0)\langle B^2\rangle \tag{2}$$

where $\alpha_Z(\mathrm{F}_{7/2}, 3, 0)$ and $\alpha_Z(\mathrm{S}_{1/2}, 0, 0)$ are the static quadratic Zeeman shift coefficients for the upper and lower clock states, with values of -17.7 mHz/µT^2 and -15.5 mHz/µT^2, respectively. The notation $\langle B^2\rangle$ indicates the root mean square (rms) value of the magnitude of the field squared, and can be separated out into parallel and perpendicular components, $\langle B^2\rangle = \langle B_{\parallel}^2\rangle + \langle B_{\perp}^2\rangle$. The quantisation axis for our E3 clock transition measurements is set in the $\hat{y}$ direction and therefore $\mathrm{B}_{\parallel}$ is also in the $\hat{y}$ direction. This results in a fractional frequency shift $\lesssim 1 \times 10^{-20}$ with an uncertainty on the shift at a comparable level.

These results can now be compared to similar measurements made in other ion traps of both endcap and linear electrode layout design. It is important to understand if similar traps produce similar ac magnetic fields, indicating that the design itself sets the scale of this effect, or if the individual tolerances of trap manufacturing cause changes in the size of the produced ac magnetic fields, to help inform future ion trap designs for high-accuracy optical frequency metrology.

5. Conclusions

The implementation of an FPGA-based control framework for the NPL Yb^+ optical clock system has transformed our ability to monitor and record to database experimental parameters and outputs, create automation routines for systematic shift analysis and system recovery, and calculate second-by-second error corrections to the output frequency of the clock. Frequency measurements over extended periods have already been used to constrain temporal variation of the fine structure constant and exclude regions of parameter space in theories beyond the Standard Model, such as those which include ultralight scalar dark matter [3, 5]. The improvement in uptimes due to autonomous fluorescence recovery algorithms has enabled longer unbroken measurement durations that will extend the sensitivity of these constraints. Additionally, the robust operation of the optical clock system has facilitated the assessment of an improved systematic uncertainty budget, for which we also verified the negligible systematic shift contribution due to the trap drive-induced ac Zeeman effect at the $\lesssim 1 \times 10^{-20}$ level, as presented in these Proceedings. Recent absolute frequency measurements against NPL's local caesium fountain as well as via satellite link to TAI, each with their associated uncertainty budget, will be reported in a separate publication, currently in preparation. Future plans for further automation across all aspects of a frequency measurement campaign aim to remove the need for human intervention, a necessary step towards implementing a future time scale based on the optical redefinition of the SI second.

Acknowledgments

The authors would like to thank Geoffrey Barwood for his helpful discussions and Maxime Favier for his careful reading of the manuscript. This work was supported by the UK government Department for Science, Innovation and Technology through the UK National Measurement System programme. The authors also acknowledge funding from the EMPIR programme for work on projects 18SIB05 ROCIT and 20FUN01 TSCAC, co-financed by the Participating States and from the European Union's Horizon 2020 research and innovation programme.

Journal of Physics: Conference Series **2889** (2024) 012044 doi:10.1088/1742-6596/2889/1/012044

References

[1] Lisdat C *et al.* 2016 *Nature Communications* **7** 12443
[2] Boulder Atomic Clock Optical Network (BACON) Collaboration 2021 *Nature* **591** 564–569
[3] Roberts B M *et al.* 2020 *New J. Phys.* **22** 093010
[4] Barontini G *et al.* 2022 *EPJ Quantum Technology* **9** 12
[5] Sherrill N *et al.* 2023 *New J. Phys.* **25** 093012
[6] Filzinger M, Dörscher S, Lange R, Klose J, Steinel M, Benkler E, Peik E, Lisdat C and Huntemann N 2023 *Phys. Rev. Lett.* **130**(25) 253001
[7] Dimarcq N *et al.* 2024 *Metrologia* **61** 012001
[8] Tofful A 2023 *Advances in performance and automation of a single ytterbium ion optical clock* Ph.D. thesis Imperial College London
[9] Nisbet-Jones P B R, King S A, Jones J M, Godun R M, Baynham C F A, Bongs K, Doležal M, Balling P and Gill P 2016 *Appl. Phys. B* **122** 57
[10] ARTIQ (Advanced Real-Time Infrastructure for Quantum physics) experimental control URL `https://m-labs.hk/experiment-control/artiq/`
[11] Baynham C F A 2019 *Frequency metrology at the* 10^{18} *level with an ytterbium ion optical clock* Ph.D. thesis University of Oxford
[12] Gan H C J, Maslennikov G, Tseng K W, Tan T R, Kaewuam R, Arnold K J, Matsukevich D and Barrett M D 2018 *Phys. Rev. A* **98** 032514
[13] Autler S and Townes C 1955 *Phys. Rev.* **100** 703

Journal of Physics: Conference Series **2889** (2024) 012045
doi:10.1088/1742-6596/2889/1/012045

Superradiant active optical atomic clocks: motivations and current challenges

Martina Matusko and Marion Delehaye

Université de Bourgogne Franche-Comté, SUPMICROTECH, CNRS, Institut FEMTO-ST, F-25000 Besançon, France

E-mail: marion.delehaye@femto-st.fr

Abstract.

Current state-of-the-art frequency standards are passive optical atomic clocks where the frequency of an optical resonator is stabilized to a narrow atomic transition. Passive clocks have achieved unprecedented stabilities of 6.6×10^{-19} over one hour of averaging time [1]. However, they face intrinsic limitations, particularly due to thermal and mechanical fluctuations of the local oscillator. To surpass the limitations of the passive clocks and go beyond the state-of-the-art, the idea of building active optical atomic clocks emerges. These clocks would be optical counterparts of hydrogen masers, with the emitted frequency defined by the atomic transition and therefore inherently stable against cavity instabilities. This paper discusses the latest developments and future prospects in the field of active optical atomic clocks.

1. Introduction

Optical atomic clocks, with their remarkable stability [1, 2] and accuracy at the 10^{-18} level [3, 2, 4], have initiated a new era in precision measurement. This exceptional level of precision paves the way for a wide array of diverse applications, ranging from probing variations in fundamental constants [5, 6, 7, 8] to the redefinition of the second [9, 10, 11, 12, 13] or advancing the field of the relativistic geodesy [14, 15, 16, 17, 18, 19, 20], where optical atomic clocks can serve as extremely accurate altimeters. The capabilities of these advanced timekeeping devices represent a significant leap forward, offering novel opportunities and insights across various scientific domains. Yet, surpassing the current state-of-the-art clocks in stability and accuracy would open even more advanced possibilities, especially in seismic mitigation studies, deep-space navigation applications and fundamental sciences.

The development of optical atomic clocks with fractional frequency stabilities below 10^{-18} achieved within an hour is for instance crucial for measuring height differences below 1 cm accuracy. This precision surpasses that of GNSS monitoring, which offers a best-case height uncertainty of typically a few cm due to atmospheric delays [21], with longer integration time. The ongoing improvements in optical clocks are expected to significantly enhance seismic and volcanic hazard mitigation studies [20]. Essential for realizing the full potential of this technology are stable and accurate on-field systems or strong fiber links between optical atomic clocks.

The improved precision of atomic clocks not only enhances terrestrial measurements but also opens up advanced possibilities in space-based navigation. The Deep Space Atomic Clock (DSAC), a mercury-based ion clock, with its demonstrated stability of 2×10^{-15} in one day of averaging time was operated between 2019 and 2021. Even though it is a microwave clock, it already offered unprecedented accuracy in one-way radiometric tracking, improving the efficiency of spacecraft operations in deep space missions [22], showing potential future improvements of space studies that would be enabled by more performant space clocks [23].

Current bounds on fundamental constants have been established by today's best optical clocks and do not show yet any clear variation. A network comprising various types of clocks, including highly-charged ion clocks, molecular, and nuclear clocks, offers a promising strategy for measuring variations in the fine structure constant (α) and the electron-to-proton mass ratio (μ) with unparalleled sensitivity [24]. The association of these diverse clocks, each sensitive to variations in different ways, is expected to significantly boost the ability to detect subtle changes in these fundamental constants.

Furthermore, atomic clocks are sensitive to ultralight scalar bosonic dark matter, which could cause oscillations in the fundamental constants, detectable through variations in clock transition frequencies [8]. This sensitivity extends across a broad range of dark matter masses and interaction strengths. Networks of clocks could potentially detect transient changes in fundamental constants induced by dark matter objects with large spatial extents, such as stable topological defects [25, 26, 27].

Additionally, an array of optical atomic clocks could serve as an innovative technique for gravitational wave detection [28, 29]. This method, contrasting with the distance variation measurements of interferometric techniques used in projects like Advanced-LIGO and eLisa, concentrates on tracking timing variations to detect the differential time dilation experienced by clocks located at different phases of a passing gravitational wave. This approach could lead to the detection of gravitational waves at various frequencies, broadening the range of perceptible astrophysical phenomena.

Current best optical clocks are based on a passive clock scheme, in which the frequency of an external local oscillator is stabilized to a narrow atomic transition. The schematic of the passive clock is depicted in Fig. 1. For classical atomic ensembles, the stability of these clocks is limited by the quantum projection noise of the measurement, given by the formula:

$$\sigma_y(\tau) = \frac{1}{SNR}\frac{\Delta\nu}{\nu_0}\sqrt{\frac{T}{N\tau}}, \tag{1}$$

where SNR is the signal-to-noise ratio, $\Delta\nu$ is the linewidth of the transition, ν_0 the transition frequency, T the interrogation time, N the number of atoms and τ the integration time. However, this ultimate stability is not fully realized even in cutting-edge optical atomic clocks, primarily due to residual instabilities from the local oscillator. Here, we will provide an overview of passive optical atomic clocks and their limitations. Following this, we will focus on the new concept of active optical atomic clocks where light emitted directly on a narrow atomic transition is harnessed as a stable frequency reference.

2. Passive optical atomic clocks

Current passive optical atomic clocks are realized by the frequency stabilization of an external optical local oscillator to an atomic transition. The operational cycle of a passive optical atomic clock [30] involves an initial process of laser cooling and preparation of the atomic state, followed by the interrogation stage, and finally destructive state read-out atom imaging, and signal processing. In the state preparation stage, atoms are typically loaded into a single-ion trap [31, 32, 33] or an optical dipole trap that follows a magneto-optical trap [34, 35, 36, 30]. To ensure frequency stability of the laser between two interrogations of the clock transition, the interrogating laser is frequency-stabilized onto a resonant mode of an ultra-stable Fabry-Perot cavity. This narrows the linewidth of the probing laser and enhances the mid-term stability by maintaining the laser close to resonance between two interrogations.

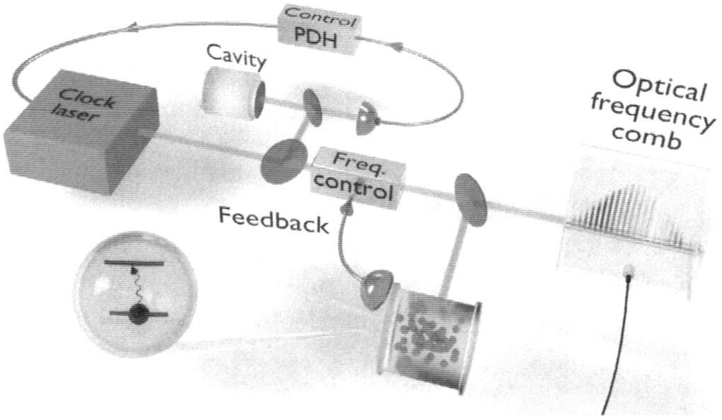

Figure 1. Scheme of a passive optical atomic clock. A clock laser, prestabilized on an ultra-stable Fabry-Perot cavity, is sent to an atomic reference, either a single ion or neutral atoms trapped in an optical lattice.

The clock frequency can be influenced by various effects stemming from the atomic environment or operational conditions. Such influences include laser-induced electric fields [37], blackbody radiation [38, 39], magnetic fields [40, 33, 41], Doppler effect [42, 43, 44, 45, 46] and other systematic shifts [34, 47, 48, 49, 50]. A comprehensive discussion of these systematic effects can be found in [30, 36]. Significant efforts have been dedicated to minimizing these shifts, resulting in the enhanced accuracy of atomic clocks that is now at the low 10^{-18} level [3, 51, 39, 52, 53].

Despite the unparalleled precision measurements made possible by passive optical atomic clocks, intrinsic limitations persist, both in terms of stability and accuracy. Even better accuracies could be obtained in principle by using highly charged ions or a nuclear transition in thorium, but limitations of the fractional frequency stabilities will be detailed in the next paragraph.

Fractional frequency stability limitations of a passive optical atomic clock

In passive clock schemes, the laser is stabilized to the atomic transition, with the laser frequency adjusted to the atomic resonance transition through a feedback loop. The destructive nature of state read-out imaging necessitates thorough preparation of the subsequent atomic ensemble at each new measurement cycle. Given the inherently complex nature of atom preparation and the necessity for repeated cycles, only a fraction (typically 20–50%) of the clock cycle allows for the actual probing of atoms.

Intermittent interrogation of the atoms during each measurement cycle leads to the Dick effect [54, 55, 56] - an aliasing of the local oscillator high-frequency noise into the low-frequency noise within the atomic resonator bandwidth, limiting the stability of the atomic clock. Efforts to mitigate the Dick effect involve increasing the duty cycle, facilitating non-destructive measurements [57], and employing more stable lasers.

One approach for minimizing the instability imposed by the Dick effect involves increasing the measurement duty cycle. A zero-dead-time optical clock based on the interleaved interrogation of two cold-atom ensembles demonstrated minimal Dick noise, achieving an exceptional fractional frequency instability assessed at $6 \times 10^{-17}/\sqrt{\tau}$ where τ is the integration time in seconds [58]. Similarly, the Dick effect can be reduced when two distinct clocks operating on the same transition are compared [59].

Enhancing the quality of optical cavities can lead to improved laser frequency noise and thus reduce the Dick effect. Presently, the most advanced optical resonators exhibit fractional frequency instabilities as low as 4×10^{-17} [60]. However, these resonators face limitations imposed by thermal Brownian noise originating from their components, particularly mirror coatings [61]. Their enhanced performance can be credited to several techniques developed to reduce thermal effects in optical cavities. One strategy for mitigating the thermal effects in the optical cavities involves the use of a long resonator operating at room temperature, simplifying the system and its operation. The achieved stability closely approached the thermal noise limit, highlighting the potential of long resonators at ambient temperatures for high precision in frequency stability [62]. Moreover, significant attention is directed towards optical coating technology to tackle Brownian noise in cavity mirrors. A notable accomplishment is a tenfold reduction in Brownian noise achieved through high-reflectivity coatings formed by directly bonding monocrystalline multilayers [63]. Another noteworthy strategy involves operating optical cavities at cryogenic temperatures to mitigate thermal noise-induced fluctuations in the cavity's optical length [64, 65, 66].

The challenges associated with the Dick effect are therefore reaching the technological limits of feasible improvements in cavity design. Hence, there is a need for new approaches. One proposed method to enhance the properties of the optical local oscillator involves employing "Spectral Hole Burning" spectroscopy [67, 68]. In this method, narrow spectral holes are imprinted into a cryogenically cooled non-linear crystal. These features are then employed for laser-frequency stabilization, where the laser frequency is actively tuned to maximize transmission through the crystal. The proposal of employing Ramsey-Bordé matter-wave interferometry within a passive-like clock scheme has emerged as another strong candidate for advancing laser stabilization [69]. This approach has demonstrated fractional frequency instability below 2×10^{-16} for short integration times. Utilizing thermal atomic beams, this method significantly outperforms the stability of other thermal atomic systems and requires minimal thermal shielding and vibration isolation, potentially surpassing the Fabry-Perot cavity in terms of insensitivity to thermal and mechanical disturbances. Demonstrated instability results, combined with advantages in low vibration sensitivity and compact design, position Ramsey-Borde optical-frequency stabilization as a highly promising approach for diverse applications. However, due to the pulsed operational nature, the standard limitations due to the Dick effect persist.

Proofs of principle for vapor-cell based optical frequency references have also been demonstrated, using either a two-photon transition in ^{87}Rb [70, 71] or dual-frequency sub-Doppler spectroscopy in ^{133}Cs [72]. These experiments continuously interrogate the atoms without the need for an optical Fabry-Perot cavity, making them insensitive to the Dick effect. However, the stability achieved so far has been limited to the 10^{-13} range at one second of integration time.

In the rest of this article, we will focus on a proposal for an active optical atomic clock that appears promising in overcoming these Dick effect-related limitations.

3. Active optical atomic clocks

3.1. Generalities

The idea of active optical clocks stems from superradiance, a quantum phenomenon initially described by Dicke in 1954 [73]. When a large number N of indistinguishable emitters are coupled to a single electromagnetic field, the constructive interference of possible decay paths between the fully excited state to the global ground state leads to an enhanced, collective emission called superradiance. Its intensity scales with N^2, while its duration is reduced by a factor N with respect to uncoupled spontaneous emission [74]. Superradiance has been observed in various experimental media, such as quantum dots [75, 76], light-harvesting complexes [77], or astrophysical phenomena [78, 79, 80]. In the case of emission in the radiofrequency, microwave,

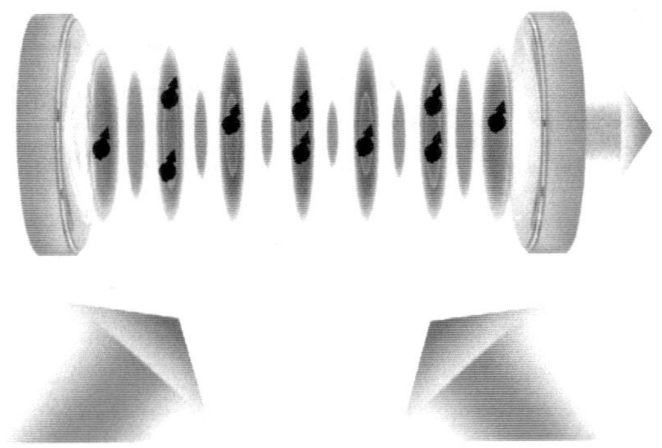

Figure 2. Scheme of an active optical atomic clock. Atoms (black) are trapped by an optical lattice at the magic wavelength (red) and coupled to the optical mode of an ultra-stable Fabry-Perot cavity (yellow).

or far-infrared domain, coupling the emitters to a common field can be achieved by placing the emitters at a distance smaller than the emission wavelength λ_0 [81, 82]. Regarding superradiant emission in the optical domain using atomic emitters, a dense sample with many atoms in a volume smaller than λ_0^3 would undergo decoherence preventing superradiant emission, so instead atoms are coupled to a single mode of an optical cavity [83], as it is depicted in Fig. 2.

In this case, the key figures are the cavity linewidth κ, the spontaneous emission rate γ and the single-atom Rabi frequency $2g$. The build-up of quantum correlations that lead to a superradiance pulse is only possible when the collective vacuum Rabi frequency $\Omega = 2g\sqrt{N}$ is much larger that κ and $\gamma_\perp = \gamma/2 + 1/T_2$, where T_2 is the typical timescale of the other decoherence mechanisms in the system. Optical superradiance is usually realized in a regime where $\gamma \ll g$, so that photons are preferably emitted in the cavity mode, and where $g \ll \kappa$ so that the probability of emitted photons escaping the cavity exceeds the probability of their re-absorption by atoms. The system is then in the so-called "bad-cavity regime" illustrated in Fig. 3 and the coherence is stored in the atomic media rather than in the light field as is the case for usual, "good-cavity" lasers.

The concept of superradiant lasers stems from the process of replenishing atoms that have transitioned back to the ground state, maintaining population inversion [83]. In these systems, the renewing rate is of paramount importance, as a too-small rate cannot maintain a population inversion, while an excessively high rate induces too much decoherence and can destroy the collective dipole of the system. Their very high potential as optical frequency references [84, 85] has paved the way for the development of active optical atomic clocks, that can be seen as an optical counterpart to the masers. Especially, it has been predicted that the N^2 scaling of the output power could lead to values in the picowatt range on a millihertz linewidth transition. The superradiant emission linewidth could be as low as $C\gamma$, where C is the cooperativity factor

$$C = \frac{(2g)^2}{\kappa\gamma}, \tag{2}$$

laying the foundations for oscillators with linewidths of a few mHz and fractional frequency instabilities in the 10^{-18} range at one second. Operation in the bad-cavity regime ensures that superradiant photons are minimally impacted by the intrinsic frequency instability of the cavity. The change of the superradiant laser output frequency due to a cavity frequency change is

Journal of Physics: Conference Series **2889** (2024) 012045 doi:10.1088/1742-6596/2889/1/012045

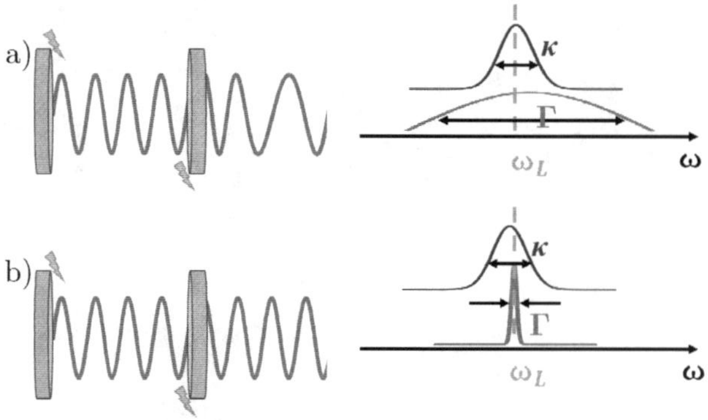

Figure 3. a) In the good-cavity regime, the laser output phase is dictated by the cavity, with fluctuations (symbolized by a lightning flash) on cavity mirrors notably affecting laser light properties. b) In the bad-cavity regime, atoms retain the phase information, substantially reducing the impact of cavity perturbations.

characterized by the cavity pulling coefficient

$$P = \mathrm{d}f/\mathrm{d}f_{\mathrm{cav}} = 2\gamma_\perp/(2\gamma_\perp + \kappa) \approx 2\gamma_\perp/\kappa \tag{3}$$

that can be in the 10^{-5} range.

Two main strategies are currently investigated to create a superradiant active optical clock: optical lattice superradiant lasers [85], and atomic beam superradiant lasers [84]. These methods are primarily differentiated by their distinct mechanisms to maintain inversion population.

3.2. Optical lattice superradiant lasers

Superradiant lasers based on atoms trapped in optical lattices stand out as a promising concept for ultra-stable oscillators.

Within this framework, atoms would be confined in an optical lattice operating at a so-called magic wavelength and operate in the Lamb-Dicke regime [86]. In this configuration, trapped atoms are continuously pumped to maintain population inversion, essential for sustained lasing. After the seminal paper for optical lattice superradiant lasers in 2009 [85], the first proof of principle for optical lattice superradiant lasers was obtained at JILA [87] using Raman dressed states of ^{87}Rb repumped to reach a steady-state. The N^2 scaling of the output power was demonstrated, and the conditions for the stable lasing operation in their effective 3-level system were explored in [88]. A scheme for a four-level active optical clock based on ^{87}Rb has also been proposed [89].

Superradiant emission directly on a relatively narrow optical transition was demonstrated on the 8 kHz-wide $^1\mathrm{S}_0 \leftrightarrow {}^3\mathrm{P}_1$ transition of ^{88}Sr at JILA [90] and at UCPH [91], and on the 375 Hz-wide $^1\mathrm{S}_0 \leftrightarrow {}^3\mathrm{P}_1$ transition of ^{40}Ca in Hamburg [92]. While atoms are trapped in a 1D optical lattice at JILA and Hamburg, at UCPH they are prepared in a Magneto-Optical Trap (MOT) inside a Fabry-Perot cavity and then released in free-fall to produce superradiance in the cavity. The JILA and the UCPH groups managed to reach a steady state regime for a few milliseconds by repumping the atoms. At JILA, emission ceases after approximately 35 photons per atom are emitted due to repump-induced heating, while at UCPH it is limited by the free-fall of the atoms that leave the cavity volume after the emission of approximately one photon per atom. Linewidths measurements were performed by UCPH group showing a subnatural linewidth of 820 Hz, Fourier-limited by the emission duration [91].

The JILA group also demonstrated pulsed superradiant emission on the $^1S_0 \leftrightarrow {}^3P_0$ clock transition of ^{87}Sr [93]. A heterodyne beatnote between superradiant pulses and a stable laser was used to extract the average frequency of each pulse and perform frequency stability measurements of the succession of pulses. A fractional frequency of $6 \times 10^{-16}\tau^{-1/2}$ (for τ larger than the cycle time of 1.1 s) has been reported [94], assessing a promising long-term frequency stability. The system however faces two challenges due to the intermittent emission: firstly, there is no phase continuity between the pulses; secondly, the short-term stability over the pulse duration has not been evaluated.

One of the main current challenges for superradiant optical clocks is therefore the realization of a true continuous operation. A prevalent solution is to use an incoherent pump [83, 85, 95], efficiently repumping atoms from the ground to the excited state during superradiant emission. This approach ensures sustained population inversion without introducing coherence between the pump and lasing mode, therefore avoiding amplified phase diffusion and intensity fluctuations. Relaxation oscillations and stable behavior of the effective 3-level superradiant laser based on dressed states of ^{87}Rb has been investigated in [88]. It has been shown that the repumping rate could be tuned to control relaxation oscillations. UCPH also investigated the conditions for stable lasing in their system. They observed strong oscillations of the output power, consistent with an excessive gain, when using an effective 3-level system, and could reach a promising stable emission regime with an effective 4-level scheme. However, the lifetime of atoms inside the cavity is anyway limited, and intense experimental efforts are currently dedicated to the issue of guiding new cold atoms to the cavity in order to compensate for atom lost during superradiant emission.

In light of these challenges, several proposals tackle directly the two issues of collective dipole continuation and atomic lifetime by investigating the feasibility of atomic beam superradiant lasers and their expected properties.

3.3. Atomic beam superradiant lasers

In atomic beam superradiant lasers, a continuous stream of atoms, pre-excited to the upper lasing state, traverses the cavity where they produce superradiance [84]. In these systems, atoms are untrapped and experience minimal light shifts, but other challenges arise. In particular, the finite transit time τ_{tr} of atoms through the cavity affects the boundaries of the continuous superradiant lasing regime [96, 97]. The relevant regime for metrological applications is obtained when $\tau_{tr} \gg \kappa/(Ng^2)$, the typical timescale necessary for the formation of the collective dipole. In this case, the output linewidth is given by $C\gamma$. The inhomogeneous broadening related to a first-order Doppler shift resulting from the atomic motion along the cavity axis leads to additional linewidth broadening.

A realistic scheme has been proposed in [98], showing that linewidths in the mHz regime were possible with a simple and robust scheme using thermal atoms. One of the main challenges of these systems is the relatively high flux of atoms that is necessary to initiate and sustain superradiance, but remarkable continuous guidance of ^{88}Sr with a high phase-space density and sub-μK radial temperature has been demonstrated recently [99, 100] and could lead to precursors of cold superradiant beam lasers.

3.4. Hybrid superradiant laser scheme

An alternative approach suggests the combination of the optical lattice laser and the atomic beam laser, dynamically transferring trapped atomic ensembles into and out of the cavity [101]. As previously mentioned, the significance of adequate transit time through the cavity persists. A more specific proposal involves introducing atoms already in the upper lasing state into the cavity while atoms from the previous ensemble are still emitting [102]. Sequentially introduced into the cavity, the inverted atomic ensembles stimulate the atoms of each subsequent ensemble to emit photons into the cavity mode primarily through stimulated emission, thereby sustaining optical phase coherence.

Extensive theoretical and experimental investigations have concentrated on pulsed and quasi-continuous superradiant emission. Nevertheless, achieving continuous lasing remains a fundamental objective with broad applications. Merging the advantages of the superradiant optical lattice laser and the superradiant atomic beam laser, the combined hybrid lasing scheme offers the potential to overcome the limitations associated with pulsed superradiant lasing, potentially paving the way for truly continuous operation.

3.5. Limitations of active optical atomic clocks

The performance of superradiant lasers may be constrained by the difficulty of maintaining uniform coupling of atoms to a common field and efficiently pumping the atoms without perturbing the system. Additionally, dipole-dipole interactions between the atoms can lead to decoherence in the collective dipole and generate phase noise, ultimately limiting the laser's frequency stability.

A particular study concentrates on dipole-dipole interactions between atoms within the cavity, identifying them as a major noise source in low-density lattices [103]. Utilizing an idealized pumping method to reduce decoherence, while disregarding light shifts from pump lasers, they found that optimal laser linewidth is achieved at moderate pump strengths and is robust against cavity fluctuations even when these significantly exceed the laser linewidth. However, superradiant decay of a highly dense atomic assembly notably broadens the laser linewidth and increases its sensitivity to cavity fluctuations.

Another recent study investigated the variation of laser linewidth on different laser parameters in both homogeneous and inhomogeneous systems [104]. This study demonstrated that the linewidth of a superradiant laser is robust to the incoherent repumping rate, but for systems with inhomogeneous atomic dephasing, both the minimum achievable linewidth of the laser and the number of intracavity photons are affected by the atoms dephasing rate.

The same study focused on an inhomogeneous system of an atomic ensemble interacting with a single cavity mode to examine the attainable frequency stability of an active optical atomic clock. It has been demonstrated that the ultimate stability of an incoherently pumped active optical atomic clock can achieve a level of stability equivalent to the quantum projection noise (QPN)-limited stability of a passive optical atomic clock at the same atom number for both types of clocks. Instabilities of $\sigma_y(\tau) \approx 3.7 \times 10^{-18}/\sqrt{\tau}$ should be reachable, with the short-term stability constrained by the shot noise of the active clock. This demonstrates that the short-term stability of an active optical atomic clock can either match or even significantly surpass the stability of an ideal, zero dead-time, QPN-limited passive optical atomic clock with the same number of atoms as the considered active clock. On the other hand, the long-term stability of active optical atomic clocks will be constrained by the phase diffusion of the superradiant signal at the cavity output.

In addition to the constrained lifetime of atoms within the cavity, resulting in pulsed emission, and losses due to atom heating caused by continuous repumping, another challenge affecting the performance of the optical lattice superradiant laser is the interference from first-order Zeeman and vector light shifts [101]. In contrast to passive clocks, where the Zeeman effect induced by a bias magnetic field can be suppressed by averaging Zeeman transitions with opposite shifts, alternating from one interrogation cycle to the next, active clocks may operate on two Zeeman transitions simultaneously. However, this potential issue can be effectively mitigated by configuring the magnetic field in a manner that keeps the splitting between these two transitions significantly smaller than the cavity linewidth, allowing the system to remain within the bad-cavity regime. Furthermore, the splitting should be considerably larger than the pumping rate, ensuring that there is no reduction in the achievable stability, as discussed in [104]. On the other hand, the loss of atoms caused by continuous repumping can be tackled by shelving atoms into a long-lived state.

Finally, unlike passive clocks, active clocks can sustain their accuracy even when the frequency of the cavity resonance marginally strays from the atomic transition. This characteristic makes active clocks potential candidates for applications outside controlled laboratory environments, where it is challenging to suppress environmental noise. This perspective would require a thorough examination of systematic effects.

4. Conclusion

In this paper we presented a comprehensive examination of the development and potential of superradiant active optical atomic clocks. These timekeepers represent a significant leap forward from the current passive optical atomic clocks. The core advantage of active clocks lies in their inherent robustness against cavity instabilities, as their emitted frequency is defined by atomic transitions rather than being reliant on external factors.

We have explored various strategies and advancements in the field, such as optical lattice superradiant lasers and atom beam lasers. These approaches have shown promising results in achieving continuous operation and overcoming the limitations of passive clocks. Notably, the ability of active clocks to remain stable even when the frequency of the cavity resonance deviates slightly from the atomic transition marks them as more adaptable to environments outside controlled laboratory settings. Furthermore, their superior stability makes them well-suited for advanced applications in space-based navigation and gravitational wave detection.

Despite the progress, several challenges remain. The difficulty of achieving continuous lasing or managing atom heating due to continuous repumping are among the primary concerns. However, the ongoing theoretical and experimental efforts in this domain indicate a promising future for these clocks. The potential of active optical atomic clocks to achieve fractional frequency instabilities in the 10^{-18} range at one second is particularly noteworthy.

In conclusion, the advancement of active optical atomic clocks holds immense promise for the fields of metrology, geodesy, and fundamental physics. Their development not only represents a technological breakthrough but also opens the door to new scientific discoveries and applications, ranging from probing fundamental constants to detecting subtle cosmic phenomena.

Acknowledgements

The authors would like to thank Clément Lacroûte, Emmanuel Klinger and Vincent Giordano for their fruitful comments and careful reading of the manuscript.

References

[1] E. Oelker, R. B. Hutson, C. J. Kennedy, L. Sonderhouse, T. Bothwell, A. Goban, D. Kedar, C. Sanner, J. M. Robinson, G. E. Marti, D. G. Matei, T. Legero, M. Giunta, R. Holzwarth, F. Riehle, U. Sterr, and J. Ye. Demonstration of 4.8×10^{-17} stability at 1 s for two independent optical clocks. *Nature Photonics*, 13(10):714–719, July 2019.

[2] W. F. McGrew, X. Zhang, R. J. Fasano, S. A. Schäffer, K. Beloy, D. Nicolodi, R. C. Brown, N. Hinkley, G. Milani, M. Schioppo, T. H. Yoon, and A. D. Ludlow. Atomic clock performance enabling geodesy below the centimetre level. *Nature*, 564(7734):87–90, November 2018.

[3] T. L. Nicholson, S. L. Campbell, R. B. Hutson, G. E. Marti, B. J. Bloom, R. L. McNally, W. Zhang, M. D. Barrett, M. S. Safronova, G. F. Strouse, W. L. Tew, and J. Ye. Systematic evaluation of an atomic clock at 2×10^{-18} total uncertainty. *Nature Communications*, 6(1), April 2015.

[4] S. M. Brewer, J. S. Chen, A. M. Hankin, E. R. Clements, C. W. Chou, D. J. Wineland, D. B. Hume, and D. R. Leibrandt. $^{27}Al^{+}$ quantum-logic clock with a systematic uncertainty below 10^{-18}. *Phys. Rev. Lett.*, 123:033201, July 2019.

[5] V. A. Dzuba and V. V. Flambaum. Atomic optical clocks and search for variation of the fine-structure constant. *Phys. Rev. A*, 61:034502, February 2000.

[6] R. M. Godun, P. B. R. Nisbet-Jones, J. M. Jones, S. A. King, L. A. M. Johnson, H. S. Margolis, K. Szymaniec, S. N. Lea, K. Bongs, and P. Gill. Frequency ratio of two optical clock transitions in $^{171}Yb^+$ and constraints on the time variation of fundamental constants. *Phys. Rev. Lett.*, 113:210801, November 2014.

[7] N. Huntemann, B. Lipphardt, C. Tamm, V. Gerginov, S. Weyers, and E. Peik. Improved limit on a temporal variation of m_p/m_e from comparisons of Yb^+ and Cs atomic clocks. *Phys. Rev. Lett.*, 113(21):210802, November 2014.

[8] M. S. Safronova. The search for variation of fundamental constants with clocks. *Annalen der Physik*, 531(5), January 2019.

[9] M. Gurov, J. J. McFerran, B. Nagórny, R. Tyumenev, Z. Xu, Y. Le Coq, R. Le Targat, P. Lemonde, J. Lodewyck, and S. Bize. Optical lattice clocks as candidates for a possible redefinition of the SI second. *IEEE Transactions on Instrumentation and Measurement*, 62(6):1568–1573, 2013.

[10] F. Riehle. Towards a redefinition of the second based on optical atomic clocks. *Comptes Rendus Physique*, 16(5):506–515, 2015. The measurement of time / La mesure du temps.

[11] F. Bregolin, G. Milani, M. Pizzocaro, B. Rauf, P. Thoumany, F. Levi, and D. Calonico. Optical lattice clocks towards the redefinition of the second. *Journal of Physics: Conference Series*, 841(1):012015, May 2017.

[12] J. Lodewyck. On a definition of the SI second with a set of optical clock transitions. *Metrologia*, 56(5):055009, September 2019.

[13] W. F. McGrew, X. Zhang, H. Leopardi, R. J. Fasano, D. Nicolodi, K. Beloy, J. Yao, J. A. Sherman, S. A. Schäffer, J. Savory, R. C. Brown, S. Römisch, C. W. Oates, T. E. Parker, T. M. Fortier, and A. D. Ludlow. Towards the optical second: verifying optical clocks at the SI limit. *Optica*, 6(4):448–454, Apr 2019.

[14] C. W. Chou, D. B. Hume, T. Rosenband, and D. J. Wineland. Optical clocks and relativity. *Science*, 329(5999):1630–1633, 2010.

[15] R. Bondarescu, A. Schärer, A. Lundgren, G. Hetényi, N. Houlié, P. Jetzer, and M. Bondarescu. Ground-based optical atomic clocks as a tool to monitor vertical surface motion. *Geophysical Journal International*, 202(3):1770–1774, July 2015.

[16] J. Flury. Relativistic geodesy. *Journal of Physics: Conference Series*, 723(1):012051, June 2016.

[17] G. Lion, I. Panet, P. Wolf, C. Guerlin, S. Bize, and P. Delva. Determination of a high spatial resolution geopotential model using atomic clock comparisons. *Journal of Geodesy*, 91(6):597–611, January 2017.

[18] T. Mehlstäubler, G. Grosche, C. Lisdat, P. Schmidt, and H. Denker. Atomic clocks for geodesy. *Reports on Progress in Physics*, 81(6):064401, April 2018.

[19] J. Grotti, S. Koller, S. Vogt, S. Häfner, U. Sterr, C. Lisdat, H. Denker, C. Voigt, L. Timmen, A. Rolland, F. N. Baynes, H. S. Margolis, M. Zampaolo, P. Thoumany, M. Pizzocaro, B. Rauf, F. Bregolin, A. Tampellini, P. Barbieri, M. Zucco, G. A. Costanzo, C. Clivati, F. Levi, and D. Calonico. Geodesy and metrology with a transportable optical clock. *Nature Physics*, 14(5):437–441, February 2018.

[20] Y. Tanaka and H. Katori. Exploring potential applications of optical lattice clocks in a plate subduction zone. *Journal of Geodesy*, 95(8), July 2021.

[21] Suelynn Choy, Sunil Bisnath, and Chris Rizos. Uncovering common misconceptions in GNSS Precise Point Positioning and its future prospect. *GPS Solutions*, 21:13–22, 2017.

[22] T. A. Ely, E. A. Burt, J. D. Prestage, J. M. Seubert, and R. L. Tjoelker. Using the Deep Space Atomic Clock for navigation and science. *IEEE Transactions on Ultrasonics, Ferroelectrics, and Frequency Control*, 65(6):950–961, 2018.

[23] Vladimir Schkolnik, Dmitry Budker, Oliver Fartmann, Victor Flambaum, Leo Hollberg, Tigran Kalaydzhyan, Shimon Kolkowitz, Markus Krutzik, Andrew Ludlow, Nathan Newbury, Christoph Pyrlik, Laura Sinclair, Yevgeny Stadnik, Ingmari Tietje, Jun Ye, and Jason Williams. Optical atomic clock aboard an earth-orbiting space station (oacess): enhancing searches for physics beyond the standard model in space. *Quantum Science and Technology*, 8(1):014003, nov 2022.

[24] G. Barontini, L. Blackburn, V. Boyer, F. Butuc-Mayer, X. Calmet, J. R. Crespo López-Urrutia, E. A. Curtis, B. Darquié, J. Dunningham, N. J. Fitch, E. M. Forgan, K. Georgiou, P. Gill, R. M. Godun, J. Goldwin, V. Guarrera, A. C. Harwood, I. R. Hill, R. J. Hendricks, M. Jeong, M. Y. H. Johnson, M. Keller, L. P. Kozhiparambil Sajith, F. Kuipers, H. S. Margolis, C. Mayo, P. Newman, A. O. Parsons, L. Prokhorov, B. I. Robertson, J. Rodewald, M. S. Safronova, B. E. Sauer, M. Schioppo, N. Sherrill, Y. V. Stadnik, K. Szymaniec, M. R. Tarbutt, R. C. Thompson, A. Tofful, J. Tunesi, A. Vecchio, Y. Wang, and S. Worm. Measuring the stability of fundamental constants with a network of clocks. *EPJ Quantum Technology*, 9(1), May 2022.

[25] A. Derevianko. Atomic clocks and dark-matter signatures. *Journal of Physics: Conference Series*, 723:012043, June 2016.

[26] P. Wcisło, P. Morzyński, M. Bober, A. Cygan, D. Lisak, R. Ciuryło, and M. Zawada. Experimental constraint on dark matter detection with optical atomic clocks. *Nature Astronomy*, 1(1), December 2016.

[27] B. M. Roberts, P. Delva, A. Al-Masoudi, A. Amy-Klein, C. Bærentsen, C. F. A. Baynham, E. Benkler, S. Bilicki, S. Bize, W. Bowden, J. Calvert, V. Cambier, E. Cantin, E. A Curtis, S. Dörscher, M. Favier, F. Frank, P. Gill, R. M. Godun, G. Grosche, C. Guo, A. Hees, I. R. Hill, R. Hobson, N. Huntemann, J. Kronjäger, S. Koke, A. Kuhl, R. Lange, T. Legero, B. Lipphardt, C. Lisdat, J. Lodewyck, O. Lopez, H. S. Margolis, H. Álvarez Martínez, F. Meynadier, F. Ozimek, E. Peik, P. E. Pottie, N. Quintin, C. Sanner, L. De Sarlo, M. Schioppo, R. Schwarz, A. Silva, U. Sterr, C. Tamm, R. Le Targat, P. Tuckey, G. Vallet, T. Waterholter, D. Xu, and P. Wolf. Search for transient variations of the fine structure constant and dark matter using fiber-linked optical atomic clocks. *New Journal of Physics*, 22(9):093010, September 2020.

[28] S. Kolkowitz, I. Pikovski, N. Langellier, M. D. Lukin, R. L. Walsworth, and J. Ye. Gravitational wave detection with optical lattice atomic clocks. *Physical Review D*, 94(12), December 2016.

[29] A. Loeb and D. Maoz. Using atomic clocks to detect gravitational waves, 2015.

[30] A. D. Ludlow, M. M. Boyd, J. Ye, E. Peik, and P. O. Schmidt. Optical atomic clocks. *Rev. Mod. Phys.*, 87:637–701, June 2015.

[31] J. Zhang, K. Deng, J. Luo, and Z. Lu. Direct laser cooling Al^+ ion optical clocks. *Chinese Physics Letters*, 34(5):050601, may 2017.

[32] M. Delehaye and C. Lacroute. Single-ion, transportable optical atomic clocks. *Journal of Modern Optics*, 65:622–639, 03 2018.

[33] S. A. King, L. J. Spieß, P. Micke, A. Wilzewski, T. Leopold, E. Benkler, R. Lange, N. Huntemann, A. Surzhykov, V. A. Yerokhin, C. López-Urrutia, R. José, and P. O. Schmidt. An optical atomic clock based on a highly charged ion. *Nature*, 611(7934):43–47, November 2022.

[34] M. Takamoto, F. Hong, R. Higashi, and H. Katori. An optical lattice clock. *Nature*, 435:321–4, June 2005.

[35] G. K. Campbell and W. D. Phillips. Ultracold atoms and precise time standards. *Philosophical Transactions of the Royal Society A: Mathematical, Physical and Engineering Sciences*, 369(1953):4078–4089, 2011.

[36] N. Poli, C. W. Oates, P. Gill, and G. M. Tino. Optical atomic clocks. *La Rivista del Nuovo Cimento*, 36(12):555–624, December 2013.

[37] B. Bloom, T. Nicholson, J. Williams, S. Campbell, M. Bishof, X. Zhang, W. Zhang, S. Bromley, and J. Ye. An optical lattice clock with accuracy and stability at the 10^{-18} level. *Nature*, 506, January 2014.

[38] S. Falke, N. Lemke, C. Grebing, B. Lipphardt, S. Weyers, V. Gerginov, N. Huntemann, C. Hagemann, A. Al-Masoudi, S. Häfner, S. Vogt, U. Sterr, and C. Lisdat. A strontium lattice clock with 3×10^{-17} inaccuracy and its frequency. *New Journal of Physics*, 16(7):073023, July 2014.

[39] I. Ushijima, M. Takamoto, M. Das, T. Ohkubo, and H. Katori. Cryogenic optical lattice clocks, 2015.

[40] X. Lu, M. Yin, T. Li, Y. Wang, and H. Chang. An evaluation of the Zeeman shift of the ^{87}Sr optical lattice clock at the National Time Service Center. *Applied Sciences*, 10(4):1440, February 2020.

[41] Zhi-Ming Tang, Yuan-Fei Wei, B. K. Sahoo, Cheng-Bin Li, Yang Yang, Yaming Zou, and Xue-Ren Huang. Suppression of black-body radiation induced Zeeman shifts in the optical clocks due to the fine-structure intramanifold resonances, 2023.

[42] J. C. Bergquist, W. M. Itano, and D. J. Wineland. Recoilless optical absorption and Doppler sidebands of a single trapped ion. *Phys. Rev. A*, 36:428–430, July 1987.

[43] F. Diedrich, J. C. Bergquist, W. M. Itano, and D. J. Wineland. Laser cooling to the zero-point energy of motion. *Phys. Rev. Lett.*, 62:403–406, January 1989.

[44] D. J. Berkeland, J. D. Miller, J. C. Bergquist, W. M. Itano, and D. J. Wineland. Minimization of ion micromotion in a Paul trap. *Journal of Applied Physics*, 83(10):5025–5033, May 1998.

[45] C. Degenhardt, T. Nazarova, C. Lisdat, H. Stoehr, U. Sterr, and F. Riehle. Influence of chirped excitation pulses in an optical clock with ultracold calcium atoms. *IEEE Transactions on Instrumentation and Measurement*, 54(2):771–775, 2005.

[46] J. Keller, T. Burgermeister, D. Kalincev, J. Kiethe, and T. E. Mehlstäubler. Evaluation of trap-induced systematic frequency shifts for a multi-ion optical clock at the 10^{-19} level. *Journal of Physics: Conference Series*, 723:012027, June 2016.

[47] X. Baillard, M. Fouché, R. Le Targat, P. G. Westergaard, A. Lecallier, Y. Le Coq, G. D. Rovera, S. Bize, and P. Lemonde. Accuracy evaluation of an optical lattice clock with bosonic atoms. *Opt. Lett.*, 32(13):1812–1814, Jul 2007.

[48] A. D. Ludlow, T. Zelevinsky, G. K. Campbell, S. Blatt, M. M. Boyd, M. H. G. de Miranda, M. J. Martin, J. W. Thomsen, S. M. Foreman, J. Ye, T. M. Fortier, J. E. Stalnaker, S. A. Diddams, Y. Le Coq, Z. W. Barber, N. Poli, N. D. Lemke, K. M. Beck, and C. W. Oates. Sr lattice clock at 1×10^{-16} fractional uncertainty by remote optical evaluation with a Ca clock. *Science*, 319(5871):1805–1808, 2008.

[49] N. D. Lemke, A. D. Ludlow, Z. W. Barber, T. M. Fortier, S. A. Diddams, Y. Jiang, S. R. Jefferts, T. P. Heavner, T. E. Parker, and C. W. Oates. Spin-1/2 optical lattice clock. *Phys. Rev. Lett.*, 103:063001, August 2009.

[50] S. Falke, H. Schnatz, J. S. R. Vellore Winfred, T. Middelmann, S. Vogt, S. Weyers, B. Lipphardt, G. Grosche, F. Riehle, U. Sterr, and C. Lisdat. The ^{87}Sr optical frequency standard at PTB. *Metrologia*, 48(5):399, September 2011.

[51] N. Huntemann, C. Sanner, B. Lipphardt, Chr. Tamm, and E. Peik. Single-ion atomic clock with 3×10^{-18} systematic uncertainty. *Phys. Rev. Lett.*, 116:063001, Feb 2016.

[52] Yao Huang, Baolin Zhang, Mengyan Zeng, Yanmei Hao, Zixiao Ma, Huaqing Zhang, Hua Guan, Zheng Chen, Miao Wang, and Kelin Gao. Liquid-nitrogen-cooled Ca^+ optical clock with systematic uncertainty of 3×10^{-18}. *Phys. Rev. Appl.*, 17:034041, Mar 2022.

[53] T. Bothwell, D. Kedar, E. Oelker, J. M. Robinson, S. L Bromley, W. L. Tew, J. Ye, and C. J. Kennedy. JILA SrI optical lattice clock with uncertainty of 2.0×10^{-18}. *Metrologia*, 56(6):065004, October 2019.

[54] G. J. Dick, J. D. Prestage, C. A. Greenhall, and L. Maleki. Local oscillator induced degradation of medium-term stability in passive atomic frequency standards. In *22nd Annual Precise Time and Time Interval (PTTI) Applications and Planning Meeting*, pages 487–508, 1990.

[55] G. Santarelli, C. Audoin, A. Makdissi, P. Laurent, G.J. Dick, and A. Clairon. Frequency stability degradation of an oscillator slaved to a periodically interrogated atomic resonator. *IEEE Transactions on Ultrasonics, Ferroelectrics, and Frequency Control*, 45(4):887–894, 1998.

[56] A. Quessada, R. P. Kovacich, I. Courtillot, A. Clairon, G. Santarelli, and P. Lemonde. The Dick effect for an optical frequency standard. *Journal of Optics B: Quantum and Semiclassical Optics*, 5(2):S150, April 2003.

[57] J. M. Robinson, M. Miklos, Yee Ming Tso, C. J. Kennedy, T. Bothwell, D. Kedar, J. K. Thompson, and J. Ye. Direct comparison of two spin squeezed optical clocks below the quantum projection noise limit, 2022.

[58] M. Schioppo, R. C. Brown, W. F. McGrew, N. Hinkley, R. J. Fasano, K. Beloy, T. H. Yoon, G. Milani, D. Nicolod, J. A. Sherman, N. B. Phillips, C. W. Oates, and A. D. Ludlow. Ultrastable optical clock with two cold-atom ensembles. *Nature Photonics*, 11(1):48–52, January 2017.

[59] T. Takano, M. Takamoto, I. Ushijima, N. Ohmae, T. Akatsuka, A. Yamaguchi, Y. Kuroishi, H. Munekane, B. Miyahara, and H. Katori. Geopotential measurements with synchronously linked optical lattice clocks. *Nature Photonics*, 10(10):662–666, October 2016.

[60] D. G. Matei, T. Legero, S. Häfner, C. Grebing, R. Weyrich, W. Zhang, L. Sonderhouse, J. M. Robinson, J. Ye, F. Riehle, and U. Sterr. 1.5 μm lasers with sub 10 mHz linewidth. *Physical Review Letters*, 118(26), June 2017.

[61] K. Numata, A. Kemery, and J. Camp. Thermal-noise limit in the frequency stabilization of lasers with rigid cavities. *Phys. Rev. Lett.*, 93:250602, December 2004.

[62] S. Häfner, S. Falke, C. Grebing, S. Vogt, T. Legero, M. Merimaa, C. Lisdat, and U. Sterr. 8×10^{-17} fractional laser frequency instability with a long room-temperature cavity. *Opt. Lett.*, 40(9):2112–2115, May 2015.

[63] G. Cole, W. Zhang, M. Martin, J. Ye, and M. Aspelmeyer. Tenfold reduction of Brownian noise in high-reflectivity optical coatings. *Nature Photonics*, 7:644–650, July 2013.

[64] T. Kessler, C. Hagemann, C. Grebing, T. Legero, U. Sterr, F. Riehle, M. Martin, L. Chen, and J. Ye. A sub-40 mHz laser based on a silicon single-crystal optical cavity. *Nature Photonics*, 6, December 2011.

[65] W. Zhang, J. M. Robinson, L. Sonderhouse, E. Oelker, C. Benko, J. L. Hall, T. Legero, D. G. Matei, F. Riehle, U. Sterr, and J. Ye. Ultrastable silicon cavity in a continuously operating closed-cycle cryostat at 4 K. *Phys. Rev. Lett.*, 119:243601, December 2017.

[66] J. M. Robinson, E. Oelker, W. R. Milner, W. Zhang, T. Legero, D. G. Matei, F. Riehle, U. Sterr, and J. Ye. Crystalline optical cavity at 4 K with thermal-noise-limited instability and ultralow drift. *Optica*, 6(2):240–243, Feb 2019.

[67] S. Cook, T. Rosenband, and D. R. Leibrandt. Laser-frequency stabilization based on steady-state spectral-hole burning in Eu^{3+} : Y_2SiO_5. *Phys. Rev. Lett.*, 114:253902, June 2015.

[68] M. J. Thorpe, L. Rippe, T. M. Fortier, M. S. Kirchner, and T. Rosenband. Frequency stabilization to 6×10^{-16} via spectral-hole burning. *Nature Photonics*, 5(11):688–693, September 2011.

[69] J. Olson, R. W. Fox, T. M. Fortier, T. F. Sheerin, R. C. Brown, H. Leopardi, R. E. Stoner, C. W. Oates, and A. D. Ludlow. Ramsey-Bordé matter-wave interferometry for laser frequency stabilization at 10^{-16} frequency instability and below. *Physical Review Letters*, 123(7), August 2019.

[70] Zachary L. Newman, Vincent Maurice, Tara Drake, Jordan R. Stone, Travis C. Briles, Daryl T. Spencer, Connor Fredrick, Qing Li, Daron Westly, B. R. Ilic, Boqiang Shen, Myoung-Gyun Suh, Ki Youl Yang, Cort Johnson, David M. S. Johnson, Leo Hollberg, Kerry J. Vahala, Kartik Srinivasan, Scott A. Diddams,

John Kitching, Scott B. Papp, and Matthew T. Hummon. Architecture for the photonic integration of an optical atomic clock. *Optica*, 6(5):680–685, May 2019.

[71] Zachary L. Newman, Vincent Maurice, Connor Fredrick, Tara Fortier, Holly Leopardi, Leo Hollberg, Scott A. Diddams, John Kitching, and Matthew T. Hummon. High-performance, compact optical standard. *Opt. Lett.*, 46(18):4702–4705, Sep 2021.

[72] Anthony Gusching, Jacques Millo, Ivan Ryger, Remy Vicarini, Moustafa Abdel Hafiz, Nicolas Passilly, and Rodolphe Boudot. Cs microcell optical reference with frequency stability in the low 10^{-13} range at 1 s. *Opt. Lett.*, 48(6):1526 – 1529, Mar 2023.

[73] R. H. Dicke. Coherence in spontaneous radiation processes. *Phys. Rev.*, 93:99–110, January 1954.

[74] M. Gross and S. Haroche. Superradiance: an essay on the theory of collective spontaneous emission. *Physics Reports*, 93(5):301–391, 1982.

[75] M. Scheibner, T. Schmidt, L. Worschec, A. Forchel, G. Bacher, T. Passow, and D. Hommel. Superradiance of quantum dots. *Nature Physics*, 3(2):106–110, February 2007.

[76] P. Tighineanu, R. S. Daveau, T. B. Lehmann, H. E. Beere, D. A. Ritchie, P. Lodahl, and S. Stobbe. Single-photon superradiance from a quantum dot. *Phys. Rev. Lett.*, 116:163604, April 2016.

[77] S. Doria, T. S. Sinclair, N. D. Klein, D. I. G. Bennett, C. Chuang, F. S. Freyria, C. P. Steiner, P. Foggi, K. A. Nelson, J. Cao, A. Aspuru-Guzik, S. Lloyd, J. R. Caram, and M. G. Bawendi. Photochemical control of exciton superradiance in light-harvesting nanotubes. *ACS Nano*, 12(5):4556–4564, 2018. PMID: 29701947.

[78] Fereshteh Rajabi and Martin Houde. Dicke's superradiance in astrophysics. I. the 21 cm line. *The Astrophysical Journal*, 826(2):216, aug 2016.

[79] D. Blas and S. J. Witte. Quenching mechanisms of photon superradiance. *Phys. Rev. D*, 102:123018, December 2020.

[80] M. Baryakhtar, M. Galanis, R. Lasenby, and O. Simon. Black hole superradiance of self-interacting scalar fields. *Phys. Rev. D*, 103:095019, May 2021.

[81] N. Skribanowitz, I. P. Herman, J. C. MacGillivray, and M. S. Feld. Observation of Dicke superradiance in optically pumped HF gas. *Phys. Rev. Lett.*, 30:309–312, February 1973.

[82] A. Kumarakrishnan and X. L. Han. Superfluorescence from optically trapped calcium atoms. *Phys. Rev. A*, 58:4153–4162, November 1998.

[83] F. Haake, M. I. Kolobov, C. Fabre, E. Giacobino, and S. Reynaud. Superradiant laser. *Phys. Rev. Lett.*, 71:995–998, August 1993.

[84] J. Chen. Active optical clock. *Chinese Science Bulletin*, 54(348 - 352), 2009.

[85] D. Meiser, Jun Ye, D. R. Carlson, and M. J. Holland. Prospects for a millihertz-linewidth laser. *Phys. Rev. Lett.*, 102:163601, April 2009.

[86] H. Katori, M. Takamoto, V. G. Pal'chikov, and V. D. Ovsiannikov. Ultrastable optical clock with neutral atoms in an engineered light shift trap. *Phys. Rev. Lett.*, 91:173005, October 2003.

[87] J. G. Bohnet, Z. Chen, J. M. Weiner, D. Meiser, M. J. Holland, and J. K. Thompson. A steady-state superradiant laser with less than one intracavity photon. *Nature*, 484(7392):78–81, April 2012.

[88] J. G. Bohnet, Z. Chen, J. M. Weiner, K. C. Cox, and J. K. Thompson. Relaxation oscillations, stability, and cavity feedback in a superradiant Raman laser. *Phys. Rev. Lett.*, 109:253602, December 2012.

[89] T. G. Zhang, Y. F. Wang, X. R. Zang, W. Zhuang, and J. B. Chen. Active optical clock based on four-level quantum system. *Chinese Science Bulletin*, 58(17):2033–2038, June 2013.

[90] M. A. Norcia and J. K. Thompson. Cold-strontium laser in the superradiant crossover regime. *Phys. Rev. X*, 6:011025, March 2016.

[91] S. L. Kristensen, E. Bohr, J. Robinson-Tait, T. Zelevinsky, J. W. Thomsen, and J. H. Müller. Subnatural linewidth superradiant lasing with cold ^{88}Sr atoms. *Phys. Rev. Lett.*, 130:223402, May 2023.

[92] T. Laske, H. Winter, and A. Hemmerich. Pulse delay time statistics in a superradiant laser with calcium atoms. *Phys. Rev. Lett.*, 123(103601), 2019.

[93] M. A. Norcia, M. N. Winchester, J. R. K. Cline, and J. K. Thompson. Superradiance on the millihertz linewidth strontium clock transition. *Science Advances*, 2(10):e1601231, 2016.

[94] M. A. Norcia, J. R. K. Cline, J. A. Muniz, J. M. Robinson, R. B. Hutson, A. Goban, G. E. Marti, J. Ye, and J. K. Thompson. Frequency measurements of superradiance from the strontium clock transition. *Phys. Rev. X*, 8:021036, May 2018.

[95] K. Henschel, J. Majer, J. Schmiedmayer, and H. Ritsch. Cavity QED with an ultracold ensemble on a chip: Prospects for strong magnetic coupling at finite temperatures. *Phys. Rev. A*, 82:033810, September 2010.

[96] D. Yu and J. Chen. Laser theory with finite atom-field interacting time. *Phys. Rev. A*, 78:013846, July 2008.

[97] B. Laburthe-Tolra, Z. Amodjee, B. Pasquiou, and M. Robert-de Saint-Vincent. Correlations and linewidth of the atomic beam continuous superradiant laser. *SciPost Phys. Core*, 6:015, 2023.

[98] H. Liu, S. B. Jäger, X. Yu, S. Touzard, A. Shankar, M. J. Holland, and T. L. Nicholson. Rugged mHz-linewidth superradiant laser driven by a hot atomic beam. *Phys. Rev. Lett.*, 125:253602, December 2020.

[99] Chun-Chia Chen, S. Bennetts, R. G. Escudero, B. Pasquiou, and F. Schreck. Continuous guided strontium beam with high phase-space density. *Phys. Rev. Appl.*, 12:044014, October 2019.

[100] Chun-Chia Chen, R. González Escudero, J. Minář, B. Pasquiou, S. Bennetts, and F. Schreck. Continuous Bose–Einstein condensation. *Nature*, 606(7915):683–687, June 2022.

[101] G. A. Kazakov and T. Schumm. Active optical frequency standards using cold atoms: perspectives and challenges, 2015.

[102] G. A. Kazakov and T. Schumm. Active optical frequency standard using sequential coupling of atomic ensembles. *Physical Review A*, 87(1), January 2013.

[103] T. Maier, S. Kraemer, L. Ostermann, and H. Ritsch. A superradiant clock laser on a magic wavelength optical lattice. *Opt. Express*, 22(11):13269–13279, June 2014.

[104] G. A. Kazakov, S. Dubey, A. Bychek, U. Sterr, M. Bober, and M. Zawada. Ultimate stability of active optical frequency standards. *Phys. Rev. A*, 106:053114, November 2022.

Journal of Physics: Conference Series **2889** (2024) 012046 doi:10.1088/1742-6596/2889/1/012046

Development of commercial fountain clocks in NIM

Kun Liu[1,2], Fasong Zheng[1,2], Weiliang Chen[1,2], Xiao Yacong[1], Shaoyang Dai[1,2], Fang Fang[1,2]*

[1]Time and Frequency Division, National Institute of Metrology, Beijing, 100029, China

[2]Key Laboratory of Time Frequency and Gravity Primary Standards, State Administration for Market Regulation

*E-mail: fangf@nim.ac.cn

Abstract. A new Rb fountain clock has been built in NIM aiming to operate semi-continuously and achieve an excellent long-term instability for time keeping. A double metal microwave cavity with a thermal expansion self-compensating mechanism is used for Ramsey interrogation to reduce the sensitivity to ambient temperature fluctuations and increase the operating temperature range. With an optimized design, the thermal coefficient of the resonance frequency has been reduced to less than 10 kHz/°C. This represents an improvement of more than an order of magnitude compared to a copper cavity. The optical system with two independent frequency stabilized lasers with automatic re-locking system is located on a 400×600 mm optical breadboard with special designed optical mounts to ensure stable output light powers, which ensure to resume operating without any realignment of optical system after transportation. The clock has been operated for more than a year without any failure. A long-term instability of 3.7×10^{-16} was achieved compared with NIM5 Cs fountain clock. By implementing feedback of the frequency difference to the Voltage Controlled Crystal Oscillator within a servo-loop, the fountain clock can generate a standard frequency output, the instability of which can enter 10^{-16} order at 40000 seconds.

1. Introduction

Since their development in 1995 [1], most fountain clocks used cesium-133(^{133}Cs) atoms as quantum reference to reproduce the definition of the second [2-17]. In order to achieve a high signal-to-noise ratio (SNR) for a higher stability, it is highly desirable for a larger number of atoms. However, a greater number of atoms results in a larger collisional frequency shift [18,

19]. Therefore, it becomes necessary to trade-off between these two factors in order to optimize the clock's performance during operation. The collisional coefficient of rubidium-87(^{87}Rb) is about one or two orders lower than that of ^{133}Cs. Moreover, the lasers used for trapping and cooling ^{87}Rb atoms exhibit higher reliability compared to those used for ^{133}Cs, ensure a robust fountain clock operation. Thus, Rb fountain clocks have gained significant interests and are used as a secondary frequency standard to calibrate TAI as well as time-keeping clocks to contribute to EAL directly [20-24], and also they are commercially available.

For a commercial clock, it is crucial to reduce its sensitive to the operating environment, including ambient temperature. For a fountain clock, both Ramsey cavity and optical system are sensitive to temperature variations. A Ramsey cavity made of oxygen-free copper (OFC) is particularly susceptible to change of ambient temperature, with a typical thermal coefficient of resonance frequency reaching up to 150 kHz/°C [3, 12]. Either a temperature-controlled system is need to stabilize the Ramsey cavity to the required temperature or the lab temperature needs to be controlled to a narrow range. To reduce temperature sensitivity, a double metal Ramsey cavity was engineered to self-compensate its thermal expansion. In addition, this cavity also serves as a vacuum chamber, resulting in a more compact vacuum system. We developed a commercial Rb fountain clock based on this. A compact and robust laser system with its frequency auto-locked is developed, which ensure a continuous laser system operation for more than a year without failure. Moreover, the fountain can be operated under two modes, one is the same as the most of the primary frequency standards which output the relative frequency offsets of the local oscillators to fountain clocks, the other is to generate a standard 100 MHz signal by feeding the frequency errors directly back to the crystal oscillators.

2. Apparatus description

Our commercial Rb fountain clock includes four units: a physical package, a laser-optical system, microwave synthesizers and an electronic control system. The physical package creates a vacuum and electromagnetic environment for atomic interacting with lasers and microwaves. The laser-optical system supplies lasers with specific frequencies, amplitudes, and polarizations to manipulate and detect atoms. The microwave synthesizers generate the signals required for clock transitions, while the electronic control system generate time sequences to control lasers and microwaves as well as data acquisition and processing.

2.1. Physical package

The physical package includes a magneto-optical trap (MOT) chamber, a state selection cavity, a detection chamber, a Ramsey cavity, and a flight tube. Except of the Ramsey cavity and the flight tube, the other parts have a similar design to NIM5 Cs fountain clock [25]. The Ramsey cavity has two ceramic windows hermetically located on the center cylindrical wall to seal the vacuum as while as to be used as microwave feed-through windows, as illustrated in figure 1(b). This design enables the cavity to concurrently act as a microwave resonator and an element of the vacuum enclosure within the physical system. Indium wires are used as a vacuum seal between the Ramsey cavity with the detection chamber and the flight tube. The rigid copper coaxial cables for microwave transmission are entirely outside the vacuum makes it possible to fine tune the cavity resonance frequency. This design is different from the traditional atomic fountain clocks' framework [26], which install a Ramsey cavity within a vacuum. The new physical system is more compact, having a less weight and easier to assemble and tune. Moreover, the microwave feedthroughs are all outside the closed flight tube reduce microwaves leaking into the interrogation zone further.

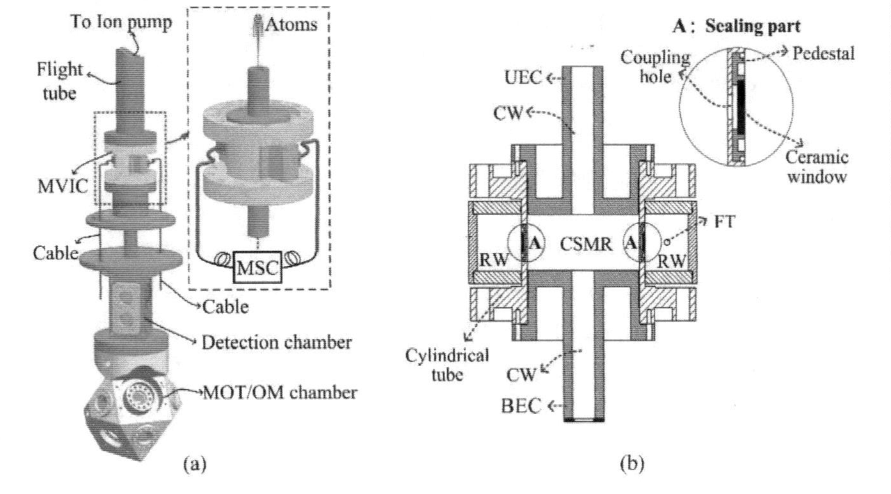

Figure 1 (a)The physical package of the commercial fountain clock; (b) vertical section through the microwave-vacuum integrated Ramsey cavity. A combination pump is on the top of the flight tube which is not shown in the figure ensures a better vacuum during the interrogation. And 3 layers of μ-metal magnetic shields surround the flight tube with a residual magnetic field of less than 1 nT. (MVIC, microwave-vacuum integrated Ramsey cavity; MSC, microwave synthesizer chain; CSMR, cylindrical microwave resonator; RW, rectangular waveguide; UEC, upper end cap; BEC, bottom end cap; FT, feed-through; CW, cut-off waveguide; CSMR, cylindrical microwave resonator)

Figure 2 shows the design of the bimetallic microwave cavity, which features low thermal sensitivity and vacuum integration [27-29]. The top and bottom end caps, as well as the cut-off waveguide, are made of oxygen free copper (OFC). The cavity itself is fabricated from titanium, with its surface coated with copper to increase electrical conductivity. The rectangular microwave feed cavity is made of OFC. Two coupling holes are symmetrically distributed in the middle of the cavity wall with a diameter of 5 mm, sealed with ceramic to ensure vacuum integrity. Indium wire is used to seal between the end caps and the cavity to minimize microwave leakage. The Ramsey cavity is made of a 99.995% pure titanium to ensure a sufficient strength and non-magnetic. With a copper coating inside, a quality factor of this cavity is about 17,000. And temperature coefficient of resonance frequency is 0.73 kHz/°C within a range of 22.48 °C to 37.48 °C, which is about 150 times smaller than that of an oxygen-free copper cavity.

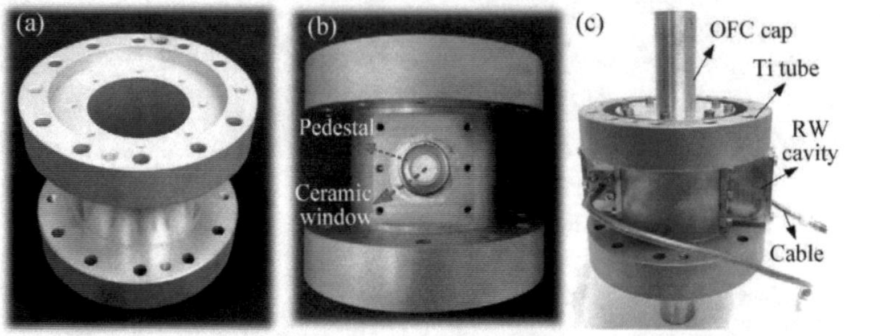

Figure 2. Photos of Ti tube after (a) the coating and (b) sealing processes, and (c) prepared bimetallic microwave cavity body

Other auxiliary components of the physical system, such as the vacuum pump, detection optical setup, geomagnetic field compensation coils, and magnetic shielding system, are consistent with previously reported designs without notable modifications and the detailed descriptions can be found in reference [22].

2.2. Laser-optical system

The design of the laser-optical system for this commercial Rb fountain clock is shown in the figure 3. This system includes a frequency stabilized master laser, a frequency stabilized repump laser, and a miniaturized optical system. The master laser utilizes commercial frequency doubling fiber laser with a built-in frequency stabilization module, providing a laser locked to the ^{87}Rb atoms two transition crossover of $|F = 2\rangle \rightarrow |F' = 1\rangle$ and $|F = 2\rangle \rightarrow |F' = 3\rangle$ with a linewidth of about 250 kHz. The repumping laser employs a DFB laser locked to the two transition crossover of $|F = 1\rangle \rightarrow |F' = 1\rangle$ and $|F = 1\rangle \rightarrow |F' = 2\rangle$. Both frequency stabilized lasers are installed on the same rack cabinet, and the lasers are transported to a 0.6 m × 0.4 m standard optical breadboard via polarization-maintaining fibers. The miniaturized optical system controls the frequency, polarization, and power of the laser to cool and manipulate atoms as well as detect them. The output power of the master laser is 1 W. After passing through the first λ/2 wave-plate/polarizing beam splitter (HP/PBS) combination to adjust laser power and the polarization direction, it is split by the two HP/PBS combinations into three beams for detection, upward and downward cooling lights, respectively. Each beam is double-passed through an AOM to shift the laser frequency while keeping the alignment. The detection beam is coupled directly into polarization-maintaining fibers after AOM and transported to a detection optical cage system mounted on the detection camber; the upward and downward cooling beams after AOM frequency shifting are split into three beams by HP/PZT combination units and coupled into six polarization-maintaining fibers respectively, and transported to the cooling light cylinder tubes mounted on the MOT chamber for upward and downward cooling lights with a power of about 20 mW for each beam. The repumping light power is about 10 mW. The beam is split into 2 beams, most of the light is transported to a fiber for MOT repump light, and a small portion is transmitted to the detection optical cage system for detection repump light.

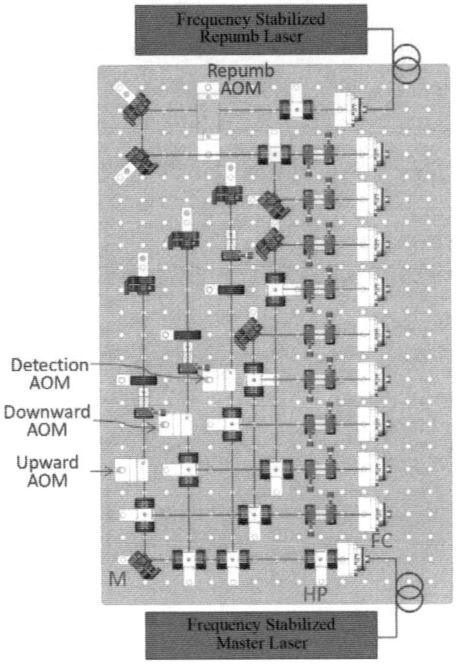

Figure 3. Compact optical system (AOM, acousto-optic modulator; FC, fiber coupler; M, mirror; HP, λ/2 wave-plate and polarization beam splitter unit; M, mirror (only one of each identical device except AOM is labeled in the figure)

To improve the stability of the optical path, the miniaturized optical system has been optimized in layout, beam height, optical path and so on, while ensuring adjustability. The beam height is fixed to 27 mm with standard optical mounts and customized adapters and bases. The upward, downward, and detection AOMs are vertically mounted, ensuring an efficiency optimization while still keeping a compact optical path.

2.3. Microwave system

The microwave system of the Rb fountain clock includes a state-selection microwave synthesizer, a Ramsey interrogation microwave synthesizer, and a microwave phase control box. The state-selection microwave synthesizer consists of a 6.8 GHz dielectric resonator oscillator (DRO) and a 34 MHz direct digital synthesizer (DDS), both referenced directly to a 100 MHz output signal from a hydrogen maser.

For commercial fountain clocks, there are two schemes for the Ramsey interrogation microwave synthesizer: one scheme uses a hydrogen maser as the local oscillator to generate the Rb clock frequency and the maser frequency bias is calculated from the error signal and put out. Most fountain clocks operate under this scheme. The second scheme employs a voltage-controlled crystal oscillator (VCXO) as the local oscillator, the error signal is used to lock the crystal oscillator frequency and generate a standard frequency output.

The first scheme is shown in figure 4(a), the interrogation microwave synthesizer is created by mixing a 6.9 GHz DRO with a 34 MHz DDS to produce a 6.934 GHz microwave signal, which is then mixed with a 100 MHz Mach-Zehnder interferometric switch [30, 31] (in pink block) to produce the 6.834 GHz microwave signal for Ramsey transitions. The switching of the interrogation microwave signal is accomplished with the Mach-Zehnder interferometric switch, while amplitude is controlled by an electronically tuned attenuator on the 34 MHz signal. The microwave phase control box splits the interrogation microwave output into two branches, matches their phases and powers, and supplies them to the physical system, providing phase and amplitude-matched microwave signals to a dual-ended Ramsey cavity. When the fountain clock is running, it utilizes square wave frequency modulation and proportional-integral(PI) locking techniques to determine the frequency deviation between the hydrogen maser and the atomic transition signal, which is then saved in a computer.

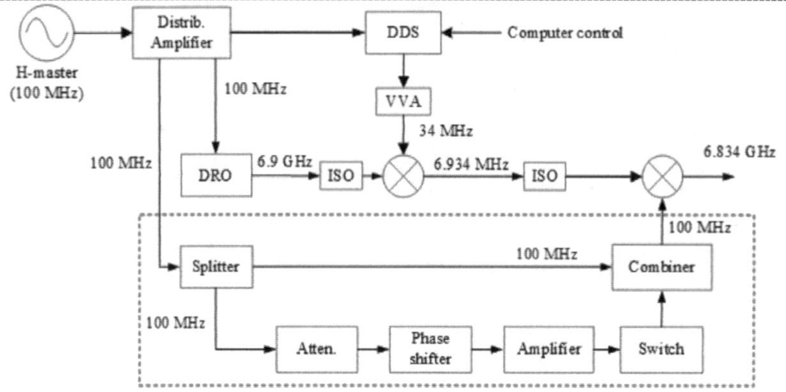

(a)

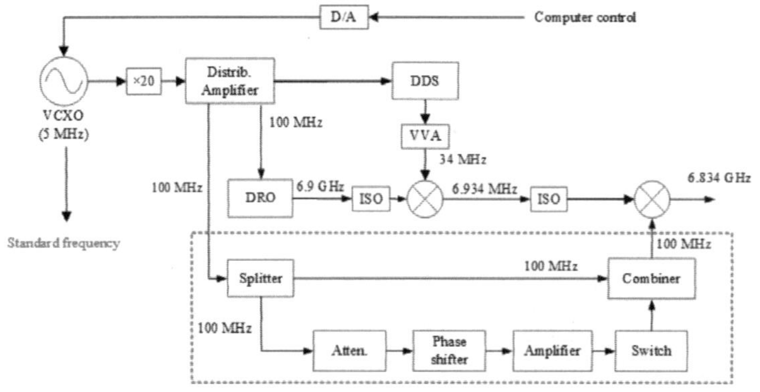

(b)

Figure 4. Two schemes of home made interrogation microwave synthesizer (a) a hydrogen maser is used as the local oscillator and the maser frequency bias is calculated from the error signal (b) a crystal oscillator (VCXO) is used as a local oscillator and its frequency locks to the Rb clock transition and generate standard 5 MHz signal. The control voltage required for VCXO using the PI algorithm, based on the obtained frequency error signal. (H-Maser, hydrogen maser; DRO, dielectric resonator oscillator; ISO, isolater; DDS, direct digital synthesizer; VVA, voltage variable attenuator; VCXO, voltage control crystal oscillator)

The second scheme is shown in figure 4(b), a 100 MHz signal multiplied from a VCXO is used as a local oscillator. In this scheme, the initial task is to identify the control voltage for the VCXO corresponding to the central peak of the Ramsey fringes. The Ramsey fringes is obtained by scanning the control voltage for the VCXO. And then, this voltage is set to the control pins of VCXO (Rakon HSO-14) using a 20-bit digital-to-analog (D/A) converter board (ADI AD5791). After that, the frequency of the DDS is alternated between 34682610.4 Hz and 34682611.4 Hz for each fountain clock cycle. The transition probabilities corresponding to these two DDS frequency values are obtained. The frequency error signal is then derived from the difference between these transition probabilities. Based on the frequency error signal, the control voltage required for VCXO using is derived from the PI algorithm. The new voltage value is then applied to the control pins of VCXO, initiating a next cycle of closed-loop locking. Notably, the PI coefficients influence the short-term stability of the output standard frequency. Optimal coefficients must be ascertained through a series of experiments. When the control voltage variation fluctuates near zero, the standard frequency (5 MHz) can be output from the second output port (first output port is connected to the synthesizer).

2.4. Electronic control system

The operation cycle of the fountain clock is approximately 1.2 seconds, managed by over 20 digital and 8 analog timing sequences. Digital sequences controls such as the operation of optomechanical switches, magnetic field switching, microwave signal feeding and cutting, and triggering frequency changes in AOM. Analog sequences generates periodic analog waveforms to control the amplitude of AOM, and the coil current variations of the MOT magnetic field and compensating magnetic fields. The industrial computer equipped with multifunctional I/O devices like PCIe-6343 from National Instruments is used to run the program. The multifunctional I/O devices include ports for high-speed acquisition of fluorescence signals as atoms fall and up to 32 digital I/O channels with a 10 MHz clock rate. The digital I/O channels can be used for digital sequences outputs but have limited driving capabilities. To drive optomechanical switches, we have added a custom driver circuit board based on AQV212 photo-relays. Additionally, there is an analog output card like PCI-6733, capable of outputting up to 8 channels of arbitrary analog voltage signals, which serve as the analog sequences. To control the current in the MOT field and compensating magnetic field coils, V-I (voltage to current) conversion modules are added after the analog sequences. Employing a LabVIEW coded control program, the computer is capable of producing both digital and analog timing sequences. Additionally, it can capture the analog signals emanating from the fluorescence detectors. This setup facilitates the implementation of closed-loop frequency control for the interrogation microwave synthesizer, ensuring precise adjustments and stable operation during the measurement process.

3. Experiments and results

In the first scheme of interrogation microwave synthesizer, the Rb fountain clock uses a 100 MHz signal from a hydrogen maser as the local oscillator to measure the frequency difference between the Rb fountain clock and the hydrogen maser signal. The Rb fountain clock maintained frequency locked for a year from May 20, 2022, to May 20, 2023 (from MJD 59719

Journal of Physics: Conference Series **2889** (2024) 012046 doi:10.1088/1742-6596/2889/1/012046

to MJD 60084) with an operational rate of 97.5% (lasers are unlocked occasionally). The lone term frequency instability is shown in figure 5.

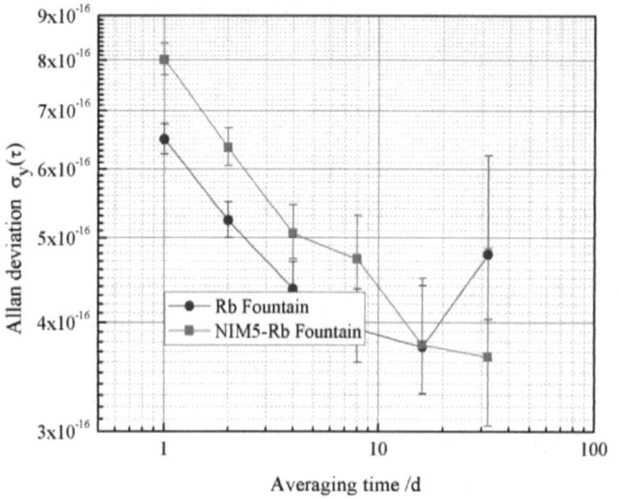

Figure 5. Long-term frequency instability of Rb fountain clocks

The blue dots represent instabilities measured directly against H-maser directly, with a daily instability of 6.5×10^{-16}, a 16 days instability of 3.8×10^{-16}, and a 32 days instability of 7.9×10^{-16}. The instability shifts up after tens days due to the drift of the H-maser. Both Rb fountain clock and the NIM5 fountain clock use the 100 MHz signal from the same hydrogen maser as a reference. By calculating the difference between the frequencies of the two fountain clocks, the drift of the hydrogen maser can be excluded, and the long-term instability is shown in figure 5 red squares, the instability is 8.0×10^{-16} in a day. And the long term frequency drift is eliminated.

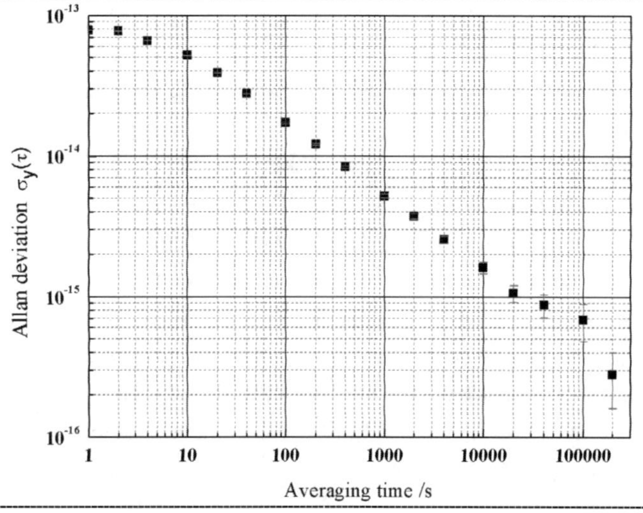

Figure 6. The instability of standard frequency output from commercial fountain clock measured by 5125A with reference of a high-quality hydrogen maser (14 days)

In the second scheme of interrogation microwave synthesizer, the frequency of VCXO is locked by fountain clock. And then, the standard frequency's instability is measured with the reference of high-quality hydrogen maser using 5125A. The results of 14 days' measurement is shown in figure 6. The instability is 2×10^{-15} at 10000 seconds, and it enter 10^{-16} order at 40000 seconds. The daily instability is about 7×10^{-16}.

4. Summary

A commercial Rb fountain clock has been built at NIM with the following notable features. First, a Ramsey cavity, unified with the vacuum chamber and whose RF can be insensitive to the ambient temperature fluctuations, significantly improve fountain clock's tolerance of ambient temperature range. Second, the miniaturized optical system, equipped with two frequency-stabilized lasers with automatic re-locking system, can ensure to resume operating without any realignment of optical path after transportation. Third, there are two schemes for the interrogation microwave synthesizer. One is for measuring relative frequency deviation of the local oscillator (hydrogen maser), the other can output standard frequency. The results indicate that the commercial Rb fountain clock can reach a long term instability of 3.7×10^{-16} in the first interrogation microwave synthesizer scheme, while reach daily instability of 7×10^{-16} in the second interrogation microwave synthesizer scheme.

References

[1] Clairon A, Laurent P, Santarelli G, Ghezali S, Lea S N and Bahoura M 1995 *IEEE Trans. Instrum. Meas.* **44** 128
[2] Wynands R and Weyers S 2005 Metrologia 42 64
[3] Cladé P, Mirandes E D, Cadoret M, Guellati-Khélifa S, Schwob C, Nez F, Julien L and Biraben F 2006 Phys. Rev. Lett. 96 033001
[4] Kumagai M, Ito H, Kajita M and Hosokawa M 2008 Metrologia 45 139
[5] Lipphardt B, Grosche G, Sterr U, Tamm C, Weyers S and Schnatz H 2009 IEEE Trans. Instrum. Meas. 58 1258
[6] Szymaniec K, Park S E, Marra G and Chałupczak W 2010 Metrologia 47 363
[7] Guéna J, Abgrall M, Rovera D, Laurent P, Chupin B, Lours M, Santarelli G, Rosenbusch P, Tobar M E, Li R, Gibble K, Clairon A and Bize S 2012 IEEE Trans. Ultrason. Ferroelectr. Freq. Control 59 391
[8] Heavner T P, Donley E A, Levi F, Costanzo G, Parker T E, Shirley J H, Ashby N, Barlow S and Jefferts S R 2014 Metrologia 51 174
[9] Du Y B, Wei R, Dong R C, Zou F and Wang Y Z 2015 Chin. Phys. B 24 070601
[10] Abgrall M, Guéna J, Lours M, Santarelli G, Tobar M E, Bize S, Grop S, Dubois B, Fluhr Ch and Giordano V 2016 IEEE Trans. Ultrason. Ferroelectr. Freq. Control, 63 1198
[11] Blinov I Y, Boiko A I, Domnin Y S, Kostromin V P, Kupalova O V and Kupalov D S 2017 Meas. Tech. 60 30
[12] Weyers S, Gerginov V, Kazda M, Rahm J, Lipphardt B, Dobrev G and Gibble K, 2018 Metrologia 55 789

[13] Pavlenko K Y, Pavlenko Y K, Belyaev A A, Blinov I Y, Khromov M N, Bize S and Lorini L 2018 Quantum Electron. 48 967

[14] Lipphardt B, Gerginov V and Weyers S 2017 IEEE Trans. Ultrason. Ferroelectr. Freq. Control 64 761

[15] Brewer S M, Chen J S, Hankin A M, Clements E R, Chou C W, Wineland D J, Hume D B and Leibrandt D R 2019 Phys. Rev. Lett. 123 03320

[16] Boudot R, Guerandel S and De Clercq E 2009 IEEE Trans. Instrum. Meas. 58 3659

[17] Beattie S, Jian B, Alcock J, Gertsvolf M, Hendricks R, Szymaniec K and Gibble K 2020 Metrologia 57 1

[18] Fertig C and Gibble K 2000 *Phys. Rev. Lett.* **85** 1622

[19] Sortais Y, Bize S, Nicolas C, Clairon A, Salomon C and Williams C 2000 *Phys. Rev. Lett.* **85** 3117

[20] Peil S, Hanssen J, Swanson T B, Taylor J and Ekstrom C R 2015 *8th Symposium on Frequency Standards and Metrology (Potsdam, GERMANY)*

[21] Cheng H N, et al. 2021 Review of Scientific Instruments **92**

[22] Chen W L, Fang F, Kun L, Zheng F S, Dai S Y, Zuo Y N and Li T C 2022 *Frontiers in Physics* **10**

[23] Kupalov D S, Baryshev V N, Blinov I Y, Boiko A I, Domnin Y S and Ivanchenko E V 2022 *Measurement Techniques* **64** 817

[24] Zhang H, et al. 2022 *IEEE Trans. Instrum. Meas.* **71**

[25] Fang F, et al. 2015 *Metrologia* **52** 454

[26] Fang F, Chen W L, Liu K, Liu N F, Han L, Li T C and Ieee 2019 URSI Asia-Pacific Radio Science Conference (AP-RASC) (New Delhi, INDIA)

[27] Zheng F S, Fang F, Chen W L, Liu K, Dai S Y, Cao S Y and Li T C 2021 Proc. Joint Conf. EFTF IEEE Int. Freq. Control Symp. (EFTF/IFCS) (United states) p 1

[28] Zheng F S, Fang F, Wang X M, Chen W L, Liu K, Dai S Y, Cao S Y, Zuo Y N and Li T C 2022 China Satellite Navigation Conference (CSNC 2022) Proceedings (Beijing, China) p 393

[29] Zheng F S, Fang F, Chen W L, Liu K, Dai S Y, Cao S Y, Zuo Y N and Li T C 2023 *IEEE Trans. Microwave Theory Tech.* **71** 1752

[30] Santarelli G, et al. 2009 IEEE Trans. Ultrason. Ferroelectr. Freq. Control **56** 1319

[31] Zhang Y, Liu K, Fang F, Liu N and Li T 2014 2014 IEEE International Frequency Control Symposium, IFCS 2014, May 19, 2014 - May 22, 2014 (Taipei, Taiwan) p 376

Acknowledgments

This work is supported by the National Basic Research Program of China (Grant No. 2021YFF0603801).

Journal of Physics: Conference Series **2889** (2024) 012047 doi:10.1088/1742-6596/2889/1/012047

Sr Optical Lattice Clock and Precision Optical Frequency Measurement at NIM

T. Yang[1], B. Lu[2], L. Zhu[2], T. Liao[2], Q. Wang[1], Y. Li[1], F. Meng[1], B. Lin[1], Y. Wang[1], F. Fang[1], Y. Lin[1], Z. Fang[1]

[1]Division of Time and Frequency Metrology, National Institute of Metrology, Beijing 100029, People's Republic of China

[2]Department of Precision Instrument, Tsinghua University, Beijing 100084, People's Republic of China

E-mail: `linyige@nim.ac.cn`

Abstract. Two Sr optical lattice clocks are being built at the National Institute of Metrology of China (NIM). Sr1 was built on NIM's Hepingli Campus. It was first evaluated in 2015 and improved later on. In 2021, the systematic shift uncertainty of Sr1 was evaluated to be 2.9×10^{-17}. Sr2 was started to be built on NIM's Changping Campus in 2017. The systematic shift uncertainty of Sr2 was evaluated to be 7.2×10^{-18} in 2022. With a new BBR optimizing process, the total BBR uncertainty of Sr2 is reduced to 4.6×10^{-18} which corresponds to the total uncertainty of Sr2 is evaluated to 6.5×10^{-18} in 2023. To compare these two Sr clocks, an optical frequency transfer fiber link with a length of 54 km was established between these two clocks. A preliminary study of generating a local time scale with Sr1 has been carried out at NIM. The time scale generated by post-processing of the comparison data between Sr1 and an H-maser demonstrated better stability than the time scale steered by UTC.

1. Introduction

Over the past two decades, cooling and trapping of alkaline-earth atoms have been widely used in ultra-high precision spectroscopy and precision tests in fundamental physics. In the field of time-frequency measurement, the best uncertainty of optical clocks based on laser cooling and trapping technology has reached 10^{-18}, or even 10^{-19} level[1–5]. According to a roadmap toward a redefinition of the second from the Consultative Committee for Time and Frequency (CCTF) of the International Bureau of Weights and Measures (International Committee for Weights and Measures CIPM), many measurements of the frequency of the Sr lattice clock have so far been performed and published, and the recommended frequency was changed by 0.2 Hz and the fractional uncertainty was reduced to 1.9×10^{-16}[6]. This uncertainty is limited by the relative uncertainty in realizing the SI second with the best primary cesium fountains[7].

2. The Sr1 optical lattice clocks at NIM

Two Sr optical lattice clocks have been constructed at NIM. NIM started the research of the strontium optical lattice clock (Sr1) in 2006[8]. The first systematic shifts evaluation of Sr1, located on the Hepingli campus, was carried out in 2015 with an uncertainty of 2.3×10^{-16}. In 2021, the time interleaved self-comparison frequency measurement stability is $\sim 1.8 \times 10^{-15} / \sqrt{\tau}$, and the systematic shift uncertainty of Sr1 was evaluated to be 2.9×10^{-17}[9].

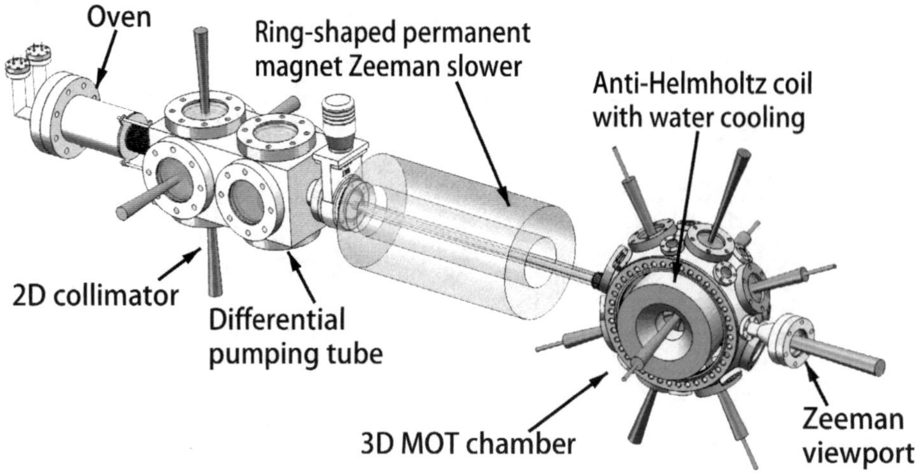

Figure 1. A sketch of the experimental setup.

3. The Sr2 optical lattice clocks at NIM

In 2017, the second strontium atomic optical lattice clock (Sr2) was started to be built on NIM's Changping campus. Based on the experience in building Sr1, many improvements were proposed for the Sr2 systems in the early stages of construction. A sketch of the experimental setup is shown in Fig. 1. A differential pumping stage is introduced between the oven and the magneto-optical trap (MOT) chamber to reduce the impact of the operation of the oven to the vacuum pressure at the MOT chamber. A ring-shaped permanent magnet Zeeman slower has been implemented in the Sr2 system, which is notable for its lack of power consumption or heat generation. A water cooling MOT coil is engaged to control the temperature gradient around the atoms precisely. In 2022, The BBR shift induced by the MOT chamber is -5105.1×10^{-18} with an uncertainty of 3.8×10^{-18}. Furthermore, the oven was maintained at a temperature of 793(50) K and the viewport's temperature was measured to be 353(20) K. The BBR frequency shift associated with these two parts is $-11.3(4.0) \times 10^{-18}$[10].

In 2023, a new passively temperature-stabilized enclosure is installed around the MOT chamber, which isolates various heat sources from the environment except for the MOT coils. To achieve an accurate temperature distribution on the MOT chamber surface, we employed computer simulations to simulate the temperature distribution within the cavity and then validated it using an infrared camera. This allowed us to pinpoint the precise locations for installing high-precision commercial platinum resistors (PT100-385) on the cavity surface. The distribution of the platinum resistance thermometer probe on the cavity is shown in Fig.2. The temperature scanner is equipped with a 10-channel multiplexer module (Fluke, 1589–2588). Under a slow scanning rate (4 s per channel), the temperature scanner has a measurement accuracy of ± 5 mK for platinum resistance thermometers (PRTs).

Throughout the entire measurement process (approximately 56 hours), the lowest temperature was 23.1195 $°C$ and the highest temperature was 23.4913 $°C$. The effective temperature measured was 23.3054 $°C$. By taking the maximum difference in all channels for each group, we obtained a plot of the measurement data as shown in Fig.3. During the measurement period, we adopted point-to-point correction using BBR frequency shift measurements between individual channels. For a single set of seven measurement channels, the maximum temperature difference was found to be 0.1635 K. The BBR frequency shift and

Journal of Physics: Conference Series **2889** (2024) 012047 doi:10.1088/1742-6596/2889/1/012047

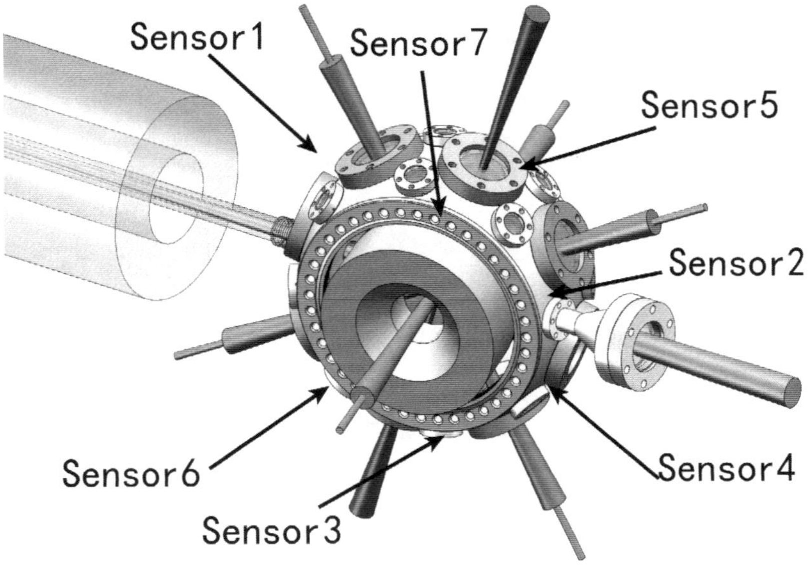

Figure 2. Layout of the sensors on the MOT chamber of Sr2.

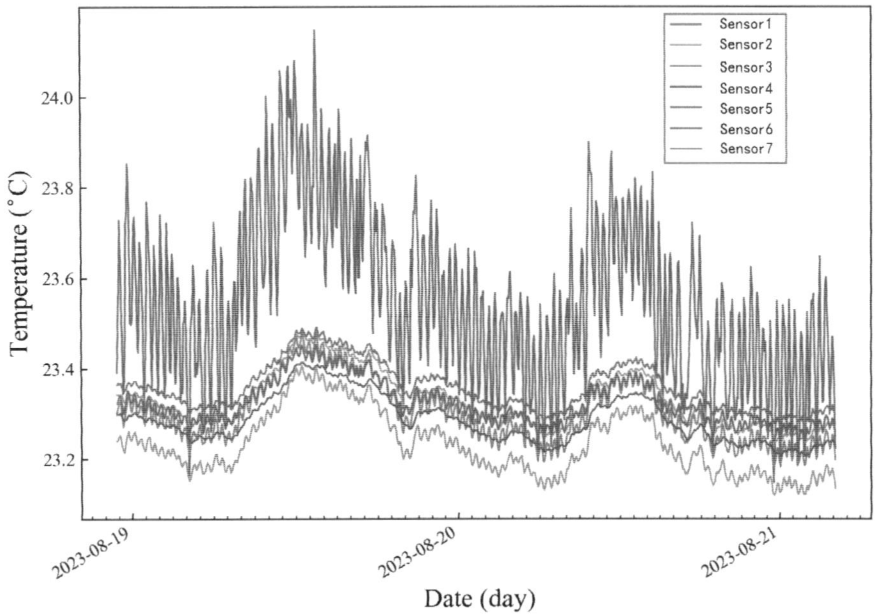

Figure 3. Temperature of 8 scanner channels.

its uncertainty as $5062.83 \pm 3.71 \times 10^{-18}$.

Based on the size of the heating Zeeman window and oven, as well as their distance from each other, we can calculate the spatial solid angle subtended by the heat source at the atoms.

Journal of Physics: Conference Series **2889** (2024) 012047 doi:10.1088/1742-6596/2889/1/012047

According to the system physical structure, as shown in Fig. 1. Given that the radius of the Zeeman window is r = 19 mm and its distance from the atom is L = 502.45 mm, we can calculate the solid angle subtended by the Zeeman window at the atomic system to be $\pi r^2/L^2 = 0.00449$ sr. The temperature monitoring for the Zeeman window was set to 85 $°C$, with an Omega thermometer accuracy of within ±10 K which allows us to calculate the BBR frequency shift caused by the Zeeman window and its uncertainty as $7.45 \pm 2.28 \times 10^{-18}$.

There is a small hole with a radius of r = 3 mm between the atomic oven and the cold atom, at a distance L = 825.65 mm from the atomic oven to the cold atom. Therefore, the solid angle subtended by the oven at the cold atom is $\pi r^2/L^2 = 4.15 \times 10^{-5}$ sr. The temperature control of the atomic oven was set to 510 $°C$, with a maximum difference between the monitored temperature and controlled temperature probe of approximately ± 50 K. The corresponding BBR shift by the oven and its uncertainty is calculated as $2.91 \pm 1.59 \times 10^{-18}$. The total BBR shift is reduced to -5073.2×10^{-18} with an uncertainty of 4.6×10^{-18}.

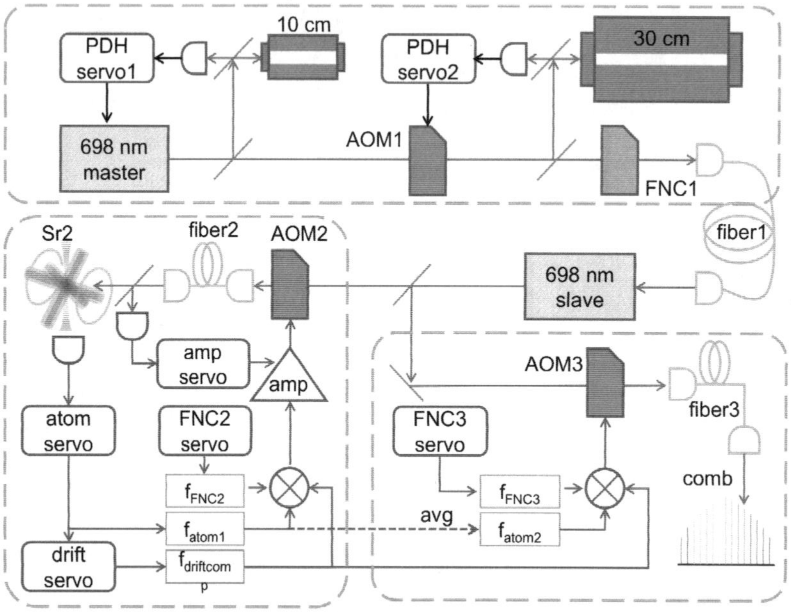

Figure 4. The Sr2 locking scheme

A homemade ultra-stable laser system is established in the Sr2 system[11]. To reduce the thermal noise influence, a 30-cm-long Ultra-low expansion glass (ULE) spacer is sandwiched by a pair of high-reflectivity fused silica mirrors. An active temperature control cover and two passive shielding covers are found in the ultra-high vacuum chamber to isolate the influence of the external environment temperature fluctuation on the ultra-stable cavity. The short-term frequency stability of the clock laser is evaluated at approximately 2.8×10^{-16}at 1 s[12], not far from the theoretical thermal noise limit of 1.4×10^{-16}[13]. As shown in Fig. 4, the 698-nm master laser is initially locked onto an internally developed 10cm pre-stabilized cavity using PDH servol, and then uses PDH servo2 to lock onto a home-made 30-cm ultra-stable cavity with AOM1 used for fast frequency feedback. The output of the 30cm super-stabilized cavity is transmitted through fiberl after fiber noise canceling by FNC1 and used as injection locking seed for the 698-nm slave laser. The output of the 698-nm slave laser is split into two parts: one is used as clock laser output, and the other is used for Sr2 atomic transition detection.

AOM2 not only serves as a frequency feedback component for the clock laser but also plays a role in addressing issues related to the clock laser's power stability. The frequency control of AOM2 includes three parts: FNC2 servo is used for fiber noise cancellation on fiber2, while the clock-locking signal and cavity drift compensation frequency signals are synthesized together to adjust the output frequency of AOM2 to achieve atomic locking. The same modulation signal is synchronized onto AOM3 to ensure that the clock laser output frequency synchronizes with the clock transition frequency.

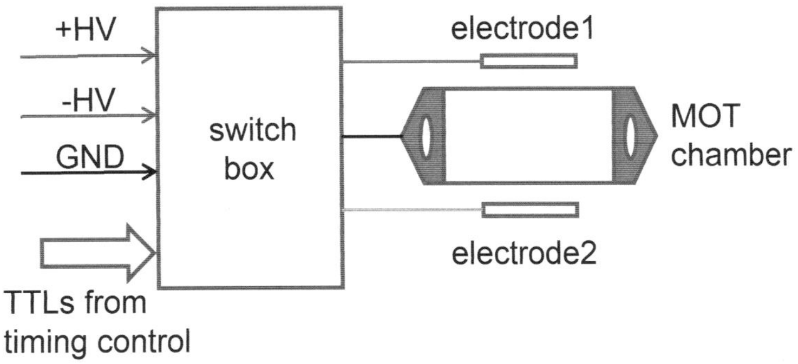

Figure 5. DC-stark operation scheme.

As shown in Table 1, apart from BBR frequency shift, various factors that affect the uncertainty of Sr2 system are measured separately, including collisional induced frequency shift, AC Stark shift, Zeeman effect, and DC-Stark effect. To measure the DC-Stark effect, a pair of static electric field electrodes are installed on the viewports near the atomic cloud of the MOT chamber, utilizing a time-sequencing program to control the direction of the electric field sequentially executing a cycle operation from $V_+ \rightarrow GND \rightarrow V_-$, as shown in Fig. 5. The frequency shift causes by the DC-stark effect is measured, allowing for the calculation of DC-Stark frequency shift being $-20.9(2) \times 10^{-18}$. Benefiting from a series of improvements based on the 1st evaluation of the Sr2 system, the systematic shift uncertainty of Sr2 is evaluated to be 6.5×10^{-18}.

4. Absolute frequency measurement of optical clock

The absolute frequencies of Sr1 still need to be traced to cesium clocks, which are the primary frequency standards according to the International System of Units (SI) definition of the second. At NIM the cesium fountains and all Hydrogen Maser (HM) clocks are located on the Changping campus. A 54-km long optical fiber link was established to connect the Hepingli campus and the Changping campus. The microwave reference frequency from the HM is transferred to the Hepingli campus via a microwave frequency noise canceling (FNC) system[14]. The beat frequency f_b between the Sr output and the corresponding comb tooth is phase-locked to a frequency synthesizer, which means the repetition rate f_r of the comb is referenced to Sr1. The f_0, f_r, and f_b are synchronously counted by a multi-channel frequency counter integrated with the comb. All these frequency measurements lead to the absolute frequency of Sr1. The HM is compared to the local coordinated universal time UTC(NIM) with a dual mixer time difference (DMTD) measurement system. UTC(NIM) has a satellite link to International Atomic Time (TAI). BIPM publishes a monthly Circular T bulletin, which gives the differences between UTC(NIM), TAI, and the best realization of the SI second by PSFS. In 2021, based on the above optical fiber link transmission technology, the absolute frequency of Sr1 is measured to

Table 1. Relative frequency corrections and uncertainties from different systematic effects of Sr2.

Systematic effect	Correction($\times 10^{-18}$)	Uncertainty($\times 10^{-18}$)
Total BBR	5073.2	4.6
Lattice scalar/tensor	11.5	2.6
Hyperpolarizability	-11.9	0.4
Lattice M1/E2	-20.1	0.8
Collisions	58.5	3.1
2nd Zeeman	105.2	0.1
Background gas collisions	10.0	1.1
DC Stark	20.9	0.2
Probe AC Stark	0	1.0
Servo error	0	1.0
Tunneling	0	1.0
Total	5247.3	6.5

be 429 228 004 229 873.07(0.13) Hz, with an uncertainty of 3.1×10^{-16} [9] which is one order of magnitude lower than the previous measurement in 2015[8].

5. Generating a local time scale with Sr1

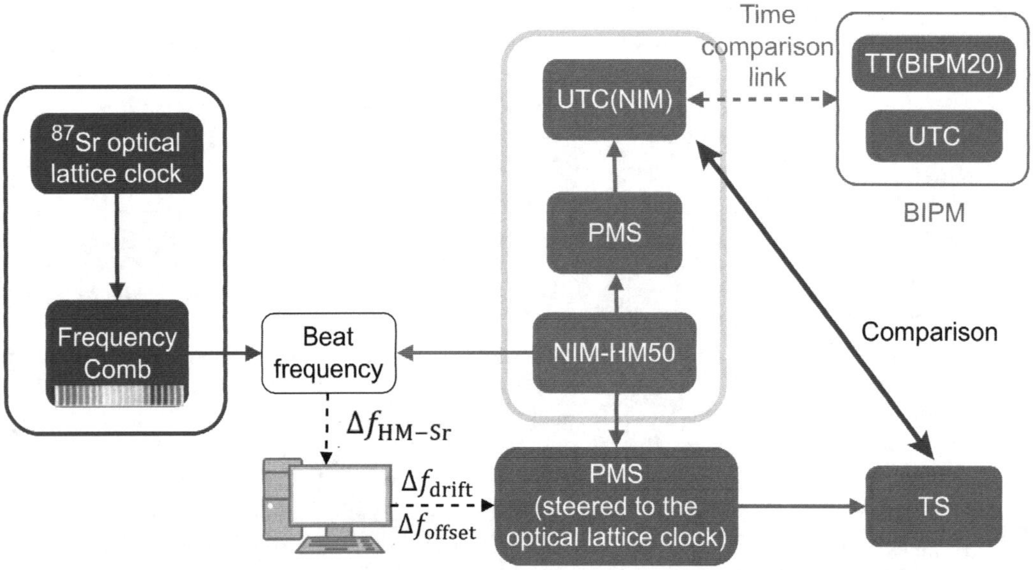

Figure 6. The schematic of steering a hydrogen maser to ^{87}Sr optical lattice clock to generate the local time scale TS. TS could be compared with UTC(NIM) and traced back to UTC and TT(BIPM20) through the time comparison link.

Optical clocks have the great potential to improve the performance of local time scales[15]. Figure 6 shows the schematic of steering a hydrogen maser (HM50) to the ^{87}Sr optical lattice clock (Sr1) to generate a local time scale TS. In this system, the HM works as a flywheel clock to ensure the continuity of the local time scale. Sr1 operates intermittently and acts as the frequency reference. The frequency of Sr1 is down-converted to the microwave frequency domain by means of an optical frequency comb, and then compared with the frequency of the hydrogen maser. During the uptime of the optical lattice clock, the fractional frequency difference Δf_{HM-Sr} between Sr1 and HM50 is derived from the beat frequency of the comb measurement. A month-long measurement campaign was conducted at NIM in 2020. Simulations have been made to evaluate the performance of the time scale steered to Sr1 with the real uptime rate of Sr1 from the modified Julian date (MJD) of 59029 to 59059. The noise model of HM50 adopted in the simulation is the same as Ref. [9, 15]. A 100 simulations show that the maximum time error is 0.91 ns, and the time error after 30 days is only 0.21 ns which demonstrates better stability than the time scale steered by UTC.

6. Conclusion

Two Sr optical clocks have been constructed on two separate campuses of NIM. In 2021, the systematic shift uncertainty of Sr1 has been evaluated to be 2.9×10^{-17}, and the absolute frequency of Sr1 was measured to be 429 228 004 229 873.07(0.13) Hz, with an uncertainty of 3.1×10^{-16}. A preliminary study of generating a local time scale with Sr1 was demonstrated in 2020, where the time error after 30 days is only 0.21 ns. In 2022, the Sr2 system was evaluated to be 7.2×10^{-18}. In 2023 a new process assists in reducing the total BBR shift of the Sr2 to -5073.2×10^{-18} with an uncertainty of 4.6×10^{-18} which corresponds to the total uncertainty of Sr2 is evaluated to 6.5×10^{-18}. In future work, the Sr2 system will undergo an absolute frequency measurement.

7. Acknowledgments

This work is supported by the National Key R&D Program of China (Nos. 2021YFF0603802, 2017YFA0304404 and 2016YFF0200201) and the National Natural Science Foundation of China (Nos. 61905231 and 91736310).

References

[1] Ushijima I, Takamoto M, Das M, Ohkubo T and Katori H 2015 *Nature Photonics* **9** 185–189 ISSN 1749-4893 URL `https://doi.org/10.1038/nphoton.2015.5`

[2] McGrew W F, Zhang X, Fasano R J, Schäffer S A, Beloy K, Nicolodi D, Brown R C, Hinkley N, Milani G, Schioppo M, Yoon T H and Ludlow A D 2018 *Nature* **564** 87–90 ISSN 1476-4687 URL `https://doi.org/10.1038/s41586-018-0738-2`

[3] Huntemann N, Sanner C, Lipphardt B, Tamm C and Peik E 2016 *Phys. Rev. Lett.* **116** 063001 publisher: American Physical Society URL `https://link.aps.org/doi/10.1103/PhysRevLett.116.063001`

[4] Bothwell T, Kedar D, Oelker E, Robinson J M, Bromley S L, Tew W L, Ye J and Kennedy C J 2019 *Metrologia* **56** 065004 ISSN 0026-1394 publisher: IOP Publishing URL `https://dx.doi.org/10.1088/1681-7575/ab4089`

[5] Brewer S M, Chen J S, Hankin A M, Clements E R, Chou C W, Wineland D J, Hume D B and Leibrandt D R 2019 *Phys. Rev. Lett.* **123** 033201 publisher: American Physical Society URL `https://link.aps.org/doi/10.1103/PhysRevLett.123.033201`

[6] Recommendation 1 (CI-2006) `https://www.bipm.org/utils/en/pdf/CIPM/CIPM2006-EN.pdf`

[7] Riehle F, Gill P, Arias F and Robertsson L 2018 *Metrologia* **55** 188 ISSN 0026-1394 publisher: IOP Publishing URL `https://dx.doi.org/10.1088/1681-7575/aaa302`

[8] Lin Y G, Wang Q, Li Y, Meng F, Lin B K, Zang E J, Sun Z, Fang F, Li T C and Fang Z J 2015 *Chinese Physics Letters* **32** 090601 ISSN 0256-307X publisher: Chinese Physical Society and IOP Publishing URL `https://dx.doi.org/10.1088/0256-307X/32/9/090601`

[9] Lin Y, Wang Q, Meng F, Cao S, Wang Y, Li Y, Sun Z, Lu B, Yang T, Lin B, Zhang A, Fang F and Fang Z 2021 *Metrologia* **58** 035010 ISSN 0026-1394 publisher: IOP Publishing URL `https://dx.doi.org/10.1088/1681-7575/abf33e`

[10] Lu B K, Sun Z, Yang T, Lin Y G, Wang Q, Li Y, Meng F, Lin B K, Li T C and Fang Z J 2022 *Chinese Physics Letters* **39** 080601 ISSN 0256-307X publisher: Chinese Physical Society and IOP Publishing Ltd URL `https://dx.doi.org/10.1088/0256-307X/39/8/080601`

[11] Zhen Sun, Tao Yang, Yige Lin, Qiang Wang, Ye Li, Bingkun Lu, Fei Meng, Shiying Cao, Baike Lin, Tianchu Li and Zhanjun Fang 2020 Advances in strontium optical lattice clocks at NIM vol 11296 p 112961H URL `https://doi.org/10.1117/12.2552695`

[12] Peik E, Schneider T and Tamm C 2005 *Journal of Physics B: Atomic, Molecular and Optical Physics* **39** 145 ISSN 0953-4075 URL `https://dx.doi.org/10.1088/0953-4075/39/1/012`

[13] Notcutt M, Ma L S, Ludlow A D, Foreman S M, Ye J and Hall J L 2006 *Phys. Rev. A* **73** 031804 publisher: American Physical Society URL `https://link.aps.org/doi/10.1103/PhysRevA.73.031804`

[14] Wang B, Gao C, Chen W L, Miao J, Zhu X, Bai Y, Zhang J W, Feng Y Y, Li T C and Wang L J 2012 *Scientific Reports* **2** 556 ISSN 2045-2322 URL `https://doi.org/10.1038/srep00556`

[15] Zhu L, Lin Y, Wang Y, Jia Z, Wang Q, Li Y, Yang T and Fang Z 2022 *Metrologia* **59** 055007 ISSN 0026-1394 publisher: IOP Publishing URL `https://dx.doi.org/10.1088/1681-7575/ac8992`

Developments to improve the stability of optical lattice clocks

S. Bize, B. Fang, Y. Le Coq, R. Le Targat, J. Lodewyck, P.-E. Pottie, H. Shang, and C. Zyskind

LNE-SYRTE, Observatoire de Paris - PSL, CNRS, Sorbonne Université,
61 avenue de l'Observatoire, 75014 Paris, France.

E-mail: `sebastien.bize@obspm.fr`

Abstract.
We present several developments aimed at improving the stability of optical lattice clocks. First, we present our developments of an optical lattice clock using neutral mercury. We mention recent advances made with the fermionic isotope ^{199}Hg. We also mention our work aimed at using bosonic isotopes, which offer the possibility to circumvent the relatively short lifetime of the upper clock state in ^{199}Hg. Second, we present our work on a non-destructive detection in a Sr optical lattice clock. We describe developments that brought the detection scheme from the classical non-destructive regime to the quantum non-destructive regime. Our detection scheme is practical and has the capability to the used beyond proving the principle. Finally, we present our work on laser stabilization using spectral hole burning in a Eu^{3+}:Y_2SiO_5 crystal at cryogenic temperatures. We describe our development of low noise interrogation based on digital IQ modulation and detection that can probe multiple spectral features simultaneously, as well as several investigations towards fluctuating environmental factors. These advances, individually or combined for example with spectral purity transfer with combs and composite clock approaches, shall bring significant progress in clock stability and accuracy.

1. Progress in the development of a Hg lattice clock

Neutral mercury Hg offers the possibility to develop an optical lattice clock with a low sensitivity to thermal radiation and to stray electric fields. Hg is about ~ 30 times less sensitive than Sr and ~ 15 times less sensitive than Yb. Highly accurate measurements of frequency ratios between Hg and other optical transitions can bring interesting contributions to fundamental physics tests and metrology. Our group, and the group at University of Tokyo and RIKEN, studied optical lattice clock based in the fermionic isotope ^{199}Hg. These developments gave two accurate and independent measurements of the frequency ratio ^{199}Hg/^{87}Sr with uncertainties of 1.8×10^{-16} [1] and 8.4×10^{-17} [2] in agreement with the 1σ error bar. Both measurements were limited by the systematic uncertainties of the ^{199}Hg systems. Our group also made direct measurement of the absolute frequency, i.e. of the ^{199}Hg/^{133}Cs(hfs) ratio, and of the ^{199}Hg/^{87}Rb(hfs). In 2017, these measurements led CIPM to recommend the $^1S_0$-$^3P_0$ in ^{199}Hg as a secondary representation of the SI second [3]. In the 2021 update of the list of recommended frequencies and secondary representations of the second [4][5], still based on the two aforementioned measurements, the ^{199}Hg clock transition has a recommended fractional uncertainty of 2.4×10^{-16}, not far from the best recommended uncertainty (1.9×10^{-16} for ^{87}Sr, ^{171}Yb, ^{171}Yb$^+$(E3) and ^{27}Al$^+$) which is determined by the uncertainty of the realisation of the SI second by cesium fountains.

Main limitations of our work with ^{199}Hg were the following. A first limit was the uncertainty to which we were able to control the optical frequency of our optical lattice light with a commercial lambdameter. A second limit was a relatively modest lattice trap depth (estimated to $\sim$ 56 recoil energy E_R) owning to the challenging UV wavelength (362.5 nm [6]) and the comparatively low polarizability (a price to pay for the low blackbody radiation sensitivity). Another limit was the short term stability that we could achieve ($> 1.2 \times 10^{-15}$ at 1s). This limit was itself linked to several factors: a limited number of atoms that could be captured in a given time and a rather short lifetime of the atomic sample in the lattice trap, which together impose a duty cycle of the probing sequence no better than $\sim$ 10% and increase the impact of the Dick effect. The reliability of our Hg clock system was also significantly limiting our capability to perform and repeat long measurement sessions, the main factor being the 254 nm light source for cooling and detecting atoms. Another limit was coming from residual background gases in our vacuum system which contribute to limit the lifetime of the atomic sample trapped in the lattice and cause a background collision shift. The impact of the latter was found quite significant in recent studies on ^{171}Yb and ^{87}Sr [7][8].

Next, we mention several developments to improve over these limits. First, we developed a system to drastically improve the control of the lattice trap frequency. Our approach was to lock the lattice light frequency to a comb in order to have virtually no limit to the level of control. We aimed to do this with the same metrological comb already used to link the Hg clock light at 1062.5 nm (282.1 THz, one fourth of the $^1S_0$-$^3P_0$ transition frequency). To do this, we relied on a 1450 nm extended cavity diode laser (ECDL) which the frequency doubled to obtain a beat note against the 725 nm light (TiSa laser) frequency doubled to generate the optical lattice trap light at 362.5 nm. 1450 nm is a wavelength where our fiber based comb have, without any modification, enough power to obtain a beat note signal. To use these beat notes and lock frequency of the 1450 nm laser comb and the TiSa laser to the doubled 1450 nm laser, we developed a locking scheme tolerant to extremely low signal-to-noise ratios while remaining sufficiently accurate, avoiding any further requirement to the comb in terms of power in the relevant spectral region. We estimated that our system can control the absolute frequency of our lattice light to an uncertainty of $\sim$ 1 kHz, which should not be a limit to the accuracy of our Hg clock down to 10^{-18} or even better [9]. Second, we improved our laser sources for cooling and detecting the atoms with the $^1S_0$-$^3P_1$ transition at 254 nm to allow for practical 2 dimensional magneto-optical trap (2D-MOT), a possibility arising from the high vapour pressure of mercury, which had not been exploited before. Our system comprises two independent generation of 254 nm light, each with a home-built ECDL at 1015 nm seeding a commercial Yb fiber amplifier and single pass doubling to 508 nm followed by a home-built cavity-enhanced frequency doubling to 254 nm [10]. One of the light source is stabilized to Doppler-free absorption spectrum from a room temperature 1 mm long Hg vapor cell. The second light source is frequency stabilized to the first one via an optical beat note at 1015 nm, providing independent frequency tuning of the two sources. One source is set to generate about 26 mW of UV power for the MOT. The other one generates up to 50 mW for the 2D-MOT. These power values are far from the maximum achievable. They are the result of a comprise to improve the lifetime of components of the doubling stages (BBO doubling crystals, mirror coatings). While the infrared and visible part of this system prove highly reliable, the 254 nm doubling stages remain the dominant impediment to unattended and daylong continuous operation of our clock. The 2D-MOT increased the cold atom loading rate by factor of 4.5 which gave us leverage to improve clock stability, investigate cold collision shift, operate at a lower trap depth or reduce the vapor pressure in the Hg source region. For instance, we were able to operate a sequence with a cycle time of 482 ms and a duty cycle of 33% giving a short term stability of $\sim 6.3 \times 10^{-16}$ at 1 s [10] with an ultra-stable laser based on a 10 cm long cavity [11][12]. We have indications that, with we same amount of UV power, the efficiency of the 2D-MOT could be further increased by optimizing the beam shape. It is reasonable to expect that multiplication by 10 or more of the loading rate would be within reach with such modification and a feasible increase of the power.

These developments and additional improvements to increase the reliability made it possible for our ^{199}Hg clock to participate to several measurement campaigns in the last years. For instance, the ^{199}Hg clock was involved in fiber link comparison campaign in May-June 2017 between SYRTE, PTB and NPL. This campaign gave stability measurements between Hg, Sr and Yb^+(E3) that was exploited to search for transient variation of the fine-structure constant and for dark matter [13]. We measured the ^{199}Hg/^{87}Sr optical frequency ratio several times during internal and fiber link campaigns in 2017, 2018, 2019, 2020 and 2023, making use of the REFIMEVE fiber network infrastructure [14]. In the last campaign, we could collect about 27 hours of data against the remote Sr clock at PTB. A typical measurement session with optimized parameters reaches a statistical uncertainty of 2×10^{-17} in 2 hours. The successive measurements of the ^{199}Hg/^{87}Sr over the years were in good agreement with each others but tend to confirm a discrepancy with the measurement made at RIKEN [2]. A preliminary analysis of the 2023 measurements gives, instead, an agreement with [2]. We are currently trying to confirm these findings and to find the origin of the change. Note that in all these measurements, uncertainties in our Hg clock dominate and our main assumption is that discrepancies and changes come from Hg. The last campaigns shall also provide valuable frequency ratio measurements against one or several Yb^+(E3), Yb^+(E2) and Yb clocks, subject to clarifying the Hg/Sr measurements.

The $^1S_0$-$^3P_0$ clock transition in the fermionic isotope ^{199}Hg is weakly allowed for electric-dipole transition due to the hyperfine mixing. This same coupling defines the spontaneous emission rate from $^3P_0$ to $^1S_0$ and the lifetime of the $^3P_0$ state. In ^{199}Hg, the estimated lifetime is rather short (~ 1.5 s) compared to, for instance, the $^3P_0$ lifetime in ^{87}Sr and ^{171}Yb. It will be an obstacle to exploit the new generation of ultra-stable lasers like for example the one reported in [15] or [16] or the one that shall result from the work presented section 3. Using bosonic isotopes of Hg provides a means to circumvent this limit. Bosonic isotopes have no nuclear spin and their $^1S_0$-$^3P_0$ is completely forbidden. The transition can be made weakly allowed by a relatively strong magnetic field as proposed in [17] and experimented in ^{88}Sr [18][19][20] and ^{174}Yb [21]. With this quenching method, the strength of the coupling can be adjusted at will and the probing time can be as long as desired to match to the coherence time of the probe laser. We made key developments to use bosonic isotopes of Hg. We developed the means to apply a pulsed and well-controlled quenching magnetic field, using coils of our MOT, an inverting switch and a controlled current source. We have checked that we can apply magnetic field pulse up to 15.5 mT without loosing or perturbing the atomic sample trapped in the lattice. We implemented a widely tunable and flexible probe laser system at 265.6 nm that preserves the low frequency noise properties of our current (flicker noise at 4×10^{-16}) and future ultra-stable light. Based on an offset phase-locked narrow line fiber laser, this system has the capability to address the $^1S_0$-$^3P_0$ clock transition of any of the natural isotopes of Hg. We optimized the beam routing and shaping of the 265.6 nm light regenerated by resonant frequency doubling in order to maximize the coupling to atoms held the lattice trap. This optimization and several other tests were made with the help of the ^{199}Hg fermionic isotope. From the maximum Rabi frequency observed with ^{199}Hg ($\sim$ 2 kHz) and the other parameters, we infer that we should reach a Rabi frequency of 50 Hz. This is quite small and the search for the transition should be challenging but feasible for the ^{198}Hg bosonic isotope for which the transition frequency is known with a 1σ uncertainty of 10.3 MHz.

Fully exploiting the potential of bosonic isotopes and of new ultra-stable lasers will require to have long probe times, largely longer than the limit imposed by the lattice lifetime observed in our current setup. This is a further motivation for us to reduce the pressure of background gases, which we already mentioned as a limit to our work with ^{199}Hg. To improve the background pressure, especially the one of Hg, we developed a home-built cryogenic pump that can be fit to our setup without major modification. The pump consists of two copper plates clamped onto a copper tube through which liquid nitrogen circulates. We estimate that this pump shall have an effective pumping speed at the level of the lattice trap of several l.s^1 for Hg. We conducted tests that showed at a significant positive impact on the vacuum level. We expect that using this device will give us a preliminary means to investigate effects of background gas collisions (frequency shifts and lattice lifetime limitation). And drastic and perennial improvement of vacuum level will then require a major redesign of our setup.

2. Non-destructive detection for Sr lattice clocks

Utilizing spin squeezing to overcome the standard quantum limit in order to enhance the clock frequency stability has first been demonstrated in microwave atomic clocks [22]. However, spin squeezing on optical clock transitions is more challenging due to the lack of differential measurement techniques directly accessing the population difference, as well as the lower number of atoms probes in optical clocks, and was accomplished only recently. A first demonstration involved coherently transferring spin-squeezing from a microwave to an optical transition on ytterbium [23]. A very recent work presents a metrological gain of 1.9 dB on a strontium optical lattice clock, relying on the cavity-assisted quantum non-demolition (QND) measurements that works on the near-resonant regime [24].

Here, we propose and realize a different QND measurement scheme in a strontium optical lattice clock, where the clock atoms are trapped and coupled with a far-detuned probe laser in a single spatial cavity mode [25], aiming at integrating the apparatus into the operational clocks. In our scheme, spin-squeezing is produced via cavity-enhanced collective coupling between an ensemble of strontium atoms and a probe laser off-resonant with the $^1S_0 - {}^1P_1$ transition at 461 nm, as depicted in the energy level of Fig. 1. For this, the number of atoms in the electronic ground state $N_\downarrow$ is acquired by a PDH-like heterodyne scheme, in which phase-modulation sidebands are simultaneously injected in two longitudinal modes of the cavity, ensuring both immunity to technical noises (laser frequency noise and mechanical vibrations of the cavity), and homogeneous longitudinal atom-cavity coupling g [25]. Our latest implementation [26] makes use of two electro-optic phase modulators in order to reach a large atom-light detuning of $\Delta = \pm$ 5.5 GHz, and to decouple the dynamics of the detection scheme from the error signal used to lock the probe beam on the cavity. These two features are critical in order to reach the low photon scattering limit characterizing quantum non-demolition measurements.

Practically, the number of atoms is deduced from a measurement of the frequency shift of the cavity modes, given by $\delta\omega_c = Ng^2/\Delta$. For the typical number of probed atoms, this frequency shift is larger than the cavity line-width, making the atomic signal move out of the linear regime (see Fig. 1). To handle this, we feedback the atomic signal to one of the modulation frequencies Ω' and record the frequency correction which is equivalent to $\delta\omega_c$ [27]. With this scheme, we have verified the scaling between $\delta\omega_c$ and $N_\downarrow$ as measured with an EMCCD camera, and measured the noise limitations in our system, including the quantum projection noises, and the detection noise (electronic noise and photon shot noise).

Besides the detection of the atomic states with QND measurements with a high signal-to-noise ratio, the preservation of the atomic coherence is also mandatory for entering the regime of metrological improvement. For this, the probe power is tuned to ensure a scattered photon rate by one atom as small as 10^{-2} γ/μs. However, the atomic coherence still suffers from dephasing with a rate of ng^2/Δ (n is the probe photon number) which comes from the interaction with the probe light. To cancel this effect, we use a photon-echo protocol where a π clock pulse is inserted between two QND measurements with equal detection time to rephase the atomic states which are initially prepared by a $\pi/2$ clock pulse. By measuring the variation of contrast of Ramsey fringes formed by two time-separated $\pi/2$ clock pulses due to the non-demolition measurements,

Journal of Physics: Conference Series **2889** (2024) 012048 doi:10.1088/1742-6596/2889/1/012048

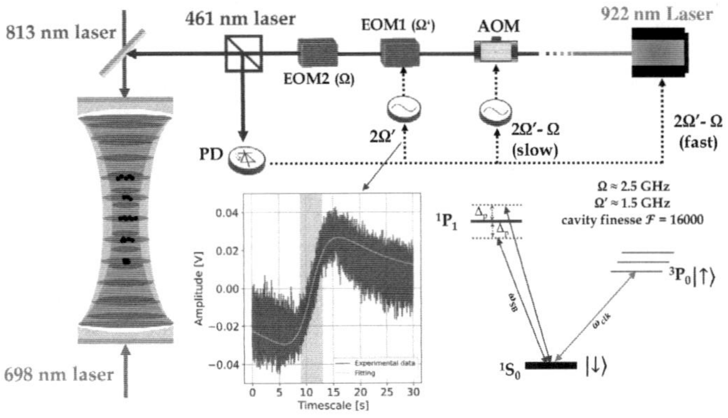

Figure 1: Quantum nondemolition measurements scheme with far-detuning probe laser. Two electro-optic modulators (EOMs) are used to phase-modulate the 461 nm probe laser and flexibly adjust the power of probe sidebands. The modulation frequency frequencies of the two EOMs are $\Omega = 2.5$ GHz and $\Omega' = 1.5$ GHz, respectively. An acoustic-optical modulator (AOM) is to match the frequency difference between the cavity resonance and probe laser. Two insets are the atomic signal and the relevant energy levels of strontium.

we observe that the photon-echo method mitigates the decoherence in our apparatus. There is still residual decoherence leading to an additional loss of the contrast by about 1.1 times at our optimal experimental conditions, and which could be further reduced with a more appropriate echo pulse [28].

With an initial preparation of the coherent spin states for around 2×10^4 atoms, we conducted two consecutive composited QND measurements to investigate the generation of the spin-squeezing, and the measurement protocol is as presented in Fig 2a. In each measurement, the transition probability which indicates the atomic states is given by $P^{(i)} = N_{\uparrow}^{(i)}/(N_{\downarrow}^{(i)} + N_{\uparrow}^{(i)})$, where i is the measurement order. Because the quantum projection noise which dominates each of the two measurements, is correlated, it can be rejected by a factor of $1/\chi_{\mathrm{QND}} = (\Delta P_{\mathrm{SQL}}/(\Delta(P_2 - \alpha P_1))^2$, where ΔP_{SQL} is the standard deviation of the transition probability in the case of standard quantum limit, and $\alpha = \mathrm{Cov}(P_1 - P_2)/(\Delta P_1)^2$ represents the conditional correlation coefficient. Generally, a smaller χ_{QND} requires a better signal-to-noise ratio which means stronger probe strength. However, we need to trade off the atomic decoherence. The Wineland parameter $\xi_w = \chi_{\mathrm{QND}} \cdot C_i/C_f^2$ can be used to represent the metrological improvement, where C_i (C_f) is the Ramsey fringe contrast before (after) the QND measurements. We investigated the ξ_w at different probe strengths, and the results are shown in Fig. 2b. At a scattered photon number of around 0.2, $\xi_w < 1$ is obtained, which infers the generation of the spin squeezing. With the further suppression of the atomic decoherence by optimizing the photon-echo protocol, reduction of the additional free-space scattered photons stemming from unwanted coupled sidebands, and enhancement of the signal-to-noise ratio by increasing the coupling efficiency of the probe, it is promising to obtain a considerable metrological improvement based on our scheme for the operational lattice clocks.

3. Laser stabilization using spectral hole burning

Given the extraordinary progress in optical lattice clocks, the need of an ultra-stable optical oscillator has become urgent. Whereas the room-temperature high-finesse Fabry-Perot cavity stabilized lasers reach their fundamental thermal noise limited fractional frequency instabilities at the low 10^{-16} around 1 s integration time [30], an improvement by at least one order of magnitude is necessary in order to demonstrate the quantum projection noise limit in optical

Journal of Physics: Conference Series **2889** (2024) 012048 doi:10.1088/1742-6596/2889/1/012048

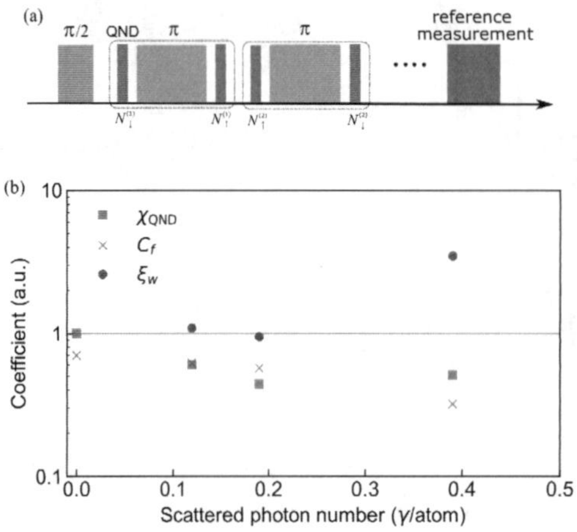

Figure 2: Observation of spin squeezing. (a) measurement protocol for observing the quantum correlations and spin squeezing. The red bar indicates the 698 nm clock laser pulse and the blue bar indicates the 461 nm QND probe pulse. The clock Rabi time is 0.3 ms. (b) at different probe strengths that are calculated from the measured photon scattering rate [29], measurement results for the quantum projection noises reduction coefficient (χ_{QND}, green-square points), the contrast of the Ramsey fringe (C_f, red-cross points), and the Wineland parameter (ξ_w, blue-circle points).

lattice clocks with the conventional cyclic operation with dead time in typical situations [31]. Several strategies to lower the thermal noise floor of the Fabry-Perot cavities have been pursued : increasing the size of the optical mode on the mirrors [15], using crystalline material with better thermal mechanical properties for the cavity components [16, 32], going to cryogenic temperatures [33], and a combination of those. However, the outcome is still limited by thermal noise, albeit under new conditions, with the price of added technological complexities, mostly related to cryogenics and vibration management.

It is in this light that we investigate the so-called 'spectral hole burning' as an alternative method to derive short-term frequency stabilities. This method is based on the spectroscopy of rare-earth ions doped in a crystalline matrix, by optically pumping and shelving these ions into metastable states to create long-lived spectral structures as frequency references. Since the homogeneous line width is related to the optical coherence time of the excited state, a narrow line laser and an appropriate choice of the dopant and the host, Europium ions doped at 0.1 at% into Yttrium orthosilicate (Eu^{3+}:Y_2SiO_5, or Eu:YSO) in our case, allow for the realization of spectral holes with full width at half maxima as low as 3 kHz at about 4 K inside a much broader inhomogeneous profile. The hyperfine splitting of the ground state of the Eu^{3+} ions gives rise to a spectral-hole life time on the order of 10 hours near 4 K. The solid-state nature of the crystal ensures a large number of ions participating in spectroscopy, which is favorable for a good signal to noise ratio. Last but not least, the fundamental limits of such a system is unknown, but preliminary estimations, including that of the thermal noise, appear to be low compared to classic Fabry-Perot based systems at room or even cryogenic temperatures.

We use heterodyne interferometry to measure and correct the frequency of a probe laser field with respect to a spectral hole structure [34, 35, 36]. The experimental setup is shown in Fig. 3(a). By the Kramers-Kronig relation, a narrow transmission spectrum of a spectral hole induces a linear dephasing of a nearly resonant laser field, with the zero crossing located at the center frequency of the spectral hole, see Fig. 3(b). We optically beat down the probe signal to the radio frequency (RF) regime, and measure the phase against that collected on a reference

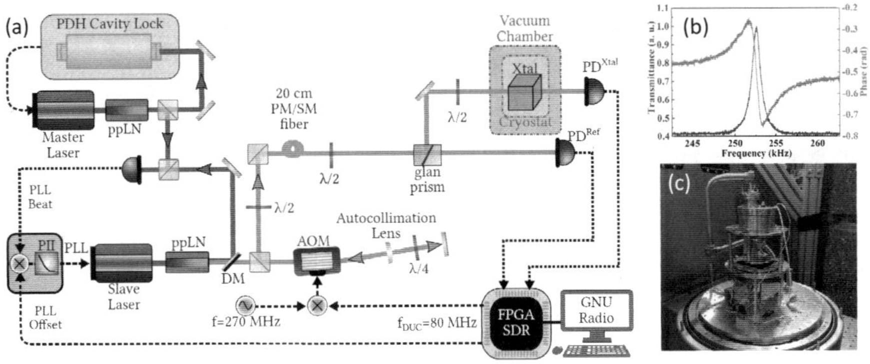

Figure 3: (a) A schematic of the experimental setup of the heterodyne interferometry. Drawing reproduced from [36]. (b) Typical transmisstance and dephasing of a narrow spectral hole. (c) The custom fit dilution stage from Pasqal-MyCryoFirm.

path, so as to obtain this dephasing that serves as an error signal for the frequency servo loop. In order to be insensitive to differential noise sources between the two paths, monitor modes can be generated and placed in a broad spectral hole to measure and subtract their contribution from the error signal. Multimode generation also constitutes a means to spread the probing optical power into several narrow spectral holes in order to increase the signal to noise ratio while keeping overburning (continuous optical pumping from the probe laser field) in check.

The technological challenge in multimode generation and detection is overcomed by using all digital signal processing, with python-based GNURadio software package coupled with a universal software-defined radio peripheral (USRP), Ettus X310 in our case, that offers phase coherent RF multiple inputs and outputs. Multiple frequency modes in time domain can be generated using inverse fast Fourrier transform of the desired spectrum centered on zero frequency. The USRP uses digital up conversion to shift the spectrum to a suitable RF signal to drive an acouto-optic modulator that imprints the multiple frequency modes onto an optical carrier. To detect the phase shift of each modes, the previously mentioned beatnote is measured by suitable photodiodes and sent to the USRP input ports, yielding both the amplitude and the phase information using IQ detection, before being digitally down converted into the base band. A narrow-band polyphase channelizer is used to isolate each frequency component before processing the phase information and generating the error and correction signal. The details of our implementation can be found in [36]. Compared to using only a single probe mode [34], this fully-digital multi-heterodyne approach allows decreasing technical noise sources by orders of magnitudes at 1 s time scale or longer and reaching a detection noise (with optical probing power compatible with hours of continuous operation) of 2×10^{-16} at 1 s, with clear prospects to reach the 10^{-17} range or even lower [36]. Our lowest reported laser frequency instability when locked onto a structure of a narrow spectral hole and a broad one is 1.7×10^{-15} at 1 s so far [35]. We expect further improvements when more narrow spectral holes are used for locking.

With the possibility of tracking the absolute frequency of the spectral holes by measuring the laser frequency, when locked, against the primary frequency standards at the SYRTE laboratory via an optical frequency comb, we have been studying several effects that could potentially impact the short-term stability, namely the sensitivity to a fluctuating DC electric field, mechanical strains (or equivalently accelerations), and the temperature of the crystal.

The measurement of the DC Stark effect is detailed in [37]. Due to the symmetry of the YSO crystal, a spectral hole essentially splits into two with positive and negative frequency shift in the presence of a homogeneous electric field. By analysing both branches for residual asymmetries, targeting a laser fractional frequency instability at the 10^{-17}, we are able to infer an upper limit for electric field fluctuations to 2.3 $V.m^{-1}$ RMS, which is experimentally within reach.

As for the sensitivity to mechanical strains, or equivalently to vibrations, we employed a cryo-compatible 3-axis translation stage in order to place or remove small masses (aluminium plates) on the crystal, while probing the transmission spectrum in the vicinity of a priorly photo-imprinted spectral hole [38, 39]. By characterizing the linear frequency shift, we established the intrinsic acceleration sensitivity to range from 1.5×10^{-11} $(m.s^{-2})^{-1}$ down to 1.3×10^{-12} $(m.s^{-2})^{-1}$ (depending on the direction of the strain and the crystal substitution site), giving an upper bound, for a target laser fractional frequency instability of 10^{-17} at 1 s, to the tolerated vibration noise to be below 10^{-5} $m.s^{-2}/\sqrt{Hz}$, which is manageable even in the presence of a pulse-tube cryocooler that allows for continuous operation without the need of regular replacement of cryogen.

Concerning the impact of the temperature of the crystal T, based on the two-phonon Raman model, one expects the resonance frequency of rare-earth dopants ν to follow a scaling law of $\nu \propto T^4$ [40]. This is in fact realized only for one of the two substitution sites in Eu:YSO. However, in both cases, the sensitivity to temperature fluctuations near 4 K largely limit the performance achieved in [35]. In the presence of an external isotropic pressure in thermal equilibrium with the crystal, however, the thermal fluctuation induced pressure change would counteract the thermal expansion and contraction of the crystal itself, giving the possibility to cancel to the first order the temperature induced frequency shift of the spectral holes for a well chosen pressure-temperature couple. Similarly to earlier work in [41], we investigated these behaviours at temperatures achievable by our pulsed-tube cryocooler. Whereas we were able to verify the cancellation of the temperature sensitivity in an environment that is coupled to an isotropic pressure [42], the technical challenges in realizing a hermetic gas cell that is robust against temperature cycling is tremendous and requires technical expertise and dedicated effort. Furthermore, this approach is only valid for timescales long-enough where thermal equilibrium is reached. We currently explore an alternative that consists of further decreasing the crystal temperature using a dilution refrigerator, see Fig. 3(c). On one hand, assuming the T^4 scaling, the temperature sensitivity, $\propto T^3$, would decrease significantly when the temperature is reduced by an order of magnitude. On the other hand, it is known that the absolute residual temperature fluctuations (in K) at dilution temperatures is lower than that at pulsed-tube temperatures. It is therefore reasonable to expect that the temperature noise induced frequency instability will be well below the target instability of 10^{-17}.

4. Conclusions

We presented developments in three different aspects of optical frequency metrology that have the potential individually, and even more so, when used together, to increase the stability of optical lattice clock beyond the 1×10^{-17} level at 1 s. To this end, one would rely on spectral purity transfer using optical frequency comb to transfer the performance of an ultra-stable laser based on spectral hole burning (discussed in sec. 3) from its specific frequency 517 THz (wavelength of 580 nm) to other relevant frequencies, in our case, the clock transition frequencies of Sr (429 THz) and Hg (4×282 THz) and the frequency of fiber networks (194 THz). It is well-established that such spectral purity transfer can be done at the 10^{-18} level [43]. Such probe lasers used in combination with ^{87}Sr or a bosonic isotope of Hg (discussed in sec. 1) in a shallow lattice trap shall permit probing time of 1 second, from which one can expect clock instabilities $\leq 10^{-17}\tau^{-1/2}$. Further improvement can be expected when making use of non-destructive detection discussed in sec. 2. In [44], we showed theoretically that a simple and practical scheme made of a Rabi interrogation pulse accompanied with 3 non-destructive detection pulses can give significant gain (7.9 dB for 2×10^4 detected atoms) when properly optimized taking into account real imperfections. Another possible use of the non-destructive detection is the so-called atomic phase lock scheme proposed in [45] and tested experimentally with a microwave transition in [46], and with an optical transition in [27]. This scheme makes it possible to extend the probing time beyond the coherence time of the probe laser. With two clocks using the same probe laser (or probe lasers with correlated phase noise via spectral purity transfer), one can also implement a compound clock scheme proposed and tested in [47] where one high-signal-to-noise clock coarsely tracks the phase of the probe laser while the other one makes a high-resolution measurement of the same phase. Like the previous one, the scheme can in principle enable a probing time for the second clock way beyond the coherence time of the laser. These few examples show the potential of the three developments reported here to reach exquisite short term stabilities, when combined together. In turn, instabilities $\leq 10^{-17}\tau^{-1/2}$ greatly facilitate investigations of systematic shifts below the 10^{-18} level. Alternatively or concurrently, these schemes can be used to reduce some biases, for instance by using them as a means to preserve an excellent stability while reducing the number of atoms and thereby collision shifts. Similarly, some of these schemes could be used to preserve an excellent stability even when using a simplified and less high-performance ultra-stable laser, in the context of technological trade-offs for transportable or industrial clocks. Developments that we described shall permit optical frequency standards and optical frequency ratio measurements with higher accuracy, that will make significant contributions towards fulfilling criteria for the redefinition of the SI second [48], towards uses of optical frequency standards in fundamental physics, and towards emerging applications like chronometric geodesy.

Acknowledgments

We acknowledge the support from SYRTE's technical services for the development of many electronic, mechanical, vacuum and computer systems involved in the present work. This paper gives a review of work done at SYRTE in the several last years on topics mentioned in the abstract. We acknowledge many essential contributions to the reviewed work from co-authors of articles that we cite in reference for the work carried out at SYRTE.

We acknowledge funding from the Ville de Paris Emergence Program, the Région Ile de France DIM C'nano and SIRTEQ, the LABEX Cluster of Excellence FIRST-TF (ANR-10-LABX-48-01) within the Program "Investissement d'Avenir" operated by the French National Research Agency (ANR), the H2020 QuantERA ERA-NET Cofund QCLOCKS, the European Commission Horizon Europe Framework Programme under the UltraStabLaserViaSHB project under the Marie Skłodowska-Curie grant agreement No GAP-101068547, and from EMPIR under project 17FUN03 USOQS, 15SIB03 OC18 and 20FUN08 NEXTLASERS. EMPIR projects are co-funded by the European Union's Horizon 2020 research and innovation program and the EMPIR participating states. We acknowledge funding from ERC Consolidator Grant AdOC (ERC-2013-CoG-617553).

References

[1] Tyumenev R, Favier M, Bilicki S, Bookjans E, Le Targat R, Lodewyck R, Nicolodi D, Le Coq Y, Abgrall M, Guéna J, De Sarlo L and Bize S 2016 *New Journal of Physics* **18** 113002 URL `http://stacks.iop.org/1367-2630/18/i=11/a=113002`

[2] Yamanaka K, Ohmae N, Ushijima I, Takamoto M and Katori H 2015 *Phys. Rev. Lett.* **114**(23) 230801 URL `https://link.aps.org/doi/10.1103/PhysRevLett.114.230801`

[3] Recommended values of standard frequencies https://www.bipm.org/en/publications/mises-en-pratique/standard-frequencies.html

[4] Recommendation CCTF PSFS 2 (2021): Updates to the CIPM list of standard frequencies https://www.bipm.org/en/committees/cc/cctf/22-_2-2021

[5] Margolis H S, Panfilo G, Petit G, Oates C, Ido T and Bize S 2024 The CIPM list "Recommended values of standard frequencies": 2021 update (*Preprint* 2401.14537) URL `http://arxiv.org/abs/2401.14537`

[6] Yi L, Mejri S, McFerran J J, Le Coq Y and Bize S 2011 *Phys. Rev. Lett.* **106** 073005

[7] McGrew W F, Zhang X, Fasano R J, Schäffer S A, Beloy K, Nicolodi D, Brown R C, Hinkley N, Milani G, Schioppo M, Yoon T H and Ludlow A D 2018 *Nature* 1 ISSN 1476-4687 URL `https://www.nature.com/articles/s41586-018-0738-2`

[8] Alves B X R, Foucault Y, Vallet G and Lodewyck J 2019 Background gas collision frequency shift on lattice-trapped strontium atoms *2019 Joint Conference of the IEEE International Frequency Control Symposium and European Frequency and Time Forum (EFTF/IFC)* pp 1–2

[9] Guo C, Favier M, Galland N, Cambier V, Álvarez-Martínez H, Lours M, De Sarlo L, Andia M, Le Targat R and Bize S 2020 *Review of Scientific Instruments* **91** 033202 ISSN 0034-6748 URL `https://aip.scitation.org/doi/10.1063/1.5140793`

[10] Guo C, Cambier V, Calvert J, Favier M, Andia M, de Sarlo L and Bize S 2023 *Phys. Rev. A* **107** 033116 URL `https://link.aps.org/doi/10.1103/PhysRevA.107.033116`

[11] Dawkins S, Chicireanu R, Petersen M, Millo J, Magalhães D, Mandache C, Le Coq Y and Bize S 2010 *Appl. Phys. B: Lasers and Optics* **99**(1) 41 ISSN 0946-2171 URL `http://dx.doi.org/10.1007/s00340-009-3830-3`

[12] Millo J, Magalhaes D V, Mandache C, Coq Y L, English E M L, Westergaard P G, Lodewyck J, Bize S, Lemonde P and Santarelli G 2009 *Phys. Rev. A* **79** 053829 URL `http://link.aps.org/abstract/PRA/v79/e053829`

[13] Roberts B M, Delva P, Al-Masoudi A, Amy-Klein A, Bærentsen C, Baynham C F A, Benkler E, Bilicki S, Bize S, Bowden W, Calvert J, Cambier V, Cantin E, Curtis E A, Dörscher S, Favier M, Frank F, Gill P, Godun R M, Grosche G, Guo C, Hees A, Hill I R, Hobson R, Huntemann N, Kronjäger J, Koke S, Kuhl A, Lange R, Legero T, Lipphardt B, Lisdat C, Lodewyck J, Lopez O, Margolis H S, Álvarez-Martínez H, Meynadier F, Ozimek F, Peik E, Pottie P E, Quintin N, Sanner C, Sarlo L D, Schioppo M, Schwarz R, Silva A, Sterr U, Tamm C, Targat R L, Tuckey P, Vallet G, Waterholter T, Xu D and Wolf P 2020 *New J. Phys.* **22** 093010 ISSN 1367-2630 URL `https://doi.org/10.1088/1367-2630/abaace`

[14] See the invited presentation of A. Amy-Klein et al., *REFIMEVE frequency and time network and applications* at the 2023 Symposium on Frequency Standards and Metrology and: URL `https://www.refimeve.fr/index.php/en/`

[15] Häfner S, Falke S, Grebing C, Vogt S, Legero T, Merimaa M, Lisdat C and Sterr U 2015 *Opt. Lett.* **40** 2112–2115 URL `http://ol.osa.org/abstract.cfm?URI=ol-40-9-2112`

[16] Matei D G, Legero T, Häfner S, Grebing C, Weyrich R, Zhang W, Sonderhouse L, Robinson J M, Ye J, Riehle F and Sterr U 2017 *Phys. Rev. Lett.* **118** 263202 URL `https://link.aps.org/doi/10.1103/PhysRevLett.118.263202`

[17] Taichenachev A V, Yudin V I, Oates C W, Hoyt C W, Barber Z W and Hollberg L 2006 *Phys. Rev. Lett.* **96** 083001 (pages 4) URL `http://link.aps.org/abstract/PRL/v96/e083001`

[18] Baillard X, Fouché M, Targat R L, Westergaard P G, Lecallier A, Coq Y L, Rovera G D, Bize S and Lemonde P 2007 *Opt. Lett.* **32** 1812 URL `https://doi.org/10.1364/OL.32.001812`

[19] Origlia S, Pramod M S, Schiller S, Singh Y, Bongs K, Schwarz R, Al-Masoudi A, Dörscher S, Herbers S, Häfner S, Sterr U and Lisdat C 2018 *Phys. Rev. A* **98** 053443

[20] Takano T, Mizushima R and Katori H 2017 *Appl. Phys. Express* **10** 072801 ISSN 1882-0778

[21] Barber Z W, Hoyt C W, Oates C W, Hollberg L, Taichenachev A V and Yudin V I 2006 *Phys. Rev. Lett.* **96**(8) 083002 URL `http://link.aps.org/doi/10.1103/PhysRevLett.96.083002`

[22] Hosten O, Engelsen N J, Krishnakumar R and Kasevich M A 2016 *Nature* **529** 505–508

[23] Pedrozo-Peñafiel E, Colombo S, Shu C, Adiyatullin A F, Li Z, Mendez E, Braverman B, Kawasaki A, Akamatsu D, Xiao Y *et al.* 2020 *Nature* **588** 414–418

[24] Robinson J M, Miklos M, Tso Y M, Kennedy C J, Bothwell T, Kedar D, Thompson J K and Ye J 2024 *Nature Physics* **20** 208–213

[25] Vallet G, Bookjans E, Eismann U, Bilicki S, Le Targat R and Lodewyck J 2017 *New Journal of Physics* **19** 083002

[26] Shang H, Marin M A C, Foucault Y, Le Targat R and Lodewyck J 2022 Quantum nondemolition detection for strontium optical lattice clock *2022 Joint Conference of the European Frequency and Time Forum and IEEE International Frequency Control Symposium (EFTF/IFCS)* (IEEE) pp 1–3

[27] Bowden W, Vianello A, Hill I R, Schioppo M and Hobson R 2020 *Physical Review X* **10** 041052

[28] Huang M Z, de la Paz J A, Mazzoni T, Ott K, Rosenbusch P, Sinatra A, Garrido Alzar C L and Reichel J 2023 *PRX Quantum* **4** 020322 ISSN 2691-3399

[29] Lodewyck J, Westergaard P G and Lemonde P 2009 *Physical Review A* **79** 061401

[30] Numata K, Kemery A and Camp J 2004 *Physical Review Letters* **93** 250602

[31] Nicholson T L, Martin M J, Williams J R, Bloom B J, Bishof M, Swallows M D, Campbell S L and Ye J 2012 *Physical Review Letters* **109** 230801

[32] Cole G D, Zhang W, Martin M J, Ye J and Aspelmeyer M 2013 *Nature Photonics* **7** 644–650 ISSN 1749-4893

[33] Robinson J M, Oelker E, Milner W R, Zhang W, Legero T, Matei D G, Riehle F, Sterr U and Ye J 2019 *Optica* **6** 240–243
[34] Gobron O, Jung K, Galland N, Predehl K, Le Targat R, Ferrier A, Goldner P, Seidelin S and Le Coq Y 2017 *Optics Express* **25** 15539–15548
[35] Galland N, Lučić N, Zhang S, Alvarez-Martinez H, Le Targat R, Ferrier A, Goldner P, Fang B, Seidelin S and Le Coq Y 2020 *Optics letters* **45** 1930–1933
[36] Lin X, Hartman M T, Zhang S, Seidelin S, Fang B and Le Coq Y 2023 *Opt. Expr.* **31** 38475
[37] Zhang S, Lučić N, Galland N, Le Targat R, Goldner P, Fang B, Seidelin S and Le Coq Y 2020 *Applied Physics Letters* **117** 221102
[38] Galland N, Lučić N, Fang B, Zhang S, Le Targat R, Ferrier A, Goldner P, Seidelin S and Le Coq Y 2020 *Phys. Rev. Applied* **13** 044022
[39] Zhang S, Galland N, Lučić N, Le Targat R, Ferrier A, Goldner P, Fang B, Le Coq Y and Seidelin S 2020 *Physical Review Research* **2** 013306
[40] Könz F, Sun Y, Thiel C W, Cone R L, Equall R W, Hutcheson R L and Macfarlane R M 2003 *Phys. Rev. B* **68** 085109
[41] Thorpe M J, Rippe L, Fortier T M, Kirchner M S and Rosenband T 2011 *Nature Photonics* **5** 688–693
[42] Zhang S, Seidelin S, Le Targat R, Goldner P, Fang B and Le Coq Y 2023 *Phys. Rev. A* **107**(1) 013518
[43] Nicolodi D, Argence B, Zhang W, Le Targat R, Santarelli G and Le Coq Y 2014 *Nat Photon* **8** 219 ISSN 1749-4893 URL `http://dx.doi.org/10.1038/nphoton.2013.361`
[44] Orenes D B, Sewell R J, Lodewyck J and Mitchell M W 2022 *Phys. Rev. Lett.* **128** 153201 URL `https://link.aps.org/doi/10.1103/PhysRevLett.128.153201`
[45] Shiga N, Mizuno M, Kido K, Phoonthong P and Okada K 2014 *New Journal of Physics* **16** 073029 URL `http://stacks.iop.org/1367-2630/16/i=7/a=073029`
[46] Kohlhaas R, Bertoldi A, Cantin E, Aspect A, Landragin A and Bouyer P 2015 *Phys. Rev. X* **5**(2) 021011 URL `http://link.aps.org/doi/10.1103/PhysRevX.5.021011`
[47] Dörscher S, Al-Masoudi A, Bober M, Schwarz R, Hobson R, Sterr U and Lisdat C 2020 *Communications Physics* **3** 1–9 ISSN 2399-3650 URL `http://www.nature.com/articles/s42005-020-00452-9`
[48] Dimarcq N, Gertsvolf M, Mileti G, Bize S, Oates C W, Peik E, Calonico D, Ido T, Tavella P, Meynadier F, Petit G, Panfilo G, Bartholomew J, Defraigne P, Donley E A, Hedekvist P O, Sesia I, Wouters M, Dubé P, Fang F, Levi F, Lodewyck J, Margolis H S, Newell D, Slyusarev S, Weyers S, Uzan J P, Yasuda M, Yu D H, Rieck C, Schnatz H, Hanado Y, Fujieda M, Pottie P E, Hanssen J, Malimon A and Ashby N 2024 *Metrologia* **61** 012001 ISSN 0026-1394 URL `https://dx.doi.org/10.1088/1681-7575/ad17d2`

Journal of Physics: Conference Series **2889** (2024) 012049 doi:10.1088/1742-6596/2889/1/012049

Duty cycle engineering for improved optical lattice clock stability

Filip Butuc-Mayer, Matthew Johnson, Chen-Hao Feng, William Bowden, Richard Hobson, Ian R Hill

National Physical Laboratory, Hampton Road, Teddington, TW11 0LW, UK

E-mail: ian.hill@npl.co.uk

Abstract.
We describe a method of duty cycle engineering in optical lattice clocks to reduce clock instability through tailoring of the Dick effect noise contribution. We apply dynamically decoupled interrogation sequences to realise an estimated improvement in single clock instability approaching a factor of two, as well as providing a route to zero dead-time operation for two interleaved clocks which are otherwise unable to reach the required duty cycle. We show through simulations and synchronous measurements of clocks NPL-Sr1 and NPL-Sr2 that a quantum projection noise limited instability $< 3 \times 10^{-17}\tau^{-1/2}$ can be realised in a near ideal zero dead-time configuration.

1. Introduction

As a tool for precision measurement, the $1/\sqrt{N}$ scaling of quantum projection noise (QPN) limited instability offered by the N-atom optical lattice clock platform [1, 2] is compelling; simply operate with an ensemble of a million cold atoms and see a 1000-fold improvement in frequency stability compared to the single atom or ion case. Alas not. Due to the pulsed nature of its implementation, i.e. the time required to repeatedly prepare atoms and detect their state which interrupts the measurement of laser frequency noise, this scaling is seldom relevant. Instead, a modest number of a few hundred atoms is typically sufficient to reach the Dick effect limited instability imposed by the aliasing of clock laser frequency noise at Fourier frequencies near synchronous to the measurement cycle [3].

To make progress, lattice clocks have leaned heavily on a complementary effort to develop frequency stabilised lasers with exceptional phase coherence [4, 5, 6]. The improved laser performance both increases the duty cycle in which the laser is measured, and decreases the noise which is present to be aliased. This approach has enabled record single-clock instabilities at $< 5{\times}10^{-17}\,\tau^{-1/2}$ [7], though Dick effect noise remains limiting. In measurements between similar clocks that share the same local oscillator (LO), the Dick effect can be circumvented through common mode rejection of synchronously sampled laser frequency noise. Here, clock comparisons have reached the underlying quantum projection noise (QPN) limit, both between distinct clock setups at $< 3.5{\times}10^{-17}\,\tau^{-1/2}$ [7], and $< 1{\times}10^{-17}\,\tau^{-1/2}$ across or between neighbouring ensembles within the same apparatus [8, 9]. These measurements prove the underlying capability of the platform and offer a powerful shortcut to a subset of precision measurements, such as probes of gravity at the mm-scale, and many body physics [10], but do not provide an improved frequency stability at the clock output. For comparisons of clocks that do not share the same LO or probe sequence, or where a phase coherent clock output is required, a different approach is needed.

2. Duty cycle engineering for improved stability

In a Ramsey interrogation scheme the atoms are put into a coherent superposition state by a $\pi/2$ pulse and the evolution of the atomic phase is compared with maximal sensitivity to the laser phase, measured after a second $\pi/2$ pulse. The pulse times, $t_\pi/2$, and in particular the dark time, t_D, between them are the moments when the laser frequency is monitored, so extending this time is desirable. There is a limit to the useful length of the probe time since only phase differences within an interval of $\pm\pi/2$ can be determined and beyond this the measurement of phase becomes ambiguous.

We can extend the useful coherence time of the probe without improvements to the frequency noise in the local oscillator by reducing the sensitivity of the interrogation. This is done by engineering the probe scheme to dynamically decouple the measurement from the laser noise to some extent, as proposed in [11]. In contrast to a normal Ramsey probe, in this type of interrogation the atomic phase evolution is occasionally interrupted by near-π probe pulses, alternating $\pm\pi/2$-phase, that rotate or 'flip' the accumulated phase. The flip pulses have an area smaller than π by amount ϵ so that they encode a part of this phase, and the sequence is ended by a final pulse of length $\epsilon/2$. In order to maintain coherence in the phase measurement for the extended probe, both the free evolution time in between flip pulses and the total integrated sensitivity should not be greater than those of the conventional Ramsey probe. Keeping the pulse separation at t_D and reducing the sensitivity level by the same factor as the probe time extension will ensure both conditions are met. In the limit of $t_\pi \ll t_D$, the probe time is extended in factors of t_D, M, by choosing $\epsilon = 2\arcsin(1/M)$.

This scheme was implemented in NPL-Sr1 [12] to extend the 334 ms Ramsey probe (t_D = 300 ms) to 679 ms ($M = 2$) and to 2.012 s ($M = 6$) without significant loss of coherence, as shown in Figure 1. The short spikes in sensitivity are due to the flip pulses and results in non-uniform sampling of laser noise, so it is beneficial to keep the pulse time as short as practical. Here we used an easily achievable t_π of 34 ms to demonstrate the technique. For $M = 6$ we estimate an improvement in single-clock instability compared to Ramsey of a factor of 1.2 for a dead-time of 650 ms and a laser with overlapping Allan deviation (OADEV) flicker frequency noise 7×10^{-17} and white frequency noise 5×10^{-17} at 1 s. This improves to a factor of 1.7 with t_π reduced to 5 ms, owing to the increased uniformity of the sensitivity throughout the probe sequence. For comparison, estimated single-clock instabilities are provided in Table 1 for two cases; a state-of-the-art room temperature 48.5 cm ULE optical cavity characterised by the noise parameters given above and with 5 ms t_π, and a 5 cm ULE cubic cavity with flicker frequency noise 7×10^{-16} and white frequency noise 5×10^{-16} at 1 s. In the latter case, t_π was reduced to 1 ms to maintain similar improvements. For each of the described conditions, the dynamical decoupling scheme offers close to a factor of two decrease in calculated single clock instability, providing benefit to both state-of-the-art lab based systems, and transportable clocks where limitations are placed on the performance of the LO.

3. Reduced sensitivity zero dead-time optical lattice clock

An ideal optical lattice clock would monitor LO noise constantly with uniform sensitivity. Zero-deadtime (ZDT) optical lattice clocks have been implemented by interleaving two atomic ensembles, to provide a composite continuous tracking of the laser phase. This was achieved in [13, 14] using two separate clock chambers operating with ^{171}Yb, which lends itself well to this application owing to the relative speed and ease of its laser cooling. With careful timing and tailoring of the $\pi/2$ pulse envelopes of each Ramsey interrogation, the ideal uniform measurement sensitivity to laser phase noise can be achieved [3]. To implement this, a duty cycle of greater

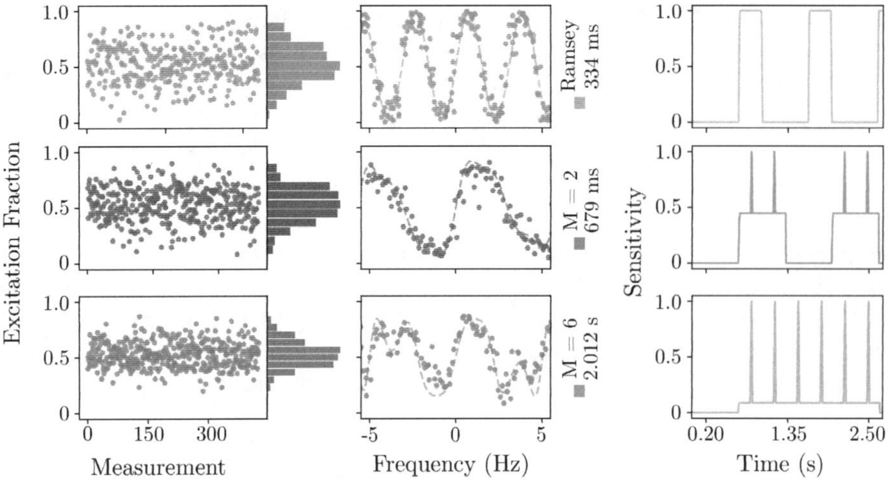

Figure 1: Experimental data from the implementation of a standard 334 ms Ramsey probe sequence, $t_\pi = 34$ ms, and two dynamically decoupled probe schemes operating with the same total sensitivity for two ($M = 2$) and six ($M = 6$) flip pulses. Left plots: excitation fraction measured while clock was locked. Centre plots: measured excitation fraction versus probe laser detuning (points) and the theoretical lineshapes (dashed lines). Right plots: theoretical sensitivities of the probe sequences as a function of time.

Table 1: Estimated single clock OADEV instability in units 1×10^{-17} for Ramsey spectroscopy with 300 ms dark-time, and dynamically decoupled spectroscopy operating with the same total sensitivity for two, and six flip pulses (denoted by $M = 2$ and $M = 6$), similar to Figure 1. Simulations were carried out for two levels of clock laser noise denoted by σ_{lab} and σ_{compact} and t_π of 5 ms and 1 ms, respectively. The contributions from QPN and Dick effect (DE) are provided along with the total instability.

	$\sigma_{\text{lab}} = (7 + 5\tau^{-1/2})$, $t_\pi = 5$ ms			$\sigma_{\text{compact}} = (70 + 50\tau^{-1/2})$, $t_\pi = 1$ ms		
Probe	Ramsey	$\text{DD}_{M=2}$	$\text{DD}_{M=6}$	Ramsey	$\text{DD}_{M=2}$	$\text{DD}_{M=6}$
QPN	1.2	1.4	1.9	2.9	3.1	4.0
DE	9.0	6.9	5.2	120.0	92.0	63.0
Total	**9.1**	**7.0**	**5.5**	**120.0**	**92.1**	**63.1**

than 50 % is required for each clock, which can be non-trivial to achieve in the commonly used species ^{87}Sr, which typically requires a preparation time greater than 0.5 s, demanding a very good LO. With dead-time difficult to reduce (in our case $\geq$ 600 ms) and the maximum coherence time for the Ramsey interrogation (in our case $\sim$ 350 ms) fixed by the laser noise, another approach is needed.

The technique of duty cycle engineering described in this paper opens new avenues for the implementation of near-ZDT (nZDT) composite clocks by enabling 50 % duty cycles to be reached somewhat arbitrarily. Here, 'near' accounts for the spikes in sensitivity which deviate from the ideal uniform case, owing to the flip pulses required in the dynamical decoupling probe scheme. We have experimentally demonstrated a 50 % duty cycle in NPL-Sr1 using a dynamically decoupled probe sequence with two flip pulses (Figure 1; row two). This method offers an easily implementable solution for existing lattice clock systems, allowing nZDT operation for interleaved systems with an unfavourable ratio of loading time to laser coherence time, including in systems with low performance local oscillators where the gap can be significant. The scheme trades off QPN through an extended clock cycle time at a given integrated probe sensitivity, for a sizeable reduction in the limiting Dick effect noise.

4. Simulations of a nZDT lattice clock

We developed numerical simulations to estimate the potential gain in stability from this method. We initially generate a noise sample that approximates the measured noise profile of our local oscillator [4], inflated slightly to account for observed short-term clock instability. We obtain a value for the average frequency offset of the laser noise measured during a single clock cycle in two stages. First, the calculated sensitivity, comprising dead-time and probe sequence, is used to estimate the measured laser frequency offset during a single clock cycle. Second, this is mapped to an excitation fraction using a calculated lineshape for the probe sequence, before finally being converted to a frequency correction, which assumes a linear frequency discriminator, and fed into the atomic servo. This correction is applied to the entire laser frequency noise sample beyond the measured period and the process repeated for the next period of laser noise. The resulting modified laser noise sample represents the output of the clock steered by the simulated clock operation. The atomic servo is implemented according to $\nu_{n+1} = \nu_n + g_1 e_n + g_2 \Delta_t \sum_{i=0}^{n} e_n$, where g_1 and g_2 are the individual servo gains, e_n the measured frequency error, Δt the time between corrections, and ν_n the laser frequency at the nth iteration of the loop.

Figure 2 shows simulations of the laser noise and resulting clock output OADEVs, along with snapshots of clock sequence sensitivity, for three probe schemes; conventional Ramsey with 600 ms dead-time, $t_D = 300$ ms and $t_\pi = 30$ ms, the same clock sequence for two interleaved clocks, and interleaved clocks each with 670 ms dynamically decoupled probe sequence with two flip pulses ($M = 2$), $t_\pi = 30$ ms. The clock noise at very short time scales is equal to that of the laser due to the bandwidth of the atomic servo. At longer time scales the noise decreases at a slope approximating $1/\tau$ before asymptoting to a $1/\sqrt{\tau}$ line set by the servo noise, roughly equal to the quadrature sum of the analytical solutions for QPN and the Dick effect noise.

Two clocks interleaved with identical Ramsey sequences have an improved instability of $7 \times 10^{-17}\tau^{-1/2}$, in comparison to $1.3 \times 10^{-16}\tau^{-1/2}$ for the single clock, yet leave some intervals of dead-time where the laser noise is not measured. In both these cases the Dick effect dominates. On the other hand, two clocks that have extended probe times afforded by the dynamical decoupling technique can monitor the laser noise without gaps, giving a combined sensitivity which is near uniform, and a simulated instability of $< 4 \times 10^{-17}\tau^{-1/2}$, dominated by QPN.

5. Experimental verification at the QPN limit

The minimised Dick effect in the nZDT lattice clock should allow for the realisation of a state-of-the-art QPN-limited instability. To verify that our systems will support this we operated the constituent clocks (NPL-Sr1 [12] and NPL-Sr2 [15]) simultaneously in a synchronous comparison, i.e., with aligned interrogations such that the sampled clock laser noise is common mode. This allows for the rejection of the local oscillator noise and reveals the underlying instability of the two systems. Clocks were compared using matched sequences with 1340 ms dead-time and $t_D = 270$ ms with $t_\pi = 100$ ms. To achieve a high degree of common mode rejection in the comparison of clocks, differential phase noise in the delivery of the clock light was minimised. This was achieved through careful attention to the path length stabilisation setups which use a combination of interferometers at both the clock and the lattice wavelengths to compensate optical paths all the way up to the atoms' position, without interruption from pulsing of the clock light during the interrogation. With atomic servo parameters consistent between clocks, we achieved a 0.996 correlation between excitation signals in each clock servo when probing the $5s^2\ ^1S_0$, $m_F = \pm 5/2$ to $5s5p\ ^3P_0$, $m_F = \pm 3/2$, $\sigma_\mp$ transitions. Being close to twenty times less

Journal of Physics: Conference Series **2889** (2024) 012049 doi:10.1088/1742-6596/2889/1/012049

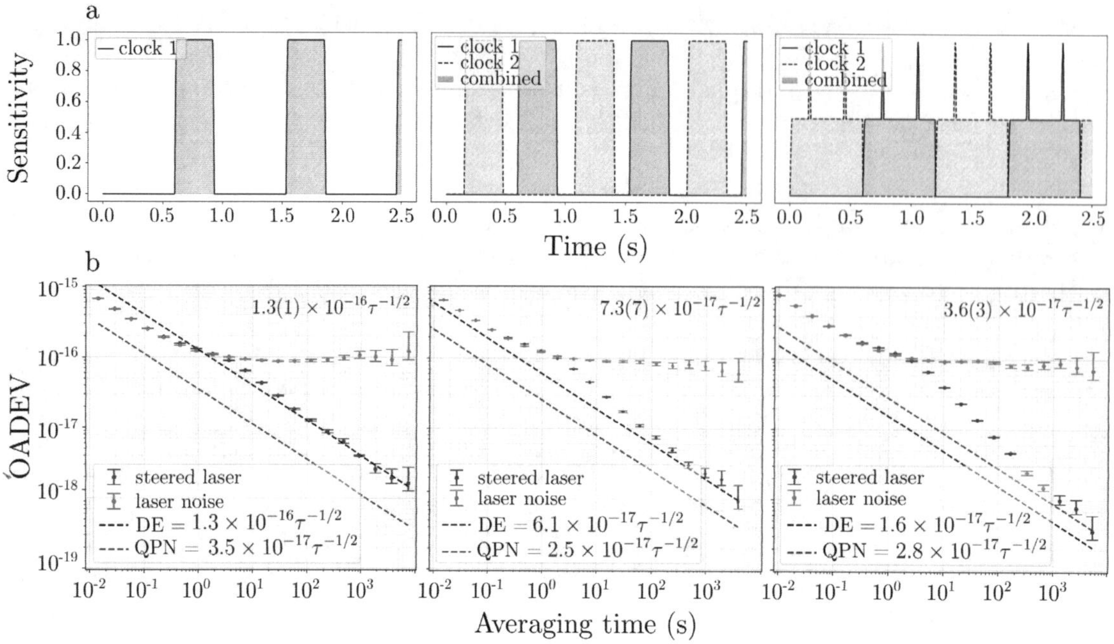

Figure 2: Simulations comparing individual and composite systems comprising lattice clocks with a nominal 600 ms dead-time and 30 ms π-pulse time: a single clock using Ramsey probes (left), two interleaved clocks with Ramsey probes (middle), and two interleaved clocks using dynamically decoupled spectroscopy to implement a nZDT composite clock, with marginally increased dead-time for each contributing clock (right). **a)** Sensitivity of the individual and combined clock sequences to laser frequency noise, with the contribution from each clock depicted by the solid and dotted lines. The composite clock with 330 ms Ramsey probes (middle) removes some dead-time while the 670 ms extended probes using dynamically decoupled pulse schemes (right) remove the dead-time completely, leaving only small spikes upwards in the sensitivity during the flip pulses and downwards during the transition between clock interrogations. The latter can be removed with careful tailoring of the opening and closing pulse envelopes. **b)** Results of numerical simulations (solid points) and analytical models (dashed lines) for comparison of long-term instability. The dashed lines are estimates of the individual contributions of the Dick effect (DE) and QPN to instability. The numbers in top right represent the total estimated asymptotic white frequency noise instabilities. Simulations are based on a laser with white frequency noise OADEV at 1 s of 8×10^{-17} and a flicker frequency floor of 9×10^{-17} and assuming an ensemble of 1000 atoms. Error bars on the OADEV points represent the 1σ confidence interval.

magnetically sensitive than transitions between the $m_F = \pm 9/2$ stretched states, they were used to suppress the contribution of differential magnetic field noise between the two independent clocks [16, 7]. The observed instability is adjusted to show the contribution of a single clock and shown in Figure 3. This corresponds to the expected QPN limit at $4 \times 10^{-17}\tau^{-1/2}$ for the comparison, similar to that achieved in [7], giving an estimated instability contributed by a single clock of $< 3 \times 10^{-17}\tau^{-1/2}$. This demonstrates an underlying performance compatible with that simulated for the nZDT QPN-limited composite clock.

Journal of Physics: Conference Series **2889** (2024) 012049
doi:10.1088/1742-6596/2889/1/012049

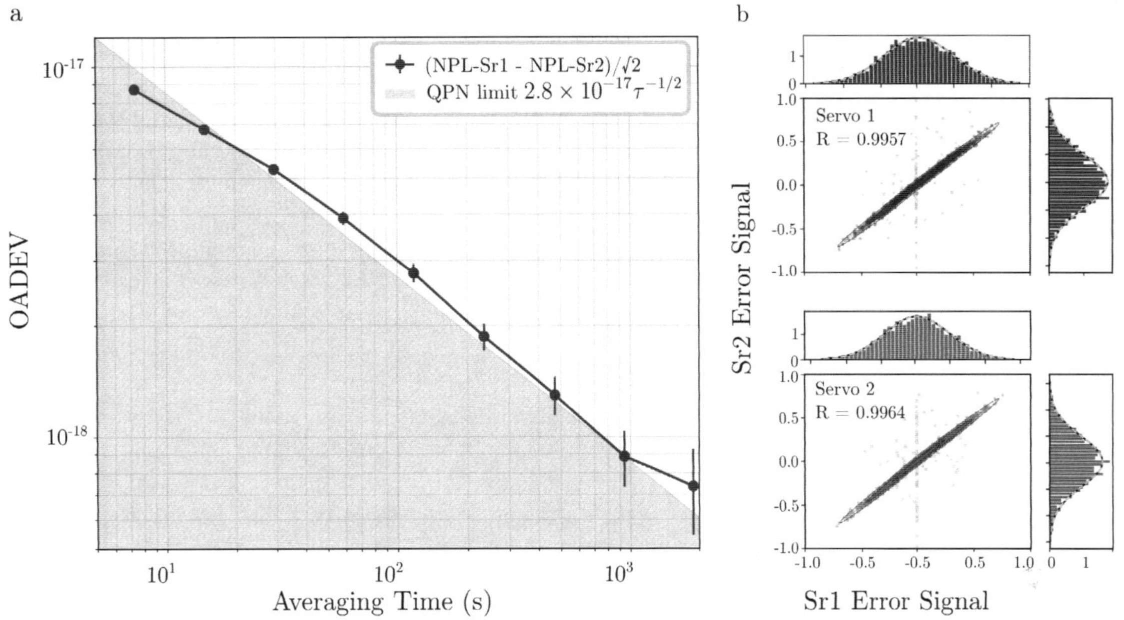

Figure 3: A synchronous comparison of clocks NPL-Sr1 and NPL-Sr2 each with dead-time 1340 ms and Ramsey probe sequence with t_D = 270 ms and t_π = 100 ms, operating on the $5s^2\ {}^1S_0$, $m_F = \pm 5/2$ to $5s5p\ {}^3P_0$, $m_F = \pm 3/2$, $\sigma_\mp$ transitions with ~4000 atoms. To remove first order sensitivity to Zeeman shifts, the comparison of clocks is based on the difference in mean frequency derived from the average of the centre frequencies in servos 1 and 2 for each clock. **a)** Overlapping Allan deviation adjusted to show the estimated single clock instability. Edge of shaded region indicates the expected limit from QPN. The first two data points sit below the indicated QPN limit as a consequence of the of atomic servo bandwidth, which is on the order of 0.1 Hz. Error bars on the points represent the 1σ confidence interval. **b)** Correlations between the error signals of the two clocks in corresponding servos for both upper and lower magnetically split clock transitions. Invalid measurements with low (missing) atom number in any of the clock cycles are plotted in grey and omitted from the histograms.

6. Conclusions

We have shown that a dynamically decoupled probe scheme can be used to modify the duty cycle of the lattice clock sequence, yielding a reduction in frequency instability when the dominant noise contribution is from the Dick effect. Numerical simulations of the modified clock cycle show that single clock instability can be reduced by a factor between 1.2 and 2 for realistic parameters of pulse duration and laser noise. The corresponding sequences were implemented in NPL-Sr1, extending probe time from 300 ms to 2 s per clock cycle without significant loss of contrast.

By applying the same technique to interleaved measurements in two clocks, each with a duty cycle increased to > 50 %, we have shown that near zero dead-time operation can be easily achieved without special apparatus, at the minor cost of a small increase in QPN. The predicted instability of the nZDT composite clock is compatible with a Dick effect limited instability of $1.6 \times 10^{-17}\tau^{-1/2}$ for realistic operating parameters, and could be improved further by shortening the t_π and tailoring the initial $\pi/2$ and closing $\epsilon/2$-pulse envelopes to engineer a composite sensitivity closer to the ideal uniform case. In experimental synchronous comparisons of clocks NPL-Sr1 and NPL-Sr2, we show that implementing this scheme in our systems could yield a QPN-limited instability at the level of $< 3 \times 10^{-17}\tau^{-1/2}$.

Journal of Physics: Conference Series **2889** (2024) 012049 doi:10.1088/1742-6596/2889/1/012049

The advantage of this solution is the minimal change required to operation with no improvements to the laser noise or to the cooling stage necessary. The relative ease of implementation makes future applications accessible. The increased stability would help in time-critical measurements, such as tidal variations in the geopotential, by allowing faster averaging of frequency noise in comparisons of frequency outputs from clocks, e.g., in the comparison of clocks between NMIs over the European dark-fibre network. In addition, the continuous phase tracking capability of the nZDT system can serve in further composite or hybrid clock schemes with clocks that share the same LO [16, 14], offering extended probe times significantly longer than that afforded by the bare laser coherence time, with implications for precision measurements including enhancing tests of time variation of the fine structure constant [17, 18, 19].

7. Acknowledgments

This work was supported by the UK government Department for Science, Innovation and Technology through the UK National Measurement System programme. The authors also acknowledge funding from the EMPIR programme for work on projects 18SIB05 ROCIT and 20FUN01 TSCAC, co-financed by the Participating States and from the European Union's Horizon 2020 research and innovation programme. Support was also received from the Science and Technology Facilities Council as part of the QSNET project ST/T00598X/1 within the Quantum Technologies for Fundamental Physics Programme.

References

[1] H. Katori, "Cold atom optical standards. Spectroscopy of strontium atoms in the Lamb-Dicke confinement," in *Proceedings of the 6th Symposium on Frequency Standards and Metrology*, P. Gill, Ed. World Scientific, Singapore, 2002.

[2] H. Katori, M. Takamoto, V. G. Pal'chikov, and V. D. Ovsiannikov, "Ultrastable Optical Clock with Neutral Atoms in an Engineered Light Shift Trap," *Physical Review Letters*, vol. 91, p. 173005, Oct 2003.

[3] G. J. Dick, "Local oscillator induced instabilities in trapped ion frequency standards," in *Proceedings of the 19th Annual Precise Time and Time Interval Systems and Applications Meeting*, 1987.

[4] M. Schioppo *et al.*, "Comparing ultrastable lasers at 7×10^{-17} fractional frequency instability through a 2220 km optical fibre network," *Nature Communications*, vol. 13, no. 1, p. 212, Jan 2022.

[5] D. G. Matei *et al.*, "1.5 μm Lasers with Sub-10 mHz Linewidth," *Physical Review Letters*, vol. 118, p. 263202, Jun 2017.

[6] J. Robinson *et al.*, "Silicon Cavity at 4 Kelvin with Thermal Noise Limited Performance," in *2018 IEEE International Frequency Control Symposium (IFCS)*, 2018, pp. 1–2.

[7] E. Oelker *et al.*, "Demonstration of 4.8×10^{-17} stability at 1 s for two independent optical clocks," *Nature Photonics*, vol. 13, pp. 714–719, 2019.

[8] T. Bothwell *et al.*, "Resolving the gravitational redshift across a millimetre-scale atomic sample," *Nature*, vol. 602, no. 7897, pp. 420–424, Feb 2022.

[9] X. Zheng, J. Dolde, M. C. Cambria, H. M. Lim, and S. Kolkowitz, "A lab-based test of the gravitational redshift with a miniature clock network," *Nature Communications*, vol. 14, no. 1, p. 4886, Aug 2023.

[10] R. B. Hutson, W. R. Milner, L. Yan, J. Ye, and C. Sanner, "Observation of millihertz-level cooperative lamb shifts in an optical atomic clock," *Science*, vol. 383, no. 6681, pp. 384–387, 2024.

[11] S. Dörscher *et al.*, "Dynamic decoupling of laser phase noise in compound atomic clocks," *Communications Physics*, vol. 3, p. 185, 2020.

[12] R. Hobson *et al.*, "A strontium optical lattice clock with 1×10^{-17} uncertainty and measurement of its absolute frequency," *Metrologia*, vol. 57, no. 6, p. 065026, oct 2020.

[13] M. Schioppo *et al.*, "Ultrastable optical clock with two cold-atom ensembles," *Nature Photonics*, vol. 11, pp. 48–52, 2017.

[14] M. E. Kim *et al.*, "Improved interspecies optical clock comparisons through differential spectroscopy," *Nature Physics*, vol. 19, no. 1, pp. 25–29, Jan 2023.

[15] W. Bowden *et al.*, "A pyramid MOT with integrated optical cavities as a cold atom platform for an optical lattice clock," *Scientific Reports*, vol. 9, no. 1, p. 11704, Aug 2019.

[16] W. Bowden, A. Vianello, I. R. Hill, M. Schioppo, and R. Hobson, "Improving the Q Factor of an Optical Atomic Clock Using Quantum Nondemolition Measurement," *Physical Review X*, vol. 10, p. 041052, Dec 2020.

[17] B. M. Roberts *et al.*, "Search for transient variations of the fine structure constant and dark matter using fiber-linked optical atomic clocks," *New Journal of Physics*, vol. 22, no. 9, p. 093010, sep 2020.

[18] R. M. Godun *et al.*, "Frequency Ratio of Two Optical Clock Transitions in $^{171}Yb^+$ and Constraints on the Time Variation of Fundamental Constants," *Physical Review Letters*, vol. 113, p. 210801, Nov 2014.

[19] N. Huntemann, B. Lipphardt, C. Tamm, V. Gerginov, S. Weyers, and E. Peik, "Improved Limit on a Temporal Variation of m_p/m_e from Comparisons of Yb^+ and Cs Atomic Clocks," *Phys. Rev. Lett.*, vol. 113, p. 210802, Nov 2014.

Journal of Physics: Conference Series **2889** (2024) 012050 doi:10.1088/1742-6596/2889/1/012050

High-accuracy multi-ion spectroscopy with mixed-species Coulomb crystals

J. Keller[1], H. N. Hausser[1], I. M. Richter[1], T. Nordmann[1], N. M. Bhatt[1], J. Kiethe[1], H. Liu[1], E. Benkler[1], B. Lipphardt[1], S. Dörscher[1], K. Stahl[1], J. Klose[1], C. Lisdat[1], M. Filzinger[1], N. Huntemann[1], E. Peik[1], and T. E. Mehlstäubler[1,2,3]

[1]Physikalisch-Technische Bundesanstalt, Bundesallee 100, 38116 Braunschweig, Germany
[2]Institut für Quantenoptik, Leibniz Universität Hannover, Welfengarten 1, 30167 Hannover, Germany
[3]Laboratory of Nano and Quantum Engineering, Leibniz Universtität Hannover, Schneiderberg 39, 30167 Hannover, Germany

E-mail: jonas.keller@ptb.de, tanja.mehlstaeubler@ptb.de

Abstract. Multi-ion optical clocks offer the possibility of overcoming the low signal-to-noise ratio of single-ion clocks, while still providing low systematic uncertainties. We present simultaneous spectroscopy of up to four $^{115}In^+$ clock ions in a linear Coulomb crystal, sympathetically cooled with $^{172}Yb^+$ ions. In first clock comparisons, we see agreement below 1×10^{-17} with results obtained using a single In^+ ion, for which we have evaluated the systematic uncertainty to be 2.5×10^{-18}. Operation with four clock ions reduces the instability from $1.6 \times 10^{-15}/\sqrt{t/(1\,\mathrm{s})}$ to $9.2 \times 10^{-16}/\sqrt{t/(1\,\mathrm{s})}$. We derive a model for decay-related dead time during state preparation, which matches the observed scaling of instability with clock ion number N, and indicates that $1/\sqrt{N}$ scaling can be achieved with the addition of a repump laser.

1 Introduction

Coulomb crystals provide a scalable approach for storing multiple clock ions in a well-controlled environment for interrogation with low systematic uncertainties. They also straightforwardly enable coupling to ions of a different species for sympathetic cooling, quantum logic operations or in-situ measurements of environmental conditions. Clocks based on Coulomb crystals have gained interest in recent years [1, 2, 3, 4, 5], as the use of multiple clock ions reduces quantum projection noise (QPN) [6], the limiting instability contribution of current single ion clocks. A given level of statistical uncertainty can thus be reached with shorter averaging times, which would otherwise need to be prohibitively long to match the decreasing systematic uncertainties of these systems. Due to its low sensitivities to electric field gradients, In^+ has been identified as one of the suitable candidates for a high-accuracy clock based on Coulomb crystals [7], with potential frequency uncertainties in the low 10^{-19} range [8]. Here, we report on spectroscopy and clock operation with multiple In^+ ions, sympathetically cooled with Yb^+ ions. We discuss site-resolved clock spectroscopy signals in section 2. In section 3, we present initial clock operation with multiple In^+ ions, compared against a Sr lattice and an Yb^+ ion clock. We investigate the scaling of the instability with clock ion number, derive a model which describes the current limitations due to dead time, and discuss our plans to overcome this limit.

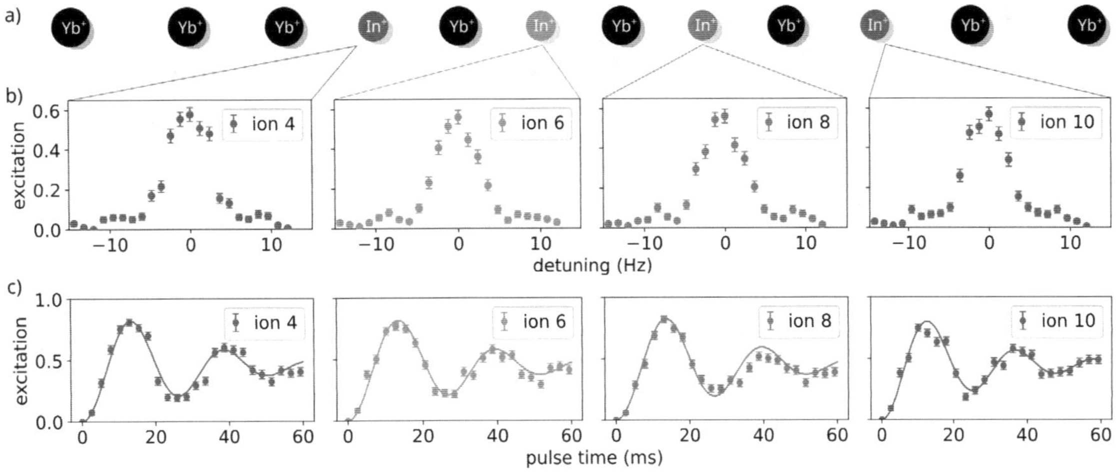

Figure 1: Spectroscopy with four In^+ ions. a) Mixed-species crystal consisting of four In^+ clock ions and eight Yb^+ cooling ions. b) Site-resolved frequency scan of the clock transition ($m_F = m_{F'} = +9/2$ component). The frequency axis offset is the same for all plots. c) Site-resolved Rabi oscillations compared to the theoretical expectation at 1.3 times the Doppler temperature for all $2N = 24$ radial modes, scaled by 0.9 to match the observed contrast. Error bars depict the quantum projection noise.

2 Spectroscopy of multiple clock ions in a mixed-species chain

We operated the clock with three different crystal compositions, containing one, two, and four In^+ clock ions, respectively. All experiments use the spectroscopy sequence described in [9], which allows independent evaluation of the individual ion signals and ensures a reproducible permutation of clock and cooling ion positions within the crystal. Denoting $^{115}In^+$ ions with "0" and $^{172}Yb^+$ with "1", we use the permutations 1011, 101011 and 111010101011 (or their mirror images, which are equivalent in terms of all relevant characteristics). These provide sufficient participation of the sympathetic cooling ions in all modes of motion. We apply Doppler cooling for 50 ms duration on the $^2S_{1/2} \leftrightarrow {}^2P_{1/2}$ transition in Yb^+, with a power corresponding to the saturation intensity I_{sat} in the first half, followed by a linear ramp down to $0.1 \times I_{\text{sat}}$. Most of the cooling light is applied along a radial direction, with projections onto both radial trap principal axes. An additional beam at 0.3 times the intensity provides cooling of the axial motion. The cooling is followed by interrogation of the clock transition. The clock laser addresses the crystal globally from the same direction as the radial cooling beam and has a waist of $w_z = 820\,\mu$m along the trap axis. Figure 1 shows measurements with the $4In^+$-$8Yb^+$ crystal, which is confined with secular frequencies (for the $^{172}Yb^+$ center-of-mass modes) of $(\omega_{\text{ax}}, \omega_{\text{rad1}}, \omega_{\text{rad2}})/2\pi = (110\,\text{kHz}, 821\,\text{kHz}, 833\,\text{kHz})$ and has an extent of $77\,\mu$m. Note that the ion numbering is reversed for iterations which use the mirrored permutation. Figure 1b shows a frequency scan over the $m_F = m_{F'} = +9/2$ Zeeman component of the $^1S_0 \leftrightarrow {}^3P_0$ clock transition, using 150 ms square pulses. All clock ions contribute with a contrast of ca. 0.6, which matches that seen in operation with a single clock ion [9].

Figure 1c shows Rabi oscillations on resonance of the same transition. The solid lines are added for a coarse temperature estimate: They use a model which takes into account spontaneous decay and thermal dephasing due to the $2N = 24$ radial modes of motion onto which the clock laser wave vector has nonzero projection (Debye-Waller effect [10]), assuming the set of motional modes along each radial principal axis to be at 1.3 times the respective Doppler temperature. The overall contrast of the model was reduced by a factor of 0.9 to fit the data, which could be caused by imperfect state preparation. For the parameters used in the frequency scans of Fig. 1b, the model reproduces the observed contrast of 0.6. The agreement of model and data would be further improved by more precise knowledge of the individual mode temperature degrees of freedom, e.g., through thermometry.

For the $1In^+$-$3Yb^+$ crystal, we have evaluated the systematic uncertainty to be 2.5×10^{-18} [9]. While a similar analysis has not yet been performed for the larger crystals, we expect most significant contributions to be independent of the ion number, with the exception of the thermal time dilation shift. The agreement with the Rabi oscillation model assuming temperatures close to the Doppler limit

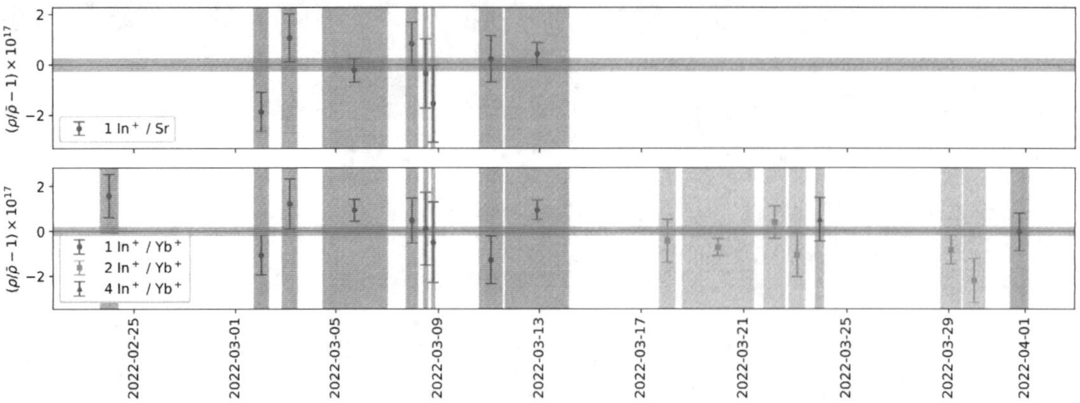

Figure 2: Optical frequency comparisons to ^{87}Sr (top) and ^{171}Yb$^+$ E3 (bottom) using different numbers of clock ions. Vertical shaded areas show the durations of the respective datasets. The green lines and shaded areas depict the weighted means and corresponding uncertainties. All shown uncertainties are purely statistical, based on the observed white frequency noise.

suggests that shifts are at the low 10^{-18} level, but more detailed analysis of the motional excitations is required to reduce the corresponding uncertainties. Lower temperatures of the relevant modes can be achieved with direct cooling of In$^+$ on the $^1S_0 \leftrightarrow {}^3P_1$ transition [11] in the intermediate cooling regime (transition linewidth on the order of the secular frequencies) [12], which is expected to reduce the time dilation shift to the low 10^{-19} range [8].

3 Multi-ion clock operation

The crystal compositions with one, two, and four In$^+$ ions have been used in optical clock operation. Figure 2 shows first measurements in which the In$^+$ clock is compared to a ^{87}Sr lattice clock [13] and an ^{171}Yb$^+$ single-ion clock based on the electric octupole (E3) transition [14]. Data before March 14th, 2022 are those used in the evaluation of [9]. The average In$^+$ / Yb$^+$ frequency ratio observed in the measurements with multiple In$^+$ ions agrees with the single-ion result to within less than 1×10^{-17}. Agreement at the 10^{-18} level will be tested in future campaigns. In the remainder, we focus instead on the observed multi-ion clock instabilities.

3.1 Decay-related dead times in state preparation

After successful excitation attempts, the ions need to be re-initialized in the ground state. The lifetime of $\tau = 195\,\text{ms}$ [15] is too short for an alternative approach which inverts the role of the two clock states in the subsequent iteration. Optical repumping can be used for deterministic initialization, but was not available during the time of the measurements presented here. Alternatively, the electronic excitation can be transferred to the state of motion and removed via laser cooling [16], but this requires additional experimental overhead and ground-state cooling. In the initial In$^+$ multi-ion clock operation, state preparation was instead achieved by repeatedly monitoring the state of the clock ions while waiting for natural decay. In the following, we investigate theoretically and experimentally the frequency instability implications of the associated dead time.

3.2 Average decay times for subsets of a clock ion ensemble

Site-resolved detection allows clock operation with a subset of n_{min} out of N clock ions, which can yield a lower instability than waiting for all clock ions to decay before every cycle. We therefore first derive the average delay until n_{min} decays have occurred.

For a two-level atom which is in the excited state at $t=0$, the excited state population follows $P_{e,1}(t) = e^{-t/\tau} = 1 - P_{g,1}(t)$. The probability to find n out of n_{exc} initially excited atoms in the ground

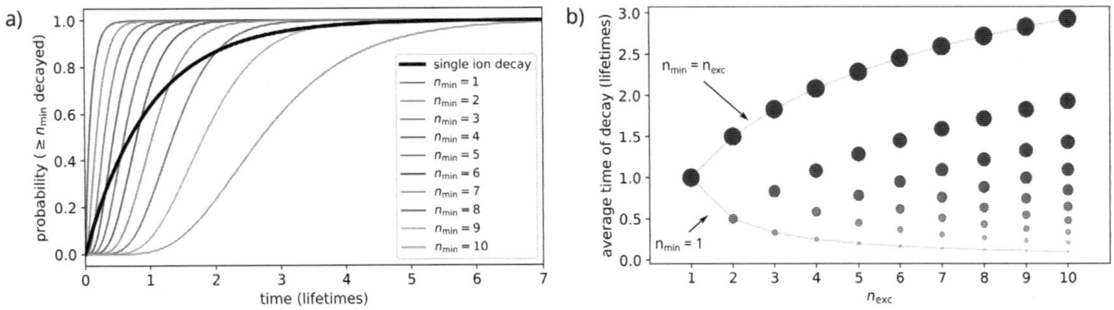

Figure 3: Decay times in ensembles of ions. a) Temporal evolution of the probability for at least $n_{\rm min}$ out of $n_{\rm exc} = 10$ ions to have decayed (see Eq. (4)). b) Average time at which the $n_{\rm min}$th ion decays (see Eq. (5)) for $n_{\rm exc} = 1 \dots 10$ and $n_{\rm min} = 1 \dots n_{\rm exc}$. The symbol sizes correspond to the respective fractions $n_{\rm min}/n_{\rm exc}$.

state is thus

$$P_g(n, n_{\rm exc}, t) = \binom{n_{\rm exc}}{n} P_{e,1}^{n_{\rm exc}-n}(t) P_{g,1}^{n}(t) \tag{1}$$

$$= \binom{n_{\rm exc}}{n} e^{-(n_{\rm exc}-n)t/\tau} \left(1 - e^{-t/\tau}\right)^n \tag{2}$$

$$= \sum_{i=0}^{n} \binom{n_{\rm exc}}{n}\binom{n}{i} (-1)^i \, e^{-(n_{\rm exc}-n+i)t/\tau} \,, \tag{3}$$

and the probability for *at least* $n_{\rm min}$ atoms to have decayed is

$$P_{g,\geq}(n_{\rm min}, n_{\rm exc}, t) = \sum_{j=n_{\rm min}}^{n_{\rm exc}} P_g(j, n_{\rm exc}, t) = \sum_{j=n_{\rm min}}^{n_{\rm exc}} \sum_{i=0}^{j} \binom{n_{\rm exc}}{j}\binom{j}{i} (-1)^i \, e^{-(n_{\rm exc}-j+i)t/\tau} \,, \tag{4}$$

as illustrated in Fig. 3a. The average time at which the $n_{\rm min}$th decay occurs is given by

$$T(n_{\rm min}, n_{\rm exc}) = \int_0^\infty t \left(\frac{\partial}{\partial t} P_{g,\geq}(n_{\rm min}, n_{\rm exc}, t)\right) dt = \tau \cdot \left(\sum_{j=n_{\rm min}}^{n_{\rm exc}} \sum_{\substack{i=0 \; (j<n_{\rm exc}) \\ i=1 \; (j=n_{\rm exc})}}^{j} \binom{n_{\rm exc}}{j}\binom{j}{i} \frac{(-1)^{(i+1)}}{n_{\rm exc} - j + i} \right) . \tag{5}$$

Figure 3b shows the average decay times (5) for $n_{\rm exc} = 1 \dots 10$ and $n_{\rm min} = 1 \dots n_{\rm exc}$. In a final step, we consider that each clock probe on average only excites a fraction $n_{\rm exc} = p_{\rm exc} N$ of the ions. By averaging over the distribution of excited ion numbers, we obtain the average dead time during clock operation:

$$T_{\rm decay}(n_{\rm min}, N, p_{\rm exc}) = \sum_{n_{\rm exc}=0}^{N} P_{\rm exc}\left(n_{\rm exc}, N\right) \, T\left(n_{\rm min} - \left(N - n_{\rm exc}\right), n_{\rm exc}\right) \,, \tag{6}$$

where $n_{\rm min}$ has been reduced by the amount $N - n_{\rm exc}$ of ions which were never excited, and we set $T(n_{\rm min}, N) = 0$ for $n_{\rm min} \leq 0$ (i.e., when the remaining number of ions ground state is enough to immediately continue with the next iteration). $P_{\rm exc}$ in (6) is another binomial distribution:

$$P_{\rm exc}(n_{\rm exc}, N) = \binom{N}{n_{\rm exc}} p_{\rm exc}^{n_{\rm exc}} \left(1 - p_{\rm exc}\right)^{(N - n_{\rm exc})} \,. \tag{7}$$

3.3 Implications for multi-ion clock instabilities

While the increased signal of a larger ion ensemble reduces QPN, the additional dead time without deterministic clock state initialization reduces the rate of data acquisition and partially counteracts this advantage. The overall effect can be expressed as

$$\sigma_y(t) = \frac{\sigma_{y,1}(t)}{\sqrt{n_{\text{min}}}} \sqrt{\frac{T_c + T_{\text{decay}}(n_{\text{min}}, N, p_{\text{exc}})}{T_c}} \,, \tag{8}$$

where $\sigma_{y,1}(t) \propto t^{-1/2}$ is the single-ion instability due to QPN at averaging time t, and T_c is the cycle time in the absence of the decay-related dead-time. Figure 4a illustrates the trade-off when choosing n_{min} for a given total number of ions for our parameters (see 3.4). The solid black line depicts the $1/\sqrt{n_{\text{min}}}$ scaling expected without additional dead time. As their number increases, it can be beneficial to operate with fewer than the total number of clock ions as long as state initialization depends on spontaneous decay. [1]

3.4 Experimental results

In clock operation with a single In^+ ion (cooled by three Yb^+ ions) and $t_p = 150\,\text{ms}$ probe pulses, we observe an average cycle time of 330 ms and an average excitation of $p_{\text{exc}} = 0.26$ at the probe detunings of $\delta = \pm 1/(2t_p)$. Of the 180 ms dead time, $p_{\text{exc}} \cdot \tau = 51\,\text{ms}$ can be attributed to decay. Most of the remaining dead time consists of Yb^+ sympathetic cooling and In^+ state detection, both of which can be subject to further optimizations. Figure 4b shows the QPN-limited instability at $t = 1\,\text{s}$ averaging time of the $^{115}In^+$ clock as determined from those of the optical frequency ratio measurements of Fig. 2. For $N = 1$, we infer $\sigma_y = 1.6 \times 10^{-15}/\sqrt{t/(1\,\text{s})}$ from the In^+ / Sr ratio, using the Sr clock instability of $2.0 \times 10^{-16}\sqrt{t/(1\,\text{s})}$ [17] (filled red circle). The blue triangles show the instabilities of the In^+ / Yb^+ (E3) frequency ratio, to which the Yb^+ single-ion clock contributes $1.0 \times 10^{-15}/\sqrt{t/(1\,\text{s})}$. The In^+ clock instabilities for $N = 2$ and $N = 4$ are derived from those data, since no comparisons to the lattice clock were available under these conditions (red open circles). With $N = 4$, the instability is reduced to $\sigma_y = 9.2(4) \times 10^{-15}/\sqrt{t/(1\,\text{s})}$. The solid lines and shaded area correspond to the theoretical expectations of Fig. 4a, adjusted to the observed instability at $N = 1$. The deviation from the ideal $\sigma_y \propto 1/\sqrt{N}$ scaling matches the expected behavior due to dead time increase. Thus, implementing a controlled clock state initialization and thereby reducing the dead time due to spontaneous decay, we expect to reach the instabilities depicted by the black line in Fig. 4.

3.5 Conclusion

In clock operation with multiple In^+ ions, the observed instabilities match the limit set by state preparation through spontaneous decay, as described by our model. To overcome this limitation in future measurements, state preparation with a repump transition ($^3P_0 \leftrightarrow {}^1P_1$) at 482 nm should be feasible [18] and will be investigated. Without any further technical improvements, this is expected to achieve the instabilities depicted by the solid black line of Fig. 4b.

Acknowledgments

We thank I. Vybornyi for helpful discussions. We acknowledge support by the projects 18SIB05 ROCIT and 20FUN01 TSCAC. These projects have received funding from the EMPIR programme co-financed by the Participating States and from the European Union's Horizon 2020 research and innovation programme. We acknowledge funding by the Deutsche Forschungsgemeinschaft (DFG) under Germany's Excellence Strategy – EXC-2123 QuantumFrontiers – 390837967 (RU B06) and through Grant No. CRC 1227 (DQ-mat, projects B02 and B03). This work has been supported by the Max-Planck-RIKEN-PTB-Center for Time, Constants and Fundamental Symmetries.

[1] Note that this treatment does not include cases in which more than n_{min} ions are available occasionally and the fact that ions on "discarded" sites might become available before the next cycle due to decay or interaction with the globally applied clock laser. The instability values for $n_{\text{min}} < N$ therefore constitute upper bounds, but neither of these effects occur in the case of $n_{\text{min}} = N$ (solid green line) tested below.

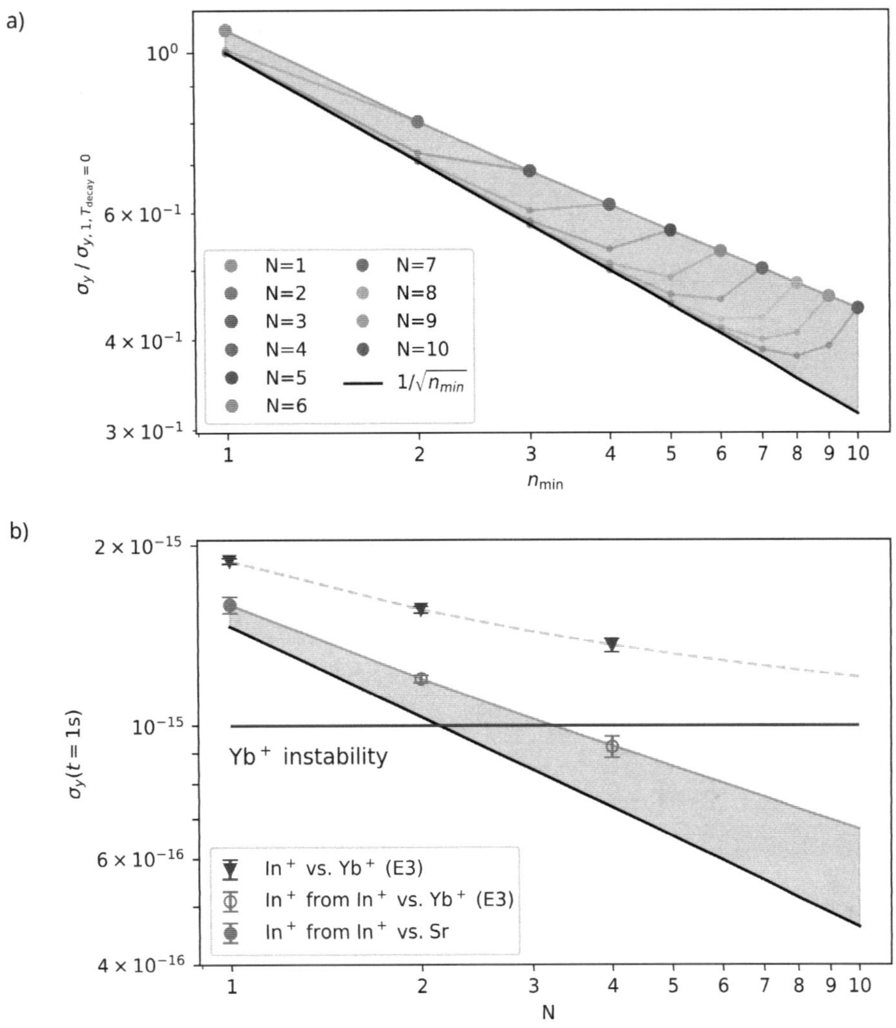

Figure 4: Multi-ion clock instabilities. a) Expected quantum projection noise limited instabilities in multi-ion clocks with atomic state initialization based on spontaneous decay (see Eq. (8)), assuming the experimental parameters of section 3.4. Each clock sequence iteration pauses until at least n_{min} clock ions are in the ground state. The different colors depict manifolds with different total numbers of clock ions N. All values are normalized to the single-ion QPN instability under the assumption of instantaneous decay. The shaded area indicates the increase with respect to the $1/\sqrt{n_{\mathrm{min}}}$ scaling in the absence of this dead time (solid black line). Note that this treatment potentially overestimates the instabilities for $n_{\mathrm{min}} < N$ (small circles), but is accurate for $n_{\mathrm{min}} = N$ (large circles) (see text). b) Experimentally observed instabilities in multi-ion operation by comparison to a Sr lattice clock and an Yb^+ (E3) single-ion clock. The solid lines and shaded area correspond to those of a). Their normalization is derived from the observed instability with a single clock ion [9]. Solid lines refer to the instability of the In^+ clock, while the dashed line shows the joint instability of the In^+ clock and Yb^+ single-ion clock. With the addition of a repump laser, the clock is expected to reach the instabilities of the solid black line.

References

[1] Arnold K, Hajiyev E, Paez E, Lee C H, Barrett M D and Bollinger J 2015 *Phys. Rev. A* **92**(3) 032108

[2] Aharon N, Spethmann N, Leroux I D, Schmidt P O and Retzker A 2019 *New J. Phys.* **21** 083040

[3] Cui K, Valencia J, Boyce K T, Clements E R, Leibrandt D R and Hume D B 2022 *Phys. Rev. Lett.* **129**(19) 193603

[4] Leibrandt D R, Porsev S G, Cheung C and Safronova M S 2022 (*Preprint* 2205.15484)

[5] Steinel M, Shao H, Filzinger M, Lipphardt B, Brinkmann M, Didier A, Mehlstäubler T E, Lindvall T, Peik E and Huntemann N 2023 *Phys. Rev. Lett.* **131**(8) 083002

[6] Itano W M, Bergquist J C, Bollinger J J, Gilligan J M, Heinzen D J, Moore F L, Raizen M G and Wineland D J 1993 *Phys. Rev. A* **47**(5) 3554–3570

[7] Herschbach N, Pyka K, Keller J and Mehlstäubler T E 2012 *Appl. Phys. B* **107** 891–906

[8] Keller J, Burgermeister T, Kalincev D, Didier A, Kulosa A P, Nordmann T, Kiethe J and Mehlstäubler T E 2019 *Phys. Rev. A* **99** 013405

[9] Hausser H N, Keller J, Nordmann T, Bhatt N M, Kiethe J, Liu H, von Boehn M, Rahm J, Weyers S, Benkler E, Lipphardt B, Dörscher S, Stahl K, Klose J, Lisdat C, Filzinger M, Huntemann N, Peik E and Mehlstäubler T E 2024 (*Preprint* 2402.16807)

[10] Wineland D J, Monroe C, Itano W M, Leibfried D and King B E and Meekhof D M 1998 *J. Res. Natl. Inst. Stand. Technol.* **103** 259

[11] Peik E, Abel J, Becker T, von Zanthier J and Walther H 1999 *Phys. Rev. A* **60**(1) 439–449

[12] Kulosa A P, Prudnikov O N, Vadlejch D, Fürst H A, Kirpichnikova A A, Taichenachev A V, Yudin V I and Mehlstäubler T E 2023 *New J. Phys.* **25** 053008

[13] Schwarz R 2022 *A cryogenic Strontium lattice clock* PhD thesis Gottfried Wilhelm Leibniz Universität Hannover

[14] Sanner C, Huntemann N, Lange R, Tamm C, Peik E, Safronova M S and Porsev S G 2019 *Nature* **567** 204–208

[15] Becker T, v Zanthier J, Nevsky A Y, Schwedes C, Skvortsov M N, Walther H and Peik E 2001 *Phys. Rev. A* **63**(5) 051802

[16] Chou C W, Kurz C, Hume D B, Plessow P N, Leibrandt D R and Leibfried D 2017 *Nature* **545** 203–207

[17] Schwarz R, Dörscher S, Al-Masoudi A, Benkler E, Legero T, Sterr U, Weyers S, Rahm J, Lipphardt B and Lisdat C 2020 *Phys. Rev. Res.* **2**(3) 033242

[18] Safronova M priv. comm.

doi:10.1088/1742-6596/2889/1/012051

Highly charged ion optical clocks

Piet O. Schmidt[1,2], Lukas J. Spieß[1], Alexander Wilzewski[1,2], Malte Wehrheim[1,2], Shuying Chen[1], Steven A. King[1,3], Peter Micke[1,4], Tobias Leopold[1,5], Nadir Khan[1,6], José R. Crespo López-Urrutia[6]

[1]Physikalisch-Technische Bundesanstalt, Braunschweig, Germany
[2]Institut für Quantenoptik, Leibniz Universität Hannover, Hannover, Germany
[3]present address: Oxford Ionics, Begbroke, UK
[4]present address: GSI Darmstadt and Friedrich-Schiller-Universität Jena, Germany
[5]present address: QUDORA Technologies, Braunschweig, Germany
[6]Max-Planck-Institut für Kernphysik, Heidelberg, Germany

E-mail: Piet.Schmidt@quantummetrology.de

Abstract. Optical atomic clocks are the most precise and accurate measurement devices, reaching fractional systematic uncertainties below one part in 10^{-18}. Their exceptional performance opens up a wide range of applications in fundamental science and technology. The extreme electronic properties of highly charged ions (HCI) make them highly sensitive probes for tests of fundamental physical theories and significantly lower sensitivity to some of the leading systematic perturbations that affect state-of-the-art optical clocks. This makes them exciting candidates for next-generation clocks. In this contribution, we will briefly review the history of optical spectroscopy of HCI and how scientific advances enabled the first coherent spectroscopy of HCI using quantum logic techniques. A first generation of optical clocks based on HCI with atom-related systematic uncertainties at the 10^{-18} level and below is demonstrated. Comparisons of the isotope shift between $^{40}Ar^{13+}$ and $^{36}Ar^{13+}$ as well as g-factor measurements with theory are interpreted as a test of quantum electrodynamic (QED) theory. This demonstrates the suitability of HCI as references for high-accuracy optical clocks and to probe for physics beyond the standard model.

1 Introduction to HCI

When electrons are removed one by one from a neutral atom, the electronic properties of the charged atoms change along the isonuclear sequence. Once a significant number of electrons are removed, the atoms are called highly charged ions (HCI). While HCI of the same isoelectronic sequence share many properties, the binding energy change dramatically. For example, for hydrogen-like HCI with charge Z, the binding energy of the last electron scales, the hyperfine splitting, and QED contributions to the binding energy scale as Z^2, Z^3, and Z^4, respectively [1]. In contrast, the polarizability and electric quadrupole moment decrease by Z^{-4} and Z^{-2}, respectively, due to the reduced size of the electron orbitals around the nucleus. At the same time, relativistic effects are enhanced, since in a semi-classical picture the electrons move faster in their smaller orbitals around the nucleus.

HCI are abundant in the universe, since matter in and around stars is highly ionized [2]. For decades, this has made HCI indispensable probes in astrophysics for temperature and magnetic fields [3], as well as x-ray astronomy [4], providing insights into, e.g., supernovae [5] and intergalactic gas [6]. Similarly, HCI have been used as probes for the temperature of plasmas on Earth, e.g. in fusion reactors [7]. In a completely different application, highly charged Sn^{12+} is employed as the source of 13.5 nm radiation in latest-generation photo-lithography systems in semiconductor industry [8, 9]. The small number of remaining electrons in H-, He-, and Li-like systems enables highly accurate *ab initio* atomic structure calculations. Together with the large QED contribution to the binding energy, comparison between experiment and calculation of, e.g. transition energies, Lamb-shift, or the ground state g-factor, provide the most stringent tests of QED theory [10–16]. HCI have also been suggested as qubit systems for quantum information processing with trapped ions [17–19], albeit not realized up to now.

The suitability of HCI as optical clock candidates was first put forward by Schiller [20]. Optical hyperfine transitions in hydrogen-like HCIs were proposed as optical clock candidates with strongly suppressed systematic frequency shifts, as well as to test the local position invariance of the electron and quark masses [20]. The low sensitivity of these HCI to external field perturbations is a direct consequence of the scaling of their electronic properties, as mentioned above. Typical systematic shifts from ac Stark shifts, such as black-body radiation (interaction of the thermal radiation field with the differential polarizability of the clock states), electric quadrupole shifts (interaction of electric quadrupole moments of the clock states with electric field gradients), or higher-order magnetic field shifts (coupling of nearby electronic levels by magnetic fields) benefit from the small polarizability, shrunk electron orbitals and sparse level structure, respectively.

Independently, HCI have been identified as promising atomic candidates with a high sensitivity of their energy levels to a change in the fine-structure constant α, since the sensitivity scales with the ionisation energy and $(Z\alpha)^2$ [21]. In particular, so-called level-crossing transitions have been predicted to be sensitive to a change in the fine-structure constant [1, 21–26] and for dark matter searches [27]. The high sensitivity is due to the very different electronic configurations and angular momentum states, which cross in the vicinity of a certain charge state and thus exhibit transitions in the optical frequency range. Many HCI species have been proposed as optical clock candidates [19, 28–53], and some of them have been summarized in [54].

To achieve any of these proposals, optical clock-type spectroscopy of HCI is necessary. The following section presents an overview of the state-of-the art in HCI optical spectroscopy prior to high-resolution coherent laser spectroscopy.

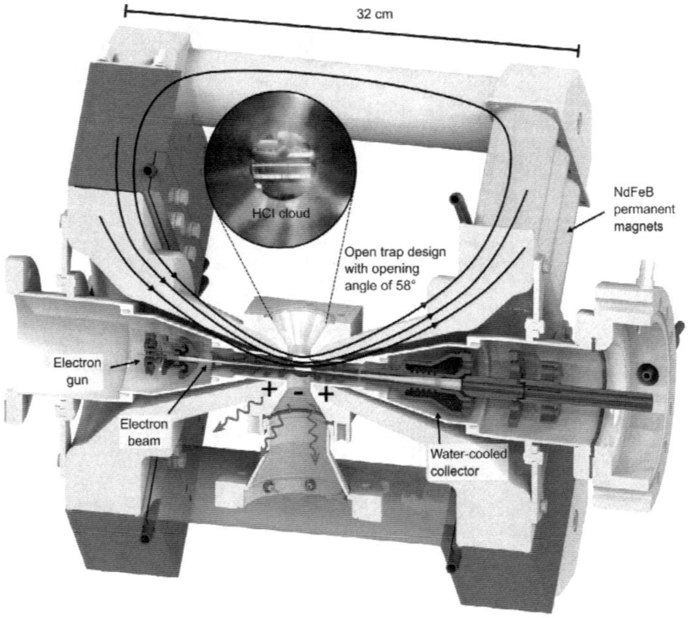

Figure 1: Schematic drawing of the Heidelberg compact electron beam ion trap (EBIT). Atoms are ionized by an electron beam, focussed by a 0.86 T strong magnetic field. It features an electron beam energy of up to 10 keV at a current of up to 80 mA. Confinement is provided by the space charge of the electron beam, the magnetic field and additional applied electric fields. After: Micke et al. [63].

2 Optical spectroscopy of HCI in electron beam ion traps

Optical transitions in HCI have been studied in storage rings [16, 55–58] and in electron beam ion traps (EBITs) [59–62], in which they are typically produced. In particular the latter have become the workhorse for HCI spectroscopy and sources of HCI for other applications. In an EBIT, electrons emitted from a thermal source (electron gun) are accelerated by a high voltage and focussed by a strong magnetic field into the target volume (see Fig. 1). The atomic species of interest is injected as a gas [63], as part of an organometallic compound [63], or through laser desorption or ablation [64]. The electron beam ionizes the atoms of interest and produces a plasma, where the distribution of charge states depends on the electron beam energy and current as well as the breeding time. Furthermore, the electron beam excites the HCI, resulting in fluorescence, which can be imaged onto high-resolution spectrometers. The ions' eV-level kinetic energy, corresponding to hundreds of thousands to millions of kelvin in temperature, result in tens of gigahertz of Doppler-broadening. The large signal-to-noise ratio and high-quality calibration of all experimental parameters enable in-EBIT spectroscopy with an uncertainty on the order of a hundred megahertz. Besides electron impact excitation, in-EBIT laser spectroscopy has been demonstrated using a pulsed laser [61], achieving a similar resolution. The most precisely measured optical transition of an HCI in an EBIT is the M1 fine-structure $^2P_{1/2} \leftrightarrow {}^2P_{3/2}$ transition in Ar^{13+} at 441 nm [61, 65–67], with a relative frequency uncertainty of 3×10^{-7}, corresponding to 200 MHz. Recently, laser spectroscopy in a Penning trap [68] has reduced this uncertainty by a factor of 30. The comparably well-known transition frequency has motivated us to

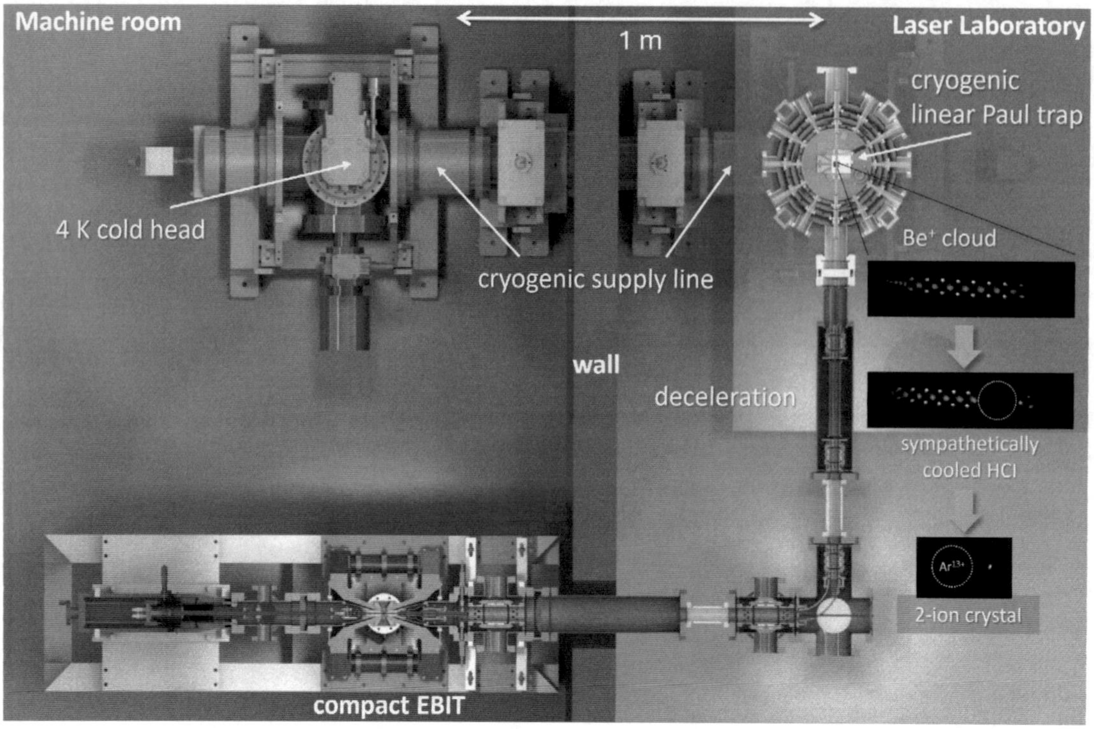

Figure 2: Overview of the setup for quantum logic spectroscopy of HCI. The inset shows a laser-cooled Coulomb crystal of trapped Be^+ ions, without and with a co-trapped Ar^{13+} HCI, and a Ar^{13+}-Be^+ two-ion crystal (top to bottom).

investigate Ar^{13+} as the first HCI candidate for high resolution coherent laser spectroscopy.
The main limitations with in-EBIT spectroscopy are the megakelvin temperatures of the plasma, paired with the general lack of suitable transitions for laser cooling, the strong inhomogeneous magnetic field, and small trapping frequencies. This results in Doppler broadening, higher-order magnetic field, and recoil shifts, respectively. All of these are detrimental to high-resolution spectroscopy. A solution to all of these issues is provided by quantum logic spectroscopy in a Paul ion trap as described in the next section.

3 Quantum logic spectroscopy of HCI

3.1 Experimental setup

Charged particles can be confined using a combination of static and radio frequency (rf) fields from a so-called Paul trap. These Paul traps [69, 70] confine ions in an almost field-free environment at megahertz secular frequencies, permitting recoil-free spectroscopy. Doppler [71, 72] and ground-state [73] laser cooling enable preparation of both external (motional) and internal (electronic) degrees of freedom in the quantum ground state. When combined with coherent laser or microwave manipulation, full control over the ion's quantum state is achieved, making them ideal candidates for applications such as quantum computing [74–77], quantum simulations [78, 79] and optical clocks [80]. Recent advances in characterizing and controlling systematic shifts in ion-based optical clocks has led to record relative systematic frequency uncertainties at a level of 10^{-18} [81–87].

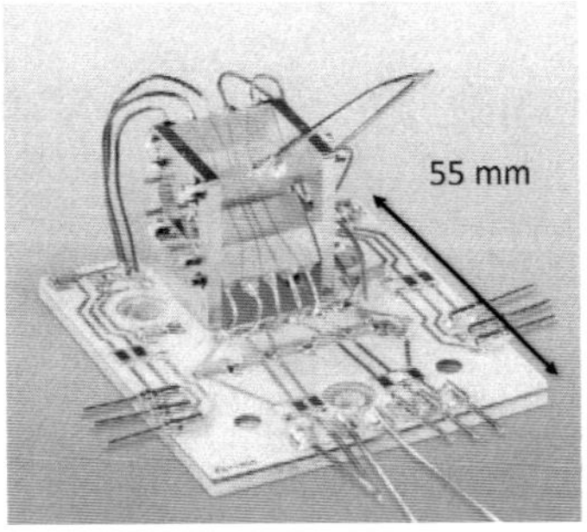

Figure 3: Picture of the PTB HCI Paul trap. It consists of gold-coated alumina blades that are laser-structured to isolate five axially segmented electrodes, providing axial confinement at secular frequencies of around 1 MHz for Be^+ at heating rates of around one motional quanta per second. Radial confinement at around 3 MHz for Be^+ is provided by an rf field applied to a pair of radial (unstructured) blade electrodes at a drive frequency of around 24 MHz. After: Leopold et al. [89].

Similarly, high accuracy optical clocks based on HCI can take advantage of the low systematic frequency shifts achievable in Paul traps. Building on the experience of the Cryogenic Paul Trap Experiment (CryPTEx) at Max-Planck-Institute für Kernphysik in Heidelberg [88], we have set up CryPTEx-PTB fulfilling all requirements for clock-type spectroscopy with HCI (see Fig. 2 for an overview of the setup). In our experiment, we employ a macroscopic linear Paul trap as shown in Fig. 3, which provides megahertz trapping frequencies and low heating rates [89]. We use Be^+ for sympathetic cooling and quantum logic spectroscopy of the HCI (see below).

HCI have a large charge transfer cross-section for collisions with background gas [90]. Thus a cryogenic vacuum can significantly prolong their lifetime in the trap. We have designed and built a low-vibration cryostat to cool the Paul trap and its environment to temperatures below 5 K [91]. From the approximately two hour's lifetime of the HCI in the trap, we estimate the vacuum pressure in the trap chamber is below 1×10^{-14} mbar. Imaging of the Be^+ ions is implemented through a single aspheric lens (22 mm optical aperture) on a remotely controllable cryogenic z-axis stage to adjust the focus around the 20 mm nominal working distance. Outside the cryostat, the near-diffraction-limited image with a magnification of three is further magnified (total of $\times 23$) and imaged with a doublet onto a charge-coupled device (CCD) camera and a photo-multiplier for internal state discrimination of the Be^+ logic ion and quantitative data analysis.

Following the pioneering experiments of Schmöger et al. [92, 93], HCI are extracted from the EBIT by applying a repulsive potential to the ions in the EBIT. Guided and focussed by ion-optical elements, the ions travel towards the cryogenic Paul trap. Ions are charge-to-mass ratio selected by their time-of-flight and slowed down to an energy of around 100 eV$/q$, where q is their charge state, using a deceleration stage [93]. The final energy loss to around 1 eV$/q$ takes place upon entering the Paul trap, which is kept at an elevated ground for this purpose. Prior to the injection of HCI, a laser-cooled crystal of tens of Be^+ ions has been prepared in the trap using laser ablation and photo-ionization [89]. The remaining kinetic energy of the HCI allows them to overcome the potential barrier of the trap which prevents the Be^+ from escaping. Using a fast electronic switch, this barrier is raised to a level sufficient for trapping HCI in the Paul trap. HCI dissipate their energy to the laser-cooled Be^+ ions in the trap until they co-crystallize. This is visible as one or several large voids in the Be^+ Coulomb crystal (see inset in Fig. 2). The re-trapping process is stochastic and is repeated every 2 s until a HCI

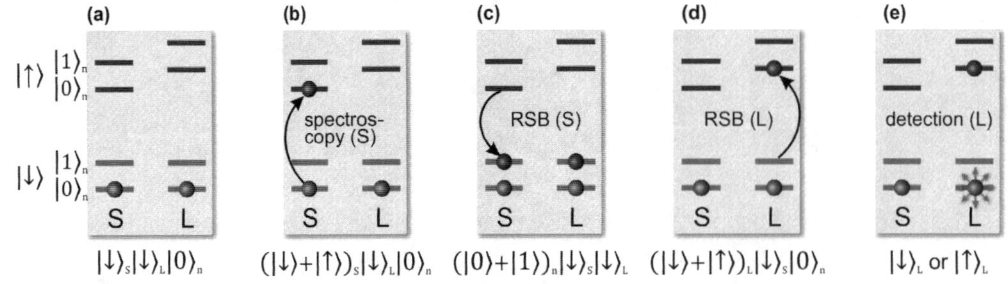

Figure 4: Basic quantum logic spectroscopy protocol. (a) Spectroscopy (index S) and logic (index L) ion are in the electronic ($|{\downarrow}\rangle_{S,L}$) and motional ($|0\rangle_n$) ground state. (b) Probe pulse on the spectroscopy ion generates an electronic superposition (normalization omitted). (c) A red-sideband (RSB) pulse on the spectroscopy ion maps the electronic into a motional superposition. (d) An RSB pulse on the logic ion maps the motional into an electronic superposition on the logic ion. (e) Internal state detection on the logic ion. After: Schmidt et al. [103].

has been crystallized. Excess Be^+ ions are then removed from the trap by applying a modulation of the radial trapping potential at the single Be^+ ion's secular frequency,while the Doppler cooling laser is turned off. This process currently requires manual interaction of the scientist, but can be automized taking advantage of the principal capability to split the ion crystal in the trap and discarding the part that does not contain the HCI [94]. This procedure prepares a 2-ion crystal containing a single Be^+ ion and a single HCI, which is the starting point for quantum logic spectroscopy.
The ions are strongly coupled due to their mutual Coulomb repulsion, resulting in two normal modes of motion in each of three mutually orthogonal directions. Manipulation of the normal mode of motion using one ion therefore directly affects the other ion and allows for sympathetic cooling. The cooling efficiency is governed by the difference in charge-to-mass ratio of the two ions and thus determines the choice of the cooling ion species. In most configurations sympathetic cooling via the logic or cooling ion becomes efficient for the axial motional modes [95–98]. A large difference though tends to decouple in particular the radial modes, resulting in modes in which the HCI (cooling ion) has a large (small) amplitude and vice versa. The modes with a small amplitude of the cooling ion (weakly coupled modes) are less efficiently cooled and involve a large amplitude of the HCI. New cooling strategies, such as algorithmic cooling, need to be employed (see below).
Interrogation of HCI is performed, e.g., using frequency-doubled extended-cavity diode lasers that are stabilized to passively stable optical cavities [99] to narrow the linewidth and for prestabilization. Long-term stability is transferred from a single crystal silicon-cavity stabilized laser [100] to the HCI clock laser using an optical frequency comb [101, 102]. The frequency comb is also used to compare our clock laser frequency to the frequency of other clock lasers.

3.2 Quantum logic spectroscopy

To perform optical spectroscopy on the HCI, the two ions are prepared in their electronic and motional ground state, as the starting point for quantum logic spectroscopy (QLS) [103, 104]. We have for the first time implemented coherent QLS on Ar^{13+} using Be^+ as the logic ion species [105]. The basic protocol is illustrated in Fig. 4. It starts with a spectroscopy pulse on Ar^{13+}, producing in general an electronic superposition of the ground and excited electronic states. A so-called red sideband (RSB) pulse converts the electronic superposition into a superposition of one of the axial modes to be in the

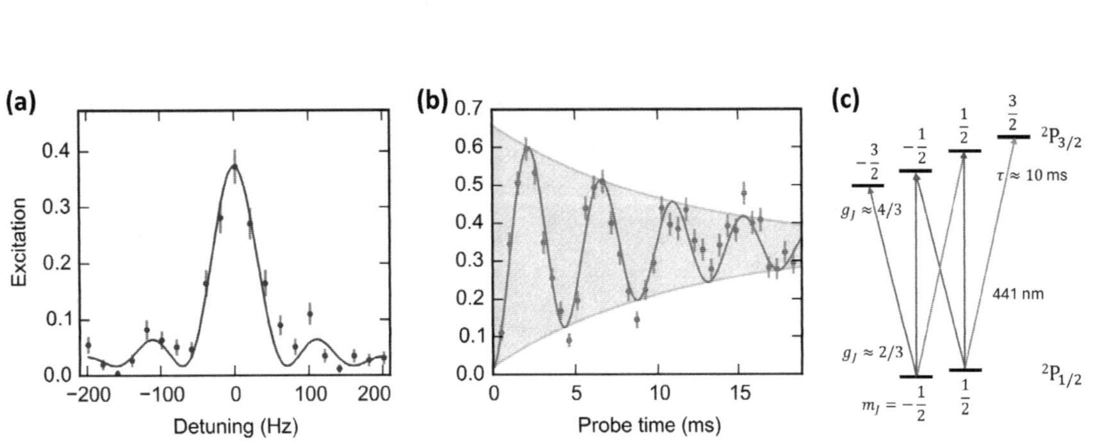

Figure 5: Coherent laser spectroscopy of Ar^{13+} using quantum logic. (a) Scan across the Ar^{13+} $^2P_{1/2} \leftrightarrow {}^2P_{3/2}$ transition. (b) Rabi flopping on the Ar^{13+} transition. The reduction in amplitude with time is consistent with the excited state lifetime of $\tau \approx 10\,\text{ms}$. (c) Relevant part of the level scheme of Ar^{13+}. After: Micke et al. [105].

ground and the first excited state. Since the motion is a shared normal mode of both ions, we can apply a similar RSB pulse on the Be^+ logic ion to convert the motional superposition back into a superposition of electronic states. Internal state discrimination on the Be^+ ion is performed by coupling its electronic ground state to a short-lived excited state scattering millions of photons if and only if it is found in the electronic ground state [106]. Through this series of laser pulses, the internal state information of the HCI after spectroscopy has been faithfully transferred to the Be^+ ion, where it is detected with $> 95\,\%$ fidelity. The entire sequence is repeated many times to suppress quantum projection noise [107].
Fig. 5(a) shows the result of a scan across the $^2P_{1/2} \leftrightarrow {}^2P_{3/2}$ fine-structure transition in Ar^{13+}, while in (b) coherent Rabi oscillations are demonstrated [105]. The observed decay in signal is dictated by the excited state lifetime of 9.97(27) s, in agreement with a previous more precise in-EBIT measurement [108]. As shown in Fig. 5(c), the ground and excited states in Ar^{13+} are split into two and four Zeeman levels, respectively. Thus, for high contrast spectroscopy initial state preparation is required. For this we implemented a quantum logic-assisted state preparation protocol, in which the internal state of Ar^{13+} is changed using motional sideband pulses, followed by ground state cooling on Be^+ [103, 105]. This final step ensures irreversibility of the state-changing laser pulse and thus provides the dissipation required for optical pumping. Through deterministic preparation of Ar^{13+} in either one of the two Zeeman ground state, all transitions between the two Zeeman manifolds can be spectroscopically investigated with near quantum projection noise-limited signal-to-noise ratio. The magnetic moment of the excited state was inferred from a measurement of the different Zeeman components and the accurately known ground-state g-factor from a Penning trap measurement [109]. It should be noted that these high-accuracy g-factor measurements are typically performed in Penning traps employing a spin detection technique with orders of magnitude longer detection time compared to the excited state lifetime of Ar^{13+}. QLS therefore enables measurements of previously inaccessible atomic parameters in HCI that can be employed, e.g., for tests of QED theory.

4 The Ar^{13+} optical clock

The Ar^{13+} ground state fine-structure transition can be turned into an optical clock by evaluating all systematic frequency shifts and by performing a frequency ratio measurement with other clocks. The

Table 1: Systematic shifts ($\Delta\nu$) and corresponding 1-σ uncertainties (σ) of the $^{40}Ar^{13+}$ optical clock. Values are given relative to the transition frequency ν. The absolute frequencies are limited by the statistical uncertainties and the total uncertainty of the previous measurement of the absolute frequency of the $^{171}Yb^{+}$ clock transition [111], which contribute approximately equally. Further systematic shifts are at the 10^{-19} level or below. After: King et al. [110].

Shift	$\Delta\nu/\nu$ (10^{-17})	σ/ν (10^{-17})
Excess micromotion	−44.3	2.2
Laser-induced a.c. Zeeman	0	0.2
Secular motion	−0.1	< 0.1
Quadrupole shift	0	< 0.1
Quadratic Zeeman	< 0.1	$\ll 0.1$
Total Ar^{13+} **clock**	−44.4	2.2
Statistics	0	10
Uncompensated paths	0	0.1
Yb^{+} $E3$ absolute frequency	0	13
Gravitational redshift	16.4	0.6
Total absolute frequency	−28	15

result of such an evaluation showing the dominant (at a level of 10^{-18}) shifts and their uncertainties is presented in Table 1 [110].

By far the largest shift and associated uncertainty comes from uncompensatable axial micromotion in the trap resulting in a time dilation shift, which has been determined using the sideband technique [112, 113]. The large micromotion is attributed to a defect in the trap, which can be mitigated by installing a new trap without this problem. Importantly, the magnitude of this shift is independent of the charge state of the HCI and would be similar for singly charged ion clocks.

The second largest uncertainty contribution is from a probe laser-induced a.c. Zeeman shift. This shift arises from non-resonant polarization components of the probe laser light coupling to the Zeeman structure of Ar^{13+} (see Fig. 5(c)), thus shifting the ground state energy level. It can be mitigated by using pure polarization in the clock laser or by implementing Stark shift compensation schemes, such as hyper Ramsey spectroscopy [114, 115].

The time dilation shift from residual secular motion contributes at a level of 10^{-18}. It is dominated by the two weakly coupled radial modes, in which the Be^{+} cooling ion has a small amplitude (see above) and sympathetic laser cooling of Ar^{13+} becomes inefficient. The slow thermalization rate would result in a large magnitude and uncertainty of the associated shift. We have therefore implemented a quantum logic-inspired cooling scheme [116], akin algorithmic cooling in nuclear magnetic resonance [117–119]. An Ar^{13+} RSB pulse on one of the weakly coupled modes removes a quantum of motion and maps it into an electronic excitation. The electronic excitation is transferred to one of the strongly coupled axial modes by an Ar^{13+} RSB pulse on this mode. Ground state cooling of the axial mode on the Be^{+} ion

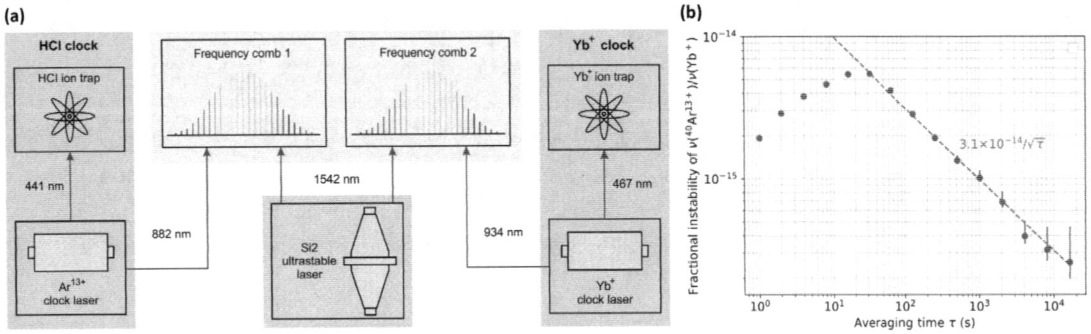

Figure 6: Frequency measurement of the Ar^{13+} optical clock. (a) Frequency comparison setup using a common reference laser at 1.5 µm and two frequency combs to transfer its stability to the respective clock lasers, which are stabilised to the two reference transitions in Ar^{13+} and $^{171}Yb^{+}$. (b) Frequency comparison between the $^{40}Ar^{13+}$ and $^{171}Yb^{+}$ E3 optical clocks with an instability of $3.1 \times 10^{-14}/\sqrt{\tau}$, with τ the averaging time in seconds. It is limited by the probe time of 15 ms, determined by the short excited state lifetime of Ar^{13+} of 10 ms. After: King et al. [110].

removes the motional excitation energy and provides the dissipation required for cooling. This way, mean phonon numbers of 0.39(15) and 0.21(8) of the two weakly coupled Ar^{13+} radial modes were achieved [116], resulting in a time dilation shift and associated uncertainty at the low 10^{-18}-level for the Ar^{13+} clock.

Other notable shifts include the electric quadrupole and the second order Zeeman shift [120]. Both shifts are strongly reduced in HCI (see scaling arguments above), which has been verified through accurate atomic structure calculations [110].

For clock operation, independent frequency stabilization servos to several Zeeman transitions are maintained. This allows elimination of the linear Zeeman and electric quadrupole shifts through averaging the frequency of these transitions. Using appropriate differences of the Zeeman transition frequencies, the ratio of ground and excited state g-factors and the excited state quadrupole moment have been determined in agreement with atomic structure calculations [110]. By probing the Ar^{13+} ion with counter propagating clock laser beams, potential linear Doppler shifts can be discovered and eliminated. The measurements did not reveal such a shift. We employed Rabi spectroscopy with a pulse duration of 15 ms, corresponding to 1.5 times the excited state lifetime, providing the smallest statistical uncertainty [121]. Spectroscopy is interleaved with "service" measurements probing for potential drifts in micromotion compensation and secular temperature.

The uncertainty budget has been verified down to a level of 1×10^{-16} through frequency comparison of the $^{40}Ar^{13+}$ clock with the $^{171}Yb^{+}$ E3 clock at PTB [82, 122] (see Fig. 6). For this, each clock's spectroscopy laser was stabilized to separate stable cavities and transfer locked using two frequency combs to the stable silicon cavity-stabilised reference laser [100]. The measurement is statistically limited by the short excited state lifetime of the excited state in Ar^{13+}. The absolute frequency of the $^{40}Ar^{13+}$ clock transition was derived from the absolute frequency of the $^{171}Yb^{+}$ clock [122] to be 679 216 462 397 957.43(11) Hz, one of the most accurately measured optical frequencies to date. It is limited by the uncertainty of the $^{171}Yb^{+}$ E3 frequency and statistics of the frequency comparison (see Tab. 1) [110].

We also measured the absolute frequency of $^{36}Ar^{13+}$ and derived the isotope shift between the two isotopes with a relative uncertainty of 6×10^{-11}. The experimental result was compared to atomic structure calculations. This way, we could verify for the first time the nuclear QED recoil effect in a many-electron system [110], which represents a test of QED theory.

In summary, the detailed evaluation of the Ar^{13+} systematic frequency uncertainties confirmed the advantageous atomic properties of HCI for highly accurate optical clocks predicted by several theory proposals [54]. The current limitation in systematic uncertainty of the demonstrated Ar^{13+} optical clock at 2.2×10^{-17} is of technical nature and can be mitigated by an upgrade to the system. The one order of magnitude larger statistical uncertainty of the Ar^{13+} clock compared to other ion clocks is dictated by the short excited state lifetime of 10 ms. Other HCI clock species with a longer excited state lifetime can overcome this limitation and become competitive or possibly superior to neutral or singly charged ion clocks. In particular the application of high accuracy HCI optical clocks for tests of fundamental physics is a very promising application.

5 Future prospects

A first potential application of HCI optical clocks is the search for new physics using isotope shift spectroscopy [54, 123–125]. The standard model predicts the frequency shift $\delta\nu_i^{A,A'}$ of a transition i between two isotopes A, A' to depend to first order on the field and the mass shift [126]:

$$\delta\nu_i^{A,A'} = F_i \delta\langle r^2\rangle_{A,A'} + k_i \frac{m_A - m_{A'}}{m_A m_{A'}}, \tag{1}$$

where $\delta\langle r^2\rangle_{A,A'}$ is the change of the squared nuclear charge radius, F_i and k_i are the field and mass shift constant, respectively, and m_A is the mass of isotope A. The typically not precisely known $\delta\langle r^2\rangle_{A,A'}$ can be eliminated from Eq. 1 by employing two different transitions, resulting in a so-called King plot, which exhibits a linear relationship between the isotope shift of one vs. a second transition [127].

A hypothetical fifth force with a coupling strength y_e to electrons and y_n to neutrons would introduce an additional term $y_e y_n X_i(A - A')$ in Eq. 1, where X_i is the overlap integral of the new force carrier with the electronic wavefunctions [125]. This additional term will break the King linearity and might be observed experimentally in high accuracy isotope shift spectroscopy. However, there are also standard model terms that lead to nonlinearties, such as higher-order field and mass shift terms, higher-order nuclear moments, nuclear polarizability, and more [128–134]. These terms are typically difficult to account for, since experimental data is more accurate than calculations of the corresponding atomic and nuclear properties. In case the electronic and nuclear properties of a transition can be factorized, a generalized King plot can be generated, in which higher-order terms are removed in a similar way as the $\delta\langle r^2\rangle_{A,A'}$ term in the original King plot by using more isotopes and transitions [135–137]. Recent experimental isotope shift measurements have reached uncertainties at the hertz level and below, exhibiting nonlinearities in the King plot analysis of Yb [138–141]. Another particularly promising isotopic sequence is that of calcium with five stable, even isotopes, for which King plots for Ca^+ between the $S_{1/2}$-$D_{3/2}$ and $S_{1/2}$-$D_{5/2}$ transitions have been generated [142–144]. However, the similarity in wavefunction of the two D-states reduces the sensitivity to new physics effects. Ca^{14+} features a fine-structure transition between the $^3P_0$ and $^3P_1$ states at 569 nm, which combined with one of the S-D transitions in Ca^+ has a large sensitivity to a possible fifth force [145]. We expect systematic and statistical uncertainties at a similar level to the ones achieved with Ar^{13+}, since charge, mass, and excited state lifetime are similar. One of the advantages of the calcium system is that it is likely to be dominated only by higher-order mass shift terms and nuclear polarizability [129], possibly enabling tighter bounds for the existence of a fifth force compared to heavier atomic systems, such as Yb^+,

Table 2: Sensitivity coefficients and transition wavelengths of selected species for a variation of the fine-structure constant α. The sensitivity coefficient K relates a relative change in α to a relative change in transition frequency: $\Delta\omega/\omega = K\Delta\alpha/\alpha$. Data taken from: Kozlov et al. [54].

System	K	λ (nm)
Sr	0.06	699
Yb^{+} E2	0.91	436
Yb^{+} E3	-6	467
Hg^{+}	-2.9	281.5
Al^{+}	0.01	267
Pr^{9+}	4.2	475
Ir^{17+} T1	-22	ca. 283
Ir^{17+} T2	145	ca. 1978
Cf^{15+}	59	ca. 775
Cf^{17+}	-48	ca. 535

which also suffer from higher-order field shifts and for which the higher-order standard model contributions are typically larger [146].

An interesting candidate for a competitive optical clock is Ni^{12+} with a $^3P_2$ ground and $^3P_0$ excited clock state at a transition wavelength of 498(3) nm [52]. We expect systematic shifts to be on the same order of magnitude as the Ar^{13+} and Ca^{14+} cases, but with much improved statistical uncertainty due to the long excited state lifetime of around 19 s, corresponding to a natural linewidth of 8 mHz. However, the large uncertainty of the calculated transition energy and the difficulty to observe such a narrow line in an EBIT, searching for the clock transition may take a long time. The availability of a logic transition to the short-lived (6.5 ms) $^3P_1$ state, which has been observed by in-EBIT spectroscopy [33], will help to reduce the time to find the transition frequency significantly using efficient search strategies [147, 148].

The stability of HCI clocks can be further improved by either trapping more than one HCI in the trap, implementing a multi-HCI optical clock [149, 150], or by increasing the transition energy towards the extreme ultra-violet regime. The latter is a feature only available to HCI due their high transition and ionization energies, but requires narrow-linewidth sources in this frequency regime [151]. This is a particularly interesting avenue, since the increase in transition energy at the same time reduces the relative uncertainty of systematic effects, which are an atomic property, independent of the transition energy.

Future tests of fundamental physics include the search for the variation of the fine-structure constant α with the largest sensitivity of any atomic system near level-crossing transitions, such as $Cf^{15+,16+,17+}$, Ir^{17+}, and Pr^{9+} [22, 23, 26, 147, 152]. The sensitivity factor of selected HCI candidates is more than an order of magnitude larger than those available from singly charged or neutral atom clocks (see Tab.2). These systems are also the most sensitive ones for a search for ultralight scalar dark matter candidates leading to an oscillation of atomic levels at the Compton frequency of the mediator particle [153] or topological dark matter [154], with characteristic transient frequency changes.

Within the framework of a grand unified theory, all gauge couplings are linked and so are their variations. The hadron masses are set by Λ_{QCD}, the gauge coupling of the strong force. If α (the gauge coupling for the electromagnetic interactions) varies, Λ_{QCD} varies as well, but much more strongly. This, for example, results in a predicted 38 times stronger relative variation of the electron-to-proton mass ratio compared to $\dot{\alpha}/\alpha$ [155–157]. A measurement sensitive to a change in the electron-to-proton mass ratio has therefore a factor of 38 larger sensitivity to new physics effects, having motivated molecular spectroscopy [158, 159] and clock comparisons involving hyperfine transitions [160–162], which are sensitive to this parameter. One problem in such clock comparisons is the high instability of microwave transitions, requiring long averaging times. This limitation can be overcome by employing HCI clocks for which hyperfine transitions are in the optical frequency regime [20].

One of the cornerstones of modern physics is Einstein's theory of relativity, which is known to be incompatible with quantum field theories. Therefore, tests of its underlying principles help to shine light on the origin of this fundamental issue. Probing for a violation of local Lorentz invariance (preferred direction in space) in the electron-photon sector using precision atomic spectroscopy is one of the many tests that can be performed. The basic idea is to compare the energy of electronic levels with different directionality with respect to a fixed background. Lorentz violation would introduce a periodic modulation of these levels as the Earth rotates and orbits around the sun [163, 164]. The electrons in HCI are strongly relativistic, resulting in large sensitivity factors for such tests [165, 166]. One of the advantages of using HCI for such a test is that it can be performed by spectroscopy in the electronic ground state, significantly simplifying the experimental protocol and extending probe times to many seconds.

Sympathetic cooling and quantum logic operations have enabled full control over all relevant degrees of freedom in HCI in Paul traps. This opened the door to fascinating new applications of HCI as high-accuracy optical clocks and for sensitive tests of fundamental physics. The large number of HCI species with different properties and features available in nature will open up many more applications in the future.

References

[1] J. C. Berengut, V. V. Flambaum, and A. Ong, "Testing spatial α-variation with optical atomic clocks based on highly charged ions", EPJ Web Conf. **57**, edited by O. Dulieu, P. Grangier, M. Leduc, and H. Perrin, 02001 (2013).

[2] J. M. Shull, B. D. Smith, and C. W. Danforth, "The baryon census in a multiphase intergalactic medium: 30% of the baryons may still be missing", ApJ **759**, 23 (2012).

[3] S. R. Habbal, M. Druckmüller, H. Morgan, I. Scholl, V. Rušin, A. Daw, J. Johnson, and M. Arndt, "Total solar eclipse observations of hot prominence shrouds", ApJ **719**, 1362–1369 (2010).

[4] P. Beiersdorfer, "Laboratory X-Ray Astrophysics", Annu. Rev. Astron. Astrophys. **41**, 343–390 (2003).

[5] Hitomi Collaboration et al., "Solar abundance ratios of the iron-peak elements in the Perseus cluster", Nature **551**, 478–480 (2017).

[6] D. Reimers, "Baryons in the Diffuse Intergalactic Medium", in *Matter in the Universe*, Vol. 14, edited by Ph. Jetzer, K. Pretzl, and R. Von Steiger (Springer Netherlands, Dordrecht, 2002), pp. 89–99.

[7] P. Beiersdorfer, "Highly charged ions in magnetic fusion plasmas: research opportunities and diagnostic necessities", J. Phys. B: At. Mol. Opt. Phys. **48**, 144017 (2015).

[8] D. C. Brandt et al., "CO2/Sn LPP EUV sources for device development and HVM", in Extreme Ultrav. EUV Lithogr. IV, Vol. 8679 (Apr. 1, 2013), pp. 396–403.

[9] I. Fomenkov et al., "Light sources for high-volume manufacturing EUV lithography: technology, performance, and power scaling", Adv. Opt. Technol. **6**, 173–186 (2017).

[10] P. Beiersdorfer, "Testing QED and atomic-nuclear interactions with high-Zions", J. Phys. B: At. Mol. Opt. Phys. **43**, 074032 (2010).

[11] J. Morgner, B. Tu, C. M. König, T. Sailer, F. Heiße, H. Bekker, B. Sikora, C. Lyu, V. A. Yerokhin, Z. Harman, J. R. Crespo López-Urrutia, C. H. Keitel, S. Sturm, and K. Blaum, "Stringent test of QED with hydrogen-like tin", Nature **622**, 53–57 (2023).

[12] T. Sailer, V. Debierre, Z. Harman, F. Heiße, C. König, J. Morgner, B. Tu, A. V. Volotka, C. H. Keitel, K. Blaum, and S. Sturm, "Measurement of the bound-electron g-factor difference in coupled ions", Nature **606**, 479–483 (2022).

[13] P. Indelicato, "QED tests with highly charged ions", J. Phys. B: At. Mol. Opt. Phys. **52**, 232001 (2019).

[14] V. M. Shabaev, A. I. Bondarev, D. A. Glazov, M. Y. Kaygorodov, Y. S. Kozhedub, I. A. Maltsev, A. V. Malyshev, R. V. Popov, I. I. Tupitsyn, and N. A. Zubova, "Stringent tests of QED using highly charged ions", Hyperfine Interact **239**, 60 (2018).

[15] S. Sturm, A. Wagner, B. Schabinger, J. Zatorski, Z. Harman, W. Quint, G. Werth, C. H. Keitel, and K. Blaum, "G Factor of Hydrogenlike $^{28}Si^{13+}$", Phys. Rev. Lett. **107**, 023002 (2011).

[16] W. Nörtershäuser, "Laser spectroscopy for QED tests in highly charged ions", Hyperfine Interact **199**, 131–140 (2011).

[17] D. J. Wineland, C. Monroe, W. M. Itano, D. Leibfried, B. E. King, and D. M. Meekhof, "Experimental issues in coherent quantum-state manipulation of trapped atomic ions", J. Res. Natl. Inst. Stand. Technol. **103**, 259–328 (1998).

[18] J. D. Gillaspy, "Highly charged ions", J. Phys. B: At. Mol. Opt. Phys. **34**, R93–R130 (2001).

[19] M. S. Safronova, V. A. Dzuba, V. V. Flambaum, U. I. Safronova, S. G. Porsev, and M. G. Kozlov, "Highly Charged Ions for Atomic Clocks, Quantum Information, and Search for α variation", Phys. Rev. Lett. **113**, 030801 (2014).

[20] S. Schiller, "Hydrogenlike Highly Charged Ions for Tests of the Time Independence of Fundamental Constants", Phys. Rev. Lett. **98**, 180801 (2007).

[21] J. Berengut, V. Dzuba, and V. Flambaum, "Enhanced Laboratory Sensitivity to Variation of the Fine-Structure Constant using Highly Charged Ions", Phys. Rev. Lett. **105**, 120801 (2010).

[22] A. Windberger, J. R. Crespo López-Urrutia, H. Bekker, N. S. Oreshkina, J. C. Berengut, V. Bock, A. Borschevsky, V. A. Dzuba, E. Eliav, Z. Harman, U. Kaldor, S. Kaul, U. I. Safronova, V. V. Flambaum, C. H. Keitel, P. O. Schmidt, J. Ullrich, and O. O. Versolato, "Identification of the Predicted 5s-4f Level Crossing Optical Lines with Applications to Metrology and Searches for the Variation of Fundamental Constants", Phys. Rev. Lett. **114**, 150801 (2015).

[23] J. C. Berengut, V. A. Dzuba, V. V. Flambaum, and A. Ong, "Optical Transitions in Highly Charged Californium Ions with High Sensitivity to Variation of the Fine-Structure Constant", Phys. Rev. Lett. **109**, 070802 (2012).

[24] J. C. Berengut, V. A. Dzuba, V. V. Flambaum, and A. Ong, "Highly charged ions with E1, M1, and E2 transitions within laser range", Phys. Rev. A **86**, 022517 (2012).

[25] J. C. Berengut, V. A. Dzuba, V. V. Flambaum, and A. Ong, "Electron-Hole Transitions in Multiply Charged Ions for Precision Laser Spectroscopy and Searching for Variations in α", Phys. Rev. Lett. **106**, 210802 (2011).

[26] S. G. Porsev, U. I. Safronova, M. S. Safronova, P. O. Schmidt, A. I. Bondarev, M. G. Kozlov, I. I. Tupitsyn, and C. Cheung, "Optical clocks based on the Cf^{15+} and Cf^{17+} ions", Phys. Rev. A **102**, 012802 (2020).

[27] M. S. Safronova, D. Budker, D. DeMille, D. F. J. Kimball, A. Derevianko, and C. W. Clark, "Search for new physics with atoms and molecules", Rev. Mod. Phys. **90**, 025008 (2018).

[28] S. O. Allehabi, V. A. Dzuba, and V. V. Flambaum, "Atomic clocks highly sensitive to the variation of the fine-structure constant based on Hf II, Hf IV, and W VI ions", Phys. Rev. A **106**, 032807 (2022).

[29] G. Barontini et al., "Measuring the stability of fundamental constants with a network of clocks", EPJ Quantum Technol. **9**, 12 (2022).

[30] K. Beloy, "Prospects of a Pb^{2+} Ion Clock", Phys. Rev. Lett. **127**, 013201 (2021).

[31] K. Beloy, V. A. Dzuba, and S. M. Brewer, "Quadruply Ionized Barium as a Candidate for a High-Accuracy Optical Clock", Phys. Rev. Lett. **125**, 173002 (2020).

[32] Jyoti, A. Chakraborty, Y.-m. Yu, J. Chen, B. Arora, and B. K. Sahoo, "Zr^{3+} ion as a prospective terahertz atomic clock", Phys. Rev. A **108**, 023115 (2023).

[33] S. Chen, Z. Zhou, J. Li, T. Zhang, C. Li, T. Shi, Y. Huang, K. Gao, and H. Guan, "Precision measurement of M1 optical clock transition in Ni^{12+}", Phys. Rev. Res. **6**, 013030 (2024).

[34] A. Derevianko, V. A. Dzuba, and V. V. Flambaum, "Highly Charged Ions as a Basis of Optical Atomic Clockwork of Exceptional Accuracy", Phys. Rev. Lett. **109**, 180801 (2012).

[35] V. A. Dzuba and V. V. Flambaum, "Cm^{15+} and bk^{16+} ion clocks with enhanced sensitivity to new physics", Phys. Rev. A **109**, L021101 (2024).

[36] V. A. Dzuba, A. Derevianko, and V. V. Flambaum, "Erratum: High-precision atomic clocks with highly charged ions: Nuclear-spin-zero f^{12}-shell ions [Phys. Rev. A 86, 054501 (2012)]", Phys. Rev. A **87**, 029906 (2013).

[37] V. A. Dzuba, A. Derevianko, and V. V. Flambaum, "High-precision atomic clocks with highly charged ions: Nuclear-spin-zero f^{12}-shell ions", Phys. Rev. A **86**, 054501 (2012).

[38] V. A. Dzuba and V. V. Flambaum, "Highly charged ions for atomic clocks and search for variation of the fine structure constant", Hyperfine Interact. **236**, 79–86 (2015).

[39] V. A. Dzuba, A. Derevianko, and V. V. Flambaum, "Ion clock and search for the variation of the fine-structure constant using optical transitions in Nd^{13+} and Sm^{15+}", Phys. Rev. A **86**, 054502 (2012).

[40] V. A. Dzuba, V. V. Flambaum, and H. Katori, "Optical clock sensitive to variations of the fine-structure constant based on the Ho^{14+} ion", Phys. Rev. A **91**, 022119 (2015).

[41] V. A. Dzuba and V. V. Flambaum, "Os^{16+} and Ir^{17+} ions as candidates for accurate optical clocks sensitive to physics beyond the standard model", Phys. Rev. A **108**, 053111 (2023).

[42] D. K. Nandy and B. K. Sahoo, "Highly charged W^{13+}, Ir^{16+}, and Pt^{17+} ions as promising optical clock candidates for probing variations of the fine-structure constant", Phys. Rev. A **94**, 032504 (2016).

[43] M. S. Safronova, V. A. Dzuba, V. V. Flambaum, U. I. Safronova, S. G. Porsev, and M. G. Kozlov, "Atomic properties of Cd-like and Sn-like ions for the development of frequency standards and search for the variation of the fine-structure constant", Phys. Rev. A **90**, 052509 (2014).

[44] M. S. Safronova, V. A. Dzuba, V. V. Flambaum, U. I. Safronova, S. G. Porsev, and M. G. Kozlov, "Highly charged Ag-like and In-like ions for the development of atomic clocks and the search for α variation", Phys. Rev. A **90**, 042513 (2014).

[45] B. Tu, R. Si, Y. Shen, J. Wang, B. Wei, C. Chen, K. Yao, and Y. Zou, "Experimental access to observing decay from extremely long-lived metastable electronic states via Penning trap spectrometry", Phys. Rev. Res. **5**, 043014 (2023).

[46] Y. Yu, D. Pan, S. Chen, B. Arora, H. Guan, K. Gao, and J. Chen, "Atomic Structure of Nd^{9+} for Highly Charged Ion Clocks", Atoms **10**, 123 (2022).

[47] Y.-m. Yu and B. K. Sahoo, "Erratum: Investigating ground-state fine-structure properties to explore suitability of boronlike S^{11+}-K^{14+} and galliumlike Nb^{10+}-Ru^{13+} ions as possible atomic clocks [Phys. Rev. A 99, 022513 (2019)]", Phys. Rev. A **104**, 069903 (2021).

[48] Y.-M. Yu, B. K. Sahoo, and B.-B. Suo, "Highly charged ion (HCI) clocks: Frontier candidates for testing variation of fine-structure constant", Front. Phys. **11**, `10.3389/fphy.2023.1104848` (2023).

[49] Y.-m. Yu and B. K. Sahoo, *A Hunt for Highly Charged Ions as Ultra-stable Optical Clock Candidates: Art of the energy level-crossing approach*, (July 26, 2023) `http://arxiv.org/abs/2307.14543`, pre-published.

[50] Y.-m. Yu and B. K. Sahoo, "Investigating ground-state fine-structure properties to explore suitability of boronlike S^{11+}-K^{14+} and galliumlike Nb^{10+}-Ru^{13+} ions as possible atomic clocks", Phys. Rev. A **99**, 022513 (2019).

[51] Y.-m. Yu and B. K. Sahoo, "Scrutinizing Al-like $^{51}V^{10+}$, $^{53}Cr^{11+}$, $^{55}Mn^{12+}$, $^{57}Fe^{13+}$, $^{59}Co^{14+}$, $^{61}Ni^{15+}$, and $^{63}Cu^{16+}$ ions for atomic clocks with uncertainties below the 10^{-19} level", Phys. Rev. A **94**, 062502 (2016).

[52] Y.-m. Yu and B. K. Sahoo, "Selected highly charged ions as prospective candidates for optical clocks with quality factors larger than 10^{15}", Phys. Rev. A **97**, 041403 (2018).

[53] V. I. Yudin, A. V. Taichenachev, and A. Derevianko, "Magnetic-Dipole Transitions in Highly Charged Ions as a Basis of Ultraprecise Optical Clocks", Phys. Rev. Lett. **113**, 233003 (2014).

[54] M. G. Kozlov, M. S. Safronova, J. R. Crespo López-Urrutia, and P. O. Schmidt, "Highly charged ions: Optical clocks and applications in fundamental physics", Rev. Mod. Phys. **90**, 045005 (2018).

[55] J. Ullmann et al., "High precision hyperfine measurements in Bismuth challenge bound-state strong-field QED", Nat Commun **8**, 15484 (2017).

[56] P. Seelig et al., "Ground State Hyperfine Splitting of Hydrogenlike $^{207}Pb^{81+}$ by Laser Excitation of a Bunched Ion Beam in the GSI Experimental Storage Ring", Phys. Rev. Lett. **81**, 4824–4827 (1998).

[57] I. Klaft, S. Borneis, T. Engel, B. Fricke, R. Grieser, G. Huber, T. Kühl, D. Marx, R. Neumann, S. Schröder, P. Seelig, and L. Völker, "Precision Laser Spectroscopy of the Ground State Hyperfine Splitting of Hydrogenlike $^{209}Bi^{82+}$", Phys. Rev. Lett. **73**, 2425–2427 (1994).

[58] M. Lochmann et al., "Observation of the hyperfine transition in lithium-like bismuth $^{209}Bi^{80+}$: Towards a test of QED in strong magnetic fields", Phys. Rev. A **90**, 030501 (2014).

[59] J. C. López-Urrutia, "The visible spectrum of highly charged ions: A window to fundamental physics", Can. J. Phys. **86**, 111–123 (2008).

[60] K. Schnorr, V. Mäckel, N. S. Oreshkina, S. Augustin, F. Brunner, Z. Harman, C. H. Keitel, J. Ullrich, and J. R. C. López-Urrutia, "Coronium in the Laboratory: Measuring the Fe XIV Green Coronal Line by Laser Spectroscopy", ApJ **776**, 121 (2013).

[61] V. Mäckel, R. Klawitter, G. Brenner, J. R. Crespo López-Urrutia, and J. Ullrich, "Laser Spectroscopy on Forbidden Transitions in Trapped Highly Charged Ar^{13+} Ions", Phys. Rev. Lett. **107**, 143002 (2011).

[62] K. Hosaka, D. N. Crosby, K. Gaarde-Widdowson, C. J. Smith, J. D. Silver, T. Kinugawa, S. Ohtani, and E. G. Myers, "Laser spectroscopy of hydrogenlike nitrogen in an electron beam ion trap", Phys. Rev. A **69**, 011802 (2004).

[63] P. Micke, S. Kühn, L. Buchauer, J. R. Harries, T. M. Bücking, K. Blaum, A. Cieluch, A. Egl, D. Hollain, S. Kraemer, T. Pfeifer, P. O. Schmidt, R. X. Schüssler, Ch. Schweiger, T. Stöhlker, S. Sturm, R. N. Wolf, S. Bernitt, and J. R. Crespo López-Urrutia, "The Heidelberg compact electron beam ion traps", Rev. Sci. Instrum. **89**, 063109 (2018).

[64] Ch. Schweiger, C. M. König, J. R. Crespo López-Urrutia, M. Door, H. Dorrer, Ch. E. Düllmann, S. Eliseev, P. Filianin, W. Huang, K. Kromer, P. Micke, M. Müller, D. Renisch, A. Rischka, R. X. Schüssler, and K. Blaum, "Production of highly charged ions of rare species by laser-induced desorption inside an electron beam ion trap", Review of Scientific Instruments **90**, 123201 (2019).

[65] R. Soria Orts, J. Crespo López-Urrutia, H. Bruhns, A. González Martínez, Z. Harman, U. Jentschura, C. Keitel, A. Lapierre, H. Tawara, I. Tupitsyn, J. Ullrich, and A. Volotka, "Zeeman splitting and g factor of the $1s^2 2s^2 2p\ ^2P_{3/2}$ and $^2P_{3/2}$ levels in Ar^{13+}", Phys. Rev. A **76**, 052501 (2007).

[66] R. S. Orts, Z. Harman, J. R. C. López-Urrutia, A. N. Artemyev, H. Bruhns, A. J. G. Martínez, U. D. Jentschura, C. H. Keitel, A. Lapierre, V. Mironov, V. M. Shabaev, H. Tawara, I. I. Tupitsyn, J. Ullrich, and A. V. Volotka, "Exploring Relativistic Many-Body Recoil Effects in Highly Charged Ions", Phys. Rev. Lett. **97**, 103002 (2006).

[67] D. J. Bieber, H. S. Margolis, P. K. Oxley, and J. D. Silver, "Studies of magnetic dipole transitions in highly charged argon and barium using an electron beam ion trap", Phys. Scr. **T73**, 64–66 (1997).

[68] A. Egl, I. Arapoglou, M. Höcker, K. König, T. Ratajczyk, T. Sailer, B. Tu, A. Weigel, K. Blaum, W. Nörtershäuser, and S. Sturm, "Application of the Continuous Stern-Gerlach Effect for Laser Spectroscopy of the $^{40}Ar^{13}$ Fine Structure in a Penning Trap", Phys. Rev. Lett. **123**, 123001 (2019).

[69] W. Paul, O. Osberghaus, and E. Fischer, "Ein Ionenkäfig", Forschungsberichte Wirtsch.-Verkehrsministerium Nordrh.-Westfal. Westdtsch. Verl. Köln Oplad. (1958).

[70] P. K. Ghosh, *Ion Traps*, 1 edition (Clarendon Press, Oxford : New York, Jan. 18, 1996), 336 pp.

[71] D. J. Wineland and W. M. Itano, "Laser cooling of atoms", Phys. Rev. A **20**, 1521 (1979).

[72] D. J. Wineland, R. E. Drullinger, and F. L. Walls, "Radiation-pressure cooling of bound resonant absorbers", Phys. Rev. Lett. **40**, 1639–1642 (1978).

[73] C. Monroe, D. M. Meekhof, B. E. King, S. R. Jefferts, W. M. Itano, D. J. Wineland, and P. Gould, "Resolved-sideband Raman cooling of a bound atom to the 3D zero-point energy", Phys. Rev. Lett. **75**, 4011–4014 (1995).

[74] J. I. Cirac and P. Zoller, "Quantum Computations with Cold Trapped Ions", Phys. Rev. Lett. **74**, 4091 (1995).

[75] C. Monroe, D. M. Meekhof, B. E. King, W. M. Itano, and D. J. Wineland, "Demonstration of a Fundamental Quantum Logic Gate", Phys. Rev. Lett. **75**, 4714 (1995).

[76] F. Schmidt-Kaler, H. Häffner, M. Riebe, S. Gulde, G. P. T. Lancaster, T. Deuschle, C. Becher, C. F. Roos, J. Eschner, and R. Blatt, "Realization of the Cirac-Zoller controlled-NOT quantum gate", Nature **422**, 408–411 (2003).

[77] S. Gulde, M. Riebe, G. P. T. Lancaster, C. Becher, J. Eschner, H. Häffner, F. Schmidt-Kaler, I. L. Chuang, and R. Blatt, "Implementation of the Deutsch-Jozsa algorithm on an ion-trap quantum computer", Nature **421**, 48–50 (2003).

[78] A. Friedenauer, H. Schmitz, J. T. Glueckert, D. Porras, and T. Schaetz, "Simulating a quantum magnet with trapped ions", Nat Phys **4**, 757–761 (2008).

[79] C. Monroe, W. C. Campbell, L.-M. Duan, Z.-X. Gong, A. V. Gorshkov, P. W. Hess, R. Islam, K. Kim, N. M. Linke, G. Pagano, P. Richerme, C. Senko, and N. Y. Yao, "Programmable quantum simulations of spin systems with trapped ions", Rev. Mod. Phys. **93**, 025001 (2021).

[80] A. D. Ludlow, M. M. Boyd, J. Ye, E. Peik, and P. O. Schmidt, "Optical atomic clocks", Rev. Mod. Phys. **87**, 637–701 (2015).

[81] S. M. Brewer, J.-S. Chen, A. M. Hankin, E. R. Clements, C. W. Chou, D. J. Wineland, D. B. Hume, and D. R. Leibrandt, "$^{27}Al^+$ Quantum-Logic Clock with a Systematic Uncertainty below 10^{-18}", Phys. Rev. Lett. **123**, 033201 (2019).

[82] N. Huntemann, M. Okhapkin, B. Lipphardt, S. Weyers, Chr. Tamm, and E. Peik, "High-Accuracy Optical Clock Based on the Octupole Transition in $^{171}Yb^+$", Phys. Rev. Lett. **108**, 090801 (2012).

[83] Y. Huang, B. Zhang, M. Zeng, Y. Hao, Z. Ma, H. Zhang, H. Guan, Z. Chen, M. Wang, and K. Gao, "Liquid-Nitrogen-Cooled Ca^+ Optical Clock with Systematic Uncertainty of $3x10^{-18}$", Phys. Rev. Appl. **17**, 034041 (2022).

[84] H. N. Hausser, J. Keller, T. Nordmann, N. M. Bhatt, J. Kiethe, H. Liu, M. von Boehn, J. Rahm, S. Weyers, E. Benkler, B. Lipphardt, S. Doerscher, K. Stahl, J. Klose, C. Lisdat, M. Filzinger, N. Huntemann, E. Peik, and T. E. Mehlstäubler, *An* $^{115}In^+$*-/sup>172Yb*$^+$ *Coulomb crystal clock with* 2.5×10^{-18} *systematic uncertainty*, (Feb. 26, 2024) `http://arxiv.org/abs/2402.16807`, pre-published.

[85] Z. Zhiqiang, K. J. Arnold, R. Kaewuam, and M. D. Barrett, "$^{176}Lu^+$ clock comparison at the 10^{-18} level via correlation spectroscopy", Sci. Adv. **9**, eadg1971 (2023).

[86] A. Tofful, C. F. A. Baynham, E. A. Curtis, A. O. Parsons, B. I. Robertson, M. Schioppo, J. Tunesi, H. S. Margolis, R. J. Hendricks, J. Whale, R. C. Thompson, and R. M. Godun, "$^{171}Yb^+$ optical clock with 2.2×10^{-18} systematic uncertainty and absolute frequency measurements", Metrologia **61**, 045001 (2024).

[87] K. Cui, S. Chao, C. Sun, S. Wang, P. Zhang, Y. Wei, J. Yuan, J. Cao, H. Shu, and X. Huang, "Evaluation of the systematic shifts of a $^{40}Ca^+$-$^{27}Al^+$ optical clock", Eur. Phys. J. D **76**, 140 (2022).

[88] M. Schwarz, O. O. Versolato, A. Windberger, F. R. Brunner, T. Ballance, S. N. Eberle, J. Ullrich, P. O. Schmidt, A. K. Hansen, A. D. Gingell, M. Drewsen, and J. R. C. López-Urrutia, "Cryogenic linear Paul trap for cold highly charged ion experiments", Rev. Sci. Instrum. **83**, 083115-083115–10 (2012).

[89] T. Leopold, S. A. King, P. Micke, A. Bautista-Salvador, J. C. Heip, C. Ospelkaus, J. R. Crespo López-Urrutia, and P. O. Schmidt, "A cryogenic radio-frequency ion trap for quantum logic spectroscopy of highly charged ions", Rev. Sci. Instrum. **90**, 073201 (2019).

[90] A. Niehaus, "A classical model for multiple-electron capture in slow collisions of highly charged ions with atoms", J. Phys. B: Atom. Mol. Phys. **19**, 2925 (1986).

[91] P. Micke, J. Stark, S. A. King, T. Leopold, T. Pfeifer, L. Schmöger, M. Schwarz, L. J. Spieß, P. O. Schmidt, and J. R. Crespo López-Urrutia, "Closed-cycle, low-vibration 4 K cryostat for ion traps and other applications", Rev. Sci. Instrum. **90**, 065104 (2019).

[92] L. Schmöger, O. O. Versolato, M. Schwarz, M. Kohnen, A. Windberger, B. Piest, S. Feuchtenbeiner, J. Pedregosa-Gutierrez, T. Leopold, P. Micke, A. K. Hansen, T. M. Baumann, M. Drewsen, J. Ullrich, P. O. Schmidt, and J. R. C. López-Urrutia, "Coulomb crystallization of highly charged ions", Science **347**, 1233–1236 (2015).

[93] L. Schmöger, M. Schwarz, T. M. Baumann, O. O. Versolato, B. Piest, T. Pfeifer, J. Ullrich, P. O. Schmidt, and J. R. C. López-Urrutia, "Deceleration, precooling, and multi-pass stopping of highly charged ions in Be^+ Coulomb crystals", Rev. Sci. Instrum. **86**, 103111 (2015).

[94] M. J. Zawierucha, T. Rehmert, J. Keller, T. E. Mehlstäubler, P. O. Schmidt, and F. Wolf, "Deterministic preparation of a dual-species two-ion crystal", Phys. Rev. A **110**, 013107 (2024).

[95] D. J. Larson, J. C. Bergquist, J. J. Bollinger, W. M. Itano, and D. J. Wineland, "Sympathetic cooling of trapped ions: A laser-cooled two-species nonneutral ion plasma", Phys. Rev. Lett. **57**, 70 (1986).

[96] M. Barrett, B. DeMarco, T. Schaetz, V. Meyer, D. Leibfried, J. Britton, J. Chiaverini, W. Itano, B. Jelenković, J. Jost, C. Langer, T. Rosenband, and D. Wineland, "Sympathetic cooling of $^9Be^+$ and $^{24}Mg^+$ for quantum logic", Phys. Rev. A **68**, 042302 (2003).

[97] J. B. Wübbena, S. Amairi, O. Mandel, and P. O. Schmidt, "Sympathetic cooling of mixed-species two-ion crystals for precision spectroscopy", Phys. Rev. A **85**, 043412 (2012).

[98] Y. Wan, F. Gebert, F. Wolf, and P. O. Schmidt, "Efficient sympathetic motional-ground-state cooling of a molecular ion", Phys. Rev. A **91**, 043425 (2015).

[99] F. Dawel, A. Wilzewski, S. Herbers, L. Pelzer, J. Kramer, M. B. Hild, K. Dietze, L. Krinner, N. C. H. Spethmann, and P. O. Schmidt, "Coherent photo-thermal noise cancellation in a dual-wavelength optical cavity for narrow-linewidth laser frequency stabilisation", Opt. Express, OE **32**, 7276–7288 (2024).

[100] D. G. Matei, T. Legero, S. Häfner, C. Grebing, R. Weyrich, W. Zhang, L. Sonderhouse, J. M. Robinson, J. Ye, F. Riehle, and U. Sterr, "1.5 μm Lasers with Sub-10 mHz Linewidth", Phys. Rev. Lett. **118**, 263202 (2017).

[101] N. Scharnhorst, J. B. Wübbena, S. Hannig, K. Jakobsen, J. Kramer, I. D. Leroux, and P. O. Schmidt, "High-bandwidth transfer of phase stability through a fiber frequency comb", Opt. Express **23**, 19771–19776 (2015).

[102] J. Stenger, H. Schnatz, C. Tamm, and H. Telle, "Ultraprecise Measurement of Optical Frequency Ratios", Phys. Rev. Lett. **88**, 073601 (2002).

[103] P. O. Schmidt, T. Rosenband, C. Langer, W. M. Itano, J. C. Bergquist, and D. J. Wineland, "Spectroscopy Using Quantum Logic", Science **309**, 749–752 (2005).

[104] D. J. Wineland, J. C. Bergquist, J. J. Bollinger, R. E. Drullinger, and W. M. Itano, "Quantum Computers and Atomic Clocks", in Proc. 6th Symp. Freq. Stand. Metrol. (2002), pp. 361–368.

[105] P. Micke, T. Leopold, S. A. King, E. Benkler, L. J. Spieß, L. Schmöger, M. Schwarz, J. R. C. López-Urrutia, and P. O. Schmidt, "Coherent laser spectroscopy of highly charged ions using quantum logic", Nature **578**, 60–65 (2020).

[106] B. Hemmerling, F. Gebert, Y. Wan, and P. O. Schmidt, "A novel, robust quantum detection scheme", New J. Phys. **14**, 023043 (2012).

[107] W. M. Itano, J. C. Bergquist, J. J. Bollinger, J. M. Gilligan, D. J. Heinzen, F. L. Moore, M. G. Raizen, and D. J. Wineland, "Quantum projection noise: Population fluctuations in two-level systems", Phys. Rev. A **47**, 3554–3570 (1993).

[108] A. Lapierre, J. Crespo López-Urrutia, J. Braun, G. Brenner, H. Bruhns, D. Fischer, A. González Martínez, V. Mironov, C. Osborne, G. Sikler, R. Soria Orts, H. Tawara, J. Ullrich, V. Shabaev, I. Tupitsyn, and A. Volotka, "Lifetime measurement of the Ar XIV $1s^2 2s^2 2p\,^2P_{3/2}{}^o$ metastable level at the Heidelberg electron-beam ion trap", Phys. Rev. A **73**, 052507 (2006).

[109] I. Arapoglou, A. Egl, M. Höcker, T. Sailer, B. Tu, A. Weigel, R. Wolf, H. Cakir, V. A. Yerokhin, N. S. Oreshkina, V. A. Agababaev, A. V. Volotka, D. V. Zinenko, D. A. Glazov, Z. Harman, C. H. Keitel, S. Sturm, and K. Blaum, "G-factor of Boronlike Argon $^{40}Ar^{13+}$", Phys. Rev. Lett. **122**, 253001 (2019).

[110] S. A. King, L. J. Spieß, P. Micke, A. Wilzewski, T. Leopold, E. Benkler, R. Lange, N. Huntemann, A. Surzhykov, V. A. Yerokhin, J. R. Crespo López-Urrutia, and P. O. Schmidt, "An optical atomic clock based on a highly charged ion", Nature **611**, 43–47 (2022).

[111] R. Lange, N. Huntemann, J. M. Rahm, C. Sanner, H. Shao, B. Lipphardt, Chr. Tamm, S. Weyers, and E. Peik, "Improved Limits for Violations of Local Position Invariance from Atomic Clock Comparisons", Phys. Rev. Lett. **126**, 011102 (2021).

[112] D. J. Berkeland, J. D. Miller, J. C. Bergquist, W. M. Itano, and D. J. Wineland, "Minimization of ion micromotion in a Paul trap", J. Appl. Phys. **83**, 5025–5033 (1998).

[113] J. Keller, H. L. Partner, T. Burgermeister, and T. E. Mehlstäubler, "Precise determination of micromotion for trapped-ion optical clocks", Journal of Applied Physics **118**, 104501 (2015).

[114] N. Huntemann, B. Lipphardt, M. Okhapkin, Chr. Tamm, E. Peik, A. V. Taichenachev, and V. I. Yudin, "Generalized Ramsey Excitation Scheme with Suppressed Light Shift", Phys. Rev. Lett. **109**, 213002 (2012).

[115] V. I. Yudin, A. V. Taichenachev, C. W. Oates, Z. W. Barber, N. D. Lemke, A. D. Ludlow, U. Sterr, Ch. Lisdat, and F. Riehle, "Hyper-Ramsey spectroscopy of optical clock transitions", Phys. Rev. A **82**, 011804 (2010).

[116] S. A. King, L. J. Spieß, P. Micke, A. Wilzewski, T. Leopold, J. R. Crespo López-Urrutia, and P. O. Schmidt, "Algorithmic Ground-State Cooling of Weakly Coupled Oscillators Using Quantum Logic", Phys. Rev. X **11**, 041049 (2021).

[117] P. O. Boykin, T. Mor, V. Roychowdhury, F. Vatan, and R. Vrijen, "Algorithmic cooling and scalable NMR quantum computers", Proc. Natl. Acad. Sci. U. S. A. **99**, 3388 (2002).

[118] G. Brassard, Y. Elias, T. Mor, and Y. Weinstein, "Prospects and limitations of algorithmic cooling", Eur. Phys. J. Plus **129**, 258 (2014).

[119] J. Baugh, O. Moussa, C. A. Ryan, A. Nayak, and R. Laflamme, "Experimental implementation of heat-bath algorithmic cooling using solid-state nuclear magnetic resonance", Nature **438**, 470–473 (2005).

[120] W. M. Itano, "External-field shifts of the $^{199}Hg^+$ optical frequency standard", J. Res. NIST **105**, 829–837 (2000).

[121] E. Peik, T. Schneider, and C. Tamm, "Laser frequency stabilization to a single ion", J. Phys. B: At. Mol. Opt. Phys. **39**, 145–158 (2006).

[122] C. Sanner, N. Huntemann, R. Lange, C. Tamm, E. Peik, M. S. Safronova, and S. G. Porsev, "Optical clock comparison for Lorentz symmetry testing", Nature **567**, 204–208 (2019).

[123] J. C. Berengut, D. Budker, C. Delaunay, V. V. Flambaum, C. Frugiuele, E. Fuchs, C. Grojean, R. Harnik, R. Ozeri, G. Perez, and Y. Soreq, "Probing New Long-Range Interactions by Isotope Shift Spectroscopy", Phys. Rev. Lett. **120**, 091801 (2018).

[124] C. Frugiuele, E. Fuchs, G. Perez, and M. Schlaffer, "Constraining new physics models with isotope shift spectroscopy", Phys. Rev. D **96**, 015011 (2017).

[125] C. Delaunay, C. Frugiuele, E. Fuchs, and Y. Soreq, "Probing new spin-independent interactions through precision spectroscopy in atoms with few electrons", Phys. Rev. D **96**, 115002 (2017).

[126] W. H. King, *Isotope Shifts in Atomic Spectra* (Springer Science & Business Media, Nov. 11, 2013), 215 pp.

[127] W. H. King, "Comments on the Article "Peculiarities of the Isotope Shift in the Samarium Spectrum"", J. Opt. Soc. Am. **53**, 638 (1963).

[128] V. V. Flambaum, I. B. Samsonov, H. B. T. Tan, and A. V. Viatkina, "Nuclear polarization effects in atoms and ions", Phys. Rev. A **103**, 032811 (2021).

[129] A. V. Viatkina, V. A. Yerokhin, and A. Surzhykov, "Calculation of isotope shifts and King-plot nonlinearities in Ca^+", Phys. Rev. A **108**, 022802 (2023).

[130] V. Debierre, N. S. Oreshkina, I. A. Valuev, Z. Harman, and C. H. Keitel, "Testing standard-model extensions with isotope shifts in few-electron ions", Phys. Rev. A **106**, 062801 (2022).

[131] R. A. Müller, V. A. Yerokhin, A. N. Artemyev, and A. Surzhykov, "Nonlinearities of King's plot and their dependence on nuclear radii", Phys. Rev. A **104**, L020802 (2021).

[132] S. O. Allehabi, V. A. Dzuba, V. V. Flambaum, and A. V. Afanasjev, "Nuclear deformation as a source of the nonlinearity of the King plot in the Yb^+ ion", Phys. Rev. A **103**, L030801 (2021).

[133] M. Tanaka and Y. Yamamoto, "Relativistic effects in the search for new intra-atomic force with isotope shifts", Progress of Theoretical and Experimental Physics **2020**, 103B02 (2020).

[134] P.-G. Reinhard, W. Nazarewicz, and R. F. Garcia Ruiz, "Beyond the charge radius: The information content of the fourth radial moment", Phys. Rev. C **101**, 021301 (2020).

[135] G. W. F. Drake, H. S. Dhindsa, and V. J. Marton, "King and second-King plots with optimized sensitivity for lithium ions", Phys. Rev. A **104**, L060801 (2021).

[136] J. C. Berengut, C. Delaunay, A. Geddes, and Y. Soreq, "Generalized King linearity and new physics searches with isotope shifts", Phys. Rev. Research **2**, 043444 (2020).

[137] K. Mikami, M. Tanaka, and Y. Yamamoto, "Probing new intra-atomic force with isotope shifts", Eur. Phys. J. C **77**, 896 (2017).

[138] K. Ono, Y. Saito, T. Ishiyama, T. Higomoto, T. Takano, Y. Takasu, Y. Yamamoto, M. Tanaka, and Y. Takahashi, "Observation of Nonlinearity of Generalized King Plot in the Search for New Boson", Phys. Rev. X **12**, 021033 (2022).

[139] J. Hur, D. P. L. Aude Craik, I. Counts, E. Knyazev, L. Caldwell, C. Leung, S. Pandey, J. C. Berengut, A. Geddes, W. Nazarewicz, P.-G. Reinhard, A. Kawasaki, H. Jeon, W. Jhe, and V. Vuletić, "Evidence of Two-Source King Plot Nonlinearity in Spectroscopic Search for New Boson", Phys. Rev. Lett. **128**, 163201 (2022).

[140] N. L. Figueroa, J. C. Berengut, V. A. Dzuba, V. V. Flambaum, D. Budker, and D. Antypas, "Precision Determination of Isotope Shifts in Ytterbium and Implications for New Physics", Phys. Rev. Lett. **128**, 073001 (2022).

[141] I. Counts, J. Hur, D. P. L. Aude Craik, H. Jeon, C. Leung, J. C. Berengut, A. Geddes, A. Kawasaki, W. Jhe, and V. Vuletić, "Evidence for Nonlinear Isotope Shift in Yb^+ Search for New Boson", Phys. Rev. Lett. **125**, 123002 (2020).

[142] T. T. Chang, B. B. Awazi, J. C. Berengut, E. Fuchs, and S. C. Doret, "Systematic-free limit on new light scalar bosons via isotope-shift spectroscopy in Ca^+", Phys. Rev. A **110**, L030801 (2024).

[143] C. Solaro, S. Meyer, K. Fisher, J. C. Berengut, E. Fuchs, and M. Drewsen, "Improved Isotope-Shift-Based Bounds on Bosons beyond the Standard Model through Measurements of the $^2D_{3/2}$-$^2D_{5/2}$ Interval in Ca^+", Phys. Rev. Lett. **125**, 123003 (2020).

[144] C. Solaro, S. Meyer, K. Fisher, J. C. Berengut, E. Fuchs, and M. Drewsen, "Erratum: Improved Isotope-Shift-Based Bounds on Bosons beyond the Standard Model through Measurements of the $^2D_{3/2}$-$^2D_{5/2}$ Interval in Ca^+ [Phys. Rev. Lett. 125, 123003 (2020)]", Phys. Rev. Lett. **127**, 029901 (2021).

[145] N.-H. Rehbehn, M. K. Rosner, H. Bekker, J. C. Berengut, P. O. Schmidt, S. A. King, P. Micke, M. F. Gu, R. Müller, A. Surzhykov, and J. R. C. López-Urrutia, "Sensitivity to new physics of isotope-shift studies using the coronal lines of highly charged calcium ions", Phys. Rev. A **103**, L040801 (2021).

[146] V. V. Flambaum, A. J. Geddes, and A. V. Viatkina, "Isotope shift, nonlinearity of King plots, and the search for new particles", Phys. Rev. A **97**, 032510 (2018).

[147] H. Bekker, A. Borschevsky, Z. Harman, C. H. Keitel, T. Pfeifer, P. O. Schmidt, J. R. C. López-Urrutia, and J. C. Berengut, "Detection of the 5p – 4f orbital crossing and its optical clock transition in Pr^{9+}", Nat Commun **10**, 5651 (2019).

[148] S. Chen, L. J. Spieß, A. Wilzewski, M. Wehrheim, K. Dietze, I. Vybornyi, K. Hammerer, J. R. C. Lopez-Urrutia, and P. O. Schmidt, *Identification of highly-forbidden optical transitions in highly charged ions*, (June 6, 2024) `http://arxiv.org/abs/2406.04015`, pre-published.

[149] N. Herschbach, K. Pyka, J. Keller, and T. E. Mehlstäubler, "Linear Paul trap design for an optical clock with Coulomb crystals", Appl. Phys. B **107**, 891–906 (2012).

[150] J. Keller, T. Burgermeister, D. Kalincev, J. Kiethe, and T. E. Mehlstäubler, "Evaluation of trap-induced systematic frequency shifts for a multi-ion optical clock at the 10^{-19} level", J. Phys. Conf. Ser. **723**, 012027 (2016).

[151] J. Nauta, J.-H. Oelmann, A. Borodin, A. Ackermann, P. Knauer, I. S. Muhammad, R. Pappenberger, T. Pfeifer, and J. R. C. López-Urrutia, "XUV frequency comb production with an astigmatism-compensated enhancement cavity", Opt. Express, OE **29**, 2624–2636 (2021).

[152] U. I. Safronova, V. V. Flambaum, and M. S. Safronova, "Transitions between the 4f-core-excited states in Ir^{16+}, Ir^{17+}, and Ir^{18+} ions for clock applications", Phys. Rev. A **92**, 022501 (2015).

[153] A. Arvanitaki, J. Huang, and K. Van Tilburg, "Searching for dilaton dark matter with atomic clocks", Phys. Rev. D **91**, 015015 (2015).

[154] A. Derevianko and M. Pospelov, "Hunting for topological dark matter with atomic clocks", Nat Phys **10**, 933–936 (2014).

[155] X. Calmet and H. Fritzsch, "The cosmological evolution of the nucleon mass and the electroweak coupling constants", Eur. Phys. J. C **24**, 639–642 (2002).

[156] X. Calmet and H. Fritzsch, "Symmetry breaking and time variation of gauge couplings", Phys. Lett. B **540**, 173–178 (2002).

[157] P. Langacker, G. Segrè, and M. J. Strassler, "Implications of gauge unification for time variation of the fine structure constant", Physics Letters B **528**, 121–128 (2002).

[158] A. Shelkovnikov, R. J. Butcher, C. Chardonnet, and A. Amy-Klein, "Stability of the Proton-to-Electron Mass Ratio", Phys. Rev. Lett. **100**, 150801 (2008).

[159] J. Kobayashi, A. Ogino, and S. Inouye, "Measurement of the variation of electron-to-proton mass ratio using ultracold molecules produced from laser-cooled atoms", Nat Commun **10**, 3771 (2019).

[160] N. Huntemann, B. Lipphardt, Chr. Tamm, V. Gerginov, S. Weyers, and E. Peik, "Improved Limit on a Temporal Variation of m_p/m_e from Comparisons of Yb^+ and Cs Atomic Clocks", Phys. Rev. Lett. **113**, 210802 (2014).

[161] R. M. Godun, P. B. R. Nisbet-Jones, J. M. Jones, S. A. King, L. A. M. Johnson, H. S. Margolis, K. Szymaniec, S. N. Lea, K. Bongs, and P. Gill, "Frequency Ratio of Two Optical Clock Transitions in $^{171}Yb^+$ and Constraints on the Time Variation of Fundamental Constants", Phys. Rev. Lett. **113**, 210801 (2014).

[162] W. F. McGrew, X. Zhang, H. Leopardi, R. J. Fasano, D. Nicolodi, K. Beloy, J. Yao, J. A. Sherman, S. A. Schäffer, J. Savory, R. C. Brown, S. Römisch, C. W. Oates, T. E. Parker, T. M. Fortier, and A. D. Ludlow, "Towards the optical second: verifying optical clocks at the SI limit", Optica, OPTICA **6**, 448–454 (2019).

[163] M. A. Hohensee, N. Leefer, D. Budker, C. Harabati, V. A. Dzuba, and V. V. Flambaum, "Limits on Violations of Lorentz Symmetry and the Einstein Equivalence Principle using Radio-Frequency Spectroscopy of Atomic Dysprosium", Phys. Rev. Lett. **111**, 050401 (2013).

[164] T. Pruttivarasin, M. Ramm, S. G. Porsev, I. I. Tupitsyn, M. S. Safronova, M. A. Hohensee, and H. Häffner, "Michelson-Morley analogue for electrons using trapped ions to test Lorentz symmetry", Nature **517**, 592–595 (2015).

[165] V. A. Dzuba, V. V. Flambaum, M. S. Safronova, S. G. Porsev, T. Pruttivarasin, M. A. Hohensee, and H. Häffner, "Strongly enhanced effects of Lorentz symmetry violation in entangled Yb^+ ions", Nat Phys **12**, 465–468 (2016).

[166] R. Shaniv, R. Ozeri, M. S. Safronova, S. G. Porsev, V. A. Dzuba, V. V. Flambaum, and H. Häffner, "New Methods for Testing Lorentz Invariance with Atomic Systems", Phys. Rev. Lett. **120**, 103202 (2018).

Journal of Physics: Conference Series **2889** (2024) 012052 doi:10.1088/1742-6596/2889/1/012052

A thermal-noise-limited fiber frequency reference with 0.1 Hz/$\sqrt{\text{Hz}}$ stability

Ya Zhang[1,2], Chathura P. Bandutunga[1,2], Terry G. McRae[1,2], Malcolm B. Gray[1] and Jong H. Chow[1,2]

[1]Centre for Gravitational Astrophysics, Australian National University, ACT 2601, Australia.
[2]OzGrav-ANU, Centre for Gravitational Astrophysics, ANU, ACT 2601, Australia.

E-mail: ya.zhang@anu.edu.au

We develop an all-passive, fiber frequency reference with a state-of-the-art short-term stability of 0.1 Hz/$\sqrt{\text{Hz}}$. We explore the impact of intrinsic fiber thermal noise and show it limits the system in the 1 Hz Fourier frequency band.

1. Introduction

Precision optical interferometry relies on the frequency stability of its optical source to reach their target sensitivity. The free-running performance of commercial laser sources often fall short for state-of-the-art instruments, necessitating stabilization to an external reference. At short integration timescales, the standard frequency reference of choice is an ultra-low-expansion (ULE) cavity. These systems deliver unrivaled short-term stability under controlled laboratory conditions, with a landmark stability reaching 0.04 Hz/$\sqrt{\text{Hz}}$ (fractional $\sim 10^{-16}$) at 1 Hz [1, 2]. However, in applications requiring field deployment or space qualification, environmental conditions become complex and challenge the stability and continuity of cavity systems.

Optical-fiber-based frequency references [3, 4] present an interesting yet less explored alternative to bulk cavities. The intrinsic alignment of optical fiber ensures uninterrupted operation even in hostile environments. Building on the structure of an armlength unbalanced interferometer, a fiber reference is highly frequency agile and, with the appropriate interrogation technique, can be also be used in a feedforward arrangement to passively remove laser frequency noise. Fiber references represent a compromise between intuitive and robust operation, and additional limitations unique to guided-wave optics [5–10]. Thus, the state-of-the-art fiber frequency reference achieves a slightly inferior performance, at 0.1 Hz/$\sqrt{\text{Hz}}$ between 1-4 kHz [11].

For both ULE cavities and fiber references, the fundamental limit in the sub-Hz frequency regime is the intrinsic thermal noise. Unlike cavities, thermal noise within optical fiber remains under-studied and less understood. Early research drew analogy between fiber thermal noise and the mirror noise seen in LIGO interferometers, where spontaneous, localized temperature fluctuations induce microscopic changes in interferometer pathlength through displacement of the coating material [12]. In the case of a fiber interferometer, these temperature variations result in the expansion and contraction of the glass as well as changes in its refractive index. These two effects were studied and modeled independently by Wanser [13] and Foster [14], both predicting a frequency independent fiber thermal noise floor.

Both theories demonstrated good experimental agreement in the high frequency regime [15–19]. However, at infrasonic frequencies, experimental measurements deviated from their predictions, as a $1/f$ type noise in excess of laser phase noise was observed in several instances [20–23]. This additional noise was found to originate from thermo-mechanical effects, where random expansion and contraction of the fiber occurred due to internal friction or structural damping [24]. This new type of thermal noise was reconciled by Duan through a universal framework utilizing the fluctuation-dissipation theorem [25]. The modeling we will present in this work is based on this prior work by Duan.

In this work, we present a fiber frequency reference reaching fundamental thermo-mechanical noise in the 400 mHz - 2 Hz band, with a frequency stability of less than 1 Hz/$\sqrt{\text{Hz}}$ above 0.3 Hz Fourier frequency. It presents first experimental results in a system where the fundamental longitudinal mechanical mode of thermo-mechanical noise is below the measurement frequency range. Using this, we verify that thermo-mechanical noise converges to the Duan model even when operating well above the first mechanical resonance frequency.

2. Experimental Configuration and Readout Stability

The experimental configuration is shown in Fig. 1. We construct two near-identical Mach-Zehnder interferometers with 15 km armlength difference. These two interferometers are separately housed in two passive, dual-layer chambers that provide thermal, mechanical and acoustic isolation from the environment. Optically, the two interferometers are used to measure the same optical source. They utilize independent photodetection and signal digitization pathways to maintain independence. The interferometric readout is handled by the homodyne variant of Digital Interferometry (DI), a code-multiplexing technique that combines phase-shifting interferometric methods with pseudo-random noise correlation techniques. In this instance, it allows us to remove all active optical components from within the interferometer itself, and to range-gate and reduce the impact of Rayleigh scattering effects.

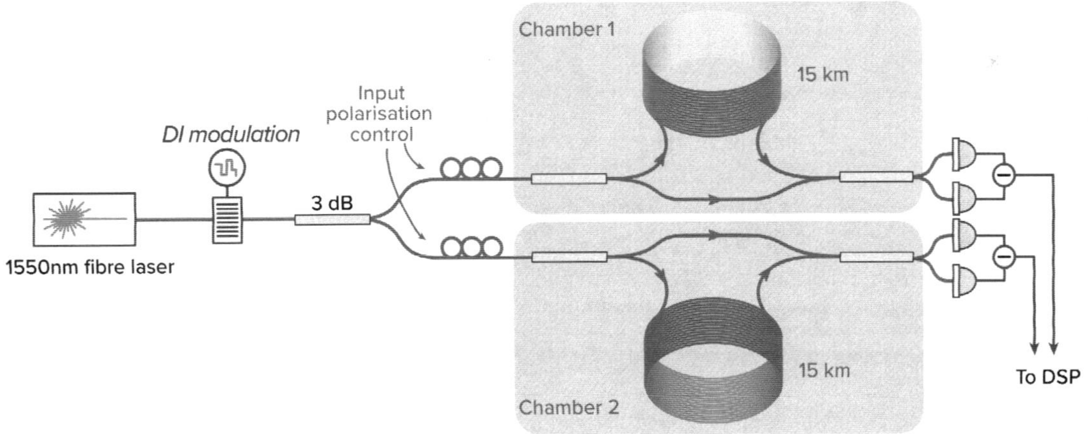

Figure 1: Optical layout of the experiment. Two near-identical Mach-Zehnder interferometers were constructed within individual isolation chambers. Each interferometer consisted of a short and long arm, with a length difference of 15 km. A fiber laser was split to interrogate both interferometers, producing two phase readouts that tracked the frequency movement of the laser.

Figure 2(a) plots a 10 minute recording of the two interferometer's frequency readout, in trace (i) and (ii). Their subtraction, shown in Fig. 2(b), subtracts the common laser frequency drift, providing an estimate of the relative stability of the fiber references.

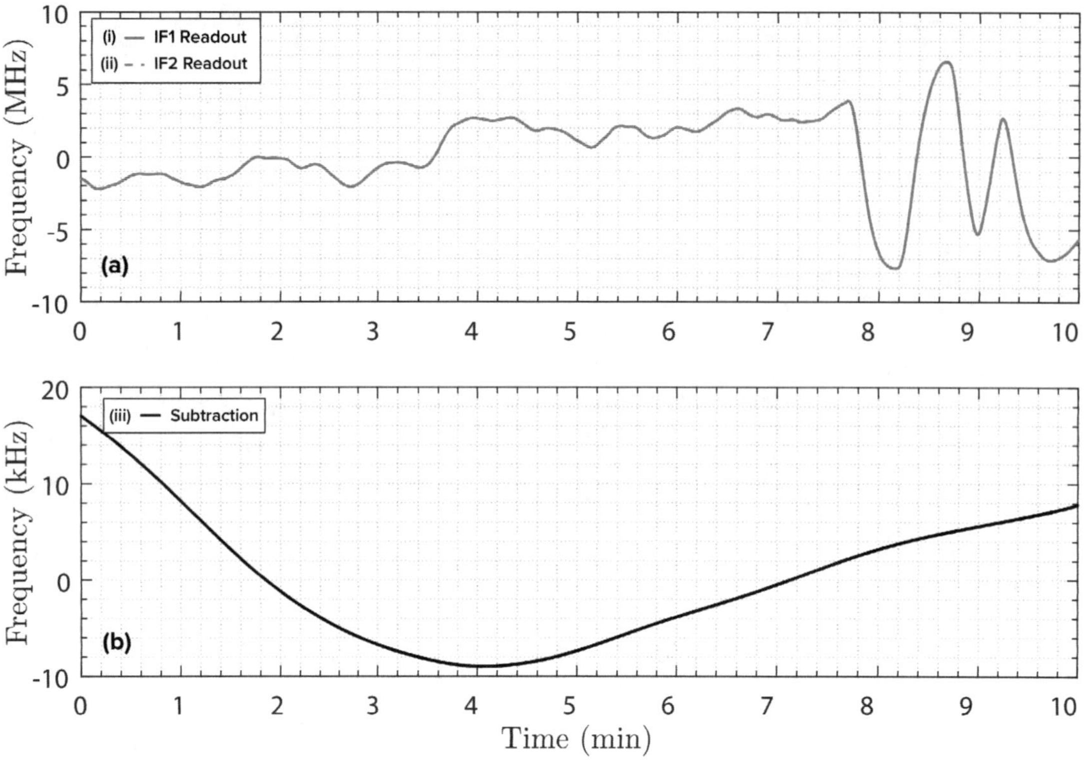

Figure 2: (a) Individual interferometer readout of the same probe laser over 10 minutes. (b) Subtraction of the above two readout traces, providing an estimate of the relative drift and readout stability of the two systems.

We compute the frequency spectral density from the time-domain measurement, shown below in Fig. 3. The computation uses Welch's method [26] with a Blackman window whose length is logarithmically shortened towards high Fourier frequencies for higher data averaging. We see in trace (b) that the fiber reference achieves a stability of 0.1 $\mathrm{Hz}/\sqrt{\mathrm{Hz}}$ above 70 Hz Fourier frequency. This performance represents the state-of-the-art sensitivity for fiber references [11, 27–29], and reaches parity with room temperature ULE systems [30–33]. Of particular interest is that between 0.4 - 2 Hz, the fiber reference reaches the intrinsic fiber thermo-mechanical noise limit, which we model and calculate in the next section.

3. Fiber Thermal Noise

In this section, we consider two mechanisms of fiber thermal noise: 1) thermo-dynamic noise, akin to the derivations presented by Wanser and Foster, and 2) thermo-mechanical noise, which fixes the low-frequency experimental discrepancy in earlier models [23, 25, 34].

3.1. Thermo-mechanical Noise

The thermo-mechanical noise estimate is derived through a normal mode expansion analysis detailed in [24] and [35], where each normal mode is treated as a harmonic oscillator to find the overall spectral density for fiber length change from structural damping. An important simplification of this calculation occurs in Eqn. 25 of [24]. This assumes the frequency range of concern is significantly below the first longitudinal resonance of the fiber, and is not entirely justified for our

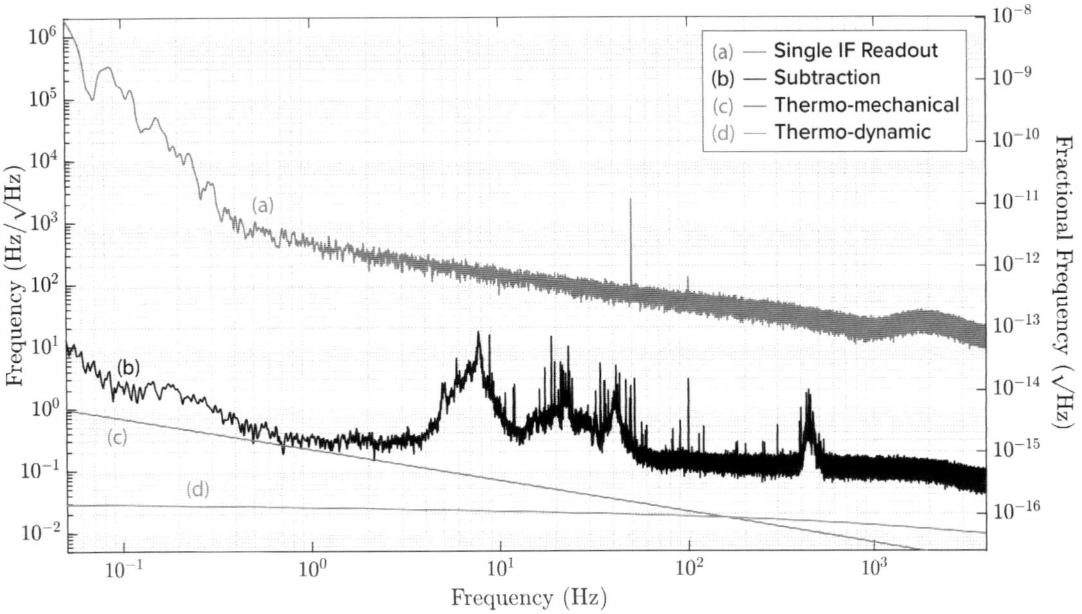

Figure 3: Frequency spectral density of (a) free-running fiber laser as measured by a single fiber interferometer, (b) the readout subtraction, representing the estimated noise floor of the fiber interferometers. Trace (c) and (d) plot the calculated thermo-mechanical and thermo-dynamic noise in optical fiber. We see that the fiber reference reaches thermo-mechanical noise from 400 mHz to 2 Hz.

system with 15 km of fiber [23]. For comparison, the corresponding first longitudinal mechanical resonance for our system is approximately 0.24 Hz. An exact calculation for our experiment would therefore require the numerical evaluation of an infinite number of higher order modes, as per Eqn. 24 in [24].

Fortunately, as the mode order increases, its impact on noise spectral density decreases exponentially. As we computed the analytical sum of 20 modes, this in fact converged to the value given by the aforementioned assumption and simplification. We further observe that the assumption still holds true for us at very low frequencies, and the $1/f$ trend continues through the mechanical resonances at higher frequencies, as shown in [24]. In addition, these resonances have not been observed experimentally, and could be caused by oversimplifications in the model such as assuming a constant, frequency independent loss angle ϕ_0. If we assume no distinct mechanical peaks and a continuation of the $1/f$ trend, we can justify applying the simplified thermo-mechanical noise estimate to all Fourier frequencies. The simplified phase spectral density for thermo-mechanical noise as given in [25] is shown in Eqn. 1:

$$\tilde{S}_\phi(f) = \sqrt{\frac{4\pi L}{\lambda^2}\frac{2k_B T n^2 \phi_0}{3E_0 A}\frac{1}{f}}\ \text{rad}/\sqrt{\text{Hz}} \tag{1}$$

We can further convert this into frequency spectral density by multiplying by the interferometer free-spectral range. We also add another factor of $\sqrt{2}$ for the incoherent noise contributions from the two interferometers, yielding Eqn 2:

$$\tilde{S}_\nu(f) = \sqrt{2}\,\frac{c}{2\pi n L}\,\tilde{S}_\phi(f)\ \ \text{Hz}/\sqrt{\text{Hz}} \tag{2}$$

 doi:10.1088/1742-6596/2889/1/012052

A glossary of the constants is provided in Tab 1. Substituting in these values into Eqn. 2, we arrive at Eqn. 3:

$$\tilde{S}_\nu(f) = \frac{26.48}{\sqrt{Lf}} \text{ Hz}/\sqrt{\text{Hz}} \tag{3}$$

Name	Symbol	Value
Boltzmann constant	k_B	1.380649×10^{-23}
Temperature	T	298 K
Mechanical loss angle	ϕ_0	0.01
Young's modulus without loss	E_0	19 GPa
Cross-sectional area of fiber	A	$\pi(125\,\mu\text{m})^2$
Optical wavelength	λ	1550 nm
Refractive index	n	1.47
Thermo-optic coefficient	$\frac{dn}{dT}$	9.488×10^{-6} K^{-1}
Linear expansion coefficient	α_L	5×10^{-7} K^{-1}
Thermal conductivity	κ	1.37 W/m/K
Power profile radius	r_0	${r_0}^2 = a^2/2$
Mode field radius	a	5.2×10^{-6} m
Thermal diffusivity	D	8.2×10^{-7} m^2/s

Table 1: Glossary of thermal constants.

Using Eqn. 3, we plot the thermo-mechanical noise estimate for our interferometer length of $L = 15$ km as in trace (c) of Fig. 3. Comparing this with our experimental noise floor of trace (b), we can identify a thermo-mechanical noise limited region of 400 mHz - 2 Hz.

3.2. Thermo-dynamic Noise

Thermo-dynamic noise in optical fiber is driven by spontaneous fiber length change and refractive index variation. The two processes are jointly related to localized temperature fluctuations, and their combined noise power spectral density $S_\phi(\omega)$ can be written as a function of the temperature power spectral density $S_{\delta T}(\omega)$ [24]:

$$S_\phi(\omega) = \frac{4\pi^2 L^2}{\lambda^2}\left(\frac{dn}{dT} + n\alpha_L\right)^2 S_{\delta T}(\omega) \tag{4}$$

Further, $S_{\delta T}(\omega)$ is derived using the fluctuation-dissipation theorem to be [24]:

$$S_{\delta T}(\omega) = \frac{k_B T^2}{4\pi^2 L\kappa}\text{Re}\left[\exp\left(\frac{i\omega r_0^2}{2D}\right)E_1\left(\frac{i\omega r_0^2}{2D}\right)\right] \tag{5}$$

where $E_1(x)$ is the special function of the exponential integral, and the other constants are specified in the glossary of Tab. 1. Substituting $S_{\delta T}(\omega)$ from Eqn. 5 into Eqn. 4 then allows us to compute the power spectral density of thermo-dynamic noise. When converting to single-sided phase spectral density, an extra factor of $\sqrt{4\pi}$ applies, resulting in Eqn. 6:

$$\tilde{S}_\phi(f) = \sqrt{4\pi}\sqrt{S_\phi(2\pi f)} \tag{6}$$

The same process is followed to convert this phase spectral density into frequency spectral density, and an extra factor of $\sqrt{2}$ is added to account for the second interferometer. The resultant thermo-dynamic noise estimate for $L = 15$ km is plotted in trace (d) of Fig. 3, intersecting with thermo-mechanical noise (trace (c)) at approximately 150 Hz. Currently, thermo-dynamic noise does not impose a limitation for our experiment in any frequency regime. It does however lie within a factor of 10 of our experimental noise floor above 150 Hz, and could become a dominant noise source if higher noise sources, such as Double Rayleigh scattering and tracking error [36, 37], are addressed.

One interesting observation from the modeling of fiber thermal noise is that the two mechanisms always intersect at the same Fourier frequency regardless of fiber length. This can be verified by equating Eqn. 1 and Eqn. 6, and we find that L is canceled out on both sides. For a transmissive, SMF-28e fiber interferometer with thermal constants, such as described in Tab. 1, the cross-over frequency occurs at ~150 Hz. Therefore in practice, these two noise sources always dominate different parts of the spectrum regardless of the optical configuration, and their relative scale remains constant.

4. Conclusion and Outlook

We have developed and characterized a fiber frequency reference with a relative frequency stability of 0.1 Hz/$\sqrt{\text{Hz}}$. The system reaches the fundamental thermal noise between 0.4 - 2 Hz, providing a valuable experimental datapoint for this less-known noise limit. Future work will focus on improving the long-timescale stability of the fiber reference, to extend the probe range of thermal noise and to meet the requirements of space-borne interferometric missions.

References

[1] Y. Y. Jiang, A. D. Ludlow, N. D. Lemke, J. A. Sherman, J. Von Stecher, R. W. Fox, L. S. Ma, A. M. Rey, and C. W. Oates. Improving the stability and accuracy of the Yb optical lattice clock. In *Proceedings of the IEEE International Frequency Control Symposium and Exposition*, pages 8–10, 2011.

[2] Y. Y. Jiang, A. D. Ludlow, N. D. Lemke, R. W. Fox, J. A. Sherman, L.-S. Ma, and C. W. Oates. Making optical atomic clocks more stable with 10^{-16}-level laser stabilization. *Nature Photonics*, 5(3):158–161, March 2011.

[3] Terry G. McRae, Silvie Ngo, Daniel A. Shaddock, Magnus T. L. Hsu, and Malcolm B. Gray. Digitally enhanced optical fiber frequency reference. *Optics Letters*, 39(7):1752–1755, April 2014.

[4] Chathura P. Bandutunga, Terry G. McRae, Ya Zhang, Malcolm B. Gray, and Jong H. Chow. Infrasonic performance of a passively stabilized, all-fiber, optical frequency reference. *Optics Express*, 28(7):9280–9287, March 2020.

[5] EP Ippen and RH Stolen. Stimulated Brillouin scattering in optical fibers. *Applied Physics Letters*, 21(11):539–541, 1972.

[6] R. G. Smith. Optical power handling capacity of low loss optical fibers as determined by stimulated raman and brillouin scattering. *Applied Optics*, 11(11):2489–2494, November 1972.

[7] R. H. Stolen and Chinlon Lin. Self-phase-modulation in silica optical fibers. *Physical Review A: Atomic, Molecular, and Optical Physics*, 17(4):1448–1453, April 1978.

[8] S. W. Lloyd, M. J. F. Digonnet, and S. Fan. Modeling coherent backscattering errors in fiber optic gyroscopes for sources of arbitrary line width. *Journal of Lightwave Technology*, 31(13):2070–2078, July 2013.

[9] Jong H. Chow, Benjamin S. Sheard, David E. McClelland, Malcolm B. Gray, and Ian C. M. Littler. Photothermal effects in passive fiber Bragg grating resonators. *Opt. Lett., OL*, 30(7):708–710, April 2005.

[10] Ian C. M. Littler, Thomas Grujic, and Benjamin J. Eggleton. Photothermal effects in fiber Bragg gratings. *Appl. Opt., AO*, 45(19):4679–4685, July 2006.

[11] Fabien Kéfélian, Haifeng Jiang, Pierre Lemonde, and Giorgio Santarelli. Ultralow-frequency-noise stabilization of a laser by locking to an optical fiber-delay line. *Optics Letters*, 34(7):914, April 2009.

[12] V. B. Braginsky, M. L. Gorodetsky, and S. P. Vyatchanin. Thermodynamical fluctuations and photo-thermal shot noise in gravitational wave antennae. *Physics Letters A*, 264(1):1–10, December 1999.

[13] K. H. Wanser. Fundamental phase noise limit in optical fibres due to temperature fluctuations. *Electronics Letters*, 28(1):53–54, January 1992.

[14] S. Foster, A. Tikhomirov, and M. Milnes. Fundamental Thermal Noise in Distributed Feedback Fiber Lasers. *IEEE Journal of Quantum Electronics*, 43(5):378–384, May 2007.

[15] S. Knudsen, A.B. Tveten, and A. Dandridge. Measurements of fundamental thermal induced phase fluctuations in the fiber of a Sagnac interferometer. *IEEE Photonics Technology Letters*, 7(1):90–92, January 1995.

[16] K. Krakenes and K. Blotekjaer. Comparison of fiber-optic Sagnac and Mach-Zehnder interferometers with respect to thermal processes in the fiber. *Journal of Lightwave Technology*, 13(4):682–686, April 1995.
[17] Alan D. Kersey. A Review of Recent Developments in Fiber Optic Sensor Technology. *Optical Fiber Technology*, 2(3):291–317, July 1996.
[18] V. Annovazzi-Lodi, S. Donati, and S. Merlo. Thermodynamic phase noise in fibre interferometers. *Optical and Quantum Electronics*, 28(1):43–49, January 1996.
[19] R. P. Moeller and W. K. Burns. Observation of thermal noise in a dynamically biased fiber-optic gyro. *Optics Letters*, 21(3):171, February 1996.
[20] Erlend Rønnekleiv. Frequency and Intensity Noise of Single Frequency Fiber Bragg Grating Lasers. *Optical Fiber Technology*, 7(3):206–235, July 2001.
[21] G. A. Cranch. Frequency noise reduction in erbium-doped fiber distributed-feedback lasers by electronic feedback. *Opt. Lett., OL*, 27(13):1114–1116, July 2002.
[22] Scott Foster, Geoffrey A. Cranch, and Alexei Tikhomirov. Experimental evidence for the thermal origin of $1/f$ frequency noise in erbium-doped fiber lasers. *Physical Review A: Atomic, Molecular, and Optical Physics*, 79(5):053802, May 2009.
[23] R. E. Bartolo, A. B. Tveten, and A. Dandridge. Thermal Phase Noise Measurements in Optical Fiber Interferometers. *IEEE Journal of Quantum Electronics*, 48(5):720–727, May 2012.
[24] Lingze Duan. General treatment of the thermal noises in optical fibers. *Physical Review A: Atomic, Molecular, and Optical Physics*, 86(2):023817, August 2012.
[25] Lingze Duan. Thermal Noise-Limited Fiber-Optic Sensing at Infrasonic Frequencies. *IEEE Journal of Quantum Electronics*, 51(2):1–6, February 2015.
[26] P. Welch. The use of fast fourier transform for the estimation of power spectra: A method based on time averaging over short, modified periodograms. *IEEE Transactions on Audio and Electroacoustics*, 15(2):70–73, 1967.
[27] Jing Dong, Yongqi Hu, Junchao Huang, Meifeng Ye, Qiuzhi Qu, Tang Li, and Liang Liu. Subhertz linewidth laser by locking to a fiber delay line. *Applied Optics*, 54(5):1152–1156, February 2015.
[28] Nicola Chiodo, Khelifa Djerroud, Ouali Acef, André Clairon, and Peter Wolf. Lasers for coherent optical satellite links with large dynamics. *Applied Optics*, 52(30):7342–7351, October 2013.
[29] Haifeng Jiang, Fabien Kéfélian, Pierre Lemonde, André Clairon, and Giorgio Santarelli. An agile laser with ultra-low frequency noise and high sweep linearity. *Optics Express*, 18(4):3284–3297, February 2010.
[30] X. Y. Zeng, Y. X. Ye, X. H. Shi, Z. Y. Wang, K. Deng, J. Zhang, and Z. H. Lu. Thermal-noise-limited higher-order mode locking of a reference cavity. *Optics Letters*, 43(8):1690–1693, April 2018.
[31] A. D. Ludlow, X. Huang, M. Notcutt, T. Zanon-Willette, S. M. Foreman, M. M. Boyd, S. Blatt, and J. Ye. Compact, thermal-noise-limited optical cavity for diode laser stabilization at 1×10^{-15}. *Optics Letters*, 32(6):641–643, March 2007.
[32] S. A. Webster, M. Oxborrow, S. Pugla, J. Millo, and P. Gill. Thermal-noise-limited optical cavity. *Physical Review A: Atomic, Molecular, and Optical Physics*, 77(3):033847, March 2008.
[33] Sebastian Häfner, Stephan Falke, Christian Grebing, Stefan Vogt, Thomas Legero, Mikko Merimaa, Christian Lisdat, and Uwe Sterr. 8×10^{-17} fractional laser frequency instability with a long room-temperature cavity. *Optics Letters*, 40(9):2112–2115, May 2015.
[34] Jing Dong, Junchao Huang, Tang Li, and Liang Liu. Observation of fundamental thermal noise in optical fibers down to infrasonic frequencies. *Applied Physics Letters*, 108(2):021108, January 2016.
[35] Kazuhiro Yamamoto, Masaki Ando, Keita Kawabe, and Kimio Tsubono. Theoretical approach to thermal noise caused by an inhomogeneously distributed loss: Physical insight by the advanced modal expansion. *Physical Review D: Particles and Fields*, 75(8):082002, April 2007.
[36] Ya Zhang, Chathura P. Bandutunga, Terry G. McRae, Malcolm B. Gray, and Jong H. Chow. Double Rayleigh scattering in a digitally enhanced, all-fiber optical frequency reference. *Optics Express*, 29(17):26319–26331, August 2021.
[37] Ya Zhang. *Development and Characterisation of a Fibre Frequency Reference.* PhD thesis, Research School of Physics, ANU College of Science, The Australian National University, 2022.

Journal of Physics: Conference Series **2889** (2024) 012053 doi:10.1088/1742-6596/2889/1/012053

A Novel Laser Power Measurement Scheme Using Rubidium Clock

Qianqian Ji[1*], Lei Han[1], Yabei Su[1], Xiaobo Xue[1], Shengkang Zhang[1] and Jun Ge[1]

[1] Science and Technology on Metrology and Calibration Laboratory, Beijing Institute of Radio Metrology and Measurement, Beijing, China

[*]E-mail: jqq202208@163.com

Abstract. To measure microwatt-level laser power, this paper presents a method based on the atomic light shift effect. A 795 nm semiconductor laser is directed into a rubidium atomic clock with an output frequency of 10 MHz. To minimize the influence of laser frequency on power measurements, an experimental setup is implemented to precisely lock the laser frequency to the wavelength meter. After stabilizing the laser frequency, we analyze its correlation with the output frequency of the rubidium atomic clock. The experimental results indicate that, during free-running, the laser undergoes a frequency drift of -90 MHz over 2 hours, which is subsequently reduced to ±0.9 MHz over 40 hours through frequency locking. Moreover, by fixing the laser wavelength at 794.99 nm, we observe an increase in the atomic clock frequency shift with rising laser power. Specifically, a laser power of 2 μW corresponds to a frequency shift of 0.13 mHz at 10 MHz. This study achieves quantum measurement of weak laser power, significantly enhancing the accuracy of low-power laser measurements.

1. Introduction

Lasers, characterized by monochromaticity and strong coherence, play a crucial role in various fields such as laser communication, laser processing, and fundamental scientific research[1]. One critical parameter is laser power, measured through various methods, including those based on photodiodes[2,3], cryogenic radiometers[4,5], photodynamic sensing, and integrating spheres[6]. However, methods for measuring low laser power often lack stability, making them susceptible to environmental noise and internal drift[7]. This inherent instability poses challenges in achieving high-precision measurements of low laser power. Precise measurements of weak laser power are essential, and few methods meet this requirement[8]. To achieve precise measurements of weak laser power at the microwatt and nanowatt levels, we utilize the light shift effect and employ atomic clocks for laser power measurement, which is traceable to the output frequency of the atomic clock.

The light shift effect is characterized by the shift in atomic energy levels within an atomic system induced by the influence of nearly resonant pumping light[9,10,11]. The magnitude of this shift is linked to the frequency and intensity of the pumping light. Research groups traditionally measure changes in light shift induced by laser power to determine the optimal laser operating power, seeking to minimize the impact of shift effects within the system[12,13,14]. In this study, however, we seek to use the light shift effect for laser power measurement.

In this paper, a 795 nm semiconductor laser and a rubidium (Rb) atomic clock are employed. The laser is directed into the rubidium clock's cell, and the atomic clock's frequency varies with the laser

Journal of Physics: Conference Series **2889** (2024) 012053 doi:10.1088/1742-6596/2889/1/012053

power. The laser frequency is locked to the wavelength meter. By tracing the measurement of laser power back to the output frequency of an atomic clock, precise measurement of weak laser power is achieved.

2. Experimental setup

The experimental design is depicted in Figure 1. A 795 nm external-cavity semiconductor laser serves as the light source. The emitted laser light traverses a half-wave plate, PBS, and a mirror before entering the Rb atomic cell in the rubidium clock. Interaction between atoms and laser induces a frequency shift in atomic energy levels due to the presence of the optical field. This light shift effect leads to a shift in both hyperfine levels of the ^{87}Rb ground state, causing a modification in the resonant frequency of atomic levels. Consequently, this induces a frequency shift in the output frequency of the atomic clock. The 53100A measured the output frequency of the rubidium atomic clock before and after the incidence of 795 nm laser. The data collected by the 53100A is processed using a computer, and the change in the atomic clock output frequency before and after the laser incidence is determined. Subsequent to this analysis, the corresponding incident laser power can be inversely deduced.

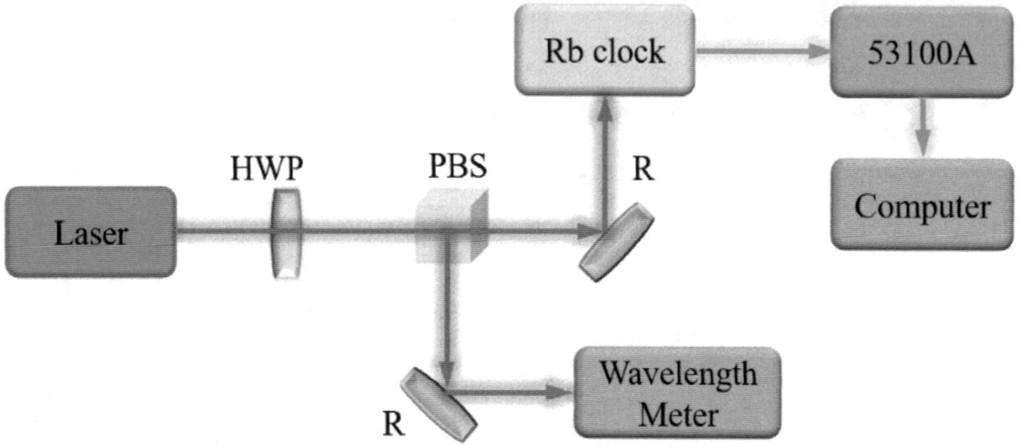

Figure 1. Schematic diagram of the experimental setup. HWP: half wave plate, PBS: polarization splitter, R: reflectors.

The frequency shift of atomic levels depends on both wavelength and laser power. Fluctuations in laser wavelength impact the atomic clock's output frequency, thereby affecting the precision of laser power measurement. Therefore, stabilizing the laser frequency is imperative. The frequency locking experimental configuration is illustrated in Figure 2. During the experiment, digital PID control of the laser frequency was employed. The laser wavelength is acquired through a wavelength meter, and the collected data is carefully processed by a computer. Subsequently, the processed data is compared with the nominal frequency to generate an error signal. This error signal serves as the input for the PID algorithm, directing the PID output to precisely regulate the laser's PZT through NI's data acquisition card. This ensures that the laser wavelength is accurately locked onto the reference frequency of the wavelength meter.

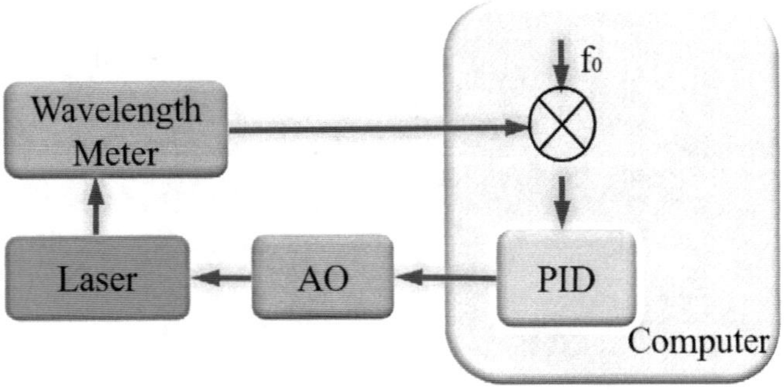

Figure 2.The principle diagram of the frequency locking.

3. Experimental results

Frequency locking of the laser was accomplished through digital PID control, and the laser's frequency drift was measured under both free-running and stabilized conditions. As depicted in Figure 3, the laser exhibited an initial frequency drift of -90 MHz after 2 hours of free operation. Subsequently, with the implementation of feedback control, the locked laser demonstrated continuous operation for 40 hours, achieving a remarkable reduction in frequency drift to within ±0.9 MHz. This performance met the stability requirements of the experimental system for laser frequency.

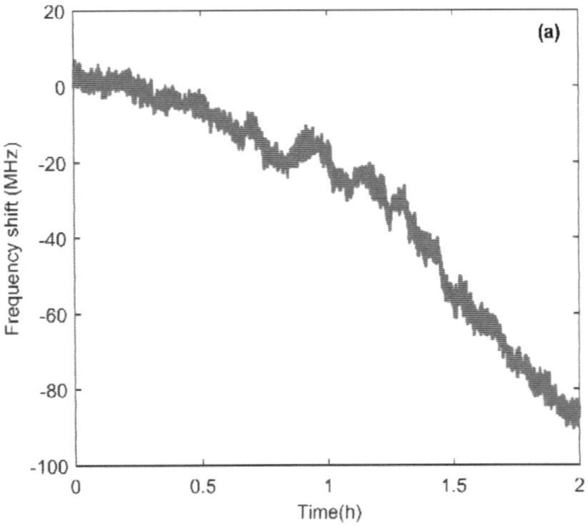

Journal of Physics: Conference Series **2889** (2024) 012053 doi:10.1088/1742-6596/2889/1/012053

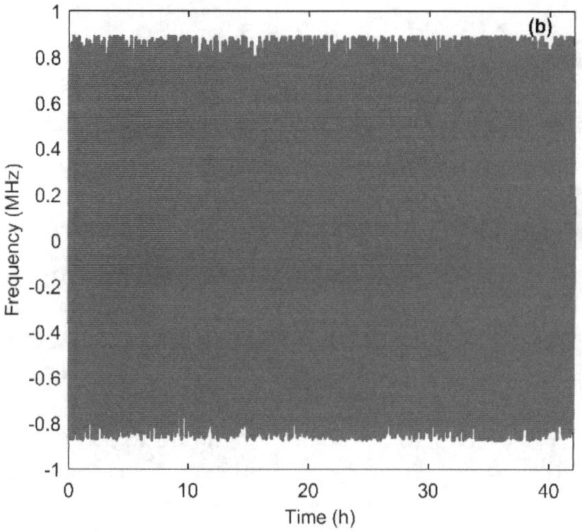

Figure 3. Experimental results of laser frequency drift. (a) laser frequency drift for 795 nm free running. (b) laser frequency drift for 795 nm PID locking.

Furthermore, to investigate the impact of laser power on the output frequency of the atomic clock, the laser output wavelength was fixed at λ=794.99 nm. The frequency shift of the atomic clock output with varying laser power is depicted in Figure 4. It is evident that a laser power of 2 μW corresponds to a frequency shift of 0.13 mHz@10MHz. The results indicate that, at lower laser power levels, the output frequency shift of the atomic clock increases with the enhancement of laser power.

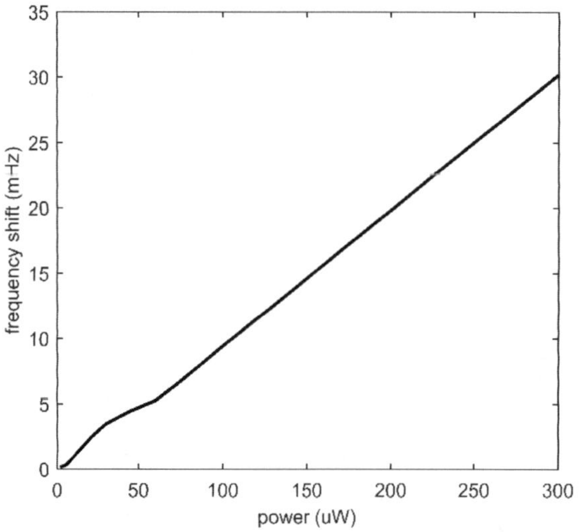

Figure 4. The Rb clock's output frequency shift varies with changes in laser power at λ =794.99 nm.

4. Conclusion

In summary, a novel method for measuring weak laser power is proposed, and the experiment is conducted to achieve quantum measurement of weak laser power. A laser frequency locking method was employed in the experiment, reducing the laser's free-running frequency from -90 MHz to a stabilized range within ±0.9 MHz. The investigation explored the impact of laser power on the output frequency of the atomic clock, revealing a notable frequency shift of 0.13 mHz@10MHz corresponding to a laser power of 2 μW. This method introduces a new approach to trace laser power to atomic frequency. Optimizing the laser beam spot size and enhancing the coupling between the laser and the atomic clock, resulting in an increased frequency shift through enhanced interaction between light and atoms. Furthermore, the experimental setup can be improved to stabilize laser power.

References

[1] Jiang, Y. Y., Ludlow, A. D., Lemke, N. D., Fox, R. W., Sherman, J. A., Ma, L. S., & Oates, C. W. (2011).*Nature Photonics*, **5(3)**, 158-161.
[2] Wang M, Kong J, Fu J, et al. 2022. *Review of Scientific Instruments*, **93(5)**.
[3] Brandt, F., Kück, S., & Grütz, A. (2010). *Mapan*, **25(1)**, 29-35.
[4] Cullen T, Aronson S, Pagano R, et al. 2022. *Optics Letters*, **47(11)**: 2746-2749.
[5] Martin J E, Fox N P, Key P J. 1985. *Metrologia*, **21(3)**: 147.
[6] Hanssen L M, Snail K A. 2002. *Handbook of Vibrational Spectroscopy*, **2**: 1175-1192.
[7] Gebhardt F G. 1976. *Applied Optics*, 1**5(6)**: 1479-1493.
[8] Peng, C., Qiao, H., Zhang, S., Luo, L., Zhou, M., & Xu, X. (2023).*IEEE Transactions on Instrumentation and Measurement.*
[9] Arditi, M., & Picque, J. L. (1975). *Journal of Physics B: Atomic and Molecular Physic*s, **8(14)**, L331.
[10] Hafiz, M. A., Vicarini, R., Passilly, N., Calosso, C. E., Maurice, V., Pollock, J. W., ... & Boudot, R. (2020). *Physical Review Applied*, **14(3)**, 034015.
[11] Shah, V., Gerginov, V., Schwindt, P. D. D., Knappe, S., Hollberg, L., & Kitching, J. (2006).*Applied Physics Letters*, **89(15)**.
[12] B. H. McGuyer, Y.-Y. Jau, and W. Happer. (2009), *Applied Physics Letters*, **94**, 251110.
[13] Li, Y., Xu, S., Chen, S., Liu, C., Wang, J., & Wang, Y. (2022). *Optics Express*, **30(24)**, 43271-43280.
[14] Fang, J., Wan, S., Chen, Y., & Li, R. (2012). *Applied Optics*, **51(31)**, 7714-7717.

Testing novel high-reflectivity mirror technologies from room-temperature to 4 K

M Kempkes, T Legero, U Sterr and D Nicolodi

Physikalisch-Technische Bundesanstalt (PTB), Bundesallee 100, Braunschweig, Germany

mona.kempkes@ptb.de

Abstract. We present a setup based on a low-vibration closed-cycle cryostat for the characterization of mirror coatings and for the direct measurement of Brownian thermal noise from room-temperature to 4 K. Using a dual frequency modulation scheme allows us to lock a laser to two eigenmodes of a high-finesse optical resonator simultaneously and to track the line splitting between the two modes. This facility will enable to investigate the temperature and power dependence of the birefringent line splitting and noise in AlGaAs/GaAs multilayer coatings across a broad temperature range.

1. Introduction

Coating thermal noise of traditional Ta_2O_5/SiO_2 multilayer high-reflectivity mirror coatings constrains the stability of state-of-the-art optical resonators and limits the performance of many precision laser experiments. The impact of coating thermal noise on the frequency stability of optical resonators can be reduced by using resonators at lower temperatures, with larger mode diameters or increased length. However, practical limitations and engineering challenges limit the improvement attainable in this way.

Coating Brownian thermal noise is proportional to the mechanical loss of the coating stack. This prompted the quest for alternative coating materials with lower mechanical losses. Only a few materials exhibiting low mechanical losses as well as low optical losses required by ultra-stable optical resonators and gravitational wave detectors have been proposed so far. Crystalline AlGaAs/GaAs multilayer coatings offer the most significant thermal noise reduction [1]. Optical resonators equipped with these coatings demonstrated performance near the predicted thermal noise floor at room-temperature [2]. Yet, crystalline coatings as one of the most promising mirror candidates for ultra-stable optical resonators with fractional frequency stabilities beyond the 10^{-17} level still hold unresolved issues.

Unlike traditional dielectric coatings, crystalline coatings exhibit significant birefringence. Their birefringence was originally explained to be due to mechanical stress embedded in the coatings during the epitaxial growth of the crystal layers [1]. From observing that the birefringence can be modified by illumination of the coatings with infrared or visible light, it is assumed that a contribution of the birefringence comes from electric fields developed in the multilayer stack coupling to the electro-optic effect in the crystal [3]. The time constants involved in the change of birefringence upon illumination are dramatically different between experiments at room- and cryogenic temperatures.

Furthermore, two experiments at 124 K and 4 K - 16 K revealed spontaneous fluctuations of the coatings birefringence and an additional noise source with hitherto unknown origin which limits the frequency stability well above the expected thermal noise floor at cryogenic temperatures [4, 5]. The

Journal of Physics: Conference Series **2889** (2024) 012054 doi:10.1088/1742-6596/2889/1/012054

sensitivity of room temperature experiments does not yet allow to confirm that these noise sources are also relevant at room temperature. The so far unidentified noise severely hinders the noise reduction expected from the low mechanical losses and challenges the potential use of crystalline coatings in cryogenic optical resonators and in the next generation of gravitational wave detectors.

A theory explaining all the different observations has not yet been formulated. Reconciling the measurement at different temperatures is particularly challenging. It cannot be excluded that the observed differences are due to specifics of individual experiments or to the properties of specific coating samples rather than to intrinsic characteristics of the crystalline coatings. An experiment where the properties of the same crystalline coating are measured over a wide range of temperature may help shine light onto the nature of the observed phenomena.

We present a low-vibration closed-cycle cryostat setup for the characterization of mirror coatings from room-temperature to 4 K focusing particularly on crystalline coatings. In this setup we use a high-finesse optical resonator with AlGaAs/GaAs crystalline mirror coatings, multiple techniques to circumvent technical noise sources related to vibrations and temperature fluctuations, and a dual frequency modulation scheme. The latter will allow us to remove the anticorrelated birefringent noise of crystalline AlGaAs/GaAs coatings by locking on the average frequency of two different polarization eigenmodes [4, 5] while at the same time tracking the birefringent line splitting. Aiming for a frequency instability for differential measurements in the 10^{-17} region, this facility will enable analyzing the birefringent line splitting and its fluctuations of AlGaAs/GaAs multilayer coatings across a broad temperature range. This might help understanding the underlying birefringent effects in crystalline coatings, gain further insight in the source of the observed excess noise and possibly identify suppression or mitigation strategies.

The same system will also offer the possibility to verify thermal noise estimates and ruling out yet unknown noise sources in other novel high-reflectivity mirror designs such as nanostructured meta-etalons [6, 7] and amorphous-Si multilayer coatings [8] over a temperature range relevant for applications in ultra-stable optical resonators and in next generation gravitational wave detectors.

2. Experimental Setup

2.1. Low-vibration closed-cycle cryostat system

Analyzing the optical behavior of mirror coatings within a temperature range relevant for applications in ultra-stable optical resonators requires a cryostat system which can cool to 4 K, provides enough space for a resonator, and introduces as little vibrations as possible to the optical system.

We are using a Cryomech PT407-RM two-stage pulse tube cryocooler with remote rotary valve option (see figure 1). In this system the pulse tube is mechanically decoupled from the vacuum chamber and the cavity's mounting station through bellows. To minimize vibrations, we rigidly connect the part above the bellows (pulse tube cooler) to the laboratory ceiling, while the part below the bellows (vacuum chamber) is mounted to an aluminum structure standing firmly on the laboratory floor.

The cooling at the sample station is realized by using helium as transfer gas in a volume between the 2nd stage of the pulse tube and the sample station. When the 2nd stage reaches a temperature of 4 K, the gaseous helium around it liquifies and builds up a bath of liquid helium at the bottom of the volume. Thermal energy at the sample station will be transferred to this helium bath which in turn starts to boil, the helium gas rises and is cooled down and liquified again by the 2nd stage. This use of helium and its liquid-gas phase change provides a high temperature stability at the sample station (temperature fluctuations below 10 mK in steady state operation at 4 K) as well as no mechanical contact to reduce vibrational transfer [9]. When operating at higher temperatures, the cooling power at the sample station can be modulated by varying the helium pressure.

We measured the acceleration at the sample station along three orthogonal directions using piezoelectric accelerometers. One of the two horizontal axis - the x-axis - was chosen to be parallel to the helium flex line which connects the cryocooler's cold head with the rotary valve. We compare the acceleration noise measured with the cryocooler on (colored lines) to the one measured with the cryocooler off (grey line) in figure 1. During cryocooler operation the acceleration was most pronounced along the x-axis (green line in figure 1). Peaks in the acceleration amplitude at the fundamental pulse tube frequency of 1.4 Hz and higher harmonics could be measured. Along the vertical axis no sharp peaks could be seen for frequencies below 20 Hz.

The cylindrical cavity is mounted rigidly along the y-axis at the cryocooler's sample station and is shielded from external influences by two radiation shields. Currently there are no further vibration isolation mechanism included, thus the stability of the system is likely to suffer from accelerations during cryocooler operation. However, as the measurements of interest in this setup are differential in nature by including two cavity eigenmodes simultaneously (see next section for details), common-mode effects like from vibrations should be strongly rejected in the measurement results.

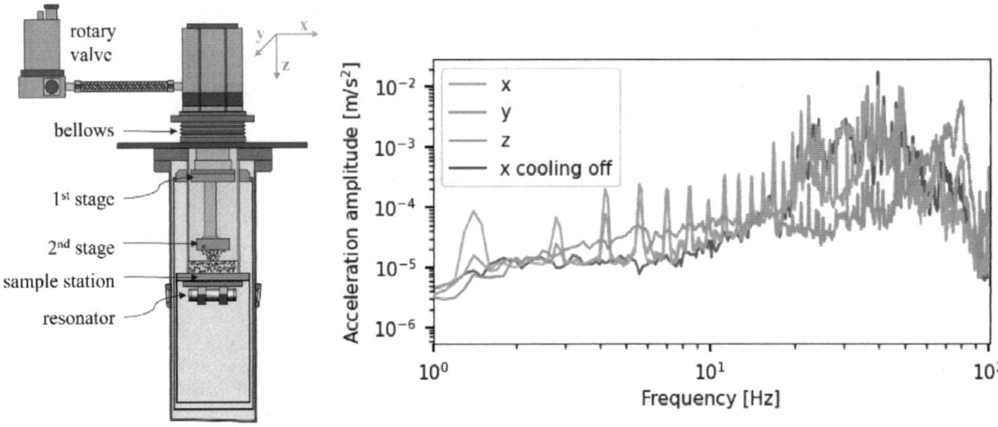

Figure 1. Left: Sketch of the cryocooler. Different to what is shown in the sketch, the cavity is aligned along the y-axis. Right: Acceleration amplitude in a resolution bandwidth of 0.1 Hz for frequencies between 1 and 200 Hz for three orthogonal directions during cryocooler operation. The x-axis is chosen to be parallel to the helium flex line connecting the rotary valve to the cold head, the z-axis is the vertical axis. The acceleration amplitude without cryocooler operation is shown for the x-axis. The y-axis baseline looks very similar to the x-axis baseline and there is no significant difference between the z-axis data shown in the plot and the z-axis baseline for frequencies below 20 Hz.

2.2. Optical setup

The central part of this setup is an optical resonator made of a 10 cm long cylindrical monocrystalline silicon spacer with silicon mirror substrates and crystalline AlGaAs/GaAs multilayer coatings. A cavity finesse of 290 000 was determined from a cavity ringdown measurement. An erbium-doped fiber laser at a wavelength of 1542 nm is locked to the fundamental HG_{00} mode of the resonator. A scheme of the experimental setup is shown in figure 2.

The laser light is frequency shifted by an acousto-optic modulator (AOM) and phase-modulated by an electro-optic modulator (EOM) before it is coupled into the silicon resonator. The AOM provides fast frequency tuning for locking the laser to the cavity. The EOM generates sidebands for the dual frequency lock described in the next section. A CCD camera and a photodetector monitor the cavity's transmission while the reflected light is used for locking the laser to the cavity's resonance frequencies.

We use a dual frequency modulation scheme, similar to the one described in [10], to simultaneously lock the frequency of the laser via the Pound-Drever-Hall (PDH) technique to the average frequency of two cavity eigenmodes and lock the rf-frequency of a local oscillator driving the EOM onto the frequency spacing of the two birefringent eigenmodes one free spectral range apart. Locking onto two cavity eigenmodes allows to remove the anticorrelated birefringent noise of crystalline coatings as described in [3, 4] while locking the rf-frequency of a local oscillator to the cavity's resonance frequency spacing opens the possibility to track the birefringent line splitting.

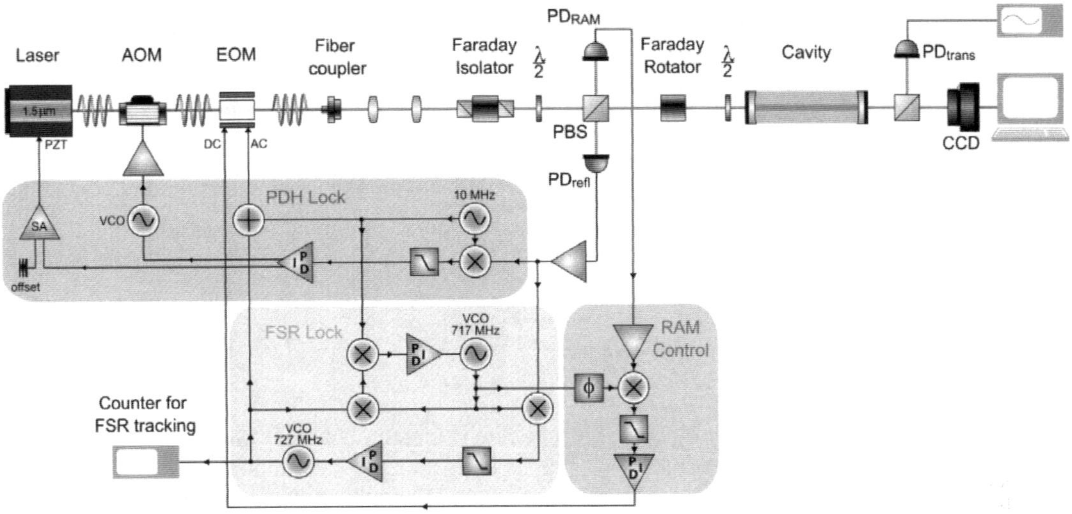

Figure 2. Scheme of the experimental setup. Light of an erbium-doped fiber laser is modulated by an acousto-optic-modulator (AOM) and electro-optic modulator (EOM) and coupled into a silicon resonator. A photodetector (PD_{refl}) picks up the light reflected off the resonator and feeds the signal to the electronics for a Pound-Drever-Hall (PDH) and a free spectral range (FSR) lock. A second photodetector (PD_{RAM}) is used to obtain a signal to compensate residual amplitude modulation (RAM). For the dual frequency modulation scheme the EOM is driven with two frequencies: $f_{\mathrm{PDH}} = 10$ MHz for the generation of PDH sidebands and a signal $f_{\mathrm{VCO}} \approx 727$ MHz from a voltage controlled oscillator (VCO) for the generation of sidebands at half of the line splitting frequency. For further details, see text.

2.2.1. Dual frequency modulation

The dual frequency modulation scheme used to simultaneously lock to the cavity's resonance frequency and the cavity's free spectral range described in [10] cannot be applied unaltered to track the birefringent line splitting $f_{\mathrm{biref}} \approx 400$ kHz because it is not possible to make the optical carrier and two modulation sidebands resonant with two different polarization eigenmodes of the cavity. Therefore, the carrier frequency ν_{carrier} of the laser is set to the center of two adjacent HG_{00} modes with different polarization (e.g. resonance frequencies of *n*-th HG_{00} mode, fast axis and *n*+1-th HG_{00} mode, slow axis) by tuning the laser's piezo input. An EOM driven at half of the cavity's free spectral range frequency minus half of the birefringent line splitting frequency $f_{\mathrm{VCO}} = f_{\mathrm{FSR/2}} - f_{\mathrm{biref/2}} \approx 727$ MHz generates two first order sidebands $\pm$ SB below and above the carrier frequency. If the carrier frequency ν_{carrier} and the modulation frequency f_{VCO} are set correctly, the first order sidebands at $\nu_{\mathrm{carrier}} \pm f_{\mathrm{VCO}}$ will be resonant with two adjacent HG_{00} modes of different polarization.

To derive error signals for the two locking schemes the EOM is not only driven with f_{VCO} but also with a second frequency $f_{\mathrm{PDH}} = 10$ MHz generating a sideband spectrum shown in figure 3. The amplitude modulation seen by a photodetector in reflection (PD_{refl} in figure 2) is demodulated with two frequencies (blue and orange shaded parts in figure 2). A 10 MHz demodulation signal (blue) is used to obtain an error signal for the PDH lock which holds the carrier frequency at the average frequency of the two HG_{00} modes of different polarization. A 717 MHz demodulation signal (orange, equals $f_{\mathrm{VCO}} - f_{\mathrm{PDH}}$) is used to obtain an error signal that locks the frequency of the local oscillator driving the EOM (and thus the sideband spacing) to the cavity's free spectral range and the birefringent line splitting.

The same frequency $f_{VCO} - f_{PDH} \approx 717$ MHz is used to demodulate the signal from the photodetector PD_{RAM}. This generates an error signal which can be used for the compensation of residual amplitude modulation (RAM) by applying an offset to the DC input of the EOM.

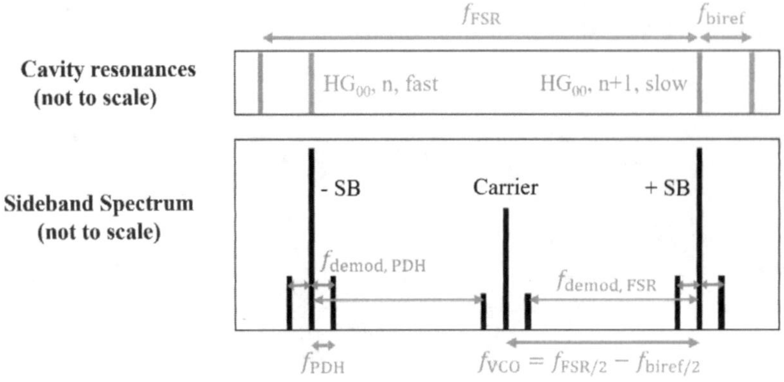

Figure 3. HG_{00} mode structure of the optical resonator and laser sideband structure. Due to the birefringent properties of crystalline coatings each fundamental eigenmode of the resonator HG_{00}, n is split into two – one resonance mode connected to the slow axis (pink) and one connected to the fast axis (green) of the crystalline coating. The carrier frequency of the laser is set to the center between HG_{00}, n, fast and HG_{00}, $n+1$, slow. An EOM modulated with f_{VCO} and f_{PDH} creates to first order a sideband structure shown in black.

2.2.2. Simulation & measurement of the FSR-lock performance

To estimate the sensitivity and performance of the FSR-lock we ran simulations in which the laser's carrier frequency $\nu_{carrier}$ is kept in the center of two adjacent HG_{00} mode resonance frequencies while the EOM's modulation frequency f_{VCO} is swept across resonance. Here, we are concentrating on two parallel polarization resonant modes.

In the conventional PDH locking, the out-of-phase component of the error signal is zero. In the FSR-locking scheme proposed here however, there is a non-zero out-of-phase component at the lock point. This introduces an error signal offset that depends on the demodulation phase. The shape of the out-of-phase component of the error signal depends on the modulation index of the modulation at f_{VCO}. For different modulation indices the shape and amplitude of the out-of-phase error signal ($\Delta\varphi = 90°$) is shown in figure 4. To minimize the effect of this component the simulation suggests an optimized modulation index of 1.925. This happens to be the setting in which the zeroth and second order sideband are nearly equal in amplitude while the first order sideband is maximal.

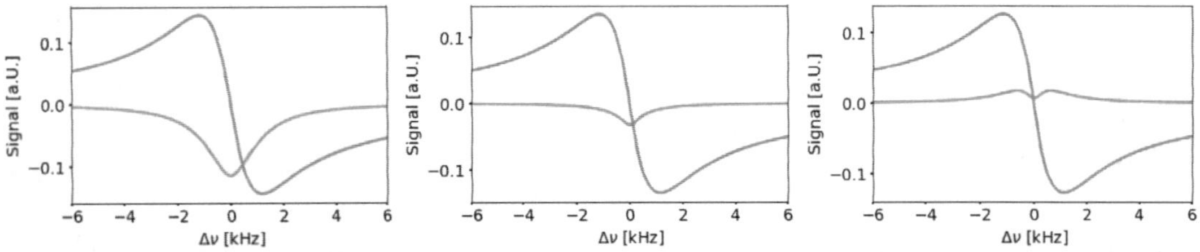

Figure 4. Simulated error signal of the FSR lock for a sweep of f_{VCO} across the resonator's mode spacing for a modulation index of 1.5 (left), 1.8 (center) and 1.94 (right) at a relative phase between modulation and demodulation signal of $\Delta\varphi = 0°$ (blue lines) and $\Delta\varphi = 90°$ (orange lines).

Given these simulation results we swept the modulation frequency f_{VCO} driving the EOM across resonance and observed the error signal for different modulation indices (figure 5). Comparing figure 4 and 5 suggests that the simulations seem to model the physical processes occurring in the FSR-lock to a reasonable degree. However, to improve the quality of the simulations the inclusion of residual amplitude modulation, a non-ideal (e.g. lossy) cavity and non-linear effects for the two modulation frequencies in the EOM would be beneficial.

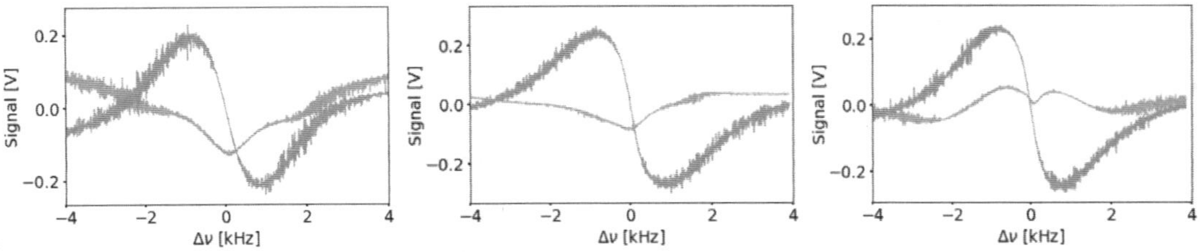

Figure 5. Measured error signal of the FSR lock for 1 dB attenuation (left), reference modulation power level (center) and 1 dB increase (right) at a relative phase between modulation and demodulation signal of $\Delta\varphi = 0°$ (blue lines) and $\Delta\varphi = 90°$ (orange lines). While the reference modulation power corresponds to a modulation index of approximately 1.8, 1 dB attenuation / increase of modulation power corresponds to a change of approximately ∓ 12 % in the modulation index.

2.2.3. Measurement results for the FSR-lock

To characterize the performance of the FSR-lock, we lock to two HG_{00} resonator modes with parallel polarization. In this configuration the laser's carrier frequency is locked to the average frequency of the two modes while the frequency $f_{FSR/2}$ of the local oscillator driving the EOM is locked onto the cavity's half FSR frequency. Probing parallel polarizations, the birefringent noise is rejected and the remaining variation of the FSR from length changes of the resonator induced by temperature fluctuations can be calculated from the laser's absolute frequency. We measure this frequency against an ultrastable silicon resonator which has a fractional frequency instability of $\approx 4 \cdot 10^{-17}$ and negligible frequency drift.

The left part in figure 6 shows the frequency deviations from the mean for the counted frequency $f_{FSR/2}$ (blue) and the carrier frequency (orange) for a room-temperature measurement lasting approximately 5 hours. The carrier frequency was divided by $\nu_{carrier} \cdot f_{FSR/2}{}^{-1}$ to account for the different influence of length changes on the optical resonance frequency and on the FSR. After subtracting the fluctuations calculated from the carrier frequency changes from $f_{FSR/2}$ we obtain a modified Allan deviation as shown on the right hand side of figure 6.

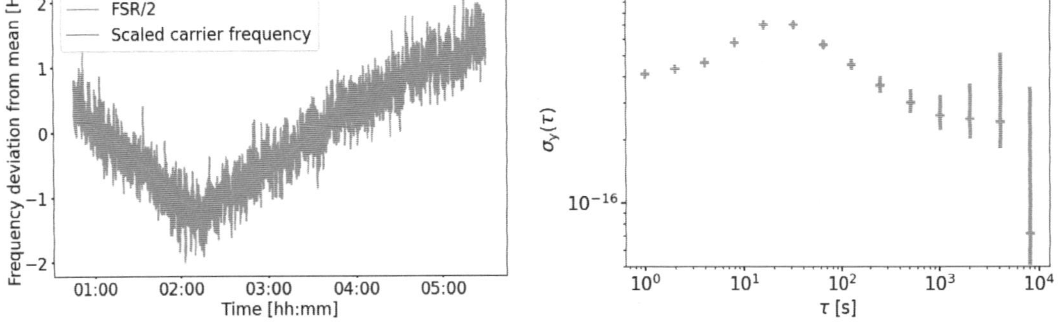

Figure 6. Left: frequency deviation from the mean for half of the FSR frequency (blue) and the carrier frequency (orange). The carrier deviation is scaled down by a factor of 258 000 to obtain the influence of the resonator's length changes on $f_{FSR/2}$. Right: modified Allan deviation $\sigma_y(\tau)$ for the fractional frequency instability of the FSR-lock referred to $\nu_{carrier}$ after correcting the influence of length changes.

The performance of the cavity's FSR measurement is currently likely to be limited by a large amount of residual amplitude modulation (RAM) which fluctuations exceed the dynamic range of our active RAM suppression control loop quite quickly. The source of the unusually large RAM is being investigated.

3. Conclusion and Outlook

We have presented our experimental setup for the characterization of mirror coatings performance from room-temperature to 4 K. Using a low-vibration closed-cycle cryostat, an optical resonator with AlGaAs/GaAs multilayer coatings, and a dual frequency modulation scheme this system allows us to lock one laser to two cavity eigenmodes simultaneously while at the same time tracking the line splitting between the two modes.

Currently we are evaluating the frequency stability of a FSR measurement at room-temperature. Reaching a fractional frequency instability of $4 \cdot 10^{-16}$ at an integration time of one second the stability of our system is probably limited by RAM and parasitic etalons. Further work will be needed to mitigate the influence of RAM, reduce technical noise sources, and reach a fractional frequency instability level of 10^{-17}, where noise of the crystalline coatings has been observed at cryogenic temperatures.

Perspectively, this experiment aims for the investigation of the temperature and power dependency of the birefringent line splitting and the recently observed excess noise in AlGaAs/GaAs multilayer coatings across a broad range of temperatures. The results possibly help identifying the origin of these noise sources and developing suppression or mitigation strategies. The same system can also be used to verify thermal noise estimates and ruling out yet unknown noise sources in other novel high-reflectivity mirror designs such as nanostructured meta-etalons [6, 7] and amorphous-Si multilayer coatings [8] over a temperature range relevant for applications in ultra-stable optical resonators and in next generation gravitational wave detectors.

Acknowledgement

We acknowledge support by the Project 20FUN08 NEXTLASERS, which has received funding from the EMPIR programme cofinanced by the Participating States and from the European Union's Horizon 2020 Research and Innovation Programme, and by the Deutsche Forschungsgemeinschaft (DFG, German Research Foundation) under Germany's Excellence Strategy–EXC-2123 QuantumFrontiers, Project-ID 390837967.

References

[1] Cole G D, Zhang W, Martin M J, Ye J and Aspelmeyer M 2013 *Nat. Phot.* **7** 644
[2] Herbers S, Häfner S, Dörscher S, Lücke T, Sterr U and Lisdat C 2022 *Opt. Lett.* **47** 5441
[3] Ma C Y, Legero T, Herbers S, Nicolodi D, Kempkes M, Riehle F, Kedar D, Robinson J M, Ye J and Sterr U 2024 *J. Phys. Conf. Ser.* this volume
[4] Yu J, Häfner S, Legero T, Herbers S, Nicolodi D, Ma C Y, Riehle F, Sterr U, Kedar D, Robinson J M, Oelker E and Ye J 2023 *Phys. Rev. X* **13** 041002
[5] Kedar D, Yu J, Oelker E, Staron A, Milner W R, Robinson J M, Legero T, Riehle F, Sterr U and Ye J 2023 *Optica* **10** 464-470
[6] Dickmann J and Kroker S 2018 *Phys. Rev. D* **98** 082003
[7] Dickmann J, Sauer S, Meyer J, Gaedtke M, Siefke T, Brückner U, Plentz J and Kroker S 2023 *Commun. Phys.* **6** 16
[8] Steinlechner J, Martin I W, Bell A S, Hough J, Fletcher M, Murray P G, Robie R, Rowan S and Schnabel R 2018 *Phys. Rev. Lett.* **120** 263602
[9] Wang C and Hartnett J G 2010 *Cryogenics* **50** 5 336-41 2010
[10] DeVoe R G and Brewer R G 1984 *Phys. Rev. A* **30** 2827

Journal of Physics: Conference Series **2889** (2024) 012055 doi:10.1088/1742-6596/2889/1/012055

Ultrastable lasers: investigations of crystalline mirrors and closed cycle cooling at 124 K

C Y Ma[1], J Yu[1], T Legero [1], S Herbers[1], D Nicolodi[1], M Kempkes[1], F Riehle[1], D Kedar[2], J M Robinson[2], J Ye[2] and U Sterr[1]

[1] Physikalisch-Technische Bundesanstalt (PTB), Bundesallee 100, Braunschweig, Germany
[2] JILA, National Institute of Standards and Technology and University of Colorado, Boulder, CO, USA

E-mail: chun.ma@ptb.de

Abstract.
We have investigated crystalline AlGaAs/GaAs optical coatings with three ultra-stable cavities operating at 4 K, 16 K, 124 K and 297 K. The response of the cavities' resonance frequencies to variations in optical power indicates non-thermal effects beyond the photo-thermo-optic effect observed in dielectric coatings. These effects are strongly dependent on the intensity of the intracavity light at 1.5 µm. When the rear side of the mirrors is illuminated with external light, we observe a prominent photo-modified birefringence for photon energies above the GaAs bandgap, which points to a possible mechanism relating our observations to the semiconductor properties of the coatings. Separately, we also present a low maintenance evolution of our 124 K silicon cavity system where the liquid nitrogen based cooling system is replaced with closed cycle cooling from a pulse-tube cryo-cooler.

1. Introduction

Lasers with long coherence time and narrow linewidth are an essential tool for quantum sensors and clocks. Since the previous 8th Symposium on Frequency Standards and Metrology in 2015, the field of ultrastable lasers has developed significantly. Ultrastable cavities and laser systems are now commercially available with fractional frequency instabilities in the mid 10^{-16} range. Ultrastable frequencies can now be distributed both locally and over long distances [1]. With femtosecond combs, frequency noise of lasers at different wavelengths can be compared, allowing frequency stability transferred from one wavelength to another [2, 3, 4] with added fluctuations well below 10^{-17} and to generate microwaves with exceptionally low phase noise [5]. This enables interrogation times of optical clocks longer than a second [6, 7]. Cavity-stabilized lasers have become very reliable and can even replace masers as flywheels to bridge downtimes of optical clocks to create a fully optical and continuous timescale [8].

For both terrestrial and space-borne transportable clocks, many transportable cavity designs and setups have been developed, the best reaching 1.6×10^{-16} instability [9]. Laser instabilities as low as 7×10^{-17} and 4×10^{-17} have been reported for room temperature and cryogenic cavities respectively [1, 10]. Recently, we have measured the stability of our two cryogenic silicon cavities (Si2 and Si5) (see section 2) and of a 48 cm long ULE cavity at 698 nm [11] with the three-cornered-hat method [12]. We obtained instabilities down to 3.2×10^{-17} for the cryogenic silicon cavity with dielectric mirrors (Si2), whereas for an otherwise identical cavity with crystalline mirrors (Si5) an instability of 3.0×10^{-17} at longer averaging times was found (see Fig. 1).

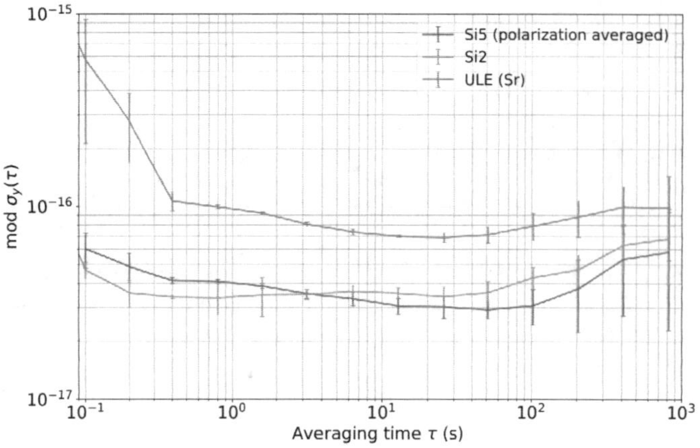

Figure 1. Fractional frequency instability, expressed as modified Allan deviation, of a cryogenic silicon cavity with dielectric mirrors (Si2), of an otherwise identical cavity with crystalline coatings (Si5), and a 48 cm long room temperature cavity ULE (Sr) with dielectric coatings at 698 nm [11], determined from a three-cornered hat analysis. The uncertainty bars indicate the standard deviation of ten separate evaluations.

Similar to large gravitational wave detectors [13, 14], the performance of these best sub-meter sized ultrastable optical resonators is currently limited by the Brownian fluctuations of the usually employed dielectric mirror coatings [15, 16]. Brownian thermal noise is related to the mechanical loss by internal friction [17] through the fluctuation-dissipation theorem [18, 19]. Thus it can be reduced by employing optical coatings with lower mechanical loss. Crystalline $Al_{0.92}Ga_{0.08}As/GaAs$ Bragg reflectors [20, 21] have appeared as alternative to conventional Ta_2O_2/SiO_2 dielectric multilayer coatings for reducing coating thermal noise. Their mechanical loss, which is about a factor of ten lower than conventional coatings, can reduce the Brownian noise, expressed in Allan deviation σ_y, by a factor of 3 [20]. At the same time, a high reflectivity can be achieved, leading to a cavity finesse exceeding 300 000 [22, 23]. Compared to dielectric coatings, these crystalline coatings exhibit much larger static birefringence on the order of $n_{\mathrm{biref}} \approx 7 \times 10^{-4}$ [16], leading to frequency splittings between the linearly polarized fast and slow eigenmodes proportional to the free spectral range (FSR) of the cavity and n_{biref}.

Our first studies at cryogenic temperatures showed novel noise contributions from the crystalline coating [15, 16]. After careful suppression of technical noise, remaining anticorrelated frequency fluctuations between the two eigenmodes that are dependent on intracavity power and mode area were observed. The anticorrelated fluctuations between the two polarization eigenmodes of this noise type relate to fluctuations of the coating birefringence, thus we refer to it as intrinsic birefringent noise. The Brownian thermal noise level was also determined and it agrees with the low mechanical loss value at room temperature. After removing the coating birefringent noise by simultaneously interrogating both modes, a remaining global noise in excess of the Brownian noise dominates the frequency instability at 124 K, 16 K and 4 K.

In addition to these intrinsic noises in AlGaAs coatings, previous studies at cryogenic temperatures have also observed an unexpected photo-induced frequency change [16] beyond the photo-thermo-optic effect observed in dielectric coatings [24] arising from the temperature change caused by the laser power absorbed in the coatings. In the following sections we will focus on expanded investigations of the coating birefringence from cryogenic to room temperature. To

Journal of Physics: Conference Series **2889** (2024) 012055
doi:10.1088/1742-6596/2889/1/012055

gain more insight into potential physical mechanisms that might be related to the semiconductor properties of AlGaAs/GaAs, we also studied the response to changes in intracavity power and to broadband illumination at different wavelengths. Our findings may help to uncover the mechanisms related to the semiconductor properties and the noise of these coatings, thus potentially enabling a reduction of that noise.

2. Experimental setup

Three optical resonators operated at 1.5 μm were used to investigate the coating birefringence. Table 1 summarizes their properties. Two resonators employ monocrystalline silicon spacers with AlGaAs coatings bonded to silicon mirror substrates [15, 16]. Based on our previous design [11], the third resonator employs a 48 cm long ULE spacer and a pair of fused silica substrates coated with AlGaAs coatings. ULE rings are attached to the back of the mirrors to compensate for CTE mismatch [25], leading to a measured CTE zero crossing at 297 K.

Similar to Si5 and Si6 [15, 16], the birefringence of the coatings in the ULE cavity results in a splitting of two polarization eigenmodes as shown in Table 1 with finesses of $1.290(2) \times 10^5$ and $1.282(2) \times 10^5$ for slow and fast axes, respectively. To study the coating-specific optical length changes, two lasers are locked independently to this ULE cavity from opposite sides as in previous studies [16]. An optical heterodyne beat signal between the two lasers is detected from the transmitted and reflected beam at one side of the cavity, forming a common-path geometry to suppress the influence of path length fluctuations.

Table 1. Summary of three investigated ultrastable cavities with AlGaAs coatings. Mirror configuration is plane concave (pl/cc) or concave - concave (cc/cc) with the indicated radius of curvature (ROC) and the beam radius on the curved (Si cavities) or on the plane mirror (ULE cavity) is indicated. For Si5 and Si6 one mirror coating stack consists of 45.5 GaAs/AlGaAs layer pairs (total thickness 11.7 μm) whereas for ULE cavity the coating stack of 38.5 pairs is 9.6 μm thick.

Cavity	Length	Temperature	Mirror configuration	FSR	Birefringent splitting	Beam radius
Si5	21 cm	124 K	cc/cc (2 m ROC)	710 MHz	200 kHz	482 μm
Si6	6 cm	4 K/16 K	cc/cc (1 m ROC)	2.5 GHz	770 kHz	294 μm
ULE	48 cm	297 K	pl/cc (1 m ROC)	310 MHz	104 kHz	495 μm

3. Photo-induced effects

It is well known that fluctuations of the intracavity power in optical cavities couple to frequency fluctuations via the photo-thermo-optic effect [24] and degrade the laser stability. Stabilizing the intracavity power is thus critical.

In addition to the photo-thermo-optic effect, for AlGaAs crystalline coatings, it has been observed that a change in intracavity optical power also couples to changes in birefringence [16]. The mechanism for this coupling is not yet fully understood. To improve our understanding of this phenomenon, we investigated the frequency change of three independent optical resonators locked to the fundamental Hermite-Gaussian HG_{00} mode when the optical power is suddenly changed.

To account for the differences in finesse and mode size of the three cavities, we normalize our measurements to the average intensity I at the AlGaAs coatings, as calculated from the intracavity power P and the $1/e^2$ intensity beam radius w

$$I = \frac{P}{\pi w^2}. \tag{1}$$

Fig. 2 shows the responses of the coating birefringence after a sudden increase of optical power. Si5 and Si6 are operated at temperatures with very low thermal expansion of silicon, thus the photo-thermal effect from the Si spacer and mirror substrates is negligible, and the induced frequency changes due to the modified birefringence are already visible in the measured frequency against a reference laser. For the ULE cavity at 297 K, the frequency difference between the fast and slow axes is calculated to remove the common photo-thermal contribution from fused silica substrates. Half of the frequency difference is shown in the figure to make the results comparable to the other measurements.

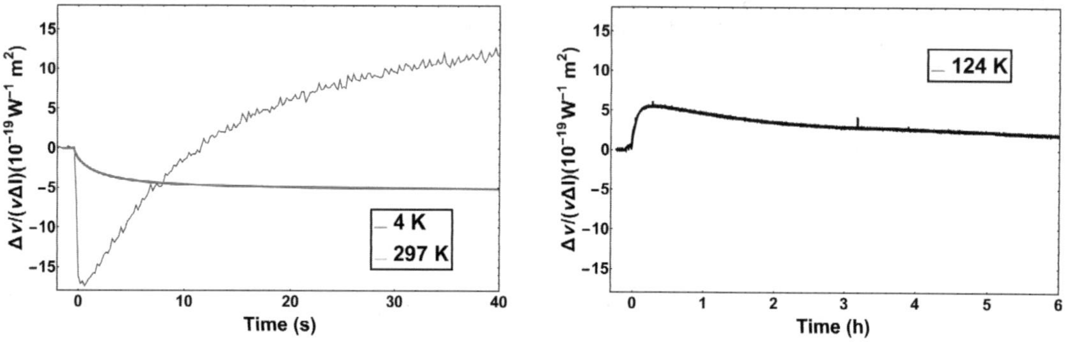

Figure 2. Fractional frequency change $\Delta\nu/\nu$ due to photo-modified birefringence after a sudden increase of intracavity power for the fast axis. The fractional frequency change is normalized by the change in intensity ΔI. For the 4 K and 297 K measurement (left), the response is in the order of seconds, whereas for 124 K (right) the response is in the order of hours.

The observed change in frequency $\Delta\nu_{\text{photo}}$ can be related to the change of coating birefringence Δn_{photo}

$$\Delta\nu_{\text{photo}} = \frac{\Delta n_{\text{photo}}}{n_{\text{biref}}}\nu_{\text{biref}} \tag{2}$$

where n_{biref} is the static birefringence of the crystalline coatings in the individual cavities [16] and ν_{biref} is the resulting line splitting. A schematic view of the intracavity light acting on the AlGaAs coating of the ULE cavity is shown in Fig. 3.

Taking the data as shown in Fig. 2, the resulting change of birefringence $\Delta n_{\text{photo}}/\Delta I$ of Si5 (124 K) is the smallest, which is followed by Si6 and the ULE cavity (297 K). The response at 124 K has a very different time scale compared to the responses at 4 and 297 K while the sign of the long term response at 297 K is different from the other two measurements.

3.1. Sensitivity to intracavity power

Fig. 4 (left) shows the line splitting as a function of the intracavity laser intensity on the mirror coatings measured with the ULE cavity. The range of intensity investigated corresponds to a transmitted power of 1.7 µW to 108 µW. Increasing intracavity intensity reduces the line splitting for the room temperature cavity. An opposite sign is observed in the 4 K and 124 K

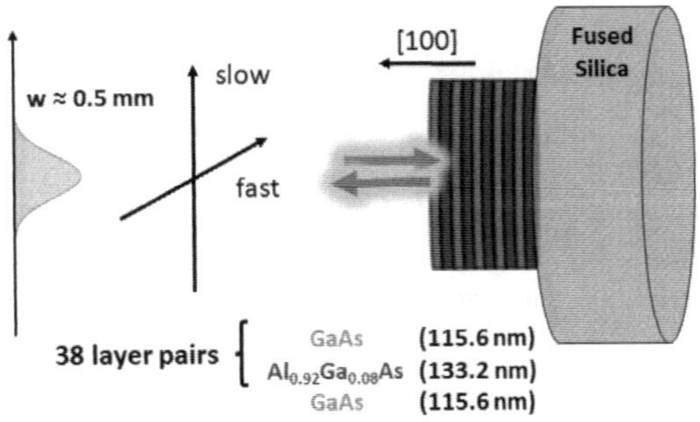

Figure 3. Schematic view of the crystalline Bragg reflector bonded to fused silica substrates of the 297 K cavity. The coating stack starts and ends with a GaAs layer.

setups as well as in recent studies at room temperature [26, 27]. This is possibly related to the fact that all setups besides our 297 K setup use "flipped" AlGaAs coatings [22], where the [1 0 0] crystalline axis points in the opposite direction. The line splitting as function of intensity of our room temperature cavity is highly nonlinear, possibly indicating a saturation at high intensity. At the highest intracavity intensity level studied, the change of mirror birefringence $\Delta n_{\mathrm{photo}}$ is about 1% of its static value n_{biref}.

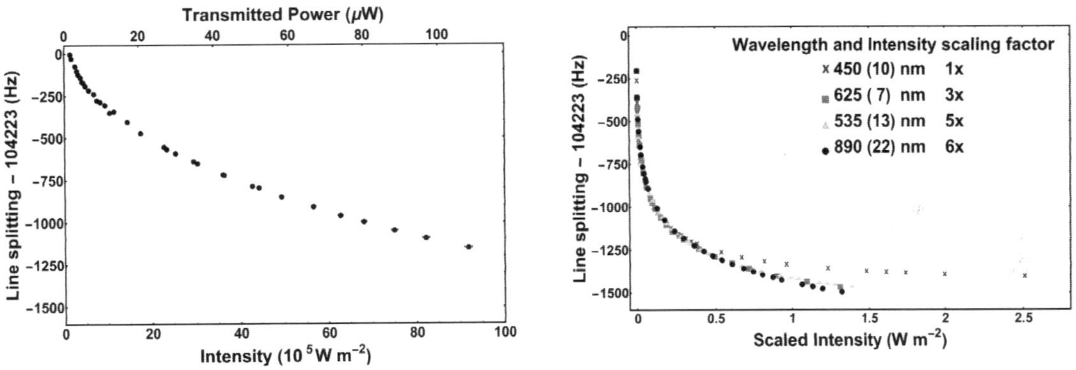

Figure 4. Birefringent splitting as a function of intracavity laser intensity (left) and external high power LED intensity (right) for the ULE cavity operating at room temperature after reaching steady state. In the left graph the corresponding power transmitted by the cavity is also indicated on the top horizontal axis. In the right graph the coating is illuminated by LEDs of four different wavelengths shorter than the GaAs bandgap at the same 1.5 μm intracavity laser intensity. The curves can be superimposed by multiplying the intensity by a wavelength-dependent scaling factor. A common frequency offset is subtracted for clarity, as indicated in the annotation.

3.2. External illumination

So far the photo-induced effect has been investigated for intracavity light at 1.5 μm. As the fused silica mirror substrates of the room temperature cavity are transparent to visible and near infrared light, we illuminate the AlGaAs coatings from outside the vacuum chamber with LEDs to further investigate the wavelength dependence of $\Delta n_{\mathrm{photo}}$ above and below the bandgap of AlGaAs. The LEDs were calibrated to determine the intensity I at the mirror coatings, accounting for divergence and the loss from the windows.

The final line splittings with respect to the LED intensities are displayed in Fig. 4 (right). The response to the intensity is also highly nonlinear as seen in the response of intracavity power changes. The responses at the different wavelengths can be described by a similar curve when their respective intensity is scaled by an empirically determined number. The smallest sensitivity is observed for blue light (see Fig. 4 right). Comparing Fig. 4 left and right panels, the sensitivity of the line splitting over intensity is much larger for external illumination with LEDs and it also shows a larger curvature. Thus the line splitting response from the cavity light cannot be superimposed with the responses from the LEDs.

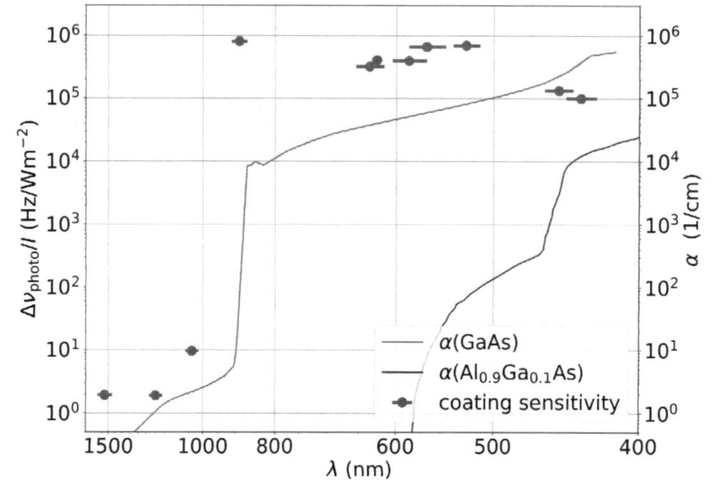

Figure 5. Wavelength dependent sensitivity $\Delta\nu_{\text{photo}}/I$ at low intensity in Hz/Wm^{-2} (green dots) of photo-modified birefringence measured with narrowband LEDs. The bandwidths of the LEDs are indicated by the horizontal bars. The uncertainty of the sensitivity is less than the marker size. The absorption coefficients α of GaAs (red) and $Al_{0.9}Ga_{0.1}As$ (blue) in cm^{-1} are included for comparison [28, 29, 30].

Fig. 5 displays the observed sensitivity of the birefringent splitting to external light in Hz/Wm^{-2} over a broader range of wavelengths between 450 nm and 1.5 µm. To account for the nonlinearity in sensitivity, the largest sensitivity of each LED is shown. The absorption profiles of GaAs and AlGaAs are included for comparison [28, 29, 30]. The sensitivity drops by five orders of magnitude across the GaAs bandgap between 890 nm to 1050 nm, as for longer wavelengths no direct photoabsorption can occur in pure GaAs. For photon energies above the GaAs bandgap we observed a slight decrease in the sensitivity compared to the increase of the absorption coefficients towards shorter wavelengths, which might be due to the variation of penetration depth with wavelength. Blue light is already absorbed in the first GaAs layer in comparison to light at 890 nm that has an absorption length beyond ten layers of GaAs. This results in charges being created in different regions.

3.3. Simple model

This wavelength-dependent behavior with correspondence to the absorption of GaAs indicates that photo-excitation of free charge carriers is likely to be the primary effect upon illumination. The carriers that are created in the first few µm thick absorption region on the back of the mirror coating stack migrate in the GaAs/AlGaAs heterostructure and finally create an electric field at the top layer. This modifies the birefringence through the electro-optic effect. For an electric field perpendicular to the mirror surface, the linear electro-optic effect in GaAs creates a birefringence Δn_{EO} with the observed orientation and a sensitivity of [31, 32]

$$\Delta n_{\mathrm{EO}} = r_{41} n_0^3 E \tag{3}$$

with the electro-optic tensor component $r_{41} = -1.33$ pm/V and the GaAs refractive index $n_0 = 3.48$. Thus, the observed maximum $\Delta n_{\mathrm{photo}}$ of 8.7×10^{-6} would require an electric field of 155 kV/m corresponding to 17 mV over the thickness of the top GaAs layer.

3.4. Temporal behavior

Previous investigation at cryogenic temperatures revealed a faster response at high final intensity [16]. To understand the mechanism of the photo-induced effect in a better way, the temporal change of line splitting due to a step in intracavity power and external light are further studied at room temperature.

Fig. 6 (left) shows three independent step responses of the splitting, when the optical power is stepped up or down to the same final level. As in Fig. 2, the fractional frequency change is normalized by the change in optical intensity, considering the signs of the quantities. Due to the nonlinear sensitivity of the line splitting with intensity (as seen in Fig. 4), the normalized frequency changes were further scaled (with scaling factors of 0.8, 1 and 1.09) to make the temporal change in frequency of all measurements overlap. The good overlap indicates that the dynamical behavior is independent on the initial condition and the step size of intensity level.

Similarly, the temporal response of the photo modified birefringence to external light is studied when LEDs at different wavelengths are switched on (Fig. 6 right). The intensities of the LEDs were adjusted to result in the same birefringent frequency change $\Delta\nu_{\mathrm{photo}} = 300$ Hz. We observe the same temporal behavior, irrespective of the LED wavelength. This observation is also in agreement with our simple model, based on initial charge creation by photoabsorption and charge migration which is followed by the modified birefringence via the electro-optic effect [15, 16, 32].

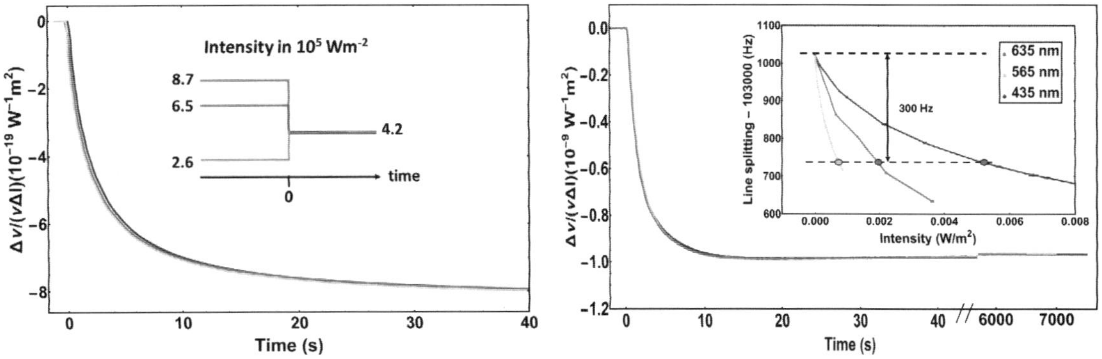

Figure 6. Temporal response of the birefringent splitting by intracavity light for different step sizes of intensities to the same final value (left). The colors of the curves correspond to the line colors of the steps in the inset. Response on switching on LEDs of different wavelengths (right). The intensity (determined from the current) is chosen to get the same $\Delta\nu_{\mathrm{photo}}$ of 300 Hz (see inset).

4. Closed cycle cooling at 124 K

Besides frequency stability, reliability and robustness are also important aspects for ultrastable lasers. In particular, the low-noise cooling of our current 124 K systems rely on cold nitrogen gas evaporated from a liquid nitrogen dewar [33, 34]. The continuous cooling of each Si cavity at 124 K requires 400 liters liquid nitrogen per week and frequent refillings. To simplify the operation and as a first step toward a transportable cryogenic system, we have replaced the liquid nitrogen based cooling system for one of our cryogenic cavity setups with a system where the cooling power is provided by a single-stage pulse-tube cryo-cooler. The cavity environment is cooled by a flow of cold helium gas that circulates between a heat exchanger connected to a pulse-tube cold-head and another heat exchanger connected to the outermost radiation shield in the vacuum chamber (Fig. 7). As in the previous setup, vacuum-isolated flexible hoses are used to isolate the cavity from vibrations of the pulse-tube cooler. The temperature of the radiation shield is actively controlled by modulating the rotation speed of the cryogenic fan used to maintain helium flow. The cavity can be kept at 124 K operating the cryo-cooler with a cooling power of roughly 35 W at 85 K. The system demonstrates sub-millikelvin outer shield temperature stability similar to the one obtained with the liquid nitrogen based cooling system. A study is underway to characterize the impact of vibrations on the stability of the cavity resonance frequency.

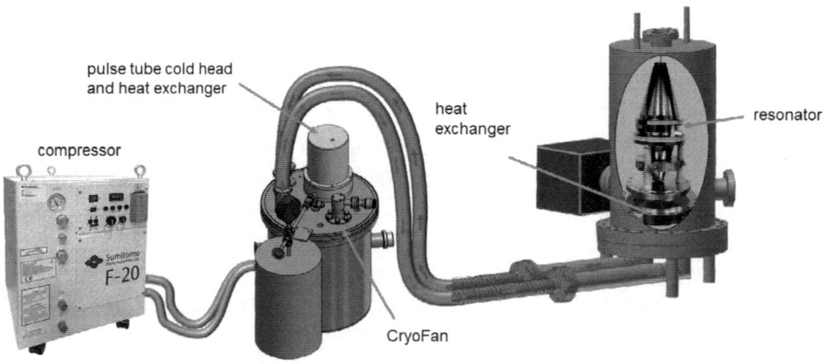

Figure 7. Setup for a low-maintenance cryogenic 124 K Si cavity (not drawn to scale).

5. Conclusions and outlook

We have investigated the photo-induced effect on the birefringence of crystalline coatings at 4 K, 16 K, 124 K and 297 K. Upon a sudden change in optical power, the original birefringence is modified at all temperatures and changes to a new stationary value. At room temperature, the change of birefringence with optical power was observed to reduce at high intracavity intensity. This non-thermal response at 297 K at different wavelength was also studied for the first time by illuminating the coating with external LED light in a broad range of wavelengths. At photon energies above the GaAs bandgap, the birefringence is modified five orders of magnitude more than at energies below the bandgap. The temporal behavior of intracavity and external light reveals the dynamic change in birefringence is dependent on the extent of the modified birefringence due to a different photo-excitation rate. We explain the observed effect with initial photoabsorption creating charge carriers. These charges migrate and create an electric field. The resulting field induces an electro-optic effect on the coatings that modifies the birefringence.

From the measured sensitivities, a relative power stabilization of 10^{-6} is required for a photo-induced instability below 1×10^{-17} at room temperature. As next step, the coating associated noise contributions with and without external light will be investigated to determine the utmost performance of AlGaAs coatings at room temperature.

Acknowledgments

We acknowledge support by the Project 20FUN08 NEXTLASERS, which has received funding from the EMPIR programme cofinanced by the Participating States and from the European Union's Horizon 2020 Research and Innovation Programme, and by the Deutsche Forschungsgemeinschaft (DFG, German Research Foundation) under Germany's Excellence Strategy–EX-2123 QuantumFrontiers (Project No. 390837967), SFB 1227 DQ-mat (Project No. 274200144). This work is partially supported by the Max Planck-RIKEN-PTB Center for Time, Constants and Fundamental Symmetries. This work is also supported by NIST, National Science Foundation QLCI OMA-2016244, AFRL, and JILA Physics Frontier Center (NSF Grant No. PHY-1734006).

References

[1] Schioppo M, Kronjäger J, Silva A, Ilieva R, Paterson J W, Baynham C F A, Bowden W, Hill I R, Hobson R, Vianello A, Dovale-Álvarez M, Williams R A, Marra G, Margolis H S, Amy-Klein A, Lopez O, Cantin E, Álvarez-Martínez H, Le Targat R, Pottie P E, Quintin N, Legero T, Häfner S, Sterr U, Schwarz R, Dörscher S, Lisdat C, Koke S, Kuhl A, Waterholter T, Benkler E and Grosche G 2022 *Nature Commun.* **13** 212

[2] Nicolodi D, Argence B, Zhang W, Le Targat R, Santarelli G and Le Coq Y 2014 *Nature Photonics* **8** 219–223

[3] Benkler E, Lipphardt B, Puppe T, Wilk R, Rohde F and Sterr U 2019 *Opt. Express* **27** 36886–36902 also see erratum [35]

[4] Li P, Rolland A, Jiang J and Fermann M E 2022 *Opt. Express* **30** 22957–22962

[5] Xie X, Bouchand R, Nicolodi D, Giunta M, Hänsel W, Lezius M, Joshi A, Datta S, Alexandre C, Lours M, Tremblin P A, Santarelli G, Holzwarth R and Le Coq Y 2017 *Nature Photonics* **11** 44–47

[6] Origlia S, Pramod M S, Schiller S, Singh Y, Bongs K, Schwarz R, Al-Masoudi A, Dörscher S, Herbers S, Häfner S, Sterr U and Lisdat C 2018 *Phys. Rev. A* **98** 053443

[7] Campbell S L, Hutson R B, Marti G E, Goban A, Darkwah Oppong N, McNally R L, Sonderhouse L, Robinson J M, Zhang W, Bloom B J and Ye J 2017 *Science* **358** 90–94

[8] Milner W R, Robinson J M, Kennedy C J, Bothwell T, Kedar D, Matei D G, Legero T, Sterr U, Riehle F, Leopardi H, Fortier T M, Sherman J A, Levine J, Yao J, Ye J and Oelker E 2019 *Phys. Rev. Lett.* **123**(17) 173201

[9] Herbers S, Häfner S, Dörscher S, Lücke T, Sterr U and Lisdat C 2022 *Opt. Lett.* **47** 5441–5444

[10] Matei D G, Legero T, Häfner S, Grebing C, Weyrich R, Zhang W, Sonderhouse L, Robinson J M, Ye J, Riehle F and Sterr U 2017 *Phys. Rev. Lett.* **118** 263202

[11] Häfner S, Falke S, Grebing C, Vogt S, Legero T, Merimaa M, Lisdat C and Sterr U 2015 *Opt. Lett.* **40** 2112–2115

[12] Yu J 2023 *Cryogenic silicon Fabry-Perot resonator with* $Al_{0.92}Ga_{0.08}As/GaAs$ *mirror coatings.* Ph.D. thesis QUEST-Leibniz-Forschungsschule der Gottfried Wilhelm Leibniz Universität Hannover

[13] Harry G M, Armandula H, Black E, Crooks D R M, Cagnoli G, Hough J, Murray P, Reid S, Rowan S, Sneddon P, Fejer M M, Route R and Penn S D 2006 *Appl. Opt.* **45** 1569–1574

[14] Granata M, Amato A, Cagnoli G, Coulon M, Degallaix J, Forest D, Mereni L, Michel C, Pinard L, Sassolas B and Teillon J 2020 *Appl. Opt.* **59** A229–A235

[15] Kedar D, Yu J, Oelker E, Staron A, Milner W R, Robinson J M, Legero T, Riehle F, Sterr U and Ye J 2023 *Optica* **10** 464–470

[16] Yu J, Häfner S, Legero T, Herbers S, Nicolodi D, Ma C Y, Riehle F, Sterr U, Kedar D, Robinson J M, Oelker E and Ye J 2023 *Phys. Rev. X* **13**(4) 041002

[17] Levin Y 1998 *Phys. Rev. D* **57** 659–663

[18] Callen H B and Welton T A 1951 *Phys. Rev.* **83** 34–40

[19] Kubo R 1966 *Rep. Prog. Phys.* **29** 255–284

[20] Cole G D, Zhang W, Martin M J, Ye J and Aspelmeyer M 2013 *Nature Photonics* **7** 644–650

[21] Cole G D, Ballmer S, Billingsley G, Cataño Lopez S B, Fejer M, Fritschel P, Gretarsson A M, Harry G M, Kedar D, Legero T, Makarem C, Penn S D, Reitze D, Steinlechner J, Sterr U, Tanioka S, Truong G W, Ye J and Yu J 2023 *Appl. Phys. Lett.* **122** 110502

Journal of Physics: Conference Series **2889** (2024) 012055 doi:10.1088/1742-6596/2889/1/012055

[22] Cole G D, Zhang W, Bjork B J, Follman D, Heu P, Deutsch C, Sonderhouse L, Robinson J, Franz C, Alexandrovski A, Notcutt M, Heckl O H, Ye J and Aspelmeyer M 2016 *Optica* **3** 647–656
[23] Kedar D, Yao Z, Ryger I, Hall J L and Ye J 2024 *Optica* **11** 58–63
[24] Farsi A, Siciliani de Cumis M, Marino F and Marin F 2012 *J. Appl. Phys.* **111** 043101
[25] Legero T, Kessler T and Sterr U 2010 *J. Opt. Soc. Am. B* **27** 914–919
[26] Zhu X Q, Cui X Y, Kong D Q, Yu H W, Jiang X, Xu P, Dai H N, Chen Y A and Pan J W 2023 *Fourteenth International Conference on Information Optics and Photonics (CIOP 2023)* vol 12935 ed Yang Y International Society for Optics and Photonics (SPIE) p 1293541
[27] Kraus B 2023 *A highly stable UV clock laser* Ph.D. thesis QUEST-Leibniz-Forschungsschule der Gottfried Wilhelm Leibniz Universität Hannover
[28] Monemar B, Shih K K and Pettit G D 1976 *J. Appl. Phys.* **47** 2604–2613
[29] Sturge M D 1962 *Phys. Rev.* **127**(3) 768–773
[30] Hutchby J A and Fudurich R L 2008 *J. Appl. Phys.* **47** 3140–3151
[31] Namba S 1961 *J. Opt. Soc. Am.* **51** 76–79
[32] Tanioka S, Vander-Hyde D, Cole G D, Penn S D and Ballmer S W 2023 *Phys. Rev. D* **107**(2) 022003
[33] Kessler T, Hagemann C, Grebing C, Legero T, Sterr U, Riehle F, Martin M J, Chen L and Ye J 2012 *Nature Photonics* **6** 687–692
[34] Hagemann C 2013 *Ultra-stable laser based on a cryogenic single-crystal silicon cavity* Ph.D. thesis Fakultät für Mathematik und Physik der Gottfried Wilhelm Leibniz Universität Hannover URL `http://edok01.tib.uni-hannover.de/edoks/e01dh13/739734164.pdf`
[35] Benkler E, Lipphardt B, Puppe T, Wilk R, Rohde F and Sterr U 2020 *Opt. Express* **28** 15023–15024

Towards a sub-kelvin cryogenic Fabry-Perot silicon cavity

Joannès Barbarat[1], Jonathan Gillot[2], Jacques Millo[2], Clément Lacroûte[1], Thomas Legero[3], Vincent Giordano[1] and Yann Kersalé[2]

[1]Université de Franche-Comté, CNRS, FEMTO-ST, 26 rue de l'Epitaphe, 25000 Besançon, France

[2]ENSMM, CNRS, FEMTO-ST, 26 rue de l'Epitaphe, 25000 Besançon, France

[3]Physikalisch-Technische Bundesanstalt, Bundesallee 100, 38116 Braunschweig, Germany

E-mail: jonathan.gillot@femto-st.fr

Abstract. We report on the development of a sub-kelvin, single-crystal silicon Fabry-Perot cavity. Operating such a cavity below 1 K should reduce the thermal noise limit of the cavity, and by this way address the current limitations of ultrastable lasers. To further decrease mechanical losses, mirrors with silicon substrates and crystalline coatings are optically contacted to the spacer, resulting in a room-temperature finesse of 220,000. To operate our cavity at sub-kelvin temperatures, we use a dilution refrigerator able to reach temperatures down to 10 mK. We have designed a mechanical mount to house our cavity in such a cryostat, with optimized heat transfer that will decrease the cooldown time for temperatures below 1 K. The estimated thermal noise is projected to be $\sim 7\times10^{-19}$ at 100 mK. However, silicon cavities with crystalline mirror coatings at cryogenic temperatures have shown birefringence correlated frequency fluctuations as well as unknown additional noise mechanisms [1, 2]. We have measured a room-temperature TEM_{00} birefringent mode splitting of about 250 kHz. Understanding and measuring these noise mechanisms will be a key to attaining fractional frequency stabilities beyond state-of-the-art.

1. Introduction:

Lasers stabilized to ultrastable optical cavities are required for a wide range of applications such as spectroscopy, gravitational waves detectors, frequency standards and tests of fundamental physics [3]. State of the art optical oscillators based on single crystal silicon cavities exhibit fractional frequency instabilities of few 10^{-17} limited by the Brownian thermal noise of the cavity coatings [4].

Silicon exhibits a very high mechanical quality factor, which increases at low temperatures [5]. There are two cancellations of the coefficient of thermal expansion (CTE), at $\sim$ 124 K and $\sim$ 17 K [6], and below 4 K the CTE asymptotically approaches zero. Several experiments have shown that fractional frequency instabilities could be attained with such cavities, at 124 K [7], 17 K [8], or below [9], along with very low frequency drifts due to the spacer crystalline structure [10, 11].

We aim to operate a silicon cavity in a yet unexplored temperature regime below 1 K. We employ crystalline coatings to try and further reduce the cavity thermal noise. Such a cavity will be a unique tool to characterize the cavity materials, including the coefficient of thermal expansion of single-crystal silicon and the behavior of crystalline AlGaAs coatings. It will also be useful to better understand the observed long-term drift of single-crystal silicon cavities, which origin is no yet well understood [12, 10, 11]. The experimental setup will help to investigate thermal and technical limitations for next generation ultrastable lasers, aiming to realize sub 10^{-17} fractional frequency instabilities. This article presents the overall setup we designed for this experiment, along with the projected fundamental and technical noises.

2. Cryocooler

Single-crystal silicon cavities have been tested at the 124 K and 17 K turning CTE cancellation points [8] and at temperatures between 1.5 K and 4 K [10]. The cavity is housed in a dilution cryocooler in order to be able to reach temperatures below 1 K. In addition to the two CTE inversion points documented since the 1970s, a third inversion point has been measured recently close to 3 K [11]. The ability to control the temperature of the spacer on the 100 mK - 4 K range will therefore be a key to the experiment, and will allow us to test this inversion point as well.

2.1. Cryocooler setup

Our cryocooler is composed of 4 main stages, shown in Fig. 1: the 50 K stage, the 4 K stage, the still stage and the mixing chamber stage (noted MX). The two first stages are based on one closed cycle ^{4}He pulse-tube cryocooler. Compression-extension cycle at $\sim$ 1.4 Hz of the gas inside "tubes" is generated with a compressor unit together with a rotating valve (three positions: open to high pressure, open to low pressure or closed). The first two stages reach temperatures of about 50 K and 3 K. Then, a Joule-Thomson stage enables temperatures around 1.3 K. To achieve lower temperatures, a dilution of ^{3}He in ^{4}He is used in a final stage. The plate connected to the mixing chamber can reach a temperature as low as a few mK.

Cylindrical aluminium shields are thermalized on each stage, except for the MX, to significantly suppress radiative heat exchange from the room temperature environment to the cavity. The available space below the mixing plate is a cylinder with a 290 mm diameter and a 300 mm height which sets the maximum dimensions for the cavity spacer. Horizontally-oriented optical access is granted through 4 pairs of AR-coated viewports. The window of the vacuum chamber is made of BK7 while thermal shields are equipped with SiO_2 windows. We estimate that the radiated infra-red power from the 300 K, 50 K and 4 K windows is below 1 µW at the MX stage, well below the available cooling power at 100 mK.

The cooldown procedure to reach the mK temperature range takes approximately 36 hours without thermal load (cavity) inside and is divided in three main phases (Fig. 1). The first one is the pulse-tube ^{4}He cryocooling procedure, of roughly 24 hours; the dilution refrigerator can then be started and the ^{3}He/^{4}He mixture is first condensed and cools down to about 0.8 K in around 6 hours at the still stage; the ^{3}He/^{4}He mixture then separates in two phases and temperatures below 100 mK are reached within roughly 6 hours.

2.2. Vibration isolation

Pulse-tube cryocoolers are known to induce vibrations that propagate to the setup through He-circulation hoses and mechanical contacts in the setup. The cavity will be mechanically decoupled from these vibrations through several stages: mechanical grounding of the rotation valve and He hoses through high masses; pneumatic isolation of the cryostat from the cryocooler head; flexible copper braids inside the cryostat that decouple the pulse-tube's vibrations from the science stages.

2.3. Temperature monitoring

The temperature of each stage is monitored by at least one sensor (see table 1). The first stage of the pulse tube (50 K) is monitored with a common PT100 sensor and only used in case

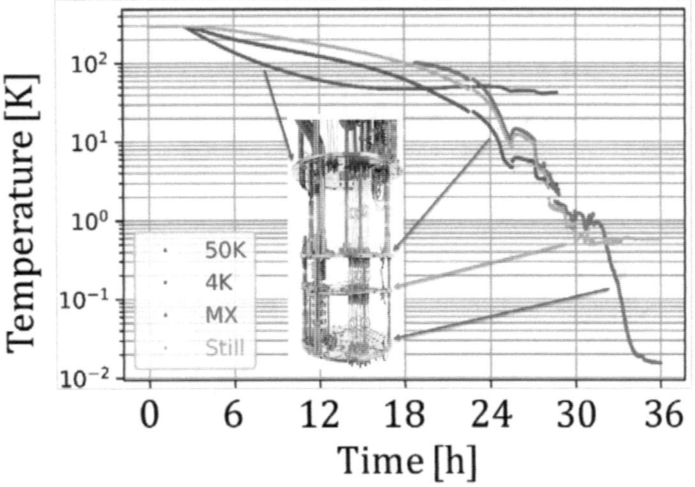

Figure 1. Cooldown of the cryostat. The insert shows the four cryocooler stages. The MX stage reaches the sub-100 mK range in 36 hours.

of cooling issues. Cernox® sensors are used for the 4 K stage and the still plate. They are calibrated for temperatures down to 0.1 K.

Two additional sensors will help monitoring the MX plate and cavity mount temperatures: a calibrated RuO_2 RX-102B and an uncalibrated RX-202A-CD, with a working range down to 0.01 K. All theses probes are monitored by a multiplexed AC resistance bridge and a temperature controller. To prevent the thermal disturbance caused by the reading of the probe, measurements are performed with a ratio of 3 seconds over 10 seconds.

Stage	type	ref.	Cal. from	to
50 K	PT100	–	310 K	20 K
4 K	Cernox®	CX-1010 (LS)	310 K	0.1 K
still	Cernox®	CX-1010 (LS)	310 K	0.1 K
MX	RuO_2	RX-102B (LS)	100 K	0.010 K
MX	RuO_2	RX-102B (LS)	100 K	0.010 K
Cavity mount	RuO_2	RX-202A-CD (LS)	uncal.	uncal.

Table 1. Temperature sensors installed in the cryostat. LS stands for LakeShore. Cal.: calibration. uncal.: uncalibrated.

2.4. Temperature measurements

We performed preliminary cooling tests for characterizing the cryocooler and temperature probes. The first one, in an unloaded configuration, achieved a temperature of 15 mK, which can be compared to the lowest temperature achieved of 7 mK without windows. During a second run we installed the cavity support and a copper part (see section 4.2) attached to the MX plate where we installed a calibrated RuO_2 RX-102B and an uncalibrated RX-202A-CD. The measured temperature of this copper block is 20 mK. The temperatures measured by the RuO_2 sensor on the copper part and the RuO_2 sensor on the mixing plate agree within 1 mK.

We tested the temperature control at the MX stage when loaded with the cavity mount. We were able to stabilize the temperature to 20 mK with fluctuations below 0.1 mK, as measured by the in-loop sensor (Fig. 2).

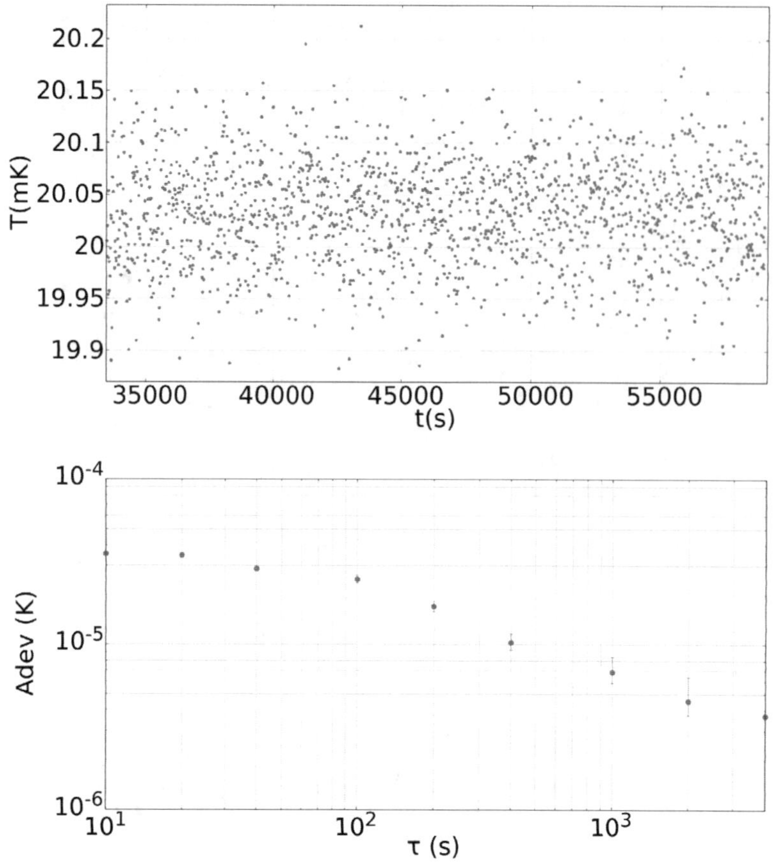

Figure 2. Top: Temperature of the MX stage when temperature stabilization is activated. This is an in-loop measurement. Bottom: Allan deviation of the temperature data.

3. Cavity

3.1. Spacer and mirrors design

The cavity is composed of a single-crystal silicon spacer and two single-crystal silicon mirror substrates. The dimensions of the cylindrical spacer are 180 mm in length and 200 mm in diameter. We chose the highest possible length given the available space in the MX chamber. The aspect ratio allows us to reduce the radial acceleration sensitivity by cancelling tilts from radial accelerations for this simple cylindrical geometry [13]. The weight of the spacer is about 13 kg. Its large thermal capacity improves the passive filtering of temperature fluctuations.

The optical axes of the spacer and the substrates are aligned to the [111] crystalline axis and include a flat for proper alignment during optical contacting. A custom contacting mask was fabricated to ensure proper centering of the mirrors and proper alignment of the crystalline axis.

The mirrors are equipped with $Al_{0.92}Ga_{0.08}As/GaAs$ crystalline coatings specified for use at 17 K. The crystalline coatings are optically contacted to the substrates. Their slow axis is indicated by a flat, which presents a misalignment of less than 6° with regards to the substrate's flat. The crystalline orientation of the coatings is most likely similar to those presented in [1], with the [100] direction of GaAs normal to the mirror surface.

3.2. Cavity mount

The cavity mount is made in duralumin in order to reduce its weight. It is attached below the MX plate by four M6 screws to minimize thermal resistance. The cavity is held horizontally on 3 contact points in order to avoid hyper static equilibrium. Further improvement will include simulations of the mount influence on the spacer vibration sensitivity and the search for possible optimal support points such as in [13].

Fig. 3 shows the cavity held in its support, attached to the MX plate. One of the three supporting spheres made of steel can be seen at the bottom-left. A second one is located in the same plane transverse to the optical axis, while the third one is at the center of the second supporting arch.

Figure 3. Single-crystal silicon cavity held in its support. The support is directly attached to the MX plate. One of the three supporting spheres is shown on the left bottom corner.

4. Cavity cooling

4.1. Cooling by radiation

As described in the previous section the cavity mount does not enable efficient cooling by heat conduction, as mechanical contacts are voluntarily minimized and the thermal conductance of the 3 contact points is expected to be negligible in the target temperature range. In this configuration, the cavity will be cooled only through the thermal radiation exchange with the surrounding thermal shield. We have performed an analytical estimation of the time required to cool the cavity from 4 K to 1 K and from 1 K to the lowest temperature, corresponding to the Joule-Thomson and dilution refrigeration steps, respectively. For this simple evaluation we assumed: i) the cavity surrounding has reached its thermal equilibrium, ii) the cavity emissivity ~ 1, iii) the cavity mass and external surface are 13 kg and 0.176 m^2 respectively, iv) the silicon heat capacity is given by $C_p = 2.5 \times 10^{-4}\ T^3$ J/kg/K and v) to simplify equation, thermal properties are considered constant within the temperature range.

In both cases, the cooling time constant is on the order of a few 100 days. We therefore aim to strongly increase the thermal conduction between the cavity and the MX plate.

Journal of Physics: Conference Series **2889** (2024) 012056 doi:10.1088/1742-6596/2889/1/012056

4.2. Optimized heat transfer to the cavity

We have constructed a new cavity mount for optimized heat transfer between the MX plate and the silicon cavity. It is made in copper and will surround the cavity. This thermal support is mechanically decoupled from the setup by using flexible copper braids to thermally contact to the MX plate (Fig. 4).

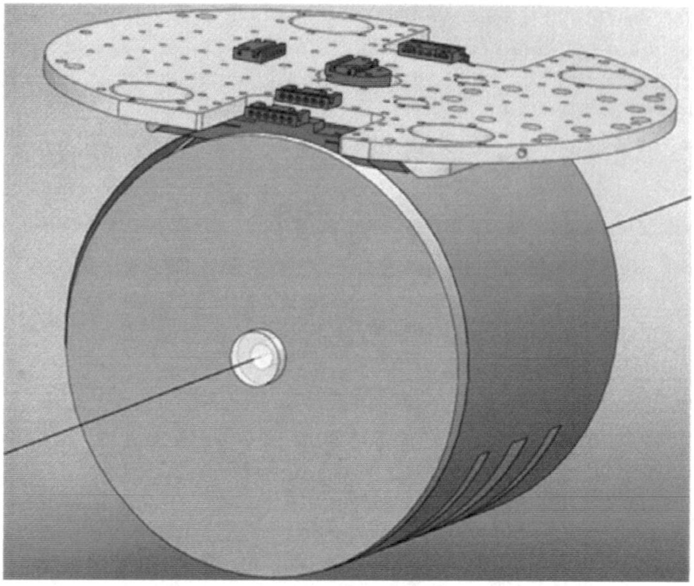

Figure 4. Cavity mount with optimized heat conduction from the cavity to the MX plate: scheme of the copper wrap and its mounting. Red parts are made of solid copper. Three connections are screwed to the MX plate and one is attached to the cavity using a copper foil wrap. Heat conduction is ensured by using soft copper breads (not shown).

5. Preliminary measurements and expected performances

5.1. Cavity finesse and birefringence splitting

The assembled cavity (Fig. 3) shows a finesse of about 220 000 in air at room temperature (Fig. 5) determined by ring-down measurement. The finesse in vacuum and at low temperatures is expected to be higher, as we face residual absorption in air at wavelengths close to 1542 nm and the mirror transmission is optimized for 17 K.

To simultaneously measure the resonance splitting caused by the coatings birefringence, we injected the cavity with linearly polarized light with a projection on both crystalline axes. This results in a beatnote signal superimposed to the typical exponential decay of the ring-down measurement. The TEM_{00} mode splitting due to the birefringence of the coatings is about 250 kHz (Fig. 5), in good agreement to the birefringence measured in other cavities using AlGaAs coatings [1, 2].

5.2. Expected performance

The expected thermal noise of this cavity is listed in table 3. We use the formula from [14, 15] for the spacer and substrates combined with the values and formula from [16] for the coatings.

Assuming peak-to-peak temperature fluctuations of 0.5 mK, the CTE needs to be below 2×10^{-15} K^{-1} to reach the 10^{-18} level. Several models have been proposed for single-crystal silicon CTE below 1 K. Lyon *et al.* proposed a 4.8×10^{-13} T^3 dependency, where T is the

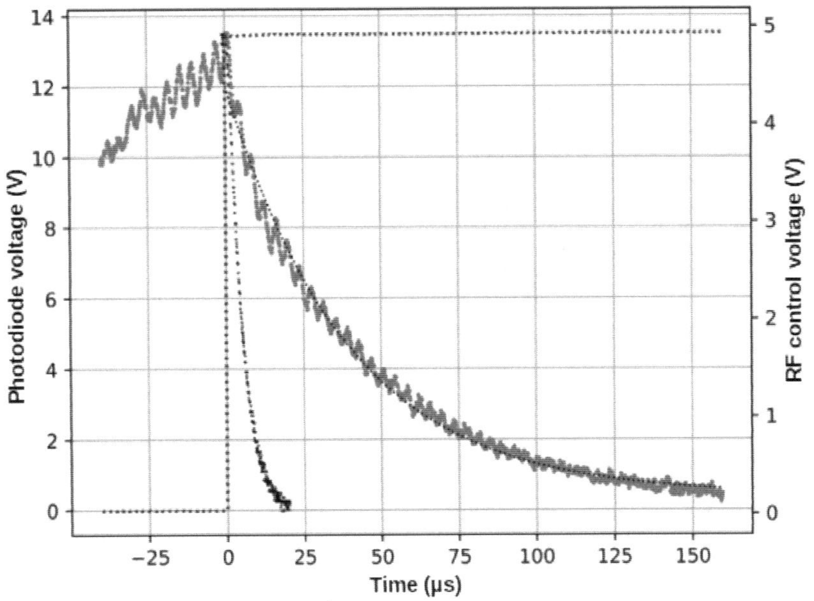

Figure 5. Finesse measurement of the single-crystal silicon cavity in air at room temperature. Red: cavity transmission signal with a superimposed beatnote of the two polarization eigenstates of the cavity. Black: photodiode response without cavity. Blue dotted line: voltage control of the RF switch that turns the light off.

temperature in K [6]. Wiens *et al.* [12] proposed a polynomial law derived from experimental values; its accuracy below 1 K is not clear, and it suggests there could be several additional CTE zero-crossings. Those models predict CTE values summarized in table 2. One can clearly see the advantage of going to the lowest possible temperature. But it is also clear that a better knowledge of the silicon CTE is required in order to accurately determine both, the operating temperature, and the temperature stability constraint.

Model	Lyon *et al.* [6]	Wiens *et al.* [12]
1 K	4.8×10^{-13} K^{-1}	1.7×10^{-13} K^{-1}
0.5 K	6×10^{-14} K^{-1}	-8.8×10^{-15} K^{-1}
0.1 K	4.8×10^{-16} K^{-1}	4.0×10^{-17} K^{-1}

Table 2. Expected coefficient of thermal expansion of single-crystal silicon at cryogenic temperatures, calculated from the model in [6] and the fit from [12].

Low sensitivity to vibrations will be a key to achieve the targeted performance in a dilution cryostat. We can have a first estimation of these effects assuming a maximum sensitivity of 10^{-10}/m s^{-2}. According to the manufacturers, the typical vibration level in our cryostat can be on the order of 10 nm displacement amplitude; some teams have measured levels of order 5×10^{-6} m s^{-2} at 1 Hz. This would lead to a contribution between 10^{-17} and 5×10^{-16} from vibrations. Table 3 shows that such vibrations would limit the performance even at 1 K. To reach the desired levels, the vibration sensitivity and the level of vibrations must be reduced by at least a factor of ten.

It will also be interesting to probe the cavity sensitivity to intra-cavity power fluctuations, which is related to coatings losses and material properties. A sensitivity to optical power was already observed in other experiments [12, 10, 11, 2, 1].

Two additional contributions have been unveiled recently for cryogenic silicon cavities with crystalline mirror coatings [1, 2]. A birefringence noise, dependent on the intra-cavity optical power and much higher than the predicted Brownian noise, can be averaged out by using the two polarization axis simultaneously. An additional, unknown mechanism dubbed "residual" or "global excess noise" that is independent of temperature or optical power was observed in two independent experiments [1, 2], and was measured to be as high as ten times the Brownian noise [1].

Table 3 summarizes the predicted contributions of thermal noise, global excess noise and thermal fluctuations to the fractional frequency stability at 1 K, 0.5 K and 0.1 K. The "Global noise" column accounts for residual excess noise described in [1] substantially different from thermal noise. It remains after averaging the dominating but anti-correlated birefringent noise of the polarization eigenmodes of the cavity and exceeds the expected Brownian thermal noise of the coatings. As this noise shows no strong dependency on temperature, we assumed an identical value of mirror displacement noise as that measured in ref. [1] at 16 K, and scaled it to our cavity length. Measuring the actual value of this global noise at sub-K temperatures will be important, as it would most probably be a major limitation.

Temperature	Thermal noise	"Global noise"	Temperature fluctuations	Total
1 K	2.2×10^{-18}	3.6×10^{-17}	2.4×10^{-16}	2.4×10^{-16}
0.5 K	1.5×10^{-18}	3.6×10^{-17}	3.0×10^{-17}	4.7×10^{-17}
0.1 K	6.9×10^{-19}	3.6×10^{-17}	2.4×10^{-19}	3.6×10^{-17}

Table 3. Predicted contributions of thermal noise, global noise and thermal fluctuations to the fractional frequency stability. The total noise is calculated from the quadratic mean of the three contributions.

6. Conclusion

We have presented a dilution cryostat as a testbed to investigate a single-crystal silicon cavity at sub-K temperatures. In this temperature range the CTE of silicon asymptotically approaches zero and the sensitivity of the cavity to temperature fluctuations decreases significantly. We have optimized the heat transfer to the cavity by copper braids which will accelerate the cooling time in the sub-K temperature range. As a next step we need to mitigate vibrations induced by the pulse-tube cryocooler.

For this cryostat we have setup a 180 mm long silicon cavity employing crystalline AlGaAs mirrors with a projected thermal noise limit in the low 10^{-18} range. The cavity room-temperature finesse is 220,000 in air and the cavity exhibits a birefringent mode splitting of 250 kHz. Our setup will allow the study of the recently described birefringent and global noise contributions in AlGaAs coatings even at temperatures below 4 K.

Acknowledgements

This work has been supported by the EIPHI Graduate School (contract "ANR-17-EURE-0002"), by the ANR-10-LABX-48-01 First-TF and ANR-11-EQPX-0033 Oscillator-IMP, and by the Région Bourgogne Franche-Comté.

This project (20FUN08 NEXTLASERS) has received funding from the EMPIR programme co-financed by the Participating States and from the European Union's Horizon 2020 research and innovation programme.

References

[1] Yu J, Häfner S, Legero T, Herbers S, Nicolodi D, Ma C Y, Riehle F, Sterr U, Kedar D, Robinson J M, Oelker E and Ye J 2023 Phys. Rev. X **13**(4) 041002
[2] Kedar D, Yu J, Oelker E, Staron A, Milner W R, Robinson J M, Legero T, Riehle F, Sterr U and Ye J 2023 Optica **10** 464–470
[3] Ludlow A D, Boyd M M, Ye J, Peik E and Schmidt P O 2015 Rev. Mod. Phys. **87**(2) 637–701
[4] Matei D G, Legero T, Häfner S, Grebing C, Weyrich R, Zhang W, Sonderhouse L, Robinson J M, Ye J, Riehle F and Sterr U 2017 Phys. Rev. Lett. **118**(26) 263202
[5] Middelmann T, Walkov A, Bartl G and Schödel R 2015 Phys. Rev. B **92**(17) 174113
[6] Lyon K G, Salinger G L, Swenson C A and White G K 2008 Journal of Applied Physics **48** 865–868
[7] Kessler T, Hagemann C, Grebing C, Legero T, Sterr U, Riehle F, Martin M J, Chen L and Ye J 2012 Nature Photonics **6** 687–692
[8] Gillot J, Hariri Y, Barbarat J, Lacroute C, Millo J and Kersalé Y 2023 Development of an ultra-stable cryogenic silicon optical cavity 2023 Joint Conf. of the European Frequency and Time Forum and IEEE International Frequency Control Symp.
[9] Zhang W, Robinson J M, Sonderhouse L, Oelker E, Benko C, Hall J L, Legero T, Matei D G, Riehle F, Sterr U and Ye J 2017 Physical Review Letters **119** 243601
[10] Robinson J M, Oelker E, Milner W R, Zhang W, Legero T, Matei D G, Riehle F, Sterr U and Ye J 2019 Optica **6** 240–243
[11] Wiens E, Kwong C J, Müller T and Schiller S 2020 Review of Scientific Instruments **91** 045112 ISSN 0034-6748
[12] Wiens E, Chen Q F, Ernsting I, Luckmann H, Rosowski U, Nevsky A and Schiller S 2014 Optics Letters **39** 3242–3245
[13] Millo J, Magalhães D V, Mandache C, Le Coq Y, English E M L, Westergaard P G, Lodewyck J, Bize S, Lemonde P and Santarelli G 2009 Phys. Rev. A **79**(5) 053829
[14] Numata K, Kemery A and Camp J 2004 Physical Review Letters **93** 250602
[15] Kessler T, Legero T and Sterr U 2012 Journal of the Optical Society of America B **29** 178–184
[16] Cole G D, Zhang W, Martin M J, Ye J and Aspelmeyer M 2013 Nature Photonics **7** 644–650

Dual-atom-interferometric gyroscope with continuous cold atomic beams

Shengzhe Wang[1,2], Zhixin Meng[1,2,3], Peiqiang Yan[1,2], Yanying Feng[1,2,*]

1. State Key Laboratory of Precision Measurement Technology and Instruments, Tsinghua University, Beijing, China
2. Department of Precision Instrument, Tsinghua University, Beijing, China
3. Beijing Institute of Aerospace Control Devices, Beijing, China.

E-mail: yyfeng@tsinghua.edu.cn

Abstract. In closed-loop dual-atom interferometers, the initial phase shift is a non-inertial phase induced by Raman pulses. This initial optical phase affects the measurements of both acceleration and rotation rate. In this study, we present a method to modulate the initial optical phase of light pulse atom interferometers (LPAIs). Employing this modulation method, we investigated the influence of the initial optical phase on a closed-loop dual-atom interferometer. The results reveal a better short-term sensitivity in the rotation rate measurement at an initial optical phase of 0. Our closed-loop dual-atom interferometer allows for decoupled measurements of rotation rate and acceleration. With a short interrogation time of $2T = 1.73\,\text{ms}$, the short-term sensitivities of the rotation rate and acceleration measurements are $2.73 \times 10^{-5}\,(\text{rad/s})/\sqrt{\text{Hz}}$ and $272\,\mu\text{g}/\sqrt{\text{Hz}}$, respectively.

1. Introduction

Light pulse atom interferometers (LPAIs) have been developed for years [1–5]. With their high sensitivity and stability when sensing acceleration [6, 7], rotation rate [4, 8, 9], gravity and gravity gradient [3, 10, 11], LPAIs show potential in inertial navigation systems (INS) [12–14], as well as scientific applications such as gravitational wave detection [15, 16] and the search for dark matter [17, 18].

The Mach-Zehnder type atom interferometers usually provide the coupling information about acceleration and rotation rate which is not suitable for field applications. Our previous work demonstrated a closed-loop dual-atom-interferometer inertial sensor with continuous cold atomic beams, which can provide measurements of acceleration and rotation rate that are decoupled [19]. In this type of closed-loop dual-atom interferometers, the initial phase shift, which is a non-inertial optical phase induced by Raman pulses, could affect the measurements of both acceleration and rotation rate.

This study presents a method to modulate the initial optical phase of Raman pulses. We observed that the initial phase influences inertial measurements in the closed-loop dual-atom-interferometer inertial sensing. We evaluated the short-term sensitivities of both rotation rate and acceleration measurement at various initial optical phases. The results show a better short-term sensitivity in rotation rate measurement at $2\phi_0 = 0$. And the results highlight the presence of certain noise within the apparatus.

2. Experimental Setup

Our work is based on two counter-propagating continuous cold ^{87}Rb beams generated by two 2D+ magneto-optical traps (2D+MOT) [20]. These beams have mean velocities of 10.96 m/s, transverse temperatures of 16.90 μK, and fluxes of about $2 \times 10^9\,\mathrm{s}^{-1}$.

The carrier frequency of the Raman laser is locked to the $F = 2 \rightarrow F' = 2$ (^{87}Rb D2 line) with a detuning of -1.2 GHz. Raman laser frequencies are generated through phase modulation of the carrier laser using an electro-optic modulator (EOM). To separate the Doppler-sensitive Raman transition from the Doppler-insensitive transition, the Raman beams are tilted at a small angle. The Raman beams define two centrosymmetric Mach-Zehnder interferometry configurations, crucial for the decoupled measurement of rotation rate and acceleration. The Raman beams have a width of 1 mm and are separated by $2L = 19$ mm (interrogation time $2T = 1.73$ ms) [19].

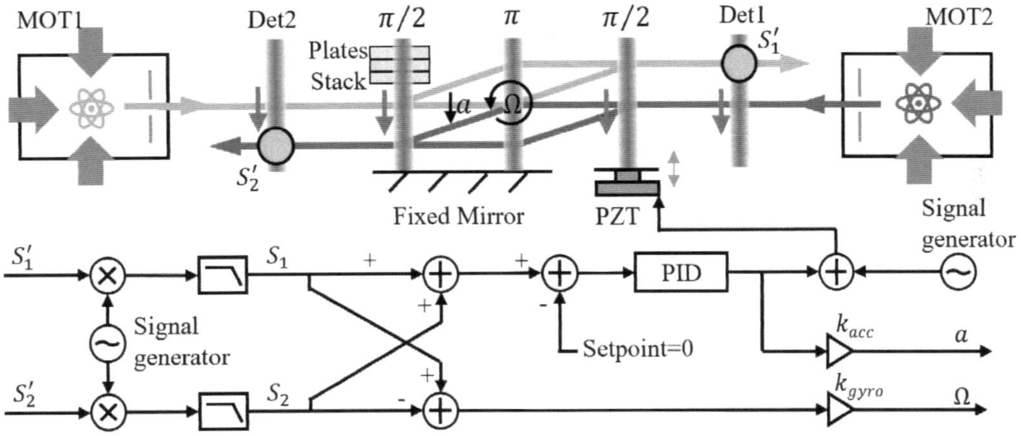

Figure 1. Atom interferometer inertial sensor based on a pair of counter-propagating continuous cold beams. One of the Raman beams is referred to a piezoelectric-transducer (PZT)-attached mirror to control and modulate the reflected optical phase shift. Sta-Pre1 and Sta-Pre2 are state-preparation lasers. Interference fringe signals S_1' andS_2' is obtained by measuring fluorescence from resonant probe lasers (Det1, Det2). Both signals are filtered and denoted as S_1 and$S_2$by lock-in amplifiers.

The interference signals S_1 andS_2 from the two interferometers encompass information about both linear acceleration a and the rotation rate Ω, and they can be modulated by the non-inertial optical phases:

$$S_1 = cos\,(\phi_a - \phi_\Omega - \phi_r + \phi_0) + n_1 \tag{1}$$
$$S_2 = cos\,(-\phi_a - \phi_\Omega + \phi_r + \phi_0) + n_2 \tag{2}$$

where ϕ_a and ϕ_Ω represent the inertial phases induced by the acceleration and rotation rate of the optical Raman beam frame, respectively. ϕ_0 and ϕ_r represent the non-inertial optical phases. ϕ_0 is the initial optical phase, that is the linear combination of the relative phase between the carrier frequency and the +6.835 GHz sideband of the Raman beam. And ϕ_r is the linear phase combination determined by the reflected optical path, while $n_{1,2}$ denote the noise and drift induced during the implementation of atom interferometers.

Since the two atom beams resonate with the Raman laser frequency $\vec{k}_{eff}$ in different directions, the interference signals S_1 and S_2 detect the linear acceleration a in opposite directions. Given the centrosymmetric nature of the interference area, S_1 and S_2 sense the rotation rate Ω in the same direction. Utilizing sum and differential signals allows for the decoupling of acceleration and rotation rate measurements, aligning them with the non-inertial optical phase:

$$S_1 + S_2 = 2cos\left(\phi_a - \phi_r\right) cos\left(\phi_\Omega - \phi_0\right) + (n_1 + n_2) \quad (3)$$

$$S_1 - S_2 = 2sin\left(\phi_a - \phi_r\right) sin\left(\phi_\Omega - \phi_0\right) + (n_1 - n_2) \quad (4)$$

The acceleration phase ϕ_a is aligned with the non-inertial optical phase ϕ_r, while the rotation rate phase ϕ_Ω is aligned with the non-inertial optical phase ϕ_0. Both the inertial phases can be extracted by phase-locking $\phi_a - \phi_r$ and $\phi_\Omega - \phi_0$.

In this work, we phase-lock $\phi_a - \phi_r$ to 0. The rotation rate is extracted from open-loop $\phi_\Omega - \phi_0$, while the acceleration is extracted from ϕ_r.

3. Results and discussion

The atom interference fringes are generated by scanning ϕ_r using a piezoelectric-transducer (PZT) attached to a retro-reflected mirror in one of the $\pi/2$ Raman pulses, as shown in Fig. 2.

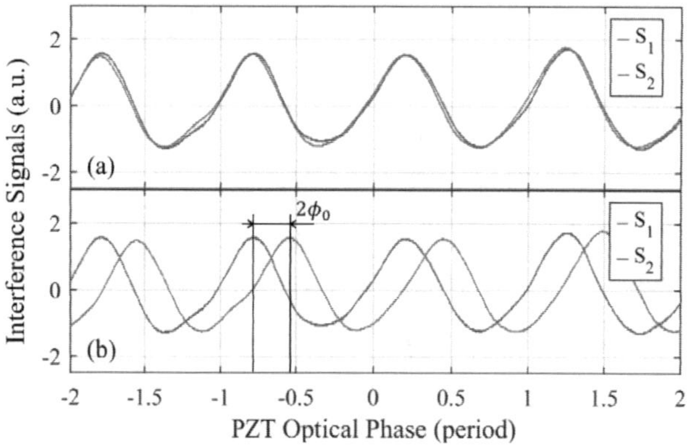

Figure 2. The atom interference fringes signal from two counter-propagating atom interferometers, with (a) $2\phi_0 = \pi/2$, and (b) $2\phi_0 = \pi/2$. The optical phase is scanned by a PZT attached to the retro-reflected mirror of the left $\pi/2$ Raman pulse in Fig. 1.

The initial optical phase ϕ_0 influences the relative phase between fringes S_1 and S_2, and it also determines the amplitude of the sum signal $S_1 + S_2$ and the differential signal $S_1 - S_2$. A stack of plates is used to adjust the initial optical phase ϕ_0 of the left $\pi/2$ Raman pulses in Fig. 1. The stack contains several plates (5 mm thick, N-BK7), each providing extra initial optical phase of 0.3680 rad (about 0.0586 periods). The dependence of the relative phase between S_1 and S_2 on the adjusted initial optical phase is shown in Fig. 3. The relative phase between S_1 and S_2 is determined by the correlation between S_1 and S_2.

When the initial optical phase $2\phi_0 = 0$, the S_1 and S_2 fringes are in phase. The sum signal $S_1 + S_2$ is used for phase-lock, and the differential signal $S_1 - S_2$ is used to calculate the rotation rate. When the initial optical phase $2\phi_0 = \pi$, the S_1 and S_2 fringes are opposite in phase. The sum signal $S_1 - S_2$ is used for phase-lock, and the differential signal $S_1 + S_2$ is used to calculate the rotation rate. If the initial optical phase $2\phi_0 = \pi/2$, both the sum signal and the differential signal can be used for phase-lock, and the other is used to calculate the rotation rate. The performance of these four operation modes is different since there can be common-mode noise, differential noise, and other noise within n_1 and n_2.

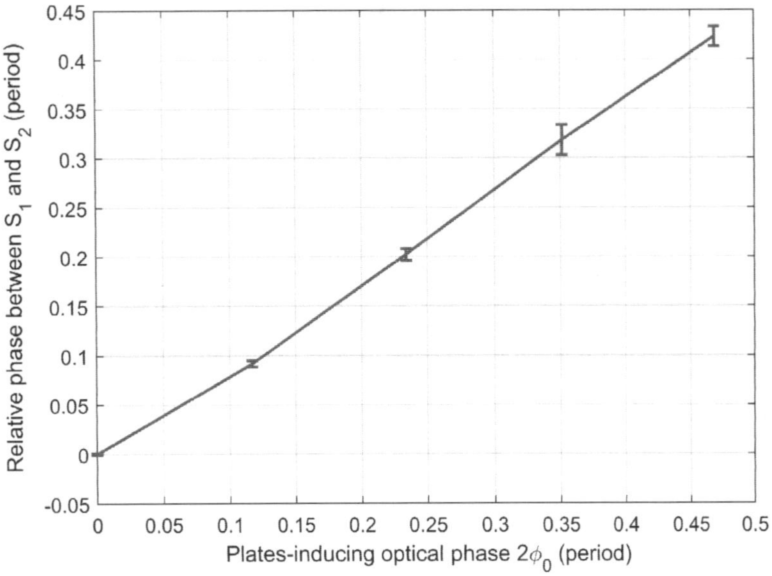

Figure 3. The relationship between plates-inducing initial optical phase $2\phi_0$ and the relative phase between S_1 and S_2.

Table 1. Short-term sensitivity in acceleration and rotation rate measurements.

Initial optical phase $2\phi_0$ (Locked signal)	Short-term sensitivity in rotation rate measurement	Short-term sensitivity in acceleration measurement
0 (S_1+S_2)	$2.73\times10^{-5}\,(\text{rad/s})/\sqrt{\text{Hz}}$	$272\,\mu\text{g}/\sqrt{\text{Hz}}$
$\pi/2$ (S_1+S_2)	$5.01\times10^{-4}\,(\text{rad/s})/\sqrt{\text{Hz}}$	$765\,\mu\text{g}/\sqrt{\text{Hz}}$
$\pi/2$ (S_1-S_2)	$1.13\times10^{-4}\,(\text{rad/s})/\sqrt{\text{Hz}}$	$435\,\mu\text{g}/\sqrt{\text{Hz}}$
π (S_1-S_2)	$6.07\times10^{-5}\,(\text{rad/s})/\sqrt{\text{Hz}}$	$651\,\mu\text{g}/\sqrt{\text{Hz}}$

We have conducted measurements using various modes discussed above. Table 1 illustrate the short-term sensitivity of rotation rate and acceleration measurements. The short-term sensitivity for rotation rate measurement is $2.73\times10^{-5}\,(\text{rad/s})/\sqrt{\text{Hz}}$, when the initial optical phase $2\phi_0 = 0$. This highlights a significant portion of common-mode noise which is effectively canceled during the differential calculation while extracting the rotation rate. The ordinary short-term sensitivity across measurements suggests that a fraction of the differential noise is canceled by the sum calculation. In our case, the short-term sensitivity performance of the acceleration measurement is constrained by the temperature drift of the PZT and polarization of the Raman laser and state-preparation laser fibers.

4. Conclusion

In this study, we propose a method to modulate the initial phase of Raman pulses, revealing its impact on the inertial measurement with the closed-loop dual-atom-interferometer inertial sensing. Short-term sensitivities for both rotation rate and acceleration measurements at various initial optical phases were assessed. The results demonstrate heightened short-term sensitivity in rotation rate measurement at $2\phi_0 = 0$, highlighting certain noise within the apparatus. Our closed-loop dual-atom interferometer allows for decoupled measurements of rotation rate and acceleration, with short-term sensitivities of $2.73 \times 10^{-5}\,(\mathrm{rad/s})/\sqrt{\mathrm{Hz}}$ and $272\,\mu\mathrm{g}/\sqrt{\mathrm{Hz}}$, respectively, achieved at a short interrogation time of $2T = 1.73\,\mathrm{ms}$. Future efforts should focus on fast initial optical phase modulation using EOM and real-time closed-loop compensation for both acceleration and rotation phases. Additionally, attention should be given to addressing temperature drift in the interference signal to enhance long-term stability.

Acknowledgments
This work was supported by the National Natural Science Foundation of China (Grant No.61473166).

References

[1] Kasevich M and Chu S 1991 *Physical Review Letters* **67** 181–184

[2] Riehle F, Kisters Th, Witte A, Helmcke J and Bordé Ch J 1991 *Physical Review Letters* **67** 177–180

[3] Peters A, Chung K Y and Chu S 2001 *Metrologia* **38** 25–61

[4] Durfee D S, Shaham Y K and Kasevich M A 2006 *Physical Review Letters* **97** 240801

[5] Geiger R, Landragin A, Merlet S and Pereira Dos Santos F 2020 *AVS Quantum Science* **2** 024702

[6] McGuinness H J, Rakholia A V and Biedermann G W 2012 *Applied Physics Letters* **100** 011106

[7] Lautier J, Volodimer L, Hardin T, Merlet S, Lours M, Pereira Dos Santos F and Landragin A 2014 *Applied Physics Letters* **105** 144102

[8] Gustavson T L, Landragin A and Kasevich M A 2000 *Classical and Quantum Gravity* **17** 2385–2398

[9] Savoie D, Altorio M, Fang B, Sidorenkov L A, Geiger R and Landragin A 2018 *Science Advances* **4** eaau7948

[10] Niebauer T M, Sasagawa G S, Faller J E, Hilt R and Klopping F 1995 *Metrologia* **32** 159–180

[11] Bidel Y, Carraz O, Charrière R, Cadoret M, Zahzam N and Bresson A 2013 *Applied Physics Letters* **102** 144107

[12] Gustavson T L, Bouyer P and Kasevich M A 1997 *Physical Review Letters* **78** 2046–2049

[13] Canuel B, Leduc F, Holleville D, Gauguet A, Fils J, Virdis A, Clairon A, Dimarcq N, Bordé Ch J, Landragin A and Bouyer P 2006 *Physical Review Letters* **97** 010402

[14] Templier S, Cheiney P, d'Armagnac De Castanet Q, Gouraud B, Porte H, Napolitano F, Bouyer P, Battelier B and Barrett B 2022 *Science Advances* **8** eadd3854

[15] Kovachy T, Asenbaum P, Overstreet C, Donnelly C A, Dickerson S M, Sugarbaker A, Hogan J M and Kasevich M A 2015 *Nature* **528** 530–533

[16] Asenbaum P, Overstreet C, Kovachy T, Brown D D, Hogan J M and Kasevich M A 2017 *Physical Review Letters* **118** 183602

[17] Graham P W, Hogan J M, Kasevich M A and Rajendran S 2013 *Physical Review Letters* **110** 171102

[18] Arvanitaki A, Graham P W, Hogan J M, Rajendran S and Van Tilburg K 2018 *Physical Review D* **97** 075020

[19] Meng Z X, Yan P Q, Wang S Z, Li X J and Feng Y Y 2022 Closed-loop dual-atom-interferometer inertial sensor with continuous cold atomic beams (*Preprint* 2210.15346)

[20] Wang S, Meng Z, , Yan P, Liu Y and Feng Y 2023 A continuous cold rubidium atomic beam with enhanced flux and tunable velocity (*Preprint* 2312.12027)

Journal of Physics: Conference Series **2889** (2024) 012058 doi:10.1088/1742-6596/2889/1/012058

Dual-species atomic interferometric sensor for simultaneous inertial measurement and clock operation

Peiqiang Yan, Weichen Jia, Shengzhe Wang, Yanying Feng

State Key Laboratory of Precision Measurement Technology and Instruments, Beijing, China
Department of Precision Instrument, Tsinghua University, Beijing, China

E-mail: yyfeng@tsinghua.edu.cn

Abstract. The atomic interferometer, serving as a precise measurement tool for multiple physical quantities, holds the potential to establish a fully quantum autonomous "position, navigation, and timing (PNT)" unit. This paper introduces a method for simultaneous inertial measurement and clock operation based on atom interferometry with dual-isotope atomic beams. The corresponding related experimental progress is also reported. A ^{87}Rb Raman-Mach-Zehnder interferometer is presented with a fringe contrast of $C = 0.05$, providing a foundational sample for further experiments.

1. Introduction

Position, navigation and timing (PNT) technologies play an essential role in various applications, such as aviation, agriculture, transportation, and scientific surveying [1]. As a crucial component of quantum sensing technology, atom interferometry is considered a key solution for autonomously providing PNT information. Following early experiments demonstrating beam or fountain clocks [2–5] and inertial sensing [6–8], atom interferometry has showcased high performance in clock [9–16], gravimeters [17–23], gravity gradiometers [24–26], accelerometers [27–29], and gyroscopes [30–34].

Light-pulse atom interferometers (LPAIs) represent a general form for precision measurements, employing pulsed light beams or microwaves to manipulate matter-wave packets with laser-cooled atomic clouds or beams. Utilizaing laser wavelength to establish inertial measurements and leverageing the oscillations between energy levels in atoms for timekeeping, it features low long-term drift and enables absolute measurement. High-precison measurements for various physical quantities, based nearly identical systematic configuration, make it a suitable platform technology for constructing an autonomous inertial navigation and timekeeping system. The inertial sensors based on LPAI, capable of simultaneously measuring acceleration and rotation, has been presented in some studies [35–40].

In this paper, we propose a scheme for a dual-spencies atomic interferometric sensor, which integrates atomic interferometric inertial sensors and an atomic clock into a unified system. The counter-propagating atomic beams both consist of two Rubidium isotopes. Specifically, the counter-propagating ^{87}Rb beams enable Mach-Zehnder interference for measuring the rotation rate and acceleration, while the ^{85}Rb beam in one direction serves as a Ramsey clock. We conducted Mach-Zehnder interference experiment based on one side of the ^{87}Rb atomic beam and obtained the interference signal. The atomic beam interacts with three Raman beams, spatially separated and modulated independently in terms of frequency or phase. The preliminary experimental results have confirmed the system's capability to perform subsequent inertial measurements, thus establishing a foundation for further experiments.

2. Sensor design

In this section, the principle and design of the simultaneous inertial measurement and clock operation system are presented. As shown in Fig. 1, two isotopes (^{87}Rb and ^{85}Rb) form dual Mach-Zehnder interferometers and a Ramsey interferometer respectively. Due to the differences in hyperfine structures and transition frequencies between the two isotopes, their interference is independent even though they are performed simultaneously in the same chamber.

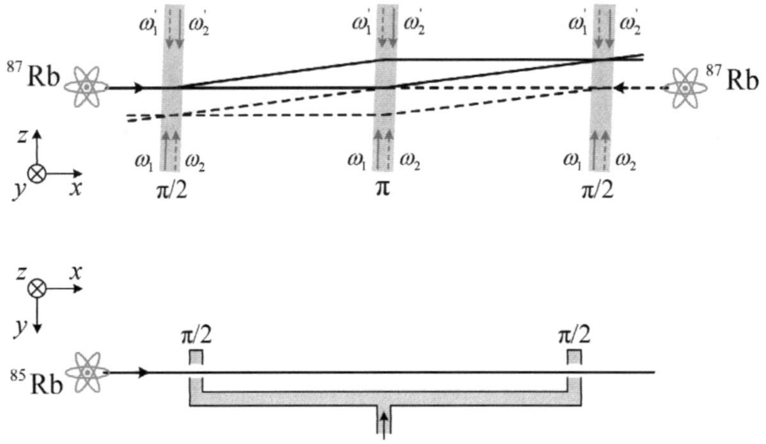

Figure 1. A schematic diagram of the interferometric configuration, including Mach-Zehnder interferometry and Ramsey interferometry in two perpendicular planes.

For inertial measurement, a pair of counter-propagating ^{87}Rb atomic beams forms dual Mach-Zehnder interferometers individually, sharing the same $\pi/2 - \pi - \pi/2$ Raman pulses. Taking the first pulse as an example, a laser comprising frequencies ω_1 and ω_2, with a small angle with the z-axis in the XOY plane, passes through the chamber and is retro-reflected by the opposite mirror as ω_1' and ω_2'. Only one combination ((ω_1, ω_2') or (ω_2, ω_1')) can drive the Raman transition due to the Doppler frequency shift applied by the small angle [32, 35, 41]. The interferometric phases from dual interferometers ϕ_1 and ϕ_2 can be extracted, both containing the rotation-induced phase shift ϕ_Ω and the acceleration-induced phase shift ϕ_a. By calculating the average and the half-difference of ϕ_1 and ϕ_2, the rotation and acceleration can be decoupled through $\phi_\Omega = -(\phi_1 + \phi_2)/2$ and $\phi_a = (\phi_1 - \phi_2)/2$ respectively.

For clock operation, the ^{85}Rb atomic beam interacts with two $\pi/2$ microwave pulses which provide a classical Ramsey sepearation resonance field. Each pulse has a Rabi phase of $\pi/2$, causing the atoms equiprobably distribute between the hyperfine states $|F=3\rangle$ and $|F=4\rangle$ of ^{85}Rb. The difference between the driving microwave frequency and the transition frequency between $|F=3\rangle$ and $|F=4\rangle$ will be revealed by the transition probability of ^{85}Rb atoms, providing an error signal for locking the local oscillator to the atomic transition.

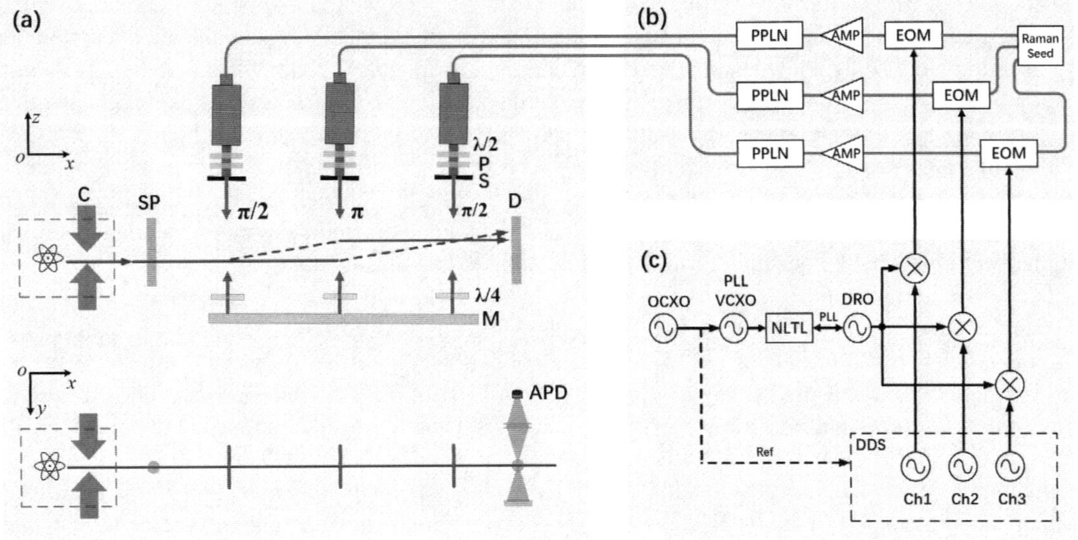

Figure 2. The schematic of a ^{87}Rb atom interferometer, illustrating the structure observed along the y-axis (above) and z-axis (below). The transversely cooled atomic beam interacts with the state preparation beam and three Raman beams subsequently, and is detected after interference. Abbreviations: C - cooling, SP - state preparation, D - detection, S - slit, P - polarizer. (b) The generation of three Raman lasers. (c) The microwave frequency synthesizer for driving the modulation of Raman lasers. It is noted that three Raman lasers are modulated individually from a Raman seed, and the frequencies and phases of modulations can be adjusted independently, providing flexible control for Raman manipulation.

3. Experimental set-up

An atom interferometer for inertial measurement is proposed, with the diagram of the interferometer illustrated in Fig. 2. For clarity in the description, we define the direction of atomic beam propagation as the x-axis and the direction of gravity as the y-axis, with the z-axis orthogonal to both.

In general, an ^{87}Rb thermal atomic beam is emitted from the atomic oven and propagates along the x-axis. Following state preparation, it interacts with three spatially separated Raman beams to undergo a Raman-Mach-Zehnder interferometer and is then detected at the end of the chamber. In the atomic oven, the Rubidium vapor is generated at a heating temperature of 155 °C, and a multi-channel array is utilized to collimate the vapor into an atomic beam. Subsequently, the atomic beam is transversely cooled by a two-dimensional optical molasses for additional collimation before entering the main chamber. A porous copper pipe with an inner diameter of 7 mm is positioned before the main chamber to absorb the diverging atoms as the beam passes through.

In the main chamber, the atoms are initially interacted with a retro-reflected linearly polarized light beam along the z-axis. This beam comprises two frequencies resonant with the $|F=2\rangle \rightarrow |F'=1\rangle$ and $|F=1\rangle \rightarrow |F'=0\rangle$ transition lines, respectively, resulting in the state preparation of atoms to the ^{87}Rb $|F=1, m_F=0\rangle$ ground state. After state preparation, a series of $\pi/2-\pi-\pi/2$ pulses interact with the atomic beam, coherently manipulating through the separation, deflection, and recombination processes successively to form a Mach-Zehnder interferometer. Each pulse consists of a pair of counter-propagating Raman beams with a frequency difference between the $|F=1\rangle$ and $|F=2\rangle$ hyperfine states. Finally, the atoms in the $|F=2\rangle$ state are induced by a retro-reflected circularly polarized probe beam propagating along the z-axis, and the fluorescence is collected and detected by an avalanche photo diode (Hamamatsu C12703-01) with a prepositive lenses group. The probe beam is resonant with the $|F=2\rangle \rightarrow |F'=3\rangle$ cycling transition to achieve as much stimulated induced fluorescence as possible. An all-fiber optical module is established to generate all lasers required above. A 1560 nm seed laser, locked by modulation transfer spectroscopy, is split into several paths. Each of them is shifted, modulated, amplified, and frequency doubled to achieve different lasers. In addition, a μ-metal magnetic shield is installed inside the main chamber, while a powered quadrupole provides a bias magnetic field of about 0.67 Gs along the z-axis.

It is worth mentioning that three Raman lasers are independently generated during the aforementioned process. Three lasers from the 1560 nm Raman seed laser are modulated by three electro-optic modulators (EOM, iXblue MPZ-LN-10). Subsequently, they are amplified and frequency doubled, ultimately resulting in three 780 nm outputs with two required frequency sidebands. A three-channel microwave frequency synthesizer for Raman EOMs is specifically built. A 6800 MHz dielectric resonator oscillator (DRO, RADITEK RDRO-A-5.0-7.9) is phase-locked to a 6800 MHz harmonic of a 100 MHz oscillator output, generated by a non-linear transmission line (NLTL). The DRO output is devided into three paths,and each is single-band mixed with one of the three programmable 35 MHz outputs from a direct digital synthesizer (DDS, AD9959), resulting in three independent 6835 MHz outputs. Both the 100 MHz oscillator and DDS are referred to an oven-controlled crystal oscillator (OCXO, Rakon HSO14). A STM32 board is utilized to drive the DDS core, and the frequency, amplitude, and phase of output waves can be programmably adjusted by modifying the control words with SPI communication. Moreover, frequency and phase scanning, as well as the acquisition of external inputs, can also be realized through the STM32. This capability allows for independent adjustments to the three Raman lasers, offering a flexible means for interference manipulation.

The Raman lasers are subsequently directed into collimators to produce three spatial-domain beams. The parallelism between the three Raman beams is strictly required with an accuracy better than 35 μrad. Moreover, they must be oriented horizontally to eliminate the effect of gravitational acceleration projected in the direction of the wave vector. An approach is proposed to adjust the collimators. Firstly, the adjustment of the middle Raman collimator (π beam in Fig. 2) is preformed. It is set to have an angle of about 0.3° to the z-axis and be parallel to the horizontal plane by coupling the reflected beam from a water surface. Then, two elongated reflecting mirrors are placed behind the opposite windows of the chamber. As the π beam strikes both mirrors, their coupling is adjusted separately, ensuring that both mirrors are perpendicular to the middle Raman collimator. Following this, the coupling of collimators on both sides ($\pi/2$ beams in Fig. 2) is individually adjusted, achieving the desired parallelism among the three collimators.

Some additional optical components are needed to perform Doppler-sensitive Mach-Zehnder interference. For each Raman beam, a set of wave plates, including $\lambda/4$ plates, $\lambda/2$ plates, and polarizers, are used to achieve orthogonal linear polarization for suppressing Doppler-free transitions. Three slits of 1mm width are added to shape each Raman beam into a narrow profile, which is helpful for enhancing signal contrast.

During the above process, the Mach-Zehnder interference of an ^{87}Rb atomic beam is realized. Both rotation rate and acceleration along the sensitive direction will be coupled in the interferometric phase. Future efforts will concentrate on employing dual atomic beam interferometers to decouple rotation and acceleration effects. Additionally, the establishment of an ^{85}Rb atomic beam-based Ramsey interferometer is planned for clock operation.

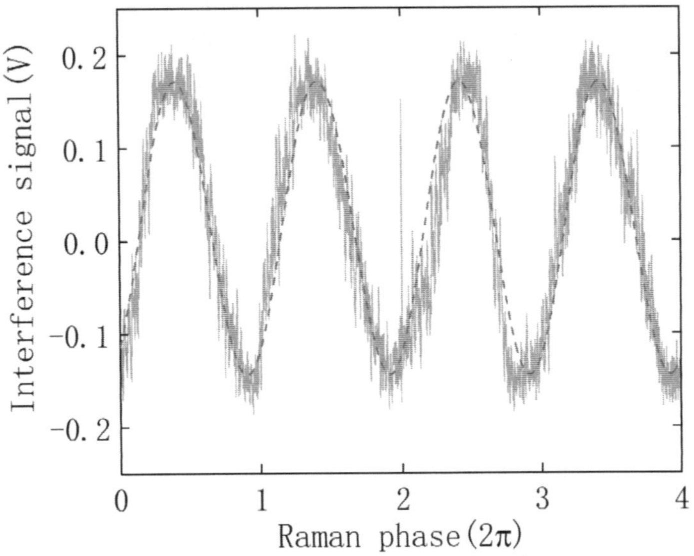

Figure 3. The Mach-Zehnder interference signal obtained by scanning the phase of the first Raman beam.

4. Results

To demonstrate the interference signals, a triangular-wave scanning is applied to the phase of the first Raman beam, resulting in the observation of fringes as depicted in Fig. 3. The noise and spurs in the fringes are primarily caused by small phase jumps occurring when rewrite the phase control words of DDS. A sine fitting analysis is performed on the signal, revealing a peak-to-peak value of approximately 0.3 V, resulting in a fringe contrast of $C = 0.05$. Subsequent experimental improvements will involve using a more advanced phase scanning source to minimize spurious noise. Moreover, it is anticipated that further narrowing of the Raman beam waist could enhance fringe amplitude, consequently improving the signal-to-noise ratio.

5. Conclusion

In this paper, we propose a scheme for establishing a full-quantum PNT unit. By employing dual atomic beams consisting of two isotopes, we integrate dual ^{87}Rb Mach-Zehnder interferometers for rotation and acceleration measurements and an ^{85}Rb Ramsey interferometer for time operation into a single vacuum chamber. This integrated setup has the capability to simultaneously obtain information on inertia and time. Additionally, we have constructed a Mach-Zehnder interferometer using the transversely cooled ^{87}Rb atomic beam. The interference fringe with a contrast of $C = 0.05$ is observed by sweeping the phase of a Raman pulse. This result provides basic technical validation for our Raman optical scheme.

Some further experiments and discussions are planned. Counter-propagating ^{87}Rb beams are going to be applied to form dual Mach-Zehnder interferometers for the decoupling of rotation and acceleration, and then the ^{85}Rb Ramsey interferometer will be established. Detection methods for separating interference signals from ^{87}Rb and ^{85}Rb will also be explored. Furthermore, a systematic test and analysis of the sensitivity and stability of inertial sensing and timing is planned to conduct in the future.

Journal of Physics: Conference Series **2889** (2024) 012058 doi:10.1088/1742-6596/2889/1/012058

References

[1] Morton Y J, van Diggelen F, Spilker Jr J J, Parkinson B W, Lo S and Gao G 2021 *Position, Navigation, and Timing Technologies in the 21st Century: Integrated Satellite Navigation, Sensor Systems, and Civil Applications, Volume 1* (John Wiley & Sons)

[2] Essen L and Parry J V 1955 *Nature* **176** 280–282

[3] Bender P L, Beaty E C and Chi A R 1958 *Physical Review Letters* **1** 311–313 ISSN 0031-9007

[4] Kasevich M A, Riis E, Chu S and DeVoe R G 1989 *Physical review letters* **63** 612

[5] Clairon A, Laurent P, Santarelli G, Ghezali S, Lea SN and Bahoura M 1995 *IEEE Transactions on instrumentation and measurement* **44** 128–131

[6] Kasevich M and Chu S 1991 *Physical Review Letters* **67** 181–184 ISSN 0031-9007

[7] Snadden M, McGuirk J, Bouyer P, Haritos K and Kasevich M 1998 *Physical Review Letters* **81** 971–974 ISSN 0031-9007, 1079-7114

[8] Gustavson T L, Bouyer P and Kasevich M A 1997 *Physical Review Letters* **78** 2046–2049 ISSN 0031-9007, 1079-7114

[9] Guéna J, Abgrall M, Clairon A and Bize S 2014 *Metrologia* **51** 108

[10] Szymaniec K, Lea S N, Gibble K, Park S E, Liu K and Głowacki P 2016 *Journal of Physics: Conference Series* **723** 012003 ISSN 1742-6588, 1742-6596

[11] Peil S, Swanson T B, Hanssen J and Taylor J 2017 *Metrologia* **54** 247–252 ISSN 0026-1394, 1681-7575

[12] Jallageas A, Devenoges L, Petersen M, Morel J, Bernier L G, Schenker D, Thomann P and Südmeyer T 2018 *Metrologia* **55** 366–385 ISSN 0026-1394, 1681-7575

[13] Weyers S, Gerginov V, Kazda M, Rahm J, Lipphardt B, Dobrev G and Gibble K 2018 *Metrologia* **55** 789

[14] Beattie S, Jian B, Alcock J, Gertsvolf M, Hendricks R, Szymaniec K and Gibble K 2020 *Metrologia* **57** 035010

[15] Weiliang C, Fang F, Kun L, Fasong Z, Shaoyang D, Yani Z and Tianchu L 2022 *Frontiers in Physics* **10** 956452 ISSN 2296-424X

[16] Hoth G W, Sherman J A, Radnaev A G, Mitchell P and Gerginov V 2023 Nist-f3, a cesium fountain frequency reference *Proceedings of the 54th Annual Precise Time and Time Interval Systems and Applications Meeting* pp 183–188

[17] Peters A, Chung K Y and Chu S 2001 *Metrologia* **38** 25–61 ISSN 0026-1394

[18] Merlet S, Bodart Q, Malossi N, Landragin A, Dos Santos F P, Gitlein O and Timmen L 2010 *Metrologia* **47** L9

[19] Bidel Y, Carraz O, Charrière R, Cadoret M, Zahzam N and Bresson A 2013 *Applied Physics Letters* **102** 144107 ISSN 0003-6951, 1077-3118

[20] Altin P A, Johnsson M T, Negnevitsky V, Dennis G R, Anderson R P, Debs J E, Szigeti S S, Hardman K S, Bennetts S, McDonald G D, Turner L D, Close J D and Robins N P 2013 *New Journal of Physics* **15** 023009 ISSN 1367-2630

[21] Hu Z K, Sun B L, Duan X C, Zhou M K, Chen L L, Zhan S, Zhang Q Z and Luo J 2013 *Physical Review A* **88** 043610 ISSN 1050-2947, 1094-1622

[22] Hauth M, Freier C, Schkolnik V, Senger A, Schmidt M and Peters A 2013 *Applied Physics B* **113** 49–55 ISSN 0946-2171, 1432-0649

[23] Wu X, Pagel Z, Malek B S, Nguyen T H, Zi F, Scheirer D S and Müller H 2019 *Science Advances* **5** eaax0800 ISSN 2375-2548

[24] McGuirk J M, Foster G T, Fixler J B, Snadden M J and Kasevich M A 2002 *Physical Review A* **65** 033608 ISSN 1050-2947, 1094-1622

[25] Sorrentino F, Bertoldi A, Bodart Q, Cacciapuoti L, De Angelis M, Lien Y H, Prevedelli M, Rosi G and Tino GM 2012 *Applied Physics Letters* **101**

[26] Biedermann G W, Wu X, Deslauriers L, Roy S, Mahadeswaraswamy C and Kasevich M A 2015 *Physical Review A* **91** 033629 ISSN 1050-2947, 1094-1622

[27] McGuinness H J, Rakholia A V and Biedermann G W 2012 *Applied Physics Letters* **100** 011106 ISSN 0003-6951, 1077-3118

[28] Lautier J, Volodimer L, Hardin T, Merlet S, Lours M, Pereira Dos Santos F and Landragin A 2014 *Applied Physics Letters* **105**

[29] Cheiney P, Fouché L, Templier S, Napolitano F, Battelier B, Bouyer P and Barrett B 2018 *Physical Review Applied* **10** 034030

[30] Gustavson T L, Landragin A and Kasevich M A 2000 *Classical and Quantum Gravity* **17** 2385–2398 ISSN 0264-9381, 1361-6382

[31] Durfee D, Shaham Y and Kasevich M 2006 *Physical review letters* **97** 240801

[32] Tackmann G, Berg P, Abend S, Schubert C, Ertmer W and Rasel E M 2014 *Comptes Rendus Physique* **15** 884–897 ISSN 16310705

[33] Berg P, Abend S, Tackmann G, Schubert C, Giese E, Schleich W P, Narducci F A, Ertmer W and Rasel E M 2015 *Physical Review Letters* **114** 063002 ISSN 0031-9007, 1079-7114

[34] Savoie D, Altorio M, Fang B, Sidorenkov L A, Geiger R and Landragin A 2018 *SCIENCE ADVANCES*

[35] Canuel B, Leduc F, Holleville D, Gauguet A, Fils J, Virdis A, Clairon A, Dimarcq N, Bordé Ch J, Landragin A and Bouyer P 2006 *Physical Review Letters* **97** 010402 ISSN 0031-9007, 1079-7114

[36] Takase K 2008 *Precision Rotation Rate Measurements with a Mobile Atom Interferometer* Ph.D. thesis Stanford University Stanford

[37] Dickerson S M, Hogan J M, Sugarbaker A, Johnson D M and Kasevich M A 2013 *Physical review letters* **111** 083001

[38] Rakholia A V, McGuinness H J and Biedermann G W 2014 *Physical Review Applied* **2** 054012 ISSN 2331-7019

[39] Wu X, Zi F, Dudley J, Bilotta R J, Canoza P and Müller H 2017 *Optica* **4** 1545 ISSN 2334-2536

[40] Gersemann M, Gebbe M, Abend S, Schubert C and Rasel E M 2020 *The European Physical Journal D* **74** 203 ISSN 1434-6060, 1434-6079

[41] Kwolek JM and Black AT 2022 *Physical Review Applied* **17** 024061

Journal of Physics: Conference Series **2889** (2024) 012059 doi:10.1088/1742-6596/2889/1/012059

Development of a laser stabilized on an ultra-stable silicon cryogenic Fabry-Perot cavity for dark matter detection

Yara Hariri[1], Jacques Millo[1], Clément Lacroûte[2], Joannès Barbarat[2], Yann Kersalé[1] and Jonathan Gillot[1]

[1]ENSMM, CNRS, FEMTO-ST, 26 rue de l'Epitaphe, 25000 Besançon, France

[2]Université de Franche-Comté, CNRS, FEMTO-ST, 26 rue de l'Epitaphe, 25000 Besançon, France

E-mail: jonathan.gillot@femto-st.fr

October 2023

Abstract. Astrophysical observations suggest the existence of an unknown kind of matter in the Universe, in the frame of the ΛCDM model. The research field of dark matter covers an energy scale going from massive objects to ultra-light scalar fields, which are the focus of the present work. It is supposed that ultra-light scalar fields affect the length of objects, whereas the speed of light stays unchanged. It follows that Fabry-Perot cavities are ideal tools for ultra-light dark matter detection since the fluctuations in the length of a cavity can be detected on the frequency of the laser stabilized to it. At FEMTO-ST, we have set up an ultra-stable silicon cavity suitable for a test of detection of ultra-light dark matter in an energy range close to 10^{-10} eV. Our 14 cm cavity is composed of two mirrors optically bonded to an ultra-rigid spacer, with each element made in single-crystal of silicon, and cooled at 17 K in order to cancel the first order thermal expansion coefficient of the silicon spacer. The projected fractional frequency stability of the laser is 3×10^{-17}, mainly limited by the thermal noise of the amorphous dielectric reflective coatings. To reach this remarkable stability, several effects have to be reduced below the thermal noise limit. While the contribution of the residual amplitude modulation is now acceptable, we are currently implementing a laser power lock with residual fluctuations lower than 3 nW and a piezoelectric-based servo loop to actively reduce the vibration noise that has to be inferior to -110 dB(m s^{-2})^2/Hz at 1 Hz. Here, we present both the status of the development of our ultra-stable laser and the mechanical response of the cavity in the presence of ultra-light dark matter.

1. Introduction

It has been over 30 years since ultra-stable lasers based on Fabry-Perot cavities have been proposed and developed [1]. With the increasing refinement of both material choices, spacer geometry, and overall design, lasers with fractional frequency instabilities of 10^{-16} and below have been developed [2], [3].

The advantages of using crystalline materials at cryogenic temperatures for both the spacer and the mirror substrates were pointed out pretty early on [4]; the combination of low mechanical loss materials, low-temperature operation, and, in some cases, the cancellation of the linear thermal expansion coefficient, ensure both a very low thermal noise and a low long-term drift. With the advent of closed-cycle cryostats, the continuous operation of such cavities was an additional advantage identified by several national metrology institutes [5], [6].

The extremely low thermal noise-induced fractional frequency instability flicker floor of such resonators is indeed unrivaled, making them the preferred choice as local oscillators for optical atomic clocks [7], [8]. A 21 cm long silicon cavity cooled to 124 K recently achieved the remarkable fractional frequency stability of $\sigma_y(\tau) = 4 \times 10^{-17}$ at 1 s [9]. On longer time scales, ultra-stable cryogenic cavities now exhibit very small frequency drifts over time [10], with a frequency drift of 3×10^{-19}/s in [3]. Additionally, ultra-stable Fabry-Perot resonators are particularly suitable for fundamental physics tests, such as ultralight dark matter detection, gravitational waves detection [11], and research on Lorentz invariance violation [12]. In particular, for ultralight dark matter detection, thanks to the interaction between the oscillating scalar field and several fundamental constants that induce fluctuations in the length of the cavity, the oscillating scalar field can then be detected by measuring the frequency fluctuations of the laser-stabilized on the cavity.

In this article, we present the ongoing work in the development of an ultra-stable single-crystal silicon cavity. We start by presenting the experimental setup and discussing the limiting effects on the stability of the cavity and the servo loops implemented to reduce them. We then discuss the mechanical response of the cavity in the presence of an ultra-light dark matter field, emphasizing the amplification of this effect due to the high mechanical quality factor of silicon when compared with ULE glass or fused silica.

2. Ultra-stable cryogenic Fabry-Perot cavity

The heart of the setup is a single-crystal silicon cavity, made of a horizontal tapered spacer and two optically contacted mirrors with single-crystal silicon substrates and amorphous dielectric coatings. The optical axis is 140 mm long, and we have measured a finesse of 78000. We attribute this rather low finesse to the surface quality of the mirror substrates. At 17 K, the predicted thermal noise of such a cavity is estimated to be at 3×10^{-17}.

To reach this level, several factors limiting the frequency stability of the laser should be compensated or stabilized. The most notable sources of noise in the optical cavity are vibrations, temperature fluctuations, and optical power fluctuations. Other instabilities originate from the residual amplitude modulation (RAM), the Doppler effect, and detection noise.

2.1. Cryogenerator

With a custom-made closed-cycle cryostat, the cavity is cooled down to ~17 K [13] to null out the first-order coefficient of thermal expansion. The cryostat that houses the Fabry-Perot cavity was designed by taking into account three constraints: a good isolation from the cryo-cooler vibrations, a range of operating temperatures from 4 K to 25 K, and a good temperature stability at the silicon temperature sensitivity turnover point at 17 K. To do so, the experimental chamber which holds the cavity inside of it, is separated from the cooling chambers, where the cryo-coolants arrive as seen in Figure 1. The whole setup enclosing the cavity is placed inside a vacuum chamber, to limit the fluctuations of the index of refraction which in return induce fluctuations of the cavity length.

Cooling is ensured by a Cryomech PT410 pulse tube and two-stage cooling, with a first stage at ~45 K, cooling a first thermal shield, and a final stage is a little below 4 K. Thermal transfer between the two chambers is performed by an OFHC copper guide. It is connected to the 4 K cooling stage using loose copper braids that provide both good thermal conductance and vibration isolation.

In the experimental chamber, the baseplate is supported by three rigid tripods. In this way, the experimental baseplate is "mechanically grounded" to the optical breadboard, which itself rests on active vibration-isolation posts. The mechanical contact between the cryocooler head which is the main source of vibrations and the experimental chamber is done only by the loose copper braids.

The experiment baseplate supports a 4 K thermal shield, itself housed inside a 45 K thermal shield. A final thermal shield is regulated with the cavity inside at its thermal expansion turnover point. To do so, four heating elements (100 Ω, 10 W resistors) associated with four temperature monitors NTC10K, are attached around a cylindrical support made of gold-coated OFHC copper, and are in contact with this final shield.

The temperature fluctuations are of 1.4×10^{-4} K at 1 s integration time, which corresponds to induced frequency fluctuation of $\sigma_y(\tau) = 3 \times 10^{-19}$ at 1 s integration time [13]. Therefore the contribution of the temperature fluctuations on the frequency stability is well below the thermal noise limit of 3×10^{-17}, which is essentially due to the thermal noise of the reflective coatings [14].

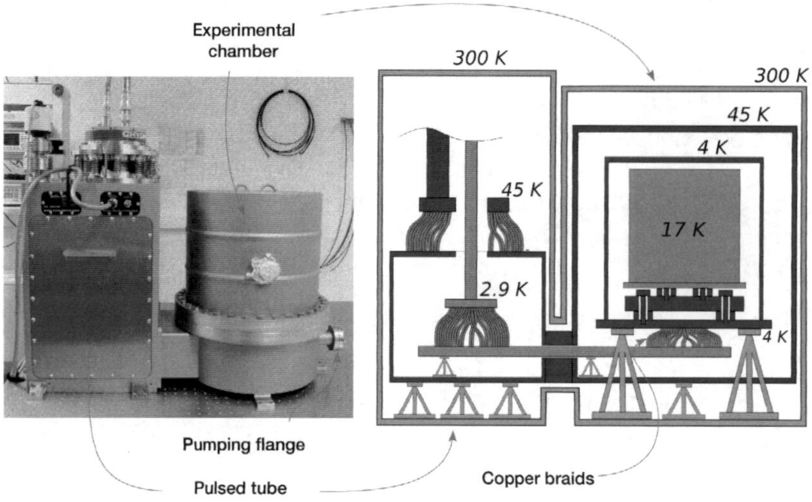

Figure 1: Setup of the two-part cooling system with a pulsed tube cryogenerator.

2.2. Optical setup

The whole experimental scheme is shown in Fig.2. The laser source emits light at 1542 nm in free space towards a 40 MHz acousto-optical modulator (AOM), serving the dual purpose of rapid frequency corrections and compensating for power fluctuations in the laser. After that, the laser beam undergoes phase modulation through an electro-optical modulator (EOM) used for the Pound-Drever-Hall (PDH) [15] frequency stabilization method, resulting in the generation of two sidebands. The modulated light is directed to a 110 MHz AOM, which will be later used to compensate for the Doppler effect before being directed toward the cavity. We lock the laser to the TEM_{00} resonance mode of the cavity using the PDH method. The cavity reflection is detected by a fast photodiode as seen in the scheme of Fig. 2. The demodulated signal is then the PDH error signal, which is fed to a numerical control loop filter that acts on a piezo-transducer for the long-term corrections, and on the 40 MHz AOM for the short-term corrections.

A drawback of this scheme is the residual amplitude modulation (RAM), which significantly affects the frequency stability of the laser by introducing an uncontrolled offset on the error signal. This effect arises primarily from the polarization mismatch between the polarization plane of the incident light and the extraordinary axis of the EOM crystal [16] and from stray etalon effects [17]. The suppression of the RAM is done by combining the stabilization of the EOM temperature near cancellation of the RAM and an active servo loop acting on the EOM DC input as shown in Figure 2 [18].

The temperature of the EOM is controlled within ±20 mK using a Peltier device. The RAM signal acquired from the in-loop photodiode (PD1) at 22.9 MHz is digitally processed. To derive the digital error signal, precise phase adjustments are made during demodulation. After being processed by a PI controller, the output correction signal is then applied to the DC port of the EOM to control its voltage. The bandwidth of the RAM servo loop is close to 8 kHz. With this technique, the RAM contribution to the fractional frequency instability is at a level of 5×10^{-19}, significantly below the thermal noise limit of the cavity at 3×10^{-17}.

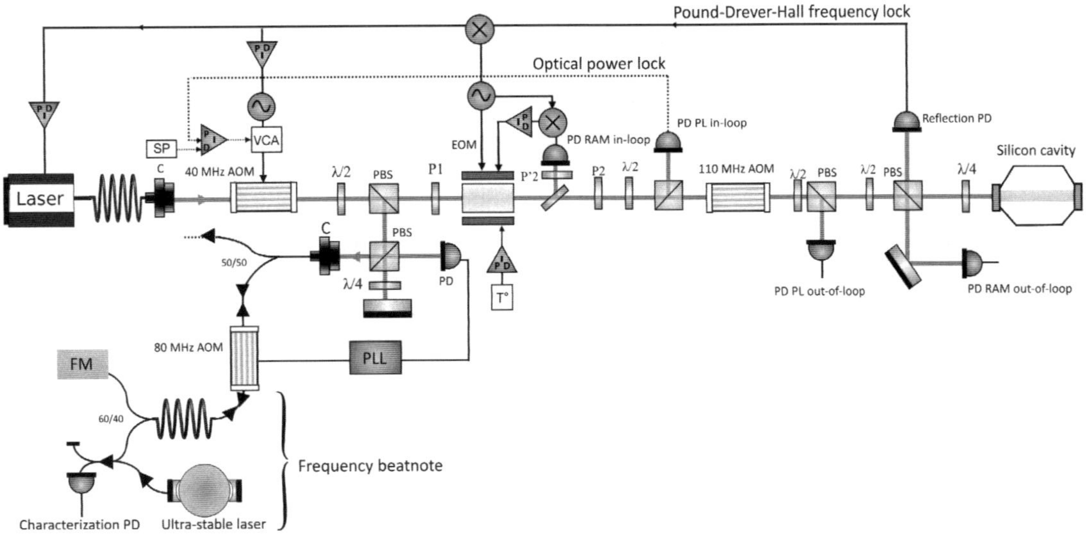

Figure 2: Set-up for stabilizing the frequency of the laser to the Fabry-Perot cavity. EOM: electro-optical modulator, AOM: acousto-optic modulator, $\lambda/2$: half-wave plate, $\lambda/4$: quarter-wave plate, PBS: polarizing beam splitter, FM: Faraday mirror, C: collimator, P: polarizer, PLL: phase-locked loop, PD: photodiode, PL: power lock, SP: setpoint, VCA: voltage-controlled attenuator.

Furthermore, the fluctuations of the laser power injected in the cavity result in temperature fluctuations of the mirrors, hence inducing length fluctuations of the cavity. In addition, fluctuations of the laser power induce temperature effects in the EOM crystal leading to an increase in the RAM effect.

In order to control these laser power fluctuations, the signal acquired from the photodiode PD PL in-loop shown in Figure 2 is compared to an adjustable voltage (SP) issued from a voltage reference, obtaining thus an error signal. The error signal is fed to a proportional-integrator (PI) controller, that delivers a correction signal to a voltage-controlled attenuator (VCA), which regulates the radio-frequency (RF) power sent to the 40 MHz AOM, thereby stabilizing the optical power. To assess the performance of the power lock in the frequency domain, we performed an analysis of the noise power spectral density (PSD) using a Fast Fourier transform analyzer. The results of both in-loop and out-of-loop measurements are presented in Figure 3. The system maintains a relative power noise (RPN) below -100 dB/Hz until 10^5 Hz, for both in-loop and out-of-loop. As seen in Figure 3, the locked regime noise level for both in-loop and out-of-loop are limited by the photodiodes noise.

To estimate the slow variations of the optical power lock in the time domain, we perform an Allan deviation measurement. The results are presented in Figure 4. The measured fractional laser-power instability from 1 s to ~100 s on the in-loop PD is around 10^{-6} whereas for the out-of-loop PD is ~ 10^{-4}. To determine if the level of the power lock is enough to limit the frequency instability caused by fluctuations of the optical power inside the cavity, it is necessary to measure the sensitivity of the laser frequency on the optical power. Preliminary measurement of this coefficient was done by modulating the optical power using the 40 MHz AOM in the setup shown in Figure 2 using a VCA and monitoring the fractional frequency of the beatnote. The sensitivity coefficient is approximately 2 Hz/μW. The optical power contribution should be below the thermal noise limit of the cavity σ_y^{th}=3×10^{-17}, which implies the necessity of stabilizing the optical power at the level of a few nanowatts. Therefore, with the power lock, the contribution of the optical power fluctuations to the fractional frequency stability of the cavity is $\sigma_y^p \approx 10^{-18}$ well below the thermal noise of the cavity.

However, the modulation of the optical power was done with an AOM positioned upstream of the EOM, and the power fluctuations within the EOM increased the RAM [18]. Further measurements on

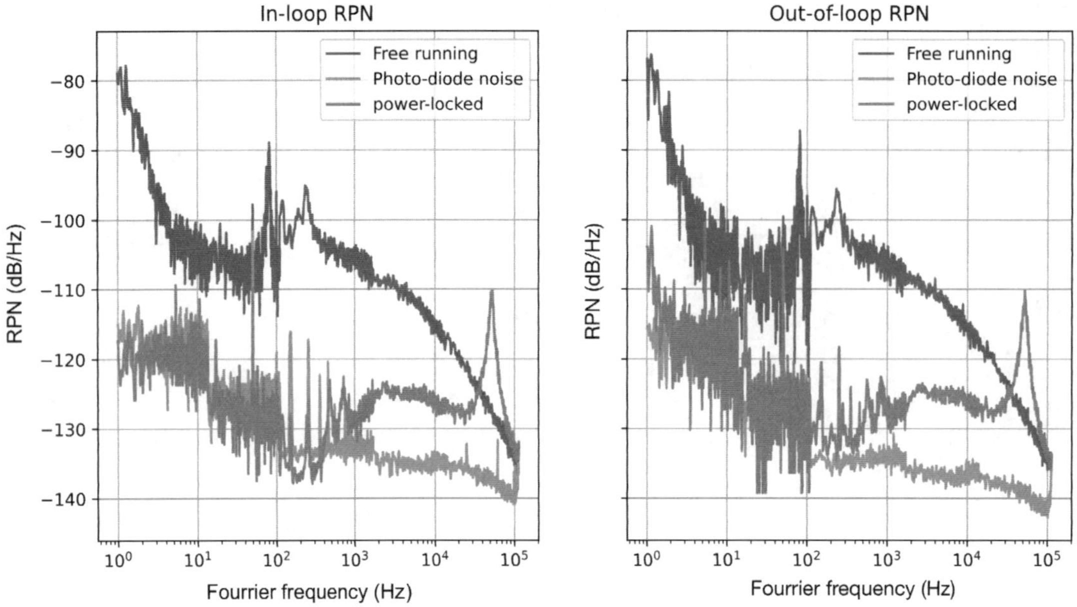

Figure 3: In-loop and out-of-loop RPN respectively.

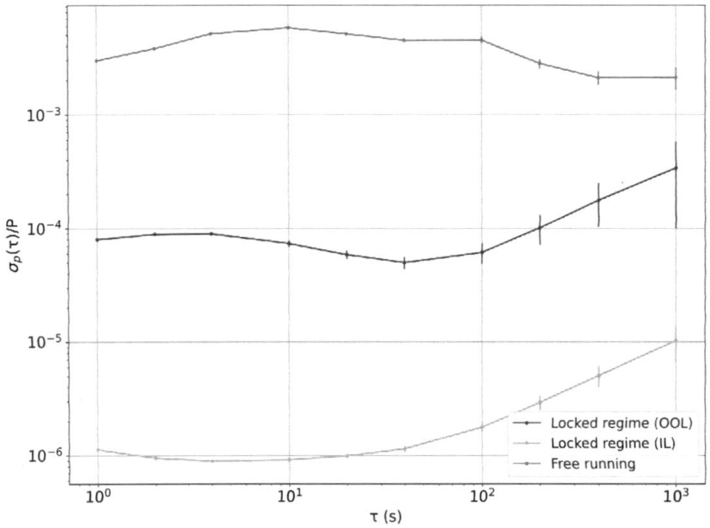

Figure 4: Relative laser-power instability. Red: free running regime, orange: in-loop locked regime, blue: out-of-loop locked regime.

this sensitivity coefficient will be conducted, excluding the influence of the RAM, by stabilizing the optical power and applying the modulation on another 110 MHz AOM placed downstream of the EOM.

2.3. Vibration sensitivity

One of the major constraints on the ultra-stable laser frequency stability is the environmental vibration noise from acoustic and seismic origins. These vibrations lead to elastic deformations of the cavity that modulate the spacer length. Vibrations-induced frequency noise on the laser locked to the resonance frequency of the cavity is modeled as:

$$S_\nu(f) = \nu_0{}^2 \sum_{i \in \{x,y,z\}} |k_i(f)|^2 \, S_{a_i}(f), \tag{1}$$

where $S_{a_i}(f)$ represents the acceleration noise PSD in each direction x, y, z with f the Fourier frequency and ν_0 the frequency of the laser. Therefore, the total frequency noise due to accelerations of the cavity is the sum of the contributions along each direction weighted by the modulus squared of the complex coefficients of sensitivity to vibration $k_i(f)$ of the cavity.

Experimental evolution of these coefficients is often obtained by measuring a transfer function by applying vibrations to the optical table supporting the cavity within its vacuum system and recording the induced frequency noise or fluctuations. So, the coefficients measured include the response of the holding support of the cavity and the vacuum chamber that shows mechanical resonances. In a first approximation, we can consider the coefficients of sensitivity to vibrations constant and independent of f below the first resonances assumed to be higher than 10 Hz.

Vibrations rising from the ground are attenuated by an active vibration isolation platform that supports an optical table with the optical setup and the cryostat. Vibrations from the cryostat are minimized by (i) the limited mechanical coupling between the pulse tube, responsible for a large source of vibrations, and the vacuum chamber and (ii) attaching the pulse tube head, which stands outside the cryostat, to a robust and rigid frame surrounding the entire setup. A seismometer with a bandwidth from 17 mHz to 100 Hz placed on the optical table measures the acceleration noise PSD shown in Figure 5 when the cryocooler is turned on. The noise floor is similar for each direction around 1 Hz with a level of about -130 dB(m s^{-2})^2/Hz. The noise strongly increases in the 10 Hz to 100 Hz decade due to the high level of acoustic perturbations. At very low frequencies, about 0.1 Hz, a difference of 40 dB is observed between the vertical z-axis and the horizontal y-axis parallel to the optical axis of the cavity. The difference is attributed to the mass distribution on the optical table that has been empirically optimized to get the lowest global vibration noise. At ~1.4 Hz and multiple frequencies, we also observe narrow lines corresponding to the harmonics of the vibration induced by the pulse tube. The acceleration PSD value in each direction at the pulsed tube frequency is between -63 and -71 dB(m s^{-2})^2/Hz.

To estimate the performance of the silicon cavity-stabilized-laser, a beatnote between the laser locked to the silicon cavity and another laser locked to a spherical ultra-low-expansion glass (ULE) cavity is detected with a fast photodiode. This laser locked to the ULE cavity achieves a frequency stability of $\sim 2 \times 10^{-15}$ at 1 s integration time [14]. The frequency noise of the stabilized laser shows a peak at ~ 1.4 Hz with a strength of ~ 4 dB Hz^2/Hz (see Figure 5).

Using the vibrations noise measured and presented in the previous paragraph we can estimate an upper bound of each $|k_i|$ coefficient value assuming that they have the same magnitude and contribute with the same phase. The estimated value is $\sim 3.4 \times 10^{-11}$/(m s^{-2}), higher by a factor ~ 7 than the sensitivity determined with finite elements modeling but this sensitivity coefficient has not been experimentally optimized. To achieve a frequency instability limited by the thermal noise of this 17 K cavity, a more accurate measurement and an optimization of these coefficients are required. The level of vibrations from the cryostat at ~ 1.4 Hz and its harmonics, has also to be strongly reduced with feedforward techniques or with an active compensation using high sensitivity seismometer.

2.4. Preliminary fractional frequency stability measurements

To determine the frequency drift of the silicon cavity, we first compensate for the drift of the ULE cavity by steering the phase of the optical signal to the phase of an ultra-stable RF reference (active hydrogen maser). The phase comparison is performed in the RF domain using an optical frequency comb as a

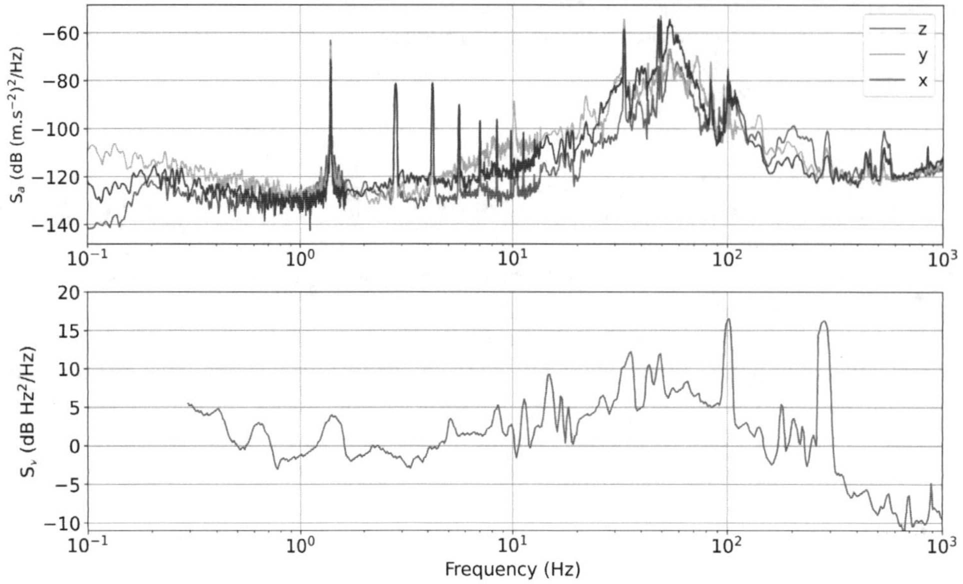

Figure 5: Up: vibration PSD measurements along the 3 directions using a seismometer placed on the optical table next to the cavity when the cryostat is turned on. z-axis is the vertical direction, y-axis is the horizontal axis along the optical axis of the cavity and x-axis is the other horizontal axis, orthogonal to the cavity axis. Down: frequency noise of the cryogenic silicon cavity stabilized laser beating with an ultra-stable laser.

divider for the optical signal [19]. We recorded the beatnote frequency shown in Figure 6 for about 3 days. Large frequency changes of $\sim 1.5 \times 10^{-13}$ to $\sim 2 \times 10^{-13}$ between the de-drifted ULE stabilized laser and the silicon stabilized laser in half a day are observed at the beginning and the end of the dataset (respectively section (1) and (3) on Figure 6). These large variations are attributed to unstable environmental conditions and an excessive sensitivity of the experimental setup. During the two days corresponding to section (2), no large frequency changes or trend is observed. We attribute this behaviour to a lower instability of the environmental parameters.

The fractional frequency drift extracted from this set of data is ranging from $\sim 7 \times 10^{-20}$/s to 4.4×10^{-18}/s depending on the considered sections. The fractional frequency stability corresponding to the full data set is reported in red on Figure 7 and is between 10^{-14} and 2×10^{-14} from few seconds integration time to $\sim 10^4$ s. No frequency drift is visible on that measurement but an upper bound of about 10^{-18}/s is compatible with this set of data and with the possible drift rate extracted from the data shown in Figure 6. This upper bound of 10^{-18}/s is compatible with the state-of-the-art cryogenic silicon cavities exhibiting fractional frequency drift of about 10^{-19}/s [20, 21].

The short term fractional frequency instability of the beatnote signal is 4×10^{-15} at 0.1 s (blue curve on Figure 7). The blue curve is computed using another measurement in order to estimate the current best performance fractional frequency stability. The progress made thus far has resulted in a frequency stability of the laser locked on the silicon cavity of 4×10^{-15} at 0.1 s integration time through the implementation of the RAM and temperature lock. The power lock at the time of these stability measurements was still not operational. However, there remain various limiting factors that degrade the stability of the laser which include vibrations, Doppler effect, and detection noise as mentioned

 IOP Publishing
doi:10.1088/1742-6596/2889/1/012059

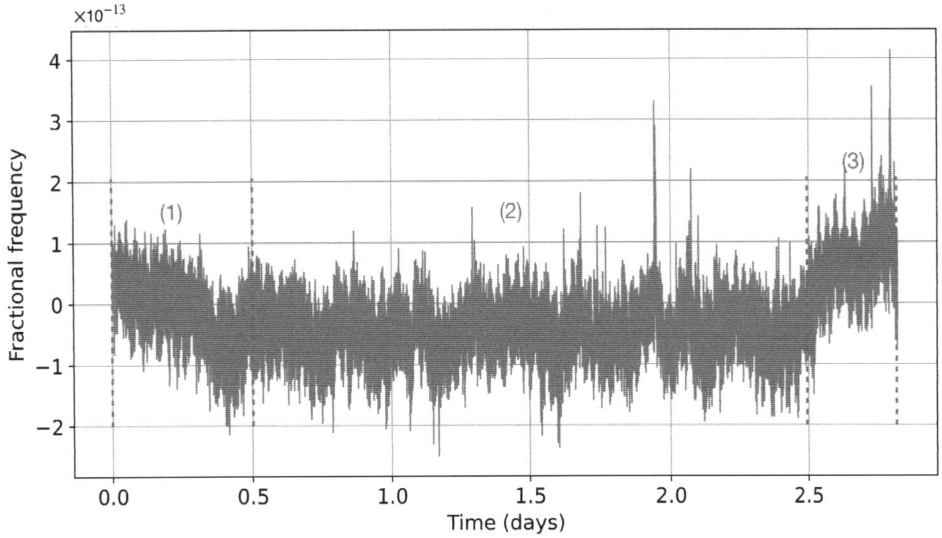

Figure 6: Fractional beatnote frequency between the laser locked to the silicon cavity and the laser locked to the spherical cavity over 3 days. The dataset is divided into 3 sections which will be useful to estimate the drift of the cavity.

previously.

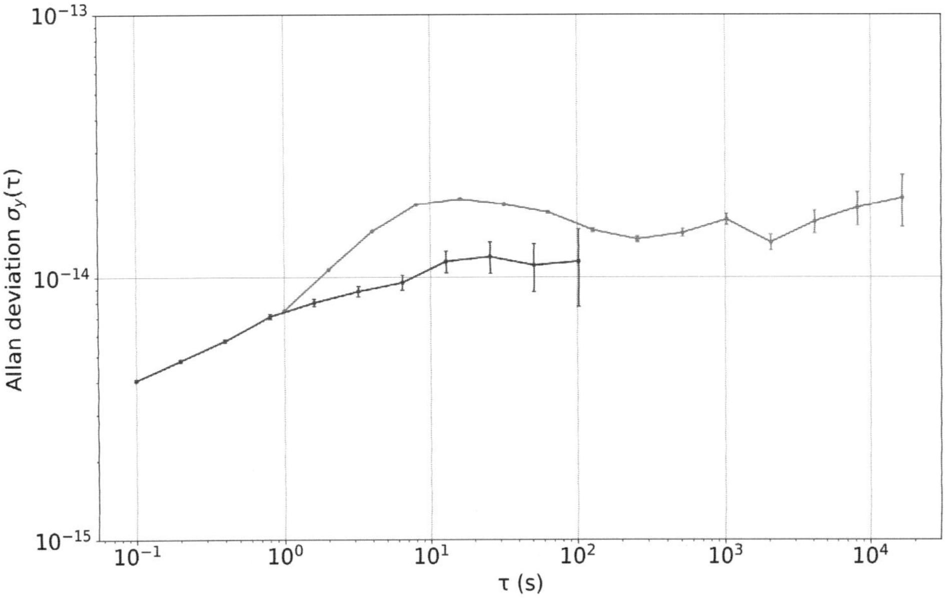

Figure 7: Allan deviation of the beatnote signal between the laser stabilized to the silicon cavity and an ULE reference laser. The red curve represents the fractional frequency instability of the signal shown in Figure 6. The blue curve is the best instability measured for the same beatnote signal.

3. Dark matter detection

Dark matter is a central question in fundamental physics and the scientific community has still not succeeded in revealing its origins and composition. This substance leaves indirect traces of its existence, and through gravitational interaction [22, 23, 24] its concentration in the Galaxy is estimated at $\rho = 0.4$ GeV/cm^3 [25], but it has never been directly detected up to now [26]. Since the constituents of dark matter can have a mass between 10^{-33} GeV and 10^{48} GeV [27], there is a plethora of models and of experiments to try to determine its nature and properties [26].

Among all the scenarios, the study of dark matter has focused a lot on the direct detection of massive constituents between 10 and 1000 GeV, WIMPs (weakly interacting massive particles), without success [27]. Another possible candidate, the ultra-light dark matter, is now the subject of increased research particularly since the discovery of the Higgs boson at the Large Hadron Collider [28, 29, 30, 31]. Since this discovery showed the existence of a scalar field to explain the mechanism giving mass to the baryonic particles, the search for dark matter in the form of a scalar field is experiencing renewed activity. Indeed, in this model, dark matter is represented in the form of an oscillating scalar field, and this oscillation is imprinted on the values of the fundamental constants of nature [32, 33, 34, 35]. Thanks to the continuous improvement of performances in time-frequency metrology, attempts to directly detect dark matter with atomic clocks or ultra-stable cavities are increasing [32, 33, 36, 37, 38, 39, 40, 41, 42, 43].

3.1. Interaction between the ultrastable cavity and a scalar field

In this theory, a scalar field ϕ of mass m_ϕ is linearly coupled to the Lagrangian of the Standard Model, with $m_\phi \ll 1$ eV. An oscillation of the scalar field will produce a variation in the value of several fundamental constants, namely the fine structure constant α, the constant Λ_3 linked to quantum chromodynamics (QCD) and the masses of the electron, of the up quark and the down quark [34]. To translate the variation of a fundamental constant C as a function of oscillations of the scalar field, we write it as $C(\phi) = C_0(1 + d_C\varphi)$ with C_0 the value of the non-perturbed constant and d_C the intensity of the coupling with the scalar field ϕ.

The current state of knowledge in astrophysics allows us to consider that dark matter is stable, self-gravitating and due to the velocity distribution in the galactic halo, this scalar field is a sum of stochastic scalar fields, giving a characteristic signature for its possible detection [44]. For simplicity, we here approximate this field in the classical form $\phi = \phi_0\cos(\omega_m t)$ where $\omega_m = m_\phi c^2/\hbar$ is the associated angular frequency at the Compton wavelength [35] and ϕ_0 is the amplitude related to the local dark matter density. In practice, the scalar field theoretically presents a spatial oscillation, but it is neglected due to the small size of the experiment [38, 34]. Thus, we expect to observe a variation of the aforementioned physical constants at this frequency, inducing a variation of the Bohr radius $a_0 = \hbar/(\alpha c m_e)$. This effect will produce a change in the frequency of clocks and variations in the length of objects. It is not possible to measure the variation in length of an object with a material tool, like a ruler, since it would be affected by this variation. On the other hand, the speed of light is not modified by the presence of dark matter. From this, and because dark matter is not sensitive to electromagnetism, photonics is the perfect tool to seek to detect these variations, for instance through ultra-stable cavities.

The cavity, made of two mirrors optically bonded to a rigid spacer, and the scalar field can be represented by a damped harmonic oscillator and a driving force [32, 45], with D the displacement of the mirrors, Q the mechanical quality factor of the resonator and ω_n the frequency of the oscillator. The right term of the equation (2) is the driving force due to the scalar field, with L_0 the length of the resonator unaffected by the scalar field and $\epsilon_L = d_e + d_{m_e}$ the coupling parameter between the scalar field and the fine structure constant and the electron mass, that the experiment aims to determine.

$$\ddot{D}(t) + \frac{\omega_n}{Q}\dot{D}(t) + \omega_n^2 D(t) = -\epsilon_L L_0 \omega_m^2 \cos(\omega_m t). \tag{2}$$

By resolving this equation, the length of the cavity is written [44] :

$$L(t) = L_0[1 - \epsilon_L((1+\alpha)\cos(\omega_m t) + \beta\sin(\omega_m t))], \tag{3}$$

and the coefficients α and β are :

$$\begin{aligned} \alpha &= \sum_{i=1}^{\infty} \frac{8}{n^2\pi^2} \frac{Q_n^2\omega_m^2(\omega_n^2 - \omega_m^2)}{Q_n^2(\omega_n^2 - \omega_m^2)^2 + \omega_m^2\omega_n^2} \\ \beta &= \sum_{i=1}^{\infty} \frac{8}{n^2\pi^2} \frac{Q_n\omega_n\omega_m^3}{Q_n^2(\omega_n^2 - \omega_m^2)^2 + \omega_m^2\omega_n^2}, \end{aligned} \tag{4}$$

with $Q_n = Q/n$ the quality factor of the mechanical resonator at the n-th harmonic [46].
From the length fluctuations in the equation (3), the laser frequency oscillations can be estimated to :

$$\frac{\delta\omega(t)}{\omega_0} = \epsilon_L[\varepsilon_c(1+\alpha)\cos(\omega_m t) + \varepsilon_s\beta\sin(\omega_m t)]. \tag{5}$$

The coefficients ε_c and ε_s depend on the optical properties of the cavity and are very close to 1 for our typical finesse and the scalar field frequency of interest (f_m > 10 kHz) [46]. The mechanical resonance frequencies are given by $f_n = c_s/2L$ depending on the size and composition of the spacer, with c_s the speed of sound in the spacer. The fundamental mode is generally a few tens of kHz.

3.2. *Effect of the scalar field on our silicon cavity*

To estimate the speed of sound in the monocrystalline silicon spacer, one can use the relation :

$$c_s = \sqrt{\frac{E}{\rho}} \tag{6}$$

where E is Young's modulus and ρ is volume density. The volume density of crystalline silicon is ρ = 2330 kg//m^3, but Young's modulus depends on the axis of the crystal. The spacers of our cavities are cut along the [111] axis which has the largest Young's modulus, i.e. E = 185 GPa. The longitudinal sound speed in the spacer is of c_s = 8910 m/s, or about 70 % greater than in the ULE glass. For our 14 cm long cavity, we will have a first resonance frequency $\nu_1 \approx$ 32 kHz. The coefficient α from the equation (4) is plotted as a function of the scalar field frequency, for the 4 first odd harmonics and for the 14 cm cavity in silicon in purple continuous line (Figure 8). The coefficient β being very sharp around the resonance, it is essentially the term in (1+α) which dominates out of resonances, in the equations (3) and (5). When $f_m \ll f_n$, α and β tend to 0, the cavity does not resonate but oscillates at the frequency of the scalar field. Conversely, when $f_m \gg f_n$, α tends towards -1 as can be seen in the insert of the Figure 8 in which the first harmonic of α as a function of f_m is enlarged. We then have $l(t) = l_0$ and the cavity adopts a constant length if f_m is not close to a resonance. Finally, at resonance, the cavity gains a factor Q_n on its sensitivity. On the lower panel of the Figure 8, the coefficient β decreases in Q/n^3 with the order of the harmonics, thus limiting the sensitivity of the instrument for high orders.

In addition, the quality factor of silicon allows for a significant increase the sensitivity. With the experiment carried out at SYRTE [46], the quality factor of ULE glass is $Q_{ULE} = 6.1 \times 10^4$ [47], while with mono-crystalline silicon, the quality factor reaches $Q_{Si} = 5 \times 10^7$, or almost 1000 times greater. As shown in the insert of the Figure 9, we therefore benefit from a much more sensitive instrument by the simple fact of using mono-crystalline silicon.

Since we also dispose of a 18 cm-long silicon cavity, we could use it for dark matter detection in the future. As seen on Figure (8), using cavities of different lengths increases the peaks of sensitivity.

The number of resonances that have to be considered depends on the bandwidth of the Pound-Drever-Hall servo loop. The piezo of the laser and the rapid correction of the AOM allows a bandwidth limited

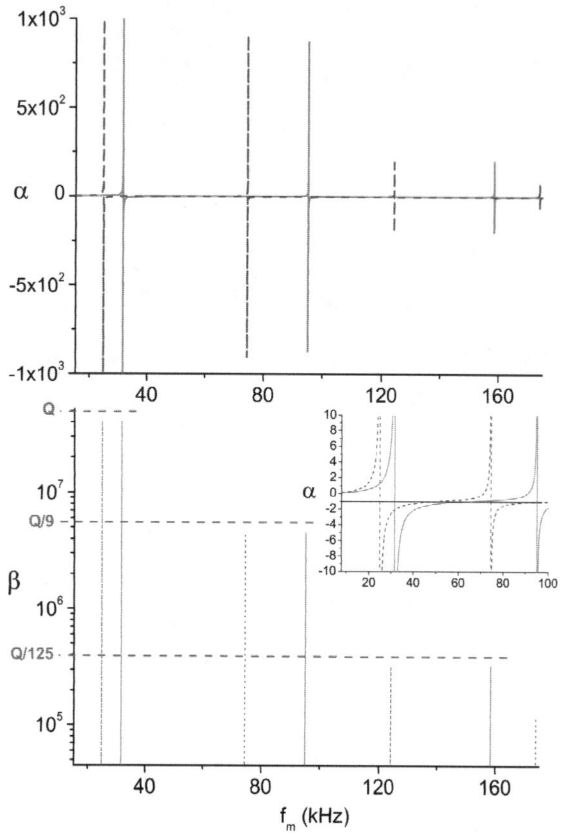

Figure 8: Plot of the α and β coefficients for our 14 cm in purple continuous line. The dashed lines are the plot of the two coefficients for an 18 cm-long silicon cavity. In the insert, zoom on the α coefficient and the first harmonic. The horizontal black line shows the value $\alpha = -1$.

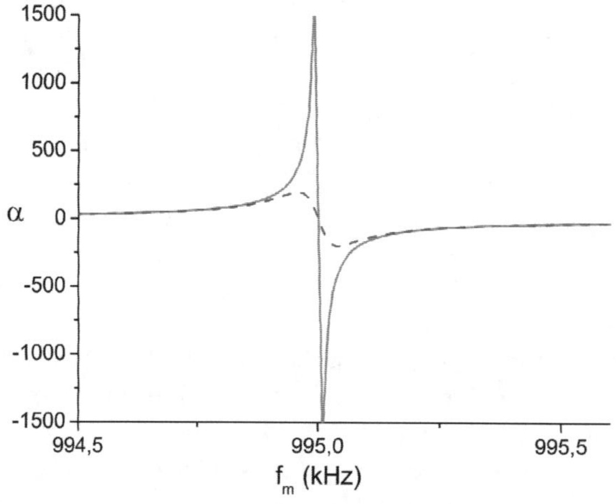

Figure 9: Plot of the α coefficient for a spacer in ULE (green dashed line) and a spacer in silicon (continuous purple line) for a resonance at the same frequency.

to about 200 kHz. However, we can significantly increase this bandwidth by using an EOM [48], which will bring it to a value between 2 MHz and 4 MHz, thus covering a much wider scalar field energy range up to $\approx 10^{-8}$ eV.

However, the ultra-stable laser alone cannot measure directly the frequency of the laser. It is necessary to do a frequency beat note between the frequency of the laser and another frequency reference to downconvert in the RF domain the frequency difference. The sensitivity of the test is different if the reference is a cavity, a maser, an atomic clock, etc [49]. Without another frequency reference, it is also possible to perform the detection test by interferometry [46]. Regardless the architecture of the ultra-light dark matter experiment we are choosing, the silicon cavity is a powerful tool to detect scalar fields or impose new constraints on the coupling parameters d_e and d_{me}.

4. Conclusion

We presented the status of the development of a 17 K cryogenic silicon ultra-stable cavity at FEMTO-ST. This is a 14-cm long single-crystal silicon cavity, made of a horizontal spacer and two optically contacted mirrors with single-crystal silicon substrates and amorphous dielectric coatings. The estimated thermal noise of such a cavity is expected to be at 3×10^{-17} and we have presented the different limitations that prevent us from reaching the state-of-the-art. While the fractional frequency instability due to the RAM is now lower than the thermal noise, we estimate that the vibrations and the optical power fluctuations are the two main remaining challenges. We already implemented a first optical power lock, stabilizing the optical power to the order of a few nanowatts. This servo loop is suitable for the RAM cancellation servo loop and is sufficient to decrease the optical power fractional frequency instability at the level of the thermal noise. Concerning the vibrations, the implementation of an active servo loop compensating them is also under investigation.

Once these limitations are overcome, we plan to set up an ultra-light dark scalar field detection test, in an energy range close to 10^{-10} eV with this ultra-stable silicon cavity. Ultra-light scalar fields are possible dark matter candidates that couple to several fundamental constants, among them the electron mass and the fine structure constant, and thus affect the length of the cavity through oscillations of the Bohr radius. With the remarkable aimed frequency stability of the laser, we hope to bring new constraints on the knowledge of the coupling parameters between fundamental constants and dark scalar fields.

Acknowledgements

This work was supported by the LABEX Cluster of Excellence FIRST-TF (ANR-10-LABX-48-01), within the Program "Investissements d'Avenir" operated by the French National Research Agency (ANR), by the EquipeX Oscillator IMP (Grant ANR 11-EQPX-0033) and EIPHI Graduate school (Grant ANR-17-EURE-0002) projects.

References

[1] Salomon C, Hils D and Hall J L 1988 *J. Opt. Soc. Am. B, JOSAB* **5** 1576–1587 ISSN 1520-8540 publisher: Optical Society of America

[2] Matei D G, Legero T, Häfner S, Grebing C, Weyrich R, Zhang W, Sonderhouse L, Robinson J M, Ye J, Riehle F and Sterr U 2017 *Phys. Rev. Lett.* **118** 263202 ISSN 0031-9007, 1079-7114

[3] Robinson J M, Oelker E, Milner W R, Zhang W, Legero T, Matei D G, Riehle F, Sterr U and Ye J 2019 *Optica* **6** 240–243 ISSN 2334-2536 publisher: Optical Society of America

[4] Sulc M 2003 *Photonics, Devices, and Systems II* ed Hrabovsky M, Senderakova D and Tomanek P (Prague, Czech Republic) p 275

[5] Zhang W, Robinson J M, Sonderhouse L, Oelker E, Benko C, Hall J L, Legero T, Matei D G, Riehle F, Sterr U and Ye J 2017 *Phys. Rev. Lett.* **119** 243601

[6] Wang W W, Chen Z A, Zhang H, Du S, Wu R, Qiao C, Kuang S and Zhang X 2023 *Frontiers in Physics* **11** 1176783 ISSN 2296-424X

[7] Nicholson T L, Campbell S L, Hutson R B, Marti G E, Bloom B J, McNally R L, Zhang W, Barrett M D, Safronova M S, Strouse G F, Tew W L and Ye J 2015 *Nature Communications* **6** 1–8 ISSN 2041-1723

[8] Brewer S, Chen J S, Hankin A, Clements E, Chou C, Wineland D, Hume D and Leibrandt D 2019 *Physical Review Letters* **123** 033201 publisher: American Physical Society

[9] Matei D, Legero T, Häfner S, Grebing C, Weyrich R, Zhang W, Sonderhouse L, Robinson J, Ye J, Riehle F and Sterr U 2017 *Physical Review Letters* **118** 263202 ISSN 0031-9007, 1079-7114
[10] Zhadnov N O, Vishnyakova G A, Kudeyarov K S, Kryuchkov D S, Khabarova K Y and Kolachevsky N N 2019 *Quantum Electronics* **49** 424 ISSN 1063-7818 publisher: IOP Publishing
[11] Drever R, Hough J, Munley A, Lee S, Spero R, Whitcomb S, Ward H, Ford G, Hereld M, Robertson N *et al.* 1981 *Laser Spectroscopy V: Proceedings of the Fifth International Conference Jasper Park Lodge, Alberta, Canada, June 29–July 3, 1981* (Springer) pp 33–40
[12] Antonini P, Okhapkin M, Göklü E and Schiller S 2005 *Phys. Rev. A* **71** 050101 publisher: American Physical Society
[13] Tetsing Talla S F 2022 *Laser stabilisé en fréquence sur une cavité Fabry-Pérot à très basse température* Theses Université Bourgogne Franche-Comté
[14] Didier A 2016 *Développement de Cavités Fabry-Perot Ultra-Stables Pour Références de Fréquence Optique de Nouvelle Génération* These de Doctorat Besançon
[15] Black E D 2001 *American Journal of Physics* **69** 79–87 ISSN 0002-9505 publisher: American Association of Physics Teachers
[16] Jin L 2021 *Optics & Laser Technology* **136** 106758 ISSN 00303992
[17] Dangpeng X, Jianjun W, Mingzhong L, Honghuan L, Rui Z, Ying D, Qinghua D, Xiaodong H, Mingzhe W, Lei D and Jun T 2010 *Opt. Express* **18** 6621–6627
[18] Gillot J, Falzon Tetsing-Talla S, Denis S, Goavec-Merou G, Millo J, Lacroûte C and Kersalé Y 2022 *Optics Express* **30** 35179 ISSN 1094-4087
[19] Gusching A, Millo J, Ryger I, Vicarini R, Abdel Hafiz M, Passilly N and Boudot R 2023 *Optics Letters* **48** 1526
[20] Robinson J M, Oelker E, Milner W R, Zhang W, Legero T, Matei D G, Riehle F, Sterr U and Ye J 2018 Crystalline optical cavity at 4 k with thermal noise limited instability and ultralow drift (*Preprint* `1812.03842`)
[21] Hagemann C, Grebing C, Lisdat C, Falke S, Legero T, Sterr U, Riehle F, Martin M J and Ye J 2014 *Optics Letters* **39** 5102 ISSN 1539-4794
[22] Zwicky F 2009 *General Relativity and Gravitation* **41** 207–224 ISSN 0001-7701, 1572-9532
[23] Rubin V C and Ford Jr W K 1970 *ApJ* **159** 379 ISSN 0004-637X, 1538-4357
[24] Rubin V C, Thonnard N and Ford Jr W K 1980 *ApJ* **238** 471 ISSN 0004-637X, 1538-4357
[25] McMillan P J 2011 *Monthly Notices of the Royal Astronomical Society* **414** 2446–2457 ISSN 0035-8711
[26] Bertone G and Tait T M P 2018 *Nature* **562** 51–56 ISSN 0028-0836, 1476-4687
[27] Safronova M S, Budker D, DeMille D, Kimball D F J, Derevianko A and Clark C W 2018 *Rev. Mod. Phys.* **90** 025008 ISSN 0034-6861, 1539-0756
[28] Higgs P W 1964 *Physical Review Letters* **13** 508–509 ISSN 0031-9007
[29] Englert F and Brout R 1964 *Physical Review Letters* **13** 321–323 ISSN 0031-9007
[30] Atlas Collaboration 2012 *Physics Letters B* **716** 1–29
[31] CMS Collaboration 2012 *Physics Letters B* **716** 30 – 61 ISSN 0370-2693
[32] Arvanitaki A, Huang J and Van Tilburg K 2015 *Physical Review D* **91** 015015 ISSN 1550-7998, 1550-2368
[33] Stadnik Y V and Flambaum V V 2015 *Phys. Rev. Lett.* **115** 201301 ISSN 0031-9007, 1079-7114
[34] Damour T and Donoghue J F 2010 *Physical Review D* **82** 084033 ISSN 1550-7998, 1550-2368
[35] Hees A, Minazzoli O, Savalle E, Stadnik Y V and Wolf P 2018 *Physical Review D* **98** 064051 ISSN 2470-0010, 2470-0029
[36] Derevianko A and Pospelov M 2014 *Nature Physics* **10** 933–936 ISSN 1745-2473, 1745-2481
[37] Van Tilburg K, Leefer N, Bougas L and Budker D 2015 *Physical Review Letters* **115** 011802 ISSN 0031-9007, 1079-7114 URL `https://link.aps.org/doi/10.1103/PhysRevLett.115.011802`
[38] Hees A, Guéna J, Abgrall M, Bize S and Wolf P 2016 *Physical Review Letters* **117** 061301 ISSN 0031-9007, 1079-7114
[39] Wcisło P, Morzyński P, Bober M, Cygan A, Lisak D, Ciuryło R and Zawada M 2017 *Nature Astronomy* **1** 0009 ISSN 2397-3366
[40] Roberts B M, Blewitt G, Dailey C, Murphy M, Pospelov M, Rollings A, Sherman J, Williams W and Derevianko A 2017 *Nature Communications* **8** 1195 ISSN 2041-1723
[41] Wcisło P, Ablewski P, Beloy K, Bilicki S, Bober M, Brown R, Fasano R, Ciuryło R, Hachisu H, Ido T, Lodewyck J, Ludlow A, McGrew W, Morzyński P, Nicolodi D, Schioppo M, Sekido M, Le Targat R, Wolf P, Zhang X, Zjawin B and Zawada M 2018 *Science Advances* **4** 4869 ISSN 2375-2548
[42] Roberts B M and Derevianko A 2021 *Universe* **7** 50 ISSN 2218-1997 arXiv: 1803.00617
[43] Wolf P, Alonso R and Blas D 2019 *Physical Review D* **99** 095019 ISSN 2470-0010, 2470-0029
[44] Savalle E 2020 *Tester la Relativité Générale avec des horloges dans l'espace, et explorer les possibilités de détection de matière noire avec des atomes froids dans l'espace et au sol* phdthesis Université Paris sciences et lettres
[45] Canuel B, Bertoldi A, Amand L, Pozzo di Borgo E, Chantrait T, Danquigny C, Dovale Álvarez M, Fang B, Freise A, Geiger R, Gillot J, Henry S, Hinderer J, Holleville D, Junca J, Lefèvre G, Merzougui M, Mielec N, Monfret T, Pelisson S, Prevedelli M, Reynaud S, Riou I, Rogister Y, Rosat S, Cormier E, Landragin A, Chaibi W, Gaffet S and Bouyer P 2018 *Sci Rep* **8** 14064 ISSN 2045-2322 publisher: Nature Publishing Group
[46] Savalle E, Hees A, Frank F, Cantin E, Pottie P E, Roberts B M, Cros L, McAllister B T and Wolf P 2021 *Phys. Rev. Lett.*

126 051301 ISSN 0031-9007, 1079-7114

[47] Millo J, Magalhães D V, Mandache C, Le Coq Y, English E M L, Westergaard P G, Lodewyck J, Bize S, Lemonde P and Santarelli G 2009 *Physical Review A* **79** 053829 ISSN 1050-2947, 1094-1622

[48] Kohlhaas R, Vanderbruggen T, Bernon S, Bertoldi A, Landragin A and Bouyer P 2012 *Optics Letters* **37** 1005–1007 ISSN 1539-4794 publisher: Optical Society of America

[49] Kennedy C J, Oelker E, Robinson J M, Bothwell T, Kedar D, Milner W R, Marti G E, Derevianko A and Ye J 2020 *Physical Review Letters* **125** 201302 ISSN 0031-9007, 1079-7114

Journal of Physics: Conference Series **2889** (2024) 012060 doi:10.1088/1742-6596/2889/1/012060

Quantum measurement limits to clock precision.

G J Milburn[1], M Kewming[2]

[1]Centre for Engineered Quantum Systems, School of Mathematics and Physics, The University of Queensland, St Lucia, Australia QLD 4072
[2]School of Physics, Trinity College Dublin, College Green, Dublin 2, Ireland

E-mail: g.milburn@uq.edu.au

Abstract. The role of projection noise, an example of quantum measurement noise, in clock design has been well understood for some time. It is based on a particular measurement protocol — Ramsey interferometry. In this paper we discuss more general quantum measurement schemes to estimate the period/frequency of a clock system based on continuous readout of a driven quantum system. The measurement needs to be chosen so as to provide a precise counting signal. We give an example based on a superconducting circuit clock.

1 Thermodynamics of clocks.

The purpose of a clock is to coordinate local coincidences, usually a recurring event and a clock count. All clocks, classical and quantum, are irreversible systems pushed away from equilibrium by access to a low entropy source of energy. This increases the free energy of the clock, $\Delta F = \Delta E - T\Delta S$, either by doing work on the clock (as in a gravity driven pendulum clock), or reducing the entropy by making measurements(discussed below)[1].

The essential components of periodic clocks are; a source of free energy, a stable oscillator, and a counter. A stable oscillation is ensured by engineering a driven, dissipative non linear oscillator to operate on a limit cycle. Small fluctuations are rapidly damped out forcing the trajectory onto the stable one-dimensional attractor of the limit cycle. Dissipation however is necessarily accompanied by fluctuations and thus all clocks are necessarily stochastic. In practice engineered artifacts will be subject to many other sources of noise. This can be mitigated by locking the limit cycle to the frequency of a naturally occurring oscillator — usually an atom — that is the same for all *local* observers. The emphasis on local is important. In order to do science we need much more than local observers. We need some way to ensure non local observers are using the same frequency standard. Fortunately relativity enables this to be achieved using optical communication schemes.

A pendulum clock illustrates these principles. Hyugens was obviously ignorant of non linear dynamics yet he was able to engineer a stable clock based on a non-linearally driven pendulum. The equations of motion for the pendulum (with unit frequency) are

$$\dot{x} = p \tag{1}$$
$$\dot{p} = -x - \Gamma p + K(x,p) + \eta(t) \tag{2}$$

where x, p are the displacement and angular momentum respectively, and we are using dimensionless units. The term $\eta(t)$ is the thermal Langevin fluctuation that necessarily accompanies viscous damping (rate Γ). The fluctuation-dissipation theorem ensures that the variance in the Langevin force is proportional to the damping rate multiplied by the local temperature. The kick function $K(x,p)$

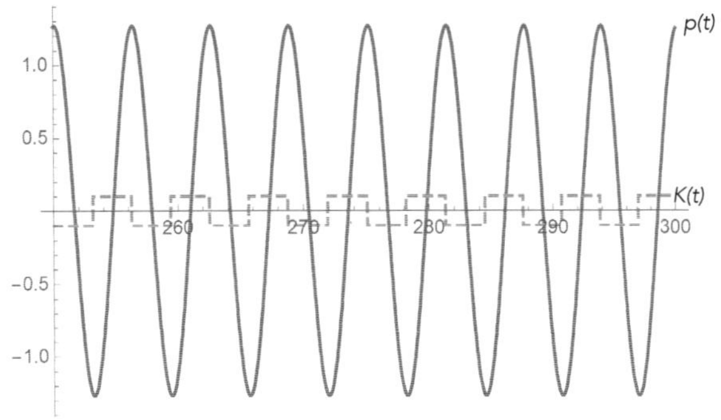

Figure 1: Angular momentum (solid) and the impulse function (dashed) versus time for the kicked pendulum with $\sin\psi_0 = 0.1$ $\Gamma = 0.1$ $\mu = 0.1$. Dimensionless units are used.

describes the nonlinear driving of the escapement mechanism in the usual over-damped limit. It is defined by

$$K(x,p) = -\mu\ \mathrm{sign}(\sin\psi_0\ x - \cos\psi_0\ p) \tag{3}$$

ψ_0 is fixed by the design of the escapement and μ has units of frequency. In Fig(1) we plot the variable $p(t)$ and the kick $K(t)$ after transients have died out. We clearly see the emergence of a stable oscillation. By counting the square-wave dynamics of the kick (i.e monitoring the escapement wheel) we have a clock.

It is a non trivial mathematical problems to determine the period of the limit cycle using centre manifold theory. It is done by parameterising the motion on the limit cycle by a phase variable, $\theta(t)$. In the absence of noise we would like this to be fixed by the parameters of the model. The period is then defined as the time take for this variable to move through 2π.

However noise causes phase diffusion and thus the time taken to move through 2π fluctuates from one cycle to the next. The general theory (phase reduction)[2] shows that the phase noise diffusion rate is slower the larger the limit cycle. For example in the normal form of a limit cycle the equations of motion near a Hopf bifurcation are

$$\dot{x} = [\mu x - \omega y - (x^2 + y^2)y]dt \tag{4}$$

$$\dot{y} = [\omega x + \mu y - (x + y^2)x]dt + \sigma dW(t) \tag{5}$$

where μ is the amplification rate. The phase variable on the limit cycle is defined by $x(t) = \sqrt{\mu}\cos(\theta(t)), \quad p(t) = \sqrt{\mu}\sin(\theta(t))$. Phase unfolding then gives

$$d\theta(t) = \omega + \frac{\sigma}{\mu}dW(t)$$

the larger the limit cycle (larger μ) the slower the phase noise. This simple result illustrates a fundamental feature of clock thermodynamics. Large limit cycles imply large heat dissipation as the work done in each cycle is dissipated as heat on each cycle. The best clocks are very dissipative machines[3]. A well know example is the laser: the harder it is driven above threshold, the smaller is the phase noise and the smaller the line-width.

We can find the statistics of the period as a first passage time problem: what is probability for time T taken for phase to change by 2π. This is given by the Wald or Inverse Gaussian distribution,

$$W(T,\alpha,\lambda) = \sqrt{\frac{\lambda}{2\pi}}T^{-3/2}\exp\left[-\frac{\lambda}{2\alpha^2 T}(T-\alpha)^2\right] \qquad t \geq 0.$$

where α, λ are positive real parameters (λ is called the spread parameter). The eman and variance are

$$\overline{T} = \alpha$$

$$\overline{\Delta T^2} = \frac{\alpha^3}{\lambda}$$

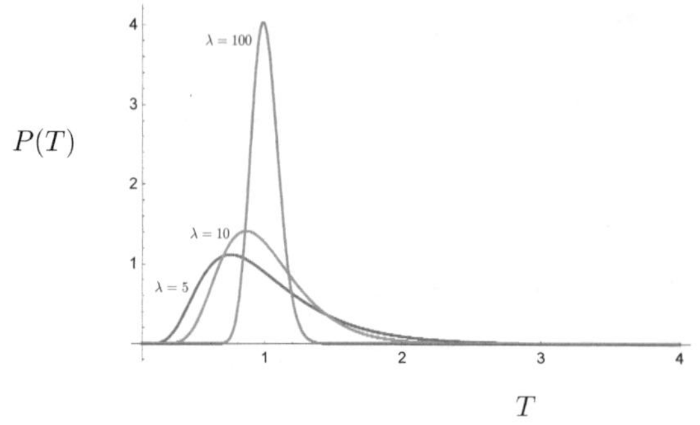

Figure 2: Mean period $\bar{T} = 1$ and three different values of the spread parameter, λ. Increasing λ is decreasing noise.

We give an example in Fig. (2). In the case of a normal form limit cycle:

$$\begin{aligned} \overline{T} &= \frac{2\pi}{\omega} \\ \overline{\Delta T^2} &= \frac{2\pi\sigma^2}{\omega^3\mu} \end{aligned}$$

fluctuations in period get smaller as the limit cycle gets bigger.

2 Quantum clocks and measurement noise.

A laser is a quantum clock, if you add a counter. It operates between a high temperature and a low temperature reservoir to maintain a population inversion. A simple model can be given in terms of a dissipative *nonlinear* oscillator[4].

Well above threshold a semi classical model is similar to the van der Pol oscillator in a rotating frame. There are two fixed points with mean field given by $\alpha_0 = 0$ and $|\alpha_0|^2 = Gn_s/\kappa$ for $G >> \kappa$. The second solution is the above threshold limit-cycle solution.

In the case of the fully quantum model we have a choice as to what we measure on the output field leaving the cavity. If we measure intensity by photon counting we see a non-oscillatory signal with a Poisson distribution of counts. Obviously we need to measure the actual field leaving the cavity to see the sable limit cycle oscillation. This may seem trivial but it reflects an important point: the parameter we seek to estimate, the period, depends on the nature of the measurement we make. Clearly we need to use heterodyne detection to get a clock signal.

The quantum theory of continuous measurement is well understood[5]. The measured homodyne current is a classical stochastic process conditioned on the mean cavity field. A full analysis is beyond the scope of this paper. However well above threshold we can use an approximate description given by Wiseman[4]. On the classical limit cycle the cavity dynamics is given by

$$\dot{\rho}_c = \Gamma\mathcal{D}[a^\dagger a]\rho + \sqrt{\eta}\mathcal{H}[a]\rho_c dW^* \tag{6}$$

the first term leaves the photon umber statistic unchanged on the limit cycle. Writing $\rho_c(t)$ as a mixture of coherent states as

$$\rho_c(t) = \int d^2\alpha P_c(\alpha, t)|\alpha\rangle\langle\alpha| \tag{7}$$

we can define $\alpha = re^{i\phi}$ for which the approximate master equation implies,

$$d\phi_c = \sqrt{\Gamma}dW \tag{8}$$

This gives the non-thermal phase diffusion rate as $\Gamma = \kappa/2\bar{n}$. This is identical to the Schalow-Townes line-width of a laser well above threshold. In this case the noise is not thermal but quantum: it originates from spontaneous emission rate κ from a cavity at zero temperature. Nonetheless a quantum thermodynamic principle holds. As the energy dissipated by cavity loss is $\propto \kappa\bar{n}$ we see that we can decrease the fluctuations in the phase by increasing $\bar{n}$ which necessarily increases the rate of energy dissipation.

Journal of Physics: Conference Series **2889** (2024) 012060 doi:10.1088/1742-6596/2889/1/012060

3 Experiment using a superconducting circuit qubit.

We now turn our focus from phase noise due to dissipation to the role of quantum measurement noise. The experiment we describe is based on a superconducting quantum circuit driven by a microwave source[6]. The particular form of the active element in the circuit is called a transmon[7]. This couples capacitively to the field in a super conducting co-planar resonator, as shown in Fig. (3) We will restrict

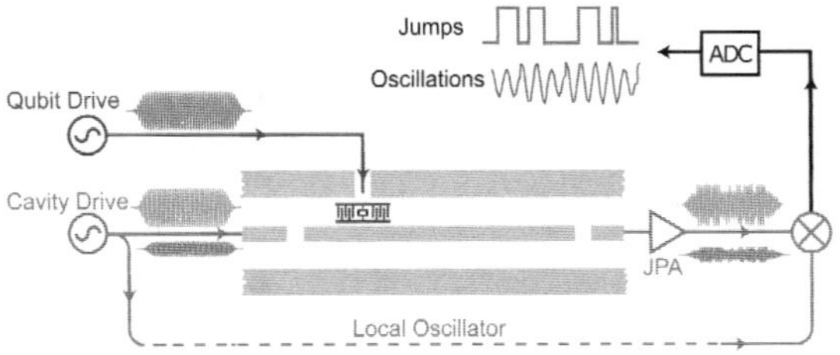

Figure 3: The experimental scheme for a transmon quantum oscillator. A transmon two level system is dispersively coupled to a superconducting microwave cavity, and the cavity and qubit are driven coherently. The output field is directed to a heterodyne detector (Josephson parametric amplifier) that produces the clock signal for counting.

the excitations of the transmon to the first two energy eigenstates — a two level system (TLS). We wish to continuously monitor the state of the qubit in the energy basis. This is done by dispersively coupling the transmon to a superconducting microwave cavity. When the cavity and qubit are driven coherently the phase of the coherent field is periodically modulated and this modulation may be monitored by subjecting the output field of the resonator to a heterodyne measurement.

Obviously using a coherent field to drive the TLS is not going to describe an autonomously driven clock. We can trace the thermodynamical origin of the driving back to the microwave source itself. Typically this is generated by the DC driving of a nonlinear circuit with a negative differential resistance (an amplifier). Amplifier thermal noise is introduced here, but at microwave frequencies in a milli-Kelvin fridge this is negligible. Our focus is on the back action noise introduced by the measurement of the transmon.

We need to make an important distinction between the unconditional dynamics of the TLS averaged over all possible measurement records and the conditional dynamics of the TLS given a history of measurement results. The first approach leads to an ensemble average while the second leads to a stochastic single-system state description. The unconditional dynamics is given by a deterministic master equation while the conditional dynamics is given by a stochastic master equation [5].

In this case the stochastic master equation is

$$\begin{aligned} d\rho_c &= -iE[a+a^\dagger, \rho_c]dt - i\Omega[\sigma_x, \rho_c]dt \\ &\quad -i\chi[a^\dagger a\sigma_z, \rho_c]dt + \gamma\mathcal{D}[\sigma_-]\rho_c + \kappa\mathcal{D}[a]\rho_c dt + \sqrt{\eta\kappa}\mathcal{H}[a]\rho_c dW(t). \end{aligned} \tag{9}$$

where χ is the dispersive coupling, γ is qubit spontaneous decay rate, κ is the cavity damping rate, and η is the homodyne detector efficiency efficiency. The final term proportional to the Wiener noise increment reflects the state updating required for conditional dynamics. We have defined the super operators $\mathcal{D}[A]$ and $\mathcal{H}[A]$ and $\mathcal{D}[A]\rho = A\rho A^\dagger - \frac{1}{2}(A^\dagger A\rho + \rho A^\dagger A)$ and $\mathcal{H}[A]\rho = A\rho + \rho A^\dagger - \mathrm{tr}(A\rho + \rho A^\dagger)\rho$. We will assume that the readout cavity is rapidly damped so that it responds quickly to the dynamics of the TLS. This means it can be adiabatically eliminated. The conditional master equation for the reduced state of the qubit (ρ_σ) becomes,

$$d\rho_\sigma = -i[H_\sigma, \rho_\sigma]dt + \gamma\mathcal{D}[\sigma_-]\rho_\sigma dt + \Gamma\mathcal{D}[\sigma_z]\rho_\sigma dt - \sqrt{\Gamma}\mathcal{H}[\sigma_z]\rho_\sigma dW(t). \tag{10}$$

Here $H_\sigma = \Omega\sigma_x + \Delta\sigma_z$ and

$$\frac{\Gamma}{\kappa} = 4\left(\frac{\chi}{\kappa}\right)^2 n_0 \quad \text{measurement dephasing rate} \tag{11}$$

$$\frac{\Delta}{\kappa} = \left(\frac{\chi}{\kappa}\right) n_0 \quad \text{effective Stark shift} \tag{12}$$

$n = |\alpha_0|^2 = 4|E|^2/\kappa$. Note that Γ can be changed by varying E.

As we are dealing with a two level system the dynamics can be described in terms of the Bloch sphere conditional dynamics.

$$\begin{aligned} dX &= -2\Delta Y dt - \gamma_2 X dt \\ dY &= 2\Delta X dt - 2\Omega Z dt - \gamma_2 Y dt \\ dZ &= 2\Omega Y dt - \gamma(1+Z)dt - 2\sqrt{\Gamma}(1-Z^2)dW \end{aligned}$$

where $X = \langle \sigma_x \rangle_c$ etc, and $\gamma_2 = \gamma/2 + 2\Gamma$ is the transverse decay rate of the conditional polarization. In the experiment the spontaneous emission rate, $\gamma << 1$, so $\gamma_2 \approx 2\Gamma$.
The unconditional dynamics is obtained by averaging over the noise. $\overline{dW} = 0$. There are two regimes, under-damped for which $\Omega > \Gamma/2$ and over-damped for which $\Omega < \Gamma/2$. The unconditional average for Z is shown in Fig. (4). The conditional dynamics corresponding to each of these regimes are shown in

Figure 4: The unconditional dynamics for $\langle \sigma_z \rangle$ for over-damped (left) and under-damped (right).

Fig. (5). In both cases, the noise is quantum not thermal. It arises from measurement back-action itself.

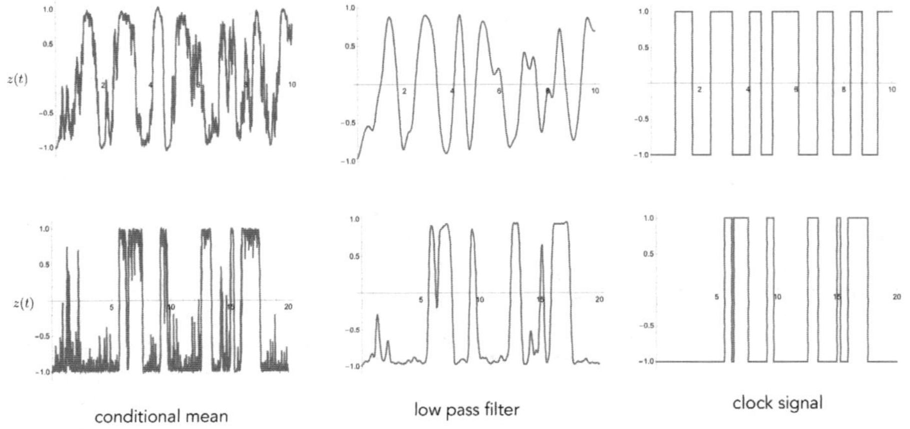

Figure 5: The conditional dynamics for $\langle \sigma_z \rangle$. Top line: under-damped, $\Omega > \Gamma/2$. Bottom line: over-damped : $\Omega < \Gamma/2$. In each row the first graph is the simulated conditional mean, the second graph is a low-pass filtered version and the third is a plot of the sign of the low-pass version that is used as a clock signal for counting. Time is in dimensionaless units.

We now define a quality measure for the period estimated from the clock signal extracted from the copnditional dynamics,

$$N_{osc} = \frac{(\mathrm{E}[T])^2}{\mathrm{Var}[T]} \tag{13}$$

We first estimate this by directly simulating the stochastic master equation. The distribution of periods in shown in Fig. (6). It is clear that decreasing measurement strength makes a better clock.
Does this distribution match a Wald distribution? To answer this we first linearise the multiplicative noise in the conditional case by using the unconditional steady state. This gives a Wald distribution

Journal of Physics: Conference Series **2889** (2024) 012060 doi:10.1088/1742-6596/2889/1/012060

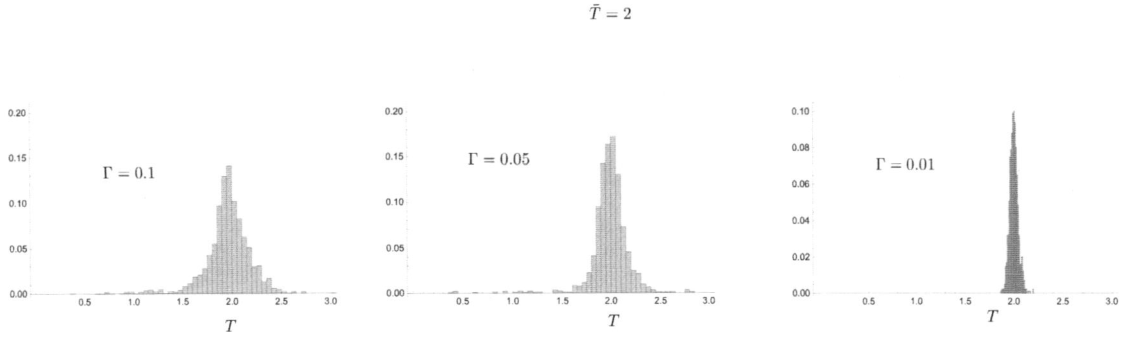

Figure 6: The probability distribution of the period extracted from the simulation of the stochastic equation.

with

$$\mathrm{E}[T] = \frac{\pi}{\Omega}, \quad \mathrm{Var}[T] = \frac{\pi\Gamma}{2\Omega^3} \tag{14}$$

Thus

$$N_{osc} = 2\pi\left(\frac{\Omega}{\Gamma}\right). \tag{15}$$

On the other hand a direct simulation gives results in Fig. (7). The results are summarised in table 1.

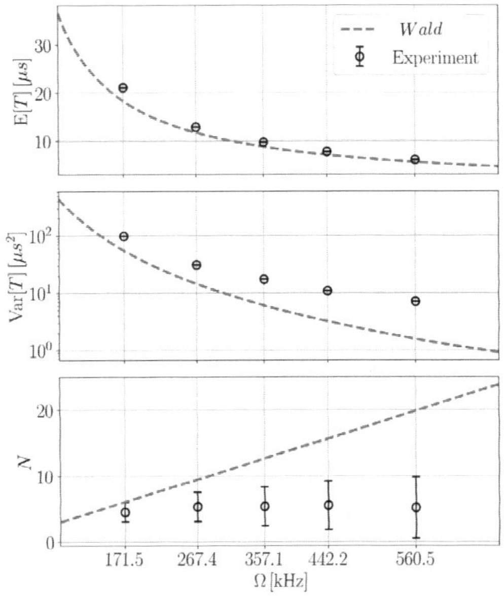

Figure 7: The clock mean period ($\mathrm{E}[T]$), variance ($\mathrm{Var}[T]$), and precision (N) as a function of qubit drive amplitude (Ω). The theoretical curves correspond to a Wald distribution. The experimental estimates of $\mathrm{E}[T]$ and $\mathrm{Var}[T]$ are computed directly from the data for each Ω. The experiment agrees with the Wald distribution only for very weak measurements. This shows that in general quantum noise due to measurement is different from the thermal phase diffusion case.

There is a better match between the Wald distribution and the Wald distribution for weak measurement.

Journal of Physics: Conference Series **2889** (2024) 012060 doi:10.1088/1742-6596/2889/1/012060

Table 1: The value of N_{osc} calculated by linearising the noise and using the Wald distribution compared to a direct numerical simulation of the stochastic Schrödinger equation is used to estimate the variance Var[T] for various values of Γ. The mean is E[T] = 2. This is compared to N_{osc} as given by fitting a Wald distribution.

Γ	numerical	Wald
0.1	56	98.7
0.05	101	197
0.01	987	1024
0.001	1861	1974

4 Quantum thermodynamic uncertainty relations.

In terms of quantum parameter estimation, the objective of a clock is to estimate the period of a system by integrating a continuous measurement record. This approach to parameter estimation was introduced by Gammelmark and Molmer [8]. In the case of period estimation this becomes a First Passage Time problem[9]. The results can be formulated in terms of a quantum kinetic uncertainty relation,

$$\frac{\mathrm{Var}[T]}{(\partial_\theta \mathrm{E}_\theta[T]|_{\theta=0})^2} \geq \frac{1}{I_Q(0)} \,. \tag{16}$$

where θ is the estimated parameter, and $I_Q(0) = I_Q(\theta = 0)$ is the quantum Fisher information (QFI). In the case considered here we get

$$I_Q(0) = \mathrm{E}[T](\overbrace{\mathcal{N}}^{classsical} + \overbrace{\mathcal{Q}}^{quantum}) \,, \tag{17}$$

where

$$\mathcal{N} = \Gamma \,, \qquad \mathcal{Q} = \frac{4\Omega^2}{\Gamma} \,, \tag{18}$$

Further details can be found in Van Vu et al. [10]. Why does the Fisher information have a quantum contribution? Quantum probabilities are given by the Born rule $P(x|\theta) = |\psi_\theta(x)|^2$ We need amplitude and phase to determine statistical distance, unlike in classical statistics where a single real number determines the probability. See Braunstein et al. [11].

In Fig. (8) we compare the experimental results with kinetic uncertainty relations (KUR)

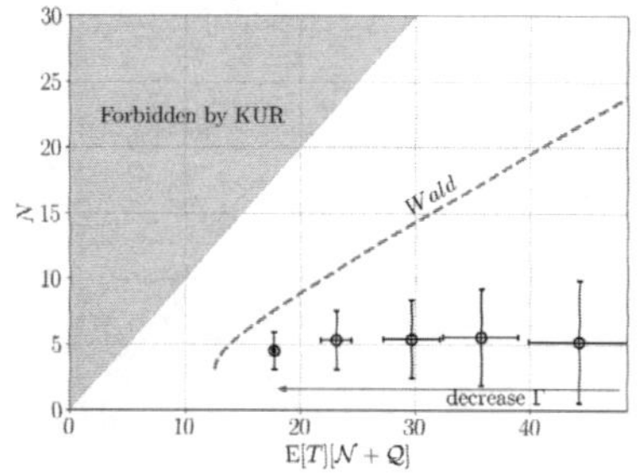

Figure 8: Precision ($N = \mathrm{E}[T]^2/\mathrm{Var}[T]$) versus the QFI. The red dashed line corresponds to the theoretical prediction from the Wald distribution, and the shaded region signifies violation of the KUR. The black circles correspond to the experimental data at each of the five values of Ω. Both theory and experiment obey the KUR.

In the over-damped regime the Wald distribution does not apply. In this case we get a non-periodic clock. Much like radiocarbon dating it is based on a pure classical Markov process.

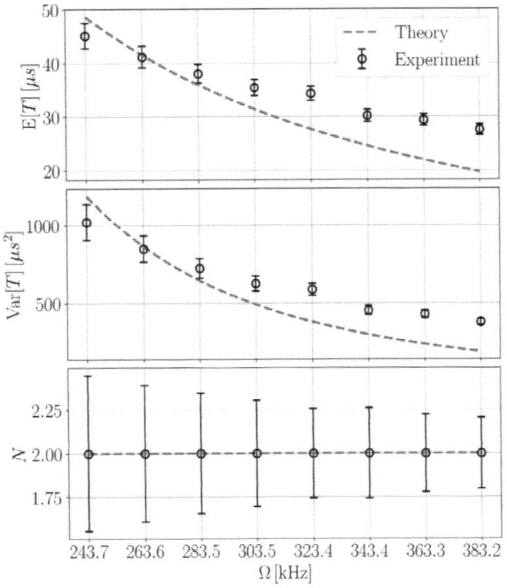

Figure 9: Jump clock period (T) statistics as a function of qubit drive amplitude (Ω). From top to bottom: mean ($\mathrm{E}[T]$), variance ($\mathrm{Var}[T]$,), and precision (N).

In the quantum jump regime we define one period to be the time from one downward transtion to the next. The Markov transition rate, μ,. can be used to calculate a FPT distribution:

$$P(T) = \mu^2 T e^{-\mu T}, \tag{19}$$

where the transition rate is $\mu = \Gamma\Omega^2/(\Gamma^2 + \Delta^2)$ Then

$$\mathrm{E}[T] = \frac{2}{\mu}, \quad \mathrm{Var}[T] = \frac{2}{\mu^2}, \quad N_{\text{jump}} = 2\,. \tag{20}$$

This is shown in Fig.(9)

5 Conclusion.

All clocks are dissipative systems pushed away from thermal equilibrium by work or measurement. At finite temperature, the fluctuation-dissipation theorem implies a limit to clock accuracy. In this paper we have considered a continuous measurement scheme (as opposed to a Ramesy scheme) to extract a clock signal. At low temperature, weak continuous measurement plays a special role in extracting a clock signal, and quantum noise (spontaneous emission, tunnelling, measurement back action) limits clock accuracy. The quantum kinetic uncertainty relations enable the quality of the clock to be predicted in terms of all quantum noise processes involved including quantum measurement noise.

Acknowledgements
We wish to thank Adil Gangat. We also acknowledge Arkady Fedorov, Eric He and Prasanna Pakkiam who did the experimental work reported here. This research project was supported by the Foundational Questions Institute Fund, a donor-advised fund of Silicon Valley Community Foundation, under Grant No. FQXi-IAF19-04. We also acknowledge support from the Australian Research Council Centre of Excellence for Engineered Quantum Systems (EQUS, CE170100009). M.J.K. acknowledges financial support from a Marie Skłodwoska-Curie Fellowship (Grant No. 101065974).

References

[1] Milburn GJ. The Thermodynamics of Clocks. Contemporary Physics. 2020;61:69.

[2] Aminzare Z, Holmes P, Srivastava V. On Phase Reduction and Time Period of Noisy Oscillators. In: 2019 IEEE 58th Conference on Decision and Control (CDC); 2019. p. 4717-22.

[3] Erker P, Mitchison MT, Silva R, Woods MP, Brunner N, Huber M. Quantum Clocks: Does Thermodynamics Limit Our Ability to Measure Time? Phys Rev X. 2017;7:31022.

[4] Wiseman HM. Stochastic quantum dynamics of a continuously monitored laser. Phys Rev A. 1993 Jun;47:5180-92. Available from: `https://link.aps.org/doi/10.1103/PhysRevA.47.5180`.

[5] Wiseman HM, Milburn GJ. Quantum Measurement and Control. Cambridge: Cambridge University Press; 2010.

[6] He X, Pakkiam P, Gangat AA, Kewming MJ, Milburn GJ, Fedorov A. Effect of Measurement Backaction on Quantum Clock Precision Studied with a Superconducting Circuit. Phys Rev Appl. 2023 Sep;20:034038. Available from: `https://link.aps.org/doi/10.1103/PhysRevApplied.20.034038`.

[7] Roth TE, Ma R, Chew WC. The Transmon Qubit for Electromagnetics Engineers: An introduction. IEEE Antennas and Propagation Magazine. 2023;65(2):8-20.

[8] Gammelmark S, Mølmer K. Fisher Information and the Quantum Cramér-Rao Sensitivity Limit of Continuous Measurements. Phys Rev Lett. 2014 Apr;112:170401. Available from: `https://link.aps.org/doi/10.1103/PhysRevLett.112.170401`.

[9] Kewming MJ, Kiely A, Campbell S, Landi GT. First Passage Times for Continuous Quantum Measurement Currents. arXiv:230807810. 2023.

[10] Van Vu T, Saito K. Thermodynamics of Precision in Markovian Open Quantum Dynamics. Phys Rev Lett. 2022 Apr;128:140602. Available from: `https://link.aps.org/doi/10.1103/PhysRevLett.128.140602`.

[11] Braunstein SL, Caves CM, Milburn GJ. Generalized Uncertainty Relations: Theory, Examples and Lorentz Invariance. Annals of Physics. 1996;247:135-73.

Journal of Physics: Conference Series **2889** (2024) 012061 doi:10.1088/1742-6596/2889/1/012061

Pure frequency-based dispersive spectroscopy

A. Cygan[1], K. Bielska[1], D. Charczun[1], G. Kowzan[1], A. Nishiyama[1,5], P. Wcisło[1], S. Wójtewicz[1], M. Zaborowski[1], P. Masłowski[1], D. Lisak[1], R. Ciuryło[1], A. J. Fleisher[2a], K. A. Gillis[2b], J. T. Hodges[2a], T. Voumard[3], T. Wilidi[3], T. Herr[3,6], V. Brasch[4]

[1]Institute of Physics, Faculty of Physics, Astronomy and Informatics, Nicolaus Copernicus University in Toruń, ul. Grudziądzka 5, 87-100 Toruń, Poland.
[2]Chemical Sciences Division[a], Sensor Science Division[b], National Institute of Standards and Technology, 100 Bureau Drive, Gaithersburg, MD 20899, USA.
[3]Deutsches Elektronen-Synchrotron DESY, Notkestr. 85, 22607 Hamburg, Germany.
[4]CSEM - Swiss Center for Electronics and Microtechnology, 2000 Neuchâtel, Switzerland.
[5]National Metrology Institute of Japan (NMIJ), National Institute of Advanced Industrial Science and Technology, 1-1-1 Umezono, Tsukuba, Ibaraki 305-8563, Japan.
[6]Physics Department, Universität Hamburg UHH, Luruper Chaussee 149, 22761 Hamburg, Germany.

E-mail: `rciurylo@umk.pl`

Abstract. Dispersion spectroscopy based on frequency measurement of the high finesse optical cavity mode shifts induced by the presence of a gas medium with absorption resonances is an attractive alternative to standard absorptive spectroscopy. The main advantage of such pure frequency-based dispersive spectroscopy technique is the unique combination of sensitivity and accuracy. High sensitivity is ensured by high finesse optical cavities providing extremely long effective optical paths. On the other hand, high accuracy is provided by frequency measurement cavity mode shift. This makes this technique free from systematic errors coming from the non-linearity of light intensity measurements, which is one of the limiting factors in typical absorptive spectroscopy. We review several realizations of dispersive spectroscopy in high finesse optical cavities from Doppler-free to broadband. These techniques are especially useful for studying weak molecular transitions applicable in testing fundamental physics, gas metrology, spectroscopic thermometry, and a new generation of spectroscopic databases.

1. Introduction

We explore the applicability of several variations of pure frequency-based spectroscopic techniques to molecular systems and their metrology. In these techniques, we use a linear phenomenon well known as mode pushing [1] in an optical cavity with an absorbing medium, exploiting the intrinsic physical connection between absorption and dispersion. It was demonstrated that mode frequency shift measurements allow the obtaining of molecular spectra with exceptional accuracy and precision.

2. Dispersive spectroscopy in high finesse optical cavity

It was shown that information about absorption resonances in optical cavities is encoded in the measured spectral shapes of cavity modes, i.e., in its width [2, 3, 4]. Use of laser tightly PDH-locked [5] to high-finesse cavity filled with absorbing gas allowed accurate scanning of narrow

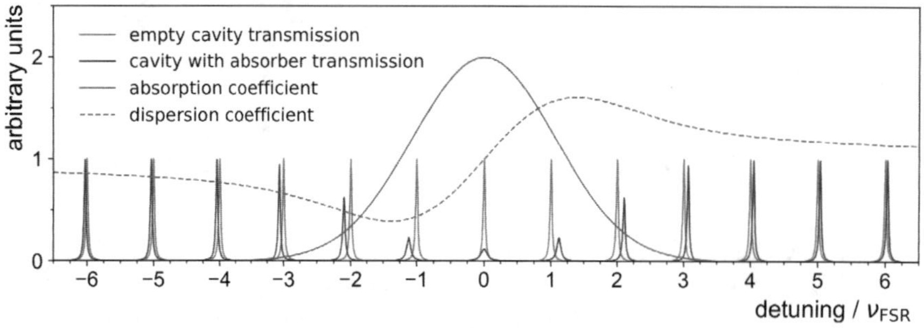

Figure 1. Optical cavity mode structure with and without absorptive and despersive gas medium. The empty cavity free specpectral range (FSR) is denoted as ν_{FSR} and the dispersive curve was shifted up for clarity.

cavity mode resonance and its frequency shift determination with uncertainties lower than 100 mHz [6] in the optical domain. It is realized by splitting a single-frequency laser light into two beams with orthogonal linear polarization, which overlaps again before entry into the optical cavity. First, a locking beam is used to tightly lock the laser to one of the cavity modes. Thanks to that, the second probe beam, which passes through a frequency shifter, can be scanned over another cavity mode unaffected by mechanical vibration, reaching sub-Hz accuracy of the relative frequency axis. The spectrum is obtained by jumping to consecutive modes and scanning through each cavity mode in several frequency steps. Determined in such a way frequency shifts of the cavity modes in the presence of absorbing gas compared to cavity mode frequencies in the case of an empty cavity versus the probe beam frequency makes a dispersion spectrum. The accuracy of determination of the cavity mode shifts should not be confused with the accuracy of frequency determination of the molecular resonance, which is typically lower even in the Doppler-free case mentioned in the next paragraph. The optical cavity length is stabilized to a separate reference laser, eliminating a slow drift of cavity mode frequencies [7]. Moreover, tuning the reference laser frequency allows the controlled shift of the comb of cavity modes and gets dispersion spectra of molecular resonances with a frequency step smaller than the cavity-free spectral range (FSR). Such an approach gave a promise for the determination of dispersion spectrum and, thanks to Kramers-Krönig relations [8] also absorption spectrum with high accuracy, see Fig. 1. This technique, called cavity mode-dispersion spectroscopy (CMDS), with its primary observable being frequency [6], is immune to bias caused by nonlinearity in detecting light intensity affecting most commonly used spectroscopic approaches. The CMDS demonstrated molecular transition intensities measurement with sub-promille relative uncertainty [9] and its traceability to primary frequency standards of the SI. Reference [10] shows how both frequency axes of dispersion spectra were linked to the primary frequency standard. Obviously, the link of frequency axes to optical atomic clocks is also possible in a similar way as it was demonstrated in the case of cavity ring-down spectroscopy (CRDS) [11]. At the same time, sub-promille relative uncertainty of line intensity measurements have been reached with calibrated CRDS [12]. It should be noted, however, that the CMDS approach is calibration-free [6, 9] while CRDS can require a sophisticated procedure to evaluate detector non-linearity and achieve equivalently high accuracy [12]. This dramatic progress in reachable accuracy of line intensity measurements enabled, for the first time, the demonstration of interlaboratory comparison of measured line intensities at sub-promille level [13]. For this study the near-infrared second overtone band (3–0) of $^{12}C^{16}O$ was chosen as a test case. Moreover, sub-promille agreement between experimental results and ab initio predictions was shown as well [13].

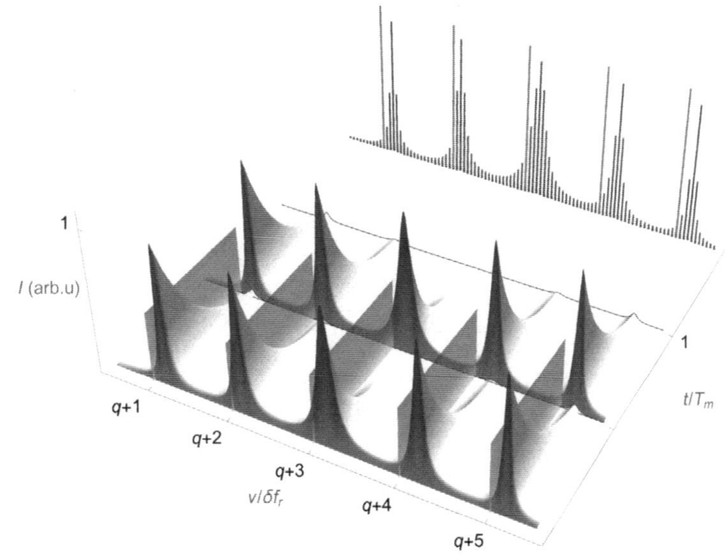

Figure 2. The time- and frequency-dependent intensity of the cavity output and LO comb beat signal is plotted in the spectral range corresponding to the down-converted radiofrequency signal. The cavity responds with ring-down decays (brown) to the alternating switching of the probe comb (red) intensity with a square-wave modulation of period T_m. Difference in repetition frequency of probe and LO comb is $\delta\nu_r$. For clarity one period of modulation is shown and the relative intensities between the probe and decay signals are not preserved. Fourier-transform power spectra of the probe comb and cavity response are shown at the top right, corresponding to the acquisition of decay signals at fixed (phase-locked) time intervals. See more details in [17].

Applicability of CMDS is not limited to Doppler broadened spectra. The CMDS was successfully used to obtain Doppler-free saturation measurements of weak molecular transitions and appeared to be superior to other techniques [14]. The accuracy of determination of molecular transition frequency was below 500 Hz, which corresponds to 3×10^{-12} relative accuracy at near-infrared (wavelength of CO transitions 1.57 μm).

The main obstacle of CMDS is the relatively long time needed to scan the probe laser through the optical mode. However, thanks to proper control of frequency and phase of probe laser exiting the optical cavity mode, it is possible to induce oscillations of light intensity [15] with a frequency corresponding to the difference between probe laser frequency and cavity mode frequency. Taking advantage of the beat of light buildup with a local oscillator (LO) precisely detuned from the cavity resonance, we demonstrated that determination of the cavity mode frequency is possible on a time scale shorter than the cavity decay time. We call this approach the cavity buildup dispersion spectroscopy (CBDS) [16].

3. Broadband dispersive spectroscopy in high finesse optical cavity

A combination of optical frequency comb (OFC) [18] as a light source and a Fourier transform spectrometer led to the realization of a broadband CMDS [19]. To get the sub-nominal resolution of Fourier transform spectra, the procedure described in [20] was used. It relies on a coincidence of optical cavity modes frequencies and nodes of instrumental sinc function of the Fourier transform spectrometer. The other way to obtain broadband high-resolution measurements of cavity modes is to use a VIPA spectrometer [21]. The broadband optical cavity mode measurements at Hz-level precision were demonstrated with a comb-based VIPA spectrometer [22], which allowed the realization of broadband CMDS as well [23].

Both detection schemes based on Fourier transform or VIPA spectrometers require mechanical moving parts, long optical paths in free space sensitive to alignment, or expensive cameras on top of the cost of the optical frequency comb. This inconvenience can be overcome by the implementation of a dual-comb detection scheme [24, 25] for CMDS [26, 23]. In one of these cases, measurements of 13 kHz-wide cavity modes with sub-nominal resolution were performed, reaching a precision of determination of their widths and position at the Hz level [23]. All broadband approaches discussed above suffer from the need for scanning through optical cavity mode. This problem could be overcome by extending the idea of CBDS [16] to the measurement of ring-down decays for many modes simultaneously, implementing a fast dual-comb detection scheme with sub-nominal resolution. Realization of these approaches is possible by beating light buildup or ringdown from many optical cavity modes excited simultaneously with a LO comb precisely detuned from the cavity resonances, see Fig. 2. We demonstrated that such an approach allows rapid measurement of the mode widths and frequencies. This technique called dual-comb cavity ring-down spectroscopy (DC-CRDS) [17], brings to comb-based CRDS, initiated in [27], the full spectral resolution offered by OFC and high-finesse cavity. In addition, DC-CRDS [17] allows us to get information about molecular resonances just from pure frequency measurement of the optical cavity mode shift which is immune to non-linearity of light intensity determination affecting absorptive techniques.

4. Conclusions

The pure frequency-based dispersive spectroscopy seems to be an attractive alternative to intensity-based measurements of cavity decay rates or light absorption, especially in studies of weak molecular transitions. Applications include reference data for a new generation of spectroscopic databases, studies of fundamental physics, gas metrology, and Doppler-width thermometry.

Acknowledgments

This work was supported by the National Science Centre, Poland, Project no. 2015/18/E/ST2/00585, 2020/39/B/ST2/00719, 2018/30/E/ST2/00864, 2019/35/B/ST2/01118, 2021/42/E/ST2/00152, and 2021/41/B/ST2/00681. The measurements were performed as a part of the research done at the National Laboratory FAMO (KL FAMO) in Toruń, Poland. KL FAMO in Toruń is supported by the subsidy of the Ministry of Education and Science.

References

[1] McFarlane R A 1964 *Phys. Rev.* **135**(3A) A543–A550

[2] Nakagawa K, Katsuda T, Shelkovnikov A, de Labachelerie M and Ohtsu M 1994 *Optics Communications* **107** 369–372 ISSN 0030-4018 URL `https://www.sciencedirect.com/science/article/pii/0030401894903492`

[3] Cygan A, Lisak D, Morzyński P, Bober M, Zawada M, Pazderski E and Ciuryło R 2013 *Opt. Express* **21** 29744–29754 URL `https://opg.optica.org/oe/abstract.cfm?URI=oe-21-24-29744`

[4] Long D A, Truong G W, van Zee R D, Plusquellic D F and Hodges J T 2014 *Applied Physics B* **114** 489–495 URL `https://doi.org/10.1007/s00340-013-5548-5`

[5] Drever R W, Hall J L, Kowalski F V, Hough J, Ford G, Munley A and Ward H 1983 *Applied Physics B* **31** 97–105 URL `https://doi.org/10.1007/BF00702605`

[6] Cygan A, Wcisło P, Wójtewicz S, Masłowski P, Hodges J T, Ciuryło R and Lisak D 2015 *Opt. Express* **23** 14472–14486 URL `https://opg.optica.org/oe/abstract.cfm?URI=oe-23-11-14472`

[7] Hodges J T, Layer H P, Miller W W and Scace G E 2004 *Review of Scientific Instruments* **75** 849–863 ISSN 0034-6748 URL `https://doi.org/10.1063/1.1666984`
[8] Peiponen K E and Saarinen J J 2009 *Reports on Progress in Physics* **72** 056401 URL `https://dx.doi.org/10.1088/0034-4885/72/5/056401`
[9] Cygan A, Wcisło P, Wójtewicz S, Kowzan G, Zaborowski M, Charczun D, Bielska K, Trawiński R S, Ciuryło R, Masłowski P and Lisak D 2019 *Opt. Express* **27** 21810–21821 URL `https://opg.optica.org/oe/abstract.cfm?URI=oe-27-15-21810`
[10] Cygan A, Wójtewicz S, Kowzan G, Zaborowski M, Wcisło P, Nawrocki J, Krehlik P, Śliwczyński L, Lipiński M, Masłowski P, Ciuryło R and Lisak D 2016 *The Journal of Chemical Physics* **144** 214202 ISSN 0021-9606 URL `https://doi.org/10.1063/1.4952651`
[11] Bielska K, Wójtewicz S, Morzyński P, Ablewski P, Cygan A, Bober M, Domysławska J, Zawada M, Ciuryło R, Masłowski P and Lisak D 2017 *Journal of Quantitative Spectroscopy and Radiative Transfer* **201** 156–160 ISSN 0022-4073 URL `https://www.sciencedirect.com/science/article/pii/S0022407317303928`
[12] Fleisher A J, Adkins E M, Reed Z D, Yi H, Long D A, Fleurbaey H M and Hodges J T 2019 *Phys. Rev. Lett.* **123**(4) 043001 URL `https://link.aps.org/doi/10.1103/PhysRevLett.123.043001`
[13] Bielska K, Kyuberis A A, Reed Z D, Li G, Cygan A, Ciuryło R, Adkins E M, Lodi L, Zobov N F, Ebert V, Lisak D, Hodges J T, Tennyson J and Polyansky O L 2022 *Phys. Rev. Lett.* **129**(4) 043002 URL `https://link.aps.org/doi/10.1103/PhysRevLett.129.043002`
[14] Bielska K, Cygan A, Konefał M, Kowzan G, Zaborowski M, Charczun D, Wójtewicz S, Wcisło P, Masłowski P, Ciuryło R and Lisak D 2021 *Opt. Express* **29** 39449–39460 URL `https://opg.optica.org/oe/abstract.cfm?URI=oe-29-24-39449`
[15] Wójtewicz S, Cygan A, Domysławska J, Bielska K, Morzyński P, Masłowski P, Ciuryło R and Lisak D 2018 *Opt. Express* **26** 5644–5654 URL `https://opg.optica.org/oe/abstract.cfm?URI=oe-26-5-5644`
[16] Cygan A, Fleisher A J, Ciuryło R, Gillis K A, Hodges J T and Lisak D 2021 *Communications Physics* **4** 14 URL `https://doi.org/10.1038/s42005-021-00517-3`
[17] Lisak D, Charczun D, Nishiyama A, Voumard T, Wildi T, Kowzan G, Brasch V, Herr T, Fleisher A J, Hodges J T *et al.* 2022 *Scientific reports* **12** 2377 URL `https://doi.org/10.1038/s41598-022-05926-0`
[18] Diddams S A, Jones D J, Ye J, Cundiff S T, Hall J L, Ranka J K, Windeler R S, Holzwarth R, Udem T and Hänsch T W 2000 *Phys. Rev. Lett.* **84**(22) 5102–5105 URL `https://link.aps.org/doi/10.1103/PhysRevLett.84.5102`
[19] Johansson A C, Rutkowski L, Filipsson A, Hausmaninger T, Zhao G, Axner O and Foltynowicz A 2018 *Opt. Express* **26** 20633–20648 URL `https://opg.optica.org/oe/abstract.cfm?URI=oe-26-16-20633`
[20] Maslowski P, Lee K F, Johansson A C, Khodabakhsh A, Kowzan G, Rutkowski L, Mills A A, Mohr C, Jiang J, Fermann M E and Foltynowicz A 2016 *Phys. Rev. A* **93**(2) 021802 URL `https://link.aps.org/doi/10.1103/PhysRevA.93.021802`
[21] Diddams S A, Hollberg L and Mbele V 2007 *Nature* **445** 627–630 URL `https://doi.org/10.1038/nature055242`
[22] Kowzan G, Charczun D, Cygan A, Trawiński R S, Lisak D and Masłowski P 2019 *Scientific Reports* **9** 8206
[23] Charczun D, Nishiyama A, Kowzan G, Cygan A, Voumard T, Wildi T, Herr T, Brasch V, Lisak D and Masłowski P 2022 *Measurement* **188** 110519 ISSN 0263-2241 URL `https://www.sciencedirect.com/science/article/pii/S0263224121013993`
[24] Schiller S 2002 *Opt. Lett.* **27** 766–768 URL `https://opg.optica.org/ol/abstract.cfm?URI=ol-27-9-766`
[25] Coddington I, Swann W C and Newbury N R 2008 *Phys. Rev. Lett.* **100**(1) 013902 URL `https://link.aps.org/doi/10.1103/PhysRevLett.100.013902`
[26] Fleisher A J, Long D A and Hodges J T 2018 *Journal of Molecular Spectroscopy* **352** 26–35 ISSN 0022-2852 URL `https://www.sciencedirect.com/science/article/pii/S0022285218301498`
[27] Thorpe M J, Moll K D, Jones R J, Safdi B and Ye J 2006 *Science* **311** 1595–1599 URL `https://www.science.org/doi/abs/10.1126/science.1123921`

Journal of Physics: Conference Series **2889** (2024) 012062 doi:10.1088/1742-6596/2889/1/012062

Time-dependent Aharonov-Bohm effect for electromagnetic and gravitational scalar potentials

RY Chiao
University of California, Merced, School of Natural Sciences, Merced, CA 95344, USA
E-mail: raymond_chiao@yahoo.com

NA Inan
Clovis Community College, 10309 N. Willow, Fresno, CA 93730 USA
University of California, Merced, School of Natural Sciences, Merced, CA 95344, USA
Department of Physics, California State University Fresno, Fresno, CA 93740-8031, USA
E-mail: ninan@ucmerced.edu

M Scheibner
University of California, Merced, School of Natural Sciences, Merced, CA 95344, USA
E-mail: mscheibner@ucmerced.edu

J Sharping
University of California, Merced, School of Natural Sciences, Merced, CA 95344, USA
E-mail: jsharping@ucmerced.edu

DA Singleton
Department of Physics, California State University Fresno, Fresno, CA 93740-8031, USA
E-mail: dougs@mail.fresnostate.edu

ME Tobar
Quantum Technologies and Dark Matter Labs, Department of Physics, University of Western Australia, Crawley, WA 6009, Australia.
E-mail: michael.tobar@uwa.edu.au

Abstract. In this article, we review a new approach to the scalar Aharonov-Bohm effect for both the electromagnetic and gravitational interaction. For both interactions a quantum system is placed in a time-dependent electromagnetic or gravitational potential, but with no force (spatial derivative of the potential) acting on the quantum system. Nevertheless, we show that the energy levels of the quantum system develop side bands which can be detected as the signature of this version of the scalar Aharonov-Bohm effect. We briefly look at the specific experimental setups required to detect the energy side bands.

1. The Time-Dependent Scalar/Electric Aharonov-Bohm effect

Prior to the seminal work by Aharonov and Bohm in reference [1] the standard view was that it was only derivatives of the potential of a particular interaction (*i.e.* the force associated with the interaction) that led to physical results, while the potentials of the interaction were a mathematical convenience. However, Aharonov and Bohm in [1] showed that the potentials associated with a given interaction had more physical reality than had previously been thought.

The best-known version of the Aharonov-Bohm (AB) effect involves the vector potential (*i.e.* $\mathbf{A}$) and magnetic field (*i.e.* $\mathbf{B} = \nabla \times \mathbf{A}$) of electromagnetism. In the idealized version of this vector/magnetic AB effect, an infinite solenoid is placed between the slits of a double-slit experiment for some quantum particle – normally electrons. The signature of this vector/magnetic AB effect is a shift of the interference pattern that depends on the magnetic flux inside the solenoid. This shift occurs despite the electrons moving entirely in a magnetic field-free region. This vector/magnetic version of the AB effect was first confirmed experimentally by Chambers [2]. A loophole-free version of the vector/magnetic AB effect using a micro-scale torus rather than an infinite solenoid was carried out in an experiment by Tonomura *et al.* [3].

There is also a scalar/electric version of the AB effect which involves the scalar potential (*i.e.* ϕ_e) and electric field (*i.e.* $\mathbf{E} = -\nabla\phi_e$). In this proceeding we focus on the scalar interaction $e\phi_e$ – the product of electric charge, e, with scalar potential, ϕ_e. *i.e.* the scalar coupling between the charge and the electric scalar potential. The original set-up for the scalar/electric AB effect is shown in Fig. 1, where a path of a charged particle is split and sent through two different, spatially separated metal tubes (Faraday cages), which are kept at different potentials while the charged particle is inside the tubes, but with the potential difference (and electric field) turned off when the charged particle is outside the tubes. When the paths of the charged particle are recombined they show a phase difference proportional to the product of the size of the potential difference and the time in the tubes. This version of the scalar/electric AB effect is harder to realize in practice due to the need to synchronize the turning off and on of the potential difference with the charged particle exiting and entering the tube. Because of this, the experimental verification of the scalar/electric AB effect is much less robust than the vector/magnetic AB effect. A well-known experimental verification of the scalar/electric AB effect was given in [4]. However, this experiment measured the effects of both the scalar electric and vector AB effect, rather than the effect of only the scalar electric AB effect. Also at some point in the experiment, the electrons did move in a region where the **E**-field was non-zero. Thus this was not a clean/pure test of the scalar/electric AB effect.

In [5] we proposed a new and cleaner version of the scalar/electric AB effect. The experimental set-up for our proposal to test the scalar electric AB effect, as given in Fig. 2, consists of a Faraday shell of radius R with a time-varying voltage. Outside R there is a time-varying $\mathbf{E}(t)$ field and time-varying scalar potential $\phi_e(t)$, while inside R there is only a time-varying potential, $\phi_e(t)$. The $\mathbf{E}$ field inside is zero. Next, we place a quantum system (*i.e.* a hydrogen-like atom like rubidium) inside this Faraday shell. In the next section we will show that this leads to the generation of additional energy levels - *i.e.* side-bands - in the quantum system. It is these energy side-bands that are the signature of this version of the scalar/electric AB effect.

It is useful to compare and contrast the two different approaches - from Fig. 1 versus Fig. 2 - to the scalar/electric AB effect. In the original set-up in Fig. 1, charged particles are spatially separated and then recombined after the potential difference has been turned on while the charges are inside the Faraday tubes, and then off when they are outside the tubes. The signature of this version of the scalar/electric AB effect is a phase difference between the two paths that the charged particles took. In the new set-up in Fig. 2, the quantum system is in one spatial location and the potential is varied with time. This exposes the quantum system to a time-varying potential but zero **E**-field. The signature of the setup in Fig. 2 is a shifting of energy levels (or more precisely the generation of energy side bands) rather than a shifting of

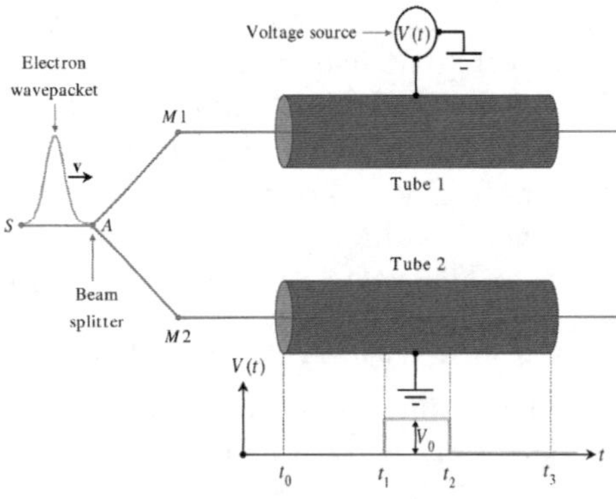

Figure 1. The originally proposed setup for the electric AB effect.

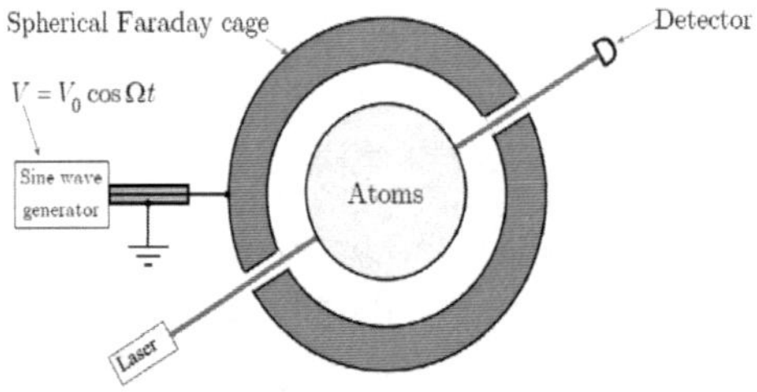

Figure 2. The basic set-up of our proposed test of the scalar electric AB effect

interference fringes. This shifting of the energy levels for the set-up in Fig. 2 is similar to what happens in the ac Stark effect [6, 7] and later we show that the mathematical analysis of the scalar/electric AB effect in Fig. 2 is almost identical to the analysis of the ac Stark effect.

2. General analysis of scalar AB effect

Here we give a general analysis for a quantum system that sees a time-varying, but spatially uniform potential. Since the potential is spatially uniform the force field associated with this potential will be zero since the force field is proportional to the spatial derivative of the potential. Thus the quantum system will see a time-varying potential but a zero force field.

To start we assume that our quantum system is described by a Hamiltonian, H_0, for which the solutions to the time-independent Schrödinger equation are known *i.e.* $H_0\Psi_i(\mathbf{x}) = E_i\Psi_i(\mathbf{x})$. We now place this quantum system in a time-varying, spatially uniform potential, $V(t)$, with a "charge" Q so that the total Hamiltonian is

$$H = H_0 + QV(t) \tag{1}$$

In section 3 we will take $V(t)$ to be the electric scalar potential, ϕ_e, and the "charge" to be the electric charge, $Q = e$, while in section 4 we will take $V(t)$ to be the gravitational scalar potential, ϕ_g, and the "charge" to be mass, $Q = m$. We take $V(t)$ to be sinusoidal of the form

$$\begin{aligned} V(t) &= 0 \quad \text{for } t < 0 \\ V(t) &= V_0 \cos\Omega t \quad \text{for } t \geq 0\,. \end{aligned} \tag{2}$$

$\frac{\Omega}{2\pi}$ is the frequency and V_0 is the amplitude. For $t < 0$ where $V(t) = 0$ the Hamiltonian is just H_0 with wave function solutions $\Psi_i(\mathbf{x})$, and energy eigenvalues E_i. The time-dependent Schrödinger equation for this new system is

$$i\hbar\frac{\partial\psi}{\partial t} = H\psi = (H_0 + QV(t))\,\psi \tag{3}$$

Hamiltonians of the form (3) with a piecewise, continuous periodic potential can be solved using Floquet's theorem. We briefly run through the solution of (3).

We begin by applying the separation-of-variables ansatz $\psi(\mathbf{x}, t) = X(\mathbf{x})T(t)$. Substituting this into (3) gives

$$i\hbar\frac{\partial\psi}{\partial t} = i\hbar X\frac{dT}{dt} = (H_0 + QV)\,XT = TH_0X + X\,(QV)\,T \tag{4}$$

Dividing by XT and moving the $QV(t)$ term to the left-hand side gives

$$-QV + i\hbar\frac{1}{T}\frac{dT}{dt} = \frac{1}{X}H_0X \tag{5}$$

Since the left-hand side of this equation only depends on t while the right-hand side only depends on $\mathbf{x}$ each side must be equal to a constant which we denote as E. This leads to the separated equations

$$-QV + i\hbar\frac{d\ln T}{dt} = E \quad \text{and} \quad H_0X = EX \tag{6}$$

The solution to the second equation in (6) is just the known solution to the time-independent Schrödinger equation namely $X = \Psi_i(\mathbf{x})$ and $E = E_i$ *i.e.* $H_0\Psi_i(\mathbf{x}) = E_i\Psi_i(\mathbf{x})$. Next we integrate the first, temporal equation in (6) over t giving

$$-Q\int V(t)dt + i\hbar\int\frac{d\ln T(t)}{dt}dt = \int E_i dt \tag{7}$$

Carrying out the integrations in (7) and solving for $T(t)$, gives

$$\begin{aligned} T(t) &= \exp\left(-\frac{i}{\hbar}E_it\right)\exp\left(-\frac{i}{\hbar}Q\int V dt\right) \\ &= \exp\left(-\frac{i}{\hbar}E_it - i\alpha\sin\Omega t\right) = \exp\left(-\frac{i}{\hbar}E_it - i\varphi(t)\right), \end{aligned} \tag{8}$$

where α is called the FM depth of modulation parameter and is defined as

$$\alpha = \frac{QV_0}{\hbar\Omega} \tag{9}$$

Multiplying $X(\mathbf{x}) = \Psi_i(\mathbf{x})$ and $T(t)$ from (8) gives the wave function, $\psi_i(\mathbf{r}, t)$, for the full Hamiltonian, $H_0 + QV(t)$ from (1) gives

$$\psi_i(\mathbf{r}, t) = \Psi_i(\mathbf{r})\exp\left(-\frac{iE_it}{\hbar} - i\varphi(t)\right). \tag{10}$$

This new wave function is the original wave function with an added AB phase factor $\exp(-i\varphi(t))$. From (8) the AB phase, $\varphi(t)$, is given by

$$\varphi(t) = \frac{Q}{\hbar}\int V(t)dt = \alpha\sin\Omega t\,. \tag{11}$$

Journal of Physics: Conference Series **2889** (2024) 012062 doi:10.1088/1742-6596/2889/1/012062

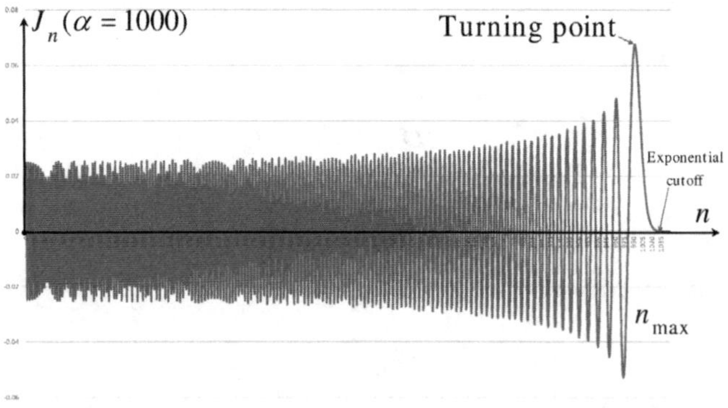

Figure 3. A plot of $J_n(\alpha)$ vs. n for the case when $\alpha = 1000$. There are side-bands in the energy $E_i^{(n)}$ which occurs up to some maximum index n given by $n_{max} \approx \alpha$. From the plot, one can see that the weighting, $J_n(\alpha)$, is largest when $n = n_{max} \approx \alpha$ and it is this state which contributes the most.

To see the emergence of the energy side-bands we exponentiate the AB phase from (11) and use the Jacobi-Anger expansion [1]

$$\exp(-i\varphi(t)) = \exp(-i\alpha \sin \Omega t) = \sum_{n=-\infty}^{\infty} (-1)^n J_n(\alpha) \exp(in\Omega t) \ . \tag{12}$$

Using (12) in (10), the wave function now reads

$$\begin{aligned} \psi_i(\mathbf{r}, t) &= \Psi_i(\mathbf{r}) \sum_{n=-\infty}^{\infty} (-1)^n J_n(\alpha) \exp(in\Omega t) \exp\left(-\frac{iE_i t}{\hbar}\right) \\ &= \Psi_i(\mathbf{r}) \sum_{n=-\infty}^{\infty} (-1)^n J_n(\alpha) \exp\left(-\frac{i\left(E_i - n\hbar\Omega\right)t}{\hbar}\right) \end{aligned} \tag{13}$$

Reading off (13) we can see that each energy level E_i is split into a series of energy side-bands, $E_i^{(n)}$, given by

$$E_i^{(n)} = E_i \pm n\hbar\Omega, \text{ with } n, \text{an integer} \tag{14}$$

Equation (14) would seem to imply that there is an infinite ladder of energy levels above and below E_i characterized by integer multiples of $\hbar\Omega$. However, this ignores the Bessel function "weighting" factor $J_n(\alpha)$ for each oscillator term from (13). If one plots this weighting factor as a function of n (see Fig. 3) one sees that beyond a maximum value of n the contributions to $\psi_i(\mathbf{r}, t)$ are exponentially suppressed. From Fig. 3 one can see that this maximum value of n is approximately equal to the value of α *i.e.*

$$n_{max} \approx \alpha \ . \tag{15}$$

Effectively one only has a finite number of energy side bands that run up to $n = n_{max} \approx \alpha$. Further from Fig. 3 one sees that the maximum weighting, $J_n(\alpha)$, occurs also at this n_{max} so

[1] The Jacobi-Anger expansion replaces the exponential of a sinusoidal function with an infinite series of the product of Bessel function $J_n(\alpha)$ and oscillatory functions of the form $\exp(in\Omega t)$.

that one can approximate the energy side-bands as consisting only of one upper and one lower side-band

$$E_i \pm n_{max}\hbar\Omega \approx E_i \pm QV_0 \ , \tag{16}$$

where in the last step in (16) we have used $n_{max} \approx \alpha = \frac{QV_0}{\hbar\Omega}$.

In conclusion, the generic scalar AB phase, $\varphi(t)$, creates energy side bands, equations (14) (16), which can be probed for via spectroscopic experiments. The dominant energy side-bands occur for $n = n_{max} \approx \alpha$. These energy side-bands are essentially the quasi-energy levels of Zeldovich [8] and studied later in more detail by Sambe [9].

3. Electric scalar AB effect

To apply the above results of section 2 to the electric/scalar AB effect we simply need to take $Q \to e$ (general charge goes to electric charge) and $V(t) \to \phi_e$ (general potential goes to electric scalar potential). With these changes, the results of section 2 can be used for our version of the scalar/electric AB effect as sketched in Fig. 2.

In reference [5] it was shown that our version of the scalar/electric AB effect from Fig. 2 was mathematically similar to the ac Stark effect (also known as the Autler-Townes effect) [6]) which also led to the generation of energy side bands as in (16). Physically the ac Stark effect is different from our set-up in Fig. 2 since in the ac Stark effect the quantum system sees a non-zero vector potential and non-zero electric field, while in the scalar/electric AB set-up from Fig. 2 the quantum system sees only a time-varying scalar potential.

The difference between the ac Stark effect and our version of the scalar/electric AB effect can be seen in how the vector and scalar potentials are coupled to the charge e. Starting from the time-independent Schrödinger equation, $i\hbar\frac{\partial}{\partial t}\Psi = \frac{\hat{\mathbf{p}}^2}{2m}\Psi$ for a particle of mass m and charge e, for the ac Stark effect, one uses minimal coupling of the vector potential $\hat{\mathbf{p}}^2 \to (\hat{\mathbf{p}} - e\mathbf{A})^2$. The vector potential in this case is taken as $\mathbf{A}(t) = \frac{E_0}{\Omega}\cos(\Omega t)\hat{\mathbf{z}}$, which leads to an electric field of $\mathbf{E} = -\frac{\partial \mathbf{A}}{\partial t} = E_0 \sin(\Omega t)\hat{\mathbf{z}}$. Following the analysis of the ac Stark effect from [7] shows that the quantum system will develop energy side-bands of the form given in (16). For the scalar/electric AB effect, the minimal coupling of the scalar potential is accomplished via $i\hbar\frac{\partial}{\partial t} \to i\hbar\frac{\partial}{\partial t} - e\phi_e(t)$ with $\phi_e(t) = V_0\cos(\Omega t)$. Then via the analysis in section 2 one also sees side-bands of the form given in (16). Now however, the quantum system sees only the scalar potential; the electric field is zero since $\mathbf{E} = -\nabla\phi_e(t) = 0$. The ac Stark effect has been observed experimentally, and we proposed in [5] that one should be able to observe the scalar/electric AB effect as similarly given in Fig. 2.

As a final comment on this version of the scalar/electric AB effect one can question this setup in Fig. 2 (and detailed in section 2) since it is possible to gauge away the scalar potential, $\phi_e(t) = V_0\cos(\Omega t)$. A general gauge transformation of the scalar and vector potential is

$$\phi_e' = \phi_e - \partial_t\lambda \quad \text{and} \quad \mathbf{A}' = \mathbf{A} + \nabla\lambda \ , \tag{17}$$

with the gauge function $\lambda(\mathbf{r}, t)$. By choosing $\lambda(t) = \frac{V_0}{\Omega}\sin(\Omega t)$ one can cancel out $\phi_e(t) = V_0\cos(\Omega t)$ so that $\phi_e' = 0$. Also since $\nabla\lambda = 0$ the new vector potential will remain zero, $\mathbf{A}' = 0$. Thus one has gauge transformed away all electromagnetic potentials so how can there be any effect? The resolution to this question is that (17) is only half of the gauge transformation. One must also transform the wave function as

$$\begin{aligned} \psi_i'(\mathbf{r}, t) &= \exp\left(-i\frac{e}{\hbar}\lambda\right)\psi_i(\mathbf{r}, t) = \exp\left(-i\frac{eV_0}{\hbar\Omega}\sin(\Omega t)\right)\Psi_i(\mathbf{r})\exp\left(-i\frac{E_i t}{\hbar}\right) \\ &= \Psi_i(\mathbf{r})\exp\left(-i\frac{E_i t}{\hbar} - i\varphi(t)\right) \ . \end{aligned} \tag{18}$$

In arriving at the results in (18) we have used results from (10) and (11). The wave function ψ' from (18), which is in the gauge where all electromagnetic potentials are zero, matches the wave function ψ from (10) which has the non-zero, sinusoidal scalar potential. The results leading to the energy side bands are gauge invariant.

4. Gravitational AB effect

In this section, we show that one can set up a gravitational version of the scalar/electric AB effect discussed above in sections 2 and 3. The gravitational AB phase can be obtained from the general formula in (11) by letting $Q \to m$ (generic charge goes to mass) and $V(t) \to \phi_g(t)$ (generic potential goes to gravitational potential. This gives the gravitational AB phase as

$$\varphi_g(t) = \frac{m}{\hbar}\int_0^t \phi_g(t')dt' \ . \tag{19}$$

For the scalar/electric AB effect, the quantum system was placed inside a Faraday cage with a sinusoidal varying potential, $\phi_e(t) = V_0\cos(\Omega t)$ - see Fig. 2. This does not work for the gravitational case, for reasons discussed below and in more detail in [10]. For the gravitational AB effect, we will instead place our quantum system in a satellite in an elliptical orbit so that the distance between the satellite and center of the Earth changes with time. Thus the gravitational potential will be of the form

$$\phi_g(t) = -\frac{GM}{r(t)} \tag{20}$$

where G is Newton's constant, M is the mass of some large body (*e.g.* the Earth) about which our quantum system will orbit, and $r(t)$ is the time-dependent distance between the satellite and one focus of the orbit *i.e.* the center of the Earth. In the scalar/electric AB effect set-up we varied the scalar potential, $\phi_e(t)$, by varying the charges oscillated onto and off of the Faraday shell *i.e.* the charge on the shell was time-dependent $e(t)$. A similar set-up for the gravitational interaction would require varying the mass rapidly *i.e.* $M \to M(t)$. This is not practical, and so we vary the radius $r \to r(t)$, by placing the quantum system in an elliptical orbit.

Since our quantum system is in orbit (free fall) in a satellite, then by the equivalence principle the quantum system will be locally in a zero gravitational field, *i.e.* the field has been transformed away by going to a free falling frame.

First we consider the orbit of the International Space Station (ISS), which is an almost circular, low Earth orbit with orbital parameters given by: (i) perigee and apogee radius from the center of the Earth are $r_p = 6.800 \times 10^6$ m and $r_a = 6.810 \times 10^6$ m, respectively, which corresponds to a perigee altitude of 400 km and apogee altitude of 410 km given that the Earth's radius is $r_E \approx 6400$ km. (ii) The period of the ISS with this apogee/perigee is about $T \approx 90$ minutes or 5400 seconds giving an angular frequency of $\Omega = \frac{2\pi}{T} = 1.0 \times 10^{-3} \frac{\text{rad}}{\text{sec}}$ (or a standard frequency of $f = 1.59 \times 10^{-4}$ Hz). The radius of the orbit as a function of time can be approximated as [2]

$$r(t) = \frac{r_p + r_a}{2} + \frac{r_p - r_a}{2}\cos(\Omega t) \equiv A + B\cos(\Omega t) \ . \tag{21}$$

Using the r_p and r_a values above, we find the A and B parameters defined in (21) become $A = 6.805 \times 10^6$ m and $B = -5.000 \times 10^3$ m. Perigee occurs at $t = 0$ and apogee at $t = \pi/\Omega$. For the chosen r_a and r_p, $A \gg B$ so one can approximate $\frac{1}{r(t)} = \frac{1}{A+B\cos(\Omega t)} \approx \frac{1}{A}\left(1 - \frac{B}{A}\cos(\Omega t)\right)$. With this the gravitational potential in (20) becomes

$$\phi_g(t) \approx -\frac{GM}{A}\left[1 - \frac{B}{A}\cos(\Omega t)\right] \ , \tag{22}$$

[2] This treatment of nearly circular orbits is essentially that found in section 9.5 of reference [11]

Inserting this in (19) gives

$$\varphi_g(t) = -\frac{GmM}{\hbar A}\int_0^t \left(1 - \frac{B}{A}\cos(\Omega t')\right) dt' \ . \tag{23}$$

The second sinusoidal term in (23) gives the gravitational AB phase. The term linear in t is packaged with the energy to give a shift of the base energy of the quantum system.

With (22) and (23) in hand we can repeat the analysis in section 2 (from (1) through (16)) to obtain the result for the gravitational AB effect.

As before, in the absence of the gravitational potential, $\phi_g(t)$, we assume that the quantum system has a known solution to the time-independent Schrödinger (*i.e.* $H_0\Psi_i(\mathbf{x}) = E_i\Psi_i(\mathbf{x})$) where H_0, Ψ_i and E_i are the Hamiltonian, wavefunction and energy eigenvalues, respectively, of the quantum system. Placing the quantum system in the potential, $\phi_g(t)$, leads to the Hamiltonian $H = H_0 + m\phi_g(t)$, with the new term being the time-dependent gravitational potential energy. The Schrödinger equation for H is analogous to (3)

$$i\hbar\frac{\partial\psi}{\partial t} = H\psi = (H_0 + m\phi_g(t))\,\psi. \tag{24}$$

At this point we solve (24) exactly using separation-of-variables with $\psi(\mathbf{x},t) = X(\mathbf{x})T(t)$. As before the spatial part of this solution is just $X = \Psi_i(\mathbf{x})$ and the energy eigenvalue is $E = E_i$. These are the given solutions to the time-independent Schrödinger equation $H_0\Psi_i(\mathbf{x}) = E_i\Psi_i(\mathbf{x})$.

Carrying out steps similar to those in section 2 we find that the time part of the separation of variables is

$$\begin{aligned} T(t) &= \exp\left(-\frac{i}{\hbar}E_i t\right)\exp\left(-\frac{i}{\hbar}m\int \phi_g dt\right) \\ &= \exp\left(-\frac{i}{\hbar}\left(E_i + \frac{GmM}{A}\right)t - i\alpha\sin\Omega t\right) \\ &= \exp\left(-\frac{i}{\hbar}\left(E_i + \frac{GmM}{A}\right)t - i\varphi_g(t)\right) \ , \end{aligned} \tag{25}$$

where $\alpha \equiv \frac{GmMB}{\hbar\Omega A}$ is the dimensionless gravitational depth of modulation parameter. To obtain a value for α we assume the satellite orbits the Earth so that $M = 5.97 \times 10^{24}$ kg. The value of A and B for the almost circular orbit are given above (22) as $A = 6.805 \times 10^6$ m and $B = -5.000\times10^3$ m, and the angular frequency given above (21) is $\Omega = \frac{2\pi}{T} = 1.0\times10^{-3}\frac{\text{rad}}{\text{sec}}$. If the quantum system is an atomic system, $m \to m_e = 9.11 \times 10^{-31}$ kg (the electron mass), and if it is a nuclear system $m \to m_N = 1.67 \times 10^{-27}$ kg (the nucleon mass). If the system for comparison is a photon, which is absorbed by a quantum system, then the energy level is the difference between two quantum states with a rest mass increase of $m_{ph} = E_{ph}/c^2$. Using these values for A, B, M, Ω and m we find $\alpha_{atomic} \approx -3.7 \times 10^{11}$ for atomic systems, $\alpha_{nuclear} \approx -6.8 \times 10^{14}$ for a nuclear system and $\alpha_{ph} \approx -2.6 f_{ph}$ for the photonic system, where f_{ph} is the photon frequency in GHz.

The $\frac{GmM}{A}$ term in (25) gives a constant shift to the unperturbed energy E_i which was not present in the scalar/electric AB effect. In the scalar/electric AB case, this constant part of the electric potential was equal to zero. It is only the $\varphi_g(t)$ term in (25) which is associated with the gravitational AB effect.

Combining the spatial part of the solution $X(\mathbf{x}) = \Psi_i(\mathbf{x})$ and the temporal part of the solution $T(t)$ from (8) gives the wave function, $\psi_i(\mathbf{r},t)$ as

$$\psi_i(\mathbf{r},t) = \Psi_i(\mathbf{r})\exp\left(-\frac{i\left(E_i + \frac{GmM}{A}\right)t}{\hbar} - i\varphi_g(t)\right) \ . \tag{26}$$

Journal of Physics: Conference Series **2889** (2024) 012062 | doi:10.1088/1742-6596/2889/1/012062

This new wave function is the original wave function with an added AB phase factor $\exp(-i\varphi_g(t))$, with the gravitational AB phase given by $\varphi_g(t) = \frac{m}{\hbar}\int \frac{GMB}{A^2}\cos(\Omega t)dt = \alpha \sin \Omega t$. Again using the Jacobi-Anger expansion to evaluate the exponentiation of $\varphi_g(t)$ gives the wave function as

$$\begin{aligned}\psi_i(\mathbf{r},t) &= \Psi_i(\mathbf{r}) \sum_{n=-\infty}^{\infty} (-1)^n J_n(\alpha) \exp(in\Omega t) \exp\left(-\frac{i\left(E_i + \frac{GmM}{A}\right)t}{\hbar}\right) \\ &= \Psi_i(\mathbf{r}) \sum_{n=-\infty}^{\infty} (-1)^n J_n(\alpha) \exp\left(-\frac{i\left(E_i + \frac{GmM}{A} - n\hbar\Omega\right)t}{\hbar}\right)\end{aligned} \tag{27}$$

The result in (27) is similar to the generic AB analysis in (13) and we can see that each energy level E_i is split into a multiplet $E_i^{(n)}$

$$E_i^{(n)} = E_i + \frac{GmM}{A} \pm n\hbar\Omega \equiv \tilde{E}_i \pm n\hbar\Omega, \quad \text{with } n, \text{an integer} . \tag{28}$$

In distinction from the generic result of (14) the result in (28) has a constant shift of $\frac{GmM}{A}$ which we have absorbed into the definition of $\tilde{E}_i$. It is the energy side bands represented by $\pm n\hbar\Omega$ which are signature for this version of the AB effect. Note, if we consider comparing photons from stable clocks in different potentials then $\tilde{E}_i = E_i(1 + \frac{GM}{c^2 A})$, which is equivalent to the energy/frequency shift of the photon due to the gravitational potential, which causes the gravitational redshift of a photon as it escapes a gravitational potential.

5. Experimental set-up and considerations

In this short section, we summarize and briefly discuss some of the details and issues associated with this new proposal to probe both the scalar/electric AB effect and the gravitational AB effect.

5.1. Experimental set-up for scalar/electric AB effect

The new version of the scalar electric AB effect, shown in Fig. 2, is cleaner than the original proposal shown in Fig. 1. We do not have to time the turn-on and turn-off of the potential difference with the charge entering or exiting the metal tubes. Also for the setup in Fig. 1, there are always fringing fields, while for the Faraday sphere setup, the electric field is zero inside the shell, modulo very small electric and magnetic fields that arise whenever there is a time variation in charges/fields.

A full discussion of the experimental details to probe this version of the scalar/electric AB effect can be found in [5]. In order to ensure that only the side bands with $n = n_{max} \approx \alpha = \frac{eV_0}{\hbar\Omega}$ dominate one needs to have $QV_0 \to eV_0 \gg \hbar\Omega$ or $\alpha \gg 1$. In [5] it was shown that this condition was easy to achieve for obtainable values of V_0 and Ω (in [5] we chose $V_0 = 0.5mV$, and $\Omega = 10^8$ s^{-1} which gave $\alpha \approx 10^4$).

Another avenue to probe the energy sidebands of (16) was to use electromagnetically induced transparency (EIT). Details of this can be found in [5].

5.2. Experimental set-up for gravitational AB effect

We briefly review some details of the experimental setups to probe the gravitational AB effect. From equation (28) it again seems that there are an infinite number of energy sidebands above and below $\tilde{E}_i$. However, using Fig. 2 and as discussed below (16) it is only the two energy side bands with $n = n_{max} \approx \alpha$ which will dominate if the condition $\alpha \gg 1$ is met. For the

ISS orbit this will be true for atomic and nuclear system, but only for the photonic system if $f_{ph} \gg 340MHz$.

In contrast to the scalar/electric AB effect, which only has one unit of charge ($Q = e$), the gravitational AB effect has several possible charges depending on the mass of the particle of the quantum state. If the quantum system in the satellite is atomic (in this case $Q = m_e$ with m_e being the electron mass) or nuclear (in this case $Q = m_N$ with m_N being the nucleon mass.) Using the numerical values for $A, B, \Omega, M, m_e, m_N$ below (25) gives $|\alpha_{atomic}| \approx 3.7\times10^{11} \approx n_{max}$, and for atomic systems, $|\alpha_{nuclear}| \approx 6.8\times10^{14} \approx n_{max}$ for nuclear systems. For a photonic system generated by a nuclear or atomic transition, the change in rest mass, m_{ph}, of the particle in the excited state depends on the energy of the transition through $E_{ph} = m_{ph}c^2 = hf_{ph}$, and for the ISS orbit $|\alpha_{ph}| \approx 2.9 \times 10^{-9} f_{ph}$, where f_{ph} is in Hz.

The issue becomes if one can observe the energy side bands given their magnitude. In [10] it was shown that one could observe the energy sidebands using the ACES (Atomic Clock Ensemble in Space) mission [12], which involves putting optical/microwave frequency clocks on the ISS. For the nuclear case reference [10] proposed using the Mössbauer effect to observe the energy sidebands. This was a temporal version of the famous Pound-Rebka experiment [14]. The Pound-Rebka experiment observed the gravitational shift of gamma rays emitted and absorbed by ^{57}Fe with the emitter and absorber being spatially separated vertically in a gravitational field. Here the emission and absorption are temporally separated due to the change of the satellite as it orbits the Earth. More experimental details about both the atomic and nuclear probes for the gravitational AB effect are in reference [10].

Data from Galileo satellites in eccentric orbits, which have been used to put the best limits on the redshift from the comparison of onboard hydrogen maser clocks with ground clocks [15, 16], may also be used to search for effects of the gravitational AB effect. When a photon excites an energy transition and is absorbed, the mass of the excited electron increases through a quantum electrodynamics virtual process, given by $\Delta m_e = \frac{hf_{ph}}{c^2}$. Thus, these eccentric orbits, which will experience a significant modulation of the gravitational potential, will in turn cause the energy levels of the hydrogen maser transition to change and hence the mass of the excited hydrogen atoms to vary. In this case the DC term in (28) is equivalent to the redshift, which was measured in [15, 16] at a precision of parts in 10^5, while the Aharonov-Bohm phase proposed here predicts extra sidebands. The orbital period of these Galileo satellite is 12.94 hours (21.5 μHz) with an eccentricity of $e = 0.162$, with a perigee and apogee radius from the center of the Earth of $r_p = 23,445km$ and $r_a = 32,510km$, respectively, so $A_G = 27,977.5$ km, and $B_G = -453.2$ km and thus the gravitational modulation parameter for the Galileo orbit, given by (9), is $\alpha_G \equiv \frac{G\Delta m_e M B_g}{\hbar\Omega A_g^2} = 1.2 \times 10^{-6} f_{ph}$ where f_{ph} is in Hz. For the hydrogen maser clock transition at 1.42 GHz, $\alpha_G = 1699$, with a predicted maximum modulation sideband at around 26.2 mHz. Note, that this orbit causes a depth in modulation, α_G, which is more than 450 times greater than the modulation index induced by the ISS orbit, which the ACES clocks experience.

5.3. Acknowledgments

MET was funded by the ARC Centre of Excellence for Engineered Quantum Systems, CE170100009, and Dark Matter Particle Physics, CE200100008

References

[1] Y. Aharonov and D. Bohm, Phys. Rev. **115**, 485 (1959).
[2] R.G. Chambers, Phys. Rev. Lett. **5**, 3 (1960).
[3] A. Tonomura *et al.*, Phys. Rev. Lett. **56**, 792 (1986).
[4] A. van Oudenaarden, M.H. Devoret, V. Yu. Nazarov, and J.E. Mooij, *Nature*, **391**, 768 (1998)
[5] R. Chiao, H. Hart, M. Scheibner, J. Sharping, N. Inan, D. Singleton, and M. Tobar Phys. Rev. A **107**, 042209 (2023).

[6] S.H. Autler and C. H. Townes, Phys. Rev., **100**, 703 (1955).
[7] N.B. Delone and V.P. Krainov, Physics-Uspekhi, **42**, 669 (1999).
[8] Y. Zel'dovich, Sov. Phys.-JETP **24**, 1006 (1967).
[9] H. Sambe, Phys. Rev. A **7**, 2203 (1973).
[10] R. Chiao, N. Inan, M. Scheibner, J. Sharping, D. Singleton, and M. Tobar, "Gravitational Aharonov-Bohm Effect", e-Print: 2311.07764 [gr-qc].
[11] D. Kleppner and R.J. Kolenkow, *An Introduction to Mechanics*, (McGraw Hill, Boston, MA 1973).
[12] L. Cacciapuoti, *et al.*, Eur. Phys. J. D, **74**, 164 (2020).
[13] R.L. Mössbauer, Zeit. Phys. A, **151**, 124 (1958).
[14] R.V. Pound and G.A. Rebka, Phys. Rev. Letts. **4**, 337 (1960).
[15] Sven Herrmann, Felix Finke, Martin Lülf, Olga Kichakova, Dirk Puetzfeld, Daniela Knickmann, Meike List, Benny Rievers, Gabriele Giorgi, Christoph Günther, Hansjörg Dittus, Roberto Prieto-Cerdeira, Florian Dilssner, Francisco Gonzalez, Erik Schönemann, Javier Ventura-Traveset, and Claus Lämmerzahl, eccentric orbit," *Phys. Rev. Lett.*, vol. 121, pp. 231102, Dec 2018.
[16] P. Delva, N. Puchades, E. Schönemann, F. Dilssner, C. Courde, S. Bertone, F. Gonzalez, A. Hees, Ch. Le Poncin-Lafitte, F. Meynadier, R. Prieto-Cerdeira, B. Sohet, J. Ventura-Traveset, and P. Wolf, *Phys. Rev. Lett.*, vol. 121, pp. 231101, Dec 2018.

AUTHOR INDEX